BROWSING SCIENCE RESEARCH AT THE FEDERAL LEVEL IN CANADA

BRIAN WILKS

Browsing Science Research at the Federal Level in Canada

History, Research Activities, and Publications

UNIVERSITY OF TORONTO PRESS
Toronto Buffalo London

Toronto Buffalo London
Printed in Canada

ISBN 0-8020-8811-2

Printed on acid-free paper

National Library of Canada Cataloguing in Publication

Wilks, Brian B.
Browsing science research at the federal level in Canada : history, research activities and publications / Brian Wilks.

Includes bibliographical references and index.
ISBN 0-8020-8811-2

1. Research – Canada. 2. Science and state – Canada – History – 20th century. 3. Federal aid to research – Canada – History – 20th century. 4. Scientific literature – Canada. 5. Government publications – Canada. I. Title.

Q180.C3W54 2004 507'.2071 C2003-906787-4

The University of Toronto Press acknowledges the financial assistance to its publishing program of the Canada Council for the Arts and the Ontario Arts Council.

University of Toronto Press acknowledges the financial support for its publishing activities of the Government of Canada through the Book Publishing Industry Development Program (BPIDP).

Contents

Acknowledgments

First, I would like to thank Virgil Duff, Executive Editor of the University of Toronto Press, and Anne Laughlin, Managing Editor at the Press for their warmth of welcome and great assistance to me. And a very large thank you, indeed, to Judy Williams for her intelligent, careful, and consistent copy-editing of the draft.

Second, I would like to thank the following people at the Meteorological Service of Canada. Morley Thomas for allowing me to read his unpublished manuscripts pertaining to the history of the Service, Roberta McCarthy, librarian, for tracing unpublished internal report series for me, and several researchers who kindly gave me information by telephone about the archiving of water data.

Third, I want to thank Dalton Robertson, a former executive editor of the *Financial Post*, for his endless reading of the draft manuscript as well as his editing of the galley proofs against the edited draft copy.

Next, a special thanks to science writer Margret Brady, whose firm grip on new developments in her field helped keep me au courant.

Finally, a huge thank you to my colleague Ismat Pasha at the Steacie Science Library for his meticulous and speedy keyboarding, which included updates and changes to the text as it took shape over the years of writing.

BROWSING SCIENCE RESEARCH AT THE FEDERAL LEVEL IN CANADA

Introduction

This book is designed to fill a gap, a large gap, in the information needed for any serious exploration of science research and politics at the federal level in Canada. And the two, science research and politics, are irretrievably intertwined because it is social, economic, and political imperatives that drive the direction and scope of the unique scientific and technical research conducted by federal agencies.

It is already impressive in total as well as range of activities. 'The federal government is a principal player in science and technology,' Statistics Canada points out, 'in which it invests almost $7 billion each year.' That is the figure for the 2000–2001 budget year as noted in their most recent annual publication entitled Federal Scientific Activities (Ottawa: Statistics Canada, Science and Technology Statistics Division, 1984–) and it would not include the extra $500 million Ottawa began talking about in mid-2001 as an addition to the billion already going annually to the universities in federal research grants. This was a very public reminder of the political imperatives behind the ups and downs of federal government-sponsored science research. At mid-2001 and still running hefty surpluses, the federal attention to science research and development spending was renewed by the urgencies of the so-called new economy and competition among the developed nations where Canada has been seen as lacking in innovation.

The almost $7 billion contrasts dramatically with the $5.4 billion of 1996–97 when there had been a decrease of 6.6% from the year before. At that point, Ottawa was cutting back everywhere it could to eliminate the federal deficit. The latest figures and the positive climate for further increases suggest another golden age for science research in Canada may be in the making.

Here is a look at who gets what, for the research areas covered by this book. These visible sciences, by department, are: Agriculture – $333 million; Environment (this includes the Canadian Wildlife Services and the Meteorological Service of Canada) – $513 million; Fisheries and Oceans – $335 million; National Research Council – $598 million; Natural Resources (including the Canadian Forest Service, the Earth Sciences Section, and the Geological Survey of Canada) – $396 million; Natural Sciences and Engineering Research Council (NSERC) – $555 million; and Industry Canada (originating as the Ministry of State for Science and Technology but formerly Industry, Science and Technology) – $427 million. These areas cover the core subjects of science research and study which generally concern a university science library. They are the subject areas of our science government documents, and monographic and serial science collections, which faculty and students use.

The areas not covered by the book fall into two groups – subject (medical), and specialized technology. In library work, the medical sciences are considered to be a separate topic, and in universities with medical faculties, their libraries and their literature are always distinct.* The specialized technology areas include the work of the Department of Defence, much of which I believe is classified, and the research of Atomic Energy of Canada Limited, the Crown Corporation. These very specialized technical topics are not in the purview of a university science library. Nevertheless, all the included and excluded areas which are designed as 'science' by Statistics Canada, along with administrative and capital expenditures, data collecting, and library costs, among others, do bring the total funding of federal science in Canada to almost the $7 billion previously mentioned.

While these figures represent handsome increases for almost all areas of research spending from the years before, and, as Statistics Canada noted, the federal government is now a principal player in science and technology, it should be kept in mind that interest in science at the federal or national level only received a heightened profile in the early 1960s after the Soviet space successes of the late 1950s. This helped to trigger debates on the role and importance of sciences in Canada as seen by the formation of the Royal Commission on Government Organization (the Glassco Commission) of the early 1960s and the Special Senate Committee on Science Policy (the Lamontagne Commission) of the late 1960s. This interest did see the creation of a Ministry of State for Science and Technology in 1971 and the creation of the granting councils, including the National Sciences

*Nevertheless, these areas have been covered as supplementary chapters available through the website of the University of Toronto Press. These chapters are: (1) Chapter 10 – Defence Research, (2) Chapter 11.1 – the Medical Research Granting Councils (this includes the Medical Research Council and its successor the Canadian Institutes of Health Research), and (3) Chapter 11.2 – Health Research (this focuses upon the laboratory research of the historic Food and Drug Directorates and the Laboratory of Hygiene). To include these chapters in the formal text would have made the monograph too large. The website is: www.utppublishing.com.

and Engineering Research Council (NSERC), separate from the National Research Council (NRC), as well as the creation of the advisory Science Council of Canada later in that decade. It was this new focus that encouraged some commentators to say that this period was the 'golden age' of science in Canada. Even then, however, politicians had, and still have, difficulty coming to terms with, on the one hand, science research funded to the university community or carried out by the National Research Council or other federal departments, and, on the other hand, applied research and innovation in partnership with industry designed to stimulate economic growth. These two themes can be noted in the change of name and absorption through time of the Ministry of State for Science and Technology to Industry Canada and the dismissal of the Science Council of Canada and its replacement with a more modest National Advisory Board on Science and Technology.

So, what kind of research is carried out by the federal departments mentioned in this book? Apart from the advisory or granting councils, which fund basic research carried out in the universities, the research consists of work that cannot (and in my view should not) be done by the private sector – agriculture and forestry research unique to Canada, geographically unique fisheries research, aquatic mammals and wildlife research, atmospheric and climatic research and services, water research and services, geological and mining research and services, and astronomical and astrophysical research and services (including the observatories). These are the topics I have covered in this book, which I started ten years ago by reading annual reports, undertaking database searches, doing basic reading in areas that were beyond my own science background (zoology and chemistry), and reading many research papers from the scientific literature that were relevant and interesting to me; hence the term 'browsing' in the title of the book.

The format that I have chosen for the text is narrative for the historical background and research activities of departments or agencies, and bibliographic, where possible, for the types of publications issued or the sources of the publications for that department or agency. The reason for this mixed format is that history and science research need the narrative format in the body of the text for continuity as well as for descriptive explanations of science. It also makes for more interesting reading. However, I think that publication lists and publication sources are more suited to the annotated format so often used by librarians.

This lengthy monograph is specifically designed to deal with the nature of the visible research done or funded by Ottawa – a gap in existing literature. As mentioned, it is the core research that one encounters on a daily basis when helping faculty and students. As science librarian at York University, it frustrated me that there was no guide to the kind and range of science research and its documentation at the federal level. So, the topics covered, the historical backgrounds discussed, and the publications described are aimed at librarians in general, science librarians, librarians in training, and the ever-needful science student. However, I think the book will also be useful to students studying the history of science in Canada.

Chapter 1.1

The Ministry of State for Science and Technology and Beyond: The Search for Science Policy

I Historical Background

The search for a federal science policy can be formally traced to the creation of a Ministry of State for Science and Technology (MOSST) in the early 1970s, and to the initiatives of the federal Conservative government in the mid-1980s and its creation of a National Advisory Board on Science and Technology, in 1991, and the abandonment of the Science Council of Canada in 1992. Then, to the decline of MOSST to its eventual name of Industry Canada, and the further reviews and strategies for science and technology by the new Liberal government in the middle to late 1990s.

The Ministry of State for Science and Technology, which was established in August, 1971, was an outgrowth of the debates in the 1960s on the role of sciences and science policy in Canada. The first official suggestion for the formation of a ministry came in 1960 from the Glassco Commission, whose reports were formally called the *Royal Commission on Government Organization: Reports* (Ottawa: Queen's Printer, 1962–63, 5 v.). The commission found a weakness in the government machinery for assisting the prime minister and cabinet with respect to science policy decisions. In Report No. 23 of Volume 4, the commission made two major points relevant to the general issue of science and politics. The first dealt with the machinery of science policy, and the second dealt with the apportionment of scientific work in Canada between the government, industrial, and academic sectors of the scientific community. Historically, the Privy Council Committee of Scientific and Industrial Research (modelled on the British Committee and created during World War I) and an Advisory Panel for Scientific Policy (an interdepartmental committee of deputy ministers created in 1949) 'had failed to function as intended,' and decisions about science tended to fall, by default, to the Treasury Board and its staff. The commission recommended that a 'central scientific bureau' be created to act as a science secretariat to the cabinet and that a national science advisory council be created that would be broadly representative of the science disciplines. In practice, a Science Secretariat was established within the Privy Council Office, in 1964, and the Science Council of Canada was established in 1966 to provide independent advice to the government.

Debate continued with the Senate Special Committee on Science Policy under the chairmanship of Senator Maurice Lamontagne. This committee began its review in late 1967, and the multi-volume report entitled *A Science Policy for Canada: Report of the Senate Special Committee on Science Policy* (Ottawa: Queen's Printer, 1970–72, 4 v.) began to appear in 1970. Volume 1 was entitled *A Critical Review: Past and Present*; Volume 2 was entitled *Targets and Strategies for the Seventies*; Volume 3 was entitled *A Government Organization for the Seventies*; and Volume 4 was entitled *Progress and Unfinished Business.* The commission stated that investment in scientific research must follow the pattern of national priorities, including a priority given to research and development (R&D) through market-oriented technologies. The commission also pointed to the need for a Ministry of Science, which, as noted above, was realized in 1971.

Other science policy initiatives were also undertaken by the Science Council of Canada, formed in 1966. These are noted in Chapter 3.

With the formation of the Ministry of State for Science and Technology in 1971, the activities of the ministry can be broken down into two phases. The first phase went from the early 1970s to the mid-1980s. This was the period when MOSST participated most actively in the development of individual subject-oriented science policies for the federal government; undertook studies pertaining to federal science developments, as well as industry and the universities; participated in task forces; and was involved in collecting data about science activities and indicators, including forecasting.

However, after the election of the Conservative government of Brian Mulroney in 1984, there were new initiatives at the federal level to develop an overall national science policy and to emphasize the links between industry, science, and technology. This new emphasis saw the merger of the Department of Regional Industrial Expansion (DRIE) with MOSST to create a new Department of Industry, Science and Technology in 1989. Then in 1991, a separate National Advisory Board on Science and Technology was created in order to study the relationships between science and technology and competitiveness in Canadian industry.

In 1993, the department was renamed Industry and Science Canada in order to serve 'the needs of consumers; industry and small business as well as the science community.' This new ministry was described in the brochure entitled *Industry and Science Canada: A New Partnership to Serve You Better* (Ottawa: Industry and Science Canada, 1993. 10 p.). After the election of the new Liberal government in late 1993, the ministry had yet another change. It became Industry Canada, which included a Secretary of State for Science, Research and Development. And, in June 1994, the federal government announced a science and technology review of federal departments and agencies. As well, the National Advisory Board of Science and Technology undertook an independent assessment of science and technology (S & T) in Canada.

However, one strategy document prepared by the Industry minister for cabinet was turned down as too weak to publish, and according to an article entitled *Canadian Research Strategy Set for Lukewarm Welcome?* (*Nature*, v. 376, 1995. p. 376), there is a 'lack of political leadership from the top of government' in science and technology.

II Publications Relating to Science Policy in Canada

Other than the work of the Science Council of Canada, described in Chapter 3, or the earlier work of the Glassco Commission and the Special Senate Committee of Maurice Lamontagne, there is not a large or organized body of literature that has come forth from the federal government or its agencies pertaining to science policy in Canada. After the formation of the Ministry of State for Science and Technology in 1971, that department's own initiatives did see the writing of a series of background papers. For example, their initiative to understand university science and technology, described in the next section, saw papers written such as *Federal Funding of University Research: Major Issues* (Ottawa: Ministry of State, Science and Technology Paper. MOSST Background Papers No. 7. 18 p.). Besides these background papers there were also unpublished internal working papers, sometimes mentioned or described in the *Annual Report: Ministry of State for Science and Technology Canada* (Ottawa: The Ministry, 1971/72–1987/88, 17 v.). Other publications pertaining to science policy have been dependent on the cycle of political changes at the federal level and have generated reviews of national science and technology, or commentaries, from the National Advisory Council on Science and Technology. These types of publications, which are elaborated on in a later section of this chapter, have titles such as *Building a Federal Science and Technology Strategy* (Ottawa: Secretariat for Science and Technology Review, Industry Canada, 1994. 16 p.), and *Spending Smarter: Report of the National Advisory Board on Science and Technology Committee on Federal Science and Technology Priorities* (Ottawa: The Committee, 1993. 40 p.).

III Activities of the Ministry of State for Science and Technology and Beyond

A MOSST Activities Affecting Government Science and Technology

In the early 1970s, the most important science policy areas identified by the cabinet for formulation and discussion were space, oceans, and northern science. Subsequently, in the 1980s, biotechnology became a focus of government interest. Other interest areas that were pursued related to the energy crisis of the 1970s, scientific manpower needs, and the transfer of technology from government to industry.

A Canadian space policy was announced in July, 1974. It emphasized the strengthening of Canadian industrial capacity to design and construct space systems, but not to develop an independent satellite launching capacity. One impetus for this initiative came from the American invitation to participate in NASA's Space Shuttle Program by developing a shuttle arm attached to a remote manipulation system. This arm was developed by Canadian industry for the National Research Council (NRC). The development of a space policy required the compilation and analysis of background information involving existing and past space programs in Canada, as well as consultation with appropriate government departments and agencies, and Canadian industrial representatives. All this was done through an Interdepartmental Committee on Space (ICS) and MOSST, in effect, contributed to policy formation via membership in that body. A Five-Year Space Program was prepared for cabinet, and the ministry did publish a series of space plans. They were: the *Canadian Space Program Plan for 1981/82–1983/84* (Ottawa: Ministry of State, Science and Technology Canada, 1981. 25 p. MOSST Background Paper No. 19); the *Canadian Space Program Plan for 1982/83–1984/85* (Ottawa: Ministry of State for Science and Technology, 1981. 14 p. MOSST Background Paper No. 20); and the *Interim Space Plan, 1985–1986* (Ottawa: Ministry of State, Science and Technology, 1985. 9 p.).

The first three-year space plan emphasized the use of space to 'contribute significantly to the attainment of social, cultural and economic goals and the economic benefits to be obtained from the creation of a strong industry to meet our needs and which is able to compete in the international marketplace. Canada would build on its existing strengths in remote sensing, the encouragement of diversity in the Canadian space industry, and the development of new and improved satellite communications.' The second plan emphasized the continuation of the existing space program, along with new initiatives such as a mobile communications satellite demonstration project (MSAT), a continuation of RADARSAT studies (an imaging radar satellite system), and continued participation in the European Remote Sensing Program. The third plan focussed upon the decision to accept US President Ronald Reagan's invitation to participate in the definition stage of the US Space Station Program. This interim plan was issued as an information package that also contained related publicity releases from other departments. They were: *MSAT: An Opportunity for Canada* (Ottawa: Department of Communications, 1985. 6 p.); *Canadian Participation in Space Station* (Ottawa: National Research Council, 1985. 14 p.); and *RADARSAT* (Ottawa: Energy, Mines and Resources Canada, 1985. 7 p.). MOSST's involvement in space policy activities came to an end in 1989 when the government announced the creation of the Canadian Space Agency, to be based in the greater Montreal area. This agency would manage Canada's civilian space program and direct expenditures of almost $3 billion to the end of the century.

Unlike the federal space policy initiatives, the cabinet-approved Oceans Policy of 1973 did not generate any publicly available publications through MOSST. However, this internally elaborated policy stated that Canada should have a world-class operational excellence in, on, or below ice-covered waters; develop a Canadian ocean industry (e.g., deep-sea mining); create an off-shore resource databank; intensify marine science and technology programs, including oceans engineering and environmental management, and ice, wind, and sea-state forecasting. As was the pattern with the space policy, MOSST worked with other departments to identify and develop a coordinated approach to matters addressed by the ocean policy. In 1976, a Panel on Ocean Management was created to deal with policy initiatives, and an internal overview report was written that reflected the federal approach to offshore mineral rights, in view of federal-provincial conflicts in this area. In 1980, MOSST wrote a separate internal report for the panel that focused on determining the requirements for an integrated ocean information system using updated surveillance technologies such as satellites. But it is interesting to note that a Canadian oceans policy was not actually published until 1986, and it was released by the Department of Fisheries and Oceans. It was entitled *Oceans Policy for Canada: A Strategy to Meet the Challenges and Opportunities on the Oceans Frontier* (Ottawa: Department of Fisheries and Oceans, 1987. 17 p.).

The government initiative on federal science and technology in the north was not as focused as it was for space, or oceans, as the government appeared to lack basic information. As a result, MOSST undertook to complete an inventory of federal science and technology expenditures in the north and created an internal database entitled *Inventory of Federal Northern Science Prospects, 1975–76*. This data was reviewed by the Interdepartmental Advisory Committee on Northern Development, but the actual review of the government's northern science objectives and policies was to be undertaken by the then-Department of Indian Affairs and Northern Development.

MOSST's interest in biotechnology began in 1980 with the publication of the background paper entitled *Biotechnology in Canada* (Ottawa: Ministry of State for Science and Technology, 1980. 64 p.). As a result of this initial process to develop a federal policy for the promotion and development of biotechnology in Canada, MOSST announced the establishment of a Task Force on Biotechnology, in June 1986, under the chairmanship of Maurice Brossard. Its report, entitled *Biotechnology: A Development Plan for Canada: Report of the Task Force on Biotechnology to the Minister of State for Science and Technology* (Ottawa: Minister of Supply and Services Canada, 1981. 51 p.), stated that this high-technology field required a strong interdisciplinary science base for its development, but the erosion of science in Canadian universities and government research establishments during the 1970s had weakened Canada's ability to respond to the challenges of biotechnology. Accordingly, Canada's current biotechnological efforts were 'characterized by a wide scattering of research activity in university and government (federal and provincial) laboratories and the absence of any major industrial activity.' As a result of the report, the National Biotechnology Advisory Committee was established in 1983 to advise the Minister of State for Science and Technology on a National Biotechnology Strategy. The government hoped to strengthen its own efforts in biotechnology through the National Research Council and the Department of Agriculture, as well as provide additional support to universities and industry. Some of the committee's early initiatives were in the form of workshops run to attempt to improve communications among industry, university, and government, and to identify the policies and resources required to facilitate commercial developments in biotechnology. Also, national surveys were undertaken to assess the growth of the biotechnology industry in Canada, and biotechnology-related expenditures and activities of the federal and provincial governments. These activities of the committee were reported in the *National Biotechnology Advisory Committee Annual Report* (Ottawa: Ministry of State for Science and Technology, v. 1, 1985–v. 5, 1991. 5 v.)

Space, oceans, the north, and biotechnology were broad policy initiatives, but during the 1970s and onward, there were other developments in science and technology that occupied MOSST and generated either reports, task force reports, or background papers. These developments occurred in energy, and because the federal government intended to increase its research and development (R&D) funding from 0.9% to 1.5% of the GNP in Canada by 1985. An increase would affect federal scientific manpower requirements, as well as the government's ability to transfer or influence research and technological developments in both the private sector and the universities.

The energy crisis of 1973 triggered studies and reviews of non-conventional or mixed energy sources, particularly nuclear energy. MOSST gave advice via the interdepartmental Nuclear Policy Review Group in studies that were not published. However, the ministry also undertook studies for the Department of Energy, Mines and Resources on the use of off-peak electricity to conserve oil. The outcome of these studies was a report published in 1984 entitled *Electric/Oil Hybrid Heating Systems* (Ottawa: Ministry of State, Science and Technology Canada, 1984. 76 p.). This report followed the federal government's National Energy Program (NEP), which established an 'off-oil' conversion program designed to promote a rapid and effective shift from oil toward gas, electricity, renewable energy, and coal. But these alternatives were not as versatile as oil for residential, commercial, and industrial space heating, nor would a shift to natural gas and all-electric space heating significantly improve the efficiency of Canada's total energy system. However, the study stated that hybrid heating systems involving the adoption, not of a new technology, but of a new operational technique – the use of electricity for space heating during periods of off-peak electric demand, and the use of oil when the demand for electricity is peaking – provided maximum end-use efficiency of both energy types.

The training and utilization of scientific manpower in the federal government, and the review of job-related issues such as aging, mobility, and rotational management opportunities, were of specific interest because of possible increased R&D funding. This resulted in an internal publication (mentioned only in the 1978/79 Annual Report of the Ministry) entitled *Recent Science Policy Initiatives and the R&D of the Scientist and Research Manager in the Public Service.* However, the broader area of interest here lay in the ability of the government to influence and have an impact on scientific research and technological developments outside the government. Two background papers elaborated on this topic. The first was entitled *Technology Transfer by Department of Communications: A Study of Eight Innovations* (Ottawa: Ministry of State Science and Technology Canada, 1980. 48 p. MOSST Background Paper No. 12). This paper studied the case histories of eight innovations which were developed with the active support of scientific and technical expertise and skills within the Department of Communications Research Centre to 'illustrate the role government laboratories can play in influencing the innovative process to foster the industrial development of the communications sector in Canada.' The study showed that under the right conditions, opportunities did exist for government and industrial laboratories to work together, and that work in government laboratories could supplement the development work being done in the private sector. The second study was entitled *R&D Policies, Planning and Programming* (Ottawa: Ministry of State for Science and Technology, 1981. 1 portfolio – MOSST Background Paper No. 13).

B MOSST Activities and Studies Affecting Industry

In order to create research and development assistance policies which would strengthen the innovative performance of Canadian industry, the federal government needed an understanding both of the implications and effects of current policies on Canadian industry and of the nature of industrial R&D. Because the government always had large R&D and procurement needs of its own, a coordinated effort on its part could have a positive impact in its approaches to industrial R&D assistance. In the studies that were carried out, there were both unpublished internal working papers written, and numerous background papers released. Later, as the government believed that its approaches to encourage industrial innovation seemed to be right, the work of the first decade was reviewed in order to further strengthen its initiatives with the private sector. In all this work, MOSST participated via interdepartmental committees.

The early work done on understanding the factors affecting the environment for innovation examined the effects of patents and licensing technology transfer and government R&D assistance and incentives to industry. The work was carried out by the Patent Law Review Committee and the Small Business Development Committee. Two internal studies were written – *The Effect of Tariffs and Taxes on Innovation in the Scientific Instrument Industry*, and *The Capital Market for Technological Innovation in Canada*. However, the first published study in this area was written by R. Grasley and entitled *Availability of Risk Capital for Technological Innovation and Invention in Canada* (Ottawa: Ministry of State, Science and Technology Canada, 1975. 80 p.). The latter was broad in scope and described the availability of capital financing for industrial innovation and the structure of Canadian capital markets.

Other studies were more focused, and the impetus for them was the need to prepare studies on industrial R&D for the first ministers' conference of February, 1978, and the Federal-Provincial Conference on Research and Industrial Research and Development scheduled for November, 1978. Several background papers emerged from this work. The first one was entitled *Industrial Research and Development in Canada* (Ottawa: Ministry of State, Science and Technology Canada, 1978. 24 p. MOSST Background Paper No. 2). This paper surveyed industrial R&D in Canada and came to the conclusion that Canadian R&D was inadequate, based on international comparisons of expenditure levels, the invisible inflows of technology, and trade performance. This inadequacy has caused industry not to be sufficiently innovative in order to maintain the strong internationally competitive stance needed to create adequate rates of growth and levels of employment.

The second study was entitled *Importation of Invisible Research and Develop-*

ment, 1974–1976 (Ottawa: Ministry of State, Science and Technology Canada, 1978. 31 p. MOSST Background Paper No. 3). This paper stated that although every nation uses the results of science and technology performed elsewhere, the magnitude of the invisible in-flows of technology, as estimated in the study, indicated that Canada's dependence on the importation of technology is quite substantial and probably greater than for most industrialized nations. The report concluded that imported R&D 'cannot fully substitute for indigenous R&D and Canada's great dependence on such imports creates a serious problem.'

The third study of this group was entitled *Performance of Canadian Manufacturing Industries by Levels of Research Intensity* (Ottawa: Ministry of State, Science and Technology Canada, 1978. 28 p. MOSST Background Paper No. 4). This paper showed that industries which do R&D have a better track record (the economic indicators used were employment, real output productivity, and prices) than those industries which do little or no R&D. These industries showing the best performance were the paper and allied industries, and the primary metals industries.

The fourth study was entitled *Canadian Trade in Technology Intensive Manufactures, 1964–1976* (Ottawa: Ministry of State, Science and Technology Canada, 1978. 26 p. MOSST Background Paper No. 5). The purpose of this study was to develop and apply a methodology to establish which manufactures were technology-intensive, compare Canada's trade position in these technology-intensive products, and assess how they had fared in international competition. In terms of trade competition, Canada's strongest technology-intensive industries in the 1970s were aircraft, followed by chemicals, and chemical products.

Further studies undertaken examined the effects of foreign ownership on Canadian R&D, as well as the impact of technological change on employment in the manufacturing industries of Canada and other countries. They resulted in the publication of two background papers. The first one was entitled *R&D in Canadian and Foreign Controlled Manufacturing Firms* (Ottawa: Ministry of State, Science and Technology Canada, 1979. 49 p. MOSST Background Paper No. 9). This paper showed that R&D in foreign-controlled firms is not only quantitatively less, but also different. It is primarily directed toward adapting foreign technology to domestic needs as opposed to exporting. 'This implies that the net gains from both imported technology and domestic R&D especially the R&D performed by foreign controlled firms have not been maximized due to insufficient opportunities for greater specialization.' The second paper was entitled *Changes in the Size and Structure of Manufacturing Employment in Industrial Countries* (Ottawa: Ministry of State, Science and Technology Canada, 1979. 20 p. MOSST Background Paper No. 10). In this paper, six industrial countries of the OECD (United States, Germany, Japan, United Kingdom, France, and Canada) were examined. The study concluded that 'a lack of employment growth was not synonymous with stagnation or a loss of competitiveness, but was rather associated with a strong shift in employment toward technology-intensive modes of production.'

Perhaps overshadowing any of these studies was the government's need to understand and publicize the impact of its own R&D procurement needs, particularly in the field of new federal military programs. In August, 1973, the government announced its 'Make-or-Buy Policy,' which was aimed at contracting out government procurement needs. An evaluation of that policy was published, in 1976, and entitled *The Make-or-Buy Policy, 1973–1975* (Ottawa: Ministry of State for Science and Technology, 1976. 44 p.). This contracting-out policy was considered a major policy success, and by 1978–9, R&D contracts awarded to the private industrial sector amounted to $107.5 million. After the government's initial successes in encouraging private-sector industrial innovation, the former wanted to further encourage industrial research and development spending so that it would reach a level of 0.75% of the GNP by 1985. To do this, MOSST reviewed various aspects of industrial R&D, including technology assessments, the role of government laboratories, government procurement policies, and the tax system and its incentives for R&D. Out of this review process came the following publication entitled *Support for Technology Development: A Summary of Federal Programs and Incentives, 1984* (Ottawa: Ministry of State, Science and Technology Canada, 1984. 32 p.).

The major publication to appear as part of this review was the *Report of the Task Force on Federal Policies and Programs for Technology Development* (Ottawa: Ministry of State, Science and Technology Canada, 1984. 69 p.). This so-called Wright Report, named after its chairman, Douglas Wright, looked at the government's industry-support programs, and how they were used, the responsiveness of the universities to the research and development needs of the private sector, and the uses of the federal laboratories. It suggested that government policies and programs aimed at technological development were not working well, or in some cases, not at all. The government needed to 'set a climate that encourages the private sector to adopt and use the most up-to-date world technologies and create new technologies when they will respond to market opportunities where Canada has a comparative advantage.'

C MOSST Activities and Studies Affecting Universities

The federal government's link to the universities is via their funding of university research programs. In 1976, the granting function of the National Research Council (NRC) passed to the Natural Sciences and Engineering Research Council

(NSERC), and two additional granting councils were created. One was for the social sciences and humanities (Social Sciences and Humanities Research Council [SSHRC]), and the other was for the health sciences (Medical Research Council). At that time, MOSST was asked to give advice on policies that affected the federal support of university research, but behind MOSST's advisory function was the government's goal to attempt to integrate university research into programs related to so-called 'national goals,' as well as to oversee that university research fulfilled Canada's international obligations. However, in order to provide advice, MOSST first needed to generate basic information about the nature of federal funding of university science and technology and university-controlled research. The ministry also had to look at new trends in scientific research, and how they affected and could be incorporated into national goals. Also, MOSST had to develop the capability to prepare forecasts on university graduates and research manpower requirements in Canada.

The basic information needed on the nature of federal funding of university research began to show up in the late 1970s in the form of internal studies and publicly available papers. For example, in 1976, MOSST became a liaison between the federal and provincial governments with the establishment of the Canadian Committee on Financing University Research and did publish two background papers on university funding which coincided with the high inflationary period in Canada in the late 1970s and early 1980s. The first paper was entitled *Federal Funding of University Research: Major Issues* (Ottawa: Ministry of State, Science and Technology Canada, 1979. 18 p. MOSST Background Paper No. 7). This paper stated that there were nine key problems that had to be addressed if university research was to make its proper contribution to Canada's science goals. These included the need to maintain and strengthen the universities' capacity to carry out research, both as a corrective for the weakening of research in the 1970s, and as part of the objective to raise the level of national research and development. As well, an assured level of real funding was necessary for longer-term plans, and for equipment and facilities that needed replacing. The second paper was entitled *A Rationale for Federal Funding of University Research* (Ottawa: Ministry of State, Science and Technology Canada, 1979. 18 p. MOSST Background Paper No. 8). This paper reviewed the structure of federal funding and the need for a more concerted effort to focus upon 'the emerging national role of R&D.'

The data gathered on university activities resulted in the release of five background papers in the autumn of 1981. The first paper in this cluster was entitled *Recent Trends in Degrees Awarded and Enrolments at Canadian Universities* (Ottawa: Ministry of State for Science and Technology, 1981. 67 p.). The purpose of this paper was to review recent trends in university enrolments and degrees in order to obtain a better understanding of the current capacity of the university system for generating highly qualified manpower. The paper estimated that 'the supply of persons with university degrees potentially available for the job market grew at about 4% per year at the undergraduate level ... at the graduate level the numbers remained almost constant' during the 1970s. The second paper was entitled *University Enrolment Projections to 2000* (Ottawa: Ministry of State, Science and Technology Canada, 1981. 60 p. MOSST Background paper No. 15). Based on the projection methods developed, this study estimated that full-time equivalent university enrolment, excluding foreign students, would peak at just over 500,000 in 1982–83, decline by 1995–96 to 429,000, then rise to 442,000 by the year 2000. The third study was entitled *The Stock of Research Trained Personnel* (Ottawa: Ministry of State, Science and Technology Canada, 1981. 22 p. MOSST Background Paper No. 16). The purpose of this paper was to provide estimates of the number of persons engaged in R&D by performing sector in the natural sciences. For the year 1978, 'the national stocks of professional R&D manpower (excluding support staff) are estimated to be 21,800 full-time equivalent person-years ...' The fourth study was entitled *Research Manpower Requirements Arising from Accelerated Expenditures on R&D* (Ottawa: Ministry of State, Science and Technology Canada, 1981. 30 p. MOSST Background Paper No. 17). The purpose of this paper was to develop several scenarios of requirements for professional R&D personnel associated with raising the level of R&D in Canada. This analysis was confined to university-trained scientists and engineers in the natural sciences. Scenarios arising from the simulations indicated there would be shortages of trained researchers particularly in the fields of science in industrial research areas, and the extent of the shortages would be a function of the size of the gross domestic expenditures on R&D in those areas. The fifth study was entitled *The Requirements for Engineering Graduates to 1985* (Ottawa: Ministry of State, Science and Technology Canada, 1981. 19 p. MOSST Background Paper No. 18). This report reviewed the trends in enrolments, degrees awarded and requirements for engineers over the 1970s and reviewed a number of factors bearing on future requirements to 1985. It said that in the 1980s, 'the university system will have to become more flexible to be able to meet the needs and preferences of students and the labour market ...' and 'industry can take steps to counteract the expected shortages through ... innovative staffing practices and more investment in existing human resources.' Simultaneously with the publication of these background papers, the ministry also released a paper describing the methodology of the HQM (Highly Qualified Manpower) demand model used for these papers. It was entitled *MOSST HQM Demand Model: Methodology* (Ottawa: Ministry of State for Science and Technology, 1981. 49 p.).

The ministry's last visible foray into the field of university research came at

Saskatoon, Saskatchewan, in 1987, when it hosted a national forum on the topic. A background paper was issued and entitled *University Research in Canada: A Background Paper for the National Forum on Post-Secondary Education* (Ottawa: Ministry of State, Science and Technology Canada, 1987. 111 p.). This provided an overview of university research and highly qualified personnel in Canada, and discussed the emerging role of universities in the transfer of knowledge, technology, and expertise. In 1988, the prime minister announced the establishment of the Canada Scholarships program to encourage students to pursue undergraduate degrees in the natural sciences, engineering, and related disciplines.

D MOSST Activities and Studies in the International Field

In the international field, MOSST had the responsibility of explaining and interpreting Canadian science and technology policies, as well as participating in cooperative agreements, either bilaterally with individual nations, or multilaterally with international bodies. Until 1976, MOSST participated at the operational level in both bilateral and multilateral activities, but after 1976, bilateral cooperation and agreements were taken over by External Affairs, and MOSST retained an advisory role only. It did, however, retain its operational functions in multinational science and technology bodies. The scope of these activities is interesting as background information, but MOSST itself published little in this area. Some of these multilateral intergovernmental activities that dealt with science policy and scientific cooperation were: (1) the Committee on Scientific and Technological Policy of the Organization for Economic Cooperation and Development (OECD), (2) the Committee of Senior Advisors to Governments on Science and Technology of the Economic Committee, and (3) the Commonwealth Science Council. MOSST also participated in United Nations conferences such as the 1979 UN Conference on Science and Technology for Development, and the UN Conference on New and Renewable Resources held at Nairobi, Kenya, in 1984.

In spite of these external activities, which involved the writing of position papers for bilateral or multilateral meetings, the only publication that was noted was an internal published MOSST report entitled *Science and Technology in the UN System: Program Elements Significant to Canadian Interests* (Ottawa: Ministry of State for Science and Technology, 1974. 178 p.).

E MOSST Data-Gathering Activities

During its science-oriented period, MOSST gathered, analysed, and published information about Canadian scientific and technological activities, based upon its policy development and advisory function. These information-gathering activities fall into two periods. The first covered the formation of the ministry to its reorganization in 1975. That period oversaw the documentation of historical data and other information. After 1975, MOSST's information-gathering activities became an annual ongoing task. All of this work was done in cooperation with the Science Statistics Centre of Statistics Canada, which took over this type of publication activity from MOSST in 1984.

The first series of four data reports, the so-called Green Books, were entitled *Scientific Activities: Federal Government Costs and Expenditures* (Ottawa: Ministry of State for Science and Technology, 1958/59–1972/73. 2 v.); and *Federal Scientific Resources: Natural and Human Sciences* (Ottawa: Ministry of State for Science and Technology, 1972/74–1973/75. 2 v.). The second series of data reports, the MOSST 'Red Books,' resulted from a recommendation of the Senate Special Committee, chaired by Senator Lamontagne, as mentioned in the introduction. The first report, entitled *Federal Science Programs* (Ottawa: Ministry of State, Science and Technology, 1977–78. 1 v.); was continued by *Federal Science Activities* (Ottawa: Ministry of State, Science and Technology, 1978/79–1982/83. 5 v.). A companion series entitled *Federal Science Expenditures and Personnel* (Ottawa: Ministry of State, Science and Technology Canada, 1977/78– 1982/83. 6 v.) was also published. The purpose of these reports was to provide members of Parliament with information on federal science expenditures when they were reviewing the government's spending plans. After a reorganization of MOSST, in the summer of 1983, these data functions were transferred to Statistics Canada's Science and Technology Statistics Division, which began the annual publication entitled *Federal Scientific Activities* (Ottawa: Statistics Canada, Science and Technology Statistics Division, 1984–). This series differed from the MOSST series because it was not intended as a companion to the government's main estimates report to Parliament, and the narrative descriptions of ongoing departmental programs were replaced by an analysis of departmental expenditures.

F MOSST Forecasting Activities

Besides the forecasting studies relating to university research and development needs, MOSST was also active in forecasting possible advances, or futures, in science and technology and assessing the effects of these advances on Canadians. There were some unpublished internal studies and reports written as well as one major set of conference proceedings.

Futures studies initially focused upon the theme 'Alternative Futures for Canada,' which was based upon separate projections pertaining to the 'Post-Industrial Society,' the 'Super-Industrial Society,' and the 'Zero-Growth Society.' These internal studies attempted to integrate the methods of econometric and sociological forecasting in an attempt to verify the hypotheses of these projections. The methodology itself was based on various editions of an internally published *Review of Technology Forecasting and Assessment Methodology*, 5th edition (Ottawa: Ministry of State for Science and Technology, 1974, 108 p.). Out of these general future studies came an internal unpublished report entitled *Canadian Post-Industrial Society: Some Quantitative Perspectives.*

MOSST's future studies culminated in the Canada Tomorrow Conference, held in Ottawa in late 1983. This conference looked at the impact of technology on women, and the impact of information technology on training education. The proceedings of the conference were published by MOSST as the *Canada Tomorrow Conference, November 6–9, 1983: Proceedings* (Ottawa: Ministry of State, Science and Technology Canada, 1984. 111 p.).

G *Science Studies and Science Policy Initiatives under the Mulroney Conservatives*

The change of government in 1984 created a flurry of new initiatives and publications which included the first written National Science and Technology Policy and the establishment of a National Advisory Board on Science and Technology in the early 1990s.

The former initiative began in 1985 at a meeting of federal, provincial, and territorial science and technology ministers in Calgary, Alberta. At that meeting, the ministers 'made a commitment to formulate a national science and technology policy with three principle objectives: (1) to strengthen private sector innovation, (2) to encourage the transfer and application of technology, and (3) to develop the long-term competitiveness of Canadian industry by supporting basic scientific research.' Then, in 1986, the Minister of State for Science and Technology wrote to selected industry, academic, and labour leaders, inviting them to take part in a national forum and to submit briefs in order to find where there was consensus as well as differences on the creation of a written science policy. The forum was organized by the Science Council of Canada and the published proceedings were entitled *The National Science and Technology Policy Forum (Winnipeg, Manitoba, 8, 9, and 10 June, 1986)* (Ottawa: Science Council of Canada, 1986. 41 p.).

The National Science and Technology Policy, which was announced in 1987, has been described in the publication entitled *Background Paper: The National Science and Technology Policy* (Ottawa: Government of Canada, 1988. 11 p.). In general, the policy called for the use of science and technology to promote economic, social, and regional development in Canada through ongoing cooperation among governments, and between the public and private sectors. In practice, the policy emphasized the use of science and technology research to strengthen Canadian competitiveness in a 'growing international economy, specialization of markets and production and increasing requirements for higher education and advanced technical skills [which] signal the emergence of a world economy that is knowledge-based.'

There were also a number of promotional publications issued around the policy initiatives of the new conservative government.

One publication by D. Doyle was entitled *Technology Venturing in Canada: A Guide to the Commercialization of the Results of Federally Funded Research in Your Community* (Ottawa: Ministry of State, Science and Technology Canada, 1986. 114 p.). This handbook was compiled to aid those who might be involved in the diffusion of technology in order to identify and exploit technology transfer opportunities through federal research laboratories. A second promotional publication was entitled *The Government of Canada's Support for Technology Development: A Summary of Federal Programs and Incentives, 1987* (Ottawa: Ministry of State, Science and Technology Canada, 1987. 40 p.). The summaries included tax incentives, financial assistance for research and development, training activities, technology and technical information support, and energy and communications programs, provided by the federal government to assist and encourage technological development in the country. A third publication entitled *Strengthening the Private Sector/University Research Partnership: The Matching Policy Rules* (Ottawa: Supply and Services Canada, 1987. 18 p.) outlined how the federal government would provide additional funds to each research council to match private-sector contributions for university research. A fourth publication entitled *Innovation: The Canadian Strategy for Science and Technology* (Ottawa: Ministry of State, Science and Technology Canada, 1987. 23 p.), which was part of an information kit, described the strategies and initiatives of the new government 'which will strengthen our scientific and technological resources, as well as improve Canada's ability to compete within the global marketplace, in the area of science and technology.' Two other related publications, describing the new strategies and initiatives, were entitled *Innovation: The Federal Microelectronics Strategy* (Ottawa: Ministry of State, Science and Technology Canada, 1987. 5 p.), and *Innovation: Canadian Manufacturing Advanced Technology Exchange (CAN-MATE)* (Ottawa: Ministry of State, Science and Technology Canada, 1987. 3 p.). The former described a strategy designed 'to stimulate the development and application of microelectronics throughout Canada,' and the latter described how 'CAN-MATE could help manufacturers maximize the benefits to be obtained by adapting existing manufacturing technologies and acquiring new technologies.'

Finally, the federal government also hosted a conference on science and technology, in Toronto, which was briefly described in the publication entitled *Na-*

tional Conference on Technology and Innovation: January 13–15, 1988, Toronto, Ontario (Ottawa: Conference Secretariat, 1987. 8 p.). These new initiatives, which emphasized industrial innovation, helped generate one unpublished internal report as well as several promotional publications and pamphlets.

The creation of the National Advisory Board on Science and Technology also saw that body generate a series of publications which emphasized strategy rather than policy. Their first statement issued to Prime Minister Brian Mulroney was entitled *Statement on Competitiveness: Report of the National Advisory Board on Science and Technology* (Ottawa: The Board, 1991. 15 p.). It pointed out that Canada's productivity was lagging behind that of all other G-7 nations because of the 'failure of our nation as a whole to adequately employ the tools of science and technology to drive innovation through the greater value-added and stronger productivity growth.' The board also defined science and technology 'as the set of skills and tools that are the cutting edge of continued economic development of modern high-wage societies.' Their second statement, issued by the board's Committee on Federal Science and Technology Priorities, was entitled *Science and Technology Innovation and National Prosperity: The Need for Canada to Change Course: Report of the National Advisory Board on Science and Technology* (Ottawa: The Board, 1991. 38 p.). This report documented 'the rapidly widening gap between the technological fitness of Canadian industry and that of its industrial country competitors,' noting that Canada ranked 17th among 23 industrialized countries of the OECD (Organization for Economic Cooperation and Development) for gross expenditure on research and development. This low expenditure, to a great degree, was blamed on the branch plant status of many of Canada's manufacturing industries, but overall 'Canadians must become more adept at applying science and technology so as to create a continuous flow of innovation and productivity growth.'

After these opening statements, the National Advisory Board began to examine the spending and role of the federal government in science and technology. One publication was entitled *Spending Smarter: Report of the National Advisory Board on Science and Technology Priorities, Committee on Federal Science and Technology Priorities* (Ottawa: The Committee, 1993. 40 p.). The committee said that 'judging by the allocation of financial resources, S & T is clearly a priority for the federal government. However, there is no evidence of an explicit rationale for the distribution of S & T spending among departments.' The committee suggested that the 'development of a knowledge-thirst society,' by research and technology diffusion, and support of 'market-driven technology development' should be the priority areas for federal science and technology programming. Their second publication, entitled *Spending Smarter: Report of the National Advisory Board on Science and Technology, Committee on Federal Science and Technology Priorities:*

Phase II (Ottawa: The Committee, 1994. 67 p.), stated, after a review of federal organizations involved in science and technology, that these organizations, or departments, should set their priorities according to those mentioned in the first report. Other studies undertaken by the Advisory Board, until the change of government in 1993, involved the examination of competitiveness in some of Canada's resource industries as well as the role of women in science and technology. These reports were entitled *Competitiveness in the Canadian Mining and Forestry Industries* (Ottawa: The Board, 1993. 38 p.), and *Report on Women in Trades, Technology, Science, and Engineering: Report of the National Advisory Board on Science and Technology, Human Resources Committee* (Ottawa: The Committee, 1993. 135 p.).

H Science Studies and Science Policy Initiatives under the Chrétien Liberals

With the change of government back to the Liberal party in late 1993, the federal government conducted another extensive federal science and technology review that included an examination of federal S & T activities, along with broad consultations with scientists, academics, and industrialists. The approach to the review was documented by the Secretariat for Science and Technology Review in the publication entitled *Building a Federal Science and Technology Strategy* (Ottawa: Secretariat for Science and Technology Review, Industry Canada, 1994. 16 p.). The related data was contained in the publication entitled *Resource Book for Science and Technology Consultations: Volume I* (Ottawa: Secretariat for Science and Technology Review, Industry Canada, 1994. 41 p.).

As well, the National Advisory Board on Science and Technology was asked to undertake an independent assessment of Canada's current science and technology strategy, or policy. It documented its findings in the report entitled *Healthy, Wealthy and Wise: A Framework for an Integrated Federal Science and Technology: Report of the National Advisory Board on Science and Technology* (Ottawa: The Board, 1995. 106 p.). This said that 'thirty years of S & T policy reviews have provided a wealth of good ideas, but these have not been effectively implemented within an integrated S & T strategy.' More specifically, the board said that setting priorities in science and technology must become better focused at the government-wide and departmental levels and that policy should aim to increase the amount of research performed outside the government. As well, the board suggested that the government should improve the business climate so that private enterprise would seek out and commercialize the results of research. Finally, it suggested that the federal fiscal restraints that began in 1994, in order to eliminate federal deficits, and were causing dislocations in the science community, were 'an opportunity to re-evaluate and re-define our fundamental goals and priorities as a nation, as well as the role of the federal government in achieving these goals. We must establish a truly results-oriented action plan to prepare Canada for the 21st century.'

As an outcome of the federal science and technology review and the input of the National Advisory Board on Science and Technology, the federal government, through Industry Canada, did announce a new science and technology strategy in a publication entitled *Science and Technology for the New Century: A Federal Strategy* (Ottawa: Industry Canada, 1996. 43 p.) This broad policy paper set out seven operating principles that reflected advice received during the federal S & T review process of 1994. These were: (1) increasing the effectiveness of federally supported research by promoting the commercialization of technologies transferred to the private sector from federal departments and agencies, (2) capturing the benefits of partnership which is 'aimed at helping funders and performers in all sectors ensure their S & T efforts are both essential and complementary to those of others,' (3) emphasizing preventive approaches and sustainable development to bolster collaboration and coordination and increased cooperation and working together on specific projects by federal departments, (4) positioning Canada competitively within emerging international regulatory standards and intellectual property regimes, (5) building information networks, which are the infrastructure of the knowledge economy, to strengthen the dissemination and use of information, (6) extending science and technology linkages internationally, and (7) promoting a stronger science culture.

One year after issuing the above publication, Industry Canada released a publication entitled *Minding Our Future: A Report on Federal Science and Technology – 1997* (Ottawa: Industry Canada, 1997. 44 p.). This reported 'highlights of activities and policy directions to illustrate how the federal effort is being performed to meet current and future challenges.' After reviewing science and technology activities, the report stated that the 'federal strategy has been a positive force for change.'

IV Departmental Sources of the Publications

As observed from the above discussions, primarily about MOSST, and the scope and nature of its activities, the ministry's interests ranged over many areas of science and technology in the form of both internal and published reports.

The internal reports, often written for interdepartmental committees, can be identified only by examining the annual reports of the ministry, which were published from 1971 to 1988. These annual reports also listed the papers, task force reports, background papers, and annual statistical reports for the particular time period when MOSST held those responsibilities. As well, MOSST did issue a separate list of publications available to the end of 1984. It was entitled *Publications* (Ottawa: Ministry of State, Science and Technology Canada, 1984. 7 p.). Because it listed only those items available at that time, it is not a complete cumulated list. Since 1984, however, with the absorption of MOSST into what is now Industry Canada, there have been no publication lists for the latter department, or for the Advisory Council on Science and Technology.

Chapter 1.2

The Natural Sciences and Engineering Research Council

I Historical Background

The Natural Sciences and Engineering Research Council (NSERC) was created after the federal government's decision in 1976 to reorganize its scholarship and research grant activities. Besides NSERC, the Medical Research Council (MRC) and the Social Sciences and Humanities Research Council (SSHRC) were formed at that time. NSERC was incorporated on 1 May 1978, and took over the management of the Program of Scholarships and Grants in Aid of Research that had been the responsibility of the National Research Council (NRC). Its goal was 'to promote and support the development and maintenance of research and the provision of highly qualified manpower in the Natural Sciences and Engineering,' according to the Natural Sciences and Engineering Research Council, *Notes to the Statement of Expenditures, March 31, 1979*. Since its inception, NSERC's history has been like a roller-coaster ride in terms of financing its activities.

Although the council began its work with an air of optimism, by 1985 the atmosphere for research became pessimistic with the change of government and reduced funding to the council. This pessimism reached its lowest level in 1986, when the council was faced with the realities and choices of reduced federal funding. The reason for the earlier optimism came from the Trudeau government, which announced its intention to increase national expenditures on R&D from 0.9% to 1.5% of the Gross National Product (GNP) within a five-year period ending in 1985. This announcement also produced hope in the Canadian scientific research community that the government, or nation, finally recognized the need for a greater national scientific research and development effort. The later pessimism occurred in January, 1986, when the NSERC budget of $289.1 million was deemed not sufficient to accommodate built-in growth of its established programs. That year, there was a reduction of 10% in new scholarships and fellowships and a cut of 4% in funds available for operating grants (the heart of NSERC's funding activities). As well, there were no allocations for new rewards for designated areas of the strategic grants program (targeted research and the general research grants to university presidents were reduced by 37%). The latter are a percentage of the total individual grants awarded to individual universities. However, in February, 1986, the federal government stabilized base funding at $311.6 million for a five-year period. The government also provided some one-time-only money.

These early ups and downs saw NSERC produce two published five-year plans and a futures projection, described in the next section, as well as a major internal program review, which was triggered by the budget restrictions of 1986/87. This program review confirmed the priority of the operating grants program, and NSERC's funding programs were reorganized into four components. They were: (1) the research base – covering operating grants (including equipment) that provide the foundation upon which the Canadian research effort in science and engineering is built, (2) scholarships and fellowships – covering university undergraduate, postgraduate, and postdoctoral activities, which are aimed at developing and securing Canada's human resource needs in the sciences, (3) targeted research – covering the strategic grants for designated research areas such as energy, oceans, environmental toxicology, food/agriculture, communications/computers, and industrial materials/processes (included in this program were the industrially oriented programs such as the industrial research fellowships, the undergraduate research awards tenable in industry, and the university-industry programs, which provide for effective university-industry interactions via government-funded industrial research chairs in Canadian universities), and (4) general research supports, which included the general research grant to university presidents, grants in aid of conferences, and other specialized grants.

In the mid-1980s, the federal government also announced new science and technology funding initiatives through the Ministry of State for Science and Technology (MOSST), mentioned in the previous chapter. Again, these initiatives were: (1) the Research Partnership Program, which supported research projects conducted jointly by a university, and (2) the Networks of Centres of Excellence program 'to promote leading-edge fundamental and long-term applied research and to provide an opportunity for the nation's best researchers to work together in support of Canada's long-term industrial competitiveness' (a conditional program that became permanently funded in 1997, was jointly administered by NSERC, the Medical Research Council, the Social Sciences and Humanities Research Council, and Industry Canada, and was described in the publication entitled *Powerful Partnerships* [Ottawa: Natural Sciences and Engineering Research Council, 1997. 10 p.]), and (3) the Matching Funds Policy. The latter was started in

1987/88 and ran until 1990/91, and allowed private-sector contributions to universities to be matched by the federal government for an annual increase of 6% to NSERC's base budget.

With these newer approaches to funding, NSERC, on its tenth anniversary, was able to take a more optimistic view of its mandate. Over its 10–year existence, new programs had been launched, the community it served had grown, and its discipline mix had undergone changes. As well, funding had increased by 73%, in net terms, from $194.6 million (1987 dollars) in 1978/79, to $338.1 million in 1987/88, according to the Natural Sciences and Engineering Research Council of Canada, *Report of the President, 1987–8*, p. 7. But, by the mid-1990s, the council received another blow in its cycle of optimism and pessimism. In the 1994 budget, the federal government announced reductions to NSERC's budget of 5%, 9.8%, and 14% over a three-year period. NSERC's initial reaction was to begin the cancellation of non-core programs such as general research grants, conference grants, etc. However, optimism returned once again in 1998, the twentieth year of NSERC's existence. In the 1998 federal budget, NSERC received $70.8 million more than had been expected and the cuts made since 1994 were cancelled or reversed. As well, the jointly administered Networks of Centres of Excellence received permanent funding in 1997, set at $47.4 million, and the federal government also announced the formation of three new Networks of Centres of Excellence. Furthermore, the government said that it would invest $182 million in the centres of excellence program over the next four years. At the same time, the federal government announced funding of $800 million for a Canada Foundation for Innovation, which was their 'novel and massive response to the pressing need for a renewal of research infrastructure in universities and research hospitals,' according to T. Brzustowski, the president of NSERC (*CONTACT* 22 [Spring, 1997], 4). Finally, the federal government announced in its 1999 budget that NSERC would receive an additional $25 million per year to its base budget. This meant that in 1999–2000, NSERC's budget would be $538.5 million.

II Publications

Because NSERC is a granting council, its publishing activities, since its inception, reflect that function. Besides its plans and projections, NSERC produces information-type publications about the council and its activities, as well as annual guides to its awards programs, and annual lists of awards granted. These are the public documents. From time to time, NSERC has also found it necessary to create task forces to study and report on problems encountered in its programs. These reports have been mentioned in the annual *Report of the President* (Ottawa: Natural Sciences and Engineering Research Council Canada, 1978/79–1991/92,

14 v.), now succeeded by *Highlights* (Natural Sciences and Engineering Research Council Canada, 1993–), but not listed anywhere for public purchase. The Report contains appendices which are a source for its funding statistics, and committee lists and memberships. NSERC also publishes a newsletter entitled *CONTACT* (Ottawa: Natural Sciences and Engineering Research Council, v. 1, 1976–). It was initially published by the National Research Council. Time-sensitive news now appears at its website <<http://www.nserc.ca>>.

III NSERC Activities

A The Long-Range Plans and Objectives

Because NSERC took over the existing scholarship and grants program from the National Research Council (NRC), the council was able to start immediately on long-term planning with a view to defining its role in supporting Canadian research and development. This was important, because cabinet requested NSERC to prepare a five-year plan covering the period 1979–80 to 1984–85, and submit it for approval. At the end of that period, NSERC also prepared a second five-year plan, and after that second five-year plan, NSERC looked 10 years into the future and produced a new futuristic strategic plan to carry the agency to the year 2000. Since then, however, NSERC has realized that the pace of change in its environment requires continuous strategy planning.

The first five-year plan, which was approved by cabinet in November, 1979, was entitled *A Five-Year Plan for the Programs of the Natural Sciences and Engineering Research Council* (Ottawa: Natural Sciences and Engineering Research Council Canada, 1979. 102 p.). In this, the Council focused on seven areas of concern. They were: (1) to improve the Canadian research environment in order to help overcome the country's shortage of highly qualified research manpower, (2) to correct the growing obsolescence of university laboratory equipment, (3) to develop a wide-ranging program of scholarships and fellowships in order to increase the national research and development effort and to overcome the problem of reduced research manpower that would come in the late 1980s with the expected volume of university teaching retirements, (4) have a five-year equipment refurbishing program that would encourage Canadian manufacturing companies into the high-technology field of instrumentation, (5) to expand targeted research on defined areas of national concern such as communications, energy, environmental toxicology, food/agriculture, and oceans, (6) to improve the grant programs in aid of the 20 areas of discipline research supported by the council, and (7) to improve the management of programs provided that staff costs did not go above 2% of the total budget.

Over the five-year period of the plan, the council thought that it could achieve its objectives, because the government did commit itself to increased funding for NSERC. For example, in 1980–81, the government announced an increase of $40 million for NSERC. By 1982–83, halfway through the plan, the council was able to say that the plan 'consists of a broadly based program that supports and funds all other present and future research and development activities in Canada, by assisting in the production of highly qualified researchers for employment by industry, government, and the universities themselves, and by fostering the production of new knowledge and advanced technology for use by industry and government,' according to the Natural Sciences and Engineering Research Council, *Report of the President, 1982–83*, p. 3.

By the end of the five-year program, the Task Force on Federal Policies and Programs for Technology Development (the Wright Report mentioned in Chapter 1.1), of the Ministry of State for Science and Technology (MOSST), confirmed NSERC's effectiveness, although NSERC, itself, said that even with government funding and awarding of $1.2 billion over the five-year period, this funding still fell short of levels anticipated by the plan. At that time, NSERC was also encountering problems brought on by reduced funding of the universities and university charges against research programs, a downturn in the economy, and a period of high inflationary pressures on awards. However, NSERC was pleased that the awards were up again to the level of 1980 funding in terms of real dollars.

At the end of the first five years, NSERC also reviewed and updated its research manpower demand study (done in 1979 and incorporated into the five-year plan) as background work for preparing a second five-year plan. This review was issued in May, 1985, under the title *Research Talent in the Natural Sciences and Engineering: Supply and Demand Projections to 1990* (Ottawa: Natural Sciences and Engineering Research Council Canada, 1985. 104 p.). This report presented conditional forecasts of the research talent required for the government's economic growth and research and development investment targets. According to the model simulations, the forecast supply of research talent (postgraduate only) 'would be inadequate to meet the requirements, even with a fairly modest growth in research and development expenditures ...' and 'there is relatively little scope for increasing the supply of postgraduates over a time span as short as the projection period to 1990.' The second five-year plan was issued the same year and entitled *Completing the Bridge to the 90's: A Second Five-Year Plan for the Programs of the Natural Sciences and Engineering Research Council* (Ottawa: Natural Sciences and Engineering Research Council Canada, 1985. 2 v.)

The second five-year plan had six major objectives and was a natural continuation of the programs of the first five years. These objectives were: (1) to expand the universities' knowledge-producing and research-training capacity, (2) to attract more young Canadians into the graduate schools of Canada, (3) to equip the university laboratories with state-of-the-art facilities and a much improved level of research infrastructure, (4) to fund our existing research facility at levels which more closely reflect the high cost of frontier science, (5) to forge closer links between the university and industrial research communities, and (6) to provide the critical mass and concentration of effort so essential for a competitive position in many of the rapidly expanding areas of research and technology.

The obstacles to the plan were seen to be the decreased research funding from the universities' own resources, along with the emergence of research challenges, including equipment that was extremely costly to acquire, and increased emphasis on teaching and teaching loads at the universities owing to increased undergraduate enrolments. Also, there was an aging professoriate. On the other hand, the opportunities seen were the greater awareness by industry and government of the vital importance of fundamental research, a greater awareness by industry and the universities of their interdependence and the need to collaborate, and the exciting new fields of research such as biotechnology and artificial intelligence to pursue.

In reality, NSERC's optimism for the next five years for research and development in Canada was dashed by a change of government. In 1986, its budget was reduced by 10%, which affected all programs. (The budget in fact did get stabilized at the previous year's level of $311.6 million as previously mentioned, with a one-time increase for that year only of $13 million. After that, the government said it would match up to 6% of the previous year's budget, provided that the equivalent amount came from the private sector – the Matching Funds Policy also mentioned above.) Hence, the government's reply to the second five-year plan was a great disappointment for NSERC. Faced with this, NSERC had to again review and adjust its activities. This review was outlined in the publication entitled *Ten Years to 2000: A Strategy Document* (Ottawa: Natural Sciences and Engineering Research Council, 1989. 24 p.)

It was a 10-year projection in which NSERC stated that 'high-technology, knowledge-based industries are playing an increasingly large part in the overall productivity and economic performance of nations.' Because of these trends, strength in science and engineering was the 'crucial competitive edge in the marketplace of world trade' at a time when there was a real decline in the 18–24-year-old university entrance age group and a levelling off of graduate enrolment in the natural sciences and engineering in Canada. This demographic trend was a challenge for NSERC to make sure that its various program areas were designed to contribute to the building of human resource strength in science and engineering. This could be done by providing training niches, and employment, in a variety of disciplines for graduate students and postdoctoral fellows, either as individuals or in teams, at universities in all regions of Canada.

By 1994, however, NSERC published another strategy document, entitled *Partnerships in Knowledge: Maximizing the Investment in University Research* (Ottawa: The Council, 1994. 12 p.). The then-president of NSERC, P. Morand, said that a new strategy was necessary because 'all government spending is under scrutiny which makes it imperative that council have a clear direction and rationale for its activities ... in addition, the way we carry out research is changing, as is the integration of our ideas and skills into the economy and society.'

Finally, in late 1999, NSERC again produced a new document, entitled *NSERC Strategy, 1999–2000* (Ottawa: The Council, 1999. 9 p.). In it the president of NSERC, T. Brzustowski, said, 'The environment in which the council operates is changing so rapidly that updating NSERC's strategy must become a continuing process.' This latest strategy emphasized the changing attitudes toward the importance of research, a new investment climate, and emergency challenges.

B Research Scholarship and Fellowship Training Programs

As noted above, one of the objectives of the first five-year plan was to increase the supply of highly trained researchers in the natural sciences and engineering, in light of the government's intention to shift 1.5% of the GNP into research and development (R&D). As well, NSERC was concerned about the 33% decline in doctoral enrolments from 1971 to 1979 and the critical shortages of research manpower that would face the country in the late 1980s because of an aging and retiring research professoriate in the universities. Although NSERC inherited the scholarship aspect of the NRC program of Scholarships and Grants in Aid of Research, the council appointed a task force to advise on the research manpower initiatives and objectives of the five-year plan. This unpublished report of the task force emphasized the inadequacy of trained personnel and programs to achieve the R&D goals and resulted in the implementation of a series of manpower scholarship programs through the period of the first five-year plan. In all, there were three ranges of scholarships available to cover the undergraduate, postgraduate, and postdoctoral levels in the universities and in industry. Over time, the basic scholarship program has remained in place, with refinements undertaken after timely reviews.

At the undergraduate level there was the program of Undergraduate Student Research Awards designed to whet the appetite of undergraduates for careers in research. This was, in effect, a summer program that applied to both universities and industry. Today, however, this undergraduate-level program is recognized as the Undergraduate Student Research Awards in Small Universities, and the Undergraduate Student Research Awards in Industry. The undergraduate awards in small universities are intended to 'stimulate the interest of undergraduate students in research in the natural sciences or engineering and to encourage them to undertake graduate studies and a research career.' The awards are granted within the students' university and the universities eligible are designated by NSERC. The industrial undergraduate student awards 'provide financial support via the host company to undergraduate students wishing to spend a period of time in an industrial research setting ... and to encourage them to undertake graduate studies and a research career.' The awards must be held in the R&D facilities of eligible Canadian industrial organizations which have met NSERC's eligibility criteria.

At the postgraduate level, NSERC created, or expanded on, programs inherited from the National Research Council (NRC). They included the Postgraduate Scholarships, which included the Industrial Postgraduate Scholarships, and the 1967 Science and Engineering Scholarships.

Today, the postgraduate programs are divided into Postgraduate Scholarships, Postgraduate Supplements, and Industrial Postgraduate Scholarships. The Postgraduate Scholarships are 'available to students who will undertake a program of study and research leading to a higher degree in one of the fields supported by NSERC.' These scholarships are divided into two categories, one for the first and second, or second and third, years of postgraduate study, and another for the third and fourth, or fourth and fifth, years of postgraduate study. Each award is tenable for a maximum duration of four years. The Postgraduate Supplements consist of the British Chevening/Athlone-Vanier Engineering Fellowship; the Canadian Forest Service Graduate Supplements; the Canadian Meteorological and Oceanographic Society Scholarship Supplement; the Canadian Space Agency Postgraduate Scholarship Supplements; the DND Postgraduate Scholarship Foundation Supplements; the Gordon M. MacNabb Scholarship Foundation Supplements; the Industry Canada Fessendan Postgraduate Scholarships; the NRCan Earth Sciences Sector Postgraduate Scholarship Supplements; and the Systematics Research Graduate Supplements. All these supplements are over and above the value of the NSERC Postgraduate Scholarships and are awarded by the group offering the supplement. The Industrial Postgraduate Scholarships are 'intended to provide highly qualified science and engineering graduates with financial support allowing them to undertake advanced studies in co-operation with interested Canadian industry, and enabling them to gain research experience in an industrial setting.' With this scholarship, NSERC grants an allocation of industrial postgraduate scholarships to universities, which in turn are responsible for nominating candidates and administering the scholarships.

At the postdoctoral level, NSERC initially offered three types of awards. They were the Postdoctoral Fellowships, the University Research Fellowships, and the Industrial Research Fellowships. Now, except for the University Research Fellowships, NSERC also offers the prestigious E.W.R. Steacie Memorial Fellowships, and administers on behalf of government departments and agencies the Visiting Fellowships in Canadian Government Laboratories, the NATO Science Fellowships, the Japanese Science and Technology Agency Fellowships, and the Japan Society for the Promotion of Science Fellowships. Finally, NSERC has also created the University Faculty Awards program (formally the Women's Faculty Award), and the Research Orientation Associateships. The broad objective of these awards has been to give postdoctoral scholars the postdoctoral research experience necessary for permanent R&D employment.

The Postdoctoral Fellowships are 'intended to secure a supply of highly qualified Canadians with leading-edge scientific and research skills, for Canadian industry, government, and universities.' They are available to persons who hold, or expect to hold, a doctorate in one of the fields of research supported by NSERC, and who intend to engage in postdoctoral research in one of these fields.

The University Research Fellowships were specifically created in 1980 to counteract the lack of faculty positions in the financially strained universities which would aggravate projected shortages of trained research personnel in the late 1980s. By 1984, 356 of these latter awards were granted. It is interesting to note that these awards also encouraged Canadian researchers to return from abroad. However, in view of the increase in the number of academic PhD positions opening up at the end of 1989, NSERC decided to terminate the university research fellowship program after the 1989 competitions. By the end of the program, over 500 fellowships had been awarded that directly contributed to the creation of approximately 400 tenure-track positions in Canadian universities.

The Industrial Research Fellowships were specifically created to assist young professional scientists to pursue an industrial career, as well as encourage companies to expand their R&D efforts. This program continues today and 'provides via the host company, financial support to recent doctoral candidates engaged in industrial research.' These fellowships have a maximum duration of two years.

The E.W.R. Steacie Memorial Fellowships are considered to be the pinnacle of the postdoctoral awards. They are granted to outstanding young research scientists in order to allow them to devote two years totally to their research activities, to enhance their career development while remaining on staff as members of a Canadian university. Approximately five awards are made each year. In 1990/91, NSERC also created the Canada Gold Medal for Science and Engineering (now called the Gerhard Herzberg Canada Gold Medal for Science and Engineering) in 'recognition of sustained and outstanding contributions to Canadian research in the natural sciences and engineering.' The Visiting Fellowships in Canadian Government Laboratories provide 'promising young scientists and engineers with an opportunity to work with research groups or leaders in Canadian government laboratories and research institutions.' The NATO Science Fellowships, which are financed by the North Atlantic Treaty Organization, provide 'NSERC with an annual grant to finance a limited number of awards to encourage international exchanges of postdoctoral fellows in the natural sciences and engineering.' The Japanese Science and Technology Agency Awards offers fellowships 'to promising young researchers from overseas to allow them to conduct research at national laboratories and certain non-profit research institutes in Japan.' NSERC is responsible for selecting Canadian nominees for these fellowships. Finally, the Japan Society for the Promotion of Science Fellowships offers postdoctoral fellowships to foreign researchers to provide 'the opportunity to conduct research in Japanese universities.' Again, NSERC is responsible for selecting Canadian nominees.

In 1991, the council launched two new programs aimed at encouraging Canadian women to undertake a research career. These programs were the Women's Faculty Awards and the Research Orientation Associateships program. The Women's Faculty Award was aimed at increasing the number of women holding faculty positions in science and engineering. In 1998, this program was renamed the University Faculty Awards program and was tenable for up to five years. The Research Orientation Associateships are aimed at helping doctoral holders who interrupted their careers for family reasons to return to their research activities.

The annual guide entitled *NSERC Scholarships and Fellowships Guide* (Ottawa: Natural Sciences and Engineering Research Council Canada, 1979–) describes the various programs of scholarships and fellowships offered by the Natural Sciences and Engineering Research Council to undergraduate and graduate students and postdoctoral fellows. They are also described at the NSERC website <<http://www.nserc.ca>>. The results of these NSERC competitions are announced in the electronic publication entitled *[Year] Scholarships and Fellowships Competition Results* through the website. Previous printed publications, elaborating on the awards, have been the *University Research Fellowship Program: A Guide for Fellows and University Administrators* (Ottawa: Natural Sciences and Engineering Research Council Canada, 1986. 22 p.), and the *Visiting Fellowships in Canadian Government Laboratories* (Ottawa: Natural Sciences and Engineering Research Council, 1981–).

C Discipline-Based Research Program

The program of operating grants for free discipline research for individuals, or groups, inherited from NRC is the cornerstone of NSERC. In fact, it is the sustaining force for all university science research and scientific human resources training. Operating grants are awarded on the recommendations made by some 24 selection committees, constituted by discipline, and composed primarily of scientists with recognized research experience from Canadian universities, industry, and government laboratories. The program supports basic or applied research, which may be of direct or indirect interest to Canadian industry, as well as research areas that may be considered to be of national interest. However, the main emphasis has always been on people rather than projects and the grants are intended to cover the pursuit of new ideas that may arise out of funded ongoing research. Grants were listed in the publication entitled *List of Grants Awarded in Aid of Research* (Ottawa: Natural Sciences and Engineering Research Council, 1979–93. 15 v.). Competition results are now listed in the electronic publication entitled *[Year] Research Grants Competition Results* through the NSERC website <<http://www.nserc.ca>>. To assist interested researchers, NSERC publishes the *NSERC Researcher's Guide* (Ottawa: The Council, 1994–). This annual guide contains information for applicants, or holders, of NSERC awards. Since 1997, this has been available as an electronic publication through the NSERC website. As well, NSERC publishes *Your NSERC Award [date]* (Ottawa: The Council, 1981–).

From time to time, individuals, or ad hoc committees, have reviewed the operating grants and made adjustments because of changing trends in research. For example, in 1983, an NSERC visiting researcher, Dr R.J. Kavanagh, assessed the relevancy and adequacy of current NSERC programs in engineering. This report, which was not published but was mentioned in the *Report of the President, 1983–84*, contained 55 recommendations, including the creation of an Advisory Committee on Engineering to foster improved communications between NSERC and the engineering community. In 1987, a Task Force on Allocations undertook an internal unpublished study of budget allocations to the various disciplines and recommended increases for engineering, computing, earth sciences, and population biology. Later, in 1991/92, a Nuclear and Particle Physics Advisory Panel was established 'to contact and report on a study of scientific opportunities and priorities in Canadian sub-atomic physics research, and to recommend a long-range plan to provide a framework for co-ordinated advancement of Canada's subatomic physics research programs over the next decade.' This review did result in the establishment of the Subatomic Physics Research Project Grants. The most recent reallocations exercise undertaken by NSERC occurred in 1998. The results were reported in the publication entitled *NSERC Report on the 1998 Reallocations Exercise* (Ottawa: National Sciences and Engineering Research Council of Canada, 1998. 30 p.), which also stated that reallocation exercises would now be undertaken every four years.

Because discipline research is the cornerstone of NSERC, the latter has been very sensitive to external factors affecting discipline research. By 1983, NSERC was concerned about the real erosion of the awards because of university charges to researchers that had previously been covered (e.g., shop charges). As well, increased undergraduate enrolments at the universities, coupled with their own shrunken funding, meant that there was more focus on teaching than on research. NSERC's concern was elaborated upon in a *Brief to the Royal Commission on the Economic Union and Development Prospects for Canada* (Ottawa: Sciences and Engineering Resarch Council Canada, 1983. 24 p.). This stated that these conditions would erode the educational and research base, and, in the long run, Canada's industrial competitiveness, particularly in the fields of high technology. These conditions, coupled with a change in government and the latter's budget decisions in 1986 which effectively reduced funding to NSERC for the foreseeable future, were a blow to discipline research funding. The council had to cut existing operating grants in 1986/87 by 4% in order to provide adequate funding for new applicants. The latter were funded again at a higher level in 1987/88, but the council said that the overall program of discipline research was underfunded by $50 million. However, with new approaches such as the Collaboration Project Grants program, mentioned below, and the overhauling of the process of allocating funds among disciplines (now, disciplines have to compete for funding increments in much the same way as individual researchers compete for individual grants), NSERC was accommodating itself to lower real government funding, which did not change again until the late 1990s.

D Targeted Research

The Strategic Grants Program of targeted research was launched by the National Research Council (NRC) in 1977. The objective of the program was to target funds to initiate or accelerate research projects of national interest. Areas of national interest were deemed to be regional science balance, national problem areas (energy, environmental toxicology, oceans, food/agriculture, biotechnology, communications, and industrial materials and processes), and interdisciplinary research. As well, the Project Research Applicable in Industry (PRAI) and the Senior Industrial Fellowships programs fell under targeted research, as they focused upon developing the university-industry interface, which was an NSERC objective.

As part of its review of targeted research in 1989, NSERC added a university-government component to the program, and this expanded grouping was renamed Research Partnerships. Today, the Research Partnerships Program of targeted research consists of Strategic Projects, Research Networks, University-Industry Projects, and Research Partnership Agreements with Canadian Government Departments and Agencies. The Strategic Projects component has targeted biotechnologies, energy efficient technologies, environmental technologies, information technologies, manufacturing and processing technologies, and materials technologies, as research areas 'targeted for acceleration by the council.' The Research Networks Program 'funds large-scale, complex research proposals that involve multi-sectorial collaborations on a common research theme and that demonstrate the added advantages of a networked approach.' The University-Industry Projects 'supports partnerships that promote high-quality university research of economic or industrial importance,' and includes within that element the Collaborative Research and Development Grants, Industrial Oriented Research Grants, Industrial Research Chairs, New Faculty Support Grants, and Chairs in the Management of Technological Change. The Research Partnership Agreements with Canadian Government Departments and Agencies consist of research partnerships administered by NSERC involving university, industry, and federal departments or agencies. Currently, there are partnership arrangements with the National Research Council, Agriculture and Agri-Food Canada, the Canadian Forest Service, and the Canadian Space Agency.

Other related targeted research programs are the Collaborative Health Research Projects and the Genomics Projects, both announced in 1999, and the International Opportunity Fund. The Collaborative Health Research Projects came out of the federal government's creation of the Canadian Institute of Health Research and the need to carry out special health-related research. The Genomic Projects program was created to provide support for genomics research that falls within NSERC's mandate. The International Opportunity Fund was created to allow Canadian scientists to participate in international research projects that are of potential benefit to Canada.

These programs have been described in the strategy document (previously mentioned) entitled *Partnerships in Knowledge: Maximizing the Investment in University Research* (Ottawa: Natural Sciences and Engineering Research Council, 1994. 12 p.), and, again, in the NSERC Researcher's Guide (Ottawa: The Council, 1994–). The latter is also available through the website <<http://www.nserc.ca>>. Applications are reviewed by NSERC selection panels or committees and those receiving grants are listed in the electronic publication entitled *[Year] Research Grants Competition Results* through the NSERC website <<http://www.nserc.ca>>.

Perhaps the largest targeted research development, however, goes back to the federal government's announcement in June, 1988, of a $240 million program to promote fundamental and long-term applied research by means of Networks of Centres of Excellence. This program, in partnership with Industry Canada, became administered by an inter-council program directorate made up of NSERC, the Medical Research Council, and the Social Sciences and Humanities Research Council (SSHRC). In 1989/90, 158 applications, involving 4,000 researchers, had been received for the program. After peer review, and recommendations to the

Ministry of State for Science and Technology, and to Cabinet, 15 Networks of Centres of Excellence across the country were approved. Some of the networks approved were the Canadian Network for Space Research centred at the University of Calgary, the Institute for Robotics and Intelligent Systems centred at PRECARN Associates, and the Canadian Aging Research Network (CARNET) centred at the University of Toronto. The program and the networks have been described in the publication entitled *Powerful Partnerships: Networks of Centres of Excellence* (Ottawa: Natural Sciences and Engineering Research Council, 1992. 16 p.). The network program was renewed by the federal government in late 1993 for an additional four years. In 1997, the program became permanently funded at $47 million. NSERC's president, T. Brzustowski, said the reason for this was the fact that 'the NCEs have turned out to be a major success; they do excellent work and they are big enough to have an impact on a national scale. The fourteen current NCEs involve about 1,000 researchers, 1,400 graduate students, and close to 500 postdoctoral fellows in 48 universities, with partners from more than 400 participating companies, 37 hospitals, and 76 government departments, ministries and agencies across the country' (*CONTACT*, v. 22 [Spring], 1997, pp. 1, 4). The details concerning applying as a network and its subsequent administration have been described in two print publications. One is entitled *Competition Announcement [1997]* (Ottawa: Networks of Centres of Excellence, 1997. 8 p.). The other is entitled *Program Guide* (Ottawa: Networks of Centres of Excellence, 1997. 22 p. This information is also available through their website <<http://www.nce.gc.ca>>.

E Equipment

Another aim of NSERC's first Five-Year Plan was to reverse the growing obsolescence of university laboratory equipment. The need for up-to-date or state-of-the-art equipment, along with funds to properly maintain that equipment, was directly linked to the human resources training programs. There was no point in producing researchers trained on obsolete equipment and unable to function in up-to-date research environments. In 1979, NSERC appointed two committees to examine the equipment needs. One was a general task force to look at the overall problem, and the other was specific to look at costly physics and astronomy equipment such as accelerators for nuclear physics. The general task force was asked by cabinet to inventory existing research equipment and give an assessment on future needs. Their unpublished report identified the magnitude of the equipment replacement problem for Canadian research to be far greater than had been estimated by NSERC. Accordingly, the council readjusted funding levels for equipment grants so that they exceeded funds allotted to the manpower training programs during the first two years of the Five-Year Plan. By 1984, 40% of NSERC's budget went for equipment replacement.

The need to allot funds for the maintenance of new equipment was also addressed by NSERC. In 1982–83, NSERC began awarding Infrastructure Support Grants to cover the costs of technicians to operate and service newly acquired equipment. It also assembled a Task Force on Research Infrastructure Needs, particularly for facilities serving groups of researchers. This was a growing problem because the universities began to charge for services to their own research communities as the former's own funding from the provinces was reduced.

The 1985 unpublished report of the task force reinforced the need for infrastructure funding. In fact, as infrastructure grants rose, NSERC's equipment grants had to be reduced to about 20% of NSERC's budget. This aggravated the council's inability to satisfy all legitimate equipment requests, and the aggravation was increased by the overall reduction of government funding in 1986. However, in 1989, the council informed the university community in an article entitled *Choices and Consequences: Reflections on the 1989 Grants Competition* (*CONTACT*, v. 14 [Summer], 1989. pp. 1–3) of its intension to gradually raise its investment in equipment grants to 10% of its program budget, or $40 million per year, starting in 1990/91. By 1994, equipment grants did reach $41 million, but with budget cuts in the mid-1990s, they were reduced to $25 million and only began climbing again in 1997/98 to $28 million.

Today, equipment grants, including maintenance and operation of the facility, consist of three categories. They are: (1) Equipment Grants – $7,001 to $150,000, (2) Major Equipment Grants – $150,001 to $325,000, and (3) Major Installation Grants – more than $325,000. As well, there is a special category of Major Facilities Access Grants to support researchers' access to major regional or national facilities. As with targeted research, the equipment grants have been described in the print *NSERC Researcher's Guide* (Ottawa: The Council, 1994–), and the information is available through the website <<http://www.nserc.ca>>.

IV Council Sources of the Publications

To date, NSERC has not produced any separate retrospective listing of its publications. Current publications, available back to 1994, are now listed for full-text display at their website entitled *Publications* <<http://www.nserc.ca/publicat.htm>>.

Chapter 1.3

The Science Council of Canada

I Historical Background

The Science Council of Canada was created in 1966 with the passing of Science Council of Canada Act by the federal government. Writing in 1975, the chair of the Science Council, R. Gaudry, said that '1966 was a year of considerable expansion in science in Canada ...' and that, 'viewed in retrospect, the mid-sixties were a golden era for science and technology, an era of enthusiasm, of rising budgets and of rising hopes.' In general, the duties of the council, as set out in the Act, were 'to assess in a comprehensive manner Canada's scientific and technological resources, requirements and potentialities and to make recommendations thereon to the Minister.' It was also expected to make reports and recommendations on the adequacy of the scientific and technological research and development being carried on in Canada, the priorities that should be assigned to research, the effective development and utilization of scientific and technological manpower, long-term planning for scientific and technological research development, the responsibilities of departments and agencies of the federal government in relationship to those of universities, private companies, and other organizations in furthering science and technology, the statistical and other information needed to provide a basis for the formulation of government policy on science and technology, and the best means of developing and exchanging information between the council and other groups. It is interesting to note that 20 years later, the Science Council viewed itself as a national think-tank with primary responsibilities to 'identify and analyze public policy issues in science and technology ... recommend policy directions ... alert Canadians to the impact of science and technology on their lives' and 'stimulate discussion of science and technology policy among governments, industry, and academic institutions.'

The council, in effect, was a crown corporation with a chairman, a vice-chairman, and 28 members, all with a specialized interest in science and technology. All members were drawn from the academic community and private industry, and specific study committees drew some of their membership from the scientific community at large. Initially, the Science Council also made use of the staff of the Science Secretariat, which had been established within the Privy Council Office in 1964, but later, before government cutbacks, had a staff of just over 60 people, until the council was disbanded in early 1992 by the government of the day. In theory, the National Advisory Board on Science and Technology, created by the prime minister in the mid-1980s, became the new advisor to the government on matters pertaining to science and technology.

Over the 25 years of its existence, the council did carry out its mandate through study committees, initially documenting the state of science and technology, including policy, in Canada and trying to unravel the connections between science and technology, R&D, and the hindrances to innovation in Canadian industry. Early on, it also examined the problems of urbanization and population growth in Canada, and in its mid-life moved on to environmental studies, the north, and the energy crisis of the 1970s. In these areas, the council tended to be driven by the political interests of the day and gave them some scientific focus. But later, the council examined the impact of computers, emerging technologies, and biotechnology on Canadians even before these developments were of general political interest or concern. All these areas of study are discussed in more detail below.

At the end of its 25 years of operation, just before it was disbanded, the last message from the chair, J.E. Halliwell, suggested that 'the Science Council has had, over its years, a significant, if intermittent and sometimes unrecognized impact on the perceptions of many of those people who influence public policy in Canada.' And the council was going to commence the writing of annual national reports that would 'assess the contributions that science and technology have made, are making, and should make toward meeting national objectives.' The first report was supposed to be published early in 1992, but the abrupt end of the council prevented this new series of publications from seeing the light of day. However, over the 25 years of its existence, the Science Council of Canada did publish a major body of works that documented Canadian science and technology and made a scientific input to related concerns, such as science policy, science and industry, urbanization, the environment, northern science, energy, computerization, and biotechnology.

II Publications

This publication output consisted of: (1) annual reports, including some annual reviews, (2) reports, including background studies, background papers on questions referred by the Minister of State for Science and Technology, statements of committees, council statements, discussion papers, manuscript reports, (3) workshop proceedings and brochures, (4) series publications, and newsletters.

A Annual Reports and Reviews

The annual reports (*Annual Report – Science Council of Canada.* Ottawa: The Council, 1967–91, 25 v.) are the source of the activities of the Science Council. Some of the reports, particularly those issued during the 1970s, contained an 'Annual Statement of the Chairman' (e.g., *Technological Sovereignty: A Strategy for Canada,* by J. Kates). The annual reviews, however, in contrast to the annual reports, were short-lived and only three were published (*Annual Review, Science Council of Canada.* Ottawa: The Council, 1980–82, 3 v.). They were an attempt at separating the annual message of the chair and the summaries of activities of the committee from the annual report into a distinctive publication.

B Reports, Background Studies, and Papers

The reports (e.g., *Winning in a World Economy: University-Industry Interaction and Economic Renewal in Canada.* Ottawa: Science Council of Canada, 84 p. 1988 Report 39) were the major policy statements which had the full sanction of the council. Technical material prepared during the course of research on a particular topic was released as background studies (e.g., Doarn, G.B. *The Peripheral Nature of Scientific and Technological Controversy in Federal Policy Formation,* 1981. 100 p. Background Study No. 46). These studies represented the views of their authors and not necessarily those of the Science Council. The background studies were also listed as special studies until the end of 1969. Background papers pertaining to questions referred by the minister were also issued from time to time (e.g., *The Optimum Size and Scope of the Canadian Space Program,* Ottawa: Science Council of Canada, 1985. 7 p.). From time to time, the council committees also issued their own statements pertaining to ongoing work (e.g., *Hard Times, Hard Choices: A Statement,* by the Industrial Policies Committee (Ottawa: Science Council of Canada, 1981. 99 p.). As well, the council itself issued statements (e.g., *A Growing Concern: Soil Degradation in Canada.* Science Council of Canada, 1986. 24 p.). The discussion papers (e.g., Buckley, P.J. *Ideas Alive: Theoretical Biology in Canada,* 1985. 30 p.) represented technical research undertaken for the council. The manuscript reports, on the other hand (e.g., Spence, J.A. *A Review of Northern European Aquaculture* (Ottawa: Science Council of Canada, 1983. 53 p.), were not considered official reports but were made available on a limited basis to provide information and elicit discussion among interested groups.

C Workshop Proceedings and Brochures

The workshop proceedings were published as a result of workshops held by the council (e.g., *National Workshop on the Status of Toxicology in Canada.* Ottawa: Science Council of Canada, 1985. 34 p.) to bring groups or people together to discuss topics of current interest to the council. Other distinct types of publication of the council were their brochures (e.g., *The Canadian Science Councillor Network*, 1985. 12 p.) which were published for wide distribution.

D Series Publications

The council also published some distinctive series publications such as the Issues in Canadian Science Policy. Three were published between 1974 and 1976 (e.g., *Issues I* [Ottawa: Science Council of Canada], 1974. 31 p.). The Perceptions series, of which only four were issued from 1975 to 1977, were related to their urbanization and population studies (e.g., Kelley, F. *Perceptions I: Population Growth and Urban Problems* [Ottawa: Science Council of Canada, 1975. 90 p.]). One other form of series publications was the council's newsletters. The first one was entitled *Agenda* (Ottawa: Science Council of Canada, v. 1, 1978–v. 5, 1982). These were published to inform the scientific reader and the general public about the emerging work of the council. This publication was replaced by *In Touch* (Ottawa: Science Council of Canada, v. 1, 1983–v. 9, 1992), which also kept people informed about the activities of the council.

III Activities of the Science Council

A Documentation of Scientific Activities in Canada

After the council's formation in 1966, a number of committees were created to undertake inventory studies that would 'take stock of activity in any desired segment of the scientific community at the present time, and would seek to identify current needs.' Others would be 'forward-looking ... enunciating broad outlines for the future development of science and technology in Canada. ...' In practice, these two types of studies tended to be mixtures, although one can discern some differences between the two. In all, there were 14 background studies written that inventoried the state of the basic sciences in Canada. As well, there were four related studies published that focused upon, or surveyed, the sciences and education and jobs in Canada, and eight forward-looking studies.

The first background study of the basic sciences to be published was written by J.H. Chapman and P.A. Forsyth and entitled *Upper Atmosphere and Space Pro-*

grams in Canada (Ottawa: Science Secretariat, Privy Council Office, 1967. 258 p. Science Council of Canada. Special Study No. 1). The study reviewed the various Canadian programs on space science, ranging from government and university to commercial activities and suggested that the 'prime objective for space technology in Canada is its application to telecommunications and survey problems ...' and that a national advisory committee in space research and development be established to reduce fragmentation in current upper atmosphere and space activities.

The second study, by a Study Group of the Canadian Association of Physicists, headed by D.C. Rose, was entitled *Physics in Canada: Survey and Outlook* (Ottawa: Science Secretariat, Privy Council Office, 1967. 385 p. Science Council of Canada. Special Study No. 2). The physics survey was broad in its interpretation of the field, which ranged from classical physics, astronomy, and astrophysics to meteorology and geophysics. The emphasis of the overall recommendations was 'that special consideration must be given to strenghtening the research effort in applied physics' and 'support for fields of research for which our terrain and location give us a natural advantage over other countries.'

The third study, by M.H. Appley and J.R. Rickwood, was entitled *Psychology in Canada* (Ottawa: Science Secretariat, Privy Council Office, 1967. 131 p. Science Council of Canada. Special Study No. 3). This study of psychology in Canada noted that psychologists with doctorates were 'mostly U.S. trained,' and American grants to psychologists in Canada provided higher levels of support. Accordingly, two of the recommendations of the study were 'the need to keep Canadian students in Canadian universities for their graduate training' and the 'need to compete with U.S. academic institutions in terms of improved research financing, reduced teaching loads, and other factors.'

The fourth study, by J.P. Bruce and D.E. Maasland, was entitled *Water Resources Research in Canada* (Ottawa: Science Secretariat, Privy Council Office, 1968. 169 p. Special Study No. 5). Here, the authors said that the three goals in which Canada's water resources were a key factor were the wise management of renewable resources, the continued growth of productivity and wealth of the country, and the development of the nation's agricultural productivity. However, in order to provide the foundations for water management, 'water resources research should be clearly identified as a field for major expansion of research work in Canada ... and the target for increased project expenditure ... should be 20% per year. ...'

The fifth study, by J.P. Tyas, was entitled *Scientific and Technical Information in Canada* (Ottawa: Science Secretariat, Privy Council Office, 1969. 2 Parts. Science Council of Canada. Special Study No. 8). The study was unique for its size and scope. Part I was the overview and summary, and Part II consisted of seven chapters, each a distinctive publication written by the sub-groups which surveyed and documented the different areas involved in scientific and technical information (STI) such as government departments and agencies, industry, universities, international organizations, and foreign countries. Included were techniques and sources, libraries, and the economics of STI.

The sixth study, by a Study Group of the Chemical Institute of Canada, was entitled *Chemistry and Chemical Engineering: A Survey of Research and Development in Canada* (Ottawa: Queen's Printer, 1969. 102 p. Science Council of Canada. Special Study No. 9). The study group noted that chemistry and related sciences 'now [account] for an estimated 45% of all expenditures on intramural industrial research and development in Canada, and therefore any planning of future research and development (R&D) effort must clearly give major consideration to this area.' Among the nine recommendations, and considered the most important, was the suggestion 'that a thorough investigation be made of the dynamics of the production and utilization of research personnel in Chemistry R&D in Canada.'

The seventh study, by B.W. Smallman and D.A. Chant, was entitled *Agricultural Science in Canada* (Ottawa: Queen's Printer, 1970. 148 p. Science Secretariat, Privy Council Office. Special Study No. 10). This pointed out that 'in its hundred years of development, agricultural science in Canada has not previously been the subject of a comprehensive study.' The authors also said that 'our report will be widely interpreted as advocating a swing back to an earlier emphasis on applied research.' Perhaps the most striking of its 25 recommendations was the suggestion that 'the federal government should reduce its responsibility for the performance of agricultural R&D in favour of increased performance by the provincial and private sectors.'

The eighth study, by R.A. Blais and C.H. Smith, was entitled *Earth Sciences Serving the Nation* (Ottawa: Information Canada, 1971. 363 p. Science Council of Canada. Special Study No. 13). After reviewing the current state of research and activities in the earth sciences, ranging from the universities to the mineral industries, the authors emphasized the need for teaching of the earth sciences in the schools as well as improved facilities for the earth sciences in the universities. Also, the report suggested that 'earth science activities should be clearly designated as an area for federal-provincial cooperation and co-ordination to meet the needs of the country.'

The ninth study, by J.H. Smith and G. Lessard, was entitled *Forest Resources Research in Canada* (Ottawa: Science Council of Canada, 1971. 204 p. Science Council of Canada. Special Study No. 14). After assessing the current status of forest resources research, the authors said that although Canada's forest resources influence, or are of direct concern to, most Canadians, 'research in how to use them wisely and more completely often appears to be too little and too late.' One of their main recommendations was that 'consideration should be given both federally and provincially to grouping all renewable resource research and management agencies in one department to provide a full range of services in agriculture, outdoor recreation, fisheries, forestry, watershed management, wildlife and wildlife grazing.'

The tenth study, by D.H. Pimlott and C.J. Kerswill, was entitled *Scientific Activities in Fisheries and Wildlife Resources* (Ottawa: Information Canada, 1971. 191 p. Science Council of Canada. Special Study No. 15). This report was considerably political and philosophical in its approach, with its beginning statement that 'fish and wildlife ... should not be sacrificed on the altar of short-term economic gain, nor should the diversity of species be destroyed in a questionable quest to feed 50 billion people in an overcrowded world which would be unfit to live in.'

The eleventh study, by R.W. Stewart and L.M. Dickie, was entitled *Ad Mare: Canada Looks to the Sea – A Study on Marine Science and Technology* (Ottawa: Science Council of Canada, 1971. 175 p. Science Council of Canada. Special Study No. 16). This attempted to forecast the levels of effort required to advance marine science and technology over the next decade in terms of people and funding. However, the authors had to create arbitrary definitions of what made up the field of marine science because 'marine science is a branch of the larger environmental science ...' and 'receives inputs from all other parts of the environment ...'

The twelfth study, by C.B. Lewis, was entitled *A Survey of Canadian Activity in Transportation R&D* (Ottawa: Science Council of Canada, 1971. 29 p. Science Council of Canada. Special Study No. 17). This published volume was the Summary Report, as the separate surveys and appendices were not published, but archived by the Science Council. The report was critical on transportation activity, for despite the size and importance of the Canadian transport sector, the author said there was no real national emphasis in transport teaching and research. Two-thirds of those surveyed thought that some form of a national transport institute was needed and the federal government should be 'initiator, co-ordinator and provider of funds for much of the research.'

The thirteenth study, by P.A. Larkin and W.J. Stephen, was entitled *From Formalin to Fortran: Some Facts and Futures about Basic Biology in Canada* (Ottawa: Information Canada, 1971. 79 p. Science Council of Canada. Special Study No. 18). This study pointed out that 'Canada is only a small contributor to the stream of biological research' and 'our great challenge is to use the world's knowledge wisely.' To do this, emphasis needed to be given 'to developing the strength of biology in the faculties of science in Canadian universities' through the catching-up in the traditional disciplines, along with new areas in biology, improved related areas such as communications, a strong national science library, and strong Canadian journals. Also, there needed to be improvement in the grant selection procedures (at that time, still under the National Research Council).

The fourteenth and last of these survey studies was by K.P. Beltzner and A.J. Coleman, and entitled *Mathematical Sciences in Canada* (Ottawa: Science Council of Canada, 1976. 369 p. Science Council of Canada. Special Study No. 37). This was a broad critical study, and the author said that 'its main thrust is towards the promotion of better liaison between the mathematicians and everybody else.' Also, 'the study contains no discussion of research publications nor of the special problems of women mathematicians.' Specific recommendations called for the improved teaching of mathematics in the elementary and secondary schools as well as a review of mathematics programs offered at the undergraduate, master's, and doctoral levels. The overall message, however, was that 'the mathematical community must be more outgoing, devoting more energy to communicating meaningfully to users of mathematics on their own terms.'

The related survey studies linked science and education, or science-educated people to jobs. One of these studies, by F. Kelly, was entitled *Prospects for Scientists and Engineers in Canada* (Ottawa: Science Council of Canada, 1971. 61 p. Science Council of Canada. Special Study No. 20). This inventoried the job prospects of scientists and engineers (at that time) and came to the conclusion that there was an oversupply of science and engineering graduates for the near future. The author suggested that these two groups should be utilized in secondary school teaching, industry, and the service industries. The background study by A.D. Boyd and A.C. Gross entitled *Education and Jobs: Career Patterns among Selected Canadian Science Graduates with International Comparisons* (Ottawa: Information Canada, 1973. 139 p. Science Council of Canada. Special Study No. 28) stated that we know little about how well scientists are utilized in various sectors and functions, but surveys undertaken did indicate that science graduates in the United States and the United Kingdom showed a closer match between undergraduate education and current work than was the case for their Canadian colleagues. Educational institutions were encouraged to be less rigid in their science study programs.

Another major study linking science and education by G.W. Orpwood and J.P. Souque was entitled *Science Education in Canadian Schools* (Ottawa: Science Council of Canada, 1984. 3v. Science Council of Canada. Background Study No. 52). This researched the state of Canadian science education and provided a 'database' for a nationwide series of conferences held to discuss questions raised by the study, and to explore future directions for science education in Canada. The outcome of this study was the report by the council entitled *Science for Every Student: Educating Canadians for Tomorrow's World* (Ottawa: Science Council of Canada, 1984. 85 p. Science Council of Canada. Report No. 36). This report provided a blueprint for an 'urgently needed renewal of science education in Canada based upon eight recommendations ranging from guaranteeing science education in every elementary school to evaluating students' progress toward all the goals of science education.' It was also an example of the forward-looking studies that sometimes emerged from the background studies. Overall, there were eight other reports issued by the council that incorporated background studies of the sciences in Canada into forward-looking reports.

The very first of these reports was entitled *A Space Program for Canada* (Ottawa: Queen's Printer, 1967. 31 p. Science Council of Canada. Report No. 1). This report was highly critical of the federal government for not having 'revealed' a coherent policy for the development and utilization of Canadian expertise on space projects. At that time, the government had agreed to a Telesat Corporation award of a contract to a US corporation to build a communications satellite for Canada in preference to using existing industrial capabilities in Canada. The second report, entitled *The Proposal for an Intense Neutron Generator: Initial Assessment and Recommendation* (Ottawa: Queen's Printer, 1967. 12 p. Science Council of Canada. Special Report No. 2), represented the federal government's efforts in the large science projects typical of that era, but it was cancelled because of opposition in the scientific community. The third report, entitled *A Major Program of Water Resources Research in Canada* (Ottawa: Queen's Printer, 1968. 37 p. Science Council of Canada. Report No. 3), encouraged the government to give priority to research on water resources, and saw the creation of the Canada Centre for Inland Waters as well as a National Advisory Committee on Water Resources Research. The fourth report, entitled *A Policy for Scientific and Technical Information Dissemination* (Ottawa: Science Council of Canada, 1969. 35 p. Science Council of Canada. Report No. 6), was pertinent because its recommendation that the National Research Council (NRC) be designated the coordinator of federal activities in the dissemination of scientific and technical information was accepted by the federal government. As a result, the NRC Library became the National Science Library of Canada. The fifth report, entitled *Canada, Science and the Oceans: A Major Program in Marine Science and Technology for Canada* (Ottawa: Information Canada, 1970. 37 p. Science Council of Canada. Report No. 10), encouraged the federal government toward the creation of a Department of the Environment, and according to the Science Council, 'constitutes a significant step towards the balanced development of our natural resources.' The sixth report, entitled *A Canadian STOL Air Transport System: A Major Program* (Ottawa: Information Canada, 1970. 33 p. Science Council of Canada. Report No. 11), promoted the development of this type of transport system and encouraged discussions between government and the industrial groups concerned in order to solve the problem of fragmentation of the industrial organizations involved in the transport industry. The seventh report, entitled *Lifelines: Some Policies for Basic Biology in Canada* (Ottawa: Information Canada, 1972. 73 p. Science Council of Canada. Report No. 17), stated that any basic research carried out in the life sciences should 'find its principal justification as an activity that supports some clearly identified practical mission such as health or food production or environmental management.' This encouragement of applied research was carried into the last of the forward-looking reports, entitled *Natural Resource Policy Issues in Canada* (Ottawa: Information Canada, 1973. 59 p. Science Council of Canada. Report No. 19). In this report, the Science Council clearly suggested a shift from basic to applied research in the field of natural resources.

B The Discussion of Science Policy

The 1960s saw a questioning of social and political organizations and their value systems, which were seen to be attached to a predominantly industrial and urban society. This re-examination also led to a worldwide reassessment of the role of science and technology in human affairs. In Canada, the Science Council, and others, thought that the Canadian scientific community could be strengthened and improved in a way so that science and technology could be applied to the achievement of national goals. The evidence of this interest in science and technology was shown by the formation of the Senate Special Committee on Science Policy, chaired by Senator Maurice Lamontagne (mentioned in chapter 1.1), the Cabinet Committee on Science Policy and Technology, and the appointment of a chief science advisor to the cabinet. This person was also the secretary of the Cabinet Committee on Science Policy and Technology.

It was in the background of renewed interest in science and technology that the Science Council issued its report, entitled *Towards a National Science Policy for Canada* (Ottawa: Queen's Printer, 1968. 56 p. Science Council of Canada. Report No. 4). The thrust of this broad report was to outline a series of goals, ranging from national prosperity and mental health and high life expectancy, to personal freedom, justice, and security for all in a 'united Canada,' around which a national science policy could evolve. At that time, the council believed that 'the ordinary Canadian was becoming less concerned with getting more things for himself and his family, and more concerned with the quality of life, the general welfare of others and especially of the poor in his community, his nation and throughout the world.' These economic and social goals, hopefully, would give an orientation to the R&D activities of the Canadian scientific community, and there would be a shift from government R&D to more R&D performed by industry and in the universities. To encourage these developments, the Science Council recommended the creation of high-level advisory committees, or boards, for each major government scientific department, or agency, which would act principally as bridges between government and the other sectors of the scientific community. This would provide the necessary links between programs in government and those in industry and the universities, and would help break down the 'solitudes,' or walls, that had traditionally separated these groups.

There were three special studies, and one report, which backed up these views. The first study, by R.W. Jackson and D.W. Henderson, was entitled *Background Studies in Science Policy: Projections of R&D Manpower and Expenditure* (Ottawa: Queen's Printer, 1969. 85 p. Science Council of Canada. Special Study No. 6). This report was made up of three sections. The first attempted to project the size of the engineering and scientific work force in Canada to the year 1978, based upon published data to 1966–67. The second section was an attempt to provide a research inflation factor based upon real costs over a ten-year period from 1954 to 1965. Here, the authors suggested a 6% per year figure for projections of research and development expenditures. The third section was a projection of Canadian total expenditures on research and development over the decade to 1978.

The second study, by J.B. Macdonald and L.P. Dugal, was entitled *The Role of the Federal Government in Support of Research in Canadian Universities* (Ottawa: Queen's Printer, 1969. 361 p. Science Council of Canada. Special Study No. 7). This was a broad study that examined the levels and conditions of financial support needed for research in Canadian universities, based on the purposes and objectives of the government and the universities that should be followed, the principles and policies that should be adopted, and the organizational structures needed. The study group made 77 recommendations, some of them rather startling. For example, recommendation no. 3 suggested that the National Research Council be divested of any research function 'so as to have as its sole responsibility the support of scientific and engineering research in universities and related institutions.' And recommendation no. 72 suggested reducing the collections mandate of the National Science Library in favour of developing 'in collaboration with Canadian university libraries[,] proposals for a co-operative acquisitions program.'

The third study, by P. Kruus, was entitled *Basic Research* (Ottawa: Information Canada, 1971. 73 p. Science Council of Canada. Special Study No. 21). This pointed out that, since basic research is neither national nor predictable, it is not possible to link it directly to national goals, except to acknowledge that it provides training for students in a 'truth-seeking atmosphere in the educational institutions,' or aid 'to achieving meaningful international communication, as it is based on objective, open information.' The report of the council was entitled *Policy Objectives for Basic Research in Canada* (Ottawa: Information Canada, 1972. 75 p. Science Council of Canada. Report No. 18). This report, unlike the previous study, said that 'we wish to reiterate that our support of the principle of greater orientation of basic research towards areas of special interest to Canada includes the clear understanding that a significant proportion of that research must continue to be conducted free from any influences external to the inner logic of science itself.'

These early discussions on science policy were given a higher profile with the establishment, in 1971, of the federal Ministry of State for Science and Technology, which was charged with developing a science policy for Canada. Also, the publication of the second volume of *A Science Policy for Canada: Report of the Senate Special Committee on Science Policy* (Ottawa: Queen's Printer, 1970–77. 4 v.), entitled *Targets and Strategies for the Seventies*, encouraged debate on science policy with its 45 specific recommendations to reform, in order to overcome, what was seen by the committee as the present state of 'accidental' science policy in Canada. This volume received critical comment from the then-chair of the Science Council, O.M. Solant, in a statement entitled *Issues in Canadian Science Policy: A Commentary on Some Aspects of Vol. 2 of the Report of the Senate Committee on Science Policy* (Ottawa: Science Council of Canada, 1972. 13 p.). In this, O.M. Solant pointed out that there was a 'profound divergence' between the two groups concerning the purpose of science and technology. 'The council, in considering the relation of science policy to the goals of the Canadian society, has concluded that science and technology should be considered as tools in the service of those goals, rather than as goals in themselves.'

The ongoing debate about science policy continued on and peaked in the mid-1970s with additional comments by the chair in the Seventh Annual Report, 1972–73, entitled *The Place of Science Policy in Canadian Governments* (Ottawa: Science Council of Canada, 1973, pp. 33–43), followed by a series of commissioned articles entitled *Issues in Canadian Science Policy*, and then a background study by R.W. Jackson entitled *Human Goals and Science Policy* (Ottawa: Science Council of Canada, 1976. 134 p. Science Council of Canada. Background Study No. 38). After the mid-1970s, there was a tendency to look not at science policy as such, but to focus more on technology and innovation, or technology policies, for the industrial sector directed toward international trade.

The chair's commentary for 1972–73 did state that there was a *de facto* science policy in Canada if one examined the actual expenditures and activities of the federal and provincial governments, since 'the act of spending public money can be legitimately be viewed as the implementation of a policy.' The chair then wondered if that *de facto* policy should be articulated into a written public policy and if the role of the council itself was to promote public debate and raise questions 'about man's use, misuse, or failure to make use of science and technology.' All these questions were further complicated by the jurisdictional and regional structure of the country and by the fact that the Science Council was viewed as a federal body not speaking for the provinces. The latter comment was reinforced in the special study by A.H. Wilson entitled *Research Councils in the Provinces: A Canadian Resource* (Ottawa: Information Canada, 1971. 115 p. Science Council of Canada. Special Study No. 19).

The first of the published series entitled *Issues in Science Policy* was released as *Issues 1* (Ottawa: Science Council of Canada, 1974. 31 p.). This publication contained a series of essays which touched on the contemporary search for more comprehensive forms of decision-making in scientific and technological areas. *Issues 2* (Ottawa: Science Council of Canada, 1976. 24 p.) contained discussions of science funding, science policy, and decision-making. *Issues 3* (Ottawa: Science Council of Canada, 1976. 24 p.) contained a single essay which emphasized northern development and the need for a balanced technology assessment system.

The final background study pertaining to science policy – *Human Goals and Science Policy* – traced how Canadian society worked in order to try to clarify the role of science policy. Attention was drawn to the role that bodies such as the Science Council could play by raising issues of social importance and how science and technology could enhance societal progress. In fact, the chair, R. Gaudry, thought that the discussion of science policy had shifted from an intellectual discussion to a political one, and the next 10 years 'may well see the moral and ethical dimensions of our support for scientific enquiry become the principal focus of public attention' with the challenges of population growth and food production, resource supplies, nuclear proliferation, and changes in the age structure of the population. In practice, however, the debate that emerged about Canada's position in changing trade patterns shifted the focus of science policy and technology to the innovative needs of Canadian industry in the international competitiveness for new markets.

This shift of view was seen in the last science policy statement made by the council entitled *Reaching for Tomorrow: Science and Technology Policy in Canada, 1991* (Ottawa: Science Council of Canada, 1992. 115 p.). This statement emphasized that nations must display science and technology in order to compete effectively in global markets. Nations unable to do this would stagnate. The last chair of the Science Council, J.E. Halliwell, said that there was now enough understanding of the structure of science and technology in Canada to renovate these institutions and policies, 'for it is the effective use of these science and technology infrastructures with managerial and organizational innovation that provides comparative advantage for a nation.' However, in Canada, there was a 'dissonance in our systems of research and development' because 'R&D investment by business is low and skewed to incremental technological advance rather than innovation. To add to this dissonance or mismatch, our attitudes, policies and programs in the vital area of technical training ... as well as in corporate workforce training ... are of the kind you might expect to find in a less developed country.'

C Science and Industry Studies

In November 1969 the Science Council created its Committee on Industrial Research and Innovation, succeeded by the Industrial Policies Committee in 1976, in order to explore and attempt to unravel and document the connection and problems which linked science and technology to the Canadian industrial environment. The committee wanted 'to enquire into matters associated with the generation and application of science and technology as they relate to the primary, secondary, and service sectors of Canadian industry.' In particular, the Science Council wanted to gain an understanding of the attitudes to science, technology, and innovation in Canada, the structure of science-based Canadian industries and its impediments and incentives to innovative activity, and the interaction of industry, universities, and government, in this innovative process. As well, the council wanted to study the problems associated with the effective utilization of highly qualified and other manpower in Canadian industry, and develop methods for technological forecasting in industry and for measuring the innovative process. This theme of innovation was constant throughout the council's history of science in industry studies, which later took a twist toward the necessity of technological sovereignty, and then coping with free trade initiatives by the government.

The first major industrial study, entitled *Innovation in a Cold Climate: The Dilemma of Canadian Manufacturing* (Ottawa: Information Canada, 1971. 49 p. Science Council of Canada. Report No. 15), was released at a time when Canada's technology-based manufacturing industries were stagnant or in decline. The council thought that these industries could prosper if innovation were made an integral part of their strategy, and the impediments to economic success in manufacturing, mainly the poor relationship between government and industry which hindered the creation of a coordinated industrial strategy, were removed. Canada could not afford to enter the post-industrial era without first developing a better industrial economy in order to have a role in the emerging global economy. The practical reason for innovation, however, was simply that new and improved goods would make Canada a more rewarding place to live and provide employment opportunities for an increasingly well-educated population. The related problem of underemployment in technology-based industries of Canadians graduating from post-secondary institutions was a concern to the Science Council, as expressed in its background study by F. Kelly entitled *Prospects for Scientists and Engineers in Canada* (Ottawa: Science Council of Canada, 1971. 61 p. Science Council of Canada. Special Study No. 20).

The theme of innovation in technology-based industries was picked up again, in 1977, by the then chair of the Science Council, J. Kates, in his theme of technological sovereignty. Kates thought that Canada must develop and control the technological capacity needed to support national sovereignty, particularly in those technologies essential to survival in the Canadian physical environment. Canada had a geographic advantage to be exploited in the fields of natural resource extraction and processing, and oceans and Arctic technologies. As well, Canada should further develop those technologies where the nation had historically shown excellence in fields such as nuclear energy, electrical transmission, and communications. According to the chair, the reasons that Canada did not have technological sovereignty were based upon the nature of the industrial structure such as reliance on resource-based industries, or mature and non-dynamic industries such as textiles that had developed over time in Canada.

The weaknesses of the industrial structure were supported by a series of studies published by the Science Council in the early 1970s. The weakness of Canadian research and development was emphasized by A.J. Cordell in the background study entitled *The Multinational Firm: Foreign Direct Investment and Canadian Science Policy* (Ottawa: Information Canada, 1971. 95 p. Science Council of Canada. Special Study No. 22); and by J.N. Britton and J.M. Gilmour in the background study entitled *The Weakest Link: A Technological Perspective on Canadian Industrial Underdevelopment* (Ottawa: Science Council of Canada, 1978. 216 p. Science Council of Canada. Background Study No. 43). The problem of innovation, again, was addressed by P.L. Bourgault in the background study entitled *Innovation and Structure of Canadian Industry* (Ottawa: Information Canada, 1972. 135 p. Science Council of Canada. Special Study No. 23). The problem of innovation and technology transfer was addressed in the council's report entitled *Technology Transfer: Government Laboratories to Manufacturing Industry* (Ottawa: Information Canada, 1975. 61 p. Science Council of Canada. Report No. 24), and in the background study by A.J. Cordell and J. M. Gilmour entitled *The Role and Function of Government Laboratories and the Transfer of Technology to the Manufacturing Sector* (Ottawa: Science Council of Canada, 1976. 397 p. Science Council of Canada. Background Study No. 35).

The Industrial Policies Committee, which was established in 1976, had a mandate to consider both the problems and solutions bearing upon the innovative process in the industrial structure in Canada. One of their major contributions was a council report entitled *Forging the Links: A Technology Policy for Canada* (Ottawa: Science Council of Canada, 1979. 72 p. Science Council of Canada. Report No. 29). This report drew attention to the need for structural change in Canadian industry and the need to encourage specialization. And, in the background study by G.P. Steed, entitled *Threshold Firms: Backing Canada's Winners*

(Ottawa: Science Council of Canada, 1982. 173 p. Science Council of Canada. Background Study No. 48), the author examined the operation of 'threshold' firms, their technology strategies, and their performance. Other background studies at this time looked at the regional nature of the economy and the need for national industrial policies. One such background study, by M.J. Jenkins, was entitled *The Challenge of Diversity: Industrial Policy in the Canadian Federation* (Ottawa: Science Council of Canada, 1983. 214 p. Science Council of Canada. Background Study No. 50). Another study, by D.L. LeRoy and P. Dufour, was entitled *Partners in Industrial Strategy: The Special Role of the Provincial Research Organizations* (Ottawa: Science Council of Canada, 1983. 146 p. Background Study No. 51).

By the mid to late 1980s, the Science Council had been pulled into the free trade debate and stated that 'we need to recognize that innovation through science and technology is an imperative for nation building and maintaining a versatile and competitive edge with our global trading partners and competitors.' The council also pointed out that the universities, as sources of innovation, must be encouraged into effective collaboration with industry because 25% of all R&D in Canada was done in the universities. In order to encourage and to stimulate greater collaboration between the universities and industry, the council undertook to publish a series of discussion papers and background studies.

The first discussion paper, by T.E. Clarke and J. Reavley, was entitled *Educating Technological Innovators and Technical Entrepreneurs at Canadian Universities* (Ottawa: Science Council of Canada, 1987. 102 p.). This paper examined the growth in the number of courses on technical innovation, or technical entrepreneurship, offered at Canadian universities. The second paper, by J.B. Barnes and G.R. Peters, was entitled *The Teaching Company Scheme: A Study of Its Application to Canada* (Ottawa: Science Council of Canada, 1987. 39 p.). This paper described a form of collaboration that had proven successful in the United Kingdom. University professors could apply for a government grant to work with companies on critical research or development projects for which the companies lacked the resources. The third paper, by W.G. Hutcheson and P. Milley, was entitled *R&D Links between Firms and Universities: Six Case Studies* (Ottawa: Science Council of Canada, 1987. 37 p.). This paper reported on the dynamics of R&D links. The fourth paper, by S. Amos, was entitled *Learning from Each Other: University-Industry Collaboration in the Continuing Education of Scientists and Engineers* (Ottawa: Science Council of Canada, 1987. 43 p.). This paper, which reported the proceedings of a workshop that discussed the growing need for updating and retraining workers, was a challenge for universities to do more in this area. Another paper, by G.A. Thom, was entitled *Employer Interaction with Public Colleges and Institutes in Canada* (Ottawa: Science Council of Canada, 1987.

53 p.). This provided background information on colleges and described the variety of links that existed between colleges and industry in Canada. These discussion papers were followed up with a council report entitled *Winning in a World Economy: University-Industry Interaction and Economic Renewal in Canada* (Ottawa: Science Council of Canada, 1988. 84 p. Science Council of Canada. Report No. 39). As well, another background study was issued by J.P. Steed, entitled *Not a Long Shot: Canadian Industrial Science and Technology Policy* (Ottawa: Science Council of Canada, 1989. 146 p. Science Council of Canada. Background Study No. 55). This study stated that science and technology are an ever-increasing force in determining success in world markets, and Canada, as a small country, must seek technological niches that have the best chance of success. This background study paralleled a council statement issued in late 1988 and entitled *Gearing Up for Global Markets: From Industry Challenge to Industry Commitment* (Ottawa: Science Council of Canada, 1988. 22 p.).

The theme of innovation in Canadian industry was visited one last time in a series of reports commissioned by [the Department of] Industry, Science and Technology to analyse the key factors driving technological innovation in 15 different business and industry sectors of the Canadian economy. In analysing these sectors – 'high-value businesses of the 1990s' – the Science Council found that 'Canadian firms, when grouped by sectors, generally perform much less R&D than their trade partners ... about one-third to one-half the rate of their leading international sectoral competitors.' The reasons given for this lower level of R&D were 'choice of business and technology strategy ... the relatively small size of Canadian firms ... the high absolute thresholds of R&D expenditure required to be a meaningful player ... and the degree of foreign ownership of Canadian firms.' The reports commissioned were *The Canadian Telecommunications Sector* (Ottawa: Science Council of Canada, 1992. 32 p. Sectoral Technology Strategy Series No. 1); *The Canadian Automotive-Parts Sector* (Ottawa: Science Council of Canada, 1992. 32 p. Sectoral Technology Strategy Series No. 2); *The Canadian Iron and Steel Sector* (Ottawa: Science Council of Canada, 1992. 29 p. Sectoral Strategy Series No. 3); *The Canadian Automotive-Vehicle Sector* (Ottawa: Science Council of Canada, 1992. 34 p. Sectoral Technology Strategy Series No. 4); *The Canadian Oil and Gas Sector* (Ottawa: Science Council of Canada, 1992. 41 p. Sectoral Technology Strategy Series No. 5); *The Canadian Petrochemicals and Resins Sector* (Ottawa: Science Council of Canada, 1992. 36 p. Sectoral Technology Strategy Series No. 6); *The Canadian Banking Sector* (Ottawa: Science Council of Canada, 1992. 44 p. Sectoral Technology Strategy Series No. 7); *The Canadian Nonferrous-Metals Sector* (Ottawa: Science Council of Canada, 1992. 36 p. Sectoral Technology Strategy Series No. 8); *The Canadian Forest Products Sector* (Ottawa: Science Council of Canada, 1992. 33 p. Sectoral Technology Strategy Series No. 9); *The Canadian Machinery Sector* (Ottawa: Science Council of Canada, 1992. 34 p. Sectoral Technology Strategy Series No. 10); *The Canadian Food and Beverage Sector* (Ottawa: Science Council of Canada, 1992. 44 p. Sectoral Technology Strategy Series No. 11); *The Canadian Electric-Power Sector* (Ottawa: Science Council of Canada, 1992. 41 p. Sectoral Technology Strategy Series No. 12); *The Canadian Consulting-Engineering Sector* (Ottawa: Science Council of Canada, 1992. 26 p. Sectoral Technology Strategy Series No. 13); *The Canadian Electronics Sector* (Ottawa: Science Council of Canada, 1992. 50 p. Sectoral Technology Strategy Series No. 14); and *The Canadian Computer Software and Services Sector* (Ottawa: Science Council of Canada, 1992. 38 p. Sectoral Technology Strategy Series No. 15).

Despite the wealth of studies that focused on the problems of Canadian industry and innovation, the council thought that Canada had fallen well behind in global competitiveness in industry because the government had not undertaken 'bold comprehensive initiatives in S&T policy to support a determined drive to competitiveness.' This was the council's last statement covering this area of its studies before it closed its doors in 1992.

D Urbanization and Population Studies

One of the early study themes of the council, which was in effect the beginning of their environmental studies, was the use and understanding of the interaction between technological developments and demographic trends. Initially, a study group on urban development, chaired by P. Dansereau, commissioned a series of studies that would focus upon urban transportation, urban information systems, the urban environment, housing, and the design of a Canadian urban model for the systems analysis of urban problems. The reason for this interest by the council in urbanization was because of its rate of acceleration in the early 1970s. The council wanted to study the way in which technological developments may have accelerated the process of urbanization and to ask questions about what role technology could play in reversing the negative aspects of that trend. The first publication that emerged was a council report entitled *Cities for Tomorrow: Some Applications of Science and Technology to Urban Development* (Ottawa: Information Canada, 1971. 67 p. Report No. 14). This stated that methodologies of systems analysis and design were needed in order to predict more precisely the consequences and spill-overs of both public and private policies and actions upon the social structure of cities. Urban transportation, housing, and recycling and waste in the urban environment were specifically pointed out as problem areas to be examined.

One year later, these urbanization studies shifted into the specific links of population growth and technology. The council thought that by the year 2000, Canada's population would reach 34 million and would have values and life styles different from the existing ones. It was assumed that there would be profound changes in people's occupations, their attitude toward leisure, where they lived, and their consumption of resources. Accordingly, there would be much greater and more diverse demands on the technology underpinning this society, and there was a need to understand which technologies were required for the future. In order to examine these potential problems, the Committee on Population, Technology and Resources decided to broaden the scope of its task and include the examination of the population's relationship to other vital resources such as energy, food, and agricultural land, as well as technology. The committee wanted to encourage a broad debate on these issues and decided to publish its findings in a series of publications called Perceptions. The aim of the series was to present a compendium of statistical and other data concerning various aspects of population growth in a format useful to the educated lay reader in order to find an audience in discussion groups, introductory university courses, interdisciplinary education, and readers in general who wanted to be knowledgeable about future options being debated by experts and policy-makers. In all, four Perceptions publications were issued.

The first, by E. Kelly, was entitled *Population Growth and Urban Problems* (Ottawa: Science Council of Canada, 1975. 95 p. Study on Population and Technology. Perceptions 1). This study stated that the concentration of population in large urban centres presented many resource, technological, and political problems which had to be urgently addressed. Perceptions of what the problems were and how they might be tackled were discussed. The second study, by L.A. Auerbach and A. Gerber, was entitled *Implications of the Changing Aging Structure of the Canadian Population* (Ottawa: Science Council of Canada, 1976. 125 p. Study on Population and Technology. Perceptions 2). This volume examined the impact of projected increases in the relative numbers of elderly and late middle-aged in Canada and of decreases in the proportion of adolescents as the 'baby-boom' children left childhood and began to look for jobs and start families. The accessibility of important technologies, transportation, communications, housing, and medicine was also discussed. The third study, by B.J. Geno and L.M. Geno, was entitled *Food Production in the Canadian Environment* (Ottawa: Science Council of Canada, 1976. 71 p. Study on Population, Technology and Resources. Perceptions 3). This volume examined the scientific and technological components involved in increasing agricultural production in Canada based upon an environmentally preferable agriculture that promoted more nutrition and less wasteful consumption. The fourth and final study of this series, by C. Beaubien and R. Tabacnik, was entitled *People and Agricultural Land* (Ottawa: Science

Council of Canada, 1977. 137 p. Study on Population, Technology and Resources. Perceptions 4). This dealt entirely with the increasingly intense and multi-faceted competition for Canada's best agricultural land and the negative influence that development can have on that land. A strong plea was made for land ownership to be considered in the context of industrial and social stewardship rather than as solely the object of private gain.

Interestingly, these concerns about food production and agricultural land were revisited again in one of the last reports of the council, entitled *Sustainable Agriculture: The Research Challenge* (Ottawa: Science Council of Canada, 1992. 46 p. Science Council of Canada. Report No. 43). This report combined the concerns of environmental degradation of farmland through current farm practices and the need to search for long-term sustainable farm practices through scientific and technical advances in order to compete effectively in global agriculture food markets.

E Environmental Studies

Perceptions 3 did give a hint of the council's direction and interest in environmental issues as we have come to think of them. The concept of environmental concerns centred on the 'conserver society,' which was the focus of a report issued in the mid-1970s, discussed below. The concept of the conserver society evolved over time but was given an additional impetus with the discussions surrounding the international energy crises, which began in 1973. The roots of the Science Council's interest in the environment, however, were part of the general growing awareness of problems facing our environment that developed in the 1960s, and many of the council's earlier reports did have a conserver orientation.

For example, the report entitled *A Major Program of Water Resources Research in Canada* (Ottawa: Queen's Printer, 1968. 56 p. Science Council of Canada. Report No. 3) called for the 'development of sound programs for the use, conservation and replenishing of resources' and the need for 'control of existing and threatened health hazards already created by the misuse of science and technology, e.g., pollution.' This latter theme was taken up again by the Science Council in the later 1970s. The previously mentioned study by R.A. Blais and C.H. Smith, entitled *Earth Sciences Serving the Nation* (Ottawa: Science Council of Canada, 1970. 363 p. Science Council of Canada. Special Study No. 138), pointed out the need for improved urban planning and the importance of natural resource development because of the fast-growing interest in the physical environment, the mounting pressures concerning landscape preservation, and pollution concerns. The report entitled *This Land Is Their Land: A Report on Fisheries and Wildlife Research in Canada* (Ottawa: Queen's Printer, 1970. 41 p. Science Council of

Canada. Report No. 9) called for making environmental improvement a new national goal and for creating an Environmental Council and a Department of Renewable Resources. The report also suggested that more emphasis be placed on the development of the environmental sciences in Canada as well as the need for long-term ecological studies. The report entitled *A Canadian STOL Air Transport System – A Major Program* (Ottawa: Information Canada, 1970. 33 p. Science Council of Canada. Report No. 11) contained a comprehensive discussion of the external costs (noise, pollution, congestion, land usage) associated with highway systems and larger aircraft systems, and their technological impact on environmental quality. The above mentioned report No. 14, entitled *Cities for Tomorrow: Some Applications of Science and Technology to Urban Development* (Ottawa: Science Council of Canada, 1971. 67 p. Science Council of Canada. Report No. 14), discussed the relationship between societal activities and the health of the environment with a particular emphasis on recycling. The theme of the health of the environment was found again in the report entitled *It Is Not Too Late – Yet: A Look at Some Pollution Problems in Canada's Natural Environment, and Identification of Some Major Concerns* (Ottawa: Information Canada, 1972. 49 p. Science Council of Canada. Report No. 16).

The term 'conserver society' was actually used for the first time in the report entitled *Natural Resource Policy Issues in Canada* (Ottawa: Information Canada, 1973. 59 p. Science Council of Canada. Report No. 19). In one of the recommendations, the report stated that Canadians must 'begin the transition from a consumer society preoccupied with resource exploitation to a conserver society engaged in more constructive endeavours.' The theme of the conserver society appeared again in the report entitled *Population, Technology and Resources* (Ottawa: Science Council of Canada, 1976. 91 p. Science Council of Canada. Report No. 25). In this report, the statement was made that 'we see Canada in the longer term becoming a conserver society using food, energy and resources frugally. ...' A proposal for a specific study on the consumer society was first made in 1973, and the major report entitled *Canada as a Conserver Society: Resource Uncertainties and the Need for New Technologies* (Ottawa: Science Council of Canada, 1977. 108 p. Science Council of Canada. Report No. 27), was issued four years later. This report stated that emerging resource restraints and environmental and social limits presented challenges for individuals at all levels of Canadian society. It also questioned the way in which our society had chosen to transform resources into commodities, and the ways Canadians had become accustomed to a way of life that involved high rates of resource use, high waste, and constantly inflated expectations. Only with concern for the future, economy of design, attention to diversity and flexibility, recognition of total costs, and respect for the regenerative capacity of the biosphere could we make a smooth transition to a conserver society.

The conserver society theme made its last appearance ten years later in the report entitled *Water 2020: Sustainable Use for Water in the 21st Century* (Ottawa: Science Council of Canada, 1988. 40 p. Science Council of Canada. Report No. 40). This report, which did not specifically use the term 'consumer society,' concluded an 18-month study of how Canada could manage its water resources more effectively in the coming decades. The report emphasized the relationship between the environment and economic development, and argued that Canadians must adopt a goal of sustainable use for water by directing science and technology to that end. After the initial flurry of the conserver society theme, the Science Council shifted its environmental interests more specifically to the environmental hazards of the workplace, the control of exposure to hazards in Canada and elsewhere, and regulatory processes necessary for the control of hazardous production in Canada.

The first report in this shift of focus was entitled *Policies and Poison: The Containment of Long-Term Hazards to Human Health in the Environment and in the Workplace* (Ottawa: Science Council of Canada, 1977. 76 p. Science Council of Canada. Report No. 28). In this report, the Science Council's Committee of Hazardous Substances of Man-Made Origin studied the decision process as it applies to the control of environmental and occupational hazards, particularly those that are injurious to the environment and long-term in their effects. Although there were over 15,000 chemicals in use in 1977, the committee focused its study upon asbestos, lead, mercury oxides of nitrogen, radiation, and vinyl chloride because these hazards were representative of the spectrum that results from industrial activity and also because they illustrated the manner in which Canadians had coped to date with the regulation of toxic substances. The background studies that contributed to the preparation of Report No. 28 were: *Canadian Law and the Control of Exposure to Hazards* by R.T. Franson and A.R. Lucas (Ottawa: Science Council of Canada, 1977. 152 p. Science Council of Canada. Background Study No. 39); *Government Regulation of the Occupational and General Environments in the United Kingdom, the United States and Sweden*, by R. Williams (Ottawa: Science Council of Canada, 1977. 155 p. Science Council of Canada. Background Study No. 40); and *Regulatory Processes and Jurisdictional Issues in the Regulation of Hazardous Products in Canada*, by G.B. Doern (Ottawa: Science Council of Canada, 1977. 201 p. Science Council of Canada. Background Study No. 41).

The occasional papers published that contributed to Report No. 28 were: *An Overview of the Canadian Mercury Problem*, by C.T. Charlebois (Ottawa: Science Council of Canada, 1977. 20 p.); *An Overview of the Vinyl Chloride Hazard in Canada*, by J. Basuk (Ottawa: Science Council of Canada, 1977. 16 p.); *An Overview of the Canadian Asbestos Problem*, by C.T. Charlebois (Ottawa: Science

Council of Canada, 1978. 20 p.); *An Overview of the Oxides of Nitrogen Problem in Canada*, by J. Basuk (Ottawa: Science Council of Canada, 1978. 48 p.); and *An Overview of the Ionizing Radiation Hazard in Canada*, by J. Basuk (Ottawa: Science Council of Canada, 1979. 225 p.).

F Northern Science and Technology Studies

The Science Council's initial interest in northern science and technology began in 1969. This was spurred on by the beginning explorations for oil and gas, which occurred before the 1973 international energy crisis, as well as attempts to demonstrate the feasibility of a commercial Northwest Passage. The council thought that Canada's efforts in using science and technology to help in the development of the north were relatively unplanned and on a small scale. What was needed was for the country to define its goals in the Arctic and then to effectively mobilize the nation's scientific and technical resources to help achieve those goals. It was this theme that became the major focus of the council's studies and continued later with the free trade developments of the 1980s and the linking of the use and marketing of unique cold-climate technologies developed in the north.

The council's first study on northern development (its concept of the north was not one that began at the 60th parallel but one that extended from the northern limits of urbanization through to the Arctic Islands) was aimed at identifying the major issues facing northern development in the mid-1970s and to provide guidelines for planning, particularly in areas where there was a concern about the pacing of development. The approach to this study was to develop a research strategy around five projects that would permit the study of science and technology in the context of the decision-making processes which the council believed would determine the future of Canada's north. In particular, the council wanted to be able to show how the decision-making process would cope with the need for new technologies, with the problem of environmental impact on the project, and with the problem of social impact. The case studies focused upon: (1) the search for oil and gas in the Arctic Islands, (2) the major mineral developments on Baffin Island, (3) the development of the Athabasca Tar Sands, (4) the exploration for oil off the coast of Labrador, and (5) the James Bay Project. As well, a general economic study was commissioned.

From these outlines, four studies were actually published, and two remained in manuscript form only. The first case study, by M. Gibbons and R. Voyer, was entitled *A Technology Assessment System: A Case Study of East Coast Offshore Petroleum Exploration* (Ottawa: Science Council of Canada, 1974. 114 p. Science Council of Canada. Background Study No. 30). The second study, by R.F. Keith and D.W. Fischer, was entitled *Northern Development and Technology Assessment Systems: A Study of Petroleum Development Programs in the Mackenzie Delta–Beaufort Sea Region and the Arctic Islands* (Ottawa: Science Council of Canada, 1976. 219 p. Science Council of Canada. Background Study No. 34). The third, and general economic study, by K.J. Rea, was entitled *The Political Economy of Northern Development* (Ottawa: Information Canada, 1976. 251 p. Science Council of Canada. Background Study No. 36). The fourth study, by R.B. Gibson, was entitled *The Strathcona Sound Mining Project: A Case Study of Decision Making* (Ottawa: Science Council of Canada, 1978. 274 p. Science Council of Canada. Background Study No. 42). The two case studies not published, but available from the Science Council in manuscript form, were *Decision Making in the North: Oil Sands Case Study*, by Canadian Resources Limited, and *Le Processus Décisionnel dans La Conception et La Réalization du Development Nordique au Canada – La Baie de James*, by E. Gourdeau.

The end result of these background studies was the publication of the report by the Science Council entitled *Northward Looking: A Strategy and a Science Policy for Northern Development* (Ottawa: Science Council of Canada, 1977. 95 p. Science Council of Canada. Report No. 26). In this report, the council advocated a northern strategy of mixed development – large-scale industrial activities, and small-scale, renewable, resource development. However, four basic principles were suggested as the guide for northern development. They were: (1) technological sovereignty, or control of the technological capacity to support national sovereignty, (2) lifestyle flexibility, (3) maintenance of the regenerative capacity of the land, and (4) the comprehensive and balanced assessment of both large- and small-scale projects in terms of their political acceptability. The report also generated a discussion paper issued by the Committee on Northern Development. It was entitled *A Northern Resource Centre: A First Step toward a University of the North* (Ottawa: Science Council of Canada, 1978. 13 p.). In this paper, the committee said that a northern resource centre would remind Canadian scientists that Canada was a northward-looking nation and this would be the first step toward the creation of a university of the north.

After this initial flurry of studies, the next impetus for additional northern studies came out of the free trade initiatives of the 1980s. The council said that this northward expansion of Canada's economic frontier had been furthered by advances in science and technology – cold climate technologies unique to Canada but having an 'enviable international reputation.' At the end of 1988, the Science Council established a Polar Science and Technology Committee to determine Canada's strengths and needs in cold climate technologies and to recommend ways to foster the growth of this emerging industrial sector. The result of the committee's work was the statement by the council entitled *Harnessing Science and*

Technology for Cold Regions (Ottawa: Science Council of Canada, 1989. 28 p.). This said that 'to use the knowledge and experience accumulated in the North as a basis to penetrate world markets required both greater self-help by the private sector and reinforcing public policies that overlays [*sic*] broad areas of industrial strategy and scientific research.'

The last northern initiative by the Science Council combined the northward expansion theme of the Canadian economy with its new technologies and the council's belief in the need to study the institutional changes required to help northeners apply science and technology to the economic development of their communities. The report entitled *Northern Science for Northern Society: Building Economic Self-Reliance* (Ottawa: Science Council of Canada, 1991. 43 p. Science Council of Canada. Report No. 41) echoed some of the thoughts of the discussion paper issued by the council just over 10 years earlier and referred to above.

G Energy Studies

As with its northern studies, the Science Council's interest in energy problems pre-dated the international energy crisis of 1973, although the latter did give a further impetus to those studies. In 1971, the Science Council created a Committee on Energy 'to study the scientific policies related to energy production, transmission, utilization and conservation.' The committee hoped to prepare a state-of-the-art survey of the energy sciences and policy issues of special interest to Canada and include an assessment of R&D capabilities and particular problems in the energy sciences.

Three years later, amid the international energy crisis, the committee issued its report, entitled *Canada's Energy Opportunities* (Ottawa: Information Canada, 1975. 135 p. Science Council of Canada. Report No. 23). The report focused upon the immediate future as well as the long-term. The immediate future was seen as a problem into the 1980s because of the inability to install the necessary productive capacities and transportation systems. As a result, Canada would not be able to satisfy demands for oil from domestic production, and would instead become a net importer of oil. Presently untapped resources to replace conventional supplies, along with conservation practices, would require a major energy research and development program in order 'to acquire and use the scientific and technical knowledge necessary for both the development of an energy policy and the achievement of its objectives.' The report then discussed energy options open to Canada for long-term research and development that would provide the nation with the widest number of options for achieving energy self-sufficiency.

The theme of energy conservation, as part of a national energy policy, was carried through in a background study by F.H. Knelman entitled *Energy Conserva-*

tion (Ottawa: Science Council of Canada, 1975. 87 p. Science Council of Canada. Background Study No. 33). The author examined both the supply and demand aspects of the Canadian situation and concluded that the country should consider adopting an active national energy conservation and efficiency program in order to produce a possible 30% energy saving by the mid-1990s. The author concluded that the major barrier to successful implementation of conservation measures was social rather than technical and this meant that the public in general must be consulted, informed, involved, and educated in the process.

The follow-up to the council's initial recommendations of the mid-1970s was an expansion of the energy committee into the Committee on Energy Scientific Policies in order to carry forward the theme of the R&D aspect of a national energy policy. In the footsteps of their first report, the expanded energy committee wrote a focused action-oriented report entitled *Roads to Energy Self-Reliance: The Necessary National Demonstrations* (Ottawa: Science Council of Canada, 1979. 200 p. Science Council of Canada. Report No. 30). The major thrust of the report was that 'amongst the vast array of actions we should, as a nation, be undertaking now to ensure adequate long-term future energy supplies, [and] it is essential to focus a substantial part of our scientific and engineering efforts on a relatively few demonstration projects.' The committee suggested that these pilot projects be designed and undertaken to assess and demonstrate the commercial and social viability of the technologies along with industries required to assure access to energy resources, and convert the latter into useable energy. In all, 11 specific demonstration projects were recommended to be pursued. These ranged from showing the technological capability for the exploration and the production of oil and gas in ice-congested waters to energy generation from solid wastes. In addition to these major projects, the committee also wanted to draw attention to alternative energy sources in the form of ideas or projects that were being unleashed in the renewable energy sector at that time. In the discussion paper by J. Passmore and R. Jackson entitled *Renewable Energy: Innovation in Action* (Ottawa: Science Council of Canada, 1984. 39 p.), the authors described about 50 ideas and systems such as wind generators, or small-scale hydro projects, that could possibly be manufactured and installed in a matter of months or a year. These systems were in contrast to developments such as nuclear stations or large-scale hydro developments that might require up to 10 years to develop.

H Computer Studies

The Science Council's interest in the implications of the widespread use of computers began in 1968 with the publication of the previously mentioned report entitled *Towards a National Science Policy for Canada* (Ottawa: Queen's Printer,

1968. 56 p. Science Council of Canada. Report No. 4). In Section 7, under specific areas for action in Canada's scientific and technological development, the report stated that 'Canada's indigenous computer industry stands as a monument to the nation's lack of entrepreneurial initiative and to the past failure to turn successful research into successful innovation.' This comment was made in the context of an awareness that the 'electronic computer may well be the basis in the 1970s of the world's third largest industry ... so the computer industry will play a major role in shaping the society of tomorrow.' Because Canada had not capitalized on its ability to design computer systems, 'the country should study the feasibility and desirability of establishing nationwide computer communications utilities as the foundation for an information transfer system.'

Three years later, the Science Council followed up on that theme by issuing a report entitled *A Trans-Canada Computer Communications Network: Phase I of a Major Program on Computers* (Ottawa: Information Canada, 1971. 41 p. Science Council of Canada. Report No. 13). This report contributed to policy discussions which ultimately led to the creation of a national computer communications network. It stated 'the development and construction of a system of interconnected computer communications networks should be planned and subsequently undertaken as a national Major Project.' The report also suggested that the responsibility for planning the networks should be assigned to the Department of Communications and, as a matter of policy, the system should span all populated regions of Canada and be entirely Canadian-owned. Two years later, however, the council came back to the idea of encouraging Canada's computer industry in the report entitled *Strategies of Development for the Canadian Computer Industry* (Ottawa: Science Council of Canada, 1973. 80 p. Science Council of Canada. Report No. 21). This report called for 'the development of a national policy with appropriate industrial strategies ... for the production of computer hardware and software, the training and certification of men and women capable of working with the new technology, and the provision of adequate standards and safeguards for ensuring that this industry serves our national interests.' As well, the council had specific recommendations such as improvements in the governmental procurement policies and in federal incentive schemes such as the application of a federal excise tax to favour domestically manufactured hardware.

By the late 1970s, with the continuous growth and use of the new information technologies in Canada, the council shifted its focus to examining the implications of such new technologies as micro-processors, word processors, and fibre optics that had contributed to the new information revolution. Accordingly, the council created a Committee on Computers and Communication: Information and Canadian Society. It issued its first report, entitled *Planning Now for an Information Society: Tomorrow Is Too Late* (Ottawa: Science Council of Canada,

1982. 77 p. Science Council of Canada. Report No. 33), which stated that Canadians faced a technology that would fundamentally change human thought and action. It was crucial for Canada to develop an industrial policy that reflected those changes. The failure to keep pace with other nations, many already more advanced in microelectronic research and development, would have a disastrous effect on Canada's ability to compete in world markets. For example, 'many sectors of our manufacturing industry would be rendered obsolete, virtually overnight. Our trade balance, already precarious, might never recover. Structural unemployment could lead to permanent joblessness, a decline in living standards and, for some, emigration.' Even if Canada did keep pace with microtechnology research and developments elsewhere in the world, the necessary restructuring of society would create formidable problems that must be solved. This report generated follow-up activities by the council such as meetings with federal officials from several government departments to discuss issues raised by the report. As well, the council organized five one-day meetings with representatives from various sectors in the Canadian economy in different regions of the country. Through these meetings, the council acted as a catalyst drawing people together who did not usually have occasion to talk with one another.

Three years later, the council published a background study based on the research done for Report No. 33. This study by A.J. Cordell was entitled *The Uneasy Eighties: The Transformation to an Information Society* (Ottawa: Science Council of Canada, 1984. 150 p. Science Council of Canada. Background Study No. 53). It consisted of a collection of essays which discussed the role and influence of information technology in society rather than simply stating the nature of the technology itself. It also included an overview of recent developments in the field of artificial intelligence.

I Biotechnology and Emerging Technology Studies

In 1983, the council began its studies of the field of biotechnology because its members thought that the development of biotechnology in Canada's resource and related industries had important implications for the country's economic future. The studies had three main objectives. They were: (1) the creation of a policy to stimulate action by the public and private sectors to encourage the advanced biological techniques in the resource-based industries, (2) the promotion of an effective communications network for biotechnologists, and (3) the establishment of a heightened awareness among decision-makers and the public of the current and potential impact of biotechnologies. The studies undertaken involved both the pulp and paper industry and agriculture, the two economic sectors considered to be of highest priority. The Biotechnology Committee that was created wanted these case studies to point out what was currently being done in the field of biotechnology in these two areas, and look for the potential developments in those areas as well the impediments to developments and how to overcome those. Finally, the committee wanted to study ways of improving research and the industrial capability of biotechnology.

Parallel to the setting up of the Biotechnology Committee, the council had also issued a report concerning biotechnology entitled *Regulating the Regulators: Science, Values, and Decisions* (Ottawa: Science Council of Canada, 1982. 106 p. Science Council of Canada. Report No. 35). This report stated that science does not exist in a void and new scientific and technical development invariably affected the public. However, the council argued that the Canadian legal system had failed to address the difficult legal and ethical questions posed by the new field of technology. The current methods of dealing with these issues of resolving science-oriented value disputes were found to be inadequate because of the relatively specialized nature of the field. The closed process of scientific deliberations worked against incorporating a wide range of scientific and public evidence.

Besides attempting to stimulate discussion concerning the legal-ethical aspects of the new technologies, and carry on with its case studies in the pulp and paper and agriculture fields, the council also undertook a series of consultations, surveys, and workshops. These were designed to bring together business and research experts in order to stimulate discussions and action pertaining to the new technologies and further heighten public awareness of the issues raised. Additional areas of study were undertaken in the fields of food processing, the neuro-sciences, medical devices, plastics and other advanced materials, and aquaculture, in order to get a better idea of where Canada stood internationally in these fields. These additional activities resulted in the publication of four discussion papers or statements. They were: *Ideas Alive: Theoretical Biology in Canada*, by P.J. Buckley (Ottawa: Science Council of Canada, 1985. 30 p.); *Advanced Industrial Materials: Canadian Perspectives and Opportunities*, by A. Elzinga and S. McCutcheon (Ottawa: Science Council of Canada, 1985. 22 p.); *Emerging Plastics Technologies and the Canadian Plastics Industry*, by F. Anderson and A. Bobkowicz (Ottawa: Science Council of Canada, 1985. 27 p.); and *Aquaculture: An Opportunity for Canadians* (Ottawa: Science Council of Canada, 1985. 28 p.).

In late 1985, the council completed its biotechnology studies and issued the report entitled *Seeds of Renewal: Biotechnology and Canada's Resource Industries* (Ottawa: Science Council of Canada, 1985. 94 p. Science Council of Canada. Report No. 38). This highlighted the opportunities that some Canadian resource industries could realize if government, universities, and industry worked together to ensure the speedy and aggressive adoption of advanced biotechnological techniques. However, the report also pointed out that Canada was not yet in a position to revitalize its resource sector using biotechnology because a number of changes had to occur first. These were: (1) that basic research in biotechnology, particularly in the universities, had to be strengthened, (2) clear direction from firms in the resource sector must be provided to the research sector, (3) links between industry and researchers in government and universities must be strengthened, and (4) the regulatory framework must be made more conducive to the development of biotechnology and legislative gaps must be filled.

After the publication of this report, the council continued to issue statements and discussion papers pertaining to the emerging technologies. Some of the statements issued were *Food Irradiation: Prospects for Canadian Technology Development* (Ottawa: Science Council of Canada, 1987. 17 p.); *A Sectoral Approach to Innovation: The Case of the Forest-Product Industries* (Ottawa: Science Council of Canada, 1987. 31 p.); and *Enabling Technologies: Springboard for a Competitive Future* (Ottawa: Science Council of Canada, 1989. 27 p.). Discussion papers issued at that time were entitled *Issues in Food Irradiation*, by S. Mills (Ottawa: Science Council of Canada, 1987. 61 p.); *Emerging Sensor Technology: The Promise and the Threat*, by S. Katz (Ottawa: Science Council of Canada, 1987. 44 p.); and *Educating Technological Innovators and Technical Entrepreneurs at Canadian Universities*, by T.E. Clarke and J. Reavley (Ottawa: Science Council of Canada, 1987. 102 p.). Besides these statements and discussion papers, the Science Council issued one more background study which further examined the new technologies emerging in the forest sector. This background study by R. Hayter was entitled *Technology and the Canadian Forest Product Industries: A Policy Perspective* (Ottawa: Science Council of Canada, 1988. 138 p. Science Council of Canada. Background Study No. 54).

IV Council Sources of the Publications

The publications of the Science Council of Canada are listed, by year, at the back of the Annual Reports as 'publications released' or 'publications in [date].' The last annual listing was for 1989–90. The publications were also listed in the background studies. Accordingly, the last complete listing was published in the background study by G.D. Steed entitled *Not a Long Shot: Canadian Industrial Science and Technology Policy* (Ottawa: Science Council of Canada, 1989. 146 p.).

When the Science Council of Canada announced its closing in 1992, it listed the final publications to be released in a newsletter inserted into their serial publication entitled *In Touch*. However, no final comprehensive list of all the council publications was ever issued.

Chapter 2

The National Research Council of Canada

I Historical Background

The National Research Council of Canada had its origin in a federal Order-in-Council in 1916 which established the Honorary Advisory Council on Scientific and Industrial Research. The impetus for its creation during wartime came from Canadian industrialists who wanted a collaboration between the federal government and Canadian universities to foster industrial research. Even back in 1916, there was a modest consensus that Canada was falling behind its industrial competitors in the fields of pure science and industrial research.

Early on, the Advisory Council recognized that the federal government must formulate some form of long-term science policy in order to provide guidelines for the council's development. A science policy was not forthcoming, but the Advisory Council went ahead on its own and made its own modest beginning with a scholarship program to encourage the pursuit of scientific research in Canada. It also created advisory committees to assist with problems of national or regional concern, or specific industrial problems.

Then, in 1919, the Cronyn Committee examined the current state of scientific and industrial research in Canada and recommended the establishment of centralized laboratories and a Bureau of Standards, along with the original Advisory Council's scholarship program and advisory committees. It was not until 1932 that these laboratories were built and became commonly known as the National Research Council of Canada. By 1939, the NRC had become a well-equipped building with a laboratory staff of about 300 working in four research divisions – Biology, Chemistry, Physics, and Mechanical Engineering. Its budget was just under $900,000, with $30,000 earmarked for scholarships, and $150,000 for extramural research grants.

With the outbreak of World War II, the National Research Council became the federal government's coordinator and mobilizer of the national science establishment for the war effort. Research focused upon atomic energy, radar, protective clothing, and nutrition, including the packaging and transportation of foods. During the war years, the staff expanded to over 3,000 and the budget ranged from $5 million to $6 million a year.

After the war, the NRC benefited from the overall economic expansion that occurred in the country. The number of research divisions increased from four to 13. It had 29 active associate committees and an expanding technical information service for industry. As well, the NRC was active in national and international bodies. Its total budget grew from over $5 million in 1945/46 to almost $13 million by 1950/51. At the same time, NRC's expenditures on scholarship and research grants grew from $290,000 to $1.5 million. Within that period, its physical plant expanded with new laboratories at the Montreal Road site, east of Ottawa, and regional laboratories on the prairies, and in the Maritimes. This postwar boom also encouraged the debate about NRC's aims, purposes, and priorities, again in the absence of a national science policy.

In general, these discussions focused upon the balance between pure, or basic, research and applied, or industrial, research. During the 1960s, under the leadership of E.R. Steacie, basic research in the universities also made great strides that were not matched in industry. Canadian industries were too small to afford research facilities, and those large enough to undertake research depended upon parent companies located outside of Canada.

In the mid-1960s, Parliament passed a new National Research Council Act (1966–67) which updated NRC's specific mandate, described below. However, in 1971, the federal government created the Ministry of State for Science and Technology, and the Science Council of Canada. That left the National Research Council as an operational organization only, with no advisory role. Then in 1974, NRC's scholarship and grant role was taken away and placed in the newly created National Science and Engineering Research Council (NSERC). Again, all these changes affecting Canada's major science research organization took place in the absence of a national science policy, leaving the National Research Council in the position of having to justify its existence and redefine itself based on the parliamentary act.

The National Research Council Act (1966–67) did establish a broad mandate for NRC as an operational body. The most important sections of the Act empowered the NRC 'to undertake, assist or promote scientific and industrial research, including, without restricting the generality of the foregoing: (1) the utilization of the national resources of Canada; (2) researches with the object of improving the technical processes and methods used in the industries of Canada; (3) researches with the view of utilizing the waste products of said industries; (4) the investigation and determination of standards and methods of measurement;(5) researches, the object of which is to improve conditions in agriculture; (6)to licence, sell, or otherwise grant or make available to others, Canadian or other patent rights ... controlled by the council; and (7) to operate and administer any astronomical observatories established or maintained by the Government of Canada.'

With the passing by Parliament of this updated mandate, and the loss of any advisory role, the National Research Council began to elaborate upon its objectives based on the mandate. In 1980, the NRC published its first five-year plan, entitled *The Urgent Investment: A Long Range Plan for the National Research Council of Canada* (Ottawa: National Research Council, 1980. 105 p.). The two objectives of the plan focused upon NRC's scientific and industrial research and its scientific and technical information programs. Within these, however, there was a strong emphasis upon assisting industrial expansion in general, and regional industrial expansion in particular. Two particular areas of research interest noted were energy and 'research of social value.'

With the renewed emphasis upon assistance to industry, NRC created new interdisciplinary research clusters such as the Biotechnology Program, which was organized around the Plant Biotechnology Institute, in Saskatoon, the Division of Biological Sciences, in Ottawa, and the new Biotechnology Research Institute, in Montreal. The program was described in the publication entitled *National Research Council Biotechnology Program* (Ottawa: National Research Council of Canada, 1989. 28 p.).

Five years later, NRC issued its second five-year plan, entitled *A Practical Perspective: The NRC Plan, 1986–1990* (Ottawa: Montreal Research Council of Canada, 1985. 64 p.). Its synopsis stated that 'encouraging and assisting Canadian industries to implement technological solutions will be the major role of the National Research Council in the coming five years.' The backdrop for this was the 'urgent economic need for Canadian industries to become more competitive. Canada now imports far more technology than it exports, relying on natural resources to maintain our standard of living.' Internally, NRC also faced reduced budgeting after 1984.

Two years later, NRC issued an update to the plan entitled *A Practical Perspective: The NRC Plan, 1986–90 – Second Update, 1987* (Ottawa: National Research Council of Canada, 1987. 22 p.). In this publication, NRC said its thrust in raising the overall technological capacity of Canadian industry and improving Canada's competitiveness had been helped by some initiatives taken by the new Conservative government in Ottawa. These initiatives were the creation of the Prime Minister's National Advisory Board on Science and Technology, the writing of a National Science and Technology Policy, the government's 'Innov Action' strategy to focus federal activities and initiatives in industrial innovation, and the government's first National Conference on Science and Technology.

On its own side, NRC established an active marketing program, which, led by a Vice-President for Marketing and Technology Transfer, coordinated marketing support and evaluated NRC programs. NRC also undertook some specific research initiatives such as the creation of an Associate Committee on Artificial Intelligence to examine potential applications and the direction of research to assist industry, the establishment of a Centre for Protein Design as part of its Biotechnology Program, and the opening of its Canadian Institute of Industrial Technology in Winnipeg. NRC also joined PRECARN (Pre-Competitive Applied Research Network), a non-profit industry-led consortium of 22 major companies to conduct advanced research in artificial intelligence and robotics. This new emphasis upon marketing NRC's services can be seen in its publication entitled *Return on Investment* (Ottawa: National Research Council, 1989. 15 p.) which described successful industrial research projects that evolved using NRC's research facilities and technological knowledge.

As the National Research Council moved into the 1990s, it began to revise its organizational structure due to the budget cuts and downsizing that had taken place since 1984. This restructuring, which had stabilized by 1996, saw the creation of two vice-president positions under the president, one for research and the other for technology and industry support. The four historic subject divisions were eliminated with the creation of additional research institutes (17 by 1996). It is important to note that one of these institutes, originally created back in 1974, is the Canada Institute for Scientific and Technical Information – CISTI – which grew out of the original NRC library and is in fact Canada's national science library and is also responsible now for the publication of NRC's research journals, which have international stature.

As well, the historic associate committees made up of external research advisors drawn from university and industry were replaced by advisory boards for the council's institutes and programs. The last listing and description of the associate committees was a publication entitled *Associate Committees, 1988* (Ottawa: National Research Council, 1988. 31 p.). At the time of its publication, there were 24 associate committees.

Another initiative of the council was the writing of a long-term plan covering the period 1990 to 1995. The plan was entitled *The Competitive Edge 1990–1995* (Ottawa: National Research Council of Canada, 1991. 35 p.) and was designed to address the looming economic and social issues facing Canada (brought about by the Free Trade Agreement of 1989). It listed three objectives: (1) 'to preserve and strengthen NRC's base of world-class research, whose leading quality and relevance continues to be solidly supported by peer review ...,' (2) 'to promote and develop strategic partnerships through roughly 200 collaborations and partnerships with individuals, firms and organizations in the private, public and university sectors ...,' and (3) 'to stimulate Canada's competitive edge through the performance of R&D, an underlying principle of the Long Range Plan.' And in 1994, the council announced the preparation of another long-range plan. This plan, published in 1996, was entitled *Vision to 2001: Science and Technology for Canada's Future* (Ottawa: National Research Council, 1996. 25 p.) and said that NRC would 'fine-tune its activities based on its successes.' In the mid-1990s, the National Research Council still employed more than 3,000 researchers, support staff, and technology advisors in its industrial assistance offices and laboratories across Canada. Its budget, in spite of cutbacks, was approximately $4 million annually. Currently, its governing body is still the council headed by a president who presides over a body of 21 members appointed by the government. This group oversees the R&D activities of the NRC.

Although the 1980s and 1990s have seen a greater emphasis placed on research assistance to industry, over the years NRC has displayed a discernible pattern of activities covering a broad range of scientific interests that are well documented. These activities have been and remain: (1) basic research in areas of national or regional importance including energy, food production, building and construction research, and transportation research, (2) research considered to have a social impact such as pollution, destruction of PCBs, or the development of new vaccines, (3) basic research for the creation of new knowledge or new applications of science in the fields of astrophysics and space research, (4) research carried out through the associate committee structure, and (5) work carried out to encourage and assist in areas such as the supervision of national research facilities, the development and maintenance of new physical standards of measurement, assistance to industry programs, and, until 1978, the program of scholarships and grants to universities. These activities also display a pattern in the types of publications generated by NRC's rather large publishing output over the years.

II Publications

The publications issued by the National Research Council can be divided into ongoing or terminated serial publications which describe the overall activities of the council, its historic divisions or the current institutes, and monographs which reflect NRC publicity about their goals carried out by the associate committees or institutes. Most recently, NRC has also published scientific conference proceedings through their publishing department and their NRC research journal affiliation. However, the vast amount of research carried out by NRC scientists in the laboratories or institutes for NRC, or for its industrial assistance programs, is reported in the open international scientific literature.

A Serial Publications

The current major publication which describes the activities of the National Research Council is its *Report to Parliament* (Ottawa: National Research Council 1993/94–). Previous titles for this report, which go back to 1918, have been *Annual Report*; *Report of the President*; and *Honorary Advisory Council for Scientific and Industrial Research Report*. A historic serial publication which also described research activities was entitled *Review of the National Research Council of Canada* (Ottawa: National Research Council, 1938/39–1967/68. 32 v.).

The research activities of the divisions have been reported in varying ranges of years. These reports are:

National Research Council. Division of Biological Sciences. *Annual Report* (Ottawa: The Division, 1981/82–1989/90. 10 v.);
National Research Council. Division of Building Research. *Research Program* (Ottawa: The Division, 1976/77–1985/86. 10 v.);
National Research Council. Division of Chemistry. *Annual Report* (Ottawa: The Division, 1986/87–1989/90. 3 v.);
National Research Council. Division of Electrical Engineering. *Bulletin* (Ottawa: The Division, 1975/76–1989/90. 15 v.);
National Research Council. Division of Energy. *Annual Report* (Ottawa: The Division, 1982/83–1985. 3 v.);
National Research Council. Division of Mechanical Engineering. *Annual Report* (National Research Council, The Division, 1986/87–1988/89. 3 v.);
National Research Council. Division of Microstructural Science. *Annual Report* (Ottawa: The Division, 1984–85. 2 v.);
National Research Council. Division of Physics. *Progress Report* (Ottawa: The Division, 1975/76–1980/81. 6 v.); and
National Research Council. Radio and Electrical Division. *Bulletin* (Ottawa: The Division, 1951/52–1971/72. 21 v.).

With the reorganization of the National Research Council primarily into institutes in the 1980s, current research activities of the council are described in detail in the institutes' annual reports in print format or at their websites. They are:

Biotechnology Research Institute. *Annual Report* (Montreal: The Institute, 1985/86–); <<http://www.nrc.ca/bri>>;

Canada Institute for Scientific and Technical Information – CISTI. *Report* (Ottawa: National Research Council, 1974/77–); <<(http://www.nrc.ca/cisti>>;

Herzberg Institute of Astrophysics. *Annual Report* (Ottawa: Herzberg Institute of Astrophysics, 1991/92–1993/94); <<http://www.nrc.ca/hia>> ;

Industrial Materials Institute. *Annual Report* (Boucherville, Quebec: The Institute, 1982/83–);

Institute for Aerospace Research. *Annual Report* (Ottawa: The Institute, 1990/91–); <<http://www.nrc.ca/iar>>;

Institute for Biodiagnostics. *Annual Report* (Winnipeg: The Institute, 1993–); <<http://www.nrc.ca/ibd>>;

Institute for Biological Sciences. *Annual Report* (Ottawa: The Institute, 1990/91–1993/94); <<http://www.nrc.ca/ibs>>;

Institute for Chemical Process and Environmental Technology. *Annual Report* (Ottawa: The Institute, 1996–); <<http://www.icpet.nrc.ca>>;

Institute for Environmental Quality. *Annual Report* (Ottawa: NRC Institute for Environmental Quality, 1991–94. (Now closed);

Institute for Information Technology. *Annual Report* (Ottawa: The Institute, 1991–); <<http://www.nrc.ca/iit>>;

Institute for Marine Biosciences. *Annual Report* (Halifax, NS: The Institute, 1990/91–); <<http://www.nrc.ca/imb>>;

Institute for Marine Dynamics. *Annual Report* (St John's, Nfld: NRC Institute for Marine Dynamics, 1990/91–); <http://www.nrc.ca/imd>>;

Institute for Mechanical Engineering. *Annual Report* (Ottawa: The Institute, 1990/91–);

Institute for Microstructural Sciences. *Activities Report* (Ottawa: The Institute, 1990/91–); <<http://www.nrc.ca/ims>>;

Institute for National Measurement Standards. *Annual Report* (Ottawa: National Research Council Canada. Institute for National Measurement Standards, 1991–); <<http://www.nrc.ca/inms>>;

Institute for Research in Construction. *Annual Review* (Ottawa: Institute for Research in Construction, 1990–); <<http://www.nrc.ca/irc/>>;

Integrated Manufacturing Technologies Institute. *Annual Report* (Ottawa: The Institute, 1996–);

Plant Biotechnology Institute. *Annual Report* (Saskatoon, Sask. The Institute, 1984/85–1993/94); <<http://www.nrc.ca/pbi>> and

Steacie Institute for Molecular Sciences. *Activities Report* (Ottawa: The Institute, 1991/92–); <<http://www.nrc.ca/sims>>.

B Monograph Publications

As mentioned above, the monograph activity of the National Research Council consists of publicity publications issued by the council or its institutes, research series issued by the latter, and monographs issued by the now-defunct associate committees. The publicity publications of the council are their planning documents, which have already been described. However, from time to time, the institutes issue monographs to explain and promote their work with industrial partners. An example of this type of publication issued by the Institute for Research in Construction was entitled *Working with the Construction Industry: Research in Action* (Ottawa: National Research Council, 1982. 48 p.).

Among the research series issued by NRC are the proceedings of the Institute for Research in Construction. One publication in this series had the title *Canada's Framework for the Regulation and Design of Buildings/Building Science Insight' 91.* Edited by R.A. Hewett and G.L. Walt (Ottawa: National Research Council, 1992. 85 p. *Proceedings* (Institute for Research in Construction [Canada]), No. 17). Another series example, this one issued by the Institute for Aerospace Research, and written by S.J. Zan, was entitled *Measurements of Wing and Fin Buffeting on the Standard Dynamic Model* (Ottawa: National Research Council, 1993. 75 p. Aeronautical Note IAR-AN-76).

During their existence, the associate committees had a considerable publishing output. For example, the Associate Committee on Hydrology was responsible for publishing the monograph entitled *Hydrology of Floods in Canada: A Guide to Planning and Design* (Ottawa: National Research Council. Associate Committee on Hydrology, 1989. 245 p.), and the Associate Committee on Geotechnical Research sponsored the *Proceedings of Avalanche Workshop, 3–5 November, 1980, Vancouver, B.C.* Edited by Canadian Avalanche Committee (Ottawa: National Research Council, Associate Committee on Geotechnical Research, 1981. 248 p. Technical Memorandum No. 133). Another example of a publication by this committee is the *Proceedings of the Workshops on Subsea Permafrost, November 18, 1985 and Pipelines in Permafrost, November 19, 1985.* Prepared by G.H. Johnston and V.R. Parameswaran (Ottawa: National Research Council, 1986. 168 p. Technical Memorandum [National Research Council of Canada. Associate Committee on Geotechnical Research], No. 139).

With the collapse of the associate committee structure, the National Research Council has become the direct publisher of research previously undertaken by the committees. An example of this direct publishing is the conference entitled *Perspectives in Comparative Endocrinology: Papers from XII International Congress of Comparative Endocrinology, Toronto, Ontario, Canada, 16–21 May, 1993.* Edited by K.G. Davey and R.E. Peter (Ottawa: National Research Council, 1994. 653 p.).

Another example is the occasional paper entitled *Methods of Modifying Habitat to Benefit the Great Lakes Ecosystem.* Edited by J.R. Kelso and J.H. Hartig (Ottawa: National Research Council, 1995. 294 p. CISTI (Canada Institute for Scientific and Technical Information) Occasional Paper No. 1).

C Research Journals

Through CISTI, the National Research Council also provides the editorial and publishing function for 13 of Canada's science research journals which span the physical and biological sciences and engineering. These journals use the volunteer efforts of an extensive network of editors and reviewers across the country and around the world and publish Canadian and international scientific research. The journals published are:

Biochemistry and Cell Biology (Ottawa: National Research Council, v. 64, 1986–);
Canadian Geotechnical Journal (Ottawa: National Research Council, v. 1, 1963–);
Canadian Journal of Botany (Ottawa: National Research Council, v. 29, 1951–);
Canadian Journal of Chemistry (Ottawa: National Research Council, v. 29, 1951–);
Canadian Journal of Civil Engineering (Ottawa: National Research Council, v. 1, 1974–);
Canadian Journal of Earth Sciences (Ottawa: National Research Council, v. 1, 1964–);
Canadian Journal of Forest Research (Ottawa: National Research Council, v. 1, 1971–);
Canadian Journal of Microbiology (Ottawa: National Research Council, v. 1, 1954–);
Canadian Journal of Physics (Ottawa: National Research Council, v. 29, 1951–);
Canadian Journal of Physiology and Pharmacology (Ottawa: National Research Council, v. 42, 1964–);
Canadian Journal of Zoology (Ottawa: National Research Council, v. 29, 1951–);
Environmental Reviews (Ottawa: National Research Council, v. 1, 1993–);
Genome (Ottawa: National Research Council, v. 29, 1987–).

Some of the active research carried out in the laboratories of NRC's institutes finds its way into these journals, and some – particularly that of international interest – is published in the overall scientific journal literature.

III Research and Activities of the National Research Council

A Basic Research of National or Regional Importance

This category of research by the NRC has included, at one time or another, alternate energy sources (including solar applications, wind energy, peat energy, hydrogen energy, and fusion energy), food and forest production, building and construction research, and transportation research.

(a) Energy Research

In the early 1970s, Canada's annual fuel bill for heating private residences was estimated at more than $1 billion, and that factor, coupled with rising oil prices and the ensuing international energy crisis, triggered energy research at the NRC. The main impetus for energy research and the development of energy guidelines came from a special interdepartmental committee set up by the Division of Building Research. In 1981, this committee was organized into a Division of Energy, which existed until 1986. Guidelines and standards involving air leakage and insulation were developed, along with other energy conservation measures. These were first to be applied in federal buildings, and then extended to other jurisdictions. Some of the specific work was done by G.P. Mitalas and entitled *Net Annual Heat Loss Factor Method for Estimating Heat Requirements of Buildings* (Ottawa: National Research Council, 1974. 22 p. Building Research Note 117); by W.G. Brown, entitled *Practical Methods for Estimating Ground Temperatures and Heat Flow between Engineering Structures and the Ground* (Ottawa: National Research Council, 1974. 53 pp); and by R.L. Quiroutte, entitled *Energy Conservation and Building Design* (Ottawa: National Research Council, 1976. 9 p. DBR Building Research Note 110).

The overall guidelines were published two years later by NRC and entitled *Measures for Energy Conservation in New Buildings, 1978* (Ottawa: National Research Council, 1978. 52 p). These guidelines covered enclosures for buildings, lighting requirements, fans and pumps, cooling and ventilation, and heating.

In the late 1970s, NRC became actively involved in research on passive solar thermal systems for heating and cooling buildings as well as on active solar residential systems using the applications of heated domestic water. NRC funded solar heating systems in 16 multi-family residential buildings in locations across Canada and specific progress reports on the project were written. For example, R.M. Hardy and Associates published a report entitled *Report on Solar Heating System Demonstration House: Calgary, Alberta* (Ottawa: National Research Council, 1977. 20 p). NRC also published some general information materials as part of a solar information series. One was entitled *Solar Heated Homes in Canada* (Ottawa: National Research Council, 1979, 24 p.). The program to date was described by A.H. Grignon in the publication *The National Research Council Solar Energy Program* (Ottawa: National Research Council, 1979, 36 p.).

By the 1980s, NRC had also built facilities for testing large industrial and commercial heating applications and participated in technical developments necessary for industry to manufacture absorber plates for solar collectors. The overall program, along with NRC's participation in it, was summarized in a review publication issued by the Division of Energy entitled *Alternative Energy Technology in Canada: NRC's R&D Program, 1975–1985* (Ottawa: National Research Council of Canada, 1986. 154 p.).

The goal of the wind energy programs at NRC was to determine whether or not wind energy development systems appropriate to Canadian conditions could be developed to make cost-effective contributions to current and future energy needs, as well as to create a visible Canadian domestic and export industry. The Atlantic Region was selected for site locations and the Atmospheric Environment Service cooperated with NRC on the initial evaluations before any technology could be developed. Some work was also contracted out to private business. One paper generated during the initial studies was written by C. Brothers and H. Arthur and entitled *Effects of Wind Speed Sampling and Averaging Intervals on Accuracy of Wind Energy Estimates* (Ottawa: National Research Council, 1985. 36 p.). Another paper was written by Resource Ventures Inc. of Charlottletown, PEI, and entitled *An Investigation and Assessment of Procedures to Measure and Process Wind Energy Conversion System Performance Data at the Atlantic Wind Test Site* (Ottawa: National Research Council, 1984. 180 p.). The technology for wind conversion systems required the development of a vertical axis wind turbine. Some work had been done at NRC in the 1970s, and was described by R.S. Range and R.J. Templin in a paper entitled *Wind Power and the Vertical Axis Wind Turbine Developed at the National Research Council (Quarterly Bulletin of the Division of Mechanical Engineering and the National Aeronautical Establishment,* v. 2, 1974. pp. 1–5, 7–14). The overall technical program was described in the regular journal literature, again by R.S. Range and R.J. Templin in a paper entitled *Vertical-Axis Wind Turbine Development in Canada* (*Proceedings of the IEE,* v. 130A, 1983. pp. 555–61).

Another energy source explored by the NRC throughout the 1970s and early 1980 was peat energy. Canadian peat deposits are the second largest in the world and contain potential energy second only to coal, but this energy source has remained virtually untapped because of technical, economic, and environmental constraints. For these resources, considerable research had to be undertaken to estimate their energy value and develop mining, dewatering, and combustion technologies.

First, a peatland inventory had to be undertaken to accurately determine the locations and extent of Canada's peatland areas and estimate, by volume, the peatland types in Canada. With this information in hand, a computerized Wetlands Registry was created. This work was described by C. Tarnocai of the Land Resource Research Institute of Agriculture Canada, in the publication entitled *Peat Resources in Canada* (Ottawa: National Research Council of Canada, 1984. 17 p.). Test trials on the mining of peat, which requires slurry pumping and then dewatering, were carried out in central Ontario by the firm Monenco Consultants Ltd, and described in the publication *Peat Slurry Pumping/Pipeline Trials on a Canadian Peatland Site* (Ottawa: National Research Council, 1987. 281 p.). The dewatering research was carried out for NRC at Polysar Ltd in Sarnia, Ontario, and described by A.L. Ayub in the paper entitled *Dewatering of a Fuel Grade Peat with Surfactants-Adsorption and Electrokinetics* (*Colloids and Surfaces*, v. 18, 1986. pp. 43–52). Finally, research into the pollution aspects of peat combustion was carried out, again by Monenco Ltd, in New Brunswick, and described in the publication Pollution *Potential of Peat Combustion Liquefaction and Gasification* (Ottawa: National Research Council, 1983. 147 p.).

Other non-conventional sources of energy studied by the National Research Council were hydrogen, and fusion energy. Hydrogen energy was, and is, a technically complex problem both from the point of view of developing large-scale electrolysers for producing hydrogen and oxygen from water, and of storing the energy released in full-scale storage cells. There were also safety problems involved in its storage and distribution as a fuel, and this concern was described by R.E. Knowlton in a paper entitled *An Investigation of the Safety Aspects in the Use of Hydrogen as a Ground Transportation Fuel* (*International Journal of Hydrogen Energy*, v. 9, 1984. pp. 129–36).

Fusion energy development at NRC was carried on in collaboration with the International Energy Agency (IEA). Canada's research has concentrated on the magnetic confinement of high temperature plasmas (the state of the nuclear fuel used when it has been ionized during the initial fusion process) and on studies of the management of tritium, one of the fusion fuels involved (tritium, which is a radioactive isotope of hydrogen, is fused with deuterium to create the energy release). This highly specialized work has been described by H.A. Baldis and D.M. Villeneuve of NRC's Division of Physics, in the article entitled *Plasma Waves in Laser Fusion Plasmas* (*Canadian Journal of Physics*, v. 64, 1986. pp. 961–68); and by W.R. Dixon and R.S. Storey, in an article entitled *Radiation Damage Effects in Tritium Implanted SI(LI) Detections (Nuclear Instruments & Methods in Physics Research*, v. B12, 1985. pp. 304–10).

(b) Food Production

Food research, including handling and production, became an NRC focus in the mid-1970s. The former involved traditional research approaches, but the latter entered the realm of high technology in the mid-1980s with NRC's new emphasis on biotechnology and genetic engineering in the search for new foods and greater food productivity.

In the mid-1970s, the Food Technology Section of the Division of Biological Sciences carried out research on the refrigeration of meats in order to develop techniques that would increase the shelf life of pre-packaged beef. NRC researchers concentrated on studying temperature and gaseous composition of storage rooms to find the optimum combinations that would both inhibit bacterial growth and preserve meat colour. Some of this work appeared in the journal literature. L.A. Both and D.S. Clark wrote a paper entitled *Studies on the Bacterial Flora of Vacuum-packed Fresh Meat* (*Canadian Journal of Microbiology*, v. 18, 1972. pp. 1761–66); and D.S. Clark and C.P. Linty wrote a paper entitled *Use of Mixtures of Carbon Dioxide and Oxygen for Extending Shelf-Life of Pre-packaged Fresh Beef* (*Canadian Institute of Food Science and Technology Journal*, v. 6, 1973. pp. 194–96).

Also in the 1970s, the Prairie Regional Laboratory became involved in food-related research that witnessed advances in understanding how legumes (peas, beans, and alfalfa) and rhyzobacteria (the bacterial symbionts that lodge in the plant roots) combine to 'fix' atmospheric nitrogen.

The first experimental work involved inducing bacteria to fix nitrogen in association with, rather than inside, the root cell. This work was described by J.J. Child in a paper entitled *Nitrogen Fixation by a Rhizobium sp. in Association with Non-leguminous Plant Cell Cultures* (*Nature*, v. 253, 1975. pp. 350–51). Subsequent studies determined the nature of the substances that plant cells transmit to cause fixation to take place. These trigger substances turned out to be pentose sugars such as arabinose and tylose. These results were then verified by removing the plant cells and inducing the fixation process by simply adding the required sugars. This work was described by F.B. Holl in the article *Host Plant Control of the Inheritance of Dinitrogen Fixation in the Pisum-Rhizobium Symbiosis* (*Euphytica*, v. 24, 1975. pp. 767–70).

By the 1980s, food research became important in NRC's biotechnology program when the Division of Biological Sciences in Ottawa carried out industrially important work with enzymes and carbohydrates. For example, the Microbial Processes Group worked with plant polysaccharides to develop enzyme-modified compounds responsible for the flavours and fragrances of many natural foods and beverages. This process used yeast cells for the biogeneration of natural acetaldehyde for the flavour esters. The work was considered particularly important because of the commercial demand for 'natural' as opposed to 'artificial' flavouring. D.W. Armstrong and H. Yamazaki described this work in the article entitled *Natural Flavours Production: A Biotechnical Approach* (*Trends in Biotechnology*, v. 4, 1986. pp. 264–68).

Another example of this type of work involved the development of novel yeast strains with increased fermentation rates and with the ability to utilize starch and dextrin substrates. In a continuing process of technology transfer with the Labatt Brewing Company, that firm used these new yeast strains to produce new products and to increase fermentation and brewhouse capacity. This work was described in the following articles, which came out of the Division of Biological Sciences. One was by H. Lee and P. Biely entitled *Utilization of Xylan by Yeasts and Its Conversion to Ethanol by* Pichea stipitis *Strains* (*Applied and Environmental Microbiology*, v. 52, 1986. pp. 320–24). Another was by L.G. Neirinck and S.C. Tsai entitled *Xylitol as a Carbon Source for Growth and Ethanol Production by* Packysolen tannophilus (*Canadian Journal of Microbiology*, v. 31, 1985. pp. 529–37).

The search for new foods can be seen in the 1990s at the Institute for Marine Biosciences with the study of the economic feasibility of aquaculture operations for some marine algae, specifically seaweeds as food. Some seaweeds have a long-standing tradition as food sources, as well as being sources of polysaccharides required by the food, pharmaceutical, and cosmetic industries. More specifically, two species have been examined. They are *Chondrus crispus* (the Irish moss) and *Palmaria palmata* (dulse). *Chondrus* has potential as a sea vegetable product for export to Japan, and *Palmaria* as a feed in American abalone aquaculture. Genetic studies of the two seaweeds have been carried out by C.J. Bird and C.A. Murphy in their paper entitled *The 18s rRNA Gene Sequences of Four Commercially Important Seaweeds* (*Journal of Applied Phycology*, v. 4, 1992. pp. 379–84). These genetic studies are important as part of fundamental and taxonomic studies at the molecular level, for, as C.J. Bird and J.P. Van Der Meer said, in their paper entitled *Systematics of Economically Important Marine Algae: A Canadian Perspective* (*Canadian Journal of Botany*, v. 71, 1993. pp. 361–69), 'it should be axiomatic that a commercially important species must be well defined by sound taxonomy so that it is recognizable to the user and predictable in its performance.'

Further genetically based taxonomic studies can also be noted in the paper by C.J. Bird entitled *A Review of Recent Taxonomic Concepts and Developments in the Gracilariaceae (Rhodophyta)* (*Journal of Applied Phycology*, v. 7, 1995. pp. 255–67); and by C.J. Bird and T. Chopin entitled *A Molecular Investigation of Polymorphism in the North Atlantic Red Algae* Chondrus crispus (*Gigartinales*) (*Phycological Research*, v. 44, 1996. pp. 69–80).

Also by the 1990s, the Plant Biotechnology Institute, which was formed in 1985, began to undertake research to identify and regulate genes useful for increasing plant productivity as well as to create genetically engineered plants potentially useful to the Canadian agriculture and forestry industries. In agriculture, wheat and barley plants have been a focus of the research. This research has involved the study of the genetic transformation of wheat and barley using microprojectile bombardment on immature zygotic (early fertilized and dividing) embryos and cell suspension cultures. This work has been carried out by the Cereal Biotechnology Group, and reported by R.N. Chibbar and K.K. Kartha in papers entitled *Microprojectile Mediated Delivery of Marker Genes and Their Expression in* In Vitro *Cultures and Immature Zygotic Embryos of Wheat and Barley* (*Journal of Cellular Biochemistry. Supplement*, v. 14E, 1990. p. 277); *Transient Expression of Chloramphenicol Acetyltransferase (CAT) Gene in Barley Cell Cultures and Immature Embryos through Microprojectile Bombardment* (*Plant Cell Reports*, v. 8, 1990. pp. 429–32); and *Transient Expression of Marker Genes in Immature Zygotic Embryos of Spring Wheat* (Triticum aestivum) *through Microprojectile Bombardment* (*Genome*, v. 34, 1991. pp. 453–60). An overview of the work in this field was reported by K.K. Kartha in the report entitled *Cereal Biotechnology Project: A Review of Activities and Progress 1986–1990* (Ottawa: National Research Council, 1991. 39 p. NRC Internal Report PBI-91–02).

The Conifer Biotechnology Group has been studying the procedures for the production and multiplication of genetically improved conifer trees, primarily spruce and pine, by the cloning of selected trees using several methods. One method used has been the development of shoots from buds of one- to five-year-old pine trees (*Pinus montiola*) cultured *in vitro* in a medium of benzyladine (a plant growth regulator). As well, hormonal spray treatments *in vitro* have been used to improve rooting and cuttings from one- to 10-year-olds from the pine, *Pinus banksea*. Another method has used suspension cultures of somatic (immature and dividing) embryos of the spruce tree *Picea glauca*, using the plant growth hormone abscisic acid (ABA) to obtain embryo maturation. This conifer research has been described by D.I. Dunstan in the report entitled *Conifer Biotechnology Project: A Review of Activities and Progress, 1986–1990* (Ottawa: National Research Council, 1991. 42 p. NRC Internal Report PBI-91–04); and in the papers by T.D. Bethune and D.I. Dunstan entitled *Racemic Abscisic Acid and Abciysl Alcohol Promote Maturation of Spruce* (Picea glauca) *Somatic Embryos* (*Plant Science*, v. 76, 1991. pp. 219–28); *Somatic Embryo Maturation from Long-Term Suspension Cultures of White Spruce* (Picea glauca) (*In Vitro Cellular and Developmental Biology – Plant*, v. 29, 1993. pp. 109–112); *Variability in Maturation and Germination from White Spruce Somatic Embryos as Affected by Age and Use of Solid or Liquid Culture* (*In Vitro Cellular and Developmental Biology – Plant*, v. 32, 1996. pp. 165–70); by

D.I. Dunstan and C.A. Bock, in their paper entitled *Abscisic Acid [(+)-ABA] Content in White Spruce Somatic Embryo Tissues Related to Concentration of Fed ABA* (*Journal of Plant Physiology*, v. 150, 1997. pp. 691–96); and by D.I. Dunstan and J.Z. Dong, in their papers entitled *Events Following ABA Treatment of Spruce Somatic Embryos* (*In Vitro Cellular and Developmental Biology – Plant*, v. 34, 1998. pp. 159–68); and *Cloning and Characterization of Six Embryogenesis-associated cDNAs from Somatic Embryos of* Picea glauca *and Their Comparative Expression during Zygotic Embrogenesis* (*Plant Molecular Biology*, v. 39, 1999. pp. 856–64).

(c) Building and Construction Research

Building and construction research was underlined as a major activity of the National Research Council through its Division of Building Research (later called the Institute for Research in Construction) in the early 1940s when NRC began to review its programs and look beyond the end of World War II. Besides being responsible for the National Building Code since 1935 and its later 'spin-offs,' NRC has been active in ongoing research of all aspects of construction such as building performance, building materials, and geotechnical research. Of particular interest has been its work in fire safety, building structures, acoustics, and noise in the urban environment.

Canada's first comprehensive building code was published in 1941 and entitled *National Building Code* (Ottawa: National Research Council, 1941. 1 v.). By 1943, an associate committee was formed to oversee updates or supplements and new editions of the code. This early history of the National Building Code has been described by H.B. Dickens and A.T. Hansen in a paper entitled *The National Building Code – Its Development and Use in Canada* (Ottawa: National Research Council. Division of Building Research, 1974. 3 p.). A second edition of the National Building Code was not issued until 1953. It was entitled *National Building Code of Canada, 1953* (National Research Council, 1953. 8 v. in 1 – looseleaf). This expanded edition covered the topics of administration, climate, use and occupancy, design, materials, general services, plumbing services, and construction safety measures. The most recent edition of the Code was published in 1995, and was entitled *The National Building Code of Canada, 1995.* 11th Edition (Ottawa: National Research Council, Institute for Research in Construction, 1995. 571 p.). Along with the new edition, NRC also published related material that clarified the new code. One of these publications was entitled *What's New in the NBC 1995* (Ottawa: National Research Council, 1996. 100 p.). Another was entitled *User's Guide – NBC 1995: Structural Commentaries (Part 4).* 1st Edition (Ottawa: National Research Council, 1996. 135 p.).

Besides the basic building code, NRC now also publishes separate related codes. They are the *National Farm Building Code of Canada, 1995*. 8th Edition (Ottawa: National Research Council, 1995. 31 p.); the *National Fire Code of Canada, 1995*. 7th Edition (Ottawa: National Research Council, 1995. 162 p.); and its related publication *What's New in the NFC 1995* (Ottawa: National Research Council, 1996. 73 p.); the *Canadian Plumbing Code of Canada, 1995*. 7th Edition (Ottawa: National Research Council, 1995. 113 p.); and the *National Housing Code of Canada, 1998 and Illustrated Guide* (Ottawa: National Research Council, 1998. 1 v.).

Fire safety research at the Institute for Research in Construction is 'directed towards reducing the loss of life and property due to fires, as well as lowering the cost of fire protection in buildings.' For example, in 1989, NRC estimated that building fires took about 500 lives in Canada each year injured over 4,000 people and destroyed over $1 billion worth of properties. Some of the specific research carried out has involved fire safety in high buildings and residential buildings, fire spread over external walls, and the fire resistance of building elements and assemblies.

The fire studies with high buildings began in the early 1970s. The increase of populations in large urban centres and the soaring rate of high-rise construction outpaced expert knowledge about what happened in them during a fire. In these buildings, it would only take a series of critical circumstances to occur together to create the conditions for catastrophic results. In 1973, NRC issued *Measures for Fire Safety in High Buildings: To Form Part of Supplement No. 3 to the National Building Code of Canada* (Ottawa: National Research Council, 1973. 127 p.). This publication was revised in 1977 and issued as *Measures for Fire Safety in High Buildings, 1977*. 2nd Edition (Ottawa: National Research Council, 1977. 74 p.). In 1985, this title was issued as a *Supplement to the National Building Code of Canada, 1985: Commentaries of Part II of the National Building Code of Canada, 1985* (Ottawa: National Research Council, 1985. 278 p.).

Overall, measures set out provisions to increase the safety of occupants in high-rise buildings that are more than 60 feet high, contain sleeping accommodation, or have a certain population density. Some of these measures included techniques such as the injection of air into major shafts and venting others, the use of ventilated and pressurized lobbies, elevators, and stair shafts, and total building pressurization with venting of the fire floor. Within buildings, NRC became specifically interested in the flammability of certain new building materials, particularly because our northern climate requires heating of buildings for several months a year. For example, with the advent of plastic foams, the traditional concept of flammability became doubtful, because plastic foams with a low flame-spread were developed at the Fire Research Laboratory of the Division of Building Research. T.T. Lie discussed this problem in a paper entitled *Contribution of*

Protected Plastic Foams to Fire Growth (Ottawa: National Research Council, 1975. 19 p. DBR Fire Study 37). In conjunction with the Underwriters Laboratories of Canada, the Fire Research Laboratory developed further fire hazard tests, which better represented fire development in a room. This meant there was a more accurate assessment of the fire hazards of different materials. Two papers representing further studies were written by M. Day and D.M. Wales. They were entitled *Carpet Flammability: An Oxygen Index Modification to the Pile Test* (*Journal of Fire and Flammability*, v. 4, 1973. pp. 165–73); and *An Experimental Procedure for Determining the Flammability of Carpets Using the Tri-Flammability Analyzer* (*Textile Research Journal*, v. 44, 1974. pp. 5–11).

The overall experience from work on fire safety with tall buildings was discussed by M. Galbraith in the paper entitled *Canadian Experience of Fire Safety in High Buildings* (*Fire Safety Journal*, v. 7, 1984. pp. 87–91).

Later, the Institute for Research in Construction was able to augment fire studies considerably with the building of a large-size burn hall and a 10-story tower for experimental burns. For example, a paper by G.T. Tamura and J.H. Klote described the results of a study of smoke movement from a large fire through elevator shafts. It was entitled *Experimental Fire Tower Studies of Elevator Pressurization Systems for Smoke Control* (*ASHRAE Transactions*, v. 93, Part 2, 1987. pp. 2235–56). At the same time, NRC also began to use computers to create computer models for assessing fire hazards in all residential buildings. This work has been described by V.R. Beck and D. Yung in their paper entitled *A Cost Effective Risk Assessment Model for Evaluating Fire Safety and Protection in Canadian Apartment Buildings* (*Journal of Fire Protection Engineering*, v. 2, 1990. pp. 65–74).

Related to fire safety in tall buildings has been NRCs work on fire hazards in other residential buildings. In the late 1980s, NRC developed a three-story facility at the National Fire Laboratory. The smaller-scale burn rooms allowed the experimental assessment of hazards such as wall assemblies, window assemblies, the spread of fire over exterior walls, the use of cavity wall with insulation for fire prevention, and the quality of air after indoor fires. Some of this research was reported in papers by J.K. Richardson and G.A. Chown entitled *Glazing in Fire Resistant Wall Assemblies* (Ottawa: National Research Council Institute for Research in Construction, 1988. 4 p.); by J.K. Richardson and I. Oleszkiewiez, in their paper entitled *Fire Tests on Window Assemblies Protected by Automatic Sprinkers* (*Fire Technology*, v. 23, 1987. pp. 115–32); and by K.K. Choc, in the paper entitled *Effects of Insulation on Post-flashover Room Fire* (*Fire Technology*, v. 23, 1987. pp. 19–25). Y. Tsuchiya has studied air quality problems in papers entitled *Volatile Organic Compounds in Indoor Air* (*Chemosphere*, v. 17, 1988. pp. 79–82); *Air Quality Problems inside a House Following a Fire* (*Journal of Fire Sciences*,

v. 10, 1992. pp. 58–71); and *Chemical Modelling of Fire Gases* (*Journal of Fire Sciences*, v. 13, 1995. pp. 214–23).

Another intensive study of fire resistance in building elements or assemblies has involved the testing of reinforced concrete columns. This was done in a joint study with the Portland Cement Association and has led to a series of research papers with T.T. Lie as the principal investigator. The first paper by T.T. Lie was entitled *Fire Resistance of Reinforced Concrete Columns: A Parameter Study* (*Journal of Fire Protection Engineering*, v. 1, 1989. pp. 49–61). Another by T.T. Lie and B. Celikkol was entitled *Method to Calculate the Fire Resistance of Circular Reinforced-Concrete Columns* (*ACI Materials Journal*, v. 88, 1991. pp. 84–91). Yet another study on this topic by T.T. Lie and D.C. Stringer was entitled *Calculation of the Fire Resistance of Steel Hollow Structural Section Columns Filled with Plain Concrete* (*Canadian Journal of Civil Engineering*, v. 21, 1994. pp. 382–85). And finally, four papers by T.T. Lie and V.K. Kodur were entitled *Thermal and Mechanical Properties of Steel-Fibre-Reinforced Concrete at Elevated Temperatures* (*Canadian Journal of Civil Engineering*, v. 23, 1996. pp. 511–17); *Fire Resistance of Circular Steel Columns Filled with Fibre-Reinforced Concrete* (*Journal of Structural Engineering – ACSE*, v. 122, 1996. pp. 776–82); *Fire Resistance of Steel Columns Filled with Bar-Reinforced Concrete* (*Journal of Structural Engineering – ASCE*, v. 122, 1996. pp. 30–36; and *Evaluation of Fire Resistance of Rectangular Steel Columns Filled with Fibre-Reinforced Concrete* (*Canadian Journal of Civil Engineering*, v. 24, 1997. pp. 339–49).

The Acoustics Section of the Institute for Research in Construction undertakes research to develop and apply noise control technology because of sound transmission within buildings and through enclosures. Research is also undertaken on sound transmission in enclosed spaces, especially spaces for speech propagation and the performing arts. In urban areas, NRC has also studied the noise problems of road traffic and the noise produced by air conditioners. In order to facilitate noise studies, the Acoustics Section has developed easily used, but sophisticated, design tools and measurement techniques such as computer software for studying building acoustics as well as predicting noise reduction. In the late 1980s, new software was also developed to provide for the full automation of standard sound measurement in the IRC reverberation chambers for sound transmission through walls and floors, impact noise from floors, sound power emission from noise sources, and sound absorption.

Within modern office buildings, the problem of noise is sound transmission between offices caused by sound transference through suspended ceilings which are used as a conduit for building services such as electricity, plumbing, or air supply ducts. The Acoustics Section has studied this problem and achieved noise reductions by the use of the 'fuzzblock' technique (a stack of absorption material to block the space above the suspended ceiling system). This research, using absorption batts, was undertaken by R.C. Halliwell and J.D. Quirt, and described in their paper entitled *Controlling Inter-office Sound Transmission through a Suspended Ceiling* (*Journal of the Acoustical Society of America*, v. 90, 1991. pp. 1446–53). The authors concluded that the fuzzwall 'provides a simple, economical means of increasing the sound insulation between offices without compromising the advantages of open plan construction.' Other materials tested for sound transmission in buildings have been drywall, cavity blocks, and concrete blocks. This work was carried out by A.C. Warnock, and reported in the papers entitled *Sound Transmission Loss Measurements through 190mm. and 140mm. Blocks with Added Drywall and through Cavity Block Walls* (Ottawa: National Research Council, 1990. 39 p. Institute of Research in Construction Internal Report No. 586); *Sound Transmission through Concrete Block Walls (Construction Canada*, v. 32, 1990. pp. 41–50); *Sound Transmission through Concrete Blocks with Attached Drywall* (*Journal of the Acoustical Society of America*, v. 90, 1991. pp. 1454–63); and *Sound Transmission through Two Kinds of Porous Concrete Blocks with Attached Drywall* (*Journal of the Acoustical Society of America*, v. 92, 1992. pp. 1452–60).

Research involving auditorium acoustics took J.S. Bradley to Europe to examine the acoustics of seven well-known European concert halls along with Roy Thomson Hall and the St Lawrence Centre in Toronto, and the National Arts Centre in Ottawa. One of the first papers to appear from this work by J.S. Bradley was entitled *A Comparison of 3 Classical Concert Halls* (*Journal of the Acoustical Society of America*, v. 89, 1991. pp. 1176–92). The three halls examined were two European halls (the Amsterdam Concertgebouw and the Vienna Grosser Musikvereinsaal) and the Boston Symphony Hall. The acoustical measurements undertaken helped to define the range of conditions expected in good concert halls. One aspect of concert-hall acoustics is the sound absorption of theatre chairs. This was studied by J.S. Bradley in the paper entitled *Predicting Theatre Chair Absorption from Reverberation Chamber Measurements* (*Journal of the Acoustical Society of America*, v. 91, 1992. pp. 1514–24). Related directly to this work is the sound absorption from occupied auditorium seats. This study was also carried out by J.S. Bradley, and reported in the paper entitled *The Sound Absorption of Occupied Auditorium Seating* (*Journal of the Acoustical Society of America*, v. 99, 1996. pp. 990–95). This study was again carried out in the reverberation chambers of IRC's Acoustics Section. Other auditorium research carried out concerned the sense of spatial impression to the listener. For this, J.S. Bradley examined 14 North American concert halls, and in the paper entitled *Comparison of Concert Hall Measurements of Spatial Impression* (*Journal of the Acoustical Society of America*, v. 96, 1994. pp. 3525–35) said that the theoretical measurements of spatial impression coincided with actual hall measurements. Finally, J.S. Bradley and

G.A. Soulodre studied and wrote *Objective Measures of Listening Envelopment* (*Journal of the Acoustical Society of America*, v. 98, 1995. pp. 2590–97). The authors concluded that listener envelopment 'depends on having strong lateral reflections arriving at the listener 80 mm. or more after direct sound.' This listening envelopment was emphasized once again in their paper entitled *Listener Envelopment: An Essential Part of Good Concert Hall Acoustics* (*Journal of the Acoustical Society of America*, v. 99, 1996. p. 22). Related work carried out by J.S. Bradley has examined the acoustics of speech. These papers have the titles *Relationships among Measures of Speech Intelligibility in Rooms* (*Journal of the Audio Engineering Society*, v. 46, 1998. pp. 396–405); and, with R.D. Reich, *On the Combined Effects of Signal-to-Noise Ratio and Room Acoustics on Speech Intelligibility* (*Journal of the Acoustical Society of America*, v. 106, 1999. pp. 1820–28).

Beyond building noise, or acoustics, NRC's Acoustics Section has also studied urban traffic noise, and noise from residential air conditioners. This type of noise can cause speech and sleep interference, as well as annoyance. J.S. Bradley and B.A. Jonah studied traffic noise in a series of papers entitled *The Effects of Site Selected Variables on Human Responses to Traffic Noise Part I: Type of Housing by Traffic Noise Level* (*Journal of Sound and Vibration*, v. 66, 1979. pp. 589–604); *The Effects of Site Selected Variables on Human Responses to Traffic Noise Part II: Road Type by Socio-Economic Status by Traffic Noise Level* (*Journal of Sound and Vibration*, v. 67, 1979. pp. 395–407); and *The Effects of Site Selected Variables on Human Responses to Traffic Noise Part III: Community Size by Socio-Economic Status by Traffic Noise Level* (*Journal of Sound and Vibration*, v. 67, 1979. pp. 409–23). Loud traffic transportation noises were the major source of noise (74%) in neighbourhoods. However, other noises can also cause annoyance. One such noise is residential air conditioners. J.S. Bradley carried out a field survey and obtained responses on the level of disturbance from the type of noise. The results were published in the paper entitled *Disturbance Caused by Residential Air Conditioner Noise* (*Journal of the Acoustical Society of America*, v. 93, 1993. pp. 1978–86), and indicated that subjective responses to air conditioner noise were significantly related to measured noise levels and 'annoyance to the noise of the neighbour's air conditioner tended to be related to the cube of the objective noise measures.'

(d) Transportation Research

Transportation research at the National Research Council covers both ground and air transportation. Since 1991, NRC's ground transportation studies have been amalgamated into the Ground Transportation Technology Program of the Institute for Mechanical Engineering (formerly the Division of Mechanical Engineering) and involve railway and highway vehicle research. To assist in these studies, the Program has national facilities located in Ottawa and at the Western Laboratory in Vancouver made up of a Curved Track Simulator (designed to measure phenomena associated with railway 'hunting' and curving behaviour); the Wheel Bearing and Brake Facility (designed for controlled full-scale reproduction of railway conditions); the Vehicle Squeeze Frame; the Vehicle Vibration Facility; and the Truck Suspension Measurement Facility. Air transportation studies since 1991 have been centred in the Institute for Aerospace Research (formerly the National Aeronautics Establishment). However, it is also interesting to note that NRC's earlier ground transportation studies were also carried out by the National Aeronautics Establishment. These studies, from the 1970s, involved car headlamps and non-rigid highway barriers.

The object of NRC's research was the study of driver-headlight interaction in order to reduce traffic accidents. Once headlights currently in use had been analysed and catalogued, this data could be used to make recommendations regarding how headlights could be better designed, focused, powered, and installed for greater highway driving safety. Some of this work at the National Aeronautical Establishment was written up in reports by R. Kulchyski entitled *Headlight Illumination Measurement in Real Traffic Situations. Part I: A Pilot Study to Evaluate the Data Acquisition and Analysis Systems* (Ottawa: National Aeronautical Establishment, LTR-ST-944, 1977. 9 p.); and by A. Harrison, entitled *View: A Headlight Research Program for Producing an Illumination Map on the Vertical Plane about the Eye (or Some Other) Axis* (Ottawa: National Aeronautical Establishment, LTR-ST-608, 1973. 149 p.). This research program was recognized by the International Commission on Lumination to be authoritative.

NRC's studies with non-rigid highway barriers were the first done on these barriers under controlled conditions. The studies used computer simulation models developed from data obtained during field tests carried out at the beginning of 1968 in order to understand what happens when an automobile hits a flexible safety barrier. The testing facilities also utilized controlled experimental conditions for checking the validity of the computer simulation studies. One of the early papers describing this work was written by H.F. Pinkney and G.L. Basso and entitled *The NAE Model of the Highway Cable Barrier for Parametric Studies of Vehicle Redirection Using Digital Simulation* (Ottawa: National Research Council, 1972. 67 p. Report 12694). A later summary of this highway barrier research was written by G.L. Basso and entitled *A Parametric Study on Highway Cable Barrier Performance and Its Effect on Vehicle Redirection Dynamics* (Ottawa: National Research Council, 1977. 64 p. Report 15885).

As mentioned above, more recent land transport research has been undertaken by the Institute for Mechanical Engineering and has focused on railway and highway vehicle research. NRC's railway studies, sometimes in collaboration with

CP Rail, have been concerned with the displacement, velocity, acceleration, and curving performance of rail vehicles in real service conditions, using equipment and techniques developed as part of their testing facilities. As well, studies are carried out on wheel/rail interactions.

Acceleration studies began in the mid-1970s, and were reported by F.B. Blader and R.J. Senn in their report entitled *Acceleration, Braking, Curving and High Speed Tests of the LRC Locomotive and Coach on CP Track: A Preliminary Report* (Ottawa: National Research Council, 1976. 20 leaves. Division of Mechanical Engineering Laboratory Laboratory Technical Report LTR-IN-338). Later, a transducer which measures very small accelerations as well as velocity and distance was built to be used in road tests. The transducer was found to be particularly suited for measuring small accelerations during braking. This latter work was described by C.M. Zwarts in the report entitled *Rail Vehicle Performance Monitor for Acceleration, Velocity and Distance* (Ottawa: National Research Council, 1983. 22 p. Mechanical Engineering Report MI-840). In the early 1980s, curving performance of railway vehicles was also studied by I.H. Mufti and R.V. Dukkipati in their papers entitled *The Creepage and Spin Qualities for NRC Curved Track Simulator* (Ottawa: National Research Council, 1980. 25 p. Division of Mechanical Engineering Laboratory Technical Report LTR-AN-41); and *On the Dynamics of a Wheelset for NRC Curved Track Simulator* (Ottawa: National Research Council, 1983. leaves 545–63). By the mid-1980s, and later, computer programs and simulation models were created for the study of rail vehicle performance. One such computer program was described by R.V. Dukkipati in the report entitled *Computer Programs for Rail Vehicle Designs* (Ottawa: National Research Council, 1984. 4 p.). Simulator models were studied by R.O. Majeed and R.V. Dukkipati in their paper entitled *CTS and Freight Car Truck Data for Tangent Track Hunting Simulation Models* (Ottawa: National Research Council, 1992. 31 p.).

The rail/wheel interaction in rail vehicle performance studies have focused upon the analysis of the mechanical, metallurgical, and lubrication parameters of fatigue in order to get a better understanding of rail/wheel wear. J. Kalousek has studied lubrication factors on rail/wheel forces in the publications entitled *Lubrication: Its Various Types and Effects on Rail/Wheel Forces and Wear* (Ottawa: National Research Council, 1986. 14 leaves. Reprinted from the *Proceedings of the Rail and Wheel Lubrication Symposium, December 1–2, 1981. Peabody Hotel, Memphis*); and *The Effect of Lubrication and Steering on Rail/Wheel Wear and Contact Fatigue* (Ottawa: National Research Council, 1988. 12 leaves. Reprinted from the *Proceedings of the Conference on the Economics and Performance of Freight Car Trucks, AAR Report R-590, October, 1983*). J. Kalousek also examined the need for a testing facility in the paper entitled *Rail/ Wheel Wear Testing Facility* (Ottawa: National Research Council, 1981. 8 p.) and pursued this idea in the paper entitled

Development of a Scale Dual Disc-on-Disc Rail/Wheel Wear Testing Facility (*Canadian Metallurgical Quarterly*, v. 21, 1982. pp. 67–72). The concept for a full-scale test facility was written up in the paper by H.G. Tucker entitled *Proposal for a Full-Scale Multipurpose Test Facility* (Ottawa: National Research Council, 1984. 1 v. Division of Mechanical Engineering Report TR-RW-002). This proposal was carried further, by the same author, in the paper entitled *Design Specifications and Acceptance for a Full-Scale Multipurpose Wheel/Rail Test Facility* (Ottawa: National Research Council, 1985. 43 p. Division of Mechanical Engineering Report TR-RW-005). Later, mathematical modelling was applied to wheel/rail contact studies in the paper by S. Sankar and R.V. Dukkipati entitled *A Finite Element Model of Railway Track and Its Application to the Wheel Flat Problem* (*Proceedings of the Institution of Mechanical Engineers, Part F. Journal of Rail and Rapid Transit*, v. 208, 1994. pp. 61–72).

Heavy land vehicle studies have been carried on by the National Research Council for over 40 years, since the paper by H.B. Sutherland, entitled *A Study of Vibration Produced in Structures by Heavy Vehicles* (*Proceedings of the Highway Research Board, December, 1956*. pp. 406–19), indicated that large axle loads imparted by heavy trucks were responsible for a significant portion of the damage suffered by pavements and bridges. Of particular interest have been the joint studies by NRC with the Forest Engineering Research Institute of Canada pertaining to logging trucks. These have ranged from economic analyses to safety concerns. One of the first studies, issued by D.G. Smith and P.P. Tse, was entitled *Logging Trucks: Comparison of Productivity and Cost* (Vancouver: Forest Engineering Research Institute of Canada, 1977. 43 p. Technical Report TR-18). Cost-effectiveness of logging trucks was also studied by J.G. Garner in the paper entitled *Wind Tunnel Tests of Devices for Reducing the Aerodynamic Drag of Logging Trucks* (Vancouver: Forest Engineering Research Institute of Canada, 1980. 13 p.). Even the stacking of logs has been analysed for cost-effectiveness, as noted in the paper by M.L. Clark and D.R. Giles entitled *Orientation of Log Butts on Highway Trucks: An Economic Analysis* (Pointe Claire, Quebec: Forest Engineering Institute of Canada, 1988. 34 p. Technical Report TR-85). By the 1990s, logging truck safety, concerning stability and braking, was being studied. One study by M. El-Gindy and J.H. Woodroof was entitled *Study of Rollover Threshold and Directional Stability of Log Hauling Trucks* (Ottawa: National Research Council, 1990. 1 v. Division of Mechanical Engineering Vehicle Dynamics Laboratory TR-VOL-002). Two other studies by S.P. Parker dealt with braking systems and were entitled *Development of Brake Tests for Off-Highway Logging Trucks* (Pointe Claire, Quebec: Forest Engineering Institute of Canada, 1993. 15 p. Technical Report TR-107); and *Evaluation of an Antilock Braking System on an Off-Highway Log-Hauling Truck* (Pointe Claire, Quebec: Forest Engineering Research Institute of Canada, 1996. 9 p. Technical Note TN-241). It is interesting to note that in 1994, NRC's Ground Transportation Technology Program was incorporated into a new Centre for Surface Transportation Technology which included a Pavement Laboratory.

Air transport research at NRC has been wide ranging, since the 1960s onward, and has covered such topics as bird hazards at airports, stability studies of vertical, or short take-off and landing aircraft, studies of the Canadian-built Twin Otter aircraft, and the use of the Falcon 20 jet as a research aircraft.

Studies on bird hazards began in 1962 at the request of Transport Canada and triggered the creation of an Associate Committee on Bird Hazards. With the growth of airplane traffic at airports, bird collisions became a serious hazard because of potentially fatal damage to aircraft and passengers through windshield shattering, wing punctures, and the destruction of jet engines. Originally, the committee had hoped to be able to use information readily available from ecological studies. They wanted to gain an understanding of why birds were attracted to airport sites and why they migrated at certain hours, and whether the committee could come up with suggestions to reduce the hazard. These studies were undertaken by the Canadian Wildlife Service and described on pp. 436–37. However, the committee also carried out its own studies based on the assumption that birds live around airports for good reasons; hence they made recommendations for modifications to airports that diminished their attractiveness to bird flocks. Generally, the committee recommended that the environs in and around airports, which offered birds food, shelter, or water, be removed. For example, ditches should be drained, protective hedges cut down, and nearby garbage dumps relocated. This work was described by H.F. Lewis, of the National Aeronautical Establishment, in an article entitled *Bird Hazards to Aircraft* (*Bulletin of the International Aviation Organization*, v. 22, 1967. pp. 6–9). Along with these studies, the committee suggested research be carried out on aircraft windshield design. One such study, by D.J. Peake, was entitled *Some Additional Low Speed Bird Impact Tests on a Modified Windshield of the CL-41 Tutor Aircraft* (Ottawa: National Research Council, 1967. 41 p. Aeronautical Report LR-488). It is interesting to note that the Associate Committee on Bird Hazards disbanded in 1976 after 14 years of work and 68 meetings.

As mentioned above, another area of air transport studies focused upon fundamental studies of basic aircraft design for vertical or short take-off and landing aircraft. In the mid-1960s, there was a general acknowledgment that the requirements for rapid, inter-urban transport would lead to the extensive use of either VTOL or STOL aircraft, depending upon economic considerations. Keeping this in mind, the Flight Research Laboratory of the National Aeronautical Establishment undertook investigations of problems relevant to both STOL and VTOL aircraft.

These investigations involved an airborne V/STOL aircraft simulator, which was described by the Flight Research Laboratory in a paper entitled *Description of an Airborne V/STOL Aircraft Simulator: Operated by the Flight Research Section, National Aeronautical Establishment* (Ottawa: National Research Council, 1966. 7 leaves). Performance and stability of V/STOL aircraft were examined by R.J. Templin in a paper entitled *A Momentum Rule for Optimum Aircraft Performance in the V/STOL Transition Regime* (Ottawa: National Research Council, 1967. 49 p.); by D.M. McGregor, in a paper entitled *A Flight Investigation of Various Stability Augmentation Systems for a Jet-Lift V/STOL Aircraft (Hawker-Siddeley P1127) Using an Airborne Simulator* (Ottawa: National Research Council, 1968. 33 p. Aeronautical Report LR-500); and by D.F. Daw and K. Lum, in a paper entitled *Description of a Four Degrees of Freedom, V/STOL Aircraft Airborne Simulator* (Ottawa: National Research Council, 1968. 30 p. Aeronautical Report LR-499).

The Flight Research Laboratory had access to two variable stability helicopters, a Bell 47 and a Bell 205, which could be used for examining the flight characteristics of V/STOL aircraft under conditions of turbulence, wind shears, and low-speed manoeuvring. This flight research was carried out by D.M. McGregor, and reported in the papers entitled *Flight Assessment of a Variable-Stability Helicopter for STOL Simulations and Evaluation of the Influence of Several Lateral-Directional Stability Derivatives* (Ottawa: National Research Council, 1969. 23 p. Aeronautical Report LR-524); and *A Flight Investigation of Lateral-Directional Handling Qualities for V/STOL Aircraft in Low Speed Manoeuvring Flight* (Ottawa: National Research Council, 1971. 137 p. Aeronautical Report LR-549). These research flights led to both further simulator development and computer-based calculations for flight evaluations, as reported by W.S. Hindson and K. Lum in the bulletin entitled *Progress in the Development of a Versatile Airborne Simulator for V/STOL Aircraft* (*Quarterly Bulletin of the Division of Mechanical Engineering and the National Aeronautical Establishment, DME/NAE 1974*. pp. 25–51); by D.G. Gould and W.S. Hindson, in their bulletin entitled *Estimates of the Stability Derivatives of a Helicopter and a V/STOL Aircraft from Flight Data* (*Quarterly Bulletin of the Division of Mechanical Engineering and the National Aeronautical Establishment, DME/NAE 1974*. pp. 1–20); and later by S.R. Sinclair and T.C. West, in their paper entitled *Handling Qualities of a Simulated STOL Aircraft in Natural and Computer-Generated Turbulence and Shears* (Ottawa: National Aeronautical Establishment, 1978. 28 p.). Finally, additional flight tests were undertaken in the early 1980s, and described by W.S. Hindson and G.H. Hardy in their paper entitled *Flight-Test Evaluation of STOL Control and Flight Director Concepts in a Power-Lift Aircraft Flying Curved Decelerating Approaches* (Ottawa: National Research Council, 1981. 87 p.).

In the 1980s, aircraft stability studies shifted from V/STOL aircraft to the Canadian-built Twin Otter. These studies involved theoretical calculations, as well as practical studies of air flows and wind measurements during flight and in landing. The first of these studies by A.M. Drummond and J.I. MacPherson was entitled *Theoretical and Measured Airflow about the Twin Otter Wing* (Ottawa: National Research Council, 1984. 33 p. National Aeronautical Establishment Aeronautical Note NAE-AN-19). J.I. MacPherson also carried out modelling studies of the Twin Otter as noted in the paper entitled *NAE Twin Otter Operations in the 1988 Eulerian Model Evaluation Field Study* (Ottawa: National Aeronautical Establishment, 1989. 22 p. Flight Research Laboratory Technical Report LTR-FR-109). And B.W. Leach and J.I. MacPherson studied airborne wind measurements for the Twin Otter, in their paper entitled *An Improved Airborne Wind Measurement Technique for the NAE Twin Otter* (Ottawa: National Aeronautical Establishment, 1989. 39 p. Aeronautical Note NAE-AN-61). Wind measurements were also studied by J.I. MacPherson in the paper entitled *Wind and Flux Calculations on the NAE Twin Otter* (Ottawa: National Aeronautical Establishment, 1990. 22 p. Flight Research Laboratory Laboratory Technical Report LTR-FR-109). As well, J.B. Croll evaluated landing dynamics of the Twin Otter, in the papers entitled *Flight Evaluation of Complex Approaches in a Twin Otter Aircraft Using a Microwave Landing System* (Ottawa: Institute for Aerospace Research, 1990. 44 p. Flight Research Laboratory Laboratory Technical Report LTR-FR-114) and *Flight Evaluation of Curved MLS Precision Approaches in a Twin Otter Aircraft. Phase II* (Ottawa: National Research Council, 1991. 18 p. Institute for Aerospace Research Aeronautical Note IAR-AN-73).

J.B. Croll also undertook landing dynamic studies for the Falcon 20 jet aircraft in papers entitled *The Development of a Falcon 20 Aircraft as a Research Facility for Curved Path MLS Precision Approaches* (Ottawa: Institute for Aerospace Research, 1994. 1 v. Flight Research Laboratory Laboratory Technical Report LTR-FR-124); and *Flight Evaluation of Curved Path MLS Precision Approaches in a Falcon 20 Aircraft* (Ottawa: Institute for Aerospace Research, 1995. 22 p. Institute for Aerospace Research Aeronautical Note IAR-AN-82).

One other area of transport studies of particular interest, because of Canada's north, has been air-cushion vehicle technology. Air-cushion vehicle technology was seen by NRC as an important area of transport technology research, because of its usefulness in the field of resource exploitation in the Canadian north. The ability of air-cushion vehicles (ACV) to transport heavy loads, while exerting low ground pressures in comparison with conventional wheeled or tracked vehicles, meant that transport links could be maintained over terrain unsuitable for road construction because of permafrost. As well, adverse effects on the ecosystem by wheeled vehicles could be avoided. The first major publication about air-cushion vehicles was published in the late 1960s by NRC. It was entitled *Air Cushion Vehicles – Their Potential for Canada* (Ottawa: National Research Council of Canada, 1969. 518 p.). Later research activity in this field was reported by A.S. Jackson, in a Division of General Mechanics journal article entitled *Canadian Activities in the Development of Heavy Air Cushion Vehicles* (*Transportation*, v. 9, no. 1, 1977. 8 p.).

B Research Pertaining to Social Objectives

Research on social objectives has had a more specific focus than research carried out for national objectives. This type of research has also formed the nucleus for applications that could be transferred to Canadian industry. The types of research carried out during recent years have been broad, ranging from biomedical research to pollution studies.

(a) Biomedical Research

A major area of biomedical research carried out by NRC has been the study of antigens (substances that stimulate the production of antibodies) in yeasts, and bacteria, for the development of vaccines against meningitis, research with protein degrading enzymes called cysteine proteases involved in degenerative wasting diseases of the body, and the study of the parathyroid hormone (PTH) and its use in bone regeneration. This range of research has involved 'industrial partnerships' with pharmaceutical companies and special groups within the Institute for Biological Sciences such as the Protein Biochemistry and Spectroscopy Group, and the Biomolecular Modelling Group.

In humans, or animals, natural immunity occurs when the body produces antibodies (serum globular proteins) in response to foreign material introduced into the body by infection or injection. A prime characteristic of this immune response is its high order of specificity (e.g. immunity to one infection, such as pneumonia, does not protect against other infections, such as tuberculosis). This specificity is due to the matching-up of active sites on antibodies and antigens. Work at NRC's Biochemistry Laboratory, in the 1970s, aimed at the isolation of specific antigens and the characterization of their determinant groups. For example, polysaccharides (complex carbohydrates) in the cell walls of a series of skin-attacking fungi were shown to be group antigens. As well, a high order of specificity was found in pepido-glycan antigens isolated from the same organisms, and the peptide part of the molecules was established as the determinant group. Some of this research was described by R.J. Fielder, C.T. Bishop, and S.F. Grappel in the article entitled *An Immunogenic Polysaccharide – Protein Conjugate* (*Journal of Immunology*, v. 105, 1970. pp. 265–67). This work helped to serve as a useful background in the development of a vaccine against meningitis.

Work with *Neisseria meningitides*, the bacteria causing meningitis, the often fatal infection of the membrane protecting the brain and spinal cord, involved working out the polysaccharide structures of six meningococcus groups, in order to isolate a protein common to all groups. This group antigen held promise as an antigen against all meningococcal types in the development of a vaccine. This work was initially reported by D.R. Bundle and H.J. Jennings in a paper entitled *Preliminary Communication: An Improved Procedure for the Isolation of the Meningococcal Polysaccharide Antigens, and the Structural Determination of the Antigen from Serogroup X* (*Carbohydrate Research*, v. 26, 1973. pp. 268–70). Later work was described by A. Bhattacharjee and H.J. Jennings in the article entitled *Structural Determination of the Sialic Acid Polysaccharide Antigens of* Neisseria meningitides *Serogroups B and C with Carbon-13 Nuclear Magnetic Resonance* (*Journal of Biological Chemistry*, v. 250, 1975. pp. 1926–32).

By the late 1970s, this research resulted in the creation of a meningitis vaccine. In keeping with NRC's pattern of encouraging Canadian industry, the Armand-Froppier Institute, at Montreal, was asked to work out the large-scale production techniques for the meningitis vaccine. This Institute, affiliated with the Université du Québec, also completed test trials with the vaccine produced from the larger production techniques. The production-quality vaccine was found to have the same wide-spectrum potency as the much smaller laboratory-scale preparations. Then, this vaccine was sent to Brazil for testing on populations where the incidence of meningitis was high.

By the 1980s and 1990s, the immunochemistry research on vaccines against *Neisseria meningitides* focused upon the creation of a new generation of sub-unit vaccines using 'chemically defined parts of the antigen derived by chemical modification or synthesis.' This research involves the study of immunochemical problems of carbohydrate antigens at the molecular and cellular level and their interaction with protein receptors such as antibodies. This approach to the creation of another generation of vaccines has been described by H.J. Jennings in the papers entitled *Capsular Polysaccharides as Vaccine Candidates* (*Current Topics in Microbiology and Immunology*, v. 150, 1989. pp. 97–127); and *The Capsular Polysaccharide of Group B* Neisseria meningitides *as a Vehicle for Vaccine Development* (*Contributions to Microbiology and Immunology*, v. 10, 1989. pp. 151–65).

One aspect of the detailed research aimed toward the creation of newer vaccines has been the need to understand the molecular basis of carbohydrate antigen recognition and binding with antibodies and the latter's recognition of receptor sites on the bacteria. Part of this work has involved the identification of the crystals of two sialic acid binding antibodies, one of which binds with the polysialic acid binding site of group B. meningococcus bacteria. This research has been described by V. Pavliak and J.R. Brisson, in their paper entitled *Structure of the Sialylated L3 Lipopolysaccharide* of Neisseria meningitides (*Journal of Biological Chemistry*, v. 268, 1993. pp. 14146–52). The possibilities opened by this work toward a new range of meningitis vaccines have been discussed by H.J. Jennings and J.R. Brisson in their paper entitled *Polysialic Acid Vaccines against Meningitis Caused by* Neisseria meningitides *and* Escherichia coli *K1*. In: *Polysialic Acid*. Edited by J. Roth and V. Rutishauser (Basel, Switzerland: Birkhauser Verlag, 1993. pp. 25–38). Further studies undertaken in order to learn more about variations in the structure of *Neisseria meningitides* have been carried out by G. Kogan and H.J. Jennings in the paper entitled *Structural Basis of the* Neisseria meningitides *Immunotypes Including the L4 and L7 Immunotypes* (*Carbohydrate Research*, v. 298, 1997. pp. 191–99); and by R.A. Pon and H.J. Jennings, in the paper entitled *N-Propionylated Group B Meningococcal Polysaccharide Mimics a Unique Bactericidal Capsular Epitope in Group B.* Neisseria meningitides (*Journal of Experimental Medicine*, v. 185, 1997. pp. 1929–38). Understanding the structure of the bacteria is vital for future vaccine design.

The study of cysteine proteases related to degenerative wasting diseases began in the Biomedical Modelling Group for the development of mathematical algorithms to aid in the computer analyses of the nucleic and protein sequences of these degrading enzymes, which catalyse or assist in the decomposition of peptide bonds in proteins. Hopefully, this type of work would lead to the rational design of inhibitors to these proteins, and would have important implications for certain human disease therapies. This work has been described by P.K. Klein and R.L. Somorjai in their paper entitled *Non-Linear Methods for Discrimination and Their Application to Classification of Protein Structures* (*Journal of Theoretical Biology*, v. 130, 1988. pp. 461–68). One particular cysteine protease that was targeted for study in detail was Cathepsin B. It occurs in lysosomes (cell organelles) in cells, and normally assists in intracellular digestion and degradation of foreign materials engulfed by phagocytes. However, if lysosomes become damaged and Cathepsin B is released into the cell, these proteases can cause neurological disorders and wasting such as muscular dystrophy. The initial modelling work with Cathepsin B through computer calculations, which could predict the effect of making changes in the structure of a selected part of the protein by the introduction of recombinant DNA methodology to create an altered protein that might show useful pharmaceutical activity, was carried out by H.E. Khouri and C. Plouffe, and described in their paper entitled *A Model to Explain the pH-Dependent*

Specificity of Cathepsin B-Catalysed Hydrolyses (*Biochemical Journal*, v. 275, 1991. pp. 751–57).

The initial biochemical work, however, focused on the purification and crystallization of human liver Cathepsin B expressed (cloned) in yeast in order to solve the structure of the human cysteine protease. This work has been described in several papers. One, by S. Hasnain and C. Huber, was entitled *Investigation of Structure-Function Relationships in Cathepsin B* (*Biological Chemistry: Hoppe-Seyler Supplement*, v. 373, 1992. pp. 413–18). Others by S. Hasnain and T. Hirama have been entitled *Characterization of Recombinant Rat Cathepsin B and Non-Glycosylated Mutants Expressed in Yeast* (*Journal of Biological Chemistry*, v. 267, 1992. pp. 4713–21); *Characterization of Cathepsin B Specificity by Site-Directed Mutagenesis* (*Biological Chemistry*, v. 268, 1993. pp. 235–40); *Crystal Structures of Recombinant Rat Cathepsin B and a Cathepsin B-inhibitor Complex: Implications for Structure-based Inhibitor Design* (*Journal of Biological Chemistry*, v. 270, 1995. pp. 5527–33); and *Characterization of the S-3 Subsite Specificity of Cathepsin B* (*Journal of Biological Chemistry*, v. 270, 1995. pp. 18036–43).

Another area of research carried out by the Protein Biochemistry and Spectroscopy Group of the Institute of Biological Chemistry has been the study of the parathyroid hormone (PTH), an amino acid hormone and with potential use in therapies for conditions such as osteoporosis (the industrial partner for this research is Allelix Biopharmaceuticals). Parathyroid hormone regulates blood calcium levels, and hence both the destruction and formation of bone tissue. This is unlike estrogen, which can stop post-menopausal bone loss, or fluoride treatment, which can stimulate bone growth but creates new bone that is fragile and fracture-prone. The goal of this research has been to determine the structure features of PTH that control formation and resorption of bone, and develop synthetic PTH analogues that can stimulate bone rebuilding and be of therapeutic use.

At the cellular level, researchers have been able to identify which amino acid fragments of the hormone bind with cell receptors to stimulate two enzymes – protein kinase C (PKC) and adenylate cyclase – that play key roles in cell-signalling pathways. Initial work in this field was carried out by B.R. Chakravarthy and B.R. Franks, and reported in their paper entitled *A Novel Method for Measuring Protein Kinase C Activity in a Native Membrane Associated State* (*Biochemical and Biophysical Research Communications*, v. 160, 1989. pp. 340–45); then, by B.R. Chakravarthy and B.R. Bussey in the paper entitled *The Direct Measurement of Protein Kinase C (PKC) Activity in Isolated-Membranes Using a Selective Peptide Substrate* (*Analytical Biochemistry*, v. 196, 1991. pp. 144–50); and finally, by B.R. Chakravarthy and J.F. Whitefield in their paper entitled *Parathyroid Hormone Stimulates Protein Kinase C but Not Adenylate Cyclase in Mouse Epidermal Keratinocytes* (*Journal of Cell Physiology*, v. 150, 1992. pp. 299–303). Further work was undertaken by L. Gagnon and H. Jouishomme and reported in their papers entitled *Protein Kinase C-Activating Domains of Parathyroid Hormone Related Protein* (*Journal of Bone and Mineral Research*, v. 8, 1993. pp. 497–503); and *Further Definition of the Protein Kinase C Activation Domain of the Parathyroid Hormone* (*Journal of Bone and Mineral Research*, v. 9, 1994. pp. 943–44).

By using artificial peptides, researchers were also able to determine which parts of the hormone were responsible for stimulating the enzymes protein kinase C and adenylyl cyclase. This work was carried out by W. Neugebauer, G.E. Willick, *et al.*, and reported in their papers entitled *Structural Elements of Human Parathyroid Hormone and Their Possible Relation to Biological Activities* (*Biochemistry*, v. 31, 1992. pp. 2056–63), and *Solution Structure and Adenylyl Cyclase Stimulating Activities of C-Terminal Truncated Human Parathyroid Hormone Analogues* (*Biochemistry*, v. 34, 1995. pp. 8835–42).

Another step in this research was to attempt to clone and 'express' fragments of PTH in another medium that would pave the way for the eventual large-scale production of PTH analogues for parathyroid hormone-based drugs. This was successfully done by W.L. Sung, D.M. Zahab, *et al.*, and reported in their papers entitled *Synthesis and Characterization of Extended and Deleted Recombinant Analogues of Parathyroid Hormone – (1-84): Correlation of Peptide Structure with Function* (*Biochemistry*, v. 29, 1990. pp. 10080–89); and *Specific Degenerate Codons Enhanced Selective Expression of Human Parathyroid Hormone in* Escherichia coli (*Journal of Biological Chemistry*, v. 266, 1991. pp. 2831–35).

With the development of fragment analogues, the latter needed to be tested in animal models as another step in the research of developing drugs to combat osteoporosis. This work was undertaken by R.H. Rixon and W.F. Whitfield and reported in their papers entitled *Parathyroid Hormone Fragments May Stimulate Bone Growth in Ovariectomized Rats by Activating Adenylyl Cyclase* (*Journal of Bone and Mineral Research*, v. 9, 1994. pp. 1179–89); and *Restoration of Severely Depleted Femoral Trabecular Bone in Ovariectomized Rats by Parathyroid Hormone – (1-34)* (*Calcified Tissue International*, v. 56, 1995. pp. 227–31).

(b) Pollution Studies

Besides the intensive pollution studies carried out by the Associate Committee on Environmental Quality, and described in section D, NRC has carried out interesting pollution research, ranging from bird poisoning by pellets, mercury pollution, and the destruction of PCBs to high technology anti-pollution research for the pulp and paper industry.

Using the process of spherical agglomeration, developed by the Division of Chemistry for oil and coal recovery, NRC devised lead-iron pellets as a birdshot alternative for regular lead birdshot, environmentally harmful to pond-feeding birds. The problem with pure leadshot was that spent shotgun pellets were eaten by bottom-feeding birds and collected in the avian gizzard where they were ground up, with the bird eventually dying of lead poisoning. The lead-iron pellet was much harder and the gizzard could not grind it up so easily. As well, the lead in conjunction with iron was far less toxic to the bird. This work was described in two publications. One, by F.W. Meadus and I.E. Puddington, was entitled *Pellet Technology* (*Canadian Journal of Chemical Engineering*, v. 46, 1968. pp. 287–88). The other, by C. Dayson and T. Maloney, was entitled *Dimensional Changes in Shotgun Barrels Caused by the Firing of Hard Metal Pellets* (Ottawa: National Research Council, 1969. 47 pp.).

Another pollution study on plant and animal life took place downstream from Parliament Hill on the Ottawa River. The Ottawa River Project, from the mid- to late 1970s, was a multi-disciplinary investigation of a three-mile stretch of the Ottawa River by groups from NRC's Divisions of Biological Sciences and Mechanical Engineering, and the University of Ottawa. The principal aim of the five-year project was to measure the distribution and quantity of persistent chemicals in the water and sediments of the river and to investigate their transport through the aquatic plant and animal forms. The initial study, over a two-year period, showed that the levels of mercury in fish populations were found to be well below the maximum allowable level of 0.5 ppm (parts per million), with plant-eating fish containing less than carnivorous types. Evidence also suggested that the age and growth of the fish, rather than its position on the food chain, determined the amount of mercury in its tissues. Interestingly, the river sediments were shown to contain over 99% of the 60-kilogram total of mercury found in the three-mile stretch of the river. There was apparently little danger of this amount of mercury suddenly being released in the water, because bacteria were the vectors of the removal process. These overall results were reported by D.C. Mortimer and A. Kudo in the journal article *Interactions between Aquatic Plants and Bed Sediments in Mercury Uptake from Flowing Water* (*Journal of Environmental Quality*, v. 4, 1975. pp. 491–95).

This work was later revisited by M. Schintu and T. Kauri and elaborated on in paper entitled *Methyl (Organic) Mercury in Inland Waters and the Ratio to Total Mercury* (*Water Research*, v. 23, 1989. pp. 699–704). With more refined methods of measurement, the authors found 'a relatively constant equilibrium between organic and inorganic mercury had been maintained in these waters over the last decade.' An overall summary of the results of the Ottawa River Project was reported by A. Kudo in the book chapter entitled *Mercury in the Ottawa River Canada.* In: *Aquatic Ecotoxicology Fundamentals, Concepts and Methodologies.*

Edited by A. Boudou and F. Ribeyre (Boca Raton, Florida: CRC Press, 1989. v. 1, pp. 202–17).

Another area of pollution studies carried out by the NRC since the 1970s has been in the field of polychlorinated biphenyls. In the fall of 1978, the National Research Council was asked by the government of Saskatchewan to provide a team of scientists to examine a large spill of these toxic PCBs, which occurred in August 1974, near Regina. The panel, organized by NRC, consisted of experts from science faculties at Canadian universities and from the council itself.

The objectives of the study were the estimation of the exact chemical nature and quantity of the spill, the precise location of the contamination, and the geography of the immediate and surrounding terrain. The panel was also asked to verify the effectiveness of tests and containment measures taken after the spill and to undertake additional monitoring and tests necessary for assessing the risk of future contamination of drinking water, ground water, and surface water in the area surrounding the spill. Finally, the committee was asked to recommend appropriate remedial measures. In January, 1980, NRC issued the report. It was entitled *The Report of the Ad Hoc Panel Convened by the National Research Council of Canada to Review the Spill of Polychlorinated Biphenyl at the Federal Pioneer Ltd. Site in Regina Saskatchewan* (Ottawa: National Research Council, 1980. 15 p.).

Two years later, the spill was written up as a case study in a two-part article in the open literature, the first by J.R. Roberts and J.A. Cherry, entitled *A Case Study of a Chemical Spill: Polychlorinated Biphenyl (PCBS). 1. History, Distribution and Surface Translocation* (*Water Resources Research*, v. 18, 1982. pp. 525–34), and the second by F.W. Schwartz and J.A. Cherry, entitled *A Case Study of a Chemical Spill: Polychlorinated Biphenyl (PCBS). 2. Hydrogeological Conditions and Contaminant Migration (Water Resources Research*, v. 18, 1982. pp. 535–45).

In the late 1980s and into the 1990s, NRC's pollution studies shifted from examining environmental problems after the fact to searching for solutions to specific types of industrial pollution, in collaboration with particular industries.

For example, in the pulp and paper industry, xylan (a lignan), which is a major component of hemicellulose found in wood, is normally broken down in the kraft pulping and bleaching process by the addition of chlorine. In this process, toxic polychlorinated aromatic phenols (organic compounds) are produced, which enter the environment. This breakdown of lignans of hemicellulose was studied at NRC's Institute for Biological Sciences and described by R.W. Thring and E. Chornet in the book chapter entitled *Production and Hydrolytic Depolymerization of Ethylene Glycol Lignan.* In: *Lignan Properties and Materials.* Edited by W.G. Glasser and S. Sarkanen (Columbus, Ohio: American Chemical Society, 1989. pp. 228–44). The same authors also analysed the compounds from the breakdown of lignan in their paper entitled *Analyses of Phenols from Lignan Depolymerization by Capillary Gas Chromatography* (*Journal of Chromatography*, v. 467, 1989. pp. 441–6).

It was also known that bacteria, which carried xylan-degrading enzymes called xylanases, were potentially useful for the breakdown of lignans into organic compounds that were soluble in the environment. Initial studies on the bacterial breakdown of lignan in the pulping process were described by P.F. Vidal and J. Bouchard in their paper entitled *The Degradation of Indalin ATR by Streptomyces: Chemical Characterization of the Water Soluble Acid Precipitable Products* (*Canadian Journal of Chemistry*, v. 67, 1989. pp. 350–55), and in the book chapter entitled *Bacterial Degradation of Kraft Lignan: Production, HPSEC and FTIR Characterization of Water Soluble Intermediates Derived from* S. Badius *and* S. Viridosporus. In: *Proceedings of the American Chemical Society: Plant Cell Wall Polymers Biogenesis and Biodegradation.* Edited by N.G. Lewis and M.G. Paice (Columbus, Ohio: American Chemical Society, 1989. pp. 529–43).

In order to use these bacterial xylanase enzymes in the industrial environment, new enzymes that showed improved catalytic stability under industrial processing conditions, such as high alkalinity and temperature, had to be genetically engineered in bacteria. The initial bacteria chosen for this work was *Bacillus circulans*. Using a technique called site-directed mutagenesis, NRC researchers were able to generate several mutant xylanases stable at high temperatures. This work has been described by W.W. Wakarchuk and R.L. Campbell in the paper entitled *Mutational and Crystallographic Analysis of the Active Site Residues of the* Bacillus circulans *Xylanase* (*Protein Science*, v. 3, 1994. pp. 467–75). In addition, researchers also targeted xylanase enzymes from some fungi as a source of thermostable mutant xylanase. This work has been described by M. Ujiie amd M. Yaguchi in their paper entitled *Low-Molecular-Weight Xylanase from* Trichoderma viride (*Applied and Environmental Microbiology*, v. 57, 1991. pp. 1860–62); and by M. Gilbert and M. Yaguchi in their paper entitled *A Comparison of Two Xylanases from the Thermophilic Fungi* Thielavia terrestris *and* Thermoascus crustaceus (*Applied Microbiology and Biotechnology*, v. 40, 1993. pp. 508–14). Both these xylanases were able 'to remove about 12% of the xylan remaining in an aspen kraft pulp. This indicated that either one of these thermophilic enzymes may be useful for enhancing the bleaching of kraft pulps.' In 1994, this xylanase engineering project was advanced enough to transfer it to an industrial partner, IOGEN Corporation, for laboratory scale testing of several NRC genetically engineered xylanases.

C Research toward New Knowledge and New Applications of Science

As early as 1919, the Cronyn Committee was set up to inquire about the state of scientific research in Canada, and one of the themes of the testimony, particularly from Dr Charles Saunders, was that the focus of government-sponsored research was aimed at success, whereas 'a scientist should be looking for light.' This conflict has been a constant theme at the National Research Council with its need to justify fundamental or pure research. During World War II, which triggered the expansion of NRC and its staffing, the war work was essentially practical, and used basic or fundamental research already known, for the creation of radar or atomic bombs. After the war, however, astrophysical research became a focus of pure research, and space science a focus of applied research.

(a) Astrophysics

Astrophysical research began after World War II with the construction of the Algonquin Park radio telescope. At that time, some work was done on the study of radio waves transmitted by the sun as well as sunspots. With the development of larger and more specialized telescopes, however, astrophysicists have extended the scope of their exploration into and beyond our own galaxy (the Milky Way). Since 1975, astrophysical research at the NRC has been reorganized into the Herzberg Institute of Astrophysics and the telescopes available to Canadian scientists are described in detail under 'National Facilities.' Currently, the research programs of the Institute range from optical astronomy, including the related fields of infrared and ultraviolet astronomy, radio astronomy, and spectroscopy, to solar terrestrial physics. The scope of the research covers our own sun and solar system and beyond to the outer reaches of the universe. This includes the study of quasars, binary systems, supernova remnants, the Galactic Plane Survey, solar flares, aurorae, and cyclical studies with comets.

NRC's astrophysical program has been active in the study of quasars since their discovery in the 1960s using the new technology of radio astronomy (the study of radio emissions-electromagnetic radiation with wave lengths from about 1 mm to many metres). Quasars are minute extragalactic objects which emit extremely high-level energies. In fact, they may be the nucleus of active galaxies. A major knowledge gap in the study of galaxies was precise data about their diameters, as this information was necessary before more theorizing about these objects could be done. Until work was undertaken at NRC, diameters of radio sources were measured by means of a radio interferometer consisting of two radio telescopes spaced several miles apart, but closely interconnected with a microwave radio link to enable the received signals at both stations to be directly intercompared. However, to resolve extremely small diameters of quasars, one would need a very long baseline of thousands of miles, an impossibility with a conventional interferometer system.

It was because of this problem that a Canadian team made up of astronomers from NRC's Radio and Electrical Engineering Division, the University of Toronto, the Department of Energy, Mines and Resources, and Queen's University at Kingston developed a new technique to determine the angular diameter of quasars. This technique removed the necessity of a direct link between the two receivers. Radio signals received from a quasar 2,000 miles away, simultaneously by NRC's 130-foot radio telescope at Algonquin Park and the Dominion Observatory's 84-foot telescope at Penticton, BC, were recorded on magnetic tapes and synchronized to a fraction of a microsecond for the duration of the observation, which might last up to an hour or more. The timing accuracy was achieved with synchronized atomic clocks at each observatory. Then, the tapes from the radio telescopes were brought back to a central laboratory for playback and intercomparison. Measurements indicated that a number of quasars were 100 light-years in diameter or less. By comparison, the Milky Way is approximately 100,000 light-years in diameter. This breakthrough was reported in the journal *Nature* by N.W. Broten and eight other team-members, typically in two letters to the editor, one entitled *Observations of Quasars Using Interferometer Baselines up to 3,074 km* (*Nature*, v. 215, 1967. p. 38); and the other entitled *Diameters of Some Quasars at a Wavelength of 66.9 cm* (*Nature*, v. 216, 1967. pp. 44–45).

Besides the use of radio astronomy to further our knowledge about quasars, the search for quasars through surveys using ultraviolet, infrared, and x-ray astronomy has been undertaken. Some of this survey work has been carried out by D. Crampton at the Dominion Astrophysical Observatory in Victoria, BC, since the early 1980s. Preliminary quasar searches recorded on blue grens (a transmission grating, prism, and lens combination) plates yielded about 30 quasar candidates per plate. These findings were reported first by D. Crampton and M. Rensing in their paper entitled *Quasar Candidates Near 3C 345* (*Publications of the Astronomical Society of the Pacific*, v. 94, 1982. pp. 440–43); secondly, by D. Crampton and P. Parmar in their paper entitled *Quasar Candidates Near A1139, A1775, and 1633+33* (*Publications of the Astronomical Society of the Pacific*, v. 95, 1983. pp. 127–32); and thirdly, by D. Crampton and D. Schade in their paper entitled *Quasar Candidates from Gren Plates* (*The Astronomical Journal*, v. 90, 1985. pp. 987–96). Further quasar studies, involving D. Crampton, were carried out in the late 1980s and early 1990s using the Canada-France-Hawaii Telescope. This work was reported by D. Crampton and A.P. Cowley in the following papers entitled *The Space Distribution of Faint Quasars from the CFHT Survey* (*Astrophysical Journal*, v. 314, 1987. pp. 129–44); *Quasar Candidates in Two Extended CFHT Fields: 1338+27 and 1639+40* (*Astronomical Journal*, v. 96, 1988. pp. 816–35); *Redshifts of Quasars in the CFHT/MMT Survey: Further Evidence for Isolated Groups* (*Astrophysical Journal*, v. 345, 1989. pp. 51–71); *Redshifts of Quasars in the CFHT/MMT Survey. III* (*Astronomical Journal*, v. 100, 1990. pp. 47–55); and

Canada-France-Hawaii Telescope Blue Gren Quasar Candidates (*Astronomical Journal*, v. 104, 1992. pp. 1706–13).

As a result of this survey activity, NRC helped to sponsor a quasar workshop in Victoria, BC, as part of its 75th anniversary in 1991. The aims of the workshop 'were to bring together the results of various quasar surveys under way (at all wavelengths), to compare them with the large-scale-structure picture emerging from galaxy surveys, and to confront them with recent theories on the origin and evolution of structure in the universe.' The proceedings of this were edited by D. Crampton and entitled *The Space Distribution of Quasars: Proceedings of a Workshop Held as Part of the 75th Anniversary Celebrations of the National Research Council of Canada, at the Dominion Astrophysical Observatory, Victoria, B.C., June 3rd to 5th, 1991* (San Francisco, California: Astronomical Society of the Pacific, 1991. 401 p. Astronomical Society of the Pacific Conference Series, v. 21).

Another type of star system that is more common than quasars is the binary star. Binary stars orbit around each other, and are held together as stellar companions by their mutual gravitational attraction. About 40% of all stars known have orbiting companions, and in many cases more than two stars are involved, as some binary systems are triples or quadruples. Most binary stars are only a few astronomical units apart (one AU is the mean distance between the earth and the sun – 149,598,000 km – and is used as a fundamental unit of distance, particularly within the solar system), but some are contact binaries and orbit each other in a common envelope of gas. Binary stars are classified into three groups. One is visual binaries – the two stars are visibly separate, and the orbit motion of one star about the other can be seen in comparative photographs. A second group is called spectroscopic binaries, because they orbit so closely together that their light blends into a single blob that even powerful telescopes can't separate. The only way to measure their orbits is through spectrographic identification of their individual Doppler shifts. As the stars orbit each other, they alternatively move toward or away from the earth. Moving toward the earth causes their spectral wavelength to decrease toward the earth – this is called a blueshift because the spectrum shifts into the blue band of the spectrum. As the orbit moves away from the earth, the spectral wavelength increases into the red band of the spectrum – this is called a redshift. A third group is called eclipsing binaries, because on rare occasions the orbit of the binary star will be almost exactly edge-on as seen from the earth, and as the stars orbit each other, one will eclipse the other as it passes between its companion and the earth. At the time of the eclipse, the combined light of the binary dims, producing a cycle of variation in light intensity called a light curve, which can be used to provide orbital and other information. Besides the classification of binary systems into visual, spectroscopic, or eclipsing, a broader classification by the historic terminology of late-type (a cool star) or early-type (a hot star) is used.

At the Herzberg Institute, G. Hill, in particular, and other researchers have studied the spectrum and light curves of late-type and early-type binaries in a series of papers. The first paper, by G. Hill and W.A. Fisher, was entitled *Studies of Late Binaries I. The Physical Parameters of 44t Bootis ABC* (*Astronomy and Astrophysics*, v. 211, 1989. pp. 81–98). The second, by G. Hill, was entitled *Studies of Late Type Binaries II. The Physical Parameters of VW Cephei* (*Astronomy and Astrophysics*, v. 218, 1989. pp. 141–51). The third paper, by D.E. Holmgren and G. Hill, was entitled *Absolute Dimensions of Early-Type Eclipsing Binary Stars III. U Ophiuchi* (*Astronomy and Astrophysics*, v. 248, 1991. pp. 129–38). The fourth paper, by G. Hill and W.A. Fisher, was entitled *Studies of Late-Type Binaries IV. The Physical Parameters of ER Vulpeculae* (*Astronomy and Astrophysics*, v. 238, 1990. pp. 145–59). And the fifth paper, by Y.S. Ann and G. Hill, was entitled *Studies of Late-Type Binaries V. 1073 Cygni* (*Astronomy and Astrophysics*, v. 265, 1992. pp. 597–604). Other papers on binary stars, published by G. Hill and R.W. Hilditch, have been *Optical Spectroscopy of the Massive X-Ray Binary SMC X-1 SK160* (*Monthly Notices of the Royal Astronomical Society*, v. 261, 1993. pp. 337–45); *Studies of Early-Type Variable Stars. 8. The Massive Binary-System CC-Cassiopeiae* (*Astronomy and Astrophysics*, v. 282, 1994. pp. 455–66); *On the Reflection Effect in 3 Sdob Binary Stars* (*Monthly Notices of the Royal Astronomical Society*, v. 279, 1996. pp. 1380–92; and *Spectroscopy and Eclipse-Mapping of the Mass-Exchanging Binary Star V361 Lyr* (*Monthly Notices of the Royal Astronomical Society*, v. 291, 1997. pp. 749–62).

The study of binary stars began in the early 19th century and the knowledge gained since that time about them, as well as about single stars, has led to speculation about the evolution of binary systems. It is accepted that stars in binary systems evolve in the same way predicated by model calculations for single stars, as long as the two components of the binary are sufficiently far apart that they do not interfere with each other. However, as stars evolve, they expand, and the components of many binaries are close enough to each other that they may begin to interfere in the early stages of evolution. For example, mass may be transferred from one component to the other, or be lost from the system. It is this interference that has led to theories about the evolution of stars in close binaries and the classification of such star systems. These theories, which are of recent origin, have been discussed in a major paper by A.H. Batten from the Dominion Astrophysical Observatory, entitled *Stellar Evolution in Binary Systems* (*Reports on Progress in Physics*, v. 58, 1995. pp. 885–928).

Stars, whether single or in binary systems, also age and die. From the moment of its birth, when an interstellar gas cloud collapses to become a star, to the moment of its death, gravity constantly works to crush a star. In fact, gravity is the driving force of stellar evolution. Based upon computer modelling, a high mass star (our sun is low mass) will pass from its time as a main sequence star to become a yellow giant and then pass on to become a red giant as each layer of its fuel – hydrogen, helium, and carbon – is exhausted and the core continues to contract. Its death signal is the formation of iron in its core, where continuing pressure turns the protons and electrons of the iron into neutrons. At that stage, there is a collapse of the remaining outer layers into the core that causes the infalling gases to heat up billions of degrees. The pressure surges and lifts the outer layers away from the star in a gigantic explosion called a supernova. Gas ejected by the supernova blast surges through the surrounding interstellar space, sweeping up and compressing other gases that it encounters. This huge growing cloud of stellar debris is called a supernova remnant. The remaining star core itself may become a neutron star, or a black hole.

For astronomers and astrophysicists, supernova remnants are an important element in studies of the energetics and dynamics of the interstellar medium, and statistical studies of them, such as their distribution, birth-rate, and energy, are of particular interest. However, such studies are not straightforward, because SNRs show a considerable range of intrinsic properties owing to the nature of their creation. Also, 'the identification of SNRs is generally made from radio surveys of the Galactic plane and is basically limited by two selection effects. First, the surface brightness of the remnant must be above the sensitivity limit of the observations and be readily distinguishable from Galactic background emission. Second, in general, the angular size of the remnant must be at least several times the resolution of the observations.' These problems were pointed out by D.A. Green, who began to study supernova remnants in the early 1980s, and published 'A Catalogue of Galactic Supernova Remnants' as an appendix to the paper entitled *Statistical Studies of Supernova Remnants* (*Monthly Notices of the Royal Astronomical Society*, v. 209, 1984. pp. 449–78). Four years later, D.A. Green published *A Revised Reference Catalogue of Galactic Supernova Remnants* (*Astronomy and Space Science*, v. 148, 1988. pp. 3–74). In the introduction, the author said that observations made over recent years had led to the identification of many new galactic supernova remnants and had clarified the nature of general objects thought to have been SNRs. The revised catalogue contained 155 galactic SNRs. Three years later, D.A. Green published the paper entitled *Limitations Imposed on Statistical Studies of Galactic Supernova Remnants by Observational Selection Effects* (*Publications of the Astronomical Society of the Pacific*, v. 103, 1991. pp. 209–20). In the appendix, Green provided *A Catalog of Galactic SNRs*, which was an update of 174 galactic SNRs to the revised catalogue of 1988.

The study of supernovae, and their remnants, has also encouraged their classification. According to S. Van Den Bergh, in the paper entitled *Classification of Supernovae and Their Remnants* (*Astrophysical Journal*, v. 327, 1988. pp. 156–63), 'a few years ago only two kinds of supernovae were generally recognized even though three distinct types of supernova remnants were known. This shows that it is necessary to study both supernovae and their remnants to gain a complete understanding of the full range of phenomena that can occur during the terminal phases of stellar evolution.' Van Den Bergh went on to discuss the three distinct types of young supernova remnants. They are oxygen-rich remnants, of which Casseopeia A is the prototype, plerionic (filled) remnants, of which the Crab Nebula is the prototype, and a third type, of which the Tycho remnant is the prototype. The latter emits a distinctive series of lines in the hydrogen spectrum called Balmer lines. However, there are other odd SNRs that may represent distinct types that do not fit into the three classifications. They may be the remains of massive binary stars.

The search for supernova remnants continues, particularly young SNRs, because the existing catalogues mentioned above are deficient in this type. In 1988, T.L. Landecker and L.A. Higgs, at the Dominion Radio Astrophysical Observatory (DRAO), at Penticton, studied a galactic object named G76.9+1.0 using the Very Large Array, or VLA, in New Mexico (the most complex radio telescope on a single site in the world, involving an array of 27 moveable dishes). This object, located near the galactic equator of our own Milky Way, had an extended non-thermal radio source, which is characteristic of new supernova remnants. However, its spectral structure indicated that it might be a shell remnant, except that it did not resemble the classical SNR shell (there was an absence of a shell at its outer limits, but the inner structure of two lobes could be interpreted as a shell). Nor did it have all the characteristics of a plerionic (filled) remnant like the Crab Nebula. In their paper, entitled *G76.9+1.0, a Supernova Remnant with Unusual Properties* (*Astronomy and Astrophysics*, v. 276, 1993. pp. 522–30), the authors came to the conclusion 'that G76.9+1.0 is probably a supernova remnant. While it does not fit readily into existing SNR categories ... it is most readily classified as a filled-centre SNR.' Finally, 'further observations of G76.9+1.0 are warranted.' Further observations were undertaken, as noted in the paper by T.L. Landecker and Y. Zheng entitled *The Unusual Supernova Remnant G76.9+1.0: The Low-Frequency Radio Spectrum* (*Astronomy and Astrophysics Supplement Series*, v. 123, 1997. pp. 199–207).

The Canadian Galactic Plane Survey (the plane is the line that can be drawn through the galactic equator of the Milky Way Galaxy) began in 1995, and is a consortium of 54 scientists from Canada and other countries, who are carrying out a program of panoramic imaging of the northern Galactic plane at several radio continuum frequencies in the H I and CO spectral lines and in the infrared bands. The Dominion Radio Astrophysical Observatory (DRAO) is contributing 408 and 1420 MHz radio continuum and H I data. Prior to the commencement of the survey, a feasibility study was undertaken. Both the feasibility study and the survey to date have been described by M. Normandeau and A.R. Taylor in the paper entitled *The Dominion Radio Astrophysical Observatory Galactic Plane Survey Pilot Project: The W3/W4/W5/HB3 Region* (Astrophysical Journal Supplement Series, v. 108, 1997. pp. 279–299).

Our own sun, and solar system, are also part of the Milky Way, located on one of disks of that spiral galaxy, and observable phenomena of the sun, such as solar flares, are studied. Solar flares are caused by magnetic fields on the sun but are triggered by sunspots. Flare particles in the earth's magnetosphere create disturbances in the earth's atmosphere, including auroral displays at the earth's poles. Overall, sunspots are the most common type of solar magnetic activity. They are large dark-appearing regions that can range in size from a few hundred to many thousands of kilometres across and last from a few days to over a month. Sunspots occur where there is a local strengthening of the sun's magnetic field which cools the area to around 4,000 degrees Kelvin (vs 6,000K of the normal photosphere – the sun's visible surface). This stronger field is believed to suppress the convection of hotter gases from lower levels, hence showing us darker regions. It is sunspots that give birth to solar flares, which are brief but bright eruptions of hot gas in the chromosphere (the sun's lower atmosphere). Sunspots and flare activity change from year to year in what is called the solar cycle, which is caused by the distortion of the sun's magnetic field owing to the sun's differential rotation on its axis. Historically, the cycle rises and falls about every 11 years, with one of the most recent cycles studied being the period 1979–81.

Awareness of the cycle allowed for preparatory work within the Solar Flare Buildup Study, which culminated with the intensive observations of solar flares during the Solar Maximum Year (1979–81). This was an important study because flares and the buildup to them were, and are still, not well understood. One of the accepted theories is that the magnetic fields near a sunspot become twisted by gas motions and then readjust, whipping the gas into new configurations, heating it and causing it to expand explosively. The Flare Buildup Study officially ended at the Symposium on Synopsis of the Solar Maximum Analysis held 2–5 July 1986 in Toulouse, France. V. Gaizauskas, of the Herzberg Institute, and Z. Svestka, summarized the highlights and progress in modelling the storage and release of pre-flare energy in their conference paper entitled *Highlights of the Flare Buildup Study* (*Solar Physics*, v. 114, 1987. pp. 389–98). One model, called the Emerging Flux Model, 'provides for the storage of pre-flare energy through the formation of a current sheet at the interface between oppositely directed magnetic fields of an existing active region and a new one growing beside it.' Another model, called the Magnetic Shear Model, 'demonstrated that magnetic energy sufficient to power solar flares is accumulated as the shear of the footprints in the magnetic arcade increases.' However, emerging flux and magnetic shear alone were not considered sufficient to trigger a flare, and perhaps 'the mutual loss of magnetic flux at sites where fragments of opposite magnetic polarity encounter one another' provided the missing link. V. Gaizauskas elaborated on the pre-flare stage of solar flares, again, in the paper entitled *Pre-Flare Activity* (*Solar Physics*, v. 121, 1989. pp. 135–52). This paper stated that 'the complexity of the pre-flare state will not be resolved without many more comprehensive multi-wavelength analyses of pre-flare periods observed jointly by spacecraft and ground-based instruments' and that 'theoretical tools are being developed to cope with a much more complicated and dynamic, hence realistic, magnetic environment. ...' In May, 1993, the Herzberg Institute played host the Second Flares 22 Workshop called Recent Observations and Theories of Solar Flares, held at the University of Ottawa. In the report of Team 1, D.M. Rust, V. Gaizauskas, and others, entitled *Preflare State* (*Solar Physics*, v. 153, 1994. pp. 1–17), the authors said that 'studies of the preflare state have progressed considerably in the past few years. X-Ray observations by the Yohkoh satellite, magnetographic observations, and theoretical studies all contributed to this progress.'

The National Research Council has been making regular determinations of solar activity, since 1946, at the 2800–MHz wavelength, and because of the consistency of data at this new wavelength, known as the Penticton 2800 MHz flux, it has become accepted worldwide as the primary index of solar activity. Since 1990, the program has been located at DRAO (Dominion Radio Astrophysical Observatory), and now uses automated flux monitors for the routine acquisition and distribution of data. This data is now available from the DRAO homepage at their website, <http://www.drao.nrc.ca>>.

Some of the exploding gas from solar flares may escape from the sun as solar particles and shoot across the inner solar system to stream down on the earth. There, the solar particles encounter the earth's own magnetic field, the magnetosphere, which slows them down as they are deflected by magnetic field lines into a spiralling motion of electric currents around those field lines as they flow around the earth. As the electrons are deflected downwards into the earth's upper atmosphere – the ionosphere – they collide with molecules of nitrogen and hydrogen, which excite these atmospheric gases and cause them to lift their own electrons into upper orbits around their nuclei. Then, as the electrons drop back to lower orbits, they emit the light we see on earth as the aurorae. An aurora can take many forms; for example, as a horizonal arc with rays radiating upwards, or as a hanging curtain exhibiting flickering colours, usually red and green. Aurorae occur at altitudes between 20 and 1,000 kilometres, but most typically at 100 kilometres within the ionosphere.

The exact process by which aurorae form is still controversial, but as A.V. Jones stated in the paper entitled *Historical Review of Great Auroras* (*Canadian Journal of Physics*, v. 70, 1992. pp. 479–87), 'theoretical studies have led to an understanding of the observed unusual spectroscopic characteristics including their bright red colour and the abnormal vibrational and rotational structure in the molecular bands observed in high-altitude forms,' and, as a result of the increasing sophistication of observing techniques, 'the phenomena associated with the great auroras have passed beyond the characterization of optical and magnetic, to include information on associated changes in the composition of the neutral upper atmosphere and effects in the ionosphere.' More specifically, the electron flows themselves during magnetic storms have been studied by D.D. Wallis and A.G. McNamara in their paper entitled *Spatial Relationship of Visual Arcs to Auroral Electrojets* (*Canadian Journal of Physics*, v. 69, 1991. pp. 1047–54). The authors noted that 'on the poleward side of the auroral arc the electron flow is generally eastward and on the equatorial side it is generally westward. At the visual auroral arc, when measurements can be made, the electron flow is generally southward.' These electron flows correspond to the electric-field strengths pointing equatorward, poleward, and westward respectively.

As mentioned above, the magnetic storms, which are the source of the aurorae, occur in the magnetosphere, which extends on the day side of the earth outwards to about 10 earth radii. Inwardly, it encounters the ionosphere where the aurorae take place. Recently, shifts of ion composition in the magnetosphere during magnetic storms have been noted and compared with magnetospheric composition measurements documented over the last 20 years, which showed that ions of terrestrial origin such as 0^+ions often dominated the hot plasma throughout the magnetosphere during magnetic storms. The source regions of these terrestrial ions have been identified as the polar ionosphere. The more recent studies have recorded ion composition measurements, using a suprathermal mass spectrometer (SMS) housed on the Japanese AKEBONO satellite. One such study, by A.W. Yau and B.A. Whalen, entitled *Auroral Ion Composition during Large Magnetic Storms* (*Canadian Journal of Physics*, v. 70, 1992. pp. 500–509), noted that ion compositions from the SMS on AKEBONO at quiet times indicated H^+ and O^+ as the dominant ion species and N^+ and O^{++} as the minor species in the high-altitude auroral ionosphere. But, during large magnetic storms there was an increased abundance of N^+ ions that may have been 'a direct result of the increase in molecular nitrogen density in the F-region and topside ionosphere due to heating in the presence of prolonged auroral activity. This implies that N^+ ions may possibly be the dominant plasma component during large very large magnetic storms.'

Other phenomena that can be seen from earth from time to time are comets. Comets are icy bodies about 10 kilometres or less in diameter, which grow huge tails as they near the sun and are particularly vaporized by its heat. Most comets orbit far beyond Pluto in a region of the solar system called the Oort Cloud and only rarely move into the inner solar system. One such comet – Comet Kohoutek – was the source of much observational work at the NRC, in 1973, at its radio observatories at Algonquin Park, and Penticton, BC.

One of the key discoveries of the observations that year was due to the work of Gerhard Herzberg, the Nobel laureate and NRC Distinguished Research Scientist, after whom the Herzberg Institute was named, and his colleague H. Lew. Their identification of the molecular ion H_2O^+, as the constituent of the comet's tail, was done as part of the Division of Physics' program of spectroscopy. The unequivocal identification of H_2O^+ in the tail of the Comet Kohoutek was the first clear identification that water was indeed a constituent of a comet's nucleus. This major discovery was first reported by G. Herzberg and H. Lew in a journal article entitled *Tentative Identification of the H_2O^+. Ion in Comet Kohoutek (Astronomy and Astrophysics*, v. 31, 1974. pp. 123–24).

Two decades later, NRC astronomers undertook observations of another comet's nucleus and came up with a startling suggestion. This comet – Comet Hyakutake – orbited the inner solar system during March of 1996, coming as close as 0.137 AU (astronomical units, or 20,494,000 kilometres) to the earth. This provided an opportunity for H.E. Mathews, using the James Clerk Maxwell Telescope, to identify hydrogen isocyanide (HNC) as well as hydrogen cyanide (HCN) from the comet's nucleus. Later that month, H.E. Mathews and other non-NRC astronomers confirmed these findings in a 'Letters to Nature' entitled *Spectroscopic Evidence for Interstellar Ices in Comet Hyakutake* (*Nature*, v. 383, 1996. pp. 418–20). The authors noted that the abundance of the relative proportion of these two compounds is very similar to that observed in dormant interstellar molecular clouds and 'quite different from the equilibrium ratio expected in the outermost solar nebula where comets are thought to form ... suggesting that interstellar gases have been incorporated into the comet's nucleus, perhaps as ices frozen onto interstellar grains' and providing 'constraints on the temperature of the nebula, and the consequent chemical processes that occurred in the region where comets formed.' The 1996 appearance of Comet Hyakutake was also fortuitous for D.C. Jewitt and H.E. Mathews in allowing them to undertake submillimetre continuum spectral studies of refractory dust particles, in order to examine the large grain dusts from the nucleus. In their paper, entitled *Submillimeter Continuum Observations of Comet Hyakutake* (*1996 B2*) (*Astronomical Journal*, v. 113, 1997. pp. 1145–50), the authors said that 'our efforts were amply rewarded with the first submillimeter spectrum and a limited map of the coma [the spherical cloud of gas and dust surrounding the nucleus of an active comet] at 800 um.'

(b) Space

An interesting historical note has been NRC's involvement in civilian space research. By the mid-1970s, Canadian scientists were becoming concerned about the nature of future research in space and the resources in money, people, and time required to accomplish future programs. In January, 1976, P.A. Forsyth presented a report to NRC that had been prepared for the Associate Committee on Space Research on that subject. It was entitled *Canadian Opportunities in Space* (Ottawa: National Research Council, 1975. 96 p.). The study covered all major areas of space research, including aeronomy and space physics, meteorology, remote sensing, and space astronomy. The information compiled served as a useful review and input for those scientists required to make decisions regarding future directions in these fields. One of its immediate outcomes was the establishment by NRC of a Space Coordination Office that hoped to be able to combine the diverse interests of the space science community in the universities, industry, and government agencies, into a coordinated national scientific program. At the government level, an Interdepartmental Committee on Space (ICS), reporting to the Minister of Communications, was formed to make recommendations concerning the appropriate use of resources and the coordination of space activities.

A major development in Canada's space program came about in 1975 with the signing of a memorandum of understanding by NRC and NASA, which provided for Canadian participation in the US space shuttle program. The memorandum itself involved Canada in the construction of a teleoperator remote manipulator system. Through NRC, the government's main objective was to create a national aerospace industry with expertise in the technology of teleoperator construction. With this expertise, teleoperator systems could also be developed for areas other than space, particularly for applications such as oceanographic research on the seafloor, and the handling of radioactive materials in the nuclear power industry.

Under the terms of the memorandum, contracts were placed with a team of Canadian industries led by SPAR Aerospace Products Limited and included CAE Electronics Limited, RCA (Canada) Limited, and Dilworth, Secord, Meagher and Associates Limited, for the development and construction of the first flight remote manipulator system for the US space shuttle as well the construction of a ground-based general-purpose manipulator system simulation facility. The simulation facility was scheduled for completion at the end of 1977, and the preliminary design review of the manipulation system itself was scheduled for completion at the end of 1976. Then, NASA and NRC would agree on a design, and work would proceed through the final design, hardware development, and manufacturing stages. This preliminary work was described by K.H. Doetsch in the NRC report entitled *The Remote Manipulator System for the Space Shuttle Orbiter* (Ottawa: National Research Council, 1977. 15 p.).

As well as directing the project, NRC also provided laboratory support. For example, a computerized video camera system was designed by the National Aeronautical Establishment, to be used to calculate the precise position and orientation of any three-dimensional body. This system was intended to be used if the task of visually guiding the arm in the featureless void of space should prove too difficult. The role of the NAE in this work was described by J.M. Trenouth and C.W. Mackenzie in an NRC report entitled *The Role of NAE in the Canadian Development and Investigation of a Space Induced Friction Anomaly* (Ottawa: National Research Council, 1987. 8 p.).

The Division of Mechanical Engineering used a combined analogue and digital computer model to simulate the motion and flexibility of the arm. This tool allowed the rapid study of alternate designs and the validation of the more comprehensive simulations conducted by SPAR Aerospace. This work was described by J.W. Goodanetz, of J.D. Kendall Consultants Limited, and R.E. Gagne and I.H. Mufti, of the Division of Mechanical Engineering, in a report entitled *SRMS Joint Parameter Sensitivity Study for 32K Payload Using the DME/JDK Hybrid Computer Model* (Ottawa: National Research Council, 1979. 66 p.).

A pilot model of the arm was delivered to NASA in late 1979, and in 1981 the Remote Manipulator System (now called the Canadarm) was flown for the first time on the second and third missions of NASA's Space Shuttle Columbia. Then in 1985, NRC did create a new Space Division in its organizational structure in order to consolidate the management of the various elements of Canada's continuing space research program. However, in 1989, the federal government announced the creation of the Space Agency to manage Canada's civilian space research program, thus ending NRC's involvement in that field.

D Research Carried Out through the Associate Committees

From 1917 to 1989, the National Research Council used associate committees as instruments to provide the opportunity to bring together experts for the study, coordination, and promotion of research on problems of national significance. When an associate committee studied a particular problem, it collected and collated pertinent information, delegated research problems, coordinated research, and suggested new avenues of research. As committees achieved their purpose they were terminated. In 1988–9, when the associate committee structure was abandoned and replaced by advisory boards for each NRC institute or major program, there were 23 active associate committees, with more than 50 attendant sub-committees involving hundreds of scientists and engineers from across Canada who volunteered their expertise and time. These committees have been listed and described in the publication entitled *Associate Committees 1988* (Ottawa: National Research Council, 1988. 23 p.). Associate committees also organized and sponsored national or international meetings which brought researchers together for seminars or symposiums.

One of the most pertinent committees ever formed was the Associate Committee on Scientific Criteria for Environmental Quality, authorized in 1970 by the Cabinet of the federal government. This newly created committee replaced the Associate Committee on Water Pollution formed in 1965, and whose terms of reference were no longer adequate or appropriate in view of the rapid evolution in the field of environmental quality. The committee had a mandate to collate and publish an integrated set of scientific criteria for environmental quality, but would not be involved in the setting of standards for pollution emissions or for the enforcement of such standards. The latter would still continue to be the function of the federal, provincial, or municipal levels of government. The published data, however, would make possible an assessment of biological hazards resulting from various forms and levels of pollution, and, from this, recommend maximum permissible levels for pollution. The types of pollutants in the reference base would include not only those which had direct effect on humans, but also those which had an indirect effect through their influence on plant and animal life. Initially, the committee decided to focus upon radiation, noise, pesticides, heavy metals, and bacteria, and several sub-committees were established – Air, Water, Heavy Metals, Pesticides and Related Compounds, Physical Energy, and Biological Phenomena.

The first published work of the Associate Committee on Scientific Criteria for Environmental Quality was a compilation by R.M. Sutton and C. Quadling entitled *Inventory of Pollution-Relevant Research in Canada* (Ottawa: National Research Council of Canada, 1972–5. 4 v. in 2). The introduction stated that 'the user will find the inventory handy and easy to use since all the information is cross-referenced alphabetically by author and subject.' In order to prepare this inventory, the authors sent out questionnaires to agencies, organizations, or individuals known to be active in pollution-relevant research. The information received was then stored in a computer and two files were generated, one for expertise and one for projects. By the end of 1989, this committee had published over 90 reports documenting environmental concerns. Major clusters of reports, which reflected various sub-committee interests, were published on topics ranging from heavy metals, water pollution, air pollution, and environmental effects of forest pesticides, to evaluating human health costs of environmental damage.

The first publication, which studied heavy metals in the environment, was entitled *Lead in the Canadian Environment* (Ottawa: National Research Council,

1973. 116 p. Associate Committee on Scientific Criteria for Environmental Quality. Report No. 4). The preface, which outlined the range of topics to be covered by the subcommittee, stated that 'metals have common characteristics in that they occur above the natural background level and that, as elements, they are not biodegradable.' Besides providing the geochemical background and industrial and domestic uses for the particular metal discussed, the publication examined the range of effects, positive and negative, and special effects of this metal. The succeeding monographs in this sub-series were entitled *Effects of Chromium in the Canadian Environment* (Ottawa: National Research Council, 1976. 168 p. Associate Committee on Scientific Criteria for Environmental Quality. Report No. 18); *Effects of Arsenic in the Canadian Environment* (Ottawa: National Research Council, 1978. 349 p. Associate Committee on Scientific Criteria for Environmental Quality. Report No. 34); *Effects of Cadmium in the Canadian Environment* (Ottawa: National Research Council, 1979. 148 p. Associate Committee on Scientific Criteria for Environmental Quality. Report No. 33); *Effects of Mercury in the Canadian Environment* (Ottawa: National Research Council, 1981. 352 p. Associate Committee on Scientific Criteria for Environmental Quality. Report No. 57); and *Aluminium in the Canadian Environment.* Edited by M. Havas and J.F. Jaworski (Ottawa: National Research Council, 1986. 331 p. Associate Committee on Scientific Criteria for Environmental Quality. NRCC Publication No. NRCC24759).

The reports on the aquatic environment were broad-ranging in their topics. The first report, written by D.M. McLean, was entitled *Detection and Inactivation of Enteroviruses in Water* (Ottawa: National Research Council, 1973. 26 p. Associate Committee on Scientific Criteria for Environmental Quality. Report No. 1). Enteroviruses are a family of just over 40 viruses, including the poleomyelitis virus, that invade the central nervous system, although their mode of entry to the body is through the gut. Contamination of the aqueous environment, both fresh water and marine, by enteroviruses may occur readily through the discharge of raw or inadequately treated sewage effluents into natural bodies of water. Possibly the major report on the aquatic environment was entitled *Acidification in the Canadian Aquatic Environment* (Ottawa: National Research Council, 1981. 369 p. Associate Committee on Scientific Criteria for Environmental Quality. NRCC Publication No. 18475). The panel, made up of five researchers who wrote the report, said that 'three critical questions must be addressed to advance our understanding of the acidification phenomenon in the Canadian aquatic environment. They are ... quantitative relationships between pollutant release ... and its impact on the aquatic environment ...; the rate of acidification ...; the total effect of acidic deposition. ...' In order to answer these questions, the panel identified and made seven recommendations for research, but also noted that 'not all the effects of acidic deposition are discussed in this document. The effects on wildlife, human health, agriculture, silviculture, and groundwater must also be examined.'

Other topics studied by panels, or individually authored, indicated current concerns about the aquatic environment. J.R. Marier, in the report entitled *The Effects of Pulp and Paper Wastes with Particular Attention to Fish and Bioassay Procedures for Assessment of Harmful Effects* (Ottawa: National Research Council, 1973. 33 p. Associate Committee on Scientific Criteria for Environmental Quality. Report No. 3), emphasized that the 'ecological health status of the receiving water is the important factor, rather than the mere reliance on the composition of wastes in the outfall pipes.' In the report by D.R. Dickson entitled *Waste Heat in the Aquatic Environment* (Ottawa: National Research Council, 1975. 39 p. Associate Committee on Scientific Criteria for Environmental Quality. NRCC Publication No. 14109), the author said that 'it is possible to estimate the probable effect [from thermal electricity generating plants] on certain species of fish ... and on certain aspects of water quality. ...'

Another interesting report was authored by J.R. Roberts and entitled *Polychlorinated Biphenyls: Biological Criteria for an Assessment of Their Effect on Environmental Quality* (Ottawa: National Research Council, 1978. 172 p. Associate Committee on Scientific Criteria for Environmental Quality. Report No. 53). PCBs were first created in 1881, and since 1930 have been used in a wide range of products such as heat transfer agents, lubricants, and plasticizers, and even low levels of contamination into inland and marine aquatic ecosystems can result in high accumulations of toxic PCBs in fish, marine mammals, and birds. In fact, PCBs 'constitute a major portion of the chlorinated hydrocarbon residues found in fish and birds of the North Atlantic, the Gulf of St. Lawrence, and the Great Lakes.' In order to accomplish some of the work for this report, five contracts were awarded to biologists and chemists at York University, in Toronto. The results of their research were five reports primarily authored by M. Katz and J. Heddle. These were entitled *Evaluation of Health Effects of Polynuclear Aromatic Hydrocarbons in Suspended Particulate Matter* (Contract No. 028-2370-1975. 33 p.); *Study of Mutagenesis (Chemical Carcinogenesis) by Polynuclear Aromatic Hydrocarbons and Related Environmental Pollutants* (Contract No. 031-530-1977. 70 p.); *Biological Effects of Polycyclic Aromatic Hydrocarbons and Related Environmental Pollutants. Phase I and Phase II* (Contract No. 7-1174-1978. 24 p. and Contract No. 7-2571-1978. 15 p.). Contracted research carried out for the Associate Committee on Scientific Criteria for Environmental Quality was listed in the periodic *Status Reports* (Ottawa: Environmental Secretariat, 1972–89), issued by the Committee, as information about its activities and programs, and its subcommittees and expert panels. The first *Status Report* appeared in July 1972, and they were occasionally issued until 1989.

Other water-related pollution titles were issued by the Subcommittee on Water and included P.A. Spear's report, entitled *Zinc in the Aquatic Environment: Chemistry, Distribution, and Toxicology* (Ottawa: National Research Council, 1981. 145 p. Associate Committee on Scientific Criteria for Environmental Quality. Report No. 50); R.A. Vollenweider and P.J. Dillon's report, entitled *The Application of the Phosphorus Loading Concept to Eutrophication Research* (Ottawa: National Research Council, 1974. 42 p. Associate Committee on Scientific Criteria for Environmental Quality. NRCC Publication No. 13690); the panel authored *Polycyclic Aromatic Hydrocarbons in the Aquatic Environment: Formation, Sources, Fate and Effects on Aquatic Biota* (Ottawa: National Research Council, 1983. 209 p. Associate Committee on Scientific Criteria for Environmental Quality. NRCC Report No. 18981); and J.A. Thompson and M.G. Sheffer's report entitled *Organotin Compounds in the Aquatic Environment: Scientific Criteria for Assessing Their Effects on Environmental Quality* (Ottawa: National Research Council, 1985. 284 p. Associate Committee on Scientific Criteria for Environmental Quality. NRCC Publication No. 22494).

The studies on air pollution focused upon photochemical air pollution (the interaction between the ultraviolet component of solar radiation and certain chemical groups such as hydrocarbons and oxides of nitrogen), and aerosols in the atmosphere. The first report by the Subcommittee on Air was entitled *Photochemical Air Pollution: Formation, Transport and Effects* (Ottawa: National Research Council, 1975. 224 p. Associate Committee on Scientific Criteria for Environmental Quality. Report No. 12). In this report, the role of radiant energy in promoting photochemical air pollution was described, and the effects of this pollution on materials, vegetation, animals, and humans described. As well, research needs were pointed out, particularly research 'under conditions representative of Canadian climate. ...' The second report by the Panel on Atmospheric Aerosols was entitled *Effects of Aerosols on Atmospheric Processes* (Ottawa: National Research Council, 1982. 208 p. Associate Committee on Scientific Criteria for Environmental Quality. NRCC Publication No. 18473). Among the recommendations, the report suggested that 'a standardized terminology to describe all aspects of airborne particles be internationally adopted ... and standardized methodologies for estimating emissions from natural sources be prepared and adopted internationally.'

Pesticide studies were undertaken by the Sub-committee on Pesticides and Related Compounds and focused upon forest pesticides. One study, by C.W. Forsberg was entitled *Bacillus Thuringiensis: Its Effects on Environmental Quality* (Ottawa: National Research Council, 1976. 134 p. Associate Committee on Scientific Criteria for Environmental Quality. Report No. 21). *Bacillus thuringiensis* is the most successful and widely used microbial insecticide, which produces several related proteins that are toxic when infested by a relatively narrow range of species, primarily those within the Order Lepidoptera. In the early 1970s, it was tested in large-scale applications against forest insects, particularly spruce budworm (*Choristoneura fumiferana* (Clem.)), which infests millions of acres of eastern Canadian forests. With this background in mind, the members of the sub-committee showed concerns about the toxic components in commercial formulations of the pesticide, its stability, and the amounts and persistence of the insecticide in the environment. They suggested that genetic studies should be undertaken on the commercial organism and that tests should be conducted to determine the risk to non-target species such as scavengers that ingested large quantities of injected larvae containing *B.t.* Another widely used pesticide against the spruce budworm is fenitrothion, used since 1969. However, in the mid-1970s, there was public concern which linked the incidence of Reye's Syndrome (a disease of children characterized by alterations to the brain structure and by fatty degeneration of the viscera, notably the liver, which is fatal if untreated) to the fenitrothion spray program. It was this concern that led the sub-committee to examine this insecticide in the report entitled *Fenitrothion: The Long Term Effects of Its Use in Forest Ecosystems – Current Status* (Ottawa: National Research Council, 1977. 18 p. Associate Committee on Scientific Criteria for Environmental Quality. Report No. 24). The panel concluded that 'operationally, the use of pesticides in Canada for the control of spruce budworm must be described as involving brute force and educated ignorance. Unless the restrictions and guidelines listed in the report of this panel are implemented, there is nothing to be gained by pressing on with additional field research.' One other forest insecticide examined was aminocarb, a broad-spectrum insecticide, which was used, in particular, against agricultural pests. Its large-scale use against the spruce budworm began in 1975, and the Panel of the Sub-committee on Pesticides and Related Compounds prepared a comprehensive critical review of the scientific literature, and presented an overview of the factors needed for developing criteria that would enable the impact of aminocarb in the Canadian environment to be assessed. This information was presented in their report entitled *Aminocarb: The Effects of Its Use on the Forest and the Human Environment* (Ottawa: National Research Council, 1982. 253 p. Associate Committee on Scientific Criteria for Environmental Quality. NRCC Publication No. 18979).

The cost of the effects of the environment on human health has also been examined in two related reports by the Associate Committee. One report, by C.A. Dennis and R. Drope, was entitled *Medicare Data: Its Use in Defining the Effects of the Environment on Health* (Ottawa: National Research Council, 1977. 164 p. Associate Committee on Scientific Criteria for Environmental Quality. NRCC Publication No. 15387). This study examined the need to define the cause-and-effect relationship of the environment on human health using population data. Using health-care data sources in Saskatchewan, the authors described 'how, by the use of various recorded parameters, an association between various environmental influences and health effect can identified.' The second report, by D. Victor, was entitled *Evaluation of Costs Associated with Human Health Impacts* (Ottawa: National Research Council, 1985. 161 p. Associate Committee on Scientific Criteria for Environmental Quality. NRCC Publication No. 24489). This study was based upon 'the need to establish the balance between the benefits gained and the costs incurred by society from the use of organic chemicals. ...' Because of this need, the author undertook to produce a comprehensive analysis of the usefulness of benefit-cost analyses for developing defensible regulatory decisions concerning organic chemicals.

As mentioned above, the associate committees also served as a vehicle for national or international meetings, and the results of these seminars or symposiums were often published through the National Research Council. For example, the Associate Committee on Hydrology's Sub-Committee on Hydraulics of Ice-Covered Rivers held a workshop in Edmonton, Alberta, in 1982. The proceedings of that workshop were edited by D.D. Andres and R. Gerard, and published as *Proceedings, Workshop on Hydraulics of Ice-Covered Rivers* (Ottawa: National Research Council, 1982. 336 p.). And, the Associate Committee on Geotechnical Research was particularly active in sponsoring muskeg research conferences in Canada. The first conference was held in 1955, and the proceedings edited by T.H. Baker and J. Curran as *Proceedings of the Eighteenth Muskeg Research Conference, 15 August, 1979* (Ottawa: National Research Council, 1980. 151 p. Associate Committee on Geotechnical Research. Technical Memorandum No. 127).

IV Activities toward Encouragement and Assistance to Research

Although the National Research Council itself carries out pure and applied research, it also has, or has been responsible for, major physical structures or mechanisms attached to it in order to assist government departments, universities, or industries in their own research pursuits. These structures, or mechanisms, include the impressive array of national facilities located across Canada, the special scientific services in Ottawa to assist NRC in its mandate as the keeper of physical standards of measurement, their industrial assistance program, and their university research support programs that passed to the newly created Natural Sciences and Engineering Council (NSERC) in 1978.

A National Facilities

The National Research Council has seven types of national facilities, established to serve researchers in government, universities, or industry, who individually, or collectively, could not be expected to provide such a facility for themselves. Generally, users may participate in the experimental activities of the facility. The facilities provided are wind tunnels, a ship laboratory and towing tank, nuclear accelerators, including the TRIUMF (Tri-University Meson Facility) Cyclotron, radio and optical telescopes, a rocket range, and a neutrino observatory. Overall, these facilities have been described in the publication entitled *National Facilities* (Ottawa: National Research Council, 1979. 47 p.).

The National Aeronautical Establishment has 18 wind tunnels varying greatly in size and capacity for generating wind speeds as high as 23 times the speed of sound (Mach 23). Tunnel facilities are used primarily by the aerospace industry for the investigation of aircraft design or other aerodynamic vehicles. However, the tunnels are also used to study in model form the effect of winds on bridges, vehicles, and tall buildings. The wind tunnel operations have been described by F.R. Thurston and R.J. Templin in a report entitled *Management and Operation of Canadian Government Wind Tunnels* (Ottawa: National Research Council of Canada, 1971. 17 p.).

The ship laboratory and towing tank, originally operated by the Division of Mechanical Engineering and now part of the Institute of Marine Dynamics, is the second largest in North America, and is used to investigate new ship hull designs under a variety of conditions. Model testing at this facility enables industries involved in marine vessel construction to avoid design error. J.T. Tothill, of the Division of Mechanical Engineering, has described this facility in an article entitled *The Ship Model Towing Tank* (*Transportation*, v. 3, 1971. 3 p.), and the laboratory, in the publication entitled *Ship Laboratory* (Ottawa: National Research Council, 1971. 3 p.).

NRC has three types of nuclear facilities. One is an electron linear accelerator capable of delivering electron impulses with energies from three to 50 million electron volts (MeV). The second accelerator is a positive ion generator (Van der Graaff), which produces beams of protons (to 4 MeV). The third accelerator, located on the grounds of the University of British Columbia, is one of the world's most powerful cyclotrons. The TRIUMF, as it is called (Tri-University Meson Facility), is an accelerator, which produces an intense beam of high-energy protons of sufficient energy (several hundred MeV) to produce secondary particles called mesons. All together, these three accelerators are used for fundamental research in nuclear physics and radiation chemistry by primarily university researchers.

NRC has both radio and optical telescopes available to the Canadian scientific community. Three of these telescopes are operated by NRC's Herzberg Institute of Astrophysics, and two in Hawaii are operated as a multi-nation consortium. The Algonquin Radio Observatory (now closed) had a 150-foot (46-metre) paraboloid reflector, and was one of the world's most powerful telescopes in the centimetre wavelength range. It was located at a remote site in Algonquin Park in Ontario. This observatory has been described in a booklet entitled *Algonquin Radio Observatory* (Ottawa: National Research Council, 1969. 20 p.). The second radio observatory is the Dominion Radio Astrophysical Observatory located near Penticton, BC, and includes a synthesis telescope for high-resolution studies of neutral-hydrogen. The Dominion Astrophysical Observatory at Victoria, BC, houses two optical telescopes, one with a 48-inch (1.2-metre) reflecting mirror, and the other with a 72-inch (1.8-metre) mirror. This facility has been described in a booklet entitled *The Dominion Astrophysical Observatory, Victoria, B.C.: A Brief Non-Technical Account of the Observatory and Its Work, and of Some Related Aspects of Astronomy* (Ottawa: National Research Council of Canada, 1973. 56 pp.). The Canada-France-Hawaii Telescope, located at the summit of Mauna Kea, the highest mountain on the island of Hawaii, has a 144-inch (3.6-metre) main mirror. This facility has been described by J.L. Locke in an article entitled *Report on Construction of the Canada-France-Hawaii Telescope* (*Journal of the Royal Astronomical Society of Canada*, v. 71, 1977. pp. 9–20). Also in Hawaii, near the summit of Mauna Kea, at Hilo, is the James Clerk Maxwell Telescope, which is a radio telescope operating at millimetre and submillimetre wavelengths. It is jointly operated by the NRC, the United Kingdom, and the Netherlands. Another international project is the Gemini 8 Metre Telescopes Project of Canada, the United States, the United Kingdom, Chile, Argentina, and Brazil. These telescopes have been described in the article entitled *Gemini North Sees the Light* (*Sky and Telescope*, v. 98, 1999. pp. 18–19), and by P.F. Roche, in the article entitled *Gemini and the Existing Telescopes* (*New Astronomy Reviews*, v. 45, 2001. pp. 3–7). The most recent descriptions of all these telescopes can be found on the HIA website: <<http://www.nrc.ca/hia>>.

NRC, through its Space Research Facility Branch, operated the Churchill Research Range at Churchill, Manitoba, for rocket-borne experiments in the upper atmosphere in the auroral zone. NRC had managed this range after 1966. As early as 1971, 51 scientific publications, which reported the launchings of 72 rockets sponsored by the NRC since 1965, had been issued. The publications were listed in the reports entitled *Space and Upper Atmosphere Research in Canada: Balloons, Rockets, and Satellites* (Ottawa: National Research Council of Canada, 1969–71. 3 v.). The reports were a five-year bibliography of over 400 scientific papers resulting from Canadian space and upper atmosphere research for the period 1966–70. The Churchill Research Range has been specifically described by R.S. Ritter in a conference paper entitled *The Churchill Research Range and Canada's Space Research Program*. In: *Proceedings of the Second International Conference on Space Engineering, 1969* (Dordrecht: Reidel, 1970. pp. 650–71).

Finally, the most recent national facility developed is the Sudbury Neutrino

Observatory (SNO), which started to run at the end of April, 1999. It is situated 2,000 metres underground in INCO's Creighton Mine near Sudbury, Ontario. The project is a Canadian, US, and UK collaboration. Because it uses heavy water, the SNO will be able to detect a number of neutrino reactions. Neutrinos were postulated in 1931 to account for the missing energy in beta decay (a weak interaction, in which an unstable atomic nucleus changes into a nucleus of the same mass number, but different proton numbers. The observatory has been described by A.L. Hallin in the paper entitled *The Sudbury Neutrino Observatory* (*Nuclear Physics A*, v. 663/664, 2000. pp. 787c–790c).

B Physical Standards of Measurement

In accordance with the mandate given to the National Research Council in Canada's Weights and Measures Act, NRC is the source of the national system of physical measurement standards that is coordinated with, and forms a part of, the accepted international system of measurement (Système International d'Unités, or SI, or metric system), based upon seven primary quantities – mass, length, time, temperature, electric current, luminous intensity, and quantity of material. All of the standards derived from these primary quantities are called secondary or working standards. The system is maintained by the Bureau International des Poids et Mesures (BIPM), which in turn derives its knowledge and expertise largely from nine or 10 major standards laboratories, of which NRC's Institute for National Measurement Standards is one. With the exception of the kilogram, there are no world primary standards, but Canada's material standards, both primary and secondary, are frequently intercompared with those of other countries, usually under the auspices of the Bureau. An example of such a comparison, written by D. Reymann and B.M. Wood, was entitled *Comparison of Josephson Voltage Standards of the National Research Council of Canada and the Bureau International des Poids et Mesures* (*Metrologia*, v. 30, 1993. pp. 109–11).

One important aspect of the Institute's laboratory work with standards is the continued research to bring about their improvement. This work involves the development of new instruments for the investigation of phenomena, which could lead to totally new improved standards. For example, the demand for precise electrical measurements in Canada is greater than that for any other physical quantity. The practical standards used for electrical current (DC) are the volt and ohm. Their standards are currently maintained with a precision of about one part in one million. As well, the range of measurement of these quantities is very wide. Standard voltage measurements run from nanovolts (one billionth of a volt) to one hundred kilovolts (100,000 volts). Resistance measurements run from microhms (one millionth of an ohm) to one hundred feraohms (100 million of millions of ohms). Measurements are most commonly made on direct current (DC) and are most accurate, but these measurements can be done in alternate current (AC). This work at NRC has been described by A.F. Dunn in an article entitled *Maintenance of a Laboratory Unit of Voltage* (*IEEE Transactions on Instrumentation and Measurement*, v. 20, 1971. pp. 2–10). The work on standards for measurement is ongoing, as noted in the paper by P. Filipski and R. Arseneau entitled *A Watt and a Volt-ampere Transfer Standard* (*IEEE Transactions on Instrumentation and Measurement*, v. IM-42, 1993. pp. 170–5).

Other areas of interesting standards work involve the need by industry for accurate measurements of light and colour, required in the production control of various industrial products. As well as maintaining the basic standards of photometry and colorimetry, NRC has also undertaken ongoing work to develop improved photometric and colorimetric standards, and techniques of measurement. This work has been described in two articles, one by C.L. Sanders, entitled *Photometry: Task Problems and Organizations* (*Applied Optics*, v. 10, 1971, pp. 2589–91), and the other by G. Wyszecki, entitled *Development of New CIE Standard Sources for Colorimetry* (*Farbe*, v. 19, 1970. pp. 43–76.). A more recent interesting article on colour itself, written by A.R. Robertson of the Photometry and Radiometry Program, was entitled *Colour Perception* (*Physics Today*, v. 45, 1992. pp. 24–29).

NRC's standards-related research is correlated with the work of the corresponding Canadian standards committees, as well as with other national standards laboratories such as the National Bureau of Standards in the United States. At international meetings, improvements to current standards and future directions in standards research are established by mutual agreement. The overall standards research work and services at NRC have been most recently described in its publication entitled *Canada's National Measurement System* (Ottawa: National Research Council, 1995. 14 p.) and its calibration services in the publication entitled *Directory of Calibration Services, 1994* (Ottawa: Institute for National Measurement Standards, 1994. 76 p.).

C Assistance toward the Development of Canadian Industry

In the early 1960s, NRC realized that after a decade of building up the research capacity at Canadian universities and post-secondary institutions, industrial-sector research remained modest. This was due to the small domestic market and the fact that new researchers were gravitating primarily into university research. With these factors in mind, NRC, in 1962, launched a program of industrial research assistance to Canadian industry that has been both expanded and refined over the years. The program has been called IRAP (Industrial Research Assistance

Program), and its mandate has been to support new industrial research projects on a cost-share basis. Companies would choose, direct, and control their own research projects and staff, and receive support for a period ranging from three to five years in annual grants allocated by NRC's Committee on Industrial Research Assistance. As well, the companies retained all titles and rights to the results of jointly funded research. These grants were applicable for commercial research in companies outside the defence industry. NRC hoped that the program would help companies, new to research, to build a research capability, as well as help companies achieve a long-term viability, particularly with the lead time required from research to final product. Initially, applications came primarily from the food and beverage, textile, metal fabricating, machinery, transportation, electrical, and chemical products industries. Over time, the latter two industries tended to become the major beneficiaries of the program, although the applications were varied. In the mid-1970s, for example, GEAC Computer Corporation Limited, of Markham, Ontario, was given support which permitted the development of a general purpose on-line database management computer system. The system, designed with particular attention to banking needs, order entry, library circulation systems, and inventory control, was particularly successful in the financial community.

Fifteen years after the introduction of the program, 822 proposals from 360 companies had received support at a total cost of $310 million, with NRC providing 40% or $124 million toward the projects. It is interesting to note that over time approximately 48% of the funds were allocated to small companies (with less than 200 employees), 16% to medium-sized enterprises, and 36% to large companies (those with more than 1,000 employees). By the mid-1990s, IRAP contributed approximately $52 million to more than 3,500 projects, a reflection of reduced government expenditures, and over 80% of the project funding supported small to medium-sized enterprises with fewer than 50 employees.

This program of support for industrial research has been modified over time. In 1975, the Pilot Industry Laboratory Program (PILP) was introduced. PILP allowed Canadian companies to negotiate contracts with NRC for the latter's research on their behalf toward the development of marketable products. Besides taking advantage of NRC's research capabilities and facilities, PILP offered companies the identification, protection, and licensing of NRC's Patent Development Office. By 1982, PILP was responsible for the management of 111 projects with industry involving a commitment of $54 million. Fifty-one of these projects were based on technologies developed in the NRC laboratories, while the remainder were derived from the expertise of other participating government departments. In 1978, in order to expand IRAP benefits to smaller firms, NRC created the mini-IRAP program. The objective of this program was to encourage small firms with limited or no technical staff to be able to solve specific technical problems by using research and development facilities external to NRC. As a by-product of this program, NRC hoped that small firms would become aware of the benefits of research and development and possibly establish their own R&D teams.

Within IRAP, NRC also administered the Scientific and Technical Employment Program (STEP), which was designed to stimulate the creation of permanent jobs in the private sector for unemployed graduates who had recognized scientific or engineering backgrounds. With this program, new scientific or technical personnel hired by firms involved in IRAP programs would receive a portion of their salary from STEP.

By 1982, NRC reorganized its industrial support programs and IRAP, STEP, and the Technical Information Service (TIS) described below were integrated into a single Industrial Research Assistance Program. At the same time, the Cooperative Projects with Industry Program (COPI), managed by five other government departments, were consolidated with NRC's Program for Industry/Laboratory Projects (PILP). The latter allowed the transfer to industry of the benefits of the technical resources of 10 federal departments and agencies involved in research and development. Then, in 1985, the expanded PILP program was consolidated into the IRAP program. Finally, in the mid-1980s NRC set up an Industry Policy Analysis (IPA) Group to undertake or commission studies on the relationship between technology and industry in Canada. Some of the studies undertaken examined the possible extension of existing IRAP programs in order to improve the acquisition of needed foreign technology by Canadian firms. At that time, the title *IRAP, NRC Industrial Research Program* (Ottawa: National Research Council, 1984. 40 p.) was published.

Out of the studies came the Technology Inflow Program (TIP) which was funded until 1991. Other services which had been developed were consolidated into the IRAP-C program, a field advisory service for confidential, on-site advice for companies with limited technical resources; the IRAP-H program of undergraduate programs which funded university or technical college students hired by firms for special projects under the supervision of IRAP technology advisors; IRAP-L for short-term projects that required contract funding for outside technical services; IRAP-M for medium-term projects to provide salary support for subcontracted personnel; and IRAP-R for long-term projects with government, university, or foreign laboratories, or Canadian companies. Over this was a network of technology advisors from NRC, provincial research organizations, technology centres, the Association of Consulting Engineers of Canada, the Canadian Construction Association, and project managers from NRC and other government departments and universities. As well, this network included NRC liaison people and External Affairs (now Foreign Affairs) representatives abroad.

These services have been described by E. Maser and J. Wolfe in the publication entitled *The IRAP Technology Network: A Canada-Wide Contact Network* (Ottawa: National Research Council, 1988. 47 p.). This publication also contains a listing (for 1988) of technology advisors, project managers, and participants in the overseas network.

In 1990–91, IRAP underwent a detailed program evaluation, with IRAP's perceived strengths judged to be the 'direct delivery by technically qualified and experienced staff; the technical and brokering advice received from IRAP staff, scientific advisors and other members of the technology network; and the availability of a variety of program elements to meet different client needs.' In 1991, an IRAP strategic plan was written in order to carry the program through to the year 2000. The plan, which was accepted by NRC, was described in the publication entitled *The Industrial Research Assistance Plan: Strategic Plan for the 1990s* (Ottawa: National Research Council, 1991. 17 p.). Under this plan, the various funding elements (IRAP-H, -L, -M, -R) mentioned above were consolidated into two new categories – technology enhancement (TE), and research, development, and adaptation (RDA). As well, there would be 'increased regional decision-making and increased selectivity to focus on firms with the ability and commitment to progress technologically.' Finally, in 1995, the technology network of people was renamed the Canadian Technology Network, with 300 advisors across the country involved.

D Assistance to University Research

At the end of World War II, an NRC Review Committee, under the chairmanship of J.A. Gray, noted that the postwar prospects of Canadian universities, in relationship to research, were in difficult times. Wartime demands had not enabled them to keep up research, maintain up-to-date equipment, or be able to handle the huge backlog of undergraduate students or the expected heavy enrolments for postgraduate work. The National Research Council thought it could help with this problem, provided that a greater amount of money was made available for scholarships, bursaries, and grants-in-aid of research. There was also the threat of a brain drain of researchers to, primarily, the United States because of higher salaries and opportunities. The latter had happened regularly before the war. The importance of fundamental research was emphasized by E.W.R. Steacie, who chaired the National Research Council during the 1950s and early 1960s. He thought that fundamental or basic research was the domain of the university and it was during his tenure that the greatest expansion of science and scientific research took place in Canadian universities.

By 1957, NCR's grants for university research had grown from $14,000, in

1917–18, to just under $4 million. At that time, the pressure from the Soviet Union's highly publicized space program (the first Sputnik was launched on October, 1957) also created pressures for renewed efforts in science research and education. As a result, the federal government increased support for NRC's university activity to $4.5 million. Then, during the decade of the 1960s, support rose from that amount to $59.3 million. However, in spite of the large increases during this period, which saw the number of universities and colleges receiving support double, increase in enrolments was such that support per graduate student increased realy only a few percent per year.

At that time, a series of surveys and studies was conducted by the Forecasting Committee of NRC in order to assess the future needs and trends of the universities for research funds and human resources in science and engineering. The first report was entitled *Expenditure on Research in Science and Engineering at Canadian Universities: Report of a Survey for the Forecasting Committee of the National Research Council* (Ottawa: National Research Council, Office of Economic Studies, 1966. 55 p.). A second study was entitled *Graduate Students at Canadian Universities in Science and Engineering* (Ottawa: National Research Council, Office of the Economic Studies on Research and Development, 1968. 393 p.). A third study, by D.H. Levine, was entitled *Profiles and Characteristics of Graduate Students Enrolled for the Doctorate in Science and Engineering at Canadian Universities* (Ottawa: National Research Council of Canada, Office of Economic Studies on Research and Development, 1968. 17 p.). These reports indicated an approximate doubling of total university research funds required from 1967/68 to only 1971/72 (from $120 million to $250 million). During the same period, the number of faculty members in departments of science and engineering was forecast to increase from 5,300 to 8,100. Finally, the annual expected output of Ph.D. graduates by 1971/72 was given at 1,000 per year.

In order to accommodate a high level of support to university research and its human resources, NRC evolved a range of grants and scholarship programs, which were fine-tuned, until 1978, when the program was separated from the council and given to NSERC (the Natural Sciences and Engineering Research Council), one of the three new granting councils created (the others being the Medical Research Council and the Social Sciences and Humanities Research Council). The majority of funds earmarked by NRC for the support of university research were peer-adjudicated grants to individuals or groups (negotiated grants), awarded on the basis of proven excellence in research and the merits of the research proposal. The proposals were reviewed by approximately 20 selection committees, constituted by discipline (e.g., animal biology, cell biology and genetics, space and astronomy), and composed primarily of scientists with recognized research experience from Canadian universities, industry, and government laboratories. The recommendations of the grant selection committees were then considered by the NRC Committee on Grants and Scholarships, which had the mandate to assess the recommendations. By 1978, these grants amounted to just under 78% of the council's regular program funds. These funds financed some of the costs of research, including equipment and the salaries of research assistants. A variation in peer-adjudicated grants was the strategic or development grants, which funded research in areas of national interest, or were used to overcome regional imbalances in science research at universities, or used to link university research to industry needs. For example, in 1978, university researchers were invited to submit proposals for research in three broad areas of national concern – energy, environmental toxicology, and oceanography. The additional funds made available were $2.4 million, but the 476 grant applications submitted totalled $18 million. Another form of grants were the PRAI grants (Project Research Applicable in Industry). These originated in 1972, and were intended to facilitate the transfer of research results from university laboratories to Canadian industry. The grants provided short-term additional support to university researchers, and would lead to new techniques, processes, or products that showed promise to be useful to Canadian industry. Finally, a special grant was created in 1963 called the E.W.R. Steacie Memorial Fellowship. It was established in memory of E.W.R. Steacie, who was president of the National Research Council from 1954 to 1962. The fellowship was awarded annually to one or two outstanding young scientists to allow them to devote themselves to full-time research for one year. The other major area of NRC's support to the sciences at universities was its peer-adjudicated scholarship and fellowship program. Beginning in 1916–17, NRC allotted $10,000 from its small government appropriation to set up a modest scholarship and fellowship program. In the first 10 years of the program, NRC awarded 344 scholarships and fellowships to 199 students. By 1978, over 2,000 scholarships and fellowships were being awarded, or accepted, for renewal on an annual basis in various categories and almost $12 million was being disbursed through this program.

NRC continues to be active in providing postdoctoral down to undergraduate opportunities in the natural sciences and engineering fields through a range of programs. These programs are the Research Associateships Program, the Women in Engineering and Science Program, the new Post-Doctoral Fellowship Program, the Summer-Employment/Co-op Programs, and the NRC-NSERC Research Partnership Program. NRC has offered research associateships since the early 1980s. The program has been designed to provide promising scientists and engineers work in NRC laboratories on research problems in fields of particular interest. In 1995–96, NRC sponsored over 200 participants. These research associateships have been described in the annual publication entitled *Research Associateships for*

Research in Science and Engineering in the Laboratories of the National Research Council of Canada (Ottawa: National Research Council, 1980–). The new Post-Doctoral Fellowship Program, announced in 1995, in cooperation with the National Sciences and Engineering Research Council (NSERC), has been designed, again, to provide promising young scientists and engineers the opportunity to work in NRC laboratories. As well, there is the NRC-NSERC Research Partnership Program, a five-year $25 million program announced, in 1996, to create strong three-way linkages between NRC's institutes, and researchers in university and the private sector, and to accelerate the transfer of research results for commercial exploitation. The Women in Engineering and Science Program has been designed to encourage Canadian women undergraduates to enter the fields of science, engineering, and mathematics in which they are under-represented. The program is open to women undergraduates in the sciences to work in their specific fields of interest during the summer period, or as a co-op term. Finally, the Summer Employment/Co-op Programs have been designed to provide undergraduate science students, in general, opportunities to work during the summer months at most of NRC's research institutes. All these programs are described on the NRC website <<http://www.nrc.ca>>.

The detailed activities of NRC's scholarship and grants programs in support of university research used to be detailed in NRC's publication entitled *Annual Report on Scholarships and Grants in Aid of Research* (Ottawa: National Research Council, 1971/72–1977/78. 7 v.). Its earlier titles were *Annual Report on University Support* (Ottawa: National Research Council, 1962/63. 1 v.), and *Annual Report on Support of University Research* (Ottawa: National Research Council of Canada, 1963/64–1971/72. 9 v.).

V Other Activities

The National Research Council administers and represents Canada in several international organizations through the mechanism of national committees that reflect Canadian interests. National Committees that are established by NRC draw their membership from industry, government, and the universities. For example, NRC is the Canadian member for the International Council of Scientific Unions for 12 of the 17 constituent unions, for six of the scientific committees, and three scientific associations.

NRC also maintains programs of cooperation and exchange of Canadian scientists with foreign scientists through foreign national science agencies. For example, Canada has scientific exchanges with the Centre National de la Recherche Scientifique (France), among others.

Finally, NRC was involved in the coordination of Canadian participation and contribution to international research projects such as the International Biological Program (IBP), the International Hydrological Decade (IHD), and the Global Atmosphere Research Program (GARP) in collaboration with the Atmospheric Environment Service of Environment Canada. These projects took place in the decade of the 1970s. An example of the Canadian work with the International Hydrological Decade was M. Palmer's research entitled *Measurements of Currents in the Great Lakes* (Ottawa: Canadian National Committee for the International Hydrological Decade, 1972. 32 p.). Some of the work done with GARP was described by F. Mesinger in *Numerical Methods Used in Atmospheric Models* (Geneva: World Meteorological Organization: International Council of Scientific Unions, 1976. 2 v.).

VI Council Sources of the Publications

The vast number of publications generated by, or through, the National Research Council have been documented in print, from 1918 to 1982.

The first cumulated listing of the NRC's publications was entitled *List of Publications of the National Research Council of Canada, 1918–1938: NRC Nos. 1–773* (Ottawa: National Research Council, 1939. 38 p.). The second listing was entitled *Publications of the National Research Council of Canada, Second Edition, 1918–1946: NRC, Nos. 1–1412* (Ottawa: National Research Council, 1947. 132 p.). The third listing was entitled *List of Publications of the National Research Council of Canada, Third Edition, 1918–1952: NRC Nos. 1–2900* (Ottawa: National Research Council, 1953. 263 p.). The fourth listing was entitled *Publications of the National Research Council of Canada, Supplement, 1953–1958* (Ottawa: National Research Council, 1959. 1 v.). The fifth listing was entitled *Publications of the National Research Council of Canada. Supplement 1958–1963* (Ottawa: National Research Council, 1964. 438 p.). After that two smaller supplements were issued – *Publications of the National Research Council of Canada, 1969, Supplement No. 1* (Ottawa: National Research Council, 1969. 1 v.), and *Publications of the National Research Council of Canada, 1972, Supplement No. 2* (Ottawa: National Research Council, 1972. 1 v.). These editions also listed the serial and irregular publications of NRC as an Appendix, or as Part II of the publication. Then, the last two cumulated print publications were issued. They were *Publications of the National Research Council of Canada, 1970–76* (Ottawa: Canada Institute for Scientific and Technical Information, 1976. 756 p.), and *Publications of the National Research Council of Canada, 1977–1981* (Ottawa: Canada Institute for Scientific and Technical Information, 1982. 783 p.). These last two lists were computer generated, using a keyword subject index and an author index, linked by NRC publication number to a listing of the publications.

Since 1982, there have been no published lists and the only access now to NRC publications is through the CISTI (Canada Institute for Scientific and Technical Information) catalogue, which is now a web-based catalogue at <<http://www.nrc.ca/cisti>>.

The major source for serial publications issued by the National Research Council was the *Union List of Scientific Serials in Canadian Libraries* (Ottawa: National Research Council, 1957–94). The 18th edition, 1994, was the last one issued in paper. It was also issued in microfiche. However, from 1994 until 1997, the Union List was issued annually in a CD-ROM format called Romulus, jointly produced by CISTI and the National Library of Canada. Romulus (Version 5) consisted of four lists: (1) The Union List of Serials in the Social Sciences and Humanities, (2) the Union List of Scientific Serials in Canadian Libraries, (3) the Union List of Canadian Newspapers, and (4) the CISTI Serials. Romulus also contained a Directory of Interlibrary Loan (policies) for over 5,000 Canadian libraries, and supported a communication software package for document delivery.

From time to time, some of NRC's divisions, or more recently the institutes, have issued publication lists. For example, the Astrophysics Branch issued *Publications of the Astrophysics Branch* (Ottawa: The Council, v. 1, nos. 1–5, 1971–75). This was followed by *Publications of the Herzberg Institute of Astrophysics* (Ottawa: National Research Council, v. 1, 1975–v. 19, 1993). The longest of the individual publication lists produced were the *Publications of the Dominion Astrophysical Observatory* (Ottawa: Government Printing Bureau, v. 1, 1919/22–v. 18, 1990). Other separate lists have been the *Publications of the Division of Mechanical Engineering and the National Aeronautical Establishment, Series 1, 1961–1964.* 4 v. Series 2, 1966– (Ottawa: National Research Council, 1961–); the *Publications of the National Aeronautical Establishment, Supplement, 1990–* (Ottawa: National Research Council, 1990–); and from the Division of Building Research (later named the Institute for Research in Construction), the *List of Publications: Division of Building Research, 1947–1971* (Ottawa: National Research Council, 1947–71. 25 v.); *List of Publications: Division of Building Research, 1972–1983* (Ottawa: National Research Council, 1973–84. 12 v.); *Supplement to the List of Publications: Division of Building Research, 1984–1985* (Ottawa: The Division, 1984–85); *List of Publications: Institute for Research in Construction, 1986* (Ottawa: Institute for Research in Construction, 1987. 1 v.); *List of IRC Publications, 1987–1988* (Ottawa: Institute for Research in Construction, 1987–88. 2 v.); *Publications Catalogue: Institute for Research in Construction, 1991–1993* (Ottawa: Institute for Research in Construction, 1991–93. 3 v.). Today, there is a pattern for published work of the institutes to be listed in their print annual reports, or in their electronic versions at their particular websites, which, again, are listed on page 78 under *Part II: Publications* (of the National Research Council).

Chapter 3

Agriculture Research

I Historical Background

Before Confederation, agricultural activities in what is now Canada could be traced back first to the non-nomadic Indian groups of eastern North America, then to the St Lawrence River settlements of New France in the early 1600s, and finally to the North European settlements and farming areas, later known as Nova Scotia, Prince Edward Island, and New Brunswick on the eastern seaboard. On the west coast, there was agricultural activity, primarily in the lower Fraser Valley and on Vancouver Island. On the prairies, the District of Assiniboia, containing the Red River settlement, was a self-sufficient agricultural unit in a more hostile climate.

After Confederation, there were the prairie settlements of the Northwest Territories (Alberta and Saskatchewan), with not only their hostile climate but also their unfamiliar soil and inadequate farming methods. It was these settlements that triggered the creation of a Department of Agriculture under the Department of Agriculture Act. However, it was not until 1884 that a Select Committee of the House of Commons recommended the establishment of experimental farms as a means to improve Canadian agriculture in general, as well as help farmers in particular in newly settled territories regarding what crops to grow and how to grow them. By 1886, the House of Commons had passed the Experimental Farm Station Act, which provided for the establishment of experimental farms across Canada.

Initially, the experimental farm system consisted of a Central Experimental Farm in Ottawa and four branch farms. These branch farms were located at Nappan, Nova Scotia, Brandon, Manitoba, Indian Head, Northwest Territories (Saskatchewan), and Agassiz, British Columbia. All of these initially focused upon soil, crops, and livestock, to help Canadian farmers. During the initial decade, some of the practical results were recommendations for soil fertility for crops, and information on subjects such as the handling of manure, fertilizers to use, seedbed preparation, and seeding. The organizational structure of these experimental farms was based upon research divisions (controlled from the Central Experimental Farm) by category – agriculture (including field crops and animal husbandry), cereals, chemistry, entomology, botany, and horticulture. This system of divisional control from Ottawa continued until 1959, at which time the Research Branch was organized.

The oldest of these divisions was the Entomology and Botany Division, which was part of the Department of Agriculture prior to the creation of the experimental farms, and studied problems with insects injurious to cereal, field crops, and orchards, but in 1909 entomology and botany became separate. The Horticulture Division studied the collecting, growing, and testing of fruits and vegetables to ascertain their potential for commercial production in the different regions of Canada. The Chemistry Division initially undertook the analysis of soils and fertilizers, and studied the milling and baking qualities of hard red spring wheat, feeding concentrates for livestock, the formulation of insecticides and fungicides, and the optimum use of agricultural wastes. The Poultry Division was organized in 1888, and initially studied the crossing of poultry strains within breeds as well as cross-breeding. There was also an interest in improving egg production. The Cereals Division focused upon variety tests for wheat, barley, and oats, and the hybridizing of wheat varieties. One of the successful outcomes of this work was the development of Marquis hard red spring wheat. The Agriculture Division dealt with research on all field crops, except cereals, and with all livestock, except poultry. Its work included the growing and comparison of all available varieties of corn, sugarbeets, rootcrops, and potatoes, for the unique Canadian climate and soils. As well, cattle herds of dairy breeds were developed and work was undertaken on developing mixtures of corn and fodder plants as an ensilage mixture for feed. Besides these major divisions, the only additional one formed after the turn of the century was the Tobacco Division, in 1908. The latter was a reflection of the commercial importance of this crop.

After Alberta and Saskatchewan became provinces in 1905, and the prairie provinces grew from 390,000 people in 1901 to 1.3 million in 1911, there were further demands on the Department of Agriculture to expand its network of experimental farms, or stations, to assist new farmers. Hence, in 1906 a new experimental station was established at Lethbridge, Alberta. And, in early 1907, another station, halfway between Calgary and Edmonton, was set up at Lacombe, Alberta. Then, a third station was opened at Fort Vermilion, 350 miles north of Edmonton. In 1909, one experimental station was opened at Rosthern, Saskatchewan, and another in the Maritimes at Charlottetown, Prince Edward

Island. In Quebec, a station was opened at Cape Rouge, near Quebec City, in an attempt to preserve the French-Canadian breed of horse and cow that came from France in the seventeenth century. Besides these full-sized stations, the Department of Agriculture also began to open up sub-stations, or sub-plots, in areas not easily served by full stations. Such arrangements came about at Kamloops, BC, Lesser Slave Lake, Alberta, Forts Resolution, Smith, and Providence in the Mackenzie District, Minto Bridge in the Yukon Territory, and finally Lake Abitibi in Quebec.

By the end of 1913, there were 18 experimental farms, or stations, across the country, and the controlling Central Experimental farm had expanded to 10 divisions along with nine entomology and plant pathology laboratories. From the beginning of World War I until the fiftieth anniversary of the Experimental Farms in 1937, their basic structure changed little, except for the removal of the Entomology Division, or Branch, as an entity of its own, possibly to administer the Destructive Insect and Pest Act of 1910. And, in 1923, a Bacteriology Division was added. Also, during this period, the first of the Experimental Farm bulletins for farmers, entitled *Seasonal Hints*, was printed every four months and distributed to 300,000 farmers. As well, information was written as press release articles for publication in newspapers.

In 1937, the structure of the experimental farms was changed to reflect a separation into basic and applied research. The basic research was regrouped into a Science Service of five divisions and the experimental farms became a service that focused upon applied science, problem solving, or experimentation. In practice, this split was rather artificial (Canadian livestock research was spread over five divisions between the pure and applied science groups), and in 1959 the Experimental Farms and the Science Service were united to become the Research Branch. Since then, this structure has been retained, but the administration of the branch has been divided between Directors General for the Eastern Region, the Central Experimental Farm, and the Western Region for overseeing the research centres, as the stations are now called. There has also been some specialty research assigned to individual stations. For example, food and animal research, land and biological resources, and plant research are the focus of the Central Experimental Farm, whereas research in northern agriculture has been assigned to the research station or centre at Beaverlodge, Alberta, Pacific agriculture to Vancouver, and horticulture to St Jean, Quebec.

Whatever the shifts in the administrative structure have been over the years, since the founding of the Experimental Farms, there has been a consistent pattern to the type of agricultural research carried out by the farms, or their successor research stations and centres. The major focus of the research has been soil research, including soil degradation, soil enrichment, and fertilizers, crop research, animal research, including dairy and beef cattle, poultry research, and pest control. This research is reflected in a range of discontinued and continuing series of general departmental publications, individual research institute or research centre (farm or station) publications, and a large output of research that appears in the open scientific literature. These publications are described below by category.

II Publications

A General Departmental Publications

Bulletin New Series No. 1, 1925–No. 181, 1935. This series, which was written for farmers, ranged from research information on insect pests in the field or the greenhouse, crop fertilizers, and soil fertility, grain and pasture crops, to the maple sugar industry or Canadian apples. An example from this series is the bulletin by A.E. Cameron entitled *The Morphology and Biology of a Canadian Cattle – Infesting Black Fly* (Ottawa: Department of Agriculture, 1922. 26 p. Bulletin, New Series, No. 5). This series was superseded by the Farmer's Bulletin.

Farmer's Bulletin, No. 1, 1936–No. 164, 1950. This series retained the themes of the Bulletin. An example from this series by F. Dimmock is entitled *Soybeans* (Ottawa: Department of Agriculture, 1948. 19 p. Farmer's Bulletin No. 149).

This series was superseded by the Publication series.

Pamphlet, No. 1, 1923–No. 168, 1935. The information in this series ranged from insect pests, the dressing and cutting lamb carcasses, and winter steer feeding in Manitoba, to publication lists. An example from this series by N. Criddle is entitled *The Western Wheat-Stem Sawfly and Its Control* (Ottawa: Department of Agriculture, 1922. 8 p. Pamphlet No. 6). This series was superseded by the Farmer's Bulletin described above.

Circular, No. 1, 1922–No. 189, 1950. The information in this series ranged from discussions on the control of insect pests, such as tree tent caterpillars in the prairie provinces, or apple maggots and their control in Quebec, to mushroom cultivation, or the washing and sterilization of farm dairy utensils. An example from this series by H. Groh is entitled *Poison Ivy* (Ottawa: Department of Agriculture, 1937. 4 p. Circular No. 120).

Technical Bulletin, No. 1, 1935–No. 77, 1950. The Technical Bulletins contained information which ranged from soil descriptions, or the economic aspects of agriculture such as costs and returns in apple production, to the marketing of perishable farm produce. An example of a technical bulletin by M.B. Davis and H. Hill is entitled *Orchard Soil Management and Apple Nutrition in Eastern Canada* (Ottawa: Department of Agriculture, 1948. 28 p. Technical Bulletin No. 65).

Miscellaneous Publications. These are individual publications not issued through any series. Some are authored, and some are not. Examples of unauthored publications are *St Mary Irrigation Project – PFRA*. Revised Edition (Ottawa: Department of Agriculture, 1968. 33 p.); and *Potato Wart Quarantine in Newfoundland* (Ottawa: Department of Agriculture, 1964. 5 p.). An example of an authored publication, by J.F. Alex, is entitled *Survey of Weeds of Cultivated Land in the Prairie Provinces* (Ottawa: Department of Agriculture, 1966. 75 p.).

Publications, No. 475, 1935–. This is the longest ongoing series of research and information publications of the department. Before 1935, 474 were published, but not numbered. These publications are substantial and cover topics that range from plant and animal research, including taxonomic studies, to practical topics, such as information on fruit production, or growing rutabagas. An example from this series by R.G. Foottit and W.R. Richards is entitled *The Genera of Aphids of Canada: Homoptera: Aphidoidea and Phylloxeridae* (Ottawa: Agriculture Canada, 1993. 766 p. Publication No. 1885).

Historical Series, No. 1, 1967–. This is a smaller, ongoing series, on particular historical aspects of the Research Branch or its precursors in the Department of Agriculture. An example from this series by G.G. Dustan is entitled *A History of the Entomology and Plant Pathology Laboratories on the Niagara Peninsula 1911–1960* (Ottawa: Department of Agriculture, 1968. 35 p. Historical Series No. 4).

B Publications of the Research Establishments of the Department

(a) Research Reports of the Research Branch

The Research Reports are the most important source available for the research activities of the research institutes and centres (stations) of the Research Branch. Until 1966, they were issued separately, but in 1967 they were consolidated into the following:

Research Report (Ottawa: Department of Agriculture, 1967–71. 5 v.);
Research Branch Report (Ottawa: Department of Agriculture, 1972–89/90. 8 v.);

Research Branch Directory of Research (Ottawa: Department of Agriculture Canada, 1990/91-1992/93. 3 v.);
Directory of Research (Ottawa: Agriculture and Agri-Food Canada, 1994–).

(b) Miscellaneous Publications

From time to time, the research institutions, or their precursors, as well as individual research stations or experimental farms, have issued their own publications which, like the overall departmental miscellaneous publications, are not issued within any series. An example of this type of publication is one by D.V. Fisher entitled *Pruning Tree Fruits and Smaller Fruits* (Summerland, BC: Agriculture Research Station, 1957. 35 p.).

(c) Serial Publications

The serial publications have tended to be short-lived and focus upon particular problems. For example, for a brief period of time, the Plant Pathology Laboratory at Winnipeg issued the *Report on Rust Development*, No. 1, 1954–No.3, 1956. However, a major publication by the Plant Research Institute in Ottawa, entitled *Canadian Plant Disease Survey*, v. 1, 1920–, has had a long ongoing publication history.

C The Open Research Literature

Much of the broad range of research done in the Department of Agriculture Research Branch does appear in the open international scientific journal literature and often in major Canadian scientific journals such as the *Canadian Journal of Zoology*, the *Canadian Journal of Animal Sciences*, the *Canadian Journal of Plant Science*, and the *Canadian Entomologist*. As well, papers are sometimes presented at conference proceedings, or appear as chapters in monographs covering particular research topics. It is in the open scientific literature that the slow and methodical 'brick by brick' increment of research knowledge applicable to Canadian agriculture and elsewhere is vividly noted.

III Research

A Soil Research

(a) Soil Surveys

Although European settlers were aware of the similarities, and differences, of the soils in the new lands they occupied, soil surveys did not begin in Canada until

1914. The first was undertaken by the Department of Chemistry of the Ontario Agricultural College at Guelph in 1914. This survey used the soil classification system developed by the US Bureau of Soils. It was a hierarchical scheme of three levels – soil province (based upon general features or surficial geology); soil series (analogous to a geological formation and using colour, origin, and weathering); and soil type (based upon textural divisions within the soil series). The purpose of the survey was to identify and map the main soil types so that agricultural development and practices could be planned according to the nature of the soils.

By 1920, all of Ontario southwest of Kingston had been surveyed. However, as faculties of agriculture with soils departments were formed at emerging universities across the country, soil surveys became strictly provincial undertakings during this early period. This also meant that these provincial soil zone maps used their own terminology and scale. National developments in the coordination of soil surveys in Canada did not come about until after World War II, when the National (later Canada) Soil Survey Committee began meeting for the first time. This committee had originally been formed by the National Committee on Agricultural Services, in the 1930s, at the suggestion of E.S. Archibald, then director of the Experimental Farm Service. At that time, the Experimental Farms had become involved in soil problems, brought on by the prairie drought and dust bowl conditions, which triggered the creation of the Prairie Farm Rehabilitation Act (PFRA).

Over the period 1945 to 1960, the committee managed to bring about a greater measure of uniformity among the provinces in all phases of soil survey work, and included the development of a Canadian taxonomic system of soil classification that was accepted for use in Canada. The first major publication, describing the Canadian taxonomic soil classification system, was entitled *The System of Soil Classification for Canada* (Ottawa: Department of Agriculture, 1970. 249 p.). This taxonomic system described the classification of soils by Order, Great Group, and Subgroup, and soil was defined as 'any unconsolidated mineral or organic layer thicker than 4 inches (10 cm.), occurring naturally on the earth's surface.' This publication was completely revised in 1987 and entitled *The Canadian System of Soil Classification.* 2nd Edition (Ottawa: Agriculture Canada, Research Branch, 1987. 164 p. Publication No. 1646). This revised edition incorporated refinements to the taxonomic system in order to overcome problems identified by soil scientists from various regions in the original soil classification scheme. For example, in 1973, a Crysolic order was proposed to classify the soils with permafrost close to the surface.

Parallel to the development of a Canadian soil classification system has been this development of a soil mapping system. 'Soil mapping is the identification, description and delineation on a map of different types of soil based on direct field observations or on indirect inferences from such sources as aerial photographs.'

Over the years, there has been a body of soil mapping principles developed that are applicable across Canada, and the Mapping System Working Group of the Expert Committee on Soil Survey (formerly the Canada Soil Survey Committee) has documented these principles in a set of proposals entitled *A Soil Mapping System for Canada: Revised* (Ottawa: Agriculture Canada, 1981. 94 p. Land Resources Research Institute Contribution No. 142). In this report, the Working Group has attempted to standardize methods by proposing logical alternatives where regional differences existed.

More recently, soil map and site description information has been computerized using the Canadian Soil Information System (CanSIS). This was developed by the Land Resources Institute of the Research Branch and is maintained by them, in cooperation with provincial research agencies and the universities. As well, Parks Canada and Environment Canada provide input. The procedures and computer programs that make up CanSIS are for the storage, manipulation, and display of data related to both soil science and natural resources studies. The stored data consists of national files of verified and analysed data on basic soil properties, as well as agronomy and resource information. There are also project files that have restricted access because the data have not been thoroughly verified or analysed. The system is described in detail in the publication by K.B. MacDonald entitled *The Canada Soil Information System (CanSIS): General User's Manual* (Ottawa: Agriculture Canada, 1984. 1 v. Land Resources Research Institute Contribution No. 83-59E).

The visible results of the Canada Soil Survey Committee and the Mapping System Working Group have been documented in one major overview of Canadian soils, and individual soil surveys carried out provincially, but overseen and published by the Land Resources Research Institute in Ottawa. The major compilation of Canadian soils has been written by J.S. Clayton and W.A. Ehrlich *et al.*, and entitled *Soils of Canada* (Ottawa: Department of Agriculture, 1977. 3 v.). This compilation was undertaken as a cooperative project of the Canada Soil Survey Committee and the then Soil Research Institute (now the Land Resources Research Institute) in Ottawa. Volume 1 is the *Soils Report*, Volume 2 is the *Inventory*, and Volume 3 is a *Glossary of Terms*. Two maps are also included, one entitled *Soils of Canada*, and the other entitled *Soil Temperature*. The *Soils Report* is both descriptive and technical and includes a discussion of soil concepts, the physiography of Canada, soil climates, vegetation, and soil descriptions. As well, a correlation between Canadian, United States, and World (FAO) soil classification systems is given. The *Inventory* provides a hierarchy of the Canadian soil classification scheme, to which are appended the particular Canadian region containing a particular soil type, and the soil climate of that region. Finally, the *Glossary* provides definitions of technical terms that are used in soil science in Canada.

Individual soil surveys, such as one by T.M. Lord and A.J. Green, entitled *Soils of the Fort St. John – Dawson Creek Area, British Columbia* (Ottawa: Agriculture Canada, 1986. 130 p. Maps, Land Resources Research Centre Contribution No. 85-27. British Columbia Soil Survey Report No. 42), are part of the program to survey, update, and publish inventories of soil resources of the regions of Canada (in this example, the Peace River area of British Columbia). This report is composed of two parts: the soil maps, and the report itself. There are four soil maps covering particular regions of the area, and the report itself gives technical descriptions and chemical analyses of selected soil profiles, and brief sections on early history, geology, climate, and vegetation. Also included is a section on land use and the capability of the land to support agriculture, forestry, big game, waterfowl, and recreation.

(b) Land Degradation

Land degradation has become an accumulative problem in all regions of the country as a result of changes in production practices since World War II. Overall organic and nutrient losses have been estimated at 50%–70% of original levels and require ongoing studies in soil enrichment, described in the next section. However, this problem has been aggravated by wind and water erosion of soil, soil compaction, and salination. More recently, acidification by industrial pollution, or by chemical fertilizers, particularly nitrogen, has furthered soil degradation. An overall assessment of land degradation in Canada has been written, by D.R. Cook and J. Damanski, in their publication entitled *An Assessment of the Degradation of Agricultural Lands in Canada* (Ottawa: Agriculture Canada, 1981. 86 p. Land Resource Research Institute Contribution No. 118). These authors have also written an updated article entitled *Soil Conservation in Canada* (*Journal of Soil Water Conservation*, v. 41, 1986. pp. 204–10). Another assessment of soil degradation has been written by M. Anderson and L. Knapik and entitled *Agricultural Land Degradation in Western Canada: A Physical and Economic Overview* (Ottawa: Agriculture Canada, 1984. 138 p.). One recent overview of the problem is in a volume, written and edited by D.F. Acton and L.J. Gregovich, entitled *The Health of Our Soils: Toward Sustainable Agriculture in Canada* (Ottawa: Agriculture and Agri-Food Canada, 1995. 138 p. Publication No. 1906E).

Although soil erosion had been observed on the prairies before the turn of the century, and southern Alberta west of Lethbridge encountered regular erosion even before 1920, it was not until the 1930s that the 'dust bowl' developed into a major problem in Manitoba, Saskatchewan, and Alberta. The history and reasons for this major period of soil erosion have been described, by C.H. Anderson, in the book entitled *A History of Soil Erosion by Wind in the Palliser*

Triangle of Western Canada (Ottawa: Department of Agriculture, 1975. 25 p. Historical Series No. 8).

When the Great Plains were first cultivated, before the turn of the century, dust mulches were recommended to reduce moisture loss from summer-fallowed wheat fields. These mulches, however, together with the cropping of large adjoining areas, low precipitation, high temperatures, strong winds, evaporation, and the loss of top soil from excessive ploughing, created ideal conditions for soil erosion. The problem, including the physics of the erosion process, was studied, over the years, at the Experimental Stations at Swift Current, Saskatchewan, and Lethbridge, Alberta. By 1935, the first edition of a continuously revised manual, by E.S. Hopkins and E.S. Palmer, entitled *Soil Drifting Control in the Prairie Provinces*, 4th Revision (Ottawa: Department of Agriculture, 1964. 59 p. Publication No. 568), was published. This manual encouraged strip-farming, the use of cover crops, and the establishment of shelter belts. As well, weed control was encouraged, using the Noble blade, developed by C.S. Noble in southern Alberta. This blade cut just below the surface, in order to slice roots, but to leave plants firmly in the soil.

Because of improved management practices implemented after the 'dust bowl' period, the return of some of the more susceptible soils to permanent rangeland, and the decrease in the severity of severe droughts, soil erosion by wind has received less attention; hence, a smaller number of research studies have been carried out.

Nevertheless, some recent work has been done using Cesium-137 in soils as a measure of erosion. For example, J.J. Kiss and E. DeJong, in their paper entitled *An Assessment of Soil Erosion in West-Central Saskatchewan Using Cesium-137* (*Canadian Journal of Soil Science*, v. 66, 1986. pp. 591–600), determined a baseline value for Cesium-137 in native non-eroded soils. From this baseline, net soil erosion deposition could be estimated in local areas. This same method was used by Y.Z. Cao and D.R. Coote in their study entitled *Using 137Cs to Investigate Net Soil Erosion at Two Soil Benchmark Sites in Quebec (Canadian Journal of Social Science*, v. 73, 1993. pp. 515–26).

Cesium studies have also been used to study soil loss by water erosion, but to a lesser degree. For example, cesium studies using simulated rainfall have been carried out in Quebec by C. Bernard and M.R. Laverdière and described in their research paper entitled *Variability of the Relationship between Cesium and Soil Losses by Water Erosion* (Geoderma, v. 52, 1992. pp. 265–77). Water erosion has also been studied in Alberta, but the conclusions drawn by J.A. Toogood, in the paper entitled *Water Erosion in Alberta* (*Journal of Soil and Water Conservation*, v. 18, 1963. pp. 238–40) from the study of rainfall characteristics and erosion indicated 'that moisture conservation rather than soil erosion by water should be the major concern of farmers and their advisors in Alberta.' On the other hand, in the Maritimes, where soil loss by water is a greater problem, various planting methods with potatoes have resulted in differences in soil loss. T.L. Chow and J.L. Daigle, in their paper entitled *Effects of Potato Cropping Practices on Water Runoff and Soil Erosion* (*Canadian Journal of Soil Science*, v. 70, 1990. pp. 137–48) indicated that contour planting on a slope rather than slope planting reduced soil losses. Another paper, by T.L. Chow and H.W. Reis, entitled *Effects of Potato Hilling on Water Runoff and Soil Erosion under Simulated Rainfall* (*Canadian Journal of Soil Science*, v. 74, 1994. pp. 453–60), indicated that the practice of planting potatoes in rows, and hilling the plants at some stage of growth, which is a universal practice, increased soil loss. The method for measuring loss by water involves measuring sediment concentration, and has been discussed by L.M. Edwards and J.R. Burney, in their paper entitled *Sediment Concentration of Interrill Runoff under Varying Soil, Ground Cover, Soil Compaction, and Freezing Regimes* (*Journal of Environmental Quality*, v. 20, 1991. pp. 403–7).

Like water erosion, there has been limited research into the problem of soil compaction owing to the use of heavy machinery in contemporary agriculture. However, some work has been done studying compaction by soil types, and the effect of compaction on vegetable production in eastern Canada and in the Maritimes. J. Duval and G.S. Raghavan, in their paper entitled *Residual Effects of Compaction and Tillage on the Soil Profile Characteristics of a Clay-Textured Soil* (*Canadian Journal of Soil Science*, v. 69, 1989. pp. 417–23), indicated that the effect of compaction was very persistent in clay soil. The actual effects of compaction on plant growth have been reported, by J.C. Tu and S.C. Tan, in their paper entitled *Soil Compaction Effect on Photosynthesis, Root-Rot Severity and Growth of White Beans* (*Canadian Journal of Soil Science*, v. 68, 1988. pp. 455–59). These authors noted that soil compaction greatly reduced the root biomass, shoot growth, and leaf area. Compaction also significantly reduced the photosynthetic rate and increased root-rot severity.

Unlike water erosion or soil compaction, the problem of salinization has been studied more intensely, because it is a major agricultural problem on the northern Great Plains region of North America, and in particular southern Alberta. The causes have been examined by M.J. Hendry and G.P. Buckland, in their paper entitled *Causes of Soil Salinization: 1. A Basin in Southern Alberta, Canada* (*Ground Water*, v. 28, 1990. pp. 385–93); and again, by M.J. Hendry and G.W. Chan, in their paper entitled *Causes of Soil Salinization: 2. A Basin in East-Central Alberta, Canada* (*Ground Water*, v. 28, 1990. pp. 544–50). In southern Alberta, the topographical and geological features of larger areas (basins with up slopes and shallow bedrock layers) causes a recharging of distant water inflows that have seeped into the bedrock and then into the salinized basin area. However, in smaller depressions in southeastern Alberta, salinization was caused by seasonal high water tables due to the ponding of surface water in the depressions during spring runoff or heavy rains.

Besides natural causes of salinization, human factors have also contributed to the salinization of soils in southern Alberta. T.D. Sommerfelt and E. Rapp, in their publication entitled *Management of Saline Soils under Irrigation* (Ottawa: Agriculture Canada, 1988. 26 p. Publication No. 1624E), have stated that 80,000 hectares of irrigated lands in the prairie provinces are saline because of seepage from irrigation canals, poor water management, poor irrigation practices, and inadequate drainage. In order to reverse this problem, the authors said that excess water must be prevented from entering saline lands, excess water must be drained from those lands, and leaching practices followed to remove salt. Besides soil improvement, the adverse effects of soil salinity can be reduced by growing crops which can tolerate varying levels of salinity.

Another human factor, which has more recently contributed to soil degradation, is contamination by industrial pollution, via atmospheric deposition, or by the application of agro-chemicals such as nitrogen fertilizers, both of which create acidic environments. As well, heavy metals such as arsenic, cadmium, chromium, and cobalt from mining, metallurgy, or fossil fuels may enter soils in great enough quantities to pose risks to plant health besides humans or animals. Sewage sludge used to enrich agricultural lands also contains heavy metals. These acidic environments have been well studied, but the relative impact of acid rain in general and nitrogen fertilizers (as nitric oxide) in particular has been difficult to assess.

Nevertheless, D.R. Coote and S.S. Singh, in their paper entitled *A Proposed Methodology for Assessing the Relative Impact of Acid Rain and Nitrogen Fertilizers on Acidity of Agricultural Soils in Canada* (*Canadian Journal of Soil Science*, v. 69, 1989. pp. 611–27), created a simple model that allowed the sources or the direct inputs of acidic precipitation to be traced through soil pathways in order to determine their potential contribution to soil acidity. The model equation developed was based upon the predictability of soil reactions to nitrogen-based fertilizers and other acidic precipitation. But the assessment of heavy metals in soils requires a knowledge of the amount of naturally occurring metals in soils within a particular region. Work in this area of contamination has been done by the Geological Survey of Canada. R.G. Ganet, in the paper entitled *The Distribution of Cadmium in 'A' Horizon Soils in the Prairies of Canada and Adjoining United States.* In: *Current Research, 1994-B* (Ottawa: Geological Survey of Canada. pp. 73–82), indicated that underlying regional geological factors affect the geochemical composition of soils through naturally occurring pollution.

Finally, the use of sludge (the solid waste resulting from treatment of domestic and industrial effluents that are released into sewers), widely applied as a soil enrichment to agricultural land, can contribute to soil contamination. Although sludge contains desirable organic matter and nutrients that are returned to the soil, the sludge also contains heavy metals and other contaminants. Sludges have been studied by M.D. Webber, and reviewed in the publication entitled *Industrial Organic Compounds in Selected Canadian Municipal Sludges and Agricultural Soils* (Ottawa: Agriculture and Agri-Food Canada, 1994. 100 p.). This review indicated that the contamination of Canadian agricultural soils with sludge contaminants is not a serious problem, but in the long run the contamination of soil by heavy metals is a concern, because metals are persistent and may affect plant, animal, and human health.

(c) Soil Enrichment

Excluding irrigation, there are three areas studied by the Research Branch pertaining to soil enrichment, and hence improved crop production. They are crop rotations, moisture conservation (this particularly applies to the prairie provinces), and the use of fertilizers.

i. Crop Rotation The study of crop rotations has received more attention from agricultural researchers than perhaps any other farm problem. The history and results of early crop rotation work in Canada has been described by P.O. Ripley in the publication entitled *Crop Rotation and Productivity* (Ottawa: Department of Agriculture, 1969. 78 p. Publication No. 1376). In Canada, the first crop rotation experiments were set up at the Central Experimental Farm in Ottawa in 1888, and, in 1911, the experiments were expanded to other areas. These experiments were not designed for extensive analyses of the results, but the longevity of these rotations has provided information not available through short-term experimentation. In these early experiments, attention was given to the length of rotations and the arrangement of crops in the rotation. Shorter rotations included cash crops such as cereals, forage seed crops, and vegetable crops. Longer rotations were for the production of coarse grains and forage for mixed livestock and farming. These early experiments were useful for an understanding about what crops worked best in different soil and climate zones across the country, and what worked best in particular crop successions, or summer fallow regimes, the effect of weed, insect, and disease control, and what rotations were more effective for soil enrichment, particularly where cash crops were involved.

Most of the crop rotation studies in Canada occur on the prairies as dryland rotations. In fact, the oldest continuous rotation in North America, called Rotation U, at Lethbridge, Alberta, has been described by S. Dubetz in the publication entitled *Ten Year Irrigated Rotation U. 1911–1980* (Ottawa: Agriculture Canada,

1983. 17 p. Contribution No. 1983-21). Because of the more than 60–year span of these 10-year rotations, S. Dubetz and M. Oosterveld, in their research paper entitled *Sixty-Six-Year Trends in Irrigated Crop Yields – Barley Wheat and Oils* (*Canadian Journal of Plant Science*, v. 59, 1979. pp. 685–89), found that yield increases, over time, could be mainly attributed to the introduction of new cultivars, and not to other factors such as climate or soil quality. For this conclusion, regression analysis of long-term data was used.

The more recent review of crop rotation research on the prairies has been written, by C.A. Campbell and R.P. Zentner, and entitled *Crop Rotation Studies on the Canadian Prairies* (Ottawa: Agriculture Canada, 1990. 133 p. Publication No. 1841E). These authors said that a review of the literature, to that date, indicated 'the paucity of information regarding the influence of cropping practices on weed, disease, and insect infestation. As well, details of soil conditions have been examined only in a few experiments.' Accordingly, 'these deficiencies demonstrate the need for a comprehensive, team-oriented approach to crop rotation research.' These same authors, in their paper entitled *Effect of Crop Rotation and Fertilization on the Quantitative Relationship between Spring Wheat Yield and Moisture Use in Southwestern Saskatchewan* (*Canadian Journal of Soil Science*, v. 68, 1988. pp. 1–16), specifically investigated the effect on crop yield of rotation, fertilization, and moisture. They found that moisture content, owing to spring rains at the time of seeding, was particularly important for fallow-seeded wheat.

ii. Moisture Conservation Like crop rotation, soil moisture availability is also an important ongoing study on the prairies. The agricultural region of the prairies has a continental climate, with semi-arid to sub-humid moisture regimes, and most of the crops grown are based on dry land farming. Accordingly, crop production depends on the ability of the soil to store water, particularly over the winter, for the next growing season. These prairie soils are classified as Brown, Dark Brown, Black, and Dark Gray, and are distributed around the driest core of southwestern Saskatchewan and southeastern Alberta. Outward from that core, moisture availability increases. According to R. De Jong and A. Bootema, in their technical bulletin entitled *Characterizing the Soil Water Regime of the Canadian Prairies* (Ottawa: Agriculture Canada, 1992. 15 p. Maps. Technical Bulletin 194C-2E. Centre for Land and Biological Resources Research Contribution No. 91-130), it is important to understand the soil water storage ability and rainfall potential for estimating crop production risk in the various soil regions of the prairies. However, current crop management techniques, along with new plant varieties, do indicate that the efficiency of water preserved is much higher than in the past, even though the amount of water conserved is similar.

Some of the crop management techniques that can be used for conserving soil moisture are stubble management (using stubble to trap snow as well as conserving other additional moisture and helping to prevent erosion by wind and water), and using herbicides for weed control. The most cautious of the management practices, however, should be summer fallow. This is discussed, by A. Bootsma and J. Pumanski, in their publication entitled *Estimated Soil Moisture Conserved by Summer-Fallowing on the Canadian Prairies* (Ottawa: Agriculture Canada, 1992. 6p. Maps. Centre for Land and Biological Resources Research Contribution No. 91-145). Summer fallowing may be detrimental to soil quality through increased erosion, nutrient loss, and salinization.

Attempts to estimate the amount of soil moisture available at seeding time have been carried out through a water budget model developed in 1979 by W. Baur and J.A. Dyer and described in their publication entitled *The Versatile Soil Moisture Budget* (Ottawa: Agriculture Canada, 1979. 52 p. Land Resource Research Centre. Technical Bulletin No. 87). This model was then used by A. Bootsma and J. Damanski in their publication entitled *Soil Moisture Available at Seeding on the Canadian Prairies* (Ottawa: Agriculture Canada, 1992. 6 p. Maps. Centre for Land and Biological Resources Research. Contribution No. 91-146). These authors used climatic data for the 1956–85 period for 'agro-ecological resource areas' (ARAs) of the prairie region to come up with the estimated soil moisture available for these regions.

iii. Fertilizers Besides the study of crop rotations and understanding the importance of soil moisture for soil enrichment and crop production, research in soil fertilization is also important. Soils differ in their structure and nature, as well as their chemical content; hence their ability to provide nutrients for crops, which also have unique nutrient needs, is variable. Historically, organic materials such as animal manures, or crop residues alone, have been used to improve soil structure, but the constant growth and removal of crops with their nutrient uptake means that soils over time will lose their productive capacity. Accordingly, there is ongoing research for other sources of plant nutrients which can maintain soil fertility. Basically, the research on artificial fertilizers focuses upon primary elements such as nitrogen, phosphorus, and potassium, along with secondary elements such as boron, manganese, copper, iron, lead, and zinc. As well, research still goes on with traditional soil additions such as lime mulches, and manures. In fact, some of the recent work has returned to these traditional soil additives, coupled with artificial fertilizers.

The earlier work in Canada with soil nutrients was undertaken to show that successive cropping and poor farming methods were causing an ongoing loss of nitrogen and organic matter from soil. In the 1920s, F.T. Shutt, a pioneer in research in this area, published a bulletin entitled *Influence of Grain Growing on the Nitrogen and Organic Matter Content of Western Soils in Canada* (Ottawa: Department of Agriculture, 1925. 15 p. Bulletin New Series No. 44), which showed that data collected over a 38-year period indicated a 40% loss of nitrogen in the soil. The proportion of loss was the same within shorter spans of that time frame, and it was caused by poor cultivation practices. Later studies, by J.L. Doughty and F.D. Cook, did modify the reasons for nitrogen loss. In their paper, entitled *Effect of Cultivation on the Organic Matter and Nitrogen of Brown Soils* (*Canadian Journal of Agricultural Science*, v. 34, 1954. pp. 406–11), the loss of organic matter and nitrogen by cultivation was shown to be directly related to bacterial activity, and the evolution of carbon dioxide, and the loss of nitrogen by the leaching of nitrates beyond the depth of root penetration when brown prairie soils are put under cultivation.

Nitrogen may also be lost when it is utilized to break down straw, for by the 1960s, farmers were using more nitrogen fertilizers to compensate for this apparent loss. However, W.S. Ferguson and B.J. Gorby suggested this need not always be the case. In their paper entitled *Effect of Straw on Availability of Nitrogen to Cereal Crops* (*Canadian Journal of Soil Science*, v. 47, 1967, pp. 39–48), the authors found, under some conditions, the incorporation of straw into the soil did not generally cause nitrogen deficiency. In some soils, however, low moisture content, coupled with a long period of winter freezing, reduced micro-organism activity associated with nitrogen utilization during the decomposition of straw, and did cause a loss of nitrogen.

Perhaps the overriding factor in soil enrichment and plant productivity with nitrogen additions is the factor of soil moisture. J.J. Lehane and W.J. Staple, in their paper entitled *Influence of Soil Texture, Depth of Soil Moisture and Storage and Rainfall Distribution on Wheat Yield in Southwestern Saskatchewan* (*Canadian Journal of Soil Science*, v. 45, 1965. pp. 207–19), again indicated the overriding importance of soil moisture at seeding time on increased wheat yields. This study was reinforced by J.B. Bole and U.J. Pittman, in their paper entitled *Spring Soil-Water, Precipitation, and Nitrogen Fertilizer: Effect on Barley Yield* (*Canadian Journal of Soil Science*, v. 60, 1980. pp. 461–69). These researchers were able to show the maximum amount of nitrogen fertilizer that should be added, according to additional moisture content (0.3 kg/ha per one millimetre addition of spring soil moisture), for maximum yield and cost efficiencies.

More recently, there has been some shift to the study of manures as an alternate source of nitrogen for soils. For example, J.F. Dormaar and C.W. Lindwall, in their paper entitled *Effectiveness of Manure and Commercial Fertilizers in Restoring Productivity of an Artificially Eroded Dark Brown Chernozemic Soil under Dryland Conditions* (*Canadian Journal of Soil Science*, v. 68, 1988. pp. 669–79), indicated that manure applications to eroded soils resulted in crop yields that were higher than on check or fertilized plots.

The term 'manure' can also be applied to the use of legume plants as 'green manure' crops to replace conventional summer fallow in the spring wheat production systems of western Canada. H.H. Janzen and J.B. Bole described this method of soil enrichment in their paper entitled *Fate of N Applied as Green Manure or Ammonium Fertilizer to Soil Subsequently Cropped with Spring Wheat at Three Sites in Western Canada* (*Canadian Journal of Soil Science*, v. 70, 1990. pp. 313–23). For two seasons, three sites were seeded with flat pea and lentil, and then seeded to spring wheat the following spring. Although the spring wheat uptake of organic nitrogen was less than wheat in plots enriched with inorganic nitrogen fertilizer, the primary advantage of 'green manure' production was the replenishment of stable organic nitrogen reserves in the soil. The reason for the reduced plant uptake of organic nitrogen may relate to the time of the incorporation of the 'green manure' into the soil. In Prince Edward Island, J.B. Sanderson and J.A. Macleod reported on this in their paper entitled *Soil Nitrate Profile and Response of Potatoes to Fertilizer N in Relation to Time of Incorporation of Lupin* (Lupinus albus) (*Canadian Journal of Soil Science*, v. 74, 1994. pp. 241–46). These researchers found that early incorporation resulted in significant fall of nitrogen mineralization and winter leaching, as compared with late incorporation.

As with nitrogen, phosphorus is also a primary element and has been studied, mostly in relationship to its effect on alfalfa yields. In the 1960s, R.L. Halstead and A.J. Maclean studied phosphorus applications on soil near Ottawa. In their paper entitled *Availability of Phosphorus in Soils Pre-Treated with Superphosphates* (*Canadian Journal of Agricultural Science*, v. 36, 1956. pp. 471–82), these researchers found that the phosphorus content of alfalfa could be increased threefold by the application of a phosphatic fertilizer to the soil. The importance of phosphorus as a fertilizer was confirmed on the prairies in other studies by E.D. Spratt and E.V. McCurdy, and described in their paper entitled *The Effect of Various Long-Term Soil Fertility Treatments on the Phosphorus Status of a Clay Chernozem* (*Canadian Journal of Soil Science*, v. 46, 1966. pp. 29–35).

More recently, phosphorus studies, like nitrogen studies, have looked at the effect of temperature and moisture on the uptake of phosphorus and its effect on crop yields. For example, J.L. Dionne and A.R. Pesant, in their paper entitled *Effect of Temperature and Moisture Regimes and Soil pH on Alfalfa Response to Phosphorus Application* (*Canadian Journal of Plant Science*, v. 69, 1989. pp. 1149–63), found that yield responses due to phosphorus were greater under cool, rather than warm, growing conditions. As well, yield responses, owing to phosphorus, increased in a wet soil environment, and phosphorus uptake also increased with soil pH. Overall, these results were confirmed by H.V. Petit and A.R. Pesant, in their paper entitled *Quality and Morphological Characteristics of Alfalfa as Affected by Soil Moisture pH and Phosphorus Fertilization* (*Canadian Journal of Plant Science*, v. 72, 1992. pp. 147–62). However, they found that the quality of alfalfa was generally improved in a water deficit environment.

The relationship between temperature, precipitation, and fertilizer use has also been studied, not only with alfalfa alone, but in combination with brome grass by W.F. Nutall and D.H. McCartney, in their paper entitled *The Effect of N, P, S Fertilizer, Temperature and Precipitation on the Yield of Brome Grass and Alfalfa Pasture Established on a Luvisolic Soil* (*Canadian Journal of Plant Science*, v. 71, 1991. pp. 1047–55). In their study, these researchers found that forage yields were positively related to total precipitation, combined with an optimum level of fertilizer and a maximum temperature of 19.7 degrees C. On the other hand, yields were reduced as the temperature increased to an average maximum of 25 degrees C.

Unlike the studies on the effect of primary elements on grain and forage crops, research on secondary elements such as magnesium, or the minor elements such as manganese, has been carried out on specialty crops such as tomatoes, or fruit-bearing trees like apples. G.M. Ward and M.J. Miller discussed the problem of magnesium deficiencies in greenhouse tomatoes in their paper entitled *Magnesium Deficiency in Greenhouse Tomatoes* (*Canadian Journal of Plant Science*, v. 49, 1969. pp. 53–59). And, much earlier, H. Hill and F.B. Johnston discussed magnesium deficiencies in their paper entitled *Magnesium Deficiency of Apple Trees in Sand Culture and in Commercial Orchards* (*Scientific Agriculture*, v. 20, 1940. pp. 516–25). Manganese deficiencies have also been found in peach and apple trees. This was reported by C.G. Woodbridge and H.R. Mcharty in the paper entitled *Manganese Deficiency in Greenhouse Peach and Apple in British Columbia* (*Scientific Agriculture*, v. 31, 1951. pp. 435–38). And, at a later date, D. Neilsen and P.B. Hoyt reported in their paper entitled *Manganese Soil Tests for Both Deficient and Toxic Levels in Apple Orchards* (*Canadian Journal of Soil Science*, v. 70, 1990. pp. 503–7) that manganese concentrations in leaves of apple trees varied from deficiencies to toxic levels, but this was difficult to associate with soil tests as manganese availability varied in soil according to depth and pH level.

B Crops Research

Crops research in the Department of Agriculture concentrates on cereals (winter durum, and spring wheats, barley, oats, canola, soybeans, and corn) and forage research, the former important domestically and for export, and the latter for cattle production. However, it should not be forgotten that departmental research also encompasses vegetables, fruit, and ornamentals (flowering plants). On a smaller and regional scale, research is also carried out on minor crops such as flax, hemp linseed, and tobacco. Canada is geographically on the northern fringe of the temperate zone, and the crops we take for granted have had to be adapted to the growing conditions of the Canadian climate. And, as previously mentioned, one major reason for the passing of the Experimental Farm Station Act, in 1886, was the fact that the newly opened areas of the west, including Manitoba and the then Northwest Territories, needed an early maturing hard red spring wheat. Wheat was first produced in Canada at the beginning of the 17th century at the French settlement of Port Royal in the Maritimes. Over the next 200 years, as settlement expanded westward, wheat varieties from western Europe and New England were planted, but these varieties were not suited to the cold winters and short growing seasons of the prairies.

(a) Cereal Crops

The earliest developments in the search for suitable varieties of Canadian wheat occurred in 1842, and have been described by L.H. Newman in the paper entitled *New Wheat Creations and Their Significance* (*Canadian Geographic*, v. 18, 1939. pp. 208–16). David Fife, a farmer near Peterborough, Ontario, received samples of Ukrainian wheat, among which was a spring variety that produced seed. This became known as Red Fife, which was rust resistant with hard kernels. By the 1880s, it was the dominant variety grown in eastern Canada. In the west, however, Red Fife was too late maturing to produce consistent crops.

In 1888, the Experimental Farms system began a wheat improvement program that consisted of crossing early-maturing wheat varieties with high-quality varieties. Crosses between Red Fife and Hard Red Calcutta from India, which ripened two to three weeks earlier than Red Fife, produced a variety called Markham, out of which came Marquis wheat. The development of Marquis wheat has been described by J.W. Morison in the publication entitled *Marquis Wheat – A Triumph of Scientific Endeavour* (*Agricultural History*, v. 34, 1960. pp. 183–88). Marquis was initially resistant to wheat stem rust, and by 1970, 11 years after its introduction, 90% of the wheat seeded to hard red spring wheat was that variety. However, as new races of rust appeared, there has been an ongoing program by the department to produce varieties of spring wheat that combine rust resistance, early maturity, quality, and yield.

For example, in 1975 Thatcher wheat, which had good rust resistance, combined with the qualities of Marquis, was developed in Minnesota and grown in Canada. Other wheats that were developed were named Reward and Selkirk, both of which were crosses from Marquis. As well, Neepawa was developed in 1969, and showed not only rust resistance but also good milling qualities. A review of these spring wheat varieties, or cultivars, was undertaken in 1984 and 1985 to determine what changes had occurred in yield traits due to the ongoing breeding program. P. Hucl and R.J. Baker, in their paper entitled *A Study of Ancestral and Modern Spring Wheats* (*Canadian Journal of Plant Science*, v. 67, 1987. pp. 87–97), planted 12 spring wheat varieties in different combinations and plot locations in Saskatchewan. These varieties included seed samples ranging from the early Red Fife to Marquis, Reward, Thatcher, Neepawa, and others. The authors found that wheat grain yields have not had consistent improvements since the 1930s up to the introduction of Thatcher. However, recent plant-breeding programs, which introduced height-reducing genes, had shown yield increases of up to 30%. This was characteristic of 'semi-dwarf cultivars.' Further studies of yield data, by T.N. McCaig and R.M. DePauw, indicated a similar pattern. In their paper entitled *Breeding Hard Red Wheat in Western Canada: Historical Trends in Yield Related Varieties* (*Canadian Journal of Plant Science*, v. 75, 1995. pp. 387–93), the authors noted that yield potential did not change significantly during the 1947–92 period and any yield increases were due to the number of kernels produced by the newer varieties, and not by an increase in kernel size. Nevertheless, the genetic breeding for rust resistant varieties with good kernels (for milling and baking), yields, and resistance to other diseases such as smut and bunt infection continues, primarily at the Agriculture Canada Research Station at Winnipeg.

It is increasingly difficult to identify new genes for resistance to stem rust in collections of common wheat (*Triticum aestivum* L.). Consequently, the 'relatives' of wheat are being assessed as potential sources of resistance. The transfer of genes for resistance is not difficult to accomplish with close wheat 'relatives,' as they have chromosomes that pair readily with those of common wheat. However, with more distantly related species having little or no chromosome homology, this is more difficult. In the first instance, P.L. Dyck and P. Bartos, in their paper entitled *Attempted Transfer of Leaf Rust Resistance from Triticum Monococcum and Durum Wheat to Hexaploid Wheat* (*Canadian Journal of Plant Science*, v. 74, 1994. pp. 733–36), described how they were able to easily transfer a single gene for leaf rust resistance, although they encountered difficulties understanding clearly the mechanism of the transfer. In the second instance, P.L. Dyck and E.R. Kuber, as described in their paper entitled *Transfer of a Gene for Stem Rust Resistance from* Aeilops caudata *to Common Wheat* (*Canada Journal of Plant Science*, v. 70, 1990. pp. 931–34), did facilitate the gene transfer by a process called induced chromosome translocation. This process is possible in plants deficient for a particular Chromosome – 5B, which normally suppresses pairing – or recombinations, between chromosomes of alien species.

The end result of successful genetic breeding of new cultivars is their registration by the Food Production and Inspection Branch of Agriculture and Agri-Food Canada, and their description in the scientific literature. For example, in 1995, T.N. McCaig and R.M. DePauw announced the development of AC Barrie Hard Spring Wheat (*Canadian Journal of Plant Science*, v. 76, 1996. pp. 337–39). This new spring wheat, which had been registered in 1994, was adapted to the Canadian prairies and combined high yield with high protein content. As well, it had resistance to leaf and stem rust, common bunt, and loose smut. It was also a shorter plant, with strong straw.

Another major wheat crop grown in Canada is durum (*Triticum turgidum*), an ancient wheat of Egypt. The kernels of durum wheat are much harder than those of bread wheat, and larger, and contain more protein. Hence, it is used for making pasta. Durum, which is well adapted to semi-arid regions of the prairies, because of its high tolerance to drought, was introduced to Canada in 1918 with two varieties (Golden Ball and Minden) imported from the United States. However, by the 1950s, durum wheat yields from these varieties dropped on account of stem rust, and a Canadian breeding program was started for the development of rust resistant strains. In 1969, a durum variety resistant to stem and leaf rusts, and loose smut, was developed at the Agriculture Research Station in Winnipeg by D. Leisle and announced as *Hercules, A New Durum Wheat* (*Canadian Journal of Plant Science*, v. 50, 1970. pp. 118–19). Hercules was also superior in yield, maturity, straw strength, and kernel size to the American varieties. This variety was initially tested in 1964 in the Durum Cooperative Test, organized, in 1947, to grow and evaluate durum cultures at 12 prairie locations. Since then it has become and remains the statutory standard of durum quality in Canada. This program has been described by T.N. McCaig and J.M. Clarke in their paper entitled *Breeding Durum Wheat in Western Canada: Historical Trends in Yield and Related Variables* (*Canadian Journal of Plant Science*, v. 75, 1975. pp. 55–60). This paper also noted that, despite the problems of increasing yields in a specific environment which involves a complex process of integrating many growth stages (e.g., germination, vegetative, reproductive, and ripening), the Canadian developed cultivars have increased yields by 25% relative to foreign-developed cultivars previously grown in Canada. The most recent durum cultivar developed at the Agriculture Research Stations at Swift Current and Regina, Saskatchewan, and named after the town of Kyle, was announced as *Kyle Durum Wheat* (*Canadian Journal of Plant Science*, v. 67, 1987. pp. 225–27). This cultivar, registered in 1984, 'combines high yield with good quality and a high level of disease resistance. It is adapted to the Brown and Dark Brown Soil zones of Alberta and Saskatchewan.' Kyle was tested in the Durum Wheat Cooperative Test from 1981 to 1983, and it out-yielded all the check cultivars.

Unlike spring or durum wheats, winter wheat is primarily soft, and milled as pastry or biscuit flour. However, there are hard red winter varieties that are milled for crackers. Other winter wheats are also grown in the Maritime provinces, Ontario, and British Columbia, as forage for cattle feed. Winter wheats must be seeded in the late summer, when they will germinate and grow about 10–15 centimetres. During the cooler shorter days of autumn, the wheat plants undergo a physiological change called vernalization that enables them to flower the following spring. This is described more fully on p. 159. If the wheat does survive the low winter temperatures, the spring temperatures cause rapid growth, the development of flowers, and seed that ripens by July or early August. Because winter wheat can make effective use of winter and spring moisture, it has higher yields than spring-seeded wheats. Winter wheat predominates in the milder climate of southwestern Ontario, where it was introduced in the late 1700s by United Empire Loyalists from New York State, but today the growth of winter wheat for flour is divided equally by Ontario, southern Alberta, and Saskatchewan.

Again, as with spring and durum wheat, there is an active breeding program for winter wheat, as well as studies on winter kill, optimum seeding times, and its use as forage. The initial breeding program for winter wheat was developed in cooperation between the Central Experimental Farm, Ottawa, the Experimental Station at Harrow, Ontario, and the Ontario Agricultural College, at Guelph, in the early 1940s. The most successful of these early developments was a variety called Frederick, which was released in 1947 and soon accounted for 90% of the Ontario crop. However, in the 1980s, newer and more effective varieties were developed at the Central Experimental Farm. These varieties were announced as *Gordon Soft White Winter Wheat* (*Canadian Journal of Plant Science*, v. 61, 1981. pp. 455–57); and *Harris Soft White Winter Wheat* (*Canadian Journal of Plant Science*, v. 66, 1986. pp. 161–63). Gordon had better yield, winter survival, and pastry quality than Frederick, but with a weaker straw. It was tested in the Ontario Cooperative Test for three years, and was deemed most suitable for growing in central and eastern Ontario. Harris, on the other hand, was short-strawed and lodge resistant, and showed early maturity with high yields. In the Ontario Cooperative Test, it was found to be most suitable for growing in southwestern and central Ontario.

Often, winter wheats have been developed for the Maritimes as a forage crop. For example, H.G. Nass developed *Lennox Winter Wheat* (*Canadian Journal of Plant Science*, v. 56, 1976. pp. 401–2) at the University of Guelph, but it was tested at the Agriculture Research Station in Charlottetown, PEI. Lennox was considered suitable mainly for feeding livestock. Another variety developed and tested, by H.G. Nass and R.B. Walton, was *Borden Winter Wheat* (*Canadian Journal of Plant Science*, v. 64, 1984. pp. 207–9). In the maritime Cooperative Wheat Test, this variety was described as a 'utility grade, medium hard red winter wheat, with winter hardiness, and high yield in the Atlantic provinces.'

In the west, breeding programs for high quality and hardy winter wheat for southern Alberta and southwestern Saskatchewan began in the 1950s at the Agriculture Research Station at Lethbridge, Alberta. In 1961, J.G. Andrews and M.N. Grant succeeded in developing Winalta, which was announced as *Note on Winalta Winter Wheat* (*Canadian Journal of Plant Science*, v. 42, 1962. pp. 206–7). This variety was high yielding and hardy, with a kernel quality approaching that of hard red spring wheat. It quickly became grown as a winter wheat for bread flour, replacing other varieties of winter wheat on the prairies.

The tolerance of prairie winter wheat to low temperatures has been studied, again at Lethbridge, by D.W.A. Roberts, in the publication entitled *Winterkill of Fall-Seeded Winter Wheat* (Ottawa: Agriculture Canada, 1984. 13 p. Technical Bulletin 1984-10). Roberts found that resistance in winter wheats to low temperatures began to develop when air temperatures dropped below 10 degrees C. This process, called hardening, requires light, carbon dioxide, and adequate plant nutrition. Fully hardened plants can withstand temperatures as low as –20 degrees C. In order to further study this hardening process (vernalization), J.B. Thomas and G.B. Schaalje, in their paper entitled *Prolonged Freezing of Dark-Hardened Seedlings for Rating and Selection of Winter Wheats for Winter Survival Ability* (*Canadian Journal of Plant Science*, v. 68, 1988. pp. 47–55), evaluated a simple artificial freezing procedure under laboratory conditions as a screening test for winter survival ability of winter wheats. The artificial procedure eliminated the difficulty of accurately measuring winter survival ability under field conditions. The physiological traits of winter survival were also studied by B.L. McIntyre and T.H. Chen in their paper entitled *Physiological Traits Associated with Winter Survival of Winter Wheats and Winter Triticals in Alberta* (*Canadian Journal of Plant Science*, v. 68, 1988. pp. 361–66). These authors evaluated water content, fresh weight, dry weight, and total soluble carbohydrate in winter wheat varieties as traits for winter survival ability.

In the milder winter climate of Ontario, the focus on winter wheat research has been the evaluation of optimum planting dates and/or seeding rates for maximum yields of winter wheat. C.J. Andrews and M.K. Pomeroy, in their paper entitled *Planting Dates and Seeding Rates for Soft White Winter Wheat in Eastern Ontario* (*Canadian Journal of Plant Science*, v. 72, 1992. pp. 391–402), used planting dates ranging from late August to late October in three counties in eastern Ontario that were considered marginal for seasonal survival of soft-white winter wheat. They found that winter survival was reduced to below 60% in later plantings, but yields were greater from higher seeding rates with later planting dates. A. Bootsma and C.J. Andrews, in their paper entitled *Estimated Optimum Seeding Dates for Winter Wheat in Ontario* (*Canadian Journal of Plant Science*, v. 73, 1992. pp. 389–96), estimated the optimum seeding period for winter wheat by studying climate data.

They developed 13 optimum seeding date (OSD) zones in Ontario. Previously, fall seeding dates had been based upon 'corn heat unit zones,' which had not been satisfactory for all areas.

When winter wheats are planted in the spring, they grow vegetatively instead of producing heads, because they lack the cold period for hardening, or vernalization. This makes them useful as a forage crop. However, if they are seeded in the autumn and survive winter injury, they are useful as spring pasture, or hay, and then may be left to mature after initial grazing to produce a grain crop. But what are their forage yields? B. Birkenkamp and J. Meers, in their paper entitled *Hay and Pasture Yields of Fall and Spring-Seeded Winter Wheat and Fall Rye* (*Canadian Journal of Plant Science*, v. 68, 1988. pp. 519–22), noted that spring seeded wheat, as well as fall rye, produced more hay and pasture than when seeded in the fall.

Like winter wheats, oats are grown in Canada as forage and as a cereal crop. Settlers to Canada, in the 17th century, brought oats with them as fodder for livestock and horses as well as a food source, but it was not until the end of the 19th century that oats were studied at the Central Experimental Farm, in Ottawa, and oat breeding developed. The earliest Canadian publication on oats was by W. Saunders, Director of the Experimental Farms, and entitled *The Yield of Spring Wheat, Barley and Oats Grown as Single Plants* (Montreal: Dawson, 1889. 1 microfiche. CIHM Microfiche Series No. 26384).

It was from this early work that Ottawa developed a strong tradition of oat improvement, and maintains the largest world collection of wild species of the genus *Avena*, which has been described, by B.R. Baum and T. Raghathy, in their publication entitled *Wild Oat Gene Pool: A Collection Maintained by the Canada Department of Agriculture, Canada Avena (CAV)*. 2nd Edition (Ottawa: Department of Agriculture, 1975. 100 p. Publication No. 1475). This collection was the result of three expeditions – the first to the Mediterranean region, in 1964, the second to Israel, in 1966, and the third to the Middle East, in 1970. Those regions are the historical source of oats. B.R. Baum has also written an additional monograph entitled *Oats: Wild and Cultivated: A Monograph of the Genus Avena L. (Poaceae)* (Ottawa: Supply and Services Canada, 1977. 403 p. Canada Department of Agriculture Monograph No. 14). This book was the outcome of studying the micro-morphological characteristics of *Avena* collections.

The oat breeding program, itself, has been centred in Winnipeg since the 1920s at the Cereal Breeding Laboratory, where rust resistance in oats has been continuously studied, along with lodging (grain falling as it ripens), and the development of early-maturing varieties. More recently, much of the work on rust resistance has been carried out by D.E. Harden and R.I. McKenzie, and described in papers such as *Inheritance of Crown Rust Resistance in 3 Accessions of* Avena sterilis (*Canadian Journal of Genetics and Cytology*, v. 22, 1980. pp. 27–34); *Complex*

Additive-Type of Resistance to Puccinia coronata *in* Avena *sterilis* (*Journal of Plant Pathology*, v. 6, 1984. pp. 135–38); and *Inheritance of Adult Plant Resistance to Crown Rust in an Accession of* Avena sterilis (*Phytopathology*, v. 74, 1984. pp. 352–53). In these studies, genes for crown rust resistance were isolated and bred into Canadian varieties, particularly *Puccinia coronata*, grown in western Canada.

The lodging of cereal plants, as they ripen, causes crop losses as well as making harvesting difficult, and has been studied since the early 1950s. In the paper by D.G. Hamilton entitled *Culm, Crown and Root Development in Oats as Related to Lodging* (*Scientific Agriculture*, v. 31, 1951. pp. 286–315), the factors affecting resistance to lodging in oats were elaborated upon in order to identify specific characteristics for plant breeders to select. Hamilton found that varieties resistant to lodging had culms (stems) of greater diameter, with more rigidity and better root systems than susceptible varieties.

The work on early maturing oat varieties has been done, not at Winnipeg, but at the Agriculture Research Station at Lacombe, Alberta. The first breeding work was done in the late 1940s, and has been ongoing. A major achievement of this program has been the development of the Cascade variety by T.H. Allen and M.L. Kaufmann, who announced it in the literature as *Cascade Oat* (*Canadian Journal of Plant Science*, v. 60, 1980. pp. 283–84). This spring oat variety, which was released in April 1979, after being grown in the Western Cooperative Oat Test from 1975 to 1978, showed superior performance in yields on black and gray soil regions of Alberta. Although Cascade was susceptible to stem and crown rust, and not recommended for rust-prone areas, it has become sown extensively throughout the prairie provinces.

Canadian oat varieties are very day-length sensitive, as the plants require exposure to long day-lengths to flower normally and assure yields in the short Canadian growing season. This problem of oat photoperiodicity was first studied by D.R. Sampson and V.D. Burrows in their paper entitled *Influence of Photoperiod Short-Day Vernalization and Cold Vernalization on Days to Heading in Avena Species and Cultivars* (*Canadian Journal of Plant Science*, v. 52, 1972. pp. 471–82). In 1964, these researchers did find a species, *Avena byzantina*, selected from Turkey, that carried a gene permitting the variety to flower normally under shorter day-lengths. The transfer of this gene to Canadian oat breeding material led to the establishment of winter oat nurseries in California, where it was possible to grow two generations of test varieties in the field per year. This work saw the development of the first daylight-insensitive oat species in Canada, which was registered in 1982. This new variety was announced, by V.D. Burrows, in a note entitled *Donald Oats* (*Canadian Journal of Plant Science*, v. 64, 1984. pp. 411–13). Since then, Burrows has produced additional daylight-insensitive species, again announced in the literature as notes entitled *A.C. Hill Oat* (*Canadian Journal of*

Plant Science, v. 72, 1992. pp. 439–41); and *A.C. Lotta Oat* (*Canadian Journal of Plant Science*, v. 72, 1992. pp. 443–45).

Besides wheat and oat crops, the other major cereal crop grown in Canada is barley, which was first introduced by Champlain in 1605 for brewing purposes. This remained the main reason for growing the crop until 1890, when American tariffs restricted barley imports for brewing purposes. After that, barley went into a period of decline, as there was no local market for feed barley. These early developments were described, by W. Saunders, in the publication entitled *Barley* (Ottawa: Department of Agriculture, 1890. Microfiche Central Experimental Farm Bulletin No. 6. CIHM Microfiche Series No. 25265).

Barley consists of two types – two-rowed and six-rowed – both with the same genus-species name of *Hordeum vulgare*. The rows refer to the arrangement of the kernels on the spike of the stem. Two-rowed barley has an outer cell layer, which is yellow and is used for malt, whereas six-rowed barley has a blue outer layer, or aleurone, which is used for feed.

The earliest breeding program in Canada was undertaken by W. Saunders in Ottawa, in 1889, and has been described by P.R. Cowan in the paper entitled *The Development of Barley Varieties in Canada* (*Agricultural Institute Review*, v. 7, 1952. pp. 23–24). Saunders crossed six-rowed and two-rowed barley types, in an attempt to combine the best qualities of both types. Another breeding program was undertaken at the Experimental Station at Charlottetown, in 1912, where a successful variety of two-rowed barley called Charlottetown 80 was developed. This has been described by M.B. Bourdon in the publication entitled *Charlottetown Research Station 1909–1984* (Charlottetown: Research Branch Agriculture Canada, 1984. 82 p. Historical Series No. 19). The breeding program at Charlottetown for feed barley has been continuous, with recent varieties developed, and announced, by J.D. Sterling and T.M. Choo in notes entitled *Albany Barley* (*Canadian Journal of Plant Science*, v. 68, 1988. pp. 1139–41); and *Iona Barley* (*Canadian Journal of Plant Science*, v. 73, 1993. pp. 1083–86). Both these varieties are 'well adapted to Eastern Canada.'

In the west, a barley breeding program began in 1923 at Brandon, Manitoba, where from 1936 to 1971 W.H. Johnston sought high-yielding two- and six-rowed barley, suitable for malting and for feed. Over that period of time, Johnston developed 13 barley cultivars. The most notable of these cultivars was Bonanza, announced by Johnson in the note entitled *Bonanza, a New Six-Rowed Malting Barley* (*Canadian Journal of Plant Science*, v. 50, 1970. pp. 608–9). Bonanza was developed for the Parkland areas of the prairie provinces as a malting barley but out-yielded field barleys in Manitoba. It was also resistant to stem rust, loose and covered smut, and other diseases. It quickly became the standard for malting barleys in Canada. In 1980, R.I. Wolfe, in a note entitled *Johnston Barley* (*Cana-*

dian Journal of Plant Science, v. 60, 1980, pp. 1431–33) said that this new variety, which had been developed in the early 1970s and tested in the Western Cooperative Barley Trials from 1977 to 1979, had been named in honour of W.H. Johnston for the latter's contribution to barley breeding. Also in 1980, R.I. Wolfe announced, in a note entitled *Bedford Barley* (*Canadian Journal of Plant Science*, v. 60, 1980. pp. 1435–37), a new six-rowed feed barley which had 17% higher yields than Bonanza in eastern and central Manitoba tests.

In Alberta, the Agriculture Research Station at Lacombe has been a centre of barley breeding for varieties suitable to the dryland and irrigated lands of Alberta and Saskatchewan. At Lacombe, the program began in 1954, and recent varieties have been announced by M.L. Kaufmann and S. Kibite in notes entitled *Diamond Barley* (*Canadian Journal of Plant Science*, v. 65, 1985. pp. 216–17), and *A.C. Lacombe Barley* (*Canadian Journal of Plant Science*, v. 73, 1993. pp. 1087–89). Both are six-rowed feed barley. The former showed yields comparable to the Johnston variety mentioned above, and the latter was recommended for black and grey wooded soils of western Canada, where it showed 'plump kernels ... adequate lodging resistance and good agronomic characteristics.'

In 1990, D. Kong and T.M. Choo began a study to determine the genetic variation and adaptation of all registered Canadian barley cultivars. In 1991 and 1992, 76 cultivars were tested at four locations across Canada (Charlottetown, Ottawa, Brandon, and Bentley, Alberta). According to the authors, these cultivars showed remarkable variation in yield, test weight, kernel weight, plant height, and maturity, and responded differently to the environments. Two-rowed cultivars showed the best stability over all the environments, and two-rowed eastern cultivars performed better than western two-rowed cultivars. However, some six-rowed cultivars performed well in both eastern and western Canada. In their paper entitled *Genetic Variation and Adaptation of 76 Canadian Barley Cultivars* (*Canadian Journal of Plant Science*, v. 74, 1994. pp. 737–44), the authors concluded that 'greater exchanges of breeding materials between the two regions are recommended in order to fully exploit the genetic potentials of new cultivars.'

When breeding new varieties of barley, the problems of rust smut, and other diseases, have to be taken into account. As well, a constant disease which causes about a 10% annual yield loss on Canadian prairies is root rot. After D.G. Hamilton took over the national barley breeding program in 1950, the search was on to find varieties resistant to root rot, which is caused by the fungus *Helminthosporium sativum*. Until then, there were no satisfactory techniques available for selecting resistant seedlings. D.G. Hamilton and R. Kugan, in their paper entitled *Reaction of Barley Varieties and Selections to Root Rot and Seedling Blight Incited by* Helminthosporium sativum *P.K. and B.* (*Canadian Journal of Plant Science*, v. 40, 1960. pp. 713–20), modified existing techniques, sterilized seed, sand, and a mixture of sand and cornmeal, containing the inoculums of the fungus at appropriate temperatures and humidities to get reliable readings of resistance to root rot. Nevertheless, the breeding programs did not succeed in producing barley varieties highly resistant to root rot, although some commercial varieties showed more resistance, or tolerance, to root rot than others. L.J. Piening, in the note entitled *Differential Yield Response of Ten Barley Cultivars to Common Root Rot* (*Canadian Journal of Plant Science*, v. 53, 1973. pp. 763–64), suggested that the increase in yield of some barley varieties, with a slight amount of root rot, was caused by the recuperative effect of the plant. In this situation, the disease either stimulated additional root growth, or the unaffected root tissues became more efficient in gathering nutrients for the plant.

More recently, the problem of root rot has been approached by varying seeding depth and space of seeding of barley seed. L.J. Duczek and L.J. Piening, in their paper entitled *Effect of Seeding Depth, Seeding Date and Seed Size* (Canadian Journal of Plant Science, v. 62, 1982. pp. 885–91), noted that the symptoms of root rot were not influenced by seed size, or seeding date, but by seeding depth. As well, L.J. Piening, as the single author of the paper entitled *Effects of Seed Spacing on Common Root Rot of Some Barley Cultivars* (*Canadian Journal of Plant Science*, v. 63, 1983. pp. 611–16), noted that increasing plant spacing by thinning did not provide a consistent reduction in rot. Since then, L.J. Duczek and G.B. Wildermuth have estimated the tolerance factor again. In their paper entitled *Assessment of Tolerance in Wheat and Barley to Common Root Rot (*Cochliobolus sativus*)* (*Canadian Journal of Plant Science*, v. 73, 1993, pp. 1177–87), the authors suggested that research on tolerance to root rot should concentrate on barley, because the four barley cultivars tested at least showed inconsistent results to root rot-tolerance, and the wheat varieties showed no tolerance at all.

Besides wheats, oats, and barley, other crops of economic importance in Canada are canola (rapeseed), soybeans, and corn. Canola, which is the source of Canadian margarines, cooking oils, and salad dressings, has been a particular success of Canadian agricultural research. Rapeseed had its origins in Asia over 3,000 years ago, where it was crushed and used as cooking oil, or as a lamp oil for illumination. From China and Japan, it made its way into Europe along the Mediterranean, where it was also used as lighting oil until the 19th century. However, it was not until the 1930s that rapeseed was introduced into Canada by the Department of Agriculture for testing on farms across the country. Until 1943, there was no commercial production. Then, the Wartime Prices and Trade Board asked J.M. Stevenson, of the Department of Agriculture, about the possibility of growing rapeseed. Seed was purchased from the United States and grown under contract and processed as marine lubricating oil. By the late 1940s and early 1950s, breeding programs to develop Canadian varieties of rapeseed, and experiments to develop edible oil products such as margarine and salad oils, were under way in a cooperative program between the Department of Agriculture and the National Research Council. The first Canadian variety, called Golden, was released in 1964.

Rapeseed oil contains erucic acid, which was shown to be linked to fat accumulation around the hearts of laboratory animals and therefore possibly posing a risk to human consumption. But in 1964, R.K. Downey, at Saskatoon, was able to isolate an erucic acid-free rapeseed variety, which was announced in a note entitled *A Selection of* Brassica campestris *L. Containing No Erucic Acid in Its Seed Oil* (*Canadian Journal of Plant Science*, v. 44, 1964. p. 295). By 1977, this had resulted in the production of Low Erucic Acid Rapeseed (LEAR) plants, and the production of the Candle Variety. The term 'canola' was adopted by the rapeseed industry to indicate cultivars low in erucic acid. The development of Candle was described, by J.R. King and A.J. Klassen, in the paper entitled *Inheritance of the Mucilage Character in* Brassica campestris *L. var. oleifera 'Candle'* (*Canadian Journal of Plant Science*, v. 62, 1982. pp. 305–9). It appeared that in the development of Candle, mucilage was introduced randomly into the seed coat and caused griming of seed crushers. It was suggested further work should be done to investigate the mode of inheritance in other *Brassica* species. By this time, canola had become a billion-dollar crop, which was only out-produced by wheat in Canada. However, the only additional work on the mucilage problem has been carried out by E.S. Eskin, as noted in the paper entitled *Effect of Variety and Geographical Location on the Incidence of Mucilage in Canola Seeds* (*Canadian Journal of Plant Science*, v. 72, 1992. pp. 1223–25). Here, the author said that 'geographical location and variety both significantly affected the mucilage levels among Candle seeds studies.'

Like rapeseed, or canola, soybean has had a recent history as a viable Canadian crop both as a valuable feed supplement for forage, because of its high protein and fat content, and as an edible crop for domestic or export consumption. The objective of soybean breeding in Canada, from its beginnings in the 1930s, has been to develop varieties that would mature sufficiently early for all seed to ripen before harvest, and to breed resistance to root and stem rot by *Phytophthora*, a fungus disease. The first of the Canadian varieties, developed and released for seeding in 1943, was Harrosoy from the Experimental Station at Harrow, Ontario. By the late 1950s, this variety had lost its resistance to root rot and a new variety Harrosoy 63 was developed, but showed resistance for only 10 years.

Because of this ongoing problem to resistance, J.H. Hass and R.I. Buzzell developed a technique to search for soybean varieties that showed tolerance rather than resistance to root rot. This approach was described in their paper entitled *New Races 5 and 6 of* Phytophthora megasperma *var. sojae and Differential*

Reactions of Soybean Cultures for Races 1 to 6 (*Phytopathology*, v. 66, 1976. pp. 1361–62). The outcome of their work was a new variety announced in a note entitled *Harcor Soybeans* (*Canadian Journal of Plant Science*, v. 56, 1976. pp. 973–74), which has a good field tolerance to root rot and a high yield potential. With the release of Harcor, severe losses in yield were prevented, but R.I. Buzzell and T.R. Anderson also continued the search for race-specific varieties of soybeans, resistant to root rot, as observed in their paper entitled *Plant Loss of Soybean Cultivars to* Phytophthora megasperma *F_1 sp.* glycenia (*Plant Disease Report*, v. 66, 1982. pp. 1146–48).

Further work on soybean breeding has focused on the search for early maturing varieties, suitable for the growing season in northern Ontario, Quebec, and Manitoba. This program has been carried out in Ottawa and based on soybean varieties obtained from Sweden, but originating from the Sakalin Islands of northern Japan. By 1982, this work began to produce results. H.D. Voldeng and J.F. Seitzer announced, in a note entitled *Maple Presto Soybeans* (*Canadian Journal of Plant Science*, v. 62, 1982. pp. 501–3), the first of a series of cultivars 'for production in areas of Canada with short, cool growing seasons.' The name Presto was chosen to denote the rapidity with which the plants develop and mature. In 1985, these researchers then announced another cultivar in a note entitled *Maple Isle Soybean* (*Canadian Journal of Plant Science*, v. 65, 1985. pp. 777–79). This variety had good yields, particularly in Prince Edward Island, from where its name was derived. Also in 1985, these two researchers announced *Maple Ridge Soybean* (*Canadian Journal of Plant Science*, v. 65, 1985. pp. 781–83). This variety was developed to replace their first cultivar, Maple Presto. As with these other soybean varieties, developed by Voldeng and Seitzer, it had moderate resistance to root rot. The most recent of the new soybean varieties to come out of Ottawa was announced by H.D. Voldeng and R.I. Guillemette in a cultivar description entitled *AC Brant Soybean* (*Canadian Journal of Plant Science*, v. 76, 1995. pp. 149–50). This variety was 'intended for production in 2,500–2,600 crop heat unit areas of Ontario and Quebec.'

The one other major cereal crop grown in Canada is corn. Corn is actually a grass known as *Zea mays*, which was indigenous to tropical Central America, but cultivated by native populations as far north as the Great Lakes. Under natural conditions it cross-pollinates, and unlike other cereals such as wheat, oats, and barley, which produce flowers during lengthening days, corn begins to flower as the days shorten. Because of this feature, corn matures late and is subject to damage from frost when grown outside the tropics. However, corn produces high yields, and initially Canadians were interested in adapting it to the shorter northern growing season for the production of silage. The development of corn varieties came about by using the process of self-pollination by corn in order to produce inbred lines from which hybrid corn varieties could be created. In Canada, inbreeding programs to develop hybrids began only in 1923, at Harrow, Ontario, and Ottawa, although this type of program had been going on in the United States since the turn of the century. By 1940, hybrid varieties had been licensed for use in Canada, and by 1944, most of the corn growing of southwestern Ontario was planted with hybrids. In 1946, the first hybrid variety from the Canadian breeding program at Harrow was called Harvic 300. By 1961, hybrids had also been developed for resistance to the European corn borer (a problem in corn-growing areas since the 1920s), root rot, and stem rot.

One aspect of early-maturing corn hybrids grown primarily for forage has been the study of the correlation between planting times of hybrids and the effect on yields at various harvesting times. At the Agriculture Research Station at Agassiz, British Columbia, N.A. Fairey, as described in a paper entitled the *Effects of Hybrid Maturity, Date of Planting, and Date of Harvesting on Growth and Development of Forage Maize* (*Canadian Journal of Plant Science*, v. 60, 1980. pp. 1367–75), studied four hybrids, which were given two planting dates in late April and late May, and three harvesting dates in September and October. After studying the yield results, Fairey noted that it was important to select hybrids of 'appropriate maturity and [to maximize] the use of the growing season to produce high-quality forage in marginal maize-growing regions. ...' This project was then carried on over a three-year period by N.A. Fairey, and the longer-term results described in the paper entitled *Yield, Quality and Development of Forage Maize as Influenced by Dates of Planting and Harvesting* (*Canadian Journal of Plant Science*, v. 63, 1983. pp. 157–68). This three-year study incorporated information such as air temperature records and seeding time of the corn in order to predict when the whole plant dry moisture (DM) content reached the maximum level for harvesting as silage.

As the development of earlier-maturing, higher-yielding corns has allowed forage corn production to expand into areas such as British Columbia, as noted above, or the southern prairies, corn grown as a grain has maintained a more geographic restriction. For example, in the Maritimes, it is restricted to the Annapolis Valley of Nova Scotia and the St John River Valley of New Brunswick, and research is carried on to identify locally adapted hybrids that will produce earlier-maturing and higher-yielding hybrids in the short growing season areas. In this particular area, R.P. White and K.B. McRae, in their paper entitled *Assessment of Yield Improvement of Short Season Corn Hybrids in Atlantic Trials* (*Canadian Journal of Plant Science*, v. 64, 1984. pp. 197–201), evaluated corn hybrid performance in Atlantic trials and noted a 16% improvement in grain corn yield in 1982 as compared with that tested in 1976. In Ontario, V.A. Dirks and E.F. Bolton analysed a 13-year corn crop rotation in order to identify and evaluate the climatic components, rather than varieties, that affected grain crop yields on

Brookston clay soil. In their paper entitled *Factors Contributing to Year-to-Year Variation in Grain Yield of Corn on Brookston Clay* (*Canadian Journal of Plant Science*, v. 61, 1981, pp. 293–305), the authors noted that four climatic factors – moisture deficit days, available moisture, corn heat units (growing areas measured by natural heating amounts available), and a mixture of the three that allows for the tasselling-silking period – affected grain corn yields under these soil conditions. However, the corn yield based on climatic factors was, overall, dependent on the level of nitrogen available as fertilizer in the soil. The problem of moisture deficiency has also been studied by L.M. Dwyer and D.W. Stewart in their paper entitled *Water Extraction Patterns and Development of Plant Water Deficits in Corn* (Zea mays) (*Canadian Journal of Plant Science*, v. 65, 1985. pp. 921–34), and by R.G. Besard and G.W. Thurtell in their paper entitled *The Interactive Effects of Increased Evaporative Demand and Soil Water on Photosynthesis in Maize* (*Canadian Journal of Plant Science*, v. 71, 1991. pp. 31–40).

(b) Forage Crops

In Canada, the one other major area of crop research concerns forage crops, which are grown and stored as hay, silage, or pellets. Or they are grown and remain in pastures as feed for grazing ruminant animals. Cultivated forages are distributed fairly evenly across Canada, with about 50% in eastern Canada and British Columbia, and the rest in the prairie provinces. In addition, the prairie provinces use almost double the amount of cultivated forage land in the native grasslands as pasture.

Alfalfa is the most important forage crop grown in Canada. Historically, alfalfa is thought to have originated in Iran or Turkey and made its way, only in 1871, into Canada from France. Around 1900, the Central Experimental Farm, at Ottawa, began to study and evaluate alfalfa, and G.H. Clark and M.O. Malt wrote a monograph entitled *Fodder and Pasture Plants* (Ottawa: Minister of Agriculture, 1913. 143 p.). However, it was not until the early 1970s that major varieties of alfalfa were developed at the Ottawa station. For example, H. Baenziger developed two varieties announced in notes as *Algonquin Alfalfa* (*Canadian Journal of Plant Science*, v. 55, 1975. pp. 1093–94); and *Angus Alfalfa* (*Canadian Journal of Plant Science*, v. 55, 1975. pp. 1095–96). Both were cold tolerant and suited to the prairie provinces. As well, they were resistant to bacterial wilt, which affected alfalfa, and both produced good forage yields.

Besides the problem of winter tolerance in the west, drysoil conditions can affect rooting conditions for the alfalfa. To overcome this problem, alfalfa plants displaying a creeping-root habit of growth were studied by B.E. Murray, and reported in the paper entitled the *Ontogeny of Adventitious Stems on Roots of*

Creeping-Rooted Alfalfa (*Canadian Journal of Botany*, v. 35, 1957. pp. 463–73). Out of this work came a series of varieties developed by D.H. Heinrichs, at the Agriculture Research Station at Swift Current, Saskatchewan. Among the varieties which were drought resistant, winter hardy, and had creeping roots, were Roamer Alfalfa (*Canadian Journal of Plant Science*, v. 47, 1967, pp. 220–21); Drylander Alfalfa (*Canadian Journal of Plant Science*, v. 51, 1971. pp. 430–32); and *Rangelander Alfalfa* (*Canadian Journal of Plant Science*, v. 59, 1979. pp. 491–92). Finally, R.B. Irvine and T. Lawrence announced *Heinrichs Alfalfa* (*Canadian Journal of Plant Science*, v. 62, 1982. pp. 805–8). This cultivar was named after D.H. Heinrichs, who 'devoted his entire professional career (1933–1976) to the breeding of high-yielding winter hardy cultivars for the prairie region.'

On the irrigated lands of southern Alberta, and Saskatchewan, qualities other than drought resistance are necessary for the growing of alfalfa. Research on varieties suitable for these areas was undertaken at the Agriculture Research Station at Lethbridge, Alberta, by M.R. Hanna. Beginning in 1958, research was undertaken to breed a high-yielding, wilt-resistant variety, and by 1971, *Kane Variety* (*Canadian Journal of Plant Science*, v. 52, 1972, pp. 116–18) was announced by M.R. Hanna. This variety also had the desired creeping root characteristic.

In order for any variety of alfalfa to survive and produce seed, alfalfa flowers must be visited by insects. This usually means bees. It is now known that bumblebees (*Bombus*) and leaf-cutter bees (*Megachile*) successfully pollinate alfalfa. The first observations that indicated the efficiency of leaf-cutter bees as pollinators were noted, in 1916, around Lethbridge and Medicine Hat, Alberta, by the Dominion Apiarist, F.W. Sladen, in the paper entitled *Pollination of Alfalfa by Bees of the Genus* Megachile: *Table of Canadian Species of the* Latimanies *Group* (*Canadian Entomologist*, v. 50, 1918. pp. 301–4). However, it was not until 1940 that R.W. Salt, at the Agriculture Research Station at Lethbridge, Alberta, encouraged the use of wild bees in alfalfa seed production in the circular entitled *Utilize Wild Bees in Alfalfa Seed Production* (Ottawa: Department of Agriculture, 1940. Mimeographed Circular No. 226). Another study, by O. Peck and J.L. Bolton, entitled *Alfalfa Seed Production in Northern Saskatchewan as Affected by Bees, with a Report on Means of Increasing the Population of Mature Bees* (*Scientific Agriculture*, v. 26, 1946. pp. 338–418), indicated that alfalfa seed yield was affected by the accidental destruction of the nests of wild, or native, bees, because of the clearing and cultivating of wooded land. They noted that in northern Saskatchewan seed yields in 1930 were generally 220 to 500 kilograms per hectare, but by 1945 seed yields had dropped to about 80 kilograms per hectare. This reduction was entirely due to the lack of pollinators.

All the later research on alfalfa pollinators was carried out at the Agriculture

Research Station at Lethbridge, Alberta. There, research on alfalfa pollinators began in the late 1940s, and by the mid-1950s, the first major paper by G.A. Hobbs and C.E. Lilly, entitled *Ecology of Species of* Megachile latraille *in the Mixed Prairie Region of Southern Alberta with Special Reference to Pollination of Alfalfa* (*Ecology*, v. 35, 1954. pp. 453–62), was published. This paper reviewed the two ecological areas of alfalfa growth in western Canada. One area is the more northern parkland, where leaf-cutting bee pollinators used the tree and aspen area for nesting. The other area was the more southern mixed prairie, which included river valleys and irrigated lands. It was this southern area that was intensely studied, and ground-nesting species of leaf-cutters, primarily *Megachile dentisaries* and *periherta*, were shown to be the most important pollinators of alfalfa. Their ground-nesting habitat was alfalfa fields, and the bees emerged as the alfalfa was coming into bloom.

One year later, G.A. Hobbs, in another paper, entitled *Alfalfa and Red Clover as Sources of Nectar and Pollen for Honey, Bumble, and Leaf-Cutter Bees (Hymenoptera: Apoidea)* (*Canadian Entomologist*, v. 89, 1957. pp. 230–35), showed that both bumble- and alfalfa leaf-cutter bees were pollinators, for both red clover and alfalfa. However, neither species was entirely satisfactory as alfalfa pollinators because native food plants bloomed profusely in their nesting habitats. Further work on the food preference of bees was carried out by G.A. Hobbs and W.O. Nummi, and described in the paper entitled *Food-Gathering Behaviour of Honey, Bumble, and Leaf-Cutter Bees (Hymenoptera: Apoidea) in Alberta* (*Canadian Entomologist*, v. 93, 1961. pp. 409–19). In this paper, the authors showed that given an equal opportunity to obtain food from red clover, alsiki clover, sweet clover, or alfalfa, leaf-cutter bees (*Megachile*) preferred the latter, and therefore were good pollinators. On the other hand, bumble-bees (*Bombus*) and honey-bees (*Apis*) preferred clovers. The reason for this could, perhaps, be attributed to the lengths of their proboscises (mouthparts).

This work led G.A. Hobbs into an intensive study of the ecology of bumble-bees in southern Alberta, which produced the following publications: *Ecology of Species of* Bombus *Latr. (Hymenoptera: Apidae) in Southern Alberta I. Subgenus* Alpinsbombus *Skor.* (*Canadian Entomologist*, v. 96, 1964. pp. 1465–70); *Ecology of Species of* Bombus *Latr. (Hymenoptera: Apidae in Southern Alberta II. Subgenus* Bombias *Robt.* (*Canadian Entomologist*, v. 97, 1965. pp. 120–28); *Ecology of Species of* Bombus *Latr. (Hymenoptera: Apidae in Southern Alberta III. Subgenus* Cullumanobombus *Vogt.* (*Canadian Entomologist*, v. 97, 1965. pp. 1293–1302); *Ecology of Species of* Bombus *Latr. IV. Subgenus* Fervidobombus *Skor.* (*Canadian Entomologist*, v. 98, 1960. pp. 33–39); *Ecology of Species of* Bombus *Latr. V. Subgenus* Subterraneobombus *Vogt.* (*Canadian Entomologist*, v. 93. 1966. pp. 288–94); *Ecology of Species of* Bombus *(Hymenoptera: Apidae) in Southern Alberta VI.*

Subgenus Pyrobombus (*Canadian Entomologist*, v. 99, 1967. pp. 1271–92; and *Ecology of Species of* Bombus *(Hymenoptera: Apidae) in Southern Alberta VII. Subgenus* Bombus (*Canadian Entomologist*, v. 100, 1968. pp. 156–64).

Apart from alfalfa, the other important legume grown in Canada for forage is sweet clover, which is a biennial legume native to temperate Europe and Asia. There are two major species, *Melilotus alba*, which is white-flowered, and *Melilotus officinalis*, which is yellow-flowered. Because of sweet clover's rapid growth and nitrogen fixing capabilities, it makes an excellent green manuring crop. However, it contains coumarin, which is an anti-coagulant, and that is a serious disadvantage for a feed. Coumarin develops in poorly cured sweet clover hay, which, if fed in large quantities to livestock, causes ruminant animals to haemorrhage.

At the Agriculture Research Station at Saskatoon, B.P. Goplen and J.E. Greenshields studied the problem of coumarin. In their paper entitled *Selection Techniques in Screening for Coumarin Deficient Sweet Clover Plants* (*Canadian Journal of Botany*, v. 34, 1956. pp. 711–19), they said that the effectiveness of selection in a breeding program for the coumarin-deficient character depended upon employing reliable sampling and testing techniques. They found that samples stored in a solution of sodium hydroxide, in darkness, did not lose their fluorescent properties for measuring coumarin by fluorescent analysis. The next step was to find out the genetic pattern of coumarin inheritance in sweet clover. In their paper entitled *The Inheritance of Coumarin in Sweet Clover* (*Canadian Journal of Botany*, v. 35, 1957. pp. 583–93), B.P. Goplen and J.E. Greenshields described their study, which established that two genes were concerned with inheritance of coumarin in the common species *Melilotus alba*. One determined the type of coumarin present, and the other determined its production in the plant. With these findings, they were then able to develop a low-coumarin sweet clover. This was announced by J.E. Greenshields in the brief paper entitled *Note on Cumino Sweet Clover* (*Canadian Journal of Plant Science*, v. 38, 1958. pp. 507–8). Cumino is a white sweet clover, but yellow-flowered sweet clover is preferred by farmers, because it is leafier, has thinner stems, and matures earlier than cumino. Accordingly, B.P. Goplen started a program to develop a coumarin-free, yellow-flowered variety. Finally, in 1981, Goplen was able to announce in a note entitled *Norgold – A Low Coumarin Yellow Blossom Sweetclover* (*Canadian Journal of Plant Science*, v. 61, 1981. pp. 1019–21), the first low-coumarin cultivar of yellow-blossom sweet clover, which was 'the culmination of a 22-year breeding program.'

Alfalfa and sweet clover may be grown alone for forage, or along with grasses, which may be used for hay, pasture, or protection as a crop cover. Grass is also used as a ground cover in the form of turf or lawns. Of the grasses used for forage, bromegrass, *Bromus inermis*, originally imported from Russia when the Experimental Farms system was first organized, became the most important, particularly in the west where researchers also began to study southern strains of grass developed in the United States. According to J.D. Smith and R.P. Knowles, in their paper entitled *Mass Selection for Resistance to* Selenophorma bromigena *in Northern and Southern Types of* Bromus inermis (*Canadian Journal of Plant Science*, v. 47, 1967. pp. 679–81), southern strains were less susceptible to leaf spot than traditional northern strains. And W.R. Childers announced the development of *Redpatch Bromegrass* (*Canadian Journal of Plant Science*, v. 46, 1966. pp. 456–57) as the first Canadian-bred, southern-type bromegrass variety to be licensed in Canada. This variety was 'recommended for hay and pasture in eastern Canada where its most popular use is in mixtures with alfalfa for hay.' This work was carried out at Saskatoon, where programs also began, in the early 1970s, for the development of a southern strain with satisfactory seed yields for western Canada. In 1989, J. Surprenant and R.P. Knowles were able to announce *Radisson Smooth Bromegrass* (*Canadian Journal of Plant Science*, v. 70, 1990. pp. 1183–85). This variety was the achievement of 15 years of interpollinating southern strains to develop a desired plant variety.

Besides bromegrass, wheatgrass (*Agropyron* sp.), which has several forms – slender wheatgrass, crested wheatgrass, intermediate wheatgrass, tall wheatgrass, western wheatgrass, and northern wheatgrass – is grown as a forage crop. Only the latter two forms are native to the prairies, but it is crested wheatgrass and intermediate wheatgrass that have received the most attention in Canada.

Again, R.P. Knowles, at the Agriculture Research Station, at Saskatoon, worked with crested wheatgrass, and described the program in a brief report entitled *Summary of Crested Wheatgrass Breeding Program, Saskatoon* (Western Grass Breeders' Work Planning Conference, 14th Annual Report, 1960. p. 48). Part of this work involved the study of fairway crested wheatgrass that had been developed in Canada in the 1930s for the harsh drylands of southeastern Alberta. Further breeding was done by R.P. Knowles, and reported in the paper entitled *Performance of Synthetics and Sib Lines in Fairway Crested Wheat Grass* Agropyron cristatum *(L.) Gaertn.* (*Crop Science*, v. 10, 1970. pp. 642–45). One of the outcomes of this work was the release of Parkway crested wheatgrass, in 1969, which was considered to be an improvement over the fairway variety in terms of yield as well as being less subject to lodging.

Intermediate wheatgrass has been studied at both the Saskatoon and Swift Current Research Stations, in Saskatchewan, since the 1970s. Its productivity was compared by T.L. Lawrence in the paper entitled *Productivity of Intermediate Wheatgrass and Bromegrass as Influenced by Length of Ley and Pattern of Seeding* (*Forage Notes*, v. 22, 1977. pp. 43–49). The outcome of this program was the release of *Clarke Intermediate Wheatgrass* (*Canadian Journal of Plant Science*, v. 61, 1981.

pp. 467–69). This variety showed drought resistance, winter hardiness, and high seed yield, which made it well adapted for hay and pasture use in Manitoba, Saskatchewan, and Alberta, where it out-yielded bromegrass on irrigated as well as dryland.

As previously mentioned, legumes and grasses may be grown in combination for hay production, or for pasture. Besides the breeding of strains of bromegrass and wheatgrass, evaluated as forages, the Agriculture Research Station at Swift Current, Saskatchewan, has also had an ongoing program to evaluate standard forages along with a broad range of other introduced or 'native' grasses, or mixtures of grasses and legumes. In these evaluations, both agronomic (in the field), and *in vitro* (in the laboratory) techniques have been used. In this work, T.L. Lawrence has been the major investigator. The first evaluations, by T.L. Lawrence and J.E. Troelsen, was entitled *An Evaluation of 15 Grass Species as Forage Crops for Southwestern Saskatchewan* (*Canadian Journal of Plant Science*, v. 44, 1964. pp. 301–10). This study used crested wheatgrass (*Agropyron cristatum*) and a species of Russian ryegrass (*Elymus junceus*) that was grown extensively in southern Saskatchewan to evaluate those species and 13 others for dry matter yield, protein content, cellulose digestibility, and winter hardiness. These evaluations indicated that seven species showed some potential as forage crops when compared with the two grasses currently in general use.

Over 10 years later, there was renewed interest in the evaluation of grasses as a forage crop. The first paper, again by T.L. Lawrence, was entitled *An Evaluation of Thirty Grass Populations as Forage Crops for Southwestern Saskatchewan* (*Canadian Journal of Plant Science*, v. 58, 1978. pp. 107–15). The grass strains evaluated included a wide range of wheat grasses, ryegrass, hybrids, and native populations. Comparisons were made for yield, nitrogen content, phosphorus content, digestibility of organic matter, and winter-hardiness. The findings concluded that one strain of ryegrass, *Elymus virginicus*, 'showed relatively good quality throughout the season and both high yield and good quality in the overwinter period.' One year later, T.L. Lawrence and F.G. Warder evaluated grasses on irrigated lands in their paper entitled *Evaluation of Twenty Grass Populations as Irrigated Hay Crops for Southwestern Saskatchewan* (*Canadian Journal of Plant Science*, v. 59, 1979. pp. 691–700). The reason given for this study was the fact that, in the past, grass species had been evaluated only for use as dryland forage. In this study, the intermediate wheatgrass, *Agropyron intermedium*, was found to be the most suitable as a hay crop on well-drained irrigated soil.

Then, another six years later, T.L. Lawrence, with D.C. Ratzlaff, again studied grasses under dryland conditions in their paper entitled *Evaluation of Fourteen Grass Populations as Forage Crops for Southwestern Saskatchewan* (*Canadian Journal *of Plant Science*, v. 65, 1985. pp. 951–57). This time, grasses and alfalfa-grass mixtures were seeded in well-spaced rows to evaluate their productivity for hay or pasture. Of the 14 populations, the wheatgrass, *Agropyron desertorum*, was found to be most suitable for hay production, and a ryegrass, *Elymus angustus*, was most suitable for pasture. The study was carried out over a period of seven years in order to evaluate yearly clippings.

Finally, T.L. Lawrence and C.D. Ratzlaff studied introduced and native grasses on dryland, in part, to update the study done in 1964 and mentioned above. In their paper entitled *Performance of Some Native and Introduced Grasses in a Semiarid Region of Western Canada* (*Canadian Journal of Plant Science*, v. 69, 1989. pp. 251–54), the authors compared some more recently studied native grasses that had undergone selection for improved dry organic matter and seed yield with introduced species. The results of the study indicated that introduced grasses generally outperformed native grasses and supported the 'breeders' reluctance to spend limited resources on native species when improvements can be made to introduce species which have an inherently higher yield potential.' This study began in 1981, and covered an eight-year period.

As noted in the above studies, breeding for improved forage quality requires a slow, long-term approach. It also requires good production management, as stated by R.E. Howarth and B.P. Goplen in their paper entitled *Improvement of Forage Quality through Production Management and Plant Breeding* (*Canadian Journal of Plant Science*, v. 63, 1983. pp. 895–902). This paper was a review of the prospects for improving forage quality. The authors pointed out that the major components of forage quality were nutrient content, digestibility, voluntary intake, and the lack of anti-nutritive factors. Because maturity has a major impact on forage quality, adequate harvest systems, and equipment, were essential for consistent production of quality forage. Although legume forages were considered to be a higher quality than grasses, grass-legume mixtures should not be overlooked for improving forage quality. And breeding factors that should be emphasized for quality forage were nutritive factors and digestible energy.

C Animal Research

Today, Canada has almost twice the number of cattle, swine, and chickens as most other countries, with the only exception being sheep. It is because Canada has vast grazing tracts and ample feed grains. As well, there has been constant encouragement by federal and provincial levels of agriculture to improve animal production through research into breeding, and the feeding and management of livestock. Livestock involved in these programs are dairy and beef cattle, swine, sheep, and poultry, but the emphasis here is on cattle and poultry.

(a) Dairy Cattle

The first attempts to improve the national dairy cattle breeds occurred in 1889, when imported purebred registered bulls and cows, such as Ayrshire, Holstein, Jersey, Polled Angus, and milking Shorthorn were made available to encourage dairymen to upgrade their herds for high milk production and good butter. However, it was not until 1930s that A. Deakin, the first animal geneticist on staff of the experimental farms, used artificial insemination, which had been developed in Russia, in order to contribute desirable individual bull characteristics into the herd. Artificial insemination overcame the limitations that a single bull had for contributing desirable characteristics to a national herd, because of the relatively few offspring produced by natural breeding. Equipment was imported from England, and the first calf was born as a result of artificial insemination at Sainte-Anne-de-la-Pocatière in 1936. This early work included the study of methods for collecting semen, inseminating cows, sterilizing equipment, and storing semen. This work has been described by A. Deakin and G.W. Muir in the publication entitled *Artificial Insemination of Dairy Cattle* (Ottawa: Department of Agriculture, Experimental Farms Service, 1942. 15 p.).

After World War II, the breeding of dairy cattle was reorganized, and research developed more rapidly. The reason for the reorganization was that breeding work with livestock is both slow and expensive. Although plant crops have a generation time of only one or two years, or even less, depending on the climate, cattle have a generation time of at least five years, although newer embryo transfer methods may shorten that time. Also, there is a difference in population sizes between crops and animals, as cereal plants can be grown in the millions, compared with large animals that can best be handled in the hundreds. The reorganization of the dairy cattle breeding that took place was done by considering all animals of any one breed (Holstein, Jersey, Guernsey, or Ayrshire), regardless of location, to be part of one herd. This, in effect, increased the population size, within each breed, from which selections could be made. Then, in the 1960s, the number of breeds was reduced to two (Holsteins and Ayrshires), to maximize their numbers, with the size of each herd varying from 50 to 80 cows. This reorganization was undertaken by C.G. Hickman, who also undertook a 12-year selection experiment, using these Holstein and Ayrshire herds, to evaluate dairy cattle selection for 180-day milk solids yield in the first two lactation periods of the animals. This particular work was described in the paper entitled *Response to Selection for 180-Day Milk Solids* (*Journal of Dairy Science*, v. 54, 1971. pp. 191–98). At the same time as the publication of these results, C.G. Hickman and D.M. Bowden, in the paper entitled *Correlated Genetic Responses of Feed Efficiency, Growth, and Body Size in Cattle Selected for Milk Solids Yield* (*Journal of Dairy Science*, v. 54, 1971.

pp. 1848–55), related genetic changes in feed efficiency, growth, and body size, to selection for milks-solids yield. They also pointed out that animals producing the most milk were the heaviest at calving, gained more weight during each lactation, and consumed more feed than those producing less milk.

In 1968, scientists from the dairy cattle program, at research stations across Canada, reorganized the cattle breeding program once again. This reorganization, which is still in effect, has been described by A.J. McAllister in the bulletin entitled the *National Cooperative Dairy Cattle Breeding Project* (Ottawa: Agriculture Canada, 1978. 46 p. Animal Research Institute Technical Bulletin No. 1). This program, based on an earlier design by C.G. Hickman and A.E. Freedman, and described in the paper entitled *New Approach to Experimental Designs for Selection Studies in Dairy Cattle and Other Species* (*Journal of Dairy Science*, v. 54, 1971. pp. 191–98), hoped to improve the lifetime productivity in Canadian dairy cattle. Both pure line selection and crossbreeding were part of the breeding process. The animal selection criteria for breeding was high yield of milk protein, and 900 Ayrshire and Holstein cattle were used from the previous program and kept at five research stations. The genetic bases of both populations were broadened by using frozen semen from desirable bulls not in the Research Branch herd. The crossbreeding program began with first-generation crossbred males and females from Holstein and Ayrshire lines which were mated to a continuously crossbred female line that started with Holstein and Ayrshire lines.

It was in the early 1980s that some of the initial results of the National Cooperative Dairy Breeding Project were published in a series of papers. T.R. Batra and A.J. McAllister, in their paper entitled *Comparison of the Heifer Reproduction Traits of the Daughters of Several Sire Groups* (*Canadian Journal of Animal Science*, v. 63, 1983. pp. 269–78), noted that reproductively crossbred Holstein heifer lines came into heat, bred, and calved earlier than Ayrshire lines. However, the Holstein lines produced fewer line calves. In another paper, by T.R. Batra and A.J. McAllister, entitled *Body Weights and Dimensions of Pureline and Crossline Heifers of the Canadian Dairy Cattle Breeding Project* (*Canadian Journal of Animal Science*, v. 63, 1983. pp. 511–22), the authors noted that the progeny of the crossbreeding lines exhibited higher body weights and dimensions. Yet, in another paper by T.R. Batra and A.J. Lee, entitled *Relationship of Reproduction Traits, Body Weight and Milk Yield in Dairy Cattle* (*Canadian Journal of Animal Science*, v. 66, 1986. pp. 53–65), the authors indicated that heritability of reproduction traits, body weight, or milk yield, was very low (zero to 26%) and declined further from first to second lactation. The relationship of growth and feed efficiency has also been studied in these pure and crossbred lines. A.J. Lee and C.Y. Lin, in their paper entitled *Growth and Feed Efficiency of Pureline and Crossline Dairy Heifers* (*Journal of Dairy Science*, v. 71, 1988. pp. 1000–10), indicated that genetic differences played a small role in weights and weight gain, but not with feed consumption or efficiency.

At a later date, A.J. Lee and A.J. McAllister searched beyond the National Cooperative Dairy Breeding Project to compare their breeding results with those found in 'industry' Holsteins. In their paper, entitled *A Comparison of Canadian Holstein Bulls with Those in the National Cooperative Dairy Breeding Project* (*Canadian Journal of Animal Science*, v. 74, 1994. pp. 217–21), the authors noted that project and industry Holstein progeny were equivalent in early first lactation milk fat and protein yield, but progeny of industry bulls were larger at all ages. On the other hand, when the project data covering a 14-year period were studied, the project bulls had comparable genetic merit for milk production traits to those of the intensely selected industry bulls.

(b) Beef Cattle

Unlike the breeding of dairy cattle in Canada, the early efforts at breeding of beef cattle involved hybridization attempts with wild bovines that had adapted through centuries of natural selection to the harsh North American climate of extreme dry summer heat and frigid blizzard conditions in the winter. These animals had also adapted to pasture lands which varied from arid to semi-forest conditions, and were clear in the summer and snow-covered in the winter. The early hybridization attempts with natural bisons and yaks have been described by A. Deakin and G.W. Muir in the publication entitled *Hybridization of Domestic Cattle, Bison and Yak* (Ottawa: Department of Agriculture, 1935. 15 p. Publication No. 479). This line of hybridization research continued until 1949. At that time, selection studies shifted, first to Shorthorn cattle at Lacombe, Alberta, and then on to a large scale also involving herds at Brandon, Manitoba, and Indian Head, and Scott, in Saskatchewan. The results of these studies were eventually reported by J.A. Neuman and G. Rahnefeld in their paper entitled *Selection Intensity and Response to Selection for Yearling Weight in Beef Cattle* (*Canadian Journal of Animal Science*, v. 53, 1973. pp. 1–12). The Shorthorns achieved annual genetic gains in yearling weight at 4.5 kilograms per animal, and established the practice of selectivity for performance in beef cattle.

In the mid-1960s, interest grew in importing cattle breeds to measure the effect of crossing European and British breeds. The initial breeds imported were Charolais and Simmental, followed by the Limousin, Maine, Anjou, and Brown Swiss breeds, and the program was initially undertaken, again, at the Agriculture Research Station at Lacombe, and then expanded to include the stations at Brandon, Manitoba, and Lethbridge, Alberta. This program first undertook the breeding of hybrid females, and their evaluation under diverse environmental conditions, to assess their lifetime reproductive potential. Hybrid males were evaluated for their feed efficiency and carcass quality. This initial work has been described by H.T. Fredeen and G.M. Weiss in their paper entitled *Lifetime Reproductive Efficiency of First Cross Beef Cows under Contrasting Environments* (Canadian Journal of Animal Science, v. 61, 1981. pp. 539–54).

Another purpose of this program was to study carcass characteristics of crosses of the English and European breeds. The calves of these crosses, covering the years 1973 to 1978, were fed and evaluated at the research stations at Lacombe and Brandon. In the paper by G.W. Rahnefeld and H.T. Fredeen, entitled *Breed of Terminal Sire Effects on Carcass Characteristics of 3-Way Cross Beef Cattle Reared at 2 Locations* (*Canadian Journal of Animal Science*, v. 63, 1983. pp. 523–50), the authors noted that the progeny produced by these crosses tended to show higher carcass weights than the purebred strains. For example, Limousin progeny 'averaged 5.2% greater rib eye area/weight than Chi [Chianina] progeny.' However, Chianina progeny had the 'highest proportions of round and the combined trait of rib, round, and long loin.' On the other hand, 'Simmental progeny had significantly heavier hide weights as a percentage of liveweight and significantly lower dressing percentages than the progeny from C [Charolais], L (Limousin] and Chi [Chianina] series.' These researchers also observed carcass characteristics by sex and age of slaughter in their paper entitled *Sex and Year Effects of Carcass Characteristics of 3-Way Cross Beef Cattle Reared at Two Stations* (*Canadian Journal of Animal Science*, v. 63, 1983. pp. 285–93). They found that when the dual criteria of live weight and degree of finish (time taken to be brought to full feed – 100% grain concentration at Brandon, versus a mix of silage and grain concentrate over a five-month period at Lacombe), rather than just age alone, were used as slaughter criteria, the Brandon calves had greater slaughter weights. There were also carcass differences between heifers and steers. For example, steers had a greater rib eye area, but heifers had a greater rib eye area per unit weight. Also, steers had a greater amount of chuck, shank, and flank, but heifers had more rib, long loin, and combined cuts of rib, round, and long loin. G.W. Rahnefeld and R.L. Cliplef also studied carcass characteristics in their paper entitled *Carcass Quality Characteristics of 3-Way Cross Beef Cattle Reared at 2 Locations* (*Canadian Journal of Animal Science*, v. 65, 1985. pp. 51–68). The carcasses were evaluated for firmness, percentage of protein found, water-holding capacity, chemical fat, colour, and marbling. When everything was taken into account, and comparisons made between the animals reared at the Lacombe and Brandon stations, the authors stated that 'the differences found between breeds of terminal sire [steers] and breed cross of dam [heifers] for the traits evaluated were not of any commercial consequence.'

Another aspect of study in the breeding program was the genotype (crossbreed) and environmental interactions on carcass traits. G.W. Rahnefeld and H.T. Fredeen, in their paper entitled *Genotype-Environment Interactions for Carcass Traits of Backcross Bulls* (*Canadian Journal of Animal Science*, v. 67, 1987. pp. 549–52), noted that genotype-environment interactions for bulls raised at the two different locations (the Lacombe location was more arid than Brandon) were 'not of any major importance in planning of breeding programs, performance tests and transfer of beef cattle with respect to carcass traits.'

Finally, the most recent study pertaining to the breeding program, and undertaken between 1981 and 1985, was not concerned with carcass characteristics but with milk yield. This study was reported by R.M. McKay and G.W. Rahnefeld in their paper entitled *Milk Yield and Composition in First-Cross and Backcross Beef Cows* (*Canadian Journal of Animal Science*, v. 74, 1994. pp. 209–16). The animals studied were the crossbreeds of English and European stocks, but all were bred to Limousin bulls. The authors noted that these 'possible genotype times [X] environment interacting were observed for milk yield and composition' and 'milk yield and composition were affected by age of cow and influenced cow and calf weight changes.'

(c) Feed and Nutritional Studies on Cattle

Cattle are ruminant animals which eat forages (grass and legumes) directly from pastures, or as preserved hay, or fermented silage, all of which are initially stored in the first portion of the stomach, called the reticulum. From there, the food is regurgitated to the mouth, where it is macerated – 'chewing the cud' – before being swallowed again into the second stomach called the rumen. Here, bacteria are able to convert the cellulose of the forage into a digestible form of carbohydrate. This is the uniqueness of ruminant animals over other stock or poultry. Ruminants are also able to use concentrated foods (supplements), in the form of grains and minerals, to balance their diet.

From the 1920s to the mid-1950s, research on cattle feeds focused upon grass silage, which is more easily kept in many parts of Canada, because it takes less storage room, and there is less wastage as compared with hay. However, grass silage also freezes in extremely cold weather, and may develop objectionable odours. This earlier research with grass silage has been described by U.S. Logan in the publication entitled *Grass Silage in Dairy Cattle Rations* (Ottawa: Department of Agriculture, 1955. 11 p. Publication No. 929). Other work, by P.E. Sylvestre and E. Mercier, looked at the use of grass silage in maintaining and fattening beef cattle under feedlot conditions. In their publication, entitled *Grass Silage for Beef*

Cattle (Ottawa: Department of Agriculture, 1960. 13 p. Publication No. 955), they suggested that grass silage was as good as, or even better than, corn silage. It was at the Melfort Experimental Farm, in central Saskatchewan, in 1954, that a long-term program on beef and pasture experiments began in order to improve the use of pasture and hay crops by beef cattle.

In the first seven-year grazing study, from 1955 to 1961, at Melfort, D.A. Cooke and S.E. Beacom, in their paper entitled *Response of Six-Year-Old Grass-Alfalfa Pastures to Nitrogen Fertilizer in Northeastern Saskatchewan* (*Canadian Journal of Plant Science*, v. 48, 1973. pp. 167–73), noted that they had been able to increase beef yields by 20 kilograms per hectare when a bromegrass-alfalfa pasture under study was fertilized with nitrogen. However, when phosphorus was also applied, the beef yields increased to 36 kilograms. Further applications of nitrogen provided another 30% to the beef yields, but the high rates of nitrogen reduced, or eliminated, the alfalfa, because of the increased growth of bromegrass in the pasture. Other work by these researchers, entitled *Comparison of Continuously Grazed Bromegrass-Alfalfa with Rotationally Grazed Crested Wheatgrass-Alfalfa, Bromegrass-Alfalfa, and Russian Wild Ryegrass* (*Canadian Journal of Animal Science*, v. 53, 1973, pp. 423–29), expanded and refined their pasture studies and resulted in increased beef yields, a longer grazing season, and greater feed efficiency. Finally, J.A. Robertson and D.A. Cook expanded on these results, as noted in their paper entitled *Comparison of Four Systems of Managing Yearling Beef Steers on Rotationally Grazed Bromegrass-Alfalfa Pasture* (*Canadian Journal of Animal Science*, v. 59, 1979. pp. 519–29). The object of this study was to optimize beef yields and minimize management problems caused by insufficient feed during mid-season. By converting grass to silage, and feeding silage during the summer, they were able to increase beef yields by 3%, and retain the same number of animals on pasture throughout the season.

Another aspect of feeds for cattle has been the fine-tuning of feed for nutritional purposes. These studies have examined the utilization of feed by dairy and beef cattle, and the effect of supplements, such as trace elements, to overcome diet deficiencies in cattle. For example, the digestibility of silages by cattle, and their conversion into energy at different points of their growth, are important for beef yields. P.M. Flipot and G. Girard, in their paper entitled *Digestibility and Performance of Dairy Bulls Fed Corn and Grass Silages with Various Sequences and Levels of Barley* (*Canadian Journal of Animal Science*, v. 72, 1992. pp. 61–69), noted that the inclusion of barley in the silage, during the last third of the growth period, provided better weight gains than using barley during the beginning third of the growth period. And when the barley was combined with grass silages, rather than corn silages, there was also greater growth. Even the silage chop length can make a difference on performance in cattle. K.A. Beauchemin and B.I. Farr noted, in their paper entitled *Effects of Alfalfa Silage Chop Length and Supplementary Long Hay on Chewing and Milk Production of Dairy Cows* (*Journal of Dairy Science*, v. 77, 1994. pp. 1326–39), that increasing the silage chop length from five millimetres to 10 promoted rumination, and increased milk production while maintaining milk fat content.

Depending on where forages are grown, cattle may need to supplement their diets with trace elements, if the soil is lacking adequate concentrations of elements. For example, Canadian soils, in many areas, are deficient in copper. In 1964, J.E. Miltimore, at the Summerland Research Station in British Columbia, injected copper into cattle grazing the pastures of low-lying, poorly drained soils. In the paper by J.E. Miltmore and W.J. Pigden, entitled *Response from Copper and Selenium with Vitamin E Injections to Cattle Pastured on Mineral and Organic Groundwater Soils* (*Canadian Journal of Animal Science*, v. 53, 1973. pp. 1–12), these researchers noted that the cattle had average daily weight gains of 200 grams in weight.

Vitamin E and selenium deficiencies are closely related and have an even more serious effect on calves, as deficiencies here are related to nutritional muscular dystrophy (NMD) that kills the calves. The source of the deficiency is the winter feeding of cows on a dry diet where little vitamin E, and selenium, a compound of vitamin E, were present. Hence, the calves born of these winter-fed cattle have the vitamin deficiency. Injections of vitamin E, and selenium, were first used in 1960 at the northern Ontario Experimental Farm at Kapuskasing. M. Hidiroglou and S.K. Ho, in their paper entitled *Metabolism of Inorganic Selenium in Rumen Bacteria* (*Canadian Journal of Physiology and Pharmacology*, v. 46, 1967. pp. 229–32), noted that the injections did cure the symptoms of NMD. However, the use of injections is expensive and an unsatisfactory method for dispensing selenium to cattle. Later on, M. Hidiroglou and J. Proux did solve this problem by using a Australian-developed method for delivering selenium to cattle. In their paper, entitled *Intruminal Selenium Pellet for Control for Nutritional Muscular Dystrophy in Cattle* (*Journal of Dairy Science*, v. 68, 1985. pp. 57–66), the authors described how they administered these pellets, so that the stomach bacteria of the rumen could interact with the selenium, and make it available to the cow's digestive system.

The actual amounts of selenium absorbed by dairy cows has been studied more recently by K.M. Koenig and W.T. Buckley, at the Agriculture Research Station, at Lethbridge, Alberta. In their paper, entitled *True Absorption of Selenium in Dairy Cows: Stable Isotope Tracer Methodology and Effect of Dietary Copper* (*Canadian Journal of Animal Science*, v. 71, 1991. pp. 175–84), these researchers measured true absorption of selenium in dairy cows by evaluating 'oral and intravenous routes of isotope administration and tracer enrichment in serum and level for determining endogenous Fecal Se.'

(d) Poultry

i. Poultry Breeding The longest ongoing poultry breeding research program at Agriculture Canada has emphasized egg production, including eggshell strength. As well, there has been an active program in breeding for disease resistance.

Before the turn of the century, commercial poultry production was concerned primarily with improving egg production (meat production was considered to be a by-product in dual-type birds). The favoured bird was the Barred Plymouth Rock. However, by the 1940s, much more research began to be undertaken as egg production moved into efficient mass production. The beginnings of this emphasis were in the early 1940s, when H.S. Guthridge and J.B. O'Neil published their paper entitled *The Relative Effect of Environment and Heredity Upon Body Measurements and Production Characteristics in Poultry. II. Period of Egg Production* (*Scientific Agriculture*, v. 22, 1942. pp. 482–91). These researchers found that egg production, and days to first egg, were influenced more by environment than heredity. However, egg weight was equally influenced by heredity and environment, and maximum body weight influenced more by heredity than by environment. Two years later, H.S. Guthridge and J.B. O'Neil, in their paper entitled *Level and Source of Protein in Poultry Production. II. As Related to Economical Production of Eggs* (*Scientific Agriculture*, v. 24, 1944. pp. 240–50), noted that Barred Plymouth Rock chickens, housed in batteries, produced 30 eggs more per bird during laying seasons than similar ones housed in floor pens.

For the next 20 years, R.S. Gowe and A. Robertson conducted extensive investigations involving research stations from British Columbia to Prince Edward Island on the genotype-environment interactions in poultry, which at one point were described, overall, in their paper entitled *Environment and Poultry Breeding Problems* (*Poultry Science*, v. 38, 1959. pp. 462–71). One of the particular studies by R.S. Gowe was entitled *Environment and Poultry Breeding Problems. 2. A Comparison of the Egg Production of 75.c. White Leghorn Strains Housed in Laying Batteries and Floor Pens* (*Poultry Science*, v. 35, 1955. pp. 430–35). This paper was published at a time when the egg production industry was starting to house laying birds in cages and it showed that some strains performed better in cages than did other strains. By the mid-1980s, most laying birds in Canada were housed in cages, but studies continued to examine the role of heredity on egg production. For example, R.W. Fairfull and R.S. Gowe, in their paper entitled *Dominance and Epitasis in Heterosis of White Leghorn Strain Crosses* (*Canadian Journal of Animal Science*, v. 67, 1987. pp. 663–80), noted that 'three-strain crosses had performance for egg production that was equal [to] or better than the two-strain cross with the highest egg production. Thus, in commerce where the level of production is of great economic importance, the testing and use of a specific three-way cross combination will usually result in a better commercial product.'

Related to egg production itself is the importance of eggshell strength as production systems have become more mechanized, and marketing systems have become more organized. Hence, the increased volume of eggshell a hen produces must resist greater 'insults' without breaking. R.M. Hamilton and K.G. Hollands, in their paper entitled *Relationship between Egg Shell Quality and Shell Breakage and Factors That Affect Shell Breakage in the Field – A Review* (*World's Poultry Science Journal*, v. 35, 1979. pp. 177–90), noted that the importance of eggshell strength was indicated by losses up to the farm gate, alone, of $66 million annually in North America (at that time).

Studies in eggshell strength began in the 1960s, at the Animal Research Centre in Ottawa. P.W. Voisey and J.R. Hunt, in their paper entitled *A Technique for Determining Approximate Fracture Propagation Rates of Egg Shells (Canadian Journal of Animal Science*, v. 44, 1964. pp. 347–50), used high-speed photography for the first proof that a crack in an egg shell starts at the point where an 'insult' is applied, and propagates outward. In their 1966 paper entitled *Physical Properties of Egg Shells. 1. Relationship of Resistance to Compression and Force at Failure of Egg Shells* (*Poultry Science*, v. 55, 1966. pp. 1398–1404), these researchers introduced materials-testing machines that precisely compared direct and indirect methods of measuring shell strength without breaking the shell. They also made some basic stress analyses, as all 'insults' are in the form of a concentrated force due to the shell curvature, and found that the fracture mechanism depended on the tensile strength of the shell at its inner surface. Another experiment by Voisey and Hunt, described in their paper entitled *Effect of Compression Speed on the Behaviour of Eggshells* (*Journal of Agricultural Research*, v. 14, 1969. pp. 40–46), proved that the faster an insulting force was applied, the greater the force required to fracture the shell. This finding was important, because measuring devices previously used did not control the rate of application of force.

The research on measuring the interrelationships between an egg's physical characteristics and its shell strength was continued by P.W. Voisey and D.C. Macdonald, as described in their paper entitled *Laboratory Measurements of Eggshell Strength. 1. An Instrument for Measuring Shell Strength by Quasi-Static Compression, Puncture, and Non-Destructive Deformation* (*Poultry Science*, v. 57, 1978. pp. 860–69). This paper described the instrument they developed that measured shell strength precisely, but destructively, and provided a standard for comparison. Because the range of shell characteristics is extremely small, and the variations of shell strength are numerous, statistical analyses of data by B.K. Thompson and R.M. Hamilton, in their paper entitled *Relationships among Various Egg Traits Relating to Shell Strength among and within Five Avian Species* (*Poultry Science*,

v. 60, 1981. pp. 2388–94), have contributed to the study of eggshell strength. And by the use of electron microscopy, as noted in the paper by I.L. Stevenson and P.W. Voisey, entitled *Scanning Electron Microscopy of Fractures in Eggshells Subjected to Puncture Test* (*Poultry Science*, v. 60, 1981. pp. 89–97), researchers have developed a clear understanding that the mechanisms governing shell strength rest in the shell's microstructures. Today, the work on eggshell strength continues at the Animal Research Centre, in Ottawa, as seen in the breeding of strains of White Leghorn chickens for eggshell quality. For example, A.A. Grunder and R.W. Fairfull, in their paper entitled *Correlations between Measures of Eggshell Quality or Percentage of Intact Eggs and Various Economic Traits* (*Poultry Science*, v. 70, 1991. pp. 1855–60), studied different strains for egg quality traits (non-destructive deformation, compression fracture strength, shell weight, shell weight per unit surface area, and specific gravity), and the percentage of intact eggs from oviposition (laying) through to sorting. This research was undertaken to achieve a better understanding of the correlations between breeding for desirable economic traits, including not compromising other traits such as the age of the bird at first egg or fertility, and hatchability of the birds.

As previously mentioned, the breeding for disease resistance is another aspect of poultry breeding program. This aspect of poultry research has been most noticeable in the attempt to eradicate two viral diseases in poultry – Marek's disease and lymphoid lukosis. Marek's disease, in chickens, was first identified at the turn of the century, in Hungary. It is caused by a virus that produces tumours on nerves, muscles, and viscera, and results in dehabilitation and death of poultry. The virus attaches itself to the feather and skin scurf (flaky epidermis) of infected birds, and is transmitted on these particles through the air. Until 1969, the only means of control was the destruction of infected flocks, the thorough cleaning of buildings and equipment, and starting all over again with non-infected chicks.

Although initial work was carried out in the United Kingdom to eradicate the virus through immunization of day-old chicks, major work on the disease was achieved in Canada, in Ottawa. Early on, J.S. Gavora and A.A. Grunder, in their paper entitled *An Assessment of the Effects of Vaccination on Genetic Resistance to Marek's Disease* (*Poultry Science*, v. 53, 1974. pp. 889–97), found that resistance to the disease had a 61% inheritability, based on previously observed variations in mortality to the disease among vaccinated strains of chicks. These variations were confirmed by J.S. Gavora and R.S. Gowe, in their paper entitled *Vaccination against Marek's Disease: Efficacy of Cell-Associated and Lyophilized Herpesvirus of Turkeys in Nine Strains of Leghorns* (*Poultry Science*, v. 56, 1977. pp. 846–53), where differences in mortality from Marek's disease among strains of both selected and control strains of vaccinated White Leghorns were noted. From these studies, J.S. Gavora and J.L. Spencer were able to design and execute breeding programs for the production of poultry strains resistant to Marek's disease, which were described in their paper entitled *Breeding for Immune Responsiveness and Disease Resistance* (*Animal Blood Groups and Biochemical Genetics*, v. 14, 1983. pp. 159–80). A later follow-up of this line of research has been described by H. Ameli and J.S. Gavora in a paper entitled *Genetic Resistance to Two Marek's Disease Viruses and its Relationship to Production Traits in Chickens* (*Canadian Journal of Animal Science*, v. 72, 1992. pp. 213–25). This research examined the genetic variability of resistance to Marek's disease, and the relationship between disease incidence and egg production traits among 23 genetic groups of chickens. The authors found correlations between disease incidence, egg production, and egg weight to be 'high and negative.'

The second viral disease that has been studied at the Animal Research Centre in Ottawa is lymphoid leukosis. Unlike Marek's disease, this virus affects the reproductive tract and is transmitted by eggs, as well as by direct contact. J.S. Gavora and L.T. Spencer, in their paper entitled *Lymphoid Leukosis Virus Infection: Effects on Production and Mortality and Consequences in Selection for High Egg Production* (*Poultry Science*, v. 59, 1980. pp. 2165–78), noted that strains of chickens selected for high egg production had a lower incidence of lymphoid leukosis than unselected strains. They also wondered if the difference between the two strains of chickens could be attributed to disease resistance, rather than to the greater propensity of increased egg production. Unlike Marek's disease, where the solution was sought by both breeding and inoculation, these researchers, in their paper entitled *Effects of Lymphoid Leukosis Virus Infection on Response to Selection.* In: *Poultry Genetics and Breeding.* Edited by W.G. Hill and J.M. Mason (Oxford: British Poultry Science, 1985. pp. 17–24), concluded that eradication of the lymphoid leukosis virus from breeding flocks was the best route to prevention.

This approach was used, starting in 1983, and four generations later, in 1986, 20% of the strains in the research flock were virus-free. However, the disease persisted in some strains, and research continued, as noted in the paper by J.S. Gavora and B. Benkel, entitled *Influence of the AVL6 Recombinant Avian Leukosis Virus Transgene on Production Traits and Infection with Avian Tumour Viruses in Chickens* (*Poultry Science*, v. 74, 1995. pp. 852–63). The AVL6 recombinant transgene induces dominant resistance to the avian leukosis virus, but causes birds to have significantly lower body weight. Thus, 'transgenesis has biological costs that have to be assessed against desirable effects of transgenes.'

ii. Poultry Nutrition The study of poultry nutrition has evolved over the years, from general diet observations or diet enrichment studies, to nutritional utilization of proteins or growth hormones, the use of alternate diets, the evaluation of energy uptake, and the pass-through of feedstuffs. After the turn of the century, some work was done in Ottawa to study the effect of feed restrictions on White Leghorn hens. These birds moulted earlier, but laid more eggs than fully fed birds. This work was picked up again, in the 1960s, at the Agriculture Research Station at Brandon, Manitoba. E.D. Walter and J.R. Aitken, in their paper entitled *Performance of Laying Hens Subjected to Restricted Feeding during the Rearing and Laying Periods* (*Poultry Science*, v. 40, 1961. pp. 345–54), found that restricting feed during the rearing period caused birds to commence laying later, but like those very early studies, found that they laid more and larger eggs than those birds on full feed. Reducing feed by 12% during the laying period effected a considerable saving in feed, but was offset by a drop in egg production. Another study, by T.M. MacIntyre and J.R. Aitken, examined reduced feed intake with high energy rations for egg-layers. In their paper entitled *The Effect of High Level of Dietary Energy and Protein on the Performance of Laying Hens* (*Poultry Science*, v. 36, 1957. pp. 1211–16), these researchers submitted both Barred Plymouth Rocks and White Leghorns to high energy rations at both normal and high protein levels. Two years of testing confirmed that high energy rations decreased feed consumption, which was a saving, but it did not increase egg production.

In more recent years, dietary protein, coupled with other influences, has been examined in the search for greater poultry productivity. One of these has been temperature and diet. F.G. Proudfoot and H.W. Hulan looked at these factors and their effect on chicken and turkey broilers. One paper, entitled *Interrelationships among Lighting Ambient Temperature, and Dietary Energy and Broiler Chicken Performance* (*Poultry Science*, v. 66, 1987. pp. 1744–49), studied 4,800 broiler chickens of one commercial strain under two types of light (incandescent vs green fluorescent), and various rates of starter to finish diets, in an environment of regulated declining temperatures. They found that lighting and ambient temperature treatments had no significant effects on 'mortality, feed conversion ratio, body weight, and monetary return.' However, the better the dietary energy input, in terms of feed, the better the mortality return. In their other paper, entitled *Effect of Ambient Temperature and Dietary Energy on the Performance of Turkey Broilers* (*Canadian Journal of Animal Science*, v. 68, 1988. pp. 257–64), these authors found that with turkeys, dietary energy requirements could be reduced in warmer environments, particularly for females.

Alternate diets have been studied, particularly in the Maritimes, because of the ready availability of fish meal. The use of fish meal, primarily as a diet for meat-type birds, was first studied in the late 1960s and early 1970s. In 1971, F.G. Proudfoot and W.F. Lemoreaux, in their paper entitled *Performance of Commercial Broiler Genotypes on Fish Meal Diets with a Charcoal Supplement* (*Poultry Science*, v. 50, 1971. pp. 1124–30), found that when the fish meal component of the diet fed to broiler birds was increased from 4% to 10%, there was a marked growth response. However, when the fish meal component was further increased to 15%, there was no growth change. And, when the fish meal component was further increased to 20%, there was a reduction in weight.

The problem of lower body weights was noted again much later in a study, by H.W. Hulan and R.G. Ackman, in their paper entitled *Omega -3 Fatty Acid Levels and Performance of Broiler Chickens Fed Redfish Meal or Redfish Oil* (*Canadian Journal of Animal Science*, v. 63, 1988. pp. 533–48). Besides lower body weights, there was a poorer conversion of feed into energy and a higher proportion of free cholesterol and phospholipids in the edible meat portions of the birds.

The ability to measure the conversion of feed into metabolizable energy by poultry began in the 1970s, at the Animal Research Institute in Ottawa. At first, I.R. Sibbard and K. Price studied the apparent metabolizable energy (AME) of feed in poultry. AME is the energy in the feed, minus the energy lost in feces, urine, and combustible gases. In their paper entitled *Variation in the Metabolizable Energy Values of Diets and Dietary Components Fed to Adult Roosters* (*Poultry Science*, v. 54, 1975. pp. 448–56), the authors learned that AME values varied with feed intake, and the AME underestimated the energy actually available to a bird because excreta contained materials that are not derived directly from the feed. Over the next several years, Sibbard refined techniques to measure the true metabolizable energy (TME) through precision feeding, and by removing body wastes from the excreta. This work, by I.R. Sibbard, was described in *The T.M.E. System of Feed Evaluation* (Ottawa: Agriculture Canada, 1983. 89 p. Research Branch Contribution 1983–20E), and quickly became the standard technique in most countries for assessing ration formulations for poultry. This basic technique for measuring TME has continued to be refined, as noted in the paper by M.A. Wolynetz and I.R. Sibbard, entitled *Standard Errors of Nitrogen-Corrected True Metabolizable Energy Estimates: Effects of Pooling Excreta Samples and Ignoring Among-Control-Bird Variation* (*Poultry Science*, v. 68, 1989. pp. 1361–67). Here, the authors suggested that 'failure to account for the among-control-bird variation in excreta energy overstates the precision of TMN-n estimates.'

D Pest Control Research

Research concerning the protection of crops and animals from the effects of insects has always been a major focus of the Research Branch. The work involves not only entomologists but also chemists and geneticists. Much of this work concentrates on the life cycles of harmful insects, and the development of chemical pesticides which increase crop productivity, but at the same time cause a cumulative toxic build-up in the environment. This problem has led to the development of alternate approaches such as research on the biological control of pests through useful insect predators. More specifically, some of the highlights of this program have been the study and control of grasshoppers (*Melanoplus*), the wheat-stem sawfly (*Cephus cinetus*), warble flies and grub (*Hypoderma*), and blackflies (*Simulium*).

(a) Crop Pest Control

The orthopteroid insects, such as grasshoppers and locusts, are the most destructive to agricultural crops, and, as the prairie grasslands were broken and cultivated for wheat production, grasshopper control became essential. Historically, early control was by a mechanical hopperdozer which swept grasshoppers into pans of tar where they were burnt. Also, poison baits of paris green (an insecticide) were used until the introduction of organochlorine insecticide sprays. Despite the problem, however, it was not until the early 1930s that a systematic study of grasshoppers began as a basis for a more scientific approach to their control. Some preliminary studies were undertaken by N. Criddle in Manitoba to identify prairie species of grasshoppers, and this work was described, along with later advances, by R.H. Handford in the paper entitled *The Identification of Nymphs of the Genus* Melanoplus *of Manitoba and Adjacent Areas* (*Scientific Agriculture*, v. 26, 1964. pp. 147–80). This paper indicated that the early instars (immature forms after hatching) were difficult to identify, but the author devised a key for the identification of the various species.

Grasshopper surveys were also begun in the 1930s in the prairie provinces in order to predict where and to what extent grasshopper outbreaks might be expected the following year. These surveys, along with their prediction maps, have become an annual undertaking, and the techniques for estimating potential grasshopper populations were improved, because of the work by R.L. Randell and M.K. Mukerji, described in their paper entitled *A Technique for Estimating Hatching of Natural Egg Populations of* Melanoplus sanguinipes (*Orthoptera: Acrididae*) (*Canadian Entomologist*, v. 106, 1974. pp. 801–12). In this paper, the authors used the number of eggs, and the air temperature, to devise a mathematical model that accounted for 96% of the variation in predicting hatching dates. These dates were especially important, because R. Pickford, in an earlier paper entitled *Survival, Fecundity and Population Growth of* Melanoplus bitituratus *(Wlk.) (Orthoptera: Acrididae) in Relation to Date of Hatching* (*Canadian Entomologist*, v. 92, 1960. pp. 1–10), noted that females, hatching eggs early in the season, produced 20% more eggs than those with a later hatching. And at the Agriculture Research Station, at Lethbridge, Alberta, D.S. Smith, in the paper entitled *Crowding of Grasshoppers. II. Continuing Effects of Crowding on Subsequent Generations of* Melanoplus sanguinipes *(Orthoptera: Acrididae)* (*Environmental Entomology*, v. 1,

1972. pp. 314–17), showed that grasshopper females, from parents raised in uncrowded cages, were 30% heavier, but produced only one-half the number of eggs as did those from crowded cages. This research explained how, when crowded, grasshopper populations can increase so rapidly.

Temperature and soil moisture have also been shown to affect hatchability. In two papers, entitled *Influence of Embryonic Development and Heat on Population Trends in Three Grasshopper Species (Orthoptera: Acrididae) in Saskatchewan* (*Canadian Entomologist*, v. 109, 1977. pp. 229–36), and *A Model for Estimating Hatch and Mortality of Grasshopper Egg Populations Based on Soil Moisture and Heat* (*Annals of the Entomological Society of America*, v. 71, 1978. pp. 183–90), M.J. Mukerji and S.H. Gage, using 16 years of data, showed that a general increase in density should occur when accumulated heat units reached 1,600 degree-days above 10 degrees C, provided that soil moisture was at, or above, 13.5%. These two papers led to a refinement of the grasshopper prediction model of 1974 to take into account estimates of hatch and mortality rates of eggs.

Because of the prediction maps, when a moderate-to-severe infestation is identified, different types of biodegradable insecticides can be introduced to the hatching area to possibly reduce crop losses. One of the most effective ways to control grasshoppers is by the use of poisoned bran, impregnated with the insecticide dimethoate. M.J. Mukerji and A. Ewen, in their paper entitled *Evaluation of Insecticide-Treated Bran Baits for Grasshopper Control in Saskatchewan (Orthoptera: Acrididae)* (*Canadian Entomologist*, v. 113, 1981. pp. 705–10), found that by spreading the bait at key points around fields, the amount of insecticide needed was reduced by 65%, compared with spray applications. As well, non-target insects such as bees remained clear of the insecticide, and residues of the insecticide on the crop were negligible. Another effective method for grasshopper control has been the use of trap-strips. In the paper by O. Olfert, entitled *Evaluation of Trap-Strips for Grasshopper (Orthoptera: Acrididae) Control in Saskatchewan (Canada)* (*Canadian Entomologist*, v. 118, 1986. pp. 133–40), weed trap-strips, 10 metres wide and sprayed with dimethoate, were found to be 20% more effective than bran bait formulations containing dimethoate. As well, field margins next to trap-strips contained 60%–66% fewer egg pods than margins of check subplots. The author also found that summer fallowing reduced egg numbers by 3.5%.

The search for alternatives to insecticides can be seen in the growing amount of research on grasshopper control using biological methods. For example, A. Ewen and M.K. Mukerji have examined the effectiveness of bran bait, sprayed with the parasitic protozoan *Nosema locustae*. In their paper, entitled *Evaluation of* Nosema locustae *(Microsporida) as a Control Agent of Grasshopper Populations in Saskatchewan* (*Journal of Invertebrate Pathology*, v. 35, 1980. pp. 295–303), the authors found that targeted grasshoppers ingest the parasite, which debilitates the host by competing for food reserves in blood and fat. And *Nosema locustae* is specific to grasshoppers and locusts, and therefore safe to use. Although it is slow to act, the parasite greatly reduces grasshopper populations.

Other biological control methods studied have involved the *Melanoplus sanguinipes* entomopoxvirus and the entomopathogen *Beauveria bassiana conidia.* O. Olfert and M.A. Erlandson, in their paper entitled *Wheat Foliage Consumption by Grasshoppers (Orthoptera: Acrididae) Infected with* Melanoplus sanguinipes *Entomopoxvirus* (*Environmental Entomology*, v. 20, 1991. pp. 1720–24) found that first instar grasshoppers, infected with this dried fungal spore powder, consumed 25% less food. Further damage potential was also reduced by initial instar mortality, and by the application of the powder at a later date to later instars, which prevented grasshopper maturation and reproduction. At the Agriculture Research Station, at Lethbridge, Alberta, G.D. Inglis and R.P. Feniuk, in their paper entitled *Mortality of Grasshoppers Exposed to* Beauveria bassiana *during Oviposition and Nymphal Emergence* (*Journal of Invertebrate Pathology*, v. 65, 1995. pp. 139–46), applied this entomopathogen in a sand oviposition (egg-laying) substrate, which caused ovipositing female grasshoppers to become infected in the intersegmental folds of their abdomens. Also, males, and newly hatched nymphs were infected. The mortality rate ranged from 80% to 100% for males after 14 days. Similar results were achieved by using lettuce and bran infected with the same entomopathogen. This study, undertaken by G.D. Inglis and D.L. Johnson, was entitled *Effect of Bait Substrate and Formulation on Infection of Grasshopper Nymphs by* Beauveria bassiana (*Biocontrol Science and Technology*, v. 6, 1996. pp. 35–50).

Crop losses to grasshoppers have been studied by S.H. Gage and M.K. Mukerji, in their paper entitled *Crop Losses Associated with Grasshoppers in Relation to Economics of Crop Production*, (*Journal of Economic Entymology*, v. 71, 1978. pp. 487–98). These authors found that economic losses caused by grasshoppers tended to occur during periods of poor crop yields due to drought. Besides wheat crop losses, some species of grasshoppers also damage native grass species. J.M. Hardman and J. Smoliak, in their paper entitled *Potential Economic Impact of Rangeland Grasshoppers (Acrididae) in Southeastern Alberta* (*Canadian Entomologist*, v. 112, 1980. pp. 277–84), studied data from 1928 to 1944 from the Manyberries area of southeastern Alberta. The data indicated that 80% of the grasshopper population was made up of 26 species, and that four of those species developed in the early spring and were grass-feeders, which caused damage to rangeland.

The other major insect pest of wheat crops on the prairies has been the wheat stem sawfly (*Cephus cinctus*), which is native to North America and lives in grasses. Its life history was first described in Canada by N. Criddle, in the paper entitled

The Wheat Stem Sawfly in Canada (Annual Report of the Entomological Society of Ontario, 52nd., 1921. pp. 18–22). The sawfly is a small bee, which lays eggs through minute holes cut by the saw-like ovipositor of the female in the hollow stem of wheat plants just below the head. After hatching, the young larvae grow rapidly and eat their way down the stem to the lowest joint, where they gnaw the inside of the stem, cutting a ring almost to the outside. Then, they spin their cocoons and remain dormant until spring, at which time they transform into the pupal stage and emerge from the wheat stems as adult sawflies and repeat the cycle. Although the sawfly has a native parasite, which has been described by W.A. Nelson and C.W. Farstad in their paper entitled *Biology of* Bracon cephi *(Gahan) (Hymenoptera: Braconidae), an Important Native Parasite of the Wheat Stem Sawfly,* Cephus cinctus *Nort. (Hymenoptera: Cephilae) in Western Canada* (*Canadian Entomologist*, v. 85, 1953. pp. 103–7), the route to controlling this insect has not been by biological control but by the development of wheat strains that have displayed resistance to this small bee.

Wheat stem sawfly damage was first reported in Canada toward the end of the 19th century, by the then Dominion Entomologist and Botanist, J. Fletcher, in the *Report of the Entomologist and Botanist.* In: *Experimental Farms Report for 1896* (Ottawa: Queen's Printer, 1897. pp. 229–31). However, it was not until serious sawfly outbreaks in the early 1930s that further research was undertaken on wheat strains that had solid stem characteristics resistant to the bee, as well as biological studies on the insect itself. At Swift Current, Saskatchewan, H.J. Kemp noticed that solid-stem Golden Ball durum wheat was more resistant to sawflies than the bread wheat strains Marquis or Reward. In particular, Kemp examined and bred Portuguese strains of wheat, obtained from New Zealand, in order to select plants that were stable for their solid-stem characteristic. After that, H.J. Kemp published the paper entitled *Studies on Solid-Stemmed Wheat Varieties in Relation to Wheat-Stem Sawfly Control* (*Scientific Agriculture*, v. 15, 1934. pp. 30–38). At the same time, C.W. Farstad carried out further biological studies of the insect which were reported in the paper entitled *Thelyotokous Parthenogenesis [asexual reproduction] in* Cephus cinctus *Nort. (Hymenoptera: Cephidae)* (*Canadian Entomologist*, v. 70, 1938. pp. 206–7). Then, further work was done by A.W. Platt on the influence of the environment on the solid-stem characteristics of wheat, and described in the paper entitled *The Influence of Some Environmental Factors on the Expression of the Solid-Stem Character in Certain Wheat Varieties* (*Scientific Agriculture*, v. 21, 1941. pp. 139–51). And A.W. Platt and R. Larsen, in their paper entitled *An Attempt to Transfer Solid-Stem from* Triticum durum *to I.* Vulgare *by Hybridization* (*Scientific Agriculture*, v. 24, 1943. pp. 214–20), studied the genetic transfer of the solid-stem characteristic. Three years later, A.W. Platt and C.W. Farstad developed uniform tests of wheat varieties for resistance to sawfly, which were reported in their paper entitled *The Reaction of Wheat Varieties to Wheat Stem Sawfly Attack* (*Scientific Agriculture*, v. 26, 1946. pp. 231–47). A uniform test was essential for the development of resistant wheat varieties.

After 1948, research on wheat-stem sawfly resistant wheat shifted from Swift Current, Saskatchewan, to Lethbridge, Alberta, where both plant geneticists and entomologists continued to study the sawfly problem and develop sawfly-resistant wheats. For example, L.A. Jacobson and C.W. Farstad, in their paper entitled *Effect of Time of Seeding Apex Wheat on Infestation and Sex Ratio of the Wheat Stem Sawfly,* Cyphus cinctus *Nort. (Hymenoptera: Cephedae)* (*Canadian Entomologist*, v. 84, 1952. pp. 90–92), noted that when infestations occurred in late-seeded fields, the build-up of infestation did not come to expectations the following year. Also noted was the fact that the 'progressive increase in the proportion of males as seeding date is delayed seems to be associated with the field behaviour of the insect.'

As previously mentioned, the study and development of strains of hard spring wheat, with resistance to sawfly, had begun in the 1930s with Portuguese strains of wheat. These, crossed with the Apex and the Thatcher varieties, and further crossing, produced the Rescue and Chinook varieties of the early 1950s. However, the most promising of these resistant varieties developed was not ready until 1981, and then announced by R.M. DePauw and D.S. McBean as *Leader Hard Red Spring Wheat* (*Canadian Journal of Plant Science*, v. 62, 1982. pp. 231–32). Leader was selected from a cross made in 1969 between the cultivar Fortuna, which had resistance to the wheat-stem sawfly, and the cultivar Chris, which had a low level of alpha-amylase (an enzyme that converts starches into sugars) activity. Besides its solid-stem characteristic, this variety also had a long seed dormancy, which meant it would not sprout in the fall before threshing.

(b) Animal Pest Control

Besides being harmful to grain crops, insects can also be injurious to livestock. Two of the major insects harmful to livestock that have been studied are warble flies, which lay grub-producing eggs on cattle, and black flies, which periodically, without warning, swarm onto cattle and inflict painful bites and impart a toxic saliva. The toxins cause symptoms of an anaphylactic-type shock, which is usually fatal to calves, or adult cattle, not conditioned to the bites.

In Canada, there are two species of warble flies that cause extensive damage. The egg-laying activities of warble flies start in May, and continue through June. Flies lay their eggs on the hairs of legs, bellies, and flanks of cattle, which agitates the animals into running and gadding. That in turn interferes with normal grazing, reduces milk flow, and sometimes causes injuries. Upon hatching, the larvae, or grubs, burrow through the skin of the animal and migrate within the body for about nine months, until they reach the animal's back, where they produce the characteristic lumps called warbles. The migrating larvae lodge either in the esophagus, causing eating difficulties, or in the spinal canal, sometimes causing paralysis. Each grub makes a breathing hole in the hide of the animal through which infection can occur, and in the spring of the following year the grubs emerge from the warbles, drop to the ground, pupate, and emerge as flies in the early summer, ready to start another cycle. Infested animals have a lower market value, because the impaired carcasses require trimming, and the damaged hides are devalued as a source of leather. K.K. Klein, at the Agriculture Research Station in Lethbridge, Alberta, wrote in the paper entitled *The Economics of Warble Fly Control* (*Canadian Farm Economist*, v. 14, 1979. pp. 20–27) that the annual packing plant losses attributed to grubs was $14 million (at that time).

In the 1920s and 1930s, the only preventive measures against warble flies were the use of creosote, or the use of insecticide rotenone, both as a wash and a spray. These preventive measures have been described in the publication by S. Hadwin entitled *Insects Affecting Livestock* (Ottawa: Department of Agriculture, 1923. 32 p. Bulletin No. 29, New Series); and by E. Hearle, in the paper entitled *Insects and Allied Parasites Injurious to Livestock and Poultry in Canada* (Ottawa: Department of Agriculture, 1939. 108 p. Publication No. 604). Rotenone, an insecticide from the root of a South American plant, is still used today on dairy animals because it is completely metabolized. However, the research for other forms of control, either insecticides, or the study of the production of antibodies within the animal to fight off grubs, has been ongoing, primarily at the Agriculture Research Station at Lethbridge, Alberta.

In the 1950s, the systemic insecticide trilene was studied. When trilene is administered, it is absorbed into the blood stream (hence systemic) to provide protection from infestation. This insecticide was studied by J. Weintraub and G.B. Rich, and described in their paper entitled *Timing the Treatment of Cattle with Trilene for Systemic Control of the Cattle Grubs* Hypoderma lineatum *(De Vill.) and* H. bovis *(L) in Alberta and British Columbia* (*Canadian Journal of Animal Science*, v. 39, 1959. pp. 50–57). Groups of 25 to 30 calves at Lethbridge, Alberta, and Kamloops, BC, were administered the insecticide orally, as either a drench or a large pill, over a six-month period from late fall to early spring. Untreated control groups averaged 30 to 80 grubs per animal, compared to treated animals which showed one to three grubs per animal. But both the drench and the pill were awkward to administer, and some animals showed minor toxic symptoms. G.B. Rich, at Kamloops, continued the work on the use of trilene in order to evaluate its use under ranch conditions. As described in the paper by Rich, entitled *Systemic Treatments for Control of Cattle Grubs Hypoderma spp. in an Isolated Range*

Herd (*Canadian Journal of Animal Science*, v. 45, 1965. pp. 165–72), between 800 and 1,300 cattle in each of six years were treated. After two treatments, the number of grubs per animal was reduced from 30 to two, and remained at that level over the six-year period. However, after treatment was suspended, the number of grubs per animal rose to over 10. An even broader regional control program was undertaken by M.A. Khan in central Alberta, involving over 5,000 treated animals per year, from 1959 to 1961. In Khan's paper, entitled *Extermination of Cattle Grubs (Hypoderma spp.) on a Regional Basis* (*Veterinary Record*, v. 83, 1968. pp. 97–101), the research involved the use of newer sprays of non-toxic systemic insecticides each autumn, which reduced the grub infestation from 14 per animal to zero in the central zone. In peripheral areas, where there was no intensive spraying, the infestations were reduced by about 75%. This level of control remained the same, even two years after the end of the experiment.

Nevertheless, G. Weintraub showed that warble fly populations could recover rapidly from insecticidal decimation, and concluded that some integrated control system was needed. One possibility was the release of sterilized male flies into the fly population, as entomologists, in general, knew from laboratory studies that sterile males breeding with females caused the latter to produce either infertile eggs or sterile males. Because of this knowledge, Weintraub developed a large-scale test totalling 4,200 square kilometres in both southern Alberta and northern Manitoba, involving a systemic insecticide, and released sterile males that had been produced in the laboratory by irradiation with cobalt-60. This pilot project has been described by S.E. Kunz and J. Weintraub in the paper entitled *A Pilot Test to Study the Use of the Sterile Insect Technique for Eradication of Cattle Grubs* (*Preventive Veterinary Medicine*, v. 2, 1984. pp. 523–27). However, there was no follow-up to this work in the literature.

As previously mentioned, the other route studied for the control of warble flies has been development of antibodies to help animals fight off the grubs.

Again, this research has been undertaken at Lethbridge, Alberta. R.H. Robertson, in the paper entitled *Antibody Production in Cattle Infected with Hypoderma ssp.* (*Canadian Journal of Zoology*, v. 58, 1980. pp. 245–51), began the study of the production of antibodies in animals infested with grubs. This work was based on field observations, which indicated that previously infested cattle were less subject to reinfestations, as well as on an earlier study by M.A. Khan and R. Connell entitled *Immunization and Parenteral Chemotherapy for the Control of Cattle Grubs* Hypodermia lineatum *(De Vill) and* H. bovis *(L.) in Cattle* (*Canadian Journal of Comparative Medicine*, v. 26, 1960. pp. 177–80).

The antigens (proteins) developed to stimulate an antibody reaction (immunization) in cattle against the grubs are Hypodermin A, B, and C. However, R.W. Baron, in the paper entitled *Cleavage of Purified Bovine Complement Component*

C-3 in Larval Hypoderma lineatum *(Diptera: Oestridae) Hypodermics* (*Journal of Medical Entomology*, v. 27, 1990), found that larval grubs were able to degrade these bovine antibodies and 'may allow the parasite to escape host defense reactions.' In order to overcome this problem, R.W. Baron and D.D. Colwell, as described in their paper entitled *Enhanced Resistance to Cattle Crub Infestation* (Hypoderma lineatum *De Vill.) in Calves Immunized with Purified Hypodermin A, B and C Plus Monophosphoryl Lipid A (MPL)* (*Veterinary Parasitology*, v. 38, 1991. pp. 185–98), used MPL to heighten the immune response in calves. The antigen-specific antibody response at maximum grub count, or infestation, was found to be significantly higher than in calves injected with the Hypodermin antigens alone, and the number of grubs through to pupation was significantly lower.

Unlike warble flies, the black fly problem to livestock is periodic, because of their sudden swarming and toxic biting attacks. Outbreaks of black fly *Simulium arcticum* Malloch pose annual threats to livestock in some areas adjacent to the Saskatchewan and Athabasca rivers in central Saskatchewan and Alberta. The major outbreaks which have caused the greatest damage to livestock have been described by F.J. Fredeen in the paper entitled *Outbreaks of the Black Fly* Simulium arcticum *Malloch in Alberta* (*Quaestiones Entomologicae*, v. 5, 1969. pp. 341–72). These occurred from 1944 to 1947, when about 1,100 farm animals were killed in an area of more than 20,000 square miles in Saskatchewan. During these outbreaks, which lasted from one to several weeks, grazing and breeding activities were interrupted, and besides fatalities there were weight losses, reductions in milk yields, and delayed calving.

Until the early 1950s, the only written account of the biology and egg-laying habits of this fly was by A.E. Cameron in the publication entitled the *Morphology and Biology of a Canadian Cattle Infesting Blackfly,* Simulium semile *Mall.* (*Diptera: Simuliidae*) (Ottawa: Department of Agriculture, 1922. 26 p. Technical Bulletin No. 5 New Series). However, in 1947, F.T. Fredeen and J.G. Rempel began new studies on the fly, which were reported in the paper entitled *Egg-Laying Habits, Overwintering Stages, and Life-Cycle of* Simulium arcticum *Mall (Diptera: Simuliidae)* (*Canadian Entomologist*, v. 83, 1951. pp. 73–74). Studies along the Saskatchewan River, over a three-year period, indicated that the black fly laid eggs over the water surface during July, and they overwintered in the river bed for spring hatching. The larvae then attached themselves to river rocks to complete their development. This long development, from hatching to the adult phase, indicated that repeated applications of a larvicide would be necessary for control.

At that time, the only insecticide that could be used was DDT for the control of the black fly. Experiments with that insecticide began in 1949, and were reported by A.P. Arnason and W.A. Brown in the paper entitled *Experiments in the Control of* Simulium arcticum *Malloch by Means of DDT in the Saskatchewan River*

(*Scientific Agriculture*, v. 29, 1949. pp. 527–37). Further experiments were carried out by F.J. Fredeen and A.P. Arnason, and reported in the paper entitled *Further Experiments with DDT in the Control of* Simulium arcticum *Mall. in the North and South Saskatchewan Rivers* (*Canadian Journal of Agricultural Science*, v. 33, 1953. pp. 379–93. In these experiments, DDT was applied to reaches of the North and South Saskatchewan River, just above rapids where black fly larvae lodged. The resulting concentration of DDT in the river varied from one part per billion to one part in four billion, which resulted in an over 80% control of black fly larvae at distances up to 60 kilometres, and 30% control as far as 150 kilometres downstream from the point of application. This control was attributed to the adsorption of DDT onto silt particles that larvae injested, which in turn caused accumulative fatal dosages. Researchers also found that insect-feeding bottom fish, such as suckers, were killed when the insecticide concentrate was heavier than water.

With the increasing evidence in the late 1940s of the persistence of DDT in nature, the use of DDT as an insecticide for black fly control was not desirable and the search was resumed for more suitable insecticides. The initial work on the search for an alternate insecticide was carried out by F.J. Fredeen, and described in the paper entitled *Tests with Single Injections of Methoxychlor Black Fly (Diptera: Simuliidae) Larvicides in Large Rivers* (*Canadian Entomologist*, v. 106, 1974. pp. 285–305). In tests carried out between 1968 and 1972, larval infestations were reduced by 75% in the Saskatchewan River with the use of single 15-minute injections of about 0.2 to 0.4 parts per million (p.p.m.) of methoxychlor. In four other tests, 'single 15-minute injections of 0.182 to 0.236 p.p.m. of methoxychlor ... provided the level of abatement throughout 23 km to 24 km sections of the river ... but not beyond 23 km and 27 km in two other sites.' Methoxychlor was chosen because its residues were known not to accumulate in food chain organisms or in fish and was readily biodegradable.

Another experiment was carried out in 1973, approximately 120 kilometres upstream from the earlier site, in order to study the effects of a single injection of methoxychlor in an uninterrupted 161-kilometre (100-mile) section of the North Saskatchewan River. The results were reported, again, by F.J. Fredeen, in the paper entitled *Effects of a Single Injection of Methoxychlor Black-Fly Larvicide on Insect Larvae in a 161-km (100 Mile) Section of the North Saskatchewan River* (*Canadian Entomologist*, v. 107, 1975. pp. 807–17). This experiment demonstrated that 'a single injection of 0.6 p.p.m. of methoxychlor for 7.5 min ... removed 100% of the black-fly larvae larger than 1 mm long. (instars 3 to 10 inclusive) from exposed artificial substrates as far as 80 km downstream from the point of injection.' However, 'within 1 to 3 weeks after the treatment, populations of most taxa in a 161-km section of the North Saskatchewan River had already surpassed those observed before the treatment.' Finally, 'experience in recent years indicates that if larvae accumulate on these substrates at the rate of 1 larva or more per 2 cm^2 of surface area per day of exposure treatment is likely required to prevent a potentially dangerous outbreak.'

Besides the development of an acceptable and successful larvicide, work has continued on monitoring the distribution of black flies, their taxonomy, and their environmental biology. Although the black fly is prominent in central Alberta and Saskatchewan, an Arctic species has also been noted with other new species in southern Alberta since the early 1980s. J.L. Shipp, of the Agriculture Research Station at Lethbridge, Alberta, stated in the paper entitled *Distribution of and Notes on Blackfly Species (Diptera: Simuliidae) Found in the Major Waterways of Southern Alberta (Canada)* (*Canadian Journal of Zoology*, v. 63, 1985. pp. 1823–28), that among 21 black fly species from 22 rivers, creeks, and temporary streams collected over the period from 1981 to 1983, one species (*Prosimulium dicum* Dyar and Shannon) was new to the province, and three other species (*Simulium verecundum* Stone and Jamnback, *Simulium vernium* Macquart, and *Stegopterna mutata* Malloch) were new to southern Alberta.

In Saskatchewan, the monitoring of black flies also continues. P.G. Mason and P.M. Kustero noted in their paper entitled *Seasonal Activity of Female Black Flies (Diptera:* Simuliidae) *in Pastures in Northeastern Saskatchewan, Canada* (*Canadian Entomologist*, v. 122, 1990. pp. 825–36) that at two locations along the Saskatchewan River system (near Prince Albert and near Choiceland), the species *Simulium luggeri* was most abundant at the Prince Albert site, and there was a mix of four species taken in traps at the Choiceland site. This distribution pattern covered the years 1986 to 1988, and it is interesting to note that in the spring of 1985, an outbreak of *Simulium luggeri* had occurred north of the Saskatchewan River. The source of the outbreak was probably the tributary Torch River, which had high winter levels in 1985.

As noted in the above papers, there is a variety of species of black flies native to Alberta and Saskatchewan. Their descriptive taxonomy has been studied and described by F.J. Fredeen in the paper entitled *Keys to the Black Flies (Simuliidae) of the Saskatchewan River in Saskatchewan* (*Quaestiones Entomologicae*, v. 17, 1981. pp. 189–210). Besides this taxonomic work, more detailed biological studies have also been carried out on the influence of temperature on black fly development. In one paper by J.L. Shipp and G.H. Whitfield, entitled *Influence of Temperature on Embryonic Development and Egg Hatching of* Simulium arcticum *Malloch IIS-10.11 (Diptera: Simuliidae)* (*International Journal of Biometeorology*, v. 32, 1988. pp. 242–46), the authors noted that lower air temperatures and lower air vapour pressures increased the mortality rates of flying adult black flies.

IV Departmental Sources of the Publications

The publications of the Department of Agriculture (now Agriculture and Agri-Food Canada) in general, and the Research Branch in particular, are well documented. The bilingual, revised publication, prepared by the Headquarters Library, entitled *Publications of the Canada Department of Agriculture, 1867–1974*. 2nd Revised Edition (Ottawa: Department of Agriculture, 1975. 341, 136 p.), provides a detailed listing of the department since its inception until the early 1970s. It is a revision, and updated edition, of *Publications of the Canada Department of Agriculture, 1867–1959*. Compiled by Ella S.G. Minter (Ottawa: Queen's Printer, 1963. 387 p.). This list includes all publications, both in English and in French, issued by the Department from 1867 to 1973. However, it does not include research papers published in the open scientific journal literature, reprints, unnumbered contributions, unpublished or restricted papers of the department, unedited manuscripts, or unprocessed papers. The list is divided into separate English and French parts, each of which is subdivided into three main sections. They are: (1) General Departmental Publications, (2) Publications of Services, Branches, Divisions, and other sections of the department, and (3) Publications of Research establishments of the department. With each section, the publications are generally grouped in the following order: (1) numbered series, (2) serials and periodicals, and (3) miscellaneous publications.

This publication, within its scope mentioned above, contains the listings of the Research Branch, scattered in all three sections, as well as those listings of the other Program Branches besides Research (they are Grains and Oilseeds Branch, Agriculture Development Branch, International Programs Branch, and the Food Production and Inspection Branch). The publication was supposed to be maintained between future revisions, or editions, by the department's annual publications list. This list is separate from the bi-monthly *Selected List of Acquisitions* (Ottawa: Library, Agriculture Canada), beginning in November/December, 1972 and published until 1980.

Another separate, but consolidated, listing of interest, published by the Headquarters Library, was the *Union List of Serials in Canada Department of Agriculture Libraries* (Ottawa: Department of Agriculture, 1973. 946 p.).

In particular, the publications of the Research Branch itself are well documented. The *Research Reports of the Research Branch* (Ottawa: Agriculture Canada, 1967–), originally published separately by each station, list the publications generated by each of the 33 major research stations of the Eastern, Central, and Western organizational regions. The latter includes all publications issued through Agriculture Canada itself, as well as all research publications published in the open scientific literature.

Chapter 4

Wildlife Research

I Historical Background

The overwhelming amount of wildlife research at the federal level is carried out by the Canadian Wildlife Service. However, the earliest wildlife research in Canada was done by the National Museum of Canada. The National Museum of Canada was originally the Museum of the Geological Survey of Canada and was housed with the latter, until 1910 when it moved to the new Victoria Memorial Museum. Among its earliest researchers were John Macoun as Botanist and Zoologist, and James Macoun, as Botanist. The museum's name was acquired in 1927 by Order-in-Council, but in the 1970s and later, because of administrative changes, it became first the National Museum of Natural Sciences (1974–89), and then the Canadian Museum of Nature (1990–). Because of the history of the major appointments to the museum, its biological work, which has focused on zoology, including palaeontology, and botany has been and continues to be descriptive, systematic, or taxonomic, in nature.

The development of wildlife research and related protective services, which is the major focus of this chapter, can also be traced back to the first decade of the 20th century, when there was a drastic decline in migratory bird populations, particularly in eastern North America. For example, the passenger pigeon, which had thrived in the 1880s, became extinct after intense hunting. As well, the great auk became extinct, and on the prairies the distinctive mammal, the prairie bison, had practically disappeared. Later, even the wolf became an endangered species, because of widespread programs to control it as a 'harmful' competitor.

The decline in abundance of migratory birds was a trans-border matter, because most game waterfowl nest in Canada, and then migrate south of the border in the autumn. In 1916, Canada and the United States signed a Migratory Birds Treaty. Although the British North America Act gave each province control of the natural resources within its boundaries, the new treaty and the resulting Migratory Birds Convention Act that Parliament passed in 1917 created a federal responsibility for the management and protection of migratory birds. A superintendent of wildlife protection was appointed in the parks branch of the Department of the Interior, and three enforcement officers were named – one each for the Maritimes, Ontario, Quebec, and the western provinces. During the years between World Wars I and II, the number of trained people and the amount of money devoted to this work increased very slowly, but a rising population, coupled with resource developments, was already showing its deteriorating effect on the environment and wildlife. After World War II, additional population increases and immigration, along with more leisure and affluence, created further stresses on the environment and wildlife as the demand for outdoor recreation increased. The effective management of wildlife, based on sound research, was necessary, and the Canadian Wildlife Service was created in 1947, as a division of the National Parks Branch of the Department of Indian Affairs and Northern Development. However, with the creation of the Department of Environment in 1971, the Canadian Wildlife Service, or CWS, was transferred to that new jurisdiction. And, in 1973, Parliament passed the Canadian Wildlife Act, which enabled the federal government to carry out wildlife research and cooperate with the provinces to undertake wildlife conservation.

The Canadian Wildlife Service has also been increasingly responsible for participating in national and international committees concerning wildlife protection. In Canada, the CWS is a member of the Committee on the Status of Endangered Wildlife in Canada (COSEWIC), which publishes *Canadian Species at Risk* (Ottawa: Committee on the Status of Endangered Wildlife in Canada – COSEWIC, 2000. 24 p.), and also available at their website <<http://www.cosewic.gc.ca>>. RENEW (Recovery of Nationally Endangered Wildlife) is a national program that brings all relevant Canadian agencies and groups together in an effort to rehabilitate endangered species. At the international level, the CWS is a participant in the Convention on International Trade in Endangered Species of Wild Fauna and Flora (CITES) and produces the annual *CITES Reports* (Ottawa: Canadian Wildlife Service, No.1, 1978–) that list Canadian activities on the trade in endangered species. Other examples of international activities are the Latin America Program (LAP) and the Western Hemisphere Shorebird Resource Network (WHSRN), which both focus on research about migratory birds common to both American continents. The CWS is also a participant in the Convention on Wetlands of International Importance Especially as Waterfowl Habitat, more commonly known as the Ramsar Convention (after the city in Iran where it was drafted in 1971 and came into force in 1975). The Canadian publications relevant to this convention are one by D.I. Gillespie and H. Boyd and entitled

Wetlands for the World: Canada's Ramsar Sites (Ottawa: Canadian Wildlife Service, 1991. 40 p.), and another entitled *Canada and the Ramsar Convention* (Ottawa: Canadian Wildlife Service and the North American Wetlands Conservation Council, 1995. 20 p.). In 1986, the Canadian Wildlife Service participated, jointly with the US Fish and Wildlife Service, in the development of the *North American Waterfowl Management Plan* (Ottawa: Canadian Wildlife Service, 1986. 19 p.). This document, which recognized that the 'recovery and perpetuation of waterfowl populations depends on restoring wetlands and associated ecosystems throughout the North American continent,' was signed by the Minister of the Environment and the Secretary of the Interior, and called for the establishment of a North American Waterfowl Management Plan Committee to monitor and update the plan. The first update was issued in 1994 and entitled *1994 Update to the North American Waterfowl Management Plan: Expanding the Commitment* (Washington, DC: US Department of the Interior; Ottawa: Canadian Wildlife Service, 1994. 43 p.). By then, Mexico had become a full partner in the plan. Finally, in 1992, Canada became a signatory to the United Nations Convention on Biological Diversity, which 'helps to raise public awareness of the importance of biodiversity.'

In order to emphasize the importance of wildlife to Canadians, the CWS has undertaken and publicized national surveys, or involved wildlife professionals and organizations in developing wildlife policies. The first survey, by F.L. Filion and S.W. James, was entitled *The Importance of Wildlife to Canadians: Highlights of the 1981 National Survey* (Ottawa: Canadian Wildlife Service, 1983. 40 p.). This survey, which questioned approximately 100,000 Canadians via Statistics Canada, revealed that 80% said that maintaining wildlife was important to them. There were also two spin-off secondary publications. The first, by F.L. Filion and A. Jacquemot, was entitled *The Importance of Wildlife to Canadians: An Executive Overview of the Recreational Significance of Wildlife* (Ottawa: Canadian Wildlife Service, 1985. 19 p.), and focused on the economic benefits of wildlife-related recreational activities. The second, by F.L. Filion and S. Parker, was entitled *The Importance of Wildlife to Canadians: Demand for Wildlife to 2001* (Ottawa: Canadian Wildlife Service, 1988. 29 p.) and suggested that population growth would likely 'impose new requirements on existing wildlife management programs while providing an even broader base of support for conservation strategies.'

A second survey by a federal-provincial task force was undertaken in 1987 and entitled *The Importance of Wildlife to Canadians in 1987: Highlights of a National Survey* (Ottawa: Canadian Wildlife Service, 1989. 45 p.). It noted that over 91% of Canadians were involved in some form of wildlife-related activity in 1987. This survey also covered recreational fishing, and the trapping of small mammals, for the first time. Spin-off publications from this survey were entitled *The Importance of Wildlife to Canadians in 1987: The Economic Significance of Wildlife-Related*

Recreational Activities (Ottawa: Canadian Wildlife Service, 1990. 40 p.); and *The Importance of Wildlife to Canadians in 1987: Trends in Participation in Wildlife-Related Activities, 1981–2006* (Ottawa: Canadian Wildlife Service, 1992. 36 p.). In 1991, another federal-provincial survey was undertaken, and the results were entitled *The Importance of Wildlife to Canadians: Highlights of the 1991 Survey* (Ottawa: Canadian Wildlife Service, 1993. 60 p.). This survey indicated that just over 90% of Canadians took part in one or more wildlife-related activities, and that this participation has kept up with population growth.

These surveys have provided input for the writing of a wildlife policy, overseen jointly by the Canadian Wildlife Service and the Wildlife Ministers' Council of Canada and entitled *A Wildlife Policy for Canada* (Ottawa: Canadian Wildlife Service, 1990. 29 p.). It goals are 'maintaining and restoring geological processes ... maintaining and restoring bio-diversity ...' and 'ensuring that all uses of wildlife are sustainable.' The CWS also has written a report, compiled by E. Zurbrigg and and entitled *Conserving Canada's Wildlife Habitats: 1990 to 1995: A Report on Canadian Wildlife Services Accomplishments for 1990 to 1993 and Major Plans for 1994–1995* (Ottawa: Canadian Wildlife Service, 1995. 84 p. Technical Report Series No. 226). As well, CWS publishes a newsletter entitled *Recovery: An Endangered Species Newsletter* (Ottawa: Canadian Wildlife Service, 1990–). This newsletter is also available at the CWS website <<http://www.cws-scf.ec.gc.ca/es/recovery/archive.html>>. All these national and international activities, however, are a background for the core work of the Canadian Wildlife Service, which is wildlife research.

The research of the CWS is broad in both scope and geographical range. Because of its early responsibility for migratory birds, considerable research has been done on their breeding biology, as well as population studies in the form of aerial and land surveys. The surveys have more recently been documented in the atlas format. Even the sport side of hunters, and the hunting of waterfowl, have been studied. It is interesting to note that the biological studies of large mammals, particularly in the Arctic, are a relatively recent activity of the Canadian Wildlife Service. Much of this work dates from the late 1950s and 1960s, as the service expanded its activities. In this area of research, the studies of the Arctic caribou herds have been particularly important. Related to the mammal studies are the wildlife management studies undertaken in the national parks, which are wildlife sanctuaries. The work of the 1970s, and onward, shows an emphasis on environmental impact studies, particularly in the Arctic, as well as increased work on the impact of contaminant residues in the environment, and their effect on wildlife. Finally, as the wetland areas of Canada, including the prairie potholes, have continued to shrink, the CWS has studied this impact on wildlife. All this wide-ranging research is documented, and published in papers, reports, and monographs published by the service. As well, its research is reported in the international research literature devoted to avian and wildlife studies.

II Publications

A Canadian Wildlife Service

(a) Annual Reports

As mentioned above, the Canadian Wildlife Service has always been part of another major government department and its activities reported in a section of the appropriate departmental annual report. From 1947 to 1949, the Service was part of the Lands, Parks and Forestry Branch of the Department of Mines and Resources; then from 1949/50 to 1952/53, as part of the Department of Resources and Development then from 1953/54 to 1965/66, as part of the Department of Northern Affairs and National Resources; then as part of the Department of Indian Affairs and Northern Development from 1966/67 to 1969/70.

In 1971, the CWS became part of Environment Canada – where it has resided ever since under various groupings such as Department of the Environment, 1971–75; Fisheries and Environment Canada, 1976/77–1978/79; and Environment Canada, 1980–. However, from 1986 to 1989, the CWS did publish an *Annual Review* (Ottawa: Canadian Wildlife Service, 1986/87–1988/89. 3 v.). This review discussed in detail the work of its Migratory Birds and Wildlife Branch, the North American Waterfowl Management Plan, the Wildlife Toxicology and Survey's Branch, the newer State of the Environment Reporting Branch, and the work of the Western and Northern Region and the Pacific and Yukon Region. Earlier, the CWS also published a review entitled *Canadian Wildlife Service '66* (Ottawa: Canadian Wildlife Service, 1967. 96 p.); and *Canadian Wildlife Service '71* (Ottawa: Canadian Wildlife Service, 1971. 87 p.). The later review, in particular, provided excellent background information on the work of the service, and accounts of some of the large mammal studies being carried out at that time. Currently, the Canadian Wildlife Service maintains an overview entitled *Focus on the Canadian Wildlife Service* at its website <<http://www.cws-scf.ec,gc/focus_e.html>>.

(b) Wildlife Management Bulletins

The actual research work, apart from that published in the international scientific journals, has always been well documented in a range of publication series of the Canadian Wildlife Service. The first of these series publication were the Wildlife

Management Bulletins that were published between 1950 and 1966. There were three series issued. Series 1 concerned mammal research, and 19 papers, or small monographs, were published. An example in this series, by W. Fuller, was entitled *Wolf Control Operations, Wood Buffalo National Park, 1951–52* (Ottawa: Canadian Wildlife Service, 1955. Canadian Wildlife Service Wildlife Management Bulletin. Series 1, No. 11). Series 2 concerned bird research, and 10 papers were published. An example from this series was I.M. Cowan's study entitled *Birds of Jasper National Park, Alberta, Canada* (Ottawa: Canadian Wildlife Service, 1955. 66 p. Canadian Wildlife Service Wildlife Management Bulletin. Series 2, No. 8). Series 3 concerned fish, and only five were published. An example from this series, by J.P. Currier and F.H. Schultz, was entitled *Investigation of Lake Trout and Common Whitefish in Waterton Lakes, Waterton National Park* (Ottawa: Canadian Wildlife Service, 1957. 41 p. Canadian Wildlife Service Wildlife Management Bulletin. Series 3, No. 5).

The Wildlife Management Bulletins were superseded by a new series of ongoing publications: the Progress Notes; the Occasional Papers; the Report Series; and the Monographs.

(c) The Progress Notes

The Progress Notes were first issued in 1966, and to date, there have been 208. These notes contain 'timely data and conclusions and are presented as a service to other wildlife biologists and agencies.' Many publications in this series are data collections. An example, by F.G. Cooch and K. Newell, is entitled *Report on 1976 Sales of the Canada Migratory Game Bird Hunting Permit, Waterfowl Harvest and Hunter Activity* (Ottawa: Canadian Wildlife Service, 1978. 8 p. Canadian Wildlife Service Progress Note No. 81).

(d) Occasional Papers

The Occasional Papers are 'reports on specific aspects of Canadian Wildlife Service projects.' The first appeared in 1961, and to date 98 have been published. An example from this series, by D.A. Benson, is entitled *A Wildlife Biologist Looks at Sampling Data Processing and Computers* (Ottawa: Canadian Wildlife Service, 1964. 16 p. Canadian Wildlife Service Occasional Paper No. 6).

(e) Report Series

The Report Series is a major publication of the CWS. It first appeared in 1966, and to date 47 have been published. They provide the complete results of Canadian Wildlife Service field research studies. An example by A.J. Gaston and

D.K. Cairns is entitled *A Natural History of Digges Sound* (Ottawa: Canadian Wildlife Service, 1985. 62 p. Canadian Wildlife Service Report Series No. 46).

(f) Monographs

Finally, the Monographs describe complete studies of wildlife species indigenous to Canada. The first one appeared in 1960 and to date only six have been published. An example of a monograph by A.J. Erskine is entitled *Buffleheads* (Ottawa: Canadian Wildlife Service, 1971. 240 p. Canadian Wildlife Service Monograph No. 4).

(g) Technical Report Series

Technical Report Series, issued in typed form by the specific regions, contain technical and scientific information of interest to a limited audience, or too extensive for inclusion in scientific journals. A example from this series by W. Threlfall and R.T. Goudie, is entitled *Marine Bird Observations at Placentia Bay and Cape St. Mary's, 1973–76* (St John's: Canadian Wildlife Service, Atlantic Region, 1986. 17 p. Canadian Wildlife Service, Atlantic Region Technical Report. Series No. 6).

(h) Hinterland Who's Who

Besides the above-listed series publications, the Canadian Wildlife Service has also issued a unique series of 89 pamphlets entitled Hinterland Who's Who. This unnumbered series, which is important for any library to have as a reference series, includes endangered species, urban visitors, feeder visitors, ducks, geese, and swans, water birds, land birds, large mammals, smaller mammals, and general wildlife information. The information in each pamphlet may include the taxonomic description of the species, its life and reproductive history, its current status, and a bibliography. An example from the general wildlife information category is entitled *The Benefits of Wildlife* (Ottawa: Canadian Wildlife Service, 1989. 4 p. Canadian Wildlife Service Hinterland Who's Who). Eighty-four of the pamphlets in this series are also available through the CWS website <<http://www.cws-scf.ec.gc.ca>>.

(i) State of the Environment Series

Most recently the Canadian Wildlife Service has been involved, jointly with Environment Canada, in a State of the Environment information series. This series includes newsletters (e.g., *State of the Environment Reporting Newsletter,*

No. 1, 1988–); a *State of the Environment Bulletin* (e.g., *Urban Water: Municipal Water Use and Waste Water Treatment* (Ottawa: Environment Canada, 1998. 4 l. SOE Bulletin No. 98–4)); and a monograph series. An example from the monograph series, by D. Noble, is entitled *Contaminants in Canadian Seabirds* (Ottawa: Environment Canada, 1990. 75 p. SOE Report No. 90–2).

(j) Miscellaneous Publications

The Canadian Wildlife Service also publishes a number of distinct miscellaneous publications. Some are monographs, such as the *Revised Atlas of Eastern Canadian Seabirds* (Ottawa: Canadian Wildlife Service, 1986–). Some pertain to migratory birds, such as the Canadian Migratory Convention Act, and the annual Canada Migratory Bird Regulations. Some are unlisted environmental impact studies, carried out in the 1970s. An example of this type of study, by G.W. Calef and D.C. Heard, is entitled *The Status of Three Tundra Wintering Herds in Northeastern Mainland Northwest Territories* (Ottawa: Canadian Wildlife Service, 1980. 22 p. ESCOM Report No. AI-44). Others are in looseleaf form, such as *Careers in Wildlife Conservation* (Ottawa: Canadian Wildlife Service, n.d. 6 p.), or in pamphlet form, such as *Focus on the Canadian Wildlife Service* (Ottawa: Canadian Wildlife Service, 1991).

B Canadian Museum of Nature

Like the Canadian Wildlife Service, the Canadian Museum of Nature issues numbered series publications, unnumbered series publications, monographs, and some co-published monographs.

(a) Publications in the Natural Sciences

This is their series, also known as the Syllogeus Series. This series began in 1983 with the publication by J.J. Dickinson entitled *The Systematics and Distributional Ecology of the Superfamily Ampeliscoidea (Amphipoda: Gammaridea) in the Northeastern Pacific Region. II: The Genera Byblis and Haploops* (Ottawa: National Museum of Natural Sciences, 1983. 38 p. Publications in Natural Sciences No. 1). This series supersedes, first, the *Publications in Biological Oceanography*, No. 1, 1970–No. 11, 1982 (e.g., Clark, A.H., *Molluscs from Baffin Bay and the Northern North Atlantic Ocean*. Ottawa: National Museum of Natural Sciences, 1974. 23 p. Publications in Biological Oceanography No. 7); second, the *Publications in Botany*, No. 1, 1969–No. 12, 1982 (e.g., Lee, P.K., *A Catalogue of the Marine Algae of the Canadian Arctic*. Ottawa: National Museum of Natural Sciences,

1980. 82 p. *Publications in Botany*, No. 9); and third, the *Publications in Zoology*, No. 1, 1969–No. 17, 1982 (e.g., Youngman, P.M., *Mammals of the Yukon Territory*. Ottawa: National Museum of Natural Sciences, 1975. 192 p. Publications in Zoology, No. 10). And these three series superseded the *Bulletins*, which were issued from 1910 to 1957 (e.g., Soper, J.W., *A Faunal Investigation of Southern Baffin Island*. Ottawa: Canada Department of Mines, 1928. 143 p. National Museum of Canada Bulletin No. 53). The latter then split from 1957 to 1966, into the untitled *Contributions to Botany*, and the *Contributions to Zoology*.

(b) Unnumbered Series

The major unnumbered series publications of the Canadian Museum of Nature are the Miscellaneous Publications, and the Bibliographia Invertebratorum. The miscellaneous publications tend to be botanical in nature (e.g., Argus, G.W. and K.M. Pryor, *Rare Vascular Plants in Canada: Our Natural Heritage*. Ottawa: Canadian Museum of Nature, 1990. 278 p.); and the Bibliographia Invertebratorum is, at present, an eight-volume series (e.g., Laubitz, D.R. and I. Sutherland, *Bibliographia Invertebratorum Aquaticorum Canadensium, Volume 1: Index Generum*. Ottawa: National Museum of Natural Sciences, 1983. 56 p.).

Other unnumbered series are the *Handbooks of Canadian Mammals* (e.g. Van Zyll de Jong, C.G., *Marsupials and Insectivores*. Ottawa: National Museum of Natural Sciences, 1983. 210 p. Handbook of Canadian Mammals, vol. 1); the *Edible Wild Plants of Canada Series* (e.g., Turner, N.J., and A.F. Szczawinski, *Wild Coffee and Tea Substitutes in Canada*. Ottawa: National Museum of Natural Sciences, 1978. 112 p.), the *Children Series* (e.g., Gordon, B., *What Is a Bird?* Ottawa: National Museum of Natural Sciences, 1986. 48 p.); and the *Natural History Notebook Series* (Ottawa: National Museum of Natural Sciences, No. 1, 1977–No. 5, 1981). This illustrated series covered living and extinct birds and animals worldwide. It is perhaps a precursor of the Hinterland Who's Who of the Canadian Wildlife Service.

(c) Monographs

The monographs and co-published monographs of the Canadian Museum of Nature include the publication by F.R. Cook entitled *Introduction to Canadian Amphibians and Reptiles* (Ottawa: Canadian Museum of Nature, 1984. 200 p.); and that by A.F. Szczawinski entitled *Edible Garden Weeds of Canada* (Ottawa and Markham, Ontario: Museum of Nature and Fitzhenry and Whiteside, 1988. 184 p.).

III Wildlife Research

A Breeding Biology Research

The breeding biology research on migratory birds is a long-term project, and includes ducks, geese, swans, waterbirds, and landbirds. Geographically, the studies range from the Atlantic to the Pacific, and to Arctic coasts, and throughout the prairie migration routes of these birds. One major study has involved the Ross' goose (*Anser rossii* Cassin). The Ross' goose is important for three reasons. The goose is North America's smallest Arctic nesting goose, and during the 1950s there was concern about the species' survival, because of the low numbers reported migrating through the Canadian prairie provinces, and low numbers on their wintering ground in California. As well, there was a lack of knowledge concerning its breeding habits, which made management of the species haphazard at best.

Studies were begun in 1963, in a nesting area of the Perry River region of the Northwest Territories that is northwest of the latter's Mackenzie-Keewatin border. The objective was to develop an understanding of the reproductive biology of the Ross' goose on its nesting grounds, to study its post-breeding biology, and to study interspecific relationships in the nesting area. This research study was carried out by J.P. Ryder and entitled *The Breeding Biology of Ross' Goose in the Perry River Region, Northwest Territories* (Ottawa: Canadian Wildlife Service, 1967. 56 p. Canadian Wildlife Service Report Series No. 3). Included in the study, which was carried out during the summers of 1963 and 1964, were descriptions of the explorations, climate, vegetation, fauna, and aboriginal peoples inhabiting the area. Specifically, the author studied the reproductive biology of the goose, which included the spring migration, nesting, the egg-laying period, incubation period, and hatching success. As well, the post-breeding biology, which covered movements, the moult, physical descriptions, including colour, sex ratio, parasitism, and predation, were studied. The outcome of these initial studies was an indication that the reproductive phase of the Ross' goose did not appear to be a limiting factor, as nesting and hatching success was high, and mortality rates for adults and young was low. Nor did the severe climate, or the biotic factors, affect markedly its survival rate. However, the author suggested that the unstable ecosystem such as late seasons and predation could cause severe annual population drops.

Ryder continued studies of the Ross' goose after 1964, with visual aerial surveys, from 1965 to 1967, of the Queen Maud Gulf Migratory Bird Sanctuary in which the Perry River was located. These population studies, undertaken to delineate the breeding range of the Ross' goose and determine the factors that defined suitable nesting habitat for the species, documented both Ross' and snow geese colonies at Kanack Lake and 34 other colonies, were reported in the research article entitled *Nesting Colonies of Ross' Goose* (*Auk*, v. 86, 1969. pp. 282–92); *Distribution and Breeding Biology of the Lesser Snow Goose in Central Arctic Canada* (*Wildfowl*, v. 22, 1971. pp. 18–28); and *Biology of Nesting Ross' Geese* (*Ardea*, v. 60, 1972. pp. 185–215). Ryder estimated that the Ross' goose population in those years was just over 30,000. Visual aerial surveys were also undertaken by R.H. Kerbes *et. al.*, in 1976, and the results reported in the research paper entitled *Ross' Goose and Lesser Snow Goose Colonies in the Central Canadian Arctic* (*Canadian Journal of Zoology*, v. 61, 1983. pp. 168–73). These surveys indicated a Ross' goose population of just over 77,000. R.H. Kerbes also carried out further aerial photograph surveys, in 1982 and 1988, to determine total numbers of nesting Ross' geese as well as snow geese. By 1988, the total number of Ross' geese alone was estimated at 188,000, indicating that the Ross' Goose had made a dramatic recovery from a total population as low as 2,000 in the 1950s. These findings were reported in Kerbes' paper entitled *Colonies and Numbers of Ross' Geese and Lesser Snow Geese in the Queen Maud Gulf Migratory Bird Sanctuary* (Ottawa: Canadian Wildlife Service, 1994. 47 p. Canadian Wildlife Service Occasional Paper No. 81).

Unlike the Ross' goose, the common eider duck, which consists of three subspecies of six breeding populations – the northern eider (*Somateria mollesima borealis*, the Hudson Bay eider (*S.m. sedentaria*), and the American eider (*S.m. dresseri*) – has been well studied, for it is not only an abundant waterfowl species, but is of potential importance for the development of an eiderdown industry for Inuit populations. Accordingly, the Canadian Wildlife Service undertook a two-year study of the breeding biology and management of the duck, near Cape Dorset on Baffin Island during the mid-1950s, in order to obtain a better understanding of the eider ducks of that region.

In the study by F.G. Cooch, entitled *The Breeding Biology and Management of the Northern Eider (*Somateria mollesima borealis*) in the Cape Dorset Area, Northwest Territories* (Ottawa: Canadian Wildlife Service, 1965. 68 p. Canadian Wildlife Service. Wildlife Management Bulletin, Series 2, No. 10), the spring migration, courtship and pairing, nesting, including egg laying, incubation and down deposition, natural predation and egg loss were studied. Also, eiderdown collection techniques and the Cape Dorset eiderdown industry were examined. The author concluded that the development of a viable industry on Cape Dorset would require habitat improvements, and the production of eider duck nesting sites of which four on the cape could be considered potential eider farm areas. As well, the author said that complete protection should be given to existing eider populations to guard against their population reduction. The author also recommended times and methods for collecting down from the nests in order to maximize the amount of down that could be collected. Finally, F.G. Cooch questioned whether down collecting as a source of income would be useful to aboriginal populations on Cape Dorset, because their populations were nomadic in nature.

Twenty years later, after this first study of the eider duck, the Canadian Wildlife Service requested researchers involved with the eider duck to submit papers, which could be published as a volume that would review the current status of eider ducks in Canada. Since 1916, when the Migratory Bird Convention was signed by the United States and Great Britain (on behalf of Canada), the eider duck had benefited from limited protection, because their numbers had been dwindling since 1880. The purpose of the paper was an attempt to give an account of the success provided by 70 years of protection. Eighteen papers were accepted for publication, and the topics covered were the eider's distribution and abundance, the races of common eiders, their distribution in winter, the ecology of eiders, and their use by people. The papers were edited by A. Reed and published with the title *Eider Ducks in Canada* (Ottawa: Canadian Wildlife Service, 1986. 176 p. Canadian Wildlife Service Report Series No. 47). Overall, the papers suggested that the success of the eider duck, after 70 years of protection, was 'patchy.' In the Maritimes and the St Lawrence Estuary, the common eider was again common, and the ducks were more numerous in the Arctic, but with problems. In the Arctic, the eiders were over-exploited in some areas, because of the indifferent taking of eiderdown from nests, or the taking of eggs and the shooting of adults that take two to four years to reach maturity, by people living in a subsistence economy. By comparison, the development of efficient and protected eider farming, as carried out in Iceland, was suggested as a possibility benefiting both ducks and some Inuit areas.

Since the mid-1980s, further information has been collected on the common eider, as well as the king eider, particularly in the area of the western Canadian Arctic. The impetus for this work was, first, the exploration for oil and gas in the Canadian Beaufort Sea, which led to public hearings in the mid-1980s to examine the potential environmental impact of off-shore oil and gas production and the effect of an oil spill in the narrow system of off-shore open water areas in the Beaufort Sea on sea ducks during spring migration. Through the Northern Oil and Gas Action Program, four years of data were collected on the movement of eiders through the western Canadian Arctic during spring migration. Second, the Inuvialuit land claim and settlement specified that the Inuvialuit had preferential rights to the allowable harvest of migratory birds in the western Canadian Arctic. Because the current data on eiders was inadequate to manage the species in a way that would ensure that its population remained stable, a series of waterfowl breeding pairs surveys was undertaken throughout most of the Inuvialuit Settlement Region. The result of the data collection was another series of papers, edited by D.L. Dickson, and entitled *King and Common Eiders of the Western Canadian*

Arctic (Ottawa: Canadian Wildlife Service, 1997. 75 p. Canadian Wildlife Service Occasional Paper No. 94). The publication consisted of seven papers, representing the data from these two groups of surveys.

Other duck studies have looked at the breeding and feeding habitats of duck species using the prairie potholes as their home range. One such study involved the common shoveler (*Anas clypeata*). The other study involved young pintails (*Anas acuta*), gadwalls (*Anas strepera*), American widgeons (*Mareca americana*), and lesser scaups (*Aythya affinis*).

The common shoveler is widespread throughout western North America. Its breeding range extends from Alaska, south in an arc, from Oregon to Utah and Wisconsin, then northwesterly back to Alaska. The ducks winter in the valleys and coastal regions of California, Mexico, Texas, and South Carolina. From their wintering grounds, the ducks migrate in the spring, back to specific breeding areas in the potholes of the prairies, which are a suitable home range. Because the breeding ecology of the species was not well known, and the concept of a distinctive home range unique, the Canadian Wildlife Service chose an area east of Calgary, Alberta, in order to study these ducks over a four-year period. The study was carried out by H.J. Poston and published with the title *Home Range and Breeding Biology of the Shoveler* (Ottawa: Canadian Wildlife Service, 1974. 49 p. Canadian Wildlife Service Report Series No. 25). Over the four-year period, in the late 1960s, the author found that shoveler breeding pairs occupied a core home area, consisting of the nesting site and peripheral ponds averaging about 76 acres. These core areas appeared to supply the basic requirements of food, nesting, and isolation. The shoveler pairs also spaced themselves in an orderly manner, and the breeding pair density depended upon the size of the habitat available. Hence it was important to preserve the prairie pothole environment for this waterfowl species at a time when wetlands were being constantly reduced, and the demand for waterfowl was increasing.

After the documentation of the breeding biology of these ducks, the author suggested that more work needed to be done on the habitat characteristics of the wetland potholes that made the latter so attractive to waterfowl. Nevertheless, some work had already been carried out by the Canadian Wildlife Service, in order to develop an understanding of the food requirements of waterfowl in the wetland potholes of the prairies. It is interesting to note that in the 1960s, not enough was known about the diet of ducks, particularly the flightless young, or ducklings, whose diet would change as they grew. In order to study duckling diets, the Canadian Wildlife Service carried out research from 1963 to 1967, again, in the pothole area east of Calgary, Alberta. The research involved young pintails, gadwalls, American widgeons, and lesser scaups, in order to determine the nutritional composition of duck foods. These ducks were chosen because they are relatively common to the prairie breeding grounds, and are important game ducks. As well, the four species use different parts of the same ponds. Pintails preferred the shallow edges, gadwalls and widgeons used the areas away from the shore, and the scaups, being diving ducks, used the deeper areas.

This study was undertaken by L.G. Sugden, and the results published with the title *Feeding Ecology of Pintail, Gadwall, American Widgeon and Lesser Scaup Ducklings* (Ottawa: Canadian Wildlife Service, 1973. 43 p. Canadian Wildlife Service Report Series No. 24). The author determined that aged up to five days, pintails ate mostly insects captured on the water surface, but after that the older ducklings ate aquatic invertebrates and plants. Gadwalls first ate surface invertebrates, then more aquatic invertebrates, and also plants. But after three weeks, they became essentially herbivorous. Widgeons, on the other hand, ate primarily plants, and young scaups ate primarily invertebrates. The diet was determined by examining the esophagus-proventriculus contents of over 100 ducklings from each species. In regard to the nutrient requirements of these ducks, the study of the composition of the duck foods indicated that few foods by themselves supplied the basic nutrients in adequate proportions. A mixture of foods was necessary to supply the ducklings with a nutritionally balanced diet, which had to include protein, amino acids, vitamins, and minerals such as calcium and phosphorus.

Besides ducks, other major studies of migratory birds have been carried out and published as monographs by the Canadian Wildlife Service. The first of these was a comprehensive study of the common snipe (*Capella gallinago*) which is one of two North American game birds (the other is the woodcock). The second of these monographs was a study of a high-Arctic seabird – the thick-billed murre (*Uria lomvia*).

The common snipe, which has nearly worldwide distribution, is a much sought after game bird because its zig-zag flight at speeds up to 100 kilometres per hour makes it a challenge for sportsmen. In the summer months, it inhabits bogs, swamps, and fens, which are beyond human habitation, but during migration and winter, it inhabits wet pastures and farmlands close to populated areas. During the 20th century, hunting helped to decimate the population of many shore birds, but the snipe tended to be saved from this fate, because of its summer habitat. However, droughts in its breeding area, or severe cold in its wintering area, did affect the stock. The United States closed the season on snipe hunting from the early 1940s to the early 1950s, and Canada provided a partial ban by reducing the bag limit during the hunting season. Like other duck studies mentioned above, it was important to study the biology of the common snipe in North America, which is a sub-species called the Wilson's snipe (*Capella gallinago delicata*), in order to know better how to manage it as a game bird.

The major study of snipes by the Canadian Wildlife Service was undertaken by L.M. Tuck, and the results were published as a monograph entitled *The Snipes: A Study of the Genus Capella* (Ottawa: Canadian Wildlife Service, 1972. 428 p. Canadian Wildlife Service Monograph Series No. 5). In this work, the background chapters provided a thorough review of the species, which included its breeding habitat, its morphology, including anatomy and physiology, and its taxonomy and geographic distribution. The actual research was carried out, primarily in Newfoundland, with additional studies undertaken in northern Manitoba, and Ontario, where the bird's breeding habitat exists. Winter studies were also undertaken in Louisiana, Bermuda, Trinidad, and Venezuela, which are the winter habitats of the Wilson's snipe.

The research also documented the snipe's spring and fall migration and its winter habitat. The study of its breeding biology included pair formation and mating behaviour, the creation of the nest, eggs and incubation, and the care and growth of the chick. As well, the feeding habits and food, the snipe's plumages, moults, and weights were covered. Finally, its population structure, parasites, and natural mortality were covered, including the hunting and management of the snipe. At the end of this major work, the author concluded that management of the snipe, in its vast breeding area of northern North America, was impractical and not necessary, 'since the succession of vegetation in transitional northern organic soils normally provides the required habitat. In the snipe's wintering range in the sub-tropics, existing agricultural practices of reclaiming marshes for pasture and irrigation techniques for raising crops such as rice and cotton provide more and more winter habitat. I am satisfied that there is now an abundant population of common snipe in North America which can be safely harvested and in great numbers.'

Unlike the common snipe, the thick-billed murre (*Uria lomvia*) is a highly specialized seabird, once regarded as one of the most numerous seabirds in the world, according to L.M. Tuck, in an earlier monograph entitled *The Murres, Their Distribution, Populations and Biology: A Study of the Genus Uria* (Ottawa: Canadian Wildlife Service, 1961. 260 p. Canadian Wildlife Service Monograph Series No. 1). At that time, the author estimated a total world population for the species at about 20 million birds, 10 million of which occurred in the western Atlantic, between the eastern Canadian Arctic and the coast of western Greenland, during the breeding season. Since then, however, substantial decreases, of up to 40%, appear to have taken place over a large portion of the thick-billed murre's range. These changes were noted by P.N. Nettleship, in the symposium paper entitled *Seabird Resources of Eastern Canada: Status, Problems and Prospects*. In: *Canada's Endangered Species and Habitats*. Edited by T. Mosquin and C. Suchal (Ottawa: Canadian Naturalist Federation, 1977. pp. 96–108. Canadian Naturalist Federation Special Publication No. 6). The species had suffered heavy mortality, both during and after its breeding season, because of direct factors such as shooting, drowning in fish nets, oil fouling, general disturbance, and indirect factors such as toxic chemical poisoning or commercial fisheries development that reduced the birds' food supply. The thick-billed murre is an example of an 'organism' that requires a narrow environmental regime and is highly vulnerable to decimation because the murre's own evolution has made it incapable of rapidly changing conditions.

Because of these concerns, the Canadian Wildlife Service undertook a major study of the thick-billed murres of Prince Leopold Island. This was part of a study program, undertaken from 1975 to 1977, on the distribution and ecology of seabirds in eastern Canada. The study was undertaken by A.J. Gaston and D.N. Nettleship, and the objective was to gather as much information as possible on the reproductive biology and ecological requirements of the thick-billed murres breeding in a single location in Lancaster Sound, the gateway to the Northwest Passage. The authors, in particular, wanted to study the aspects of ecology and behaviour which might allow researchers to predict the effects of environmental changes, particularly those associated with human activities, and to use that knowledge to evaluate methods and provide guidelines for future researchers.

This research was published as a monograph, and entitled *The Thick-Billed Murres of Prince Leopold Island* (Ottawa: Canadian Wildlife Service, 1981. 350 p. Canadian Wildlife Service Monograph Series No. 6). This examined the general behaviour of breeding and non-breeding birds, and aggression within the colony. As well, breeding behaviour such as the time of breeding, egg-laying, hatching, and the incubation period, were discussed, along with the development of the young and their feeding rates, and their diet. Then, the adults were studied for size and physical condition and their diet and feeding areas. This was considered to be an important base-line document that would make it comparatively easy for investigators 20, 40, or 100 years from now to make comparisons between the state of some of the biological systems of Lancaster Sound in the 21st century and how they were faring in 1975–77. Since publishing this monograph, A.J. Gaston, and others, have undertaken further studies of thick-billed murres. These studies are, first, by A.J. Gaston, entitled *Migration of Juvenile Thick-Billed Murres* (Uria lomvia) *through Hudson Strait (Canada) in 1980* (*Canadian Field Naturalist*, v. 96, 1982. pp. 30–34); second, by A.J. Gaston and D.N. Nettleship, entitled *Factors Determining Seasonal Changes in Attendance at Colonies of the Thick-Billed Murre* Uria lomvia (*Auk*, v. 99, 1982. pp. 468–73); third, by A.J. Gaston, entitled *The Growth of the Thick-Billed Murre* (Uria lomvia) *Chicks at Colonies in Hudson Strait (Canada): Inter-Colony and Intra-Colony Variation* (*Canadian Journal of Zoology*, v. 61, 1983. pp. 2465–75); fourth, by A.J. Gaston, entitled *Energy Invested in*

*Reproduction by Thick-Billed Murres (*Uria lomvia*)* (*Auk*, v. 102, 1985. pp. 447–58); fifth, by A.J. Gaston, entitled *The Diet of Thick-Billed Murre (*Uria lomvia*) Chicks in the Eastern Canadian Arctic* (*Auk*, v. 102, 1985. pp. 727–43); sixth, by A.J. Gaston and L.N. de Forest, entitled *Monitoring Thick-Billed Murre Populations at Colonies in Northern Hudson Bay, 1972–92* (Ottawa: Canadian Wildlife Service, 1993. 16 p. Canadian Wildlife Service Occasional Paper No. 80); seventh, by L.N. de Forest and A.J. Gaston, entitled *The Effect of Age on Timing of Breeding and Reproductive Success in the Thick-Billed Murre* (*Ecology*, v. 77, 1996. pp. 1501–11); and an eighth, by G.M. Donaldson and A.J. Gaston, entitled *Winter Distributions of Thick-Billed Murres from the Eastern Canadian Arctic and Western Greenland in Relation to Age and Time of Year* (Ottawa: Canadian Wildlife Service, 1997. 26 p. Canadian Wildlife Service Occasional Paper No. 96).

B Hunting Research

An important part of the research on migratory birds is the study of the impact of hunting and the attitude of hunters themselves. The information gathered from this work assists in the management, development, and conservation of migratory birds, based upon the Canada Migratory Convention Act of 1917. As previously mentioned, Great Britain (on behalf of Canada) signed the Migratory Birds Convention, which was implemented in Canada by the passing of the 1917 Act. The objective and purpose of the Convention, the Act itself, and regulations made pursuant to the Act is the preservation of migratory birds by protecting them during their nesting season, and when flying to and from their breeding grounds, through the establishment of annual hunting season dates and bag and possession limits. These regulations are published annually as part of the Canada Migratory Birds Convention Act, 1994 and entitled *Migratory Bird Regulations: (Schedule 1 only) C.R.C, ch. 1035* (Ottawa: Public Works and Government Services Canada, 199[6]. 52 p.). This publication lists all the regulations pertaining to the hunting of migratory game birds in Canada during the most recent hunting season. It includes a complete summary of hunting zones, dates, and bag limits for each province and territory. They are also posted on the CWS website <<http://www.cws-scf.ec.gc.ca>>. The last consolidation of the regulations was issued as *Departmental Consolidation, Migratory Birds Regulations: C.R.C., c. 1035* (Ottawa: Canadian Wildlife Service, 1997. 49 p.)

Enforcement is carried out by regional Canadian Wildlife Service enforcement coordinators, who advise federal (RCMP) and provincial enforcement agencies on the Migratory Birds Convention Act and Regulations. These coordinators instruct provincial enforcement officers – the people who have actual contact with hunters – on the purpose of the Act and Regulations, bird identification, and enforcement methods. Birds under protection of the Act are listed in the CWS publication entitled *List of Birds Protected in Canada under the Migratory Birds Convention Act* (Ottawa: Canadian Wildlife Service, 1980. 41 p. Canadian Wildlife Service Occasional Paper No. 1). This publication 'lists all species of migratory bird included in the Convention and found in Canada, the continental United States and adjacent waters to a distance of 320 kilometres from land or half the distance to a neighbouring country whichever is the least.' The listing is by taxonomic grouping of game and non-game birds, and includes each bird's scientific name. As well, the status of each bird is given (CN – native breeds; V – regular visitors; A – accidental; does not breed; I – introduced; and E – extinct). Species believed to be endangered are marked by an asterisk (*), and may be subject to additional Canadian import/export controls under the Convention on International Trade in Endangered Species of Wild Fauna and Flora (CITES).

The documentation of the annual hunting season is described in the annual publication entitled *Migratory Game Birds Harvested in Canada during the [19–] Hunting Season* (Ottawa: Canadian Wildlife Service, 1987–). Previous annual publications were entitled *Migratory Birds Killed in Canada during the [19–] Season* (Ottawa: Canadian Wildlife Service, 1978–1986); *Species of Waterfowl Killed in Canada During the [19– and 19–] Seasons* (Ottawa: Canadian Wildlife Service, 1968–70; 1976–77); and *Species of Waterfowl and Age and Sex Ratios of Ducks Harvested in Canada during the [19–] Season* (Ottawa: Canadian Wildlife Service, 1971–75). It is interesting to note that the systematic collection of information about the 'harvest' began nearly 50 years after the signing of the 1916 Convention. Until then, decisions about the appropriate bag limit were made without extensive information, or the hunters involved.

The source that triggers this information is the Migratory Bird Hunting Permit, which hunters have been obliged to purchase since 1966. The list of permit purchasers has been computerized since 1968, and it generates a mailing list each year to selected permit holders from the previous year. This mailing list is used for the National Harvest Survey and the Species Composition Survey, both of which are used for the generation of the annual published statistics described above. The permits and surveys have been described by D.A. Benson in the paper entitled *The Migratory Game Permit and Related Surveys* (Ottawa: Canadian Wildlife Service, 1971. 15 p. Canadian Wildlife Service Occasional Paper No. 11). Comments on the analysis of survey results have been written by A.R. Sen in a series of research papers entitled *Relative Efficiency of the Sampling Systems in the Canadian Waterfowl Harvest Survey* (*Biometrics*, v. 26, 1970. pp. 315–26); *Increased Precision in the Canadian Waterfowl Harvest Surveys Through Successive Sampling* (*Journal of Wildlife Management*, v. 35, 1971. pp. 664–68); and *Response*

Errors in Canadian Waterfowl Surveys (*Journal of Wildlife Management*, v. 37, 1973. pp. 485–91). The surveys have also been a source of information on specific species, as described in the paper by H. Boyd, entitled *Mortality Rates of Hudson Bay Snow Geese, 1967–74* (Ottawa: Canadian Wildlife Service, 1976. 4 p. Canadian Wildlife Service Progress Note No. 61).

The survey has also been a source of information for the study of hunters themselves. An overview of this type of study has been described in a series of papers, edited by H. Boyd and G.H. Finney, entitled *Migratory Game Bird Hunters and Hunting in Canada* (Ottawa: Canadian Wildlife Service, 1978. 127 p. Canadian Wildlife Service Report Series No. 43). One paper in this collection, by F.L. Filion, entitled *Demographic and Socio-Economic Characteristics of Holders of Canada Migratory Game Bird Hunting permits* (*Ibid.* pp. 42–51), stated that 'hunters tended to be younger and better educated than the total male population ... and ... hunters over-represented blue-collar occupational groups and residents in rural and small urban centres.' Another publication, pertaining to hunters, has also been written by S.A. Parker and F.L. Filion, and entitled *Human Dimensions of Migratory Game-Bird Hunting in Canada* (Ottawa: Canadian Wildlife Service, 1984. 34 p. Canadian Wildlife Service Occasional Paper No. 51). The importance of the wording of the survey has been commented on by F.L. Filion in the paper entitled *Importance of Question Wording and Response Burden in Hunter Surveys* (*Journal of Wildlife Management*, v. 45, 1981. pp. 873–82). The author said that variations in the question statement provided significant differences in factual answers, and the requirement for more detailed activity reports generated more accuracy on hunter activity.

Hunter performance, itself, was first studied by H. Boyd over a two-year period, and described in the paper entitled *Observations on Duck Hunting in Eastern Canada in 1968 and 1969* (Ottawa: Canadian Wildlife Service, 1971. 24 p. Canadian Wildlife Service Occasional Paper No. 12). Three aspects of hunting were examined: first, the effects of some characteristics of ducks themselves, including species, body size, and flock size, upon their vulnerability to hunting; second, the effects of the time of year, and local weather, upon the success of hunters; and third, the effects of some hunting practices (such as the use of decoys, dogs, and the number of hunters in a party) upon performance.

Documented observations indicated that hunters fired at most ducks considered to be in range, and fired most readily at large ducks, regardless of their species status. The average shots fired, per single duck, was 5.5 shots and the ratio of birds brought down, to birds fired at, fell with the increasing size of the flock. Finally, duck hunters tended to shoot in pairs, and those with dogs fired more often and bagged more ducks than did hunters without dogs. Based upon these observations, the author suggested that 'population models for use in managing waterfowl hunting should not be restricted to ... any single popular species, because none is of paramount importance and none is typical.'

Additional studies on hunting performance were undertaken by G.S. Hochbaum and C.J. Walters, and the findings were entitled *Components of Hunting Mortality in Ducks: A Management Analysis* (Ottawa: Canadian Wildlife Service, 1984. 27 p. Canadian Wildlife Service Occasional Paper No. 52). This study evaluated waterfowl hunting, and kill, on the Delta Marsh, Manitoba, which is along the south shore of Lake Manitoba, and one of the major waterfowl harvest areas on the Canadian prairies. The topics evaluated were the vulnerability of different species in relation to hunter effort, determinants of bag size, determinants of hunter effort, effects of weather on kill, and whether hunters were selective. Aerial surveys and experimental hunts were conducted, and kill statistics gathered, to evaluate the data against a predator-prey model. Field data showed that duck vulnerability was inversely proportional to hunting effort, and that ducks were most wary when the hunting effort was greatest. Hunters were most numerous when duck densities were high. However, hunters did not shoot selectively and crippling losses (lost birds) totalled about one-third of those birds shot. The weather, and time of the harvest, had little influence on that kill.

C Mammal Research

Large mammal research undertaken by the Canadian Wildlife Service has included the study of polar bears, grizzly bears, muskoxen, mountain sheep, wolves, and caribou. Much of this work was carried out in the decade of the 1960s, but not exclusively.

Research on polar bears (*Ursus maritimus* Phipps) was begun in 1961, although T.H. Manning and A.H. Macpherson, in their paper entitled *The Mammals of Banks Island* (Montreal: Arctic Institute of North America, 1958. 74 p. Arctic Institute of North America Technical Paper No. 2), began the Canadian studies of these Arctic mammals, which are well adapted to their environment because of thick winter coats, made up of glossy guard hairs and dense underfur, and thick fat layers beneath the skin, all of which protect them against the cold air. As well, their short furry ears are highly vascularized, with a heavy network of blood vessels that keep the ears warm. Also, the soles of their feet are commonly bordered with dense fur to insulate them.

The early stages of the Canadian Wildlife Service research project, carried out by C.R. Harington, were concerned with the general ecology and life history of the polar bear. These internal study reports were entitled *Polar Bear Study – Clyde Area, Northwest Territories, 1961* (Ottawa: Canadian Wildlife Service, 1962.

42 p.); *Polar Bear Study – Resolute Bay Area, Northwest Territories, 1962* (Ottawa: Canadian Wildlife Service, 1962. 38 p.); *Polar Bear Study: East Coast of Baffin Island, 1961* (*Arctic Circular*, v. 15, 1963. pp. 21–24); *Polar Bear Study – Southampton Island, Northwest Territories, 1962* (Ottawa: Canadian Wildlife Service, 1963. 32 p.); *Polar Bear Study – Southampton Island, Northwest Territories, 1963* (Ottawa: Canadian Wildlife Service, 1964. 57 p.); *Polar Bear Study – Banks Island, Northwest Territories, 1963* (Ottawa: Canadian Wildlife Service, 1964. 33 p.). These reports were summarized by the author in the paper entitled *Polar Bears and Their Present Status* (*Canadian Audubon*, v. 26, 1964. pp. 4–11). These studies indicated that polar bears prefer areas with suitable combinations of pack-ice, which provides a hunting platform and protective cover, open water where seals are able to reach the surface and are abundant as a food source, and land which provides denning areas, cover, and supplementary food, when seals are not available or when plant food is required. Hence, polar bears are considered to be maritime rather than marine mammals. They are native to the Arctic sea coast.

Their denning habits have also been studied and described in detail by C.R. Harington in the paper entitled *Denning Habits of the Polar Bear (*Ursus maritimus *Phipps)* (Ottawa: Canadian Wildlife Service, 1968. 30 p. Canadian Wildlife Service Report Series No. 5). This paper incorporated the general Canadian ecology studies carried out from 1961 to 1964. Because polar bears are circumpolar, their common denning areas are northern Greenland, Svalbard, the archipelagos north of the Russian mainland, and the islands of the Canadian archipelago. It is in the autumn that polar bears transport themselves on broken pack-ice to suitable denning areas. Harington noticed that site factors exert an important influence on the location of polar bear dens. The dens are usually excavated in the thick snowbanks that develop on leeward slopes of coastal hills and valley sides, and are characterized by entranceways leading to one or more rooms. Although polar bears of both sexes and various ages occupy dens, pregnant females seem to be most regular in this habit, usually denning between October and April. The cubs are usually born in early December, and are mature enough to leave the dens and travel with their mothers down to the sea ice in the spring. Most recently, the reproductive performance of polar bears has been studied by A.E. Derocher and I. Stirling in their paper entitled *Age Specific Reproductive Performance of Female Polar Bears (*Ursus maritimus*)* (*Journal of Zoology* [London], v. 234, 1994. pp. 527–36). These researchers noted that reproductive performance declined because of aging-related processes that impaired the accumulation of fat stores necessary for reproduction.

In hunting and feeding habits, I. Stirling and P.B. Latour noted, in their paper entitled *Comparative Hunting Abilities of Polar Bear Cubs of Different Ages* (*Canadian Journal of Zoology*, v. 56, 1978. pp. 1768–72), that cubs stay with their mothers for up to two and a half years, although some cubs become independent at one and a half years of age, and begin an independent pattern of hunting by taking young seals in lairs in the spring, and hunting at seal holes as the summer progresses. By the autumn, the bears begin to patrol the coastal waters, and their diet may include lemmings, eider ducks, and late summer vegetation such as seaweed, lichen, and mosses.

Although it is known that polar bears commonly live to 20 years of age, maturing patterns through age determination up to the age of six years were carried out by T.H. Manning, and described in the paper entitled *Age Determination in the Polar Bear (*Ursus maritimus *Phipps)* (Ottawa: Department of Northern Affairs and National Resources, 1964. 12 p. Canadian Wildlife Service Occasional Paper No. 5). In this paper, skull criteria taken from 487 skulls were used to determine maturity. It was noted that sexual maturity is reached by males early in their fourth year, and by females possibly early in their third year. In another paper by the same author, skull criteria were also used to study the possibility of geographic variations in the polar bear. In this paper, entitled *Geographical Variation in the Polar Bear (*Ursus maritimus *Phipps)* (Ottawa: Canadian Wildlife Service, 1971. 27 p. Canadian Wildlife Service Report Series No. 23), the author took 17 measurements on 628 skulls segregated by sex, age, and geography (Spitzbergen, east Greenland, Canada, west Greenland, Alaska north and south, Novaya Zemlya, and eastern Siberia). An increase in skull size from east to west was indicated, but the variation in measurements was too small to positively suggest distinctive variations (sub-species) in polar bear populations.

Although polar bears are not in immediate danger of extinction, they are no longer secure in their Arctic wilderness, because of human encroachment and depletion. It was not until 1965, at the First International Scientific Meeting on the Polar Bear at Fairbanks, Alaska, that the northern countries of the Arctic Archipelago expressed concern about maintaining the bears. And, in 1974, an International Agreement on the Conservation of Polar Bears was signed in Oslo, Norway, which committed the signatory nations to continuing polar bear research and management and protection of their ecosystem. The studies, begun in the early 1960s, in Canada, have been continued by I. Sterling and H.P. Kiliaan as indicated in their paper entitled *Population Ecology Studies of the Polar Bear in Northern Labrador* (Ottawa: Canadian Wildlife Service, 1980. 19 p. Canadian Wildlife Service Occasional Paper No. 42). A good overview of knowledge about the polar bear can be found in the pamphlet publication entitled *Polar Bear* (Ottawa: Canadian Wildlife Service, 1983. 4 p. Canadian Wildlife Service. Hinterland Who's Who).

There were no extensive studies on the grizzly bear (*Ursus arctos horribilis* Ord.), until the 1960s in Canada or in the United States. The Canadian program was undertaken in 1964 by the Canadian Wildlife Service and completed in the early 1970s. The reason for the research was concern over the continuing reduction of grizzly bear numbers and distribution, and the need for sound ecological data on which management decisions could be based. According to R.L. Peterson, in the paper entitled *A Well Preserved Grizzly Bear Skull Recovered from Late Glacial Deposit Near Lake Simcoe, Ontario* (*Nature*, v. 208, 1965. pp. 1233–34), the bear originally ranged as far east as Ontario, and was common on the prairies. However, the rapid increase in resource exploitation and human settlement in North America has resulted in the extermination of the species from much of its original range.

The Canadian study hoped to obtain detailed information on the ecology of the ecotypical grizzly bear living in the mountains of northern British Columbia, south and central Yukon, the Mackenzie Mountains of the Northwest Territories, and central Alaska. Toward this end, 53 grizzlies were captured and marked in the Kluane Game Sanctuary in southwestern Yukon Territory, an area of over 1,000 square kilometres. Studies on the drugs used for bear capture were discussed in the paper by A.M. Pearson and R.M. Bradley entitled *Evaluation of Phencyclidine Hydrochloride and Other Drugs for Immobilizing Grizzly and Black Bears* (*Journal of Wildlife Management*, v. 32, 1968. pp. 532–37). However, to expand the study, it was mandatory for hunters in other parts of the Yukon to surrender the skulls of killed grizzlies and earlier data was taken from statistics of animals killed in the Yukon Territory and the Mackenzie Mountains during the early 1960s.

The data collected from live and other specimen material indicated that the population showed a high percentage of adult animals, low reproduction rates with an average litter size of 1.7 cubs, late maturation of females, and a minimum of three years between litters. Age determination studies, based on tooth measurements, were described by K.R. Mundy and W.A. Fuller in their paper entitled *Age Determination in the Grizzly Bear* (*Journal of Wildlife Management*, v. 28, 1964. pp. 863–66).

The overall study of the grizzly bear in the Yukon was undertaken by A.M. Pearson, and described in the report entitled *The Northern Interior Grizzly Bear* Ursus arctos *L.* (Ottawa: Canadian Wildlife Service, 1975. 86p. Canadian Wildlife Service Report Series No. 34). In the report, the author concluded that 'the problem of management of grizzlies becomes more a matter of managing humans intruding into grizzly habitat as it is a true wilderness animal that can survive only in relatively undisturbed areas.' An overall description of the grizzly bear, including its life cycle, has been given in the pamphlet publication entitled *Grizzly* (Ottawa: Canadian Wildlife Service, 1984. 3 p. Canadian Wildlife Service. Hinterland Who's Who).

Another large Arctic mammal is the muskox (*Ovis moschatus* subsp.), which bears a superficial resemblance to the North American bison. Its appearance is characterized by its humped shoulders and long black coat, and by the shortness of its legs. The coat is remarkable, in that it allows the muskox to face the Arctic blizzards without shelter, as the outer layer of long black hair protects the thick inner coat of soft fine wool that completely covers the animal, except for the lips and nostrils. These physical characteristics are important for the maintenance of body temperature in the Arctic environment, where, up to eight months of the year, the temperature does not rise above −18 degrees centigrade. The study of the connection between body temperature and physical characteristics has been carried out by L. Irving and J. Krog, in their paper entitled *Body Temperature of Arctic and Subarctic Birds and Mammals* (*Journal of Applied Physiology*, v. 6, 1954. pp. 667–80). The Canadian Wildlife Service did not initiate studies of muskoxen until the 1950s, although Canadian herds were given complete protection by the federal government in 1917, because their numbers had become alarmingly low. However, A.H. Macpherson, in the paper entitled *On the Abundance and Distribution of Certain Mammals in the Western Canadian Arctic Islands in 1958–59* (*Arctic Circular*, v. 14, 1961. pp. 1–17), and J.S. Tener, in the papers entitled *The Distribution of Muskoxen in Canada* (*Journal of Mammalogy*, v. 39, 1958. pp. 398–408) and *The Queen Elizabeth Islands Game Survey, 1961* (Ottawa: Canadian Wildlife Service, 1963. 50 p. Canadian Wildlife Service Occasional Paper No. 4), helped suggest that their population had risen to about 8,500 on the islands, because of their protected status. About 1,500 were on the mainland. Today, the Canadian Wildlife Service estimates that the muskox population is now about 45,000, and most are found on the Arctic islands, especially Banks and Victoria Islands. About 14,000 of that total are located on the mainland north of Great Bear Lake, around Queen Maud Gulf, and in northeastern Keewatin.

The muskoxen are found in small herds in low-lying coastal and island plains, or river valleys. Here, soil has accumulated, the melted snow is held in the permafrost, and ground-hugging plants grow quickly during the summer days and provide forage. With the melting of the snow cover and the appearance of new vegetation, muskoxen graze on mountain sorrel, blue grass, and new willow growth. And, as the summer progresses, other plant species are eaten. By September, when snow covers the ground, muskoxen must dig craters through the snow to find the willows and grasses that form the major part of their diet. These eating patterns were described by J.S. Tener, in the paper entitled *A Preliminary Study of the Musk-oxen of Fosheim Peninsula, Ellesmere Island* (Ottawa: Canadian Wildlife Service, 1954. 34 p. Canadian Wildlife Service Wildlife Management Bulletin, Series 1, No. 9). A later detailed study of muskox biology covered its physical characteristics, life history, herd characteristics and behaviour, and diseases. In its life history, mating will occur in the late summer, with calves born in the following late spring. It is at that time that the large winter herds break up into smaller summer herds, when their activity is primarily eating and resting. Unlike the caribou, muskoxen do not migrate. This CWS study was undertaken by J.S. Tener and published as a monograph entitled *Muskoxen in Canada: A Biological and Taxonomic Review* (Ottawa: Canadian Wildlife Service, 1965. 166 p. Canadian Wildlife Service Monograph No. 2). A brief survey of the muskoxen has also been published under the title *Muskoxen* (Ottawa: Canadian Wildlife Service, 1984. 4 p. Hinterland Who's Who).

Unlike the muskox, which is unique to the Arctic, the Rocky Mountain bighorn sheep (*Ovis canadensis canadensis* Shaw) is well adapted to the extremes of elevation and temperature of North America's mountain rangelands. The extremes vary from the snow-fields of the Canadian Rockies to below sea level on the desert floor of Death Valley, California, and their total range extends from Alaska all the way to northern Mexico. However, in spite of their adaptive ability, these bighorn sheep have been subject to distinctive 'die-offs.' The most recent of these die-offs, or severe population losses, occurred in 1966, and became the focus of a Canadian Wildlife Service study requested by the National Parks Branch of the Department of Northern Affairs and National Resources. Before then, there had been five die-offs in the national parks, between 1936 and 1950. As well, similar die-offs had occurred on provincial ranges in the 1940s in Alberta and in the 1960s in British Columbia. These die-offs have been described by R.A. Demarchi and D.A. Demarchi in their paper entitled *Status of the Rocky Mountain Bighorn* (*Wildlife Review*, v. 4, 1967. pp. 10–14), and by J.G. Stelfox in the paper entitled *Bighorn Sheep in the Canadian Rockies: A History 1800–1970* (*Canadian Field Naturalist*, v. 85, 1971. pp. 101–22).

The objectives of the CWS were 'to compare the importance and interrelationships of the ecological factors as exhibited on sheep winter ranges in Jasper, Banff, and Waterton during the period 1967–71.' Some of the ecological factors were range conditions, summer and winter weather, ungulate population dynamics, population trends, range grazing, carrying capacity, and population self-regulation. The study was undertaken by J.G. Stelfox and published with the title *Range Ecology of Rocky Mountain Bighorn Sheep* (Ottawa: Canadian Wildlife Service, 1976. 50 p. Canadian Wildlife Service Report Series No. 39). After five years of intensive study, the author concluded that sheep populations increased until winter ranges became overgrazed, at which time 'both endoparasite burdens and lung worm infections increased.' Malnutrition and severe winter weather, combined with heavily infected lungs, created a pneumonia-lungworm disease and caused a rapid mortality of about 75 per cent of the sheep population. This disease aspect of the die-off had been described by J.G. Stelfox, a year earlier, in a progress report entitled *Diseases and Parasites of Rocky Mountain Sheep in the National Parks of Western Canada, 1966 to 1972* (Edmonton: Canadian Wildlife Service, 1974. 149 p.). An overall summary of the ecology of the Rocky Mountain Bighorn Sheep has been provided in the pamphlet entitled *Mountain Sheep* (Ottawa: Canadian Wildlife Service, 1980. 3 p. Hinterland Who's Who).

Polar bears, grizzly bears, muskoxen, and mountain sheep have never been maligned, as has the wolf (*Canis lupus*). Until 100 years ago, this badly maligned animal was more widely distributed than any other mammal of historic times, as it inhabited large areas of North America, Europe, and Asia. In North America today, wolves still live in large areas, but their range has been greatly reduced owing to changes in their habitat and man's effort to exterminate them. They remain quite common in lightly settled areas of Canada, from Labrador to British Columbia, and in parts of the Yukon. In Canada, there are primarily three subspecies that exist – the *Canis lupus occidentalis* of the Mackenzie District of the Northwest Territories, the eastern wolf (*Canis lupus lycaon*) in Ontario and Quebec, and the Hudson Bay wolf (*Canis lupus hudsonicus*) on the mainland tundra of the Arctic.

Wolves have a highly developed pattern of social behaviour, centring on a dominant male and female, and live and hunt in packs. Their breeding time is in March and April, and litters usually consist of five or six pups, which by the autumn are hunting with the pack in their individual territories. The wolves' chief prey are large mammals such as deer, moose, caribou, elk, bison, and muskox. Their hunting and feeding patterns were intensely studied, between 1960 and 1968, in the Thelon Game Sanctuary, 400 miles northeast of Yellowknife, in the Northwest Territories, by the Canadian Wildlife Service. The Thelon Game Sanctuary is an area through which the summer and autumn barren-ground caribou migration passes. The basis for the study was the fact that in the 1950s, the CWS recommended to federal, provincial, and territorial government agencies, responsible for renewable resources, that wolf populations be reduced in the Northwest Territories in order to conserve the dwindling stocks of barren-ground caribou. The recommendation was based upon caribou studies undertaken by A.W. Banfield, and published with the title *Preliminary Investigation of the Barren-Ground Caribou* (Ottawa: Canadian Wildlife Service, 1954. 2 parts. 79 p. and 112 p. Canadian Wildlife Service Management Bulletin Series 1, Nos. 10A and 10B). Although the knowledge of the effects of wolves upon the barren-ground caribou was inadequate, the Canadian Wildlife Service went ahead with the destruction of wolves in the Northwest Territories.

By 1961, the number of wolves taken annually had declined and the costs of the operation had increased. Accordingly, the Canadian Wildlife Service wanted to fill in their knowledge gap of the wolf as predator of barren-ground caribou. The study was undertaken by E. Kuryt, and published with the title *Food Habits and Ecology of Wolves of Barren-Ground Caribou Range in the Northwest Territories* (Ottawa: Canadian Wildlife Service, 1972. 36 p. Canadian Wildlife Service Report Series No. 21). This study showed that tundra wolves may migrate up to 200 miles, and their movements are associated with those of the barren-ground caribou. During the winter, wolves eat only caribou, but during the spring and summer, they vary their diet by eating small rodents, birds, eggs, and fish. Wolves also tend to prey on caribou calves, as well as older caribou. The author estimated that a wild wolf would eat about 23 caribou annually, and the wolf numbers at that time were not a serious cause of caribou mortality.

Since this study was carried out, wolves have become more acceptable as part of our wilderness heritage. More recently, they were focus of a symposium, held in 1981, entitled *Wolves in Canada and Alaska: Their Status, Biology and Management: Proceedings of the Wolf Symposium Held in Edmonton, Alberta, 12–14 May, 1981* (Ottawa: Canadian Wildlife Service, 1983. 135 p. Canadian Wildlife Service Report Series No. 45). Wolf ecology has also been summarized in the pamphlet publication entitled *Wolf* (Ottawa: Canadian Wildlife Service, 1984. 3 p. Hinterland Who's Who).

The major prey of the wolf, the caribou, has received intensive study by the Canadian Wildlife Service. The caribou (*Rangifer tarandus*) is one of Canada's most widely distributed large mammals and is found in western Ontario and British Columbia, near the US-Canada border, to as far north as Ellesmere Island in the Arctic Archipelago. The southern species is the Woodland Caribou (*Rangifer tarandus*), which inhabits the northern boreal forest. The far northern species is the Peary Caribou (*Rangifer tarandus pearyi*), which is found only in the islands of the Canadian Arctic. The most common subspecies, however, is the barren-ground caribou, which spends much or all of the year on the tundra, from Alaska to Baffin Island. The Canadian form (*Rangifer tarandus groenlandicus*) lives east of the Mackenzie River, and the Alaskan form (*Rangifer tarandus granti*) lives west of the Mackenzie River. Eight hundred thousand of these barren-ground caribou live in herds, which migrate seasonally from the tundra to the sparsely treed northern conifer forests – the taiga. From west to east, these herds are the Porcupine Herd, the Bluenose Herd, the Bathurst Herd, the Beverly Herd, and the Kamanuriak Herd. Other barren-ground caribou, numbering about 120,000, live in smaller herds and spend their entire year in the tundra. However, it is the migrating barren-ground caribou that have received more careful study than all but a very few groups of wild animals. As previously mentioned, it was the 1954 Banfield study that indicated the decline in the numbers of barren-ground caribou, although there had been undocumented awareness of the decline before that time. Accordingly, this study stimulated additional research and documentation about the barren-ground caribou by the Canadian Wildlife Service.

The first major study published was written by J.P. Kelsall, and entitled *The Migratory Barren-Ground Caribou of Canada* (Ottawa: Canadian Wildlife Service, 1968. 339 p. 24 maps. Canadian Wildlife Service Monograph No. 3). In this monograph, the barren-ground caribou habitat, food habits, migration and movement, population dynamics, physiology, breeding biology, and growth were documented. It is interesting to note that the annual migratory pattern is a fixed cycle. Most caribou spend their winter in the taiga, with the bulls venturing the farthest south, and the cows and juveniles remaining near the tree-line. In the spring, the herds regather for the migration to the calving grounds and their summer range on the tundra. About 90% of adult cows produce calves annually, which are born in late May or early June. However, many calves die each year owing to adverse weather and predation. In July, the herds begin to move south again and arrive at the tree-line of the taiga in September. In late October, mating takes place, and the cycle begins again. As previously mentioned, wolves are the natural predator of the caribou, but it has a positive culling effect, as the aged are killed along with the weak young. However, as Kelsall points out, man, because of his historic dependence on the caribou, is the greatest predator of the caribou.

Even with the number of studies of the barren-ground caribou already undertaken, more research was still required. For example, recruitment (birth) rates, mortality rates, range relationships, and the year-to-year variability of these factors had not been sufficiently understood. Accordingly, the Canadian Wildlife Service designed a research program, in 1964, in the hope of filling in these knowledge gaps. This program focused upon the Kamanuriak Herd, whose population was considered sufficiently discrete from adjacent populations to be considered distinct. Four studies were undertaken. The first study, by G.R. Parker, was entitled *Biology of the Kaminuriak Population of Barren-Ground Caribou. Part 1: Total Numbers, Mortality, Recruitment and Seasonal Distribution* (Ottawa: Canadian Wildlife Service, 1972. 95 p. Canadian Wildlife Service Report Series No. 20). In this study, five aerial linear transect surveys provided population estimates, distribution, and births in the winter of 1967–68, as well as in the calving area in June, 1968. Mortality rates were estimated from recorded caribou kills as well as from the estimation of wolf populations and the projection of their average kills. A separate, but related, publication by F.L. Miller and E. Broughton, entitled *Calf Mortality on the Calving Ground of Kaminuriak Caribou* (Ottawa: Canadian Wildlife Service, 1974. 26 p. Canadian Wildlife Service Report Series No. 26), was also published at this time. In this paper, the cause of mortality among newborn calves from the Kamanuriak population was studied in the summer of 1970, and the authors suggested that the importance of wolf predation on young caribou had not been fully realized.

The second report in the study series was undertaken by F.L. Miller and entitled *Biology of the Kamanuriak Population of Barren-Ground Caribou. Part 2:*

Dentition as an Indicator of Age and Sex; Composition and Socialization of the Population (Ottawa: Canadian Wildlife Service, 1974. 88 p. Canadian Wildlife Service Report Series No. 31). In this study, CWS biologists collected (killed) 999 barren-ground caribou each year from 1966 to 1968. The age composition was estimated from dental measurements, and the age data was used to determine the effects of birth and mortality on the population, and the sexes and ages of animals grouping together at different times of the year. The author suggested that post-calving grouping allowed for the regrouping of the winter herds.

The third study, by D.R. Miller, was entitled *Biology of the Kaminuriak Population of Barren-Ground Caribou. Part 3: Taiga Winter Range Relationships and Diet* (Ottawa: Canadian Wildlife Service, 1976. 42 p. Canadian Wildlife Service. Report Series No. 30). In this study, the winter range of barren-ground caribou in northwestern Manitoba was studied to learn what potential forage was available, and how weather, in particular snow conditions, affected the foraging of caribou. Snow conditions, especially snow depth and crust hardiness, explained the sudden changes observed in caribou diet. The fourth study, by J.C. Dauphiné, Jr., was entitled *Biology of the Kaminuriak Population of Barren-Ground Caribou. Part 4: Growth, Reproduction and Energy Resources* (Ottawa: Canadian Wildlife Service, 1976. 71 p. Canadian Wildlife Service Report Series No. 38). This study examined the body growth, seasonal energy (fat) reserves, and productivity of the Kaminuriak Herd.

Since this major series of studies of barren-ground caribou, research has continued. For example, D.C. Thomas has studied eating and digestive patterns in two caribou species, in the papers entitled *Digestibility of Plants in Ruminal Fluids of Barren-Ground Caribou (*Rangifer tarandus groenlandicus (*Arctic*, v. 34, 1981. pp. 321–24); *Rumin Contents and Habitat Selection of Peary Caribou (*Rangifer tarandus pearyi*) in Winter* (*Arctic*, v. 15, 1983. pp. 97–106); and *Fire–Caribou Winter Range Relationships in Northern Canada* (*Rangifer*, v. 16, 1996. pp. 57–67). The latter paper correlated lichen abundance peaks after forest fires, and how they affected caribou forage patterns. On the other hand, F.L. Miller has studied caribou population variations on Bathurst Island, and its satellite islands, as well as inter-island swimming, in the papers entitled *Estimating Bathurst Island (Northwest Territories, Canada) Peary Caribou and Muskox Populations* (*Arctic*, v. 44, 1991. pp. 57–62); and *Inter-Island Water Crossing by Peary Caribou, South-Central Queen Elizabeth Islands* (*Arctic*, v. 48, 1995. pp. 8–12).

D National Parks Research

Besides its own research agenda on mammals, the Canadian Wildlife Service also conducts ecological research, including mammal studies in the national parks of Canada, at the request of the National Parks Branch (now Parks Canada). These usually involve the documentation of wildlife resources through inventories, short-term studies on specific problems, or management advice. Although the national parks are wildlife sanctuaries, their open access to visitors and commercial development has not left the parks in ecological equilibrium.

Wildlife inventories, important in themselves for the documentation of wildlife in a given area, are also an important tool for recording the state of wildlife at a given time, against which future trends can be observed and evaluated. Toward that end, the Canadian Wildlife Service has undertaken wildlife inventories in both Jasper National Park and Waterton Lakes National Park, in Western Canada. Jasper National Park is the largest mountain sanctuary in Canada, with an area of 4,200 square miles. It lies entirely on the eastern watershed of the Rocky Mountains, and the Continental Divide delineates its western boundary. Because of its regional physiography, the vegetation, habitats, seasons, and climate vary markedly with the altitude. Hence, mammal distribution is more complex in the mountains than on the plains, and the great differences in elevation indicate a marked variation in living conditions and the need for adaptation.

From the early 1800s, there had been numerous expeditions by trading parties, and naturalists had visited or passed through the Jasper Park area, which was established in 1907 as a wildlife sanctuary. Although these groups did some specimen collecting that was documented in the journal literature, or in collections, much of the inventory of the park habitat was not done until the 1950s and 1960s. The first wildlife inventory undertaken in Jasper National Park was done by I.M. Cowan and entitled *Birds of Jasper National Park, Alberta, Canada* (Ottawa: Canadian Wildlife Service, 1955. 66 p. map. Canadian Wildlife Service, Wildlife Management Bulletin, Series 2, No. 8). In this inventory, the author discussed the occasional earlier documentation of Jasper bird life, reaching back to 1815. The author then identified, in an annotated list, 195 species of birds, 'for which there seemed to be conclusive evidence of occurrence within the present boundaries of Jasper Park.' In most cases, these occurrences were substantiated by specimens, most of them in the National Museum of Canada. In the listing, birds were assigned to three categories – summer residents, permanent residents, and transients.

The next major inventory was carried out by J.D. Soper, after retirement from the Canadian Wildlife Service. This inventory was entitled *The Mammals of Jasper National Park, Alberta* (Ottawa: Canadian Wildlife Service, 1970. 80 p. Canadian Wildlife Service Report Series No. 10). In this publication, 52 forms of mammals native to the park were listed, with direct references to any published field results. Included in the listing is the status, habitat, reproduction, and general remarks about the wildlife described.

J.D. Soper also carried out one other mammal inventory in a western park for the Canadian Wildlife Service. This was entitled *The Mammals of Waterton Lakes National Park, Alberta* (Ottawa: Canadian Wildlife Service, 1973. 57 p. Canadian Wildlife Service Report Series No. 23). Waterton Lakes National Park is unique for two reasons. It was among the first of the national parks created in 1895, and, unlike Jasper or Banff National Parks, it had remained little disturbed, except by southern Alberta residents, as there was no railroad to the park, no good roads, and no advertising. The first documentation of park wildlife was done in 1874, when work was undertaken on the international boundary in that area, but it was not until the early 1960s that the author undertook a systematic study of park wildlife, covering about 70 square miles. J.D. Soper identified 55 mammals inhabiting the park and described them by order and genera. Included in the description was their status, habitat, reproduction, and general comments. The descriptions followed the same pattern as his work on Jasper.

Dissimilar to this classic inventory work in the Rocky Mountain parks was a form of inventory undertaken in La Maurice National Park, Quebec by J.L. DesGranges and entitled *Avian Community Structure of Six Forest Stands in La Maurice National Park, Quebec* (Ottawa: Canadian Wildlife Service, 1980. 33 p. Canadian Wildlife Service Occasional Paper No. 41). The author, who carried out this work for CWS, as well as for an M.Sc. degree at the University of Alberta, was interested in documenting the close ties between six types of forest stands, and their vegetation, and the structure of the bird communities that they sheltered. Over a period of one year, the author studied avian communities in six habitats (white spruce, white birch, sugar maple–yellow birch stand, mixed stand, spruce stand, and fir stand) at various successional stages. This study helped to clarify the stability and ecological succession of avian communities in the park, which is part of the temperate cold forest zone. The diversity of the avian communities was directly affected by the diversity of the plant communities, and a total of 30 different species of birds were observed and documented.

Besides the inventory studies described above, the Canadian Wildlife Service has also been involved in management studies. These are important, because resource exploitation, and tourism, have accelerated the settlement of the once-wild lands of western Canada, and have reduced the natural range of mammals such as the grizzly bear (*Ursus arctos* L). To assure the maintenance of wild stocks of animals such as grizzlies, the national parks are the only areas of adequate size where the preservation of the species and habitat is a primary objective. In the west, there are seven national parks that support populations of grizzly bears. They are Jasper, Banff, and Waterton, on the eastern slope of the Rocky Mountains in Alberta, Yoho and Kootenay on the western slope of the Rockies in British Columbia, and Glacier and Mount Revelstoke in the Selkirk Mountains of British Columbia. It is also the objective of the national parks to provide an opportunity for the public to enjoy these wilderness areas. However, in some parks, these two ideals clash, particularly when grizzly bears are brought into close proximity with people, thus increasing the risk of attack.

It is because of this problem that the Canadian Wildlife Service was asked to study the life cycle, including population dynamics and density, reproduction, longevity, mortality factors, habitat, foods, movement, and attacks on, or threats to, humans by grizzly bears, in order to come up with a management plan for these mammals in the mountain parks. The study was undertaken by K.R. Mundy and D.R. Flook and entitled *Background for Managing Grizzly Bears in the National Parks of Canada* (Ottawa: Canadian Wildlife Service, 1973. 35 p. Canadian Wildlife Service Report Series No. 22). In the report, the authors listed 12 recommendations for the management of grizzly bears. These recommendations ranged from the location of campgrounds outside favourite grizzly habitat, and assigning special wilderness status to certain areas of grizzly range, to installing concrete 'bear-proof' garbage bins wherever garbage receptacles were needed, including garbage pick-up every evening.

Other park studies have been more specialized, but with applications important to wildlife management in other national parks. One such study has concerned the problem of livestock grazing in national park areas and the conflict with the preservation of park fauna and flora. Another study involved the effects of snow conditions on wildlife populations. The first study involved the Riding Mountain National Park, which had been established in 1930, and the only park where cattle grazing was still permitted. Since then, however, the national parks' policy had shifted to considering that livestock grazing was undesirable in those preserves. To that end, the National Parks Service, in the early 1960s, asked the Canadian Wildlife Service to investigate all aspects of cattle grazing in the park. The primary objective was to determine if livestock grazing was adversely affecting the native flora, or fauna, and if so, to study possible methods of reducing this conflict. However, the study limited itself to food and feeding relationships of cattle and elk (*Cervus canadensis*) in the park.

Elk had been resident in the area since pre-settlement times, and were the most abundant large mammal in the park. Because grasses are an important elk food, and elk may have suffered from food shortages in areas grazed by cattle, the specific objective of the study was to determine if elk and cattle were competing for a limited food supply. In the report written by D.A. Blood, and entitled *Range Relationships of Elk and Cattle in Riding Mountain National Park, Manitoba* (Ottawa: Canadian Wildlife Service, 1966. 62 p. Canadian Wildlife Service. Wildlife Management Bulletin Series 1, No. 19), the prevailing levels of elk and cattle in the park, and forage competition between those two species, did not appear to threaten the welfare of either. Nevertheless, although cattle grazing had not materially altered forage production at Riding Mountain, it had brought about severe changes in the species composition of many areas. Such alterations were 'incompatible with National Parks Policy and all parks should be left in their natural state unimpaired for the use and enjoyment of future generations.'

As mentioned above, the second specialized study was concerned with the study of snow conditions and their effect on animal populations. The reason for this study was based upon the fact that there was a problem of population changes between moose (*Alces alces*) and white-tailed deer (*Odocoileus virginianus*) in Fundy National Park in the Maritimes. By the 1940s, the moose population had diminished to the point at which hunting was no longer permitted. However, deer populations, which up until that time had been greatly reduced, had expanded to a remarkable degree. Deer are chronic carriers of a parasitic nematode worm that is lethal to moose and one of the reasons for the decline of the moose. However, in the Fundy National Park, moose survived in the high country area, where there was a near absence of deer in the winter and a scarcity in the summer. Because this separation was beneficial to the moose, the Canadian Wildlife Service wanted to examine the factors which allowed moose to persist in high numbers on high ground, as well as the factors which excluded deer. Because snow conditions appeared to be a major factor, an examination of snow morphology seemed to be a logical point of attack. J.P. Kelsall and W.H. Prescott undertook the study, which was published as a paper entitled *Moose and Deer Behaviour in Snow* (Ottawa: Canadian Wildlife Service, 1971. 27 p. Canadian Wildlife Service Report Series No. 15). The authors indicated that earlier Russian studies showed that deep snow, particularly of long duration, was a principal factor limiting the range of wintering deer. Their own studies indicated that snow depths, on the high altitudes frequented by wintering moose in Fundy National Park, were consistently greater than on low-altitude deer ranges. This would account for the almost complete withdrawal of deer to low elevations with the onset of winter. Because moose are seldom restricted by snow depths in their movement, they would prefer the higher altitudes, which separated them from deer and the parasitic nematodes. In fact, observations at the park indicated that when moose did stay in low altitudes, they usually became infected with the parasitic nematode and died the following spring.

E.S. Telfer and J.P. Kelsall later used snow morphology and conditions to study the physical adaptations for snow-coping by mammals such the moose (*Alces alces*), caribou (*Rangifer tarandus*), wapiti (*Cervus canadensis*), white-tailed deer (*Odocoileus virginianus*), dall sheep (*Ovis dalli*), bighorn sheep (*Ovus canadensis*), pronghorn antelope (*Antilocopra americana*), and bison (*Bison bison*). Those mammals, with greater adaptations in chest heights, and foot-loading ability on snow, coped better in snowy regions. In their paper entitled *Adaptation of Some Large*

North American Mammals for Survival in Snow (*Ecology*, v. 65, 1984. pp. 1828–34), the authors said that the predation factor on these mammals may have been a factor influencing evolution of ungulate ability to cope with snow.

Special management problems were also the reason for two studies in the Wood Buffalo National Park, which straddles the border between northern Alberta and the Northwest Territories, south of Great Slave Lake. The first of these involved the discovery of tuberculosis in the bison of Wood Buffalo National Park, in the years 1947 to 1949. The Canadian Wildlife Service was asked to formulate a management plan for the infected herd. The second study involved research on the population dynamics, including breeding success and mortality, of whooping cranes on their nesting grounds in the park. These nesting grounds were only discovered in 1954.

Wood Buffalo National Park was created in 1922, and originally encompassed an area of 10,500 square miles to provide a protective environment for the remnant herd of wood bison that inhabited the area. When the park was created, the herd numbered about 1,500 animals. These wood buffalo were thought to be a sub-species (*Bison bison athabasca*) of the plains bison (*Bison bison bison*). However, in the late 1920s, just over 6,000 plains bison were introduced to the park, which was expanded at that time to support the additional population. W.A. Fuller, in the paper entitled *Aerial Census of Northern Bison in Wood Buffalo National Park and Vicinity* (*Journal of Wildlife Management*, v. 14, 1950. pp. 445–51), estimated the then-current population to be 10,000 to 12,000. Most of these animals were a hybrid species with genetic characteristics of the plains bison. Fuller also carried out the biology and management studies involving the wood buffalo. This was entitled *The Biology and Management of the Bison of Wood Buffalo National Park* (Ottawa: Canadian Wildlife Service, 1966. 52 p. Wildlife Management Bulletin Series 1, No. 16). Fuller studied the physical range of the bison, their reproductive characteristics, and mortality factors, in order to suggest management procedures for this mammal. Fuller said that the reasons for managing any wildlife resource were for aesthetic and historical values, its value as a source of food, and the control of disease. The effect of the tuberculosis on a basically healthy herd was to stabilize its population, and avoid over-population. And control of the disease had to be justified on economic grounds, because of significant wastage of the meat as a food source. Control of the disease could be carried out by selective slaughters. It is interesting to observe that 20 years later, as noted in the Environment Canada Annual Report, 1971–72, the Canadian Wildlife Service did initiate a vaccination program into the herds as a preventive measure against anthrax, which had affected bison populations to the north at Hook Lake, Northwest Territories.

The study of whooping cranes (*Grus americana*) is unique, because this large white bird with black wing tips has long been in danger of extinction. In 1920, there were no more than 50 birds in existence, and by 1941 that number had dropped to 15. By 1945, steps were taken to protect the birds during migration, to learn as much as possible about their habits, and to search out their nesting grounds. The first work was carried out by R.P. Allen, and described in the paper entitled *The Whooping Crane* (*National Audubon Society Research Report* No. 3, 1952. pp. 1–240). The discovery and reporting of their breeding grounds was described, again, by R.P. Allen in the paper entitled *A Report on the Whooping Crane's Northern Breeding Grounds* (*National Audubon Society Research Report* No. 3. Supplement, 1956. pp. 1–60).

Whooping cranes are migratory, and spend the winter in the Aransus Refuge on the Gulf Coast of Texas. As was previously mentioned, their breeding grounds were discovered in Wood Buffalo National Park. The habitat of the nesting grounds in the area of the Saas River is marshland, interspersed with numerous potholes that are generally shallow with soft bottoms, as described by N.S. Novakawski in the paper entitled *The Day We Rescued a Whooping Crane* (*Audubon*, v. 67, 1965. pp. 230–33). The dominant emergent vegetation of the potholes is the bulrush (*Scirpus validus*), which forms the major component of their nests. The potholes, themselves, are separated by land areas on which grow black spruce (*Picea mariana*), tamarack (*Larix laricina*), and willow. Novakowski observed these nesting areas over a 10–year period.

These observations were, in effect, an annual chronology of whooping crane population changes, and of events on the breeding grounds from 1954 to 1965. These observations were described by N.S. Novakowski in the report entitled *Whooping Crane Population Dynamics on the Nesting Grounds, Wood Buffalo National Park, Northwest Territories, Canada* (Ottawa: Canadian Wildlife Service, 1966. 26 p. Canadian Wildlife Service Report Series No. 1). In this report, the author noted that the birds produced 32 young between 1954 and 1965 in the Saas River area, and 29 young in other breeding areas at the park. However, the adult population, based upon the number of birds returning each autumn to the Aransas Refuge in Texas, was only an increase of 11 birds. The author suggested that the juvenile birds were more subject to mortality than the adults, and this resulted in the total population remaining fairly static. The high mortality seemed to occur the year following a successful breeding season, and so long as this mortality couldn't be reduced, the whooping crane had little hope of reaching population levels at which it would be out of danger. Nevertheless, since 1965 their population has increased. E. Kuyt, in the paper entitled *Whooping Crane, Grus americana Home Range and Breeding Range Expansion in Wood Buffalo National Park, 1970–1991* (*Canadian Field Naturalist*, v. 107, 1993. pp. 1–12), observed that breeding pairs of whooping cranes in the Wood Buffalo National

Park had increased from 15 in 1970 to 33 in 1991. As well, during this 21-year period, the migrating Wood Buffalo National Park–Aransas National Wildlife Refuge population had more than doubled from 57 birds to 132. The author also noted that, since 1983, there had been a 'remarkable increase in the numbers of juveniles and surviving subadults, and a consequent increase in the breeding population.'

Attempts to build a capture flock by artificially incubating eggs taken from whooping crane nests in Wood Buffalo National Park were started in 1967 by the Canadian Wildlife Service and the US Bureau of Fisheries and Wildlife. By 1970, 26 eggs had been collected, and the captured juvenile flock numbered 22. This artificial flock has been described in the pamphlet publication entitled *Whooping Crane* (Canadian Wildlife Service, rev. 1993. 4 p. Hinterland Who's Who).

E Wildlife Surveys

As has been noted above, wildlife research is long-term in nature in order to design programs for wildlife management and protection. Part of that research involves wildlife surveys that provide appropriate base data on the size and shifts in populations of wildlife, and provide researchers with signals about the status of particular species at a given time. Surveys are observational, and may be aerial, or land-based, or a mixture of both techniques.

Aerial surveys have been used by the Canadian Wildlife Service, primarily since the end of World War II. Work carried out north of 60 degrees latitude involves large mammals, primarily caribou, but also includes muskoxen and bison, as previously noted on research carried out on these mammals. South of 60 degrees latitude, aerial surveys have concentrated upon migratory waterfowl of the prairie pothole region (during the 1950s) but have since expanded to the Arctic and coastal regions of Canada. Aerial surveys are useful for obtaining population estimates over large areas, as it is not possible to obtain equally large-scale coverage by ground surveys. Ground parties cannot move more rapidly than the animals they wish to count. Also, the cost of ground surveys covering thousands of square miles would be prohibitive. On the other hand, the weaknesses of aerial surveys are error factors involved in spotting animals from the air, and statistical errors that occur in population projections. The pros and cons of aerial survey work have been documented by A.A. Benson in the paper entitled *Use of Aerial Surveys by the Canadian Wildlife Surveys, 1963* (35 p. Canadian Wildlife Service Occasional Paper No. 3).

Besides the caribou and muskoxen surveys that have contributed to the study of those large Arctic mammals, the Queen Elizabeth Islands Game Survey of 1961 stands out as a major aerial survey. Eighteen islands of the Queen Elizabeth group, at the extreme northern edge of the Arctic Archipelago, were surveyed from the air to determine the species distribution and number of the larger mammals and birds found there. Three reasons were given by the Canadian Wildlife Service for undertaking the survey. First, the area, which comprised a total of 165,000 square miles, was one of the last in Canada about which little quantitative information was available on wildlife resources. Second, 'the dwindling stocks of wildlife used by Eskimos in southern regions made necessary an examination of the islands for possible relocation of those Eskimos still desirous of living off the land.' The third reason given for the survey was that adequate information about game resources was essential, in order to prevent undesirable destruction of the game on the habitat by increased human activity on these islands. The survey covered 121,000 square miles, with 17,500 linear miles of flying done at a height of 500 feet at a speed of 85 miles per hour. Transects were one quarter of a mile to half a mile wide, for recording the survey. This survey was organized by J.S. Tener, and described in the paper entitled *Queen Elizabeth Islands Game Survey, 1961* (Ottawa: Canadian Wildlife Service, 1963. 50 p. Canadian Wildlife Service Occasional Paper No. 4). The use of transects is a standard method for delineating the terrain for the counting of animals, but, as mentioned above, there is an error rate that has to be taken into account.

Other survey methods, which combine aerial surveys with high-level photography, are also useful for estimating population sizes. This combination method was used for estimating lesser snow goose populations in the Arctic. Again, the reasons given for this survey were the changes in land use in the Arctic that had led to changes in the distribution of the lesser snow goose (*Anser caerulescens caerulescens*) in the eastern Canadian Arctic. In order to survey this population, the Canadian Wildlife Service used, under contract, a small twin-engined aircraft equipped for photography, and manned by a pilot and cameraman-navigator under the supervision of CWS researchers. This photographic survey, which used large-format vertical photography, was undertaken in June, 1973. Sixteen nesting colonies were photographed in the south and west Hudson Bay region, the southern lowlands of Southampton Island in the bay, and the southwestern coastal plain of Baffin Island.

To achieve total photographic coverage of a given nesting colony, flying patterns, along predetermined parallel lines over the area at the desired height, were exposed to a series of photographs. Forward and lateral overlap gave a continuous coverage. The photographs were then analysed by film frames with the help of binocular microscopes in order to calculate population estimates. This survey was undertaken by R.H. Kerbes, and the report was entitled *The Nesting Population of Lesser Snow Geese in the Eastern Canadian Arctic: A Photographic Inventory of June 1973* (Ottawa: Canadian Wildlife Service, 1975. 47 p. Canadian Wildlife Service

Report Series No. 35). At that time, lesser snow goose populations of adults in spring were estimated to be 56,000. However, the total number of nesting adults was estimated at 170,000. By 1995, the total number of nesting adults had increased to almost 490,000. These estimates, based on leg-band recovery data and aerial surveys, were reported in the paper by R.H. Kerbes and V.V. Baranyak entitled *Estimated Size of the Western Canadian Arctic and Wrangel Island Lesser Snow Goose Populations on Their Breeding and Wintering Grounds.* In: *Distribution, Survival, and Numbers of Lesser Snow Geese of the Western Canadian Arctic and Wrangel Island, Russia.* Edited by R.H. Kerbes and K.M. Meeres (Ottawa: Canadian Wildlife Service, 1999. pp. 25–38).

Although aerial survey methods are necessary over large land areas, other non-aerial methods can be used when populations are more concentrated. For example, for the study of thick-billed murres of Prince Leopold Island, the population size and status were determined through the use of study plots, within selected study colonies. Within the study plots, two methods of counting were used. One method provided a precise record of the number of breeding pairs on the study plots, and required up to six weeks to complete. The other provided a record of mean numbers of birds on the study plots and required ten days to complete. These survey methods have been described by T.R. Birkhead and D.N. Nettleship in the paper entitled *Census Methods for Murres, Uria Species: A Unified Approach, 1980* (Canadian Wildlife Service, 1980. 25 p. Canadian Wildlife Service Occasional Paper No. 43). In this paper, the authors said that counting errors do arise, because weather factors such as high seas, wind, rain, and fog can depress counts. As well, human error has to be factored in, because census observers differ in their counting abilities.

Another form of survey undertaken by the Canadian Wildlife Service is the habitat inventory, which is different from an inventory of wildlife populations. Such a survey, involving migratory birds and mammals, was undertaken in the early 1970s by the Special Habitat Evaluation Group of the Canadian Wildlife Service, in order to carry out wildlife assessments in areas within and adjacent to the proposed Norman Wells and Mackenzie Valley Pipeline routes in the Canadian Arctic. For this inventory, seven wildlife species, or groups, were selected for study 'on the basis of social values and feasibility of a short-term inventory and the susceptibility of the species to disturbance and aesthetic values.' The groups selected were the Arctic fox, beaver and muskrat, Peary caribou for their range, Porcupine caribou for herd movements, dall sheep, grizzly bear, moose, and waterfowl. The goal of the inventory was the production of a series of maps that would define the relative importance of landscapes, or habitat units. Included with the map series would be related information that was also gathered during the study. The methods used were aerial surveys by helicopter of denning sites.

But, once a potential site was located, the helicopter landed, and the site was investigated, photographs were taken, and brief descriptions of the site written down. In order to accomplish the survey, the pipe corridor and adjacent area was divided into 33 transects, or habitat units. These transects varied in size from one degree longitude, north/south, up to four degrees latitude, east/west. Both a winter and summer survey were undertaken.

The end result of this major study was a seven-volume atlas entitled *Atlas of Wildlife Habitat Inventory Maps for Environmental-Social Program, Northern Pipelines* (Ottawa: Canadian Wildlife Service, 1973. 7 v.). The introduction to each species habitat atlas described the objectives, methods used, and specific habitat requirements. The habitat unit descriptions were the unit, or transect maps, which showed that particular species' habitat areas.

The use of transects, or survey plots, can also be seen in the annual Co-operative Breeding Bird Survey (BBS), which began in 1966. This survey was based upon the Breeding Bird Survey, developed in the United States in 1964. In the Co-operative Breeding Bird Survey, the sampling unit is the degree-block of latitude and longitude. In the United States, the degree blocks are two degrees by two degrees, but in Canada, they are one degree by one degree, according to A.J. Erskin in the report entitled *The First Ten Years of the Co-operative Breeding Bird Survey in Canada* (Ottawa: Canadian Wildlife Survey, 1978. 61 p. Canadian Wildlife Service Report Series No. 42). This gives a plot area of 111 kilometres, north-south, with a width narrowing from about 79 kilometres, at 45 degrees latitude, to about 47.5 kilometres, at 65 degrees latitude north. In effect, the whole territory of Canada was broken into these blocks, or plots, but only in the inhabited areas of Canada – the south to central blocks, east to west – have been surveyed.

The sampling intensity, using volunteers, is governed primarily by the availability and distribution of qualified observers, and all-weather roads. Each survey is made up of 50 sampling stations, spaced at 0.5 miles (0.8 kilometres) along a road, and each route is covered on one day, in June, under acceptable weather conditions, starting 30 minutes before local sunrise, and spending three minutes at each of 50 stops. All birds heard, regardless of distance, and all seen within 0.25 miles (0.4 kilometres), are recorded on standard forms, together with the weather conditions at the start and end of the survey. Volunteers are recruited by appeals through naturalist newsletters, individual letters, or by personal contact. The year-to-year changes are collated and published in annual surveys entitled *The Co-operative Breeding Bird Survey* (Ottawa: Canadian wildlife Service, 1966–. Canadian Wildlife Service Progress Notes Nos. 15, 21 etc.). The main objectives of the BBS are to detect and measure year-to-year and long-term changes in populations of birds, particularly the smaller land birds. This information is also posted at the CWS website <<http://www.cws-scf.ec.gc.ca>>, under the heading *Canadian Bird Trends Database*. The review of the first 10 years of the survey has documented the effects of severely cold springs in reducing the numbers of small insectivorous birds in the Maritimes, in 1967 and 1974, and in central Ontario and Quebec, in 1974, as well as the year-to-year fluctuations in the number of seed-eating birds that occur apparently in response to changes in the crops of tree seeds. Also, changes in the Maritimes have been correlated with variations in the regime of forest spraying against the spruce budworm, first with DDT, and later with the highly toxic phosphamedon.

The Co-operative Breeding Bird Survey data has also been used in the study of birds of the northern conifer forest, usually called the boreal region of North America. This study by A.J. Erskine, entitled *Birds in Boreal Canada* (Ottawa: Canadian Wildlife Service, 1977. 73 p. Canadian Wildlife Service Report Series No. 41), was not a survey, or check list, but 'rather it attempts to suggest the groupings and densities of birds that occur in various biotic and geographic situations within the boreal region' and not within any particular provincial boundary.

Over a 20-year period, the author carried out individual bird census studies across the boreal region. A.J. Erskine published these individual studies in the journal literature, and some were entitled *Weedy Pond, Spruce-Hemlock Wood, and Second-Growth Spruce-Maple Bush. Census 34* (*Audubon Field Notes*, v. 11, 1957. pp. 456–57); *Birds Observed in North-Central Alberta, Summer, 1964* (*Blue Jay*, v. 26, 1968. pp. 24–31); *Sedge – Sweet Gale Marsh. Census 53* (*Audubon Field Notes*, v. 22, 1968. pp. 702–4), *Bird Communities in and around Cape Breton Wetlands* (*Canadian Field Naturalist*, v. 85, 1971. pp. 129–40); *Off into the Wet Green Yonder: Birds and Plants of a Boreal Bog* (*Blue Jay*, v. 32, 1974. pp. 33–37); and *Birds in the Fort Nelson Lowlands of Northeastern British Columbia* (*Syesis*, v. 9, 1976. pp. 1–11). According to the author, these census indicated that the impact of man on boreal birds has been less than in regions long settled by man, and that forestry practices in mature or old growth forests were a threat to their existence.

The immediate visible application of surveys is the summarizing of data into atlases. Besides the *Atlas of Wildlife Habitat Inventory Maps* mentioned earlier, the Canadian Wildlife Service has published two major atlases, based on surveys of pelagic seabirds of the Atlantic and Pacific seacoasts of Canada. Unlike most of the surveys already described, these atlases use data collected from seaboard observations, rather than aerial observations, because it is impractical for technical reasons to combine both types of data into single maps. However, aerial data may be used to supplement shipboard observations in the atlas descriptions. The first major atlas of Atlantic coast seabirds published by the Canadian Wildlife Service was entitled *Atlas of Eastern Canadian Seabirds* (Ottawa: Canadian Wildlife Service, 1975, 1 v.). This summarized the data of two bird distribution surveys, done in the late 1960s by R.G. Brown and D.N. Nettleship. The publication of this atlas was timely, because it provided basic information on seabird distributions for use in the environmental assessments required for offshore hydrocarbon exploration which began in the mid-1970s. Since the publication of that atlas, additional surveys by the CWS, and by consultant companies working for the oil industry, generated a much larger amount of data than was used for the original atlas. Hence, a new atlas was deemed necessary. A revised edition, which incorporated the additional data gathered until the end of 1983, began publication in 1986. This new series by R.G.B. Brown was entitled *Revised Atlas of Eastern Canada Seabirds I. Shipboard Surveys* (Ottawa: Canadian Wildlife Service, 1986–). It combines the old and new shipboard data of observations made in 10-minute watches in good visibility from a ship moving at not less than four knots per hour (74 kilometres per hour). The average number of birds indicated per 10-minute watch refers to the numbers of birds per linear kilometre. A computerized program plots these averages as numbers, or symbols. The distributions themselves are plotted onto two maps, divided at 60 degrees north latitude, but with some overlap for an Atlantic map and an Arctic map. The data on the Atlantic map refers to block, or plot sizes, 30 minutes north by one degree west, but because of the converging longitudes farther north, the Arctic blocks, or plots, are one degree north by two degrees west. The map scale itself is five degrees north equals 300 nautical miles or about 555 kilometres. The maps themselves have been divided into seasonal observations; January–March (winter); April–June (spring, early summer); July–August (late summer); and September–October (early winter). Further breakdowns are by species counts (e.g., thick-billed murres – *Uria lomvia*). The atlas also contains commentaries on 'The Oceanography of the Northwest Atlantic, Western Canada Arctic,' and 'Conservation and Management.'

An atlas of pelagic birds of the Pacific Coast is the most recent atlas published by the Canadian Wildlife Service. Its authors are K.H. Morgan and K. Vermeer, and it is entitled *Atlas of the Pelagic Birds of Western Canada* (Ottawa: Canadian Wildlife Service, 1991. 72 p. Canadian Wildlife Service Occasional Paper No. 72). Pelagic bird surveys off British Columbia were first initiated in 1981 on an 'opportunity' basis, from ships of the Institute of Ocean Sciences at Sydney, the Pacific Biological Station at Nanaimo, and the Defence Research Establishment Pacific at Esquimalt. The reasons for undertaking surveys at this time was to obtain a better knowledge of the numbers and species of pelagic birds coming to the coastal areas of British Columbia, and to collect baseline data for assessing the potential effects of environmental perturbations on pelagic bird populations. By 1989, enough census data had been collected to support the development of a west coast atlas. Altogether, 45 pelagic bird surveys were conducted along the

British Columbia coast, some as far seaward as the 200-mile (320-kilometre) territorial limit, between 1980 and 1990. The surveys closely followed the methods used by R.G. Brown for the east coast pelagic bird surveys, and atlas, described above. Overall, the transect, or plot data, from 5,355 transects, were analysed and broken down by season. The distances traversed were 20,675 kilometres of ocean in a range, from south to north about 48 degrees to 55 degrees latitude, and east-west from 124 degrees to 138 degrees longitude. The seasonal distributions of 20 species were mapped, along with an additional 12 rare, or accidental, species. Finally, the authors noted that the waters over the continental shelf (shallower than 200 metres) 'generally supported the highest pelagic bird density (all species combined) as well as minimal numbers of most species.'

The Canadian Wildlife Service has also participated in the publication of one other major atlas by R.W. Campbell *et al.* entitled *The Birds of British Columbia* (Vancouver: UBC Press, 1997–). This is a projected four-volume set. To date, three volumes have been published. They are: v. 1. *Nonpasserines: Introduction and Loons through Waterfowl*; v. 2. *Nonpasserines: Diurnal Birds of Prey through Woodpeckers*; and v. 3. *Passerines: Flycatchers through Vireos*. The fourth projected volume is *Passerines: Wood Warblers through Old World Finches*. The atlas, which brought together the work of volunteers, as well as researchers, consists of descriptions with illustrations along with maps showing the range of individual species.

F Environmental Impact Studies

The federal government accepted the necessity of environmental impact studies in the early 1970s because of the impetus to Arctic oil and gas exploration given by the international oil crisis of that decade. Other than the Mackenzie Valley Pipeline routes, where work was often undertaken by Fisheries Canada, the other area of focus for environmental impact studies involved the proposed Arctic Island Gas Pipeline Route. In the early 1970s, major gas fields were discovered on Melville, Ellef Ringnes, and King Christian Islands, in the Canadian High Arctic. These discoveries led to the consideration of various alternatives for moving the gas to southern markets, foremost of which was the construction of a 48-inch pipeline to connect the islands with the eastern Keewatin mainland, down through the Boothia Peninsula. In anticipation of an application by industry to build the proposed pipeline, the Arctic Islands Pipeline Program (AIPP) was launched, unofficially in 1973, and expanded in 1975, under the aegis of the AIPP Study Board. The environmental impact studies on wildlife in the high Arctic were carried out under the Environmental Social Program, Northern Pipelines, of the Government of Canada, by the Canadian Wildlife Service, and did provide a further impetus to the other unrelated mammalian and bird studies that were under way at that time, in order to increase the knowledge needed for management of northern wildlife. The studies involved caribou, muskoxen, and polar bear ecology, and migratory bird surveys and studies in the eastern Arctic, or Keewatin District.

Studies of habitats of caribou (*Rangifer tarandus pearyi* and *Rangifer tarandus groenlandicus*) and muskoxen (*Ovibos moschatus*) along the proposed pipeline route, between Resolute Bay and Spence Bay, Northwest Territories, were carried out in 1975 and 1976. These were expanded, in 1976, to include the Prince of Wales Island and northern Boothia Peninsula. The remainder of Boothia Peninsula, and the northern District of Keewatin, were examined in 1977. Prior to this study, knowledge was limited regarding what constituted important seasonal caribou and muskox habitats. Examples of earlier work were papers such as *The Caribou of Boothia Peninsula* (Ottawa: Canadian Wildlife Service, 1960. 16 p. Unpublished Report No. 822); J.P. Kelsall's 1968 monograph, previously mentioned, entitled *The Migratory Barren-Ground Caribou of Canada* (Ottawa: Canadian Wildlife Service, 1968. 340 p. Canadian Wildlife Service Monograph No. 3); and the report by D.C. Thomas entitled *Population Estimates of Barren-Ground Caribou March to May, 1967* (Ottawa: Canadian Wildlife Service, 1969. 44 p. Canadian Wildlife Service Report No. 9).

The study documented that caribou on the Prince of Wales Island, Somerset Island, and Boothia Peninsula showed a preference for low moisture to moderately moist land, and moderately to poorly vegetated habitats, in both winter and summer. Major caribou wintering grounds were found in Somerset Island and eastern Boothia Peninsula, while the Prince of Wales Island, and western and central Boothia, appeared to be mainly used as a summer range. The seasonal distribution implied east-west migrations of these caribou populations – both inter-island (across Peel Sound), and insular (across the interior highlands of Boothia Peninsula). Caribou in the southern regions of the study area tended to select upland barrens for winter use, and more productive lowlands in summer. Competition for range by caribou and muskoxen was found to be minimal. In the study, muskoxen were found in substantial numbers on Prince of Wales Island. Their preferred habitats were well-vegetated lowlands, and their seasonal movements were localized. Based on the information obtained by R.H. Russell and E.J. Edmonds, in their report entitled *Caribou and Muskoxen Habitat Studies: AIPP Report 1978* (Ottawa: Canadian Wildlife Service, 1979. 140 p. ESCOM Report No. AI-26), the authors said that the proposed primary pipeline route skirted through several critical wintering grounds of caribou. As well, portions of the north-central uplands of Boothia Peninsula were used intensely by caribou in both winter and summer, and there might also be a calving ground in the region. The authors suggested an alternate route from Bathurst Island to the Prince of Wales

Island and on to Boothia Peninsula that would bypass inland caribou ranges, and precautions should be taken to avoid excessive destruction of muskox habitat.

Additional studies pertaining to the wintering caribou herds were also undertaken, in 1976 and 1977, to obtain a better knowledge of their size and distribution in the northeastern mainland of the Keewatin District. Information on caribou occupying the vast areas north of Chesterfield Inlet and Wager Bay was virtually non-existent at the beginning of these studies. The region was divided into two study areas. The Lorillard area, occupying 65,000 square kilometres, was bounded by Chesterfield Inlet on the south, Hudson Bay to the east, and Wager Bay on the north. The Wager Bay area, bounded by the Gulf of Boothia and the Arrowsmith River to the north, Hudson Bay to the east, and Wager Bay to the south, covered approximately 57,000 square kilometres. Reconnaissance flights to determine general caribou density and distribution were undertaken north of Chesterfield Inlet in June, 1976, and north of Wager Bay in July, 1976. Population estimates were based on statistical formulae, using theoretical transit lines established for the calculations. The Lorillard Herd was estimated at about 4,000 adults, and the Wager Herd was estimated at about 10,000. The Lorillard Herd, which showed a distinct calving area, had a productivity ratio of 56 calves for every 100 cows. On the other hand, the Wager Herd, where calving occurred earlier, and north of the study area, had a calving ratio of 60 calves for every 100 cows. The initial study of these herds was undertaken by D.C. Heard and G.W. Calef, and documented in the paper entitled *Numbers, Distribution and Productivity of Caribou in Northeastern Keewatin, N.W.T. in 1976* (Ottawa: Canadian Wildlife Service, 1981. 27 p. ESCOM Report No. AI-39).

Additional work by G.W. Calef and D.C. Heard tended to reinforce those population estimates of the Lorillard and Wager Herds. As well, the calving areas of both groups were confirmed. The calving area of the Lorillard Herd was a 17,000-square-kilometre area of highlands at the headwaters of the Lorillard River, and that of the Wager Herd was a 5,000-square-kilometre area north and west of Wager Bay, near Curtis Lake. An additional survey of a herd on the Melville Peninsula completed the documentation of the caribou herds of the eastern Keewatin. This Melville Herd was estimated at 52,000. This additional work was presented at the Second International Reindeer/Caribou Symposium, Rorus, Norway, in 1979, and afterwards published with the title *The Status of Three Tundra Wintering Caribou Herds in Northeastern Mainland, Northwest Territories* (Ottawa: Canadian Wildlife Service, 1980. 22 p. ESCOM Report AI-44).

The following year, the Lorillard and Wager herds were studied again, but the estimated size of the former was placed at 1,400, and that of the latter herd at 2,976. These estimates were one-tenth and one-third of the 1976 estimates. According to J.L. Donaldson, in the report entitled *Population and Recruitment*

Estimates for the Lorillard and Wager Caribou Herds in 1977 (Ottawa: Canadian Wildlife Service, 1981. 50 p. ESCOM Report No. AI-43), 'the changes in population estimates of the Wager and Lorillard herds from 1976 to 1977 cannot be accounted for in terms of recruitment or mortality. The herds may have shifted their entire distributions or may calve over a larger area than the survey suggests.' This year-to-year difference indicated that greater knowledge of the movements and distributions of the Lorillard and Wager Herds was needed to reduce the impact of potential pipeline development. For example, the summer movements and distribution of the Lorillard Herd remained unknown, and neither herd had been surveyed between September and June.

The other large mammal that occupied the Arctic Archipelago areas through which the possible Arctic gas pipeline would go was the polar bear. Preliminary studies on environmental impact were carried out, from 1970 through 1976, north of Spence Bay on the Boothia Peninsula, and included the large islands and peninsula of Melville Island, Byam Martin Island, Bathurst Island, Little Cornwallis Island, Cornwallis Island, Grinnell Peninsula, Prince of Wales Island, and Somerset Island. In order to do a preliminary assessment, information on the population ecology, distribution, abundance, seasonal movements, numbers of discrete sub-populations affected, and location of important denning, feeding, and summer retreat areas had to be acquired. As well, researchers had to try to identify important areas, or times in the annual cycle of the polar bear, that might warrant protection from, or modification of, possible construction or operational activities related to the pipeline. I. Sterling and R.E. Schweinsburg undertook these preliminary studies, and the report was entitled *Polar Bear Population Ecology – Arctic Islands Pipeline Route: AIPP Preliminary Report 1977* (Ottawa: Canadian Wildlife Service, 1978. 71 p. ESCOM Report No. AI-15). These preliminary studies indicated that polar bears in the central and high Arctic showed a high degree of 'fidelity' to winter and summer feeding areas, and some captured and tagged bears showed a strong degree of fidelity to specific areas. As well, maternity denning in the study area appeared to occur over a widespread area. The authors concluded that the pipeline activities in land areas would not appear to have serious consequences, but the use of ice-breakers to extend the open water season in Lancaster Sound and Barrow Strait would cause significant changes in the distribution of polar bears in that area, and possibly affect their mortality and birth rates as well.

The largest of these environmental impact investigations involved the study of migratory bird populations, as the whole route for the Polar Gas Pipeline stretched from the Manitoba border, in the south, through the Districts of Franklin and Keewatin and the Boothia Peninsula, to the northern end of Somerset Island. Hence, the length of the study was about 1680 kilometres, in a 32-kilometre-wide corridor, or a total area of roughly 54,000 square kilometres. The lack of published information pertaining to migratory bird distribution, and habitat types, along the proposed Polar Gas Pipeline route indicated the need for additional knowledge before an impact assessment could be made. Studies, including the breeding requirements, began in 1976, and priority was given to ducks, geese, and swans, because of their economic importance as game birds, or because of Canada's obligations under the Migratory Birds Convention Act. Four aerial surveys were executed during the 1976 field study, using standard aerial survey techniques. The priority birds were recorded as to species, number, location, age, and sex, whenever possible, using a Cessna 185 aircraft equipped with wheel-skis or floats and flying at an altitude of 30–45 metres above ground. The area of the survey was broken down into five workable zones, which related to geographic-vegetation areas such as the forest-tundra transition zone, the low shrubs–tundra complex, the low and mid-Arctic semi-desert, the Victoria Island willow-tundra complex, and the low and polar desert zone.

These surveys were undertaken and documented by R.W. Zdan and D.B. Brackett in their paper entitled *Migratory Bird Population Surveys in the District of Keewatin and Somerset, 1976: AIPP Preliminary Report 1977* (Ottawa: Canadian Wildlife Service, 1978. 111 p. ESCOM Report No. AI-18). In this preliminary survey, the authors noted that three regions reoccurred as important for all migratory groups. These included the generally well vegetated areas south of the Thelon River, the Hayes River delta, and the surrounding area including Franklin Lake and the Rasmussen Basin lowlands, which included the delta of the Murchison and Inglis Rivers. Migratory bird abundance distribution, and use of available habitat, were documented. In total, 59 bird species were noted. However, the migratory bird abundance had yet to be studied for the Boothia Peninsula and Somerset Island study areas.

The second part of the study, carried out in 1977, was confined to the District of Keewatin, because of budgetary restraints. This study corridor was about 100 kilometres wide, stretched from the Manitoba border to Spence Bay, and consisted of four aerial surveys within the restricted area. In the final report by D.L. Allen and T.H. Hogg entitled *Bird Studies in the Keewatin District: AIPP Report 1978* (Ottawa: Canadian Wildlife Service, 1979. 127 p. ESCOM Report No. AI-27), the authors recommended that the proposed pipeline right-of-way, and ancillary facilities, be kept away from areas important to birds, construction and maintenance activities be scheduled to avoid important areas to birds, and engineering designs that minimized noise be used. The authors also listed areas for further study, such as aerial bird surveys of the Boothia Peninsula and Somerset Island, to assess the impact of the Polar Gas Pipeline in that region, an aerial survey for peregrine falcon and gyrfalcon nests, along the pipeline corridor, as well as extensive ground surveys to determine the distribution and habitat preferences of passerines (perchers) and shorebirds.

G Toxicology Research

Continual federal research into pesticide effects on wildlife in Canada has been conducted only since 1963, in contrast with American research that dates back to the 1940s. It is a specific form of environmental impact research, which documents damage already done to wildlife and the environment. According to the Canadian Wildlife Service, the lack of ongoing research was due to the fact that occasional Canadian studies, done in the 1940s and 1950s, showed no serious effects on wildlife. However, by the 1960s, research in Europe and the United States finally showed that immediate wildlife kills, caused by pesticides, was just the 'tip of the iceberg.' That research showed a magnification of persistent residues in the food chain of wildlife that produced either direct mortality, or reduced productivity in populations over time long after actual pesticide use.

The early synthetic pesticides were a group of organic compounds called organochlorides, which are also known as chlorinated hydrocarbons. These are compounds made up of carbon, hydrogen, and chlorine and include DDT (dichlorodiphenyl-trichloroethane), DDD (dichlorodiphenyl-dichloroethylene, and others such as dieldrin, heptachlor, chlordane, mirex, and toxaphene. Besides these organochlorides, which were widely used to control insect pests, there are chlorinated industrial chemicals which differ from pesticides in that they are not directly applied, but are both intentionally and accidently discharged into the environment. These organochlorines are the polychlorinated biphenyls (PCBs), and chlorinated dioxins and furans. PCBs have had a wide range of use in hydraulic fluids, lubricants, and closed-circuit heat transfer systems. Dioxins and furans have been used in the wood industry for the bleaching of wood pulp and in fuel additions. Finally, there is another group, the heavy metals such as mercury, used in pulp mills, or organic mercury derivatives, applied to grain seed to control cereal diseases, such as smut and bunt, and cadmium and lead, which are used in thermal power plants, or lead by itself, which is used as a fuel additive, or found in lead-shot, which affects the health of wildlife.

The earliest work by the Canadian Wildlife Service was an attempt to document the immediate effects on wildlife of DDT substitutes in the control of the spruce budworm (*Chloristoneura fumiferana*) in New Brunswick. Aerial spraying of New Brunswick forests to control the spruce budworm, begun in the early 1950s, had produced, so it seemed, no obvious effects on wildlife. In 1964, however, an organophosphate called phosphamidon, about which little was known, was being used as a substitute. In order to gain a greater knowledge about the toxic effects of this pesticide, the Canadian Wildlife Service undertook studies, between 1965 and 1968, to test the effects of this pesticide on bird populations. The results would also be of interest to entomologists. The initial experimental design included nine plots, each about 21 square miles (about 13,000 acres or 5,200 hectares), in an area of heavy budworm infestation, lying between the southwest Miramichi River and the Gaspereace River in the Cains River watershed. The objective of the program was to test the effects of varying concentrates of spray emissions of phosphamidon from aircraft, early and late in the season, on bird populations. The experiments were organized by C.D. Fowle, and the final published results were entitled *Effects of Phosphamidon on Forest Birds in New Brunswick* (Ottawa: Canadian Wildlife Service, 1972. 25 p. Canadian Wildlife Service Report Series No. 16). The study concluded that aerial spraying of phosphamidon, at rates of up to 0.5 pounds per acre, and spray which had a high proportion of fine evenly dispersed droplets, resulted in substantial mortality of forest birds. Also, birds could pick up lethal, or debilitating, dosages from sprayed vegetation, through their feet, or be poisoned by eating sprayed food soon after spraying. Much later, this type of work was moved into the laboratory, where small songbirds were exposed to varying concentrations of fenitrothion (a more acceptable organic pesticide forest spraying) in chambers, in order to measure pesticide residue concentrations and effects. This work was reported by P. Mineau and K.M. Sundaram in the paper entitled *An Improved Method to Study the Impact of Pesticide Sprays on Small Song Birds* (*Journal of Environmental Science and Health. Part B. Pesticides Food Contaminants and Agricultural Wastes*, v. 25, 1990. pp. 105–36). The authors said that 'significant depression in body weights and brain acetylcholinesterase levels were noted only for the highest exposure group. Fenitrothion residues in blood were detectable only at the highest exposure level, and at the two higher levels. Carcass and feather residues were much higher than those in blood and liver and were detectable at all exposure levels.'

The residual effects of organic compounds were first studied by the Canadian Wildlife Service in grain-growing districts of Alberta where the use of mercury-treated seed was common. Organic mercury derivatives were introduced, in the 1920s, for application to wheat, barley, and oat seeds, in order to control such cereal diseases as smut and bunt, as they proved highly effective in controlling various seed-borne diseases while being relatively harmless to plants. However, the use of alkyl-mercury compounds was undesirable for wildlife, because the mercury was picked up from the uncovered treated grain by seed-eating birds and mammals, and passed on to their predators. One hundred and ninety-two penned pheasants (*Phasianus colchicus*), including control groups, were studied for the effects of a feeding diet involving 100%, 50%, and 25% treated grain in the feed ration. The study was carried out by N. Fimreite and entitled *Effects of Dietary*

Methylmercury on Ring-necked Pheasants: With Special Reference to Reproduction (Ottawa: Canadian Wildlife Service, 1971. 39 p. Canadian Wildlife Service Occasional Paper No. 9). Results indicated that no weight reduction in adult birds could be ascribed to the mercury compound, but strong adverse effects on reproduction were found. These effects were reduced hatchability, reduced egg production, and a large number of shell-less eggs. The author concluded 'that hatchability, at least, can be affected among wild birds where they have access to treated grain.'

More recently, the lasting effects of banned organochlorides on birds have been studied in orchard areas of British Columbia. Over a two-year period, during 1990 and 1991, bird eggs were collected from orchard and non-orchard areas of the fruit-growing region of the Okanagan Valley of central British Columbia. Although DDT, and other organochlorines, had not been used since the early 1970s, J.E. Elliot and P.A. Martin, in their paper entitled *Organochlorines and Reproductive Success of Bird in Orchard and Non-Orchard Areas of Central British Columbia, Canada, 1990–91* (Archives of Environmental Contamination and Toxicology, v. 26, 1994. pp. 435–43), indicated that levels of DDT and DDE persisted in the eggs of orchard birds, but was not deleterious to their reproductive abilities.

The most thorough studies of pesticide residues and contaminants have been undertaken on seabirds and colonial birds of the Great Lakes, and Canada's eastern, western, and Arctic coastlines. Since 1968, the Canadian Wildlife Service has analysed levels of organochlorines and mercury in eggs, and tissues, of 24 species of marine birds collected from the Atlantic, Pacific, and Arctic coasts. The CWS also analysed a small sample of east coast birds for lead and cadmium. This monitoring program has provided an indication of the relative contamination of the Maritime regions of Canada. Based on contaminants in seabird eggs, the St Lawrence River estuary, and nearby gulf, were among the most contaminated maritime regions of Canada, followed by the Bay of Fundy, the Strait of Georgia, and the west coast of Vancouver Island. Residues among the Arctic Islands, in the marine environment of Newfoundland, and along the coast of northern British Columbia were relatively low. Overall, the residues of most contaminants, particularly PCBs, DDE, and dieldrin, have declined everywhere, although the decline in the St Lawrence estuary is not statistically significant. In general, the trend in levels of contaminants has followed the patterns of use in North America. DDT and dieldrin have shown the most significant declines, as both these compounds have been banned in Canada and the United States since the early 1970s. Declines in other compounds such as chlordane show persistent low levels, probably due to restrictions imposed in 1978. The multi-year data on pesticide residues was analysed by D. Noble, and the ensuing report was entitled *Contami-*

nants in Canadian Seabirds (Ottawa: Canadian Wildlife Service, 1990. 75 p. State of the Environment Report Series No. 90–2). Since 1990, additional documentation of organochlorine residues and heavy metals has been noted in papers by J.E. Elliott and A.M. Scheuhammer, entitled *Heavy Metal and Metallothionein Concentrations in Atlantic Canadian Seabirds* (*Archives of Environmental Contamination and Toxicology*, v. 22, 1992. pp. 63–73); by M.B. Gebauer and D.V. Weseloh, entitled *Accumulation of Organic Contaminants on Sentinel Mallards Utilizing Confined Disposal Facilities at Hamilton Harbour, Lake Ontario, Canada* (*Archives of Environmental Contamination and Toxicology*, v. 25, 1993. pp. 234–43); by L. Champoux, entitled *PCBs, Dioxins and Furans in Hooded Merganser (*Lophodytes cucullatus*), Common Merganser (*Margus merganser*) and Mink (*Mustela vison*) Collected along the St. Maurice River near La Tuque, Quebec* (*Environmental Pollution*, v. 92, 1996. pp. 147–53; and by J.E. Elliott and L.K. Wilson, entitled *Bald Eagle Mortality and Chlorinated Hydrocarbon Contaminants in Livers from British Columbia, Canada, 1989–1994* (*Environmental Pollution*, v. 94, 1996. pp. 9–18).

In 1970, a more geographically restricted study of pesticide residues began on colonies of herring gulls (*Larus aegentatus*) throughout the Great Lakes. In this study, eggs had been collected from colonies and analysed as a means of assessing trends in levels of organochloride contaminants. By 1986, 41 individual contaminants had been determined by laboratory analysis, and older samples of gull eggs were retrieved from the specimen bank for reanalysis because of method improvements. This data has been published by the Canadian Wildlife Service with the title *An Atlas of Contaminants in Eggs of Fish Eating Colonial Birds of the Great Lakes (1970–1988)* (Ottawa: Canadian Wildlife Service, 1992. 2 v.). Although the situation regarding organochlorine contamination of the environment has improved since the late 1960s, collections of avian prey species such as peregrine falcon indicated that chemical 'hot spots' were still present in southern Ontario and may have a significant effect on the recovery of the peregrine falcon. Data on organochlorine residue levels in eggs and tissue from the peregrine falcon collected in Canada, over the years 1965–1987, were documented by D.B. Peakall and D.G. Noble in their paper entitled *Environmental Contaminants in Canadian Peregrine Falcons,* Falco peregrinus: *A Toxicological Assessment* (*Canadian Field Naturalist*, v. 104, 1990. pp. 244–54). Based on this information, A. Baril and J.E. Elliot in their paper entitled *Residue Levels of Environmental Contaminants in Prey Species of the Peregrine Falcon;* Falco peregrinus, *in Canada* (*Canadian Field Naturalist*, v. 104, 1990. pp. 273–84), suggested that when planning releases of captive bred peregrines, consideration should be given to potential contamination problems.

H Birds as Hazards in the Environment

Although humans have a major impact on wildlife and their environment, the Canadian Wildlife Service also studies the management of wildlife that affect human activities. The grizzly bear studies in the western national parks as previously described are an example of this type of research. However, more of this type of research has involved migratory birds that create hazards to human activities. Two types of hazards that have been studied by the CWS involve birds as hazards to aircraft, and birds as hazards to fruit crops. In 1963, the Canadian Wildlife Service began work with the National Research Council, the Department of Transport and the Department of Defence, airline services, and engine manufacturers to study the technical problem of bird hazards. On its own, however, the Canadian Wildlife Service set out to study habitat techniques that could be used around airports to reduce the bird hazard problem, radar techniques for forecasting bird hazards, and the types of bird movement revealed by radar.

Biological studies at a number of major airports showed that many species of birds used airport areas for feeding, resting, and even nesting, for a variety of ecological reasons which could be altered. For example, garbage dumps, and other attractive food supplies, could be moved away from airports, and small bodies of water, which attracted aquatic and other birds, could be drained, or filled, to prevent birds from using them. As well, nesting and perching sites could be eliminated or reduced. Additional techniques to control birds at airports could involve mechanical means such as pyrotechnic devices (shotgun shells that fire a small explosive projectile), or distress calls using amplifiers and loudspeakers. Radar studies were combined with the use of motion picture photography of radar scopes in order to provide permanent records for study. Examination of film records indicated that the information available about bird migration, based on visual observations, was incomplete. The radar studies showed that during either the spring or autumn, migrating birds generally begin their migratory flight when conditions are favourable, and backtrack when weather conditions are less than favourable, thus making it easier to forecast when major movements will occur in the spring and autumn. The end result of these biological and radar studies was the publication of a collection of papers by Canadian Wildlife Service entitled *Studies of Bird Hazards to Aircraft* (Ottawa: Canadian Wildlife Service, 1971. 105 p. Canadian Wildlife Service. Report Series No. 14). In regard to the migratory bird hazard, the CWS suggested that the bird migration times should be handled by civil and military aircraft in the same manner as thunderstorm warnings. That is, flight patterns should be adjusted to take into account the bird migration hazard.

Of specific concern was the migration of the eastern populations of the lesser snow and blue goose which winters along the coast of the Gulf of Mexico, and in the spring migrates, in a northwesterly direction, through the United States and into southern Canada. When migrating from the plains in southern Manitoba to the coasts of James and Hudson Bay to their nesting area, the geese fly in a northeasterly direction, in large flocks, across the Terminal Control Area of Winnipeg International Airport. This is an area with a radius of 30 nautical miles (50 kilometres). Because an aircraft was struck by snow geese in 1969, about 12 nautical miles northeast of Winnipeg, the Canadian Wildlife Service was asked to monitor the migration, and design and test techniques for determining the height of migratory bird flocks that could be used in air traffic control. H. Blokpoel, a CWS specialist on radar studies of bird movements, undertook the study and the findings were entitled *Migration of Lesser Snow and Blue Geese in Spring across Southern Manitoba. Part I: Distribution, Chronology, Directions, Numbers, Heights and Speed* (Ottawa: Information Canada, 1974. 30p. Canadian Wildlife Service Report Series No. 28). For this study, the spring 1970 and 1971 migrations were monitored in the vicinity of the Winnipeg International Airport, using both long-range and short-range surveillance radars to detect the flocks. Visual observations were also used. Analysis of these observations provided information on the staging areas, routes, chronology, directions, numbers, ground speeds, heights, and air speeds of the migratory geese. However, research still had to be carried out, in cooperation with the National Research Council, to develop an inexpensive height-finding radar that could be used in air traffic control operations.

Ironically, the study of bird hazards to fruit crops, which was localized to the fruit-growing area of the Niagara Peninsula of southern Ontario, was far more complex in practice than the study of birds as hazards to aircraft. The Niagara Peninsula is a major fruit-growing region, which also supports a substantial wine-growing industry. But, the taste and visual factors that make fruit attractive to humans also make fruit attractive to birds.

It is the common robin that is the major pest to fruit in the Peninsula, and the damage caused by this and other birds such as starlings, and grackles, can have serious economic consequences. Cherries and grapes are the fruit that suffer the most damage, with cherry losses sometimes as high as 25%, and grape losses at 10% of the crop. In the study undertaken by R.G. Brown, entitled *Bird Damage to Fruit Crops in the Niagara Peninsula* (Ottawa: Canadian Wildlife Service, 1974. 57 p. Canadian Wildlife Service Report Series No. 27), the author noted that early cherries were taken by robins (*Turdus migratorius*) and grackles (*Quiscalus quiscala*), and later cherries by starlings (*Sturnus vulgaris*). Grapes were taken mainly by robins, with some damage done by Baltimore orioles (*Icterus galbula*). There was also a strong preference by robins and starlings for the reddest cherries and the blackest grapes. The author examined various possibilities of damage control, which ranged from 'spoil' crops to distract the birds, to the use of nets, or physical means, but not the control of birds themselves, which is against the mandate of the Canadian Wildlife Service.

I Wetland Studies

Although small wetlands on the prairies and elsewhere across Canada were long recognized as the breeding grounds for North American waterfowl, it was not until the 1960s that the federal government tabled in Parliament the National Wildlife Policy and Program, which included a call for the preservation of suitable wetland habitats by acquisition, lease, or agreement, in sizes sufficient to support the well-being and perpetuation of waterfowl species, including balancing hunting needs. Besides being essential for the maintenance of fish and wildlife species, wetlands are extremely productive as an ecosystem because they serve as nutrient traps, water purifiers, and erosion and flood control mechanisms. As well, wetlands help to maintain ground water levels and water storage areas. Yet, despite the essential value of this unique ecosystem, wetlands, such as marshes, have more commonly been regarded as unexploited wastelands and obstacles to development by people. According to Wildlife Habitat Canada, in their publication entitled *The Status of Wildlife Habitat in Canada: Problems, Issues and Opportunities* (Ottawa: Wildlife Habitat Canada, 1986. 72 p.), drainage of wetlands has had a very serious impact on wildlife habitat, and has affected many species. Since the days of early settlement, more than 65% of Maritime salt marshes, 70% of southern Ontario and St Lawrence valley wetlands, 40% of prairie wetlands, and up to 70% of Pacific estuary wetlands have been converted to other uses, mainly for agriculture and urban expansion. Besides, changes in existing wetlands, as well as jurisdictional problems and insufficient habitat protection measures, have all hindered wetland management.

Wildlife Habitat Canada was established by Environment Canada and the Canadian Wildlife Service in the mid-1980s as an independent, non-profit foundation whose objective was to develop cooperative programs with provincial, territorial, and federal government agencies, to arrest the continuous loss of habitat of wildlife. Besides the above-mentioned publication, the foundation has other publications which have had input from the Canadian Wildlife Service, Environment Canada, or the North American Wetlands Conservation Council. One publication by M.J. Bardecki was entitled *Wetland Evaluation: Methodology, Development and Pilot Area Selection* (Ottawa: Environment Canada, 1987. 114 p.). Other publications have been *Wetlands Are Not Wastelands. Interim Report* (Ottawa: Canadian Wildlife Service, 1988. 12 p.); *Wetland Evaluation Guide:*

Final Report of the Wetlands Are Not Wastelands Project (Ottawa: North American Wetlands Conservation Council (Canada), 1992. 121 p.); and *Saving Species: Building Habitat into Endangered Species Conservation in Canada* (Ottawa: Wildlife Habitat Canada, 1995. 23 p.). Additional information about Wildlife Habitat Canada, and their publications, can be found at their website <<http://www.whc.org>>.

The loss of wetlands has led to the constant acquisition of wetlands across Canada, and their study by the federal government since the late 1960s. For example, in 1972–73, the Canadian Wildlife Service reported that it had bought 1,777 acres of wetlands that year, which brought the total of national wildlife areas up to 37,121 acres (15,026 hectares). The cost was $6 million. In 1974, the CWS reached an agreement with the Manitoba government to rehabilitate the Delta marsh in southern Manitoba. As well, the 402-acre (162-hectare) Hahn Marsh, on Lake Erie, and the 620-acre (251-hectare) Dover Marsh, on Lake St Clair, Ontario, were purchased. With the land acquisitions program that year, the total acreage acquired for waterfowl habitat was 42,500 acres (17,200 hectares), for a cost of $8 million to date. In 1975, the 80-acre (32-hectare) Marshall–Stevenson property, at the mouth of the Little Qualicum River, on Vancouver Island, was acquired. In 1979, a new National Wildlife Area at Long Point, one of the most fragile and biologically vulnerable parts of the Lake Erie shore, was established on land given by the Long Point Company and the US Nature Conservancy. After a decade of acquisitions, the Canadian Wildlife Service had created 45 national wildlife areas which comprised about 124,000 acres (50,000 hectares). And, in 1981, as previously mentioned in this chapter, Canada became a signatory to the international RAMSAR Convention 'to stem the progressive loss of wetlands now and in the future.'

Besides acquiring wetlands, the Canadian Wildlife Service has also created just under 100 migratory bird sanctuaries, which total over 115,000 square kilometres. These sanctuaries range in size from the one-acre Christie Islet of British Columbia, to the 63,750-square-kilometre Queen Maud Gulf Migratory Bird Sanctuary in the Northwest Territories. The ability to create wildlife sanctuaries, or reserves, was an outcome of the 1973 Canadian Wildlife Act, which authorized the establishment of National Wildlife Areas and Migratory Bird Sanctuaries through donations and leases from government agencies or private citizens. The most recent international emphasis on habitat reserves saw the creation of Canada's first hemispheric shorebird reserve for the protection of migrating shorebirds at Shepody Bay, New Brunswick, in 1987. The bay attracts more southward migrating birds than any other place in North America. At the same time, Canada and Surinam signed an international agreement to create a hemispheric shorebird reserve in Surinam, and to 'twin' that reserve with the one on Shepody Bay. The two reserves are directly linked by the semi-palmated sandpapers (*Calidris pusilla*), which fly each fall non-stop for three to four days over the 4,000-kilometre distance between the two sites. The documentation of potential or actual bird sanctuaries, or reserves, in Canada has been undertaken and described by R.I. Morris *et. al.* in the publications entitled *Potential Western Hemisphere Shorebird Reserve Network Sites for Migrant Shorebirds in Canada* (Ottawa: Canadian Wildlife Service, 1991. 98 p. Canadian Wildlife Service Technical Report Series No. 144); and *Potential Western Hemisphere Shorebird Reserve Network Sites for Shorebirds in Canada*, 2nd Edition (Ottawa: Canadian Wildlife Service, 1995. 104 p. Canadian Wildlife Service Technical Report Series No. 227).

This systematic acquisition of wetlands by the Canadian Wildlife Service, and ongoing publicity, can be set against a background of studies of wetlands in Canada, ranging from the prairies, where 62% of North American ducks are produced, and where 98% of Canada's wheat is produced, to the Maritime provinces. In 1967, the Canadian Wildlife Service organized a seminar entitled *Saskatoon Wetlands: Transactions of a Seminar on Small Water Areas in the Prairie Pothole Region, held 20 to 22 February 1967, to mark the opening of the Prairie Migratory Bird Research Centre in Saskatoon, Saskatchewan* (Ottawa: Canadian Wildlife Service, 1969. 262 p. Canadian Wildlife Service Report Series No. 6). This seminar followed a House of Commons statement by the then-minister of Indian Affairs and Northern Development a year earlier, that 'suitable wetlands habitat in amounts sufficient to support desired populations of ducks and geese will be preserved.' This statement came with the awareness that the conflict with agricultural interests over small wetland areas needed to be resolved through education and a change in philosophy on the use of land – a shift from the single-use approach to a multi-use approach. The seminar reviewed the significance of small wetlands, the knowledge of and the research needs for understanding small water areas, the relationship between waterfowl and small water areas, and the assessment of breeding populations.

One of the participants in this seminar was J.B. Millar, a wildlife biologist with the Canadian Wildlife Service, who, over a 10–year period, studied approximately 100 wetlands in the grassland and aspen parkland regions of Saskatchewan. In 1972, Millar completed studies on wetland classification and published the paper entitled *Wetland Classification in Western Canada: A Guide to Marshes and Shallow Open Water Wetlands in the Grasslands and Parklands of the Prairie Provinces* (Ottawa: Canadian Wildlife Service, 1976. 38 p. Canadian Wildlife Service Report Series No. 37). In this paper, the author said that 'land planning is essential if we hope to guide future developments in a way which minimizes further unnecessary and irreversible destruction of our dwindling natural sources' and that 'the first step in planning land use is to classify the kinds of natural habitats which still exist,' so that it is possible to interpret the potential of different types of habitat for producing such valuable renewable resources as wildlife and water. The classification was based upon vegetation, as well as physical characteristics. Additional work on wetland classification has been carried out by R.D. Rowsell and M. Woo, and described in their paper entitled *Hydrological Classification of Canadian Prairie Wetlands and Prediction of Wetland Inundation in Response to Climatic Variability* (Ottawa: Canadian Wildlife Service, 1993. 24p. Canadian Wildlife Service Occasional Paper No. 79).

At about the same time, in northeastern Alberta and northwestern Saskatchewan, the Canadian Wildlife Service was concerned about the ecological effects on the Peace River–Athabasca Lake Delta, caused by the altered water regime, or flow, following the completion of the W.A.C. Bennett Dam, on the Peace River. This 1,700-square-mile delta is situated in northeastern Alberta, adjacent to the western extremity of Lake Athabasca. The delta is largely contained within the Wood Buffalo National Park and is important as a bison habitat area, and as a waterfowl breeding, moulting, and staging area. As well, muskrat trapping and fishing in the delta were traditional sources of income for about 1,500 aboriginals. Under natural conditions, the Peace River annually reached a flood stage during late June and early July, and acted as a hydrological dam in the outflow channels of Lake Athabasca. Consequently, the level of the lake rose rapidly, by at least several feet, and briefly flooded the delta, resulting in recharging of delta lakes and basins, deposition of silt, plant seeds, nutrients, and the flushing of the products of plant decomposition. The vegetation pattern and animal life in the delta developed and adapted in response to a fluctuating water regime. However, since the control of the Peace River through the Bennett Dam began in 1967, the annual flood has not recurred, and water levels of Lake Athabasca, and the shallow delta lakes, have fallen by several feet. The area of the lake has also decreased. As well, many of the basins on the flood plain have completely dried out. H.J. Dirschl, in the publication entitled *Evaluation of Ecological Effects of Recent Low Water Levels in the Peace Athabasca Delta* (Ottawa: Canadian Wildlife Service, 1972. 28 p. Canadian Wildlife Service Occasional Paper No. 13), stated that 'if the present water regime continued indefinitely, the observed plant succession will convert most of the exposed lake and marsh bottoms into willow thickets and reedgrass meadows.'

The plant succession in the Peace Athabasca Delta is one of modified invasion due to dynamic changes in the ecosystem. However, wetlands and other natural landscapes have, from the earliest days of European settlement, suffered from the problem of plants invading natural areas in North America, especially European species. These types of invasive plants are considered to be detrimental, because they move into a habitat and reproduce so aggressively that they displace some of the original components of the vegetative community. In extreme infestations, there may be a loss of most of the original vegetation which would have supported wildlife that fed, or reproduced, within the plant community. For example, when a plant community is overtaken by purple loosestrife (*Lythrum salicaria*), which was introduced to North America in the early 1800s, many of the animals are displaced along with their host plants, as muskrats cannot eat it as food, and birds such as grebes, or terns, will not nest in it. D.J. White and E. Haber, in their report entitled *Invasive Plants of Natural Habitats in Canada: An Integrated Review of Wetland and Upland Species and Legislation Governing Their Control* (Ottawa: Canadian Wildlife Service, 1993. 121 p.), stated that 'some would argue that in the long run it is futile to try to control invasive species for a variety of reasons ranging from the fact that for the most part they are well established or as native diseases and predators respond to the new species, populations of the invasive plant decline to acceptable levels.'

Most recently, wetland studies have examined the problems encountered with acidic and other pollution in existing wetlands and its effects on waterfowl populations. One such study by G.R. Parker and M.J. Petrie was entitled *Waterfowl Distribution Relative to Wetland Acidity* (*Journal of Wildlife Management*, v. 56, 1992. pp. 268–74). In this study, the authors measured the influence of water acidity on the presence of fish and the abundance of aquatic invertebrates, to better understand how these variables influenced the distribution of waterfowl broods on unmanaged freshwater wetlands in southern New Brunswick. The authors' conclusion from the study indicated that the presence of fish and invertebrates affected wetland selection by broods of black ducks when they were young, but broods tended to choose wetlands based on criteria other than fish or invertebrate abundance as they aged and plants became more important in their diets.

The problem of pollution in wetlands has also been studied by R.W. Daury and F.E. Schwab in their paper entitled *Blood Lead Concentrations of Waterfowl from Unhunted and Heavily Hunted Marshes of Nova Scotia and Prince Edward Island, Canada* (*Journal of Wildlife Diseases*, v. 29, 1993. pp. 577–81). Although researchers knew that blood level concentrations in juvenile black ducks (*Anas rubripes*) and other waterfowl affected by ingested lead shot was known (100 parts per billion as a threshold level), the authors found that over-wintering black ducks in Sullivans Pond, Dartmouth, Nova Scotia, contained elevated blood level concentrations often greater than 100 parts per million, and the source of those elevated concentrations was considered to be automobile emissions. On a more positive note, A. Hanson and C. Ellingwood began in 1991 to monitor the Sackville Waterfowl Park, a 19-hectare shallow freshwater wetland created by reflooding a saltwater marsh that was drained 300 years ago. The purpose of the park is to provide wetland and wildlife reviewing opportunities to tourists and residents. In their paper entitled *Sackville Waterfowl Park, Sackville, New Brunswick,*

Canada. Baseline Characterization, towards Long-Term Monitoring (*Hydrobiologia*, Nos. 279/280, 1994. pp. 521–24), the authors said that the wetland was supporting high densities of waterfowl and provided an opportunity for long-term monitoring for understanding breeding waterfowl populations in the wetland environment.

J Other Studies (Canadian Museum of Nature)

As mentioned in the introduction, the biological work of the museum, under its various names and changes, has been overwhelmingly descriptive and taxonomic in nature. One of the earliest studies by J.M. Macoun was entitled *The Flora of Canada* (Ottawa: Government Printing Bureau, 1917. 14 p. National Museum of Canada. Bulletin No. 26). Another, by J.H. McDunnough, was entitled *A Generic Revision of North American Agrotid Moths* (Ottawa: F.A. Acland, Printer to the King, 1928. 78 p. National Museum of Canada. Bulletin No. 55). In the 1930s, P.A. Taverner wrote the first of the 'classics' with the title *Birds of Canada* (Ottawa: J.O. Patenaude, 1934. 445 p.). More recently, a taxonomic work by D.E. McAllister was entitled *Poisonous and Venomous Fishes of Canada* (Ottawa: National Museum of Canada, 1968. 11 p. National Museum of Canada Natural History Papers). Besides listing, with pictures, fishes in Canada that were known to be poisonous to humans, this publication also described the site and nature of the poison. Twelve fishes were listed. At this time, D.E. McAllister also published a major paleontological work. It was entitled *The Evolution of Branchiostegals, and Associated Opercular, Gular, and Hyoid Bones, and the Classification of Teleostome Fishes Living and Fossil* (Ottawa: Queen's Printer, 1968. 239 p.). It was also during the 1960s that A.E. Porsild published a major botanical publication entitled *Illustrated Flora of the Canadian Arctic Archipelago* (Ottawa: Queen's Printer, 1964. 218 p.). However, in 1978, H.J. Scoggan published the definitive botanical publication entitled *The Flora of Canada* (Ottawa: National Museum of Natural Sciences, 1978. Parts 1–4. 1711 p. Publications in Botany No. 7).

Although most of the wildlife studies done by the National Museum of Canada have been specialized in nature, two studies do stand out as major contributions to the understanding of wildlife in Canada. One study by A.W. Banfield was entitled *The Mammals of Canada* (Toronto: University of Toronto Press for the National Museum of Natural Sciences, National Museum of Canada, 1974. 438 p.). The idea of this publication went back to the 1920s, when coloured illustrations of Canadian mammals by A. Brooks were undertaken by the museum. However, it was not until the early 1960s that the author was able to focus on this publication. A.W. Banfield said that 'much of the work is based on my own observations while studying mammals in every province and territory of Canada.' The purpose of the book was to 'provide a popular account of the mammals of Canada for students, lay readers, and professional biologists.' The book lists 196 species of recent mammals, known to have occurred since historic times in Canada, or in the coastal waters. The overall listing is by order, but the whole work is prefaced with the classification scheme for the Class Mammalia. Each specific listing provides a description of the mammal, habits, habitat, reproduction, economic status, North American distribution, and Canadian distribution. Bibliographic references to the mammal are also given.

The other major study of the National Museum was undertaken by W.E. Godfrey and entitled *The Birds of Canada*, revised edition (Ottawa: National Museums of Canada, 1986. 595 p.). This revised edition of the 1966 publication 'includes all bird species known by the writer to occur or to have occurred, in Canada and its coastal waters within a limit of 320 km off shore, up to December 1984, a total of 578 species.' The general breakdown is by order and family. Each species is described by measurements, including wing, tail, exposed culmen (beak), and tarsus, field marks, voice, habitat, nesting, range in the Western hemisphere and range in Canada, species numbers (common, fairly common, and subspecies). Included are colour illustrations of 498 species, as well as some black and white drawings, and distribution maps.

IV Departmental Sources of the Publications

The most current listing of the publications of the Canadian Wildlife Service is at their website <<http://www.cws.scf.ec.gc.ca>>. From time to time, however, the Canadian Wildlife Service has also issued printed lists. For example, the early publication series of the Canadian Wildlife Service – the Wildlife Management Bulletins, published from 1950 to 1966 – are listed at the end of Report No. 1 – *Whooping Crane Population Dynamics on the Nesting Grounds, Wood Buffalo National Park, Northwest Territories*, published in 1966. The publications issued since the early 1950s – the Progress Notes, Occasional papers, Reports, Monographs, and Miscellaneous Publications – are listed in a typed publication, issued in June, 1994, by the Canadian Wildlife Service, and entitled *Canadian Wildlife Service Publications*.

There is a listing of some publications of the Canadian Museum of Nature at their website <<http://www.nature.ca>>. A printed list of the publications of the Canadian Museum of Nature is in a typed publication entitled *Scientific and Technical Publications: 1993 Catalogue*. The publications are listed by series. However, a 1991 *Publications: Canadian Museum of Nature* also lists monographs not indicated in the 1993 list.

Chapter 5

Fisheries Research

I Historical Background

In 1893, E.E. Prince, a Scottish zoologist, was appointed as Commissioner and General Inspector of Fisheries for the federal government. He was initially concerned about the lack of knowledge of Canada's aquatic resources, the rumours about the depletion of certain stocks (e.g. the early extinction of the St Lawrence walrus, the decline of the Great Lakes sturgeon, and New England and Ontario salmon), and the lack of opportunity for Canadian scientists who wished to study aquatic resources.

In 1897, the British Association for the Advancement of Science met in Toronto and sponsored a Canadian Committee to advance the need for a marine station on the Atlantic coast. As well, the Royal Society of Canada pressed the government to take action. Prince chaired the Canadian Committee and on 20 April 1898 it presented a memorandum to the minister requesting the establishment of a marine station on the Atlantic coast to be administered by a Board of Management made up of members from the Department of Marine and Fisheries, and universities supporting the petition. The proposal was accepted by Parliament, and Prince, who was the government's appointment, also became chairman of the new Board of Management of the Marine Biological Station of Canada. The latter became the Biological Board of Canada in 1912 and was renamed the Fisheries Research Board of Canada in 1937.

Until 1973, the board was directly responsible for providing facilities for original research on aquatic animals and plants. It was also mandated to solve problems faced by governments in fisheries management and by the fishing industry itself in the catching, processing, and marketing of fish and fish products. As well, the Fisheries Research Board provided research staff and advice for Canada's international fishing agreements and commissions. Over the years, these have included the North Pacific Fur Seal Commission; the International Pacific Halibut Commission; the International Salmon Fisheries Commission; the International Commission for the Northwest Atlantic Fisheries; the Great Lakes Fishing Commission; the International North Pacific Fisheries Commission; the International Joint Commission; the Inter-American Tropical Tuna Commission; the International Commission for the Conservation of Atlantic Tunas; the Permanent International Council for the Exploration of the Sea; and the Northwest Atlantic Fisheries Organization.

The first research station established to help the board carry out its mandate was portable, but in 1908, a permanent Atlantic Station was built at St Andrews, New Brunswick. As well, a permanent Pacific Station was built at Nanaimo, British Columbia. In the early years, there was also a station on Georgian Bay which functioned from 1904 to 1917 only. This pattern of decentralized regional laboratories, with no central laboratory in Ottawa, except for administrative and political control in federal offices, became the pattern for fisheries research in Canada. Today, that research is carried out in 11 establishments designated as stations, laboratories, institutes, or centres, and located in the Maritimes, Quebec, Ontario, Manitoba, and British Columbia.

In 1973, after 75 years of managing Canada's fisheries research, the Fisheries Research Board of Canada became an Advisory Board only, and the direct management of the stations and fisheries research was assumed by the Fisheries and Marine Service of the Department of the Environment. Then, in 1979, the Department of Fisheries and Oceans officially came into being following the proclamation of the Department of Fisheries and Oceans Act.

This new department was basically the previous Fisheries and Marine Service elevated to a higher profile. The objectives of the department included 'the comprehensive husbandry and management of Canada's fisheries and resource base ... the best use of fisheries resources [and] ... the acquisition of the necessary knowledge base pertaining to ... the management of fishery resources and its aquatic habitat.' In 1985/86, after the election of the Mulroney Conservative government, the department was reorganized and its operations grouped into four programs – Science, Atlantic Fisheries, Pacific and Freshwater Fisheries, and Inspection. In 1994, there was some further reorganization, with the groupings becoming, again, science, as well as fisheries operations, inspection, and international activities.

The science program is divided into three sub-activities – biological science, physical and chemical sciences, and hydrology, which is the responsibility of the Canadian Hydrographic Service. The latter is discussed in Chapter 9, 'Water Research and Services.' The biological sciences are responsible for research on fish, invertebrate marine mammals, marine plants, and ecosystems. This includes stock assessments, the dynamics of marine and freshwater ecosystems, aquaculture research, and the prevention of the spread of communicable diseases. The physical and chemical sciences are responsible for, among other things, the distribution and behaviour of organic and inorganic materials and their impact in fish, and the study of pollutants in the ecosystem, as well as the physical oceanographic research program, the latter discussed in the chapter on water research.

In 1985–86, there was also realignment of research responsibilities among some of the research stations with the establishment of National Centres of Disciplinary Expertise in existing departmental facilities. These Centres of Disciplinary Expertise are: (1) the Northwest Atlantic Fisheries Centre in St John's, Newfoundland (Atlantic resource assessment and survey methodology), (2) the Bedford Institute of Oceanography in Dartmouth, Nova Scotia (biological oceanography and marine contaminants and toxicology), (3) the Maurice Lamontagne Institute in Ste Flavie, Quebec (parasitology), and (4) the Pacific Biological Station in Nanaimo and the West Vancouver Laboratory in West Vancouver, British Columbia (genetics and biotechnology for aquaculture). Again, in 1994, there was also an overall realignment of programs by region, each headed by a Regional Director General. These regions are: Newfoundland Region, headquartered in St John's, Newfoundland; Scotia-Fundy Region in Halifax, Nova Scotia; Gulf Region in Moncton, New Brunswick; Quebec Region in Quebec City; Central and Arctic Region in Winnipeg, Manitoba; and Pacific Region in Vancouver, British Columbia.

Whatever the political and administrative realignments have been from time to time involving fisheries research in Canada, fisheries studies in practice can be broken down into approximately eight areas. They are: (1) taxonomy and general biology studies, (2) the distribution and movement of fish stocks, (3) dynamics of fish stocks, (4) the artificial propagation of fish including aquaculture, (5) the study and management of marine mammals, (6) environmental pollution studies, (7) fishing gear for the effective harvesting of the resource, and (8) the handling and storage of fish. Over the years, this work has given Canada a high profile in fisheries research in the world and generated a wealth of fisheries-related publications in many forms.

II Publications

A Journal Literature

(a) Contributions

The first research publications concerning Canadian fisheries were entitled *Contributions to Canadian Biology* and appeared as a supplement to the 32nd Annual

Report of the Department of Marine and Fisheries, Fisheries Branch in 1901. There were 10 unnumbered volumes between 1901 and 1925. After that date, the numbered volumes began and were called *Contributions to Canadian Biology and Fisheries*. One to five were undivided, but starting with volume 6, 1930–31, to volume 8, 1933–34, the Contributions were split into Series A: General Papers, Series B: Experimental Papers, and Series C: Industrial Papers. In 1934, the Contributions evolved into the *Journal of the Biological Board of Canada*, volume 1, 1934–35, to volume 3, 1936–37. With volume 4, 1938–40, the latter was renamed the *Journal of the Fisheries Research Board of Canada* and over the years became internationally known for the quality of its research publications. When the non-advisory functions of the Fisheries Research Board ended in 1973, the journal fell into the Department of the Environment. In 1979, when the Department of Fisheries and Oceans came into existence, the journal was renamed the *Canadian Journal of Fisheries and Aquatic Sciences.*

Besides publishing regular research papers, from time to time the journal publishes special issues, or supplements, which are usually papers presented at symposia, sponsored by the department, and edited by Canadian fisheries scientists. For example, in 1974, Canada hosted the 19th Triennial Congress of the International Association of Limnology in Winnipeg, Manitoba. To celebrate this occasion, the journal devoted its May, 1974, issue to a special collection of 20 research papers on Canadian limnology (freshwater studies). An example of a paper presented at this congress by P.F. Elson was entitled *Impact of Recent Economic Growth and Industrial Development on the Ecology of Northwest Miramichi Atlantic Salmon (*Salmo salar*)* (*Journal of the Fisheries Research Board of Canada*, v. 31, 1974. pp. 521–44). And, in 1997, the journal published *Selected Proceedings of the Symposium on the Biology and Ecology of Northwest Atlantic Cod.* Edited by J.S. Campbell and P. Schwinghammer (*Canadian Journal of Fisheries and Aquatic Sciences*, v. 54, 1997. Supplement 1. 235 p.). The editors selected 13 papers 'for their original contributions to our knowledge of the life history, physiology, population dynamics, habitat, and interspecific interactions of Atlantic cod' for publication as a journal supplement. One of the papers in this supplement by C.M. Herbinger and R.W. Doyle was entitled *Family Relationships and Effective Population Size in a Natural Cohort of Atlantic Cod (*Gadus morhua*) Larvae* (*Canadian Journal of Fisheries and Aquatic Sciences*, v. 54, 1997. Supplement 1. pp. 11–18).

(b) Festschriften

Another type of special issue, or supplement, is the festschrift, honouring major Canadian fisheries research scientists. For example, the October 1971 issue of the journal was dedicated to Edith and Cyril Berkeley, a husband-and-wife team who, over the years, made major contributions to polychaete (marine worm) studies. This issue listed their 92 published research papers and contained 30 articles by others on related polychaete research. An example of a paper from this festschrift by L. Margolis was entitled *Polychaetes as Intermediate Hosts of Helminth Parasites of Vertebrates: A Review* (*Journal of the Fisheries Research Board of Canada*, v. 28, 1971. pp. 1385–92).

(c) Perspectives and Book Reviews

In 1974, in order to broaden its readership, the journal initiated two new sections, Perspectives and Book Reviews. Perspectives were defined as 'essays of opinion or hypotheses that are of broad Canadian and international interest in fisheries management, ocean science, and the aquatic environment.' An example of this type of essay, in 1974, by P.A. Larkin was entitled *Play It Again Sam – An Essay on Salmon Enhancement* (*Journal of the Fisheries Research Board of Canada*, v. 31, 1974. pp. 1433–59). In 1975, Letters to the Editor first made their appearance. Besides the journal, there are 15 other types of publications of fisheries interest that have been published by the board or department over the years.

(d) Bulletins

The first of these publications was the Bulletin series that had its origin during World War I in the need to find new food sources. The first bulletin by A.G. Huntsman was entitled *Histories of New Food Fishes. I. The Canadian Plaice* (Biological Board of Canada Bulletin No. 1, 1918. 32 pp.). In the ensuing years, a major objective of the bulletins has been to produce comprehensive reviews, in monograph format, of the state of knowledge on specific fisheries subjects. These monographs have achieved national and international acclaim.

Other publication types not as well known as the journal literature or bulletins, but important as stepping stones in filling knowledge gaps have been:

(e) Canadian Fauna Series

The Canadian Fauna Series was intended to provide a means of ready identification of local marine fauna and flora. Only five pertaining to Canadian Atlantic fauna were published between 1921 and 1948. An example from this series by A.L. Treadwell was entitled *Canadian Atlantic Fauna: 9. Annelida, 9b, Polychaeta* (Toronto: Printed by the University of Toronto Press for the Fisheries Research Board of Canada, 1948. 69 p.) Six pertaining to Canadian Pacific fauna were published between 1937 and 1955. An example from this series by G.W. Wailes was entitled *Canadian Pacific Fauna: 1. Protozoa, 1e. Mastigophora* (Toronto: Printed by the University of Toronto Press for the Fisheries Research Board of Canada, 1939. 45 p.)

(f) Progress Reports

These were intended as brief reports to be distributed to people interested in keeping in touch with particular research being done. The first Progress Reports of the Pacific Coast Stations were issued in 1929, and by 1962, when this series ended, 114 reports had been published. An example from this series by B.E. March was entitled *The Use of Condensed Herring Solubles in Turkey Poult Rations* (Progress Report of the Pacific Coast Station No. 113, 1959. pp. 3–5). Progress Reports of the Atlantic Coast Stations began in 1931 and 73 were published by the time this series concluded in 1967. Progress Reports from the London, Ontario, Biological Station and Technical Unit consisted of two issues only for the years 1959–61.

(g) Circulars

Circulars began publication in 1935 at the Atlantic Fisheries Experimental Station. Like the Progress Reports, they aimed at putting the board's wealth of information in a format suitable for wide use. For example, one circular by E. Hess was entitled *Improvements in Cannery Equipment and Cannery Sanitation* (Halifax: Technical Station of the Fisheries Research Board of Canada, 1940. Circular [Old Series] No. 16. 3 p.). At their peak, 11 stations issued circulars, all of which were published by the individual stations (unlike the journal or bulletins, which were controlled editorially from the journal's office in Ottawa). By the early 1970s, only the Halifax Laboratory and the Pacific Biological Station, at Nanaimo, still issued circulars, and by 1979 this series ceased to be published altogether.

(h) Canadian Industry Reports of Fisheries and Aquatic Sciences

The Canadian Industry Reports of Fisheries and Aquatic Sciences began to make their appearance in 1978. They contain the results of research and development which are useful to industry for either immediate or future applications. Topics concerned are fisheries management, technology and development, ocean sciences and aquatic environment relevant to Canada. An example for this series by G.V. Hurley and H.H. Stone is entitled *The Dogfish Scourge: Protecting Fishing Gear from Shark Attack* (Ottawa: Department of Fisheries and Oceans, 1987. Canadian Industry Report of Fisheries and Aquatic Sciences No. 180. 34 p.).

(i) Canadian Technical Reports of Fisheries and Aquatic Sciences

The Canadian Technical Reports of Fisheries and Aquatic Sciences contain scientific and technical information that is of sufficient importance to be preserved, but is not appropriate for primary scientific publication. The technical reports first appeared in 1967. An example from this series by T.J. Hara is entitled *Aquaculture in Japan: Potential, Relevance to Canadians* (Ottawa: Department of Fisheries and Oceans 1988. 36 p. Canadian Technical Report of Fisheries and Aquatic Sciences No. 1574).

(j) Canadian Manuscript Reports of Fisheries and Aquatic Sciences

The Canadian Manuscript Reports of Fisheries and Aquatic Sciences are similar to the technical reports, but they deal primarily with national and regional problems, and distribution is generally restricted to institutions or individuals located in particular regions of Canada. Unlike the technical reports, the manuscript reports have a longer history and first appeared in 1937. An example from this series by C.W. Haegele and J.F. Schweigert is entitled *Herring Stock Estimates from Spawn Surveys for the North Coast of British Columbia in 1986* (Ottawa: Department of Fisheries and Oceans, 1986. 28 p. Canadian Manuscript Report of Fisheries and Aquatic Sciences No. 1976).

(k) The Canadian Data Reports of Fisheries and Aquatic Sciences

The Canadian Data Reports of Fisheries and Aquatic Sciences contain data compilations where little or no analysis of the data is included. These compilations, which began publication in 1977, have usually been prepared in support of other journal publication reports. An example from this series by B. Irwin and C. Caverhill is entitled *Primary Productivity on the Labrador Shelf during June and July, 1984* (Ottawa: Department of Fisheries and Oceans, 1984. 162 p. Canadian Data Report of Fisheries and Aquatic Sciences No. 577).

(l) CAFSAC (Canadian Atlantic Fisheries Scientific Advisory Committee) Documents

The CAFSAC (Canadian Atlantic Fisheries Scientific Advisory Committee) Documents, which began publication in 1977, provide an archive of data compilations where little or no analysis is included. These have usually been prepared in support of other journal publications, or reports. The CAFSAC Documents consist primarily of stock assessment, and some related biological information on Atlantic fish stocks. These documents began their appearance in 1980, along with the Annual Report of the Committee. An example from this series by A.B. Stasko was entitled *Lobster Larvae on the Scotian Shelf* (Dartmouth, NS, and Ottawa: Department of Fisheries and Oceans 1977. 10 p. CAFSAC Research Document No. 77/31).

B Other Publications

Finally, there are six other numbered publication series in which research activities or individual researcher's views are published.

(a) Studies Series

From the earliest years on, Canada's fisheries scientists have published papers that appear in non–Department of Fisheries and Oceans publications. Reprints of these articles have always been collected for use in station libraries, or for distribution to a limited number of exchange organizations. Beginning in 1919, these research articles were listed in the Studies Series for information only. An example from this series by L.E. Burridge and K. Haya is entitled *Seasonal Lethality of Pentachlorophenol to Juvenile Atlantic Salmon* (*Bulletin of Environmental Contamination and Toxicology*, v. 45, 1990. pp. 888–92. Department of Fisheries and Oceans. Studies No. 4606).

(b) Interpretive Articles

From time to time, fisheries scientists are called upon to write interpretive articles for trade magazines, or give conference presentations of their personal views. These types of articles or presentations have been listed as Interpretive Articles since 1966. For example, the first article listed, by H.R. Squires, entitled *Research in Newfoundland on Marine Invertebrates*, appeared in the *Newfoundland Journal of Commerce*, v. 33, 1966. pp. 23–31.

(c) Canadian Special Publications of Fisheries and Aquatic Sciences

The third type of publication is the Special Publication. In the late 1960s, several unnumbered publications were produced for various reasons but did not fit into any established series. This prompted a new series where specialized monographs are priced for sale. An example from this series was one written by R.R. Reeves and E. Mitchell and entitled *History of the White Whale (*Delphinapterus leucas*) Exploitation in Eastern Hudson Bay and James Bay* (Ottawa: Supply and Services, 1987. 45 p. Canadian Special Publication of Fisheries and Aquatic Sciences No. 95).

(d) Canadian Bulletin of Fisheries and Aquatic Sciences

A fourth type of publication, which is also a monograph series, is the Canadian Bulletin of Fisheries and Aquatic Sciences. These are major research monographs. An example from this series by T.G. Smith and D.J. St Aubin is entitled *Advances in Research on the Beluga Whale,* Delphinapterus leucas*: Introduction and Overview* (Ottawa: Supply and Services Canada, 1990. 206 p. Canadian Bulletin of Fisheries and Aquatic Sciences No. 224).

(e) Economic and Commercial Analysis Reports

The fifth type of publication, which is fisheries related but does not pertain to research, is the Economic and Commercial Analysis Report. These numbered reports, which began to appear in 1990, contain analyses of trends, studies of government policies, marketing programs, and trade and tariffs, all of which are not considered appropriate for primary publication. An example from this series by the BC Salmon Farmers Association is entitled *Current Developments in World Salmon Markets: Implications for the Canadian Salmon Farming Industry* (Ottawa: Department of Fisheries and Oceans, 1990. 115 p. Economic and Commercial Analysis Report No. 46).

(f) Translation Series

The last numbered publications of the department occur in the Translation Series. The Translations are not research work of the department, but translations of foreign-language articles that can be applied to Canadian research. The early translations were done by board employees, but most work is done now in the federal government's Translation Bureau and available from the Translation Services, Canada Institute for Scientific and Technical Information (CISTI), in Ottawa. Translations were first listed in 1958 and have been indexed on an annual basis since 1965. An example from this series by E. Parjala was entitled *Reduction of Loading Effects of Fish Farming – a Literature Review* (Translated from Finnish. *Vesitalous,* v. 5, 1984. pp. 11–16. Translation No. 5252).

C Unnumbered Publications

The unnumbered publications of the Board, or its departmental successors, cover fisheries information for the non-scientist as well as the Annual Reports of the Board or Department.

(a) Underwater World Factsheets

In 1980, Fisheries and Oceans began publishing a series entitled Underwater World Factsheets. Each factsheet is about eight pages long and describes, along with illustrations, the life history, geographic distribution, utilization, and population status of fish, shellfish, and other marine organisms. Each contains a useful bibliography. To date over 70 factsheets have been issued. An example of a publication from this series by A.W. White is entitled *Red Tides* (Ottawa: Department of Fisheries and Oceans, 1980. 8 p.).

(b) Annual Reports and Reviews

The first Annual Report of the Biological Stations of Canada appeared in 1914, but separate reports disappeared until 1929–30 (between these years, the reports appeared as appendices, or condensations, in the department's Fisheries Branch Report). By the early 1960s, the Annual Report had grown into a major document and in 1964 it split into a smaller Annual Report and a more detailed biennial Review of the Fisheries Research Board of Canada. The latter continued until 1971–72. After the board became advisory only, reports on fisheries research became one section of the departmental annual reports (first Forestry and Fisheries, then the Environment) until 1979/80, when the Department of Fisheries and Oceans was organized. Its formal title is *Annual Report: Department of Fisheries and Oceans* (Ottawa: Department of Fisheries and Oceans, 1980–). Finally, after the reorganization of the latter into a reduced number of reporting regions, regional reviews began to appear. For example, the Pacific Region, which had previously published an *Annual Review of Activities: Institute of Ocean Sciences* (Sidney, BC: Institute of Ocean Sciences, 1981–85. 5 v.), issued a new report entitled *Science Review: Department of Fisheries and Oceans, Pacific Region* (Sidney, BC: Department of Fisheries and Oceans, 1986/87–). Both central and regional activities, along with some publications, can also be found through the Fisheries and Oceans website <<http://www.ncr.dfo.ca/>>.

III Fisheries Research Activities in Canada

A Taxonomic and Biological Studies

The collecting and identifying of living, or fossil, animals or plants began seriously in the 19th century, but the impetus for Canadian studies of aquatic vertebrates, invertebrates, and plants came about with the establishment of biological stations by the Biological Board in the early 20th century. Over the years, these studies, which cover three oceans and Canada's inland waters, have contributed markedly to our general and taxonomic knowledge of marine species. However, the catalogue of information about new species is always incomplete, and traditional descriptive classification has been supplemented by physiological and biochemical studies (blood, muscle, enzymes, etc.).

Early work identified particular marine groups. For example, on the Pacific Coast, the team of Edith and Cyril Berkeley, previously mentioned, became well known for their taxonomic work identifying marine worms (Polychaeta). A small sample of their publications, which are still considered authoritative, included *Notes on Polychaeta from the Coast of Western Canada. Part 1. Spionidae (Annals and Magazine of Natural History.* Series 10, v. 18, 1936. pp. 468–77); *Notes on Polychaeta from the Coast of Western Canada. Part 2. Syllidae* (*Annals and Magazine of Natural History*. Series 11, v. 1, 1938. pp. 33–49); *Notes on Polychaeta from the Coast of Western Canada. Part 3. Further Notes on Syllidae and Some Observations on Other Polychaeta. Errantia* (*Annals and Magazine of Natural History*. Series 11, v. 12, 1945. pp. 316–35); *Notes on Polychaeta from the Coast of Western Canada. Part 4. Polychaeta Sedentaria* (*Annals and Magazine of Natural History.* Series 12, v. 3, 1950. pp. 50–69); *North Pacific Polychaeta, Chiefly from the West Coast of Vancouver Island, Alaska, and Bering Sea* (*Canadian Journal of Research, Section D*, v. 20, 1942. pp. 183–208); *Polychaeta from the Western Canadian Arctic Region* (*Canadian Journal of Research, Section D*, v. 22, 1944. pp. 1–5); and *Polychaeta of the Western Canadian Arctic* (*Journal of the Fisheries Research Board of Canada*, v. 15, 1958. pp. 801–4).

In the freshwater fisheries, E.M. Walker's studies on the dragonfly genera, during summers spent on Georgian Bay, were a major contribution to the study of that aquatic insect. His initial work was entitled *Notes on the Odonata of the Vicinity of Go-Home Bay, Georgian Bay, Ontario* (*Contributions to Canadian Biology*, 1911–14 [volume] 1915. Fasc.II, Freshwater Fish and Lake Biology. pp. 53–94).

In the Arctic and the northwest Atlantic, D.V. Vladykov began his intensive studies of the fish of those regions. His study of Arctic fishes began in the early 1930s and was entitled *Biological and Oceanographic Conditions in Hudson Bay 9. Fishes from the Hudson Bay Region (except the Coregonidae)* (*Contributions to Canadian Biology and Fisheries* (New Series) v. 8, Series A. General Papers No. 2, 1933. pp. 13–61). His first study on northwest Atlantic fishes was entitled *Taxonomic Characters of the Eastern North American Chars (*Salvelinus *and* Crestivomer*)* (*Journal of the Fisheries Research Board of Canada*, v. 11, 1954. pp. 904–32). This latter publication was the beginning study on the Arctic char, research on which was picked up again in the late 1970s.

The Arctic char became a focus of biological studies because it inhabited both lakes (the landlocked species), as well as rivers, and the Arctic Ocean (the anadramous species), all of which were relatively untouched by man. Two studies by L. Johnson were intensive and covered a time span of 20 years. One study was entitled *Ecology of Arctic Populations of Lake Trout:* Salvelinus namaycush, *Lake Whitefish,* Coregonus clupeaformis, *Arctic Char,* S. alpinus *and Associated Species in Unexploited Lakes of the Canadian Northwest Territories* (*Journal of the Fisheries Research Board of Canada*, v. 33, 1976. pp. 2459–88). In this paper, the available data on fish populations of 35 lakes in the Northwest Territories were examined because these lakes were comparatively young, geologically, and developed in isolation. Although the lakes varied in size from the Great Bear and Great Slave Lakes down to tundra lakes, the winter climate was common to all and the fish populations lived in equilibrium with their vegetative environment. Later, Johnson returned to this theme in the paper entitled *Homeostatic Characteristics of Single Species Fish Stocks in Arctic Lakes* (*Canadian Journal of Fisheries and Aquatic Sciences*, v. 40, 1983. pp. 987–1024). In this study, involving only Arctic char, these static lakes served as control models because of their lack of environmental disturbance. The species remained in balance within its age range, and size, and functioned with a low energy dissipation and a low stock replacement. This natural balancing has been termed homeokinesis.

Later, Johnson came back to these observations of char equilibrium in the environment at an International Symposium on Arctic Char, held in Winnipeg, Manitoba, in 1981, and sponsored by the Department of Fisheries and Oceans, and the University of Manitoba. This symposium brought together fisheries research scientists from northern countries. These people were specialists on the biology of the Arctic char and presented 43 papers, which were edited by L. Johnson and B. Burns and published under the title *Biology of the Arctic Char: Proceedings of the International Symposium on Arctic Char* (Winnipeg: University of Manitoba Press, 1984. 584 pp.). Johnson's own contribution to the symposium was entitled *Char and Man: The Philosophy of Limited Interaction* (pp. 523–32). In this paper, the author emphasized the theme of biological equilibrium by stating 'we need regulation of energy input to an approximately constant level, we need to eliminate further environmental degradation and thirdly we need severe limitations on the application of technology to resource harvesting. Only if some degree of stability can be attained in these areas can we hope to proceed to a state of least fluctuation, energy epipartioning and the redevelopment of organization.' The uniqueness of the Arctic char, as an ecological study of equilibrium, was brought forth again in Johnson's later papers entitled *Population Structure, Ecological Segregation and Reproduction in Non-Anadromous Arctic Char,* Salvelinus alpinus *(L.), in Four Unexploited Lakes in the Canadian High Arctic* (*Journal of Fish Biology*,

v. 28, 1991. pp. 123–48); *Systems for Survival of Ecology, Morals and Arctic Char (*Salvelinus alpinus*)* (*Nordic Journal of Freshwater Research*, no. 71, 1995. pp. 9–22); and *Action Principles as Determinants of Ecosystem Structure: The Autonomous Lake as a Reference System* (*Ecology*, v. 76, 1995. pp. 1741–57).

These intense taxonomic and biological studies by Canadian fisheries scientists of marine species which inhabit Canadian waters have resulted in the publication of internationally acclaimed monographs by the board, or its successor departments. Examples of three major studies are one by J.L. Hart entitled *Pacific Fishes of Canada* (Ottawa: Fisheries Research Board of Canada, 1973. 740 p. Fisheries Research Board of Canada. Bulletin 180). This publication was an updating of the publications by W.A. Clemens and G.V. Wilby entitled *Fisheries of the Pacific Coast of Canada* (Ottawa: Fisheries Research Board of Canada, 1946. 368 p.); and *Fisheries of the Pacific Coast of Canada.* 2nd Edition (Ottawa: Fisheries Research Board of Canada, 1961). Hart's publication included the fishes found in salt water off the Canadian coast and provided descriptions of 325 kinds of fish (54 more than Clemens's 1961 edition). A second major study by W.B. Scott and M.G. Scott is entitled *Atlantic Fishes of Canada* (Toronto: Published by the University of Toronto Press in cooperation with the Minister of Fisheries and Oceans and the Canadian Government Publishing Centre, Supplies and Services Canada, 1988. 731 p. Canadian Bulletin of Fisheries and Aquatic Sciences No. 219). This publication was intended as a replacement volume for one by A.H. Leim and W.B. Scott entitled *Fishes of the Atlantic Coast of Canada* (Ottawa: Queen's Printer, 1966. 485 p. Fisheries Research Board of Canada Bulletin No. 155), and it covered the biology, distribution, and economics of fishes occurring off the Canadian Atlantic coast. Finally, a third major study by H.J. Squires is entitled *Decapod Crustacea of the Atlantic Coast of Canada* (Ottawa: Department of Fisheries and Oceans, 1990. 532 p. Canadian Bulletin of Fisheries and Aquatic Sciences No. 221). In this volume, 89 species of decapod crustaceans, from the Atlantic coast of Canada, have been described (including details of mouthparts and other appendages, and original drawings). Brief notes on the biology of the fishery have also been provided.

B Distribution and Movement of Fish Stocks

For fish which may be of potential or actual commercial importance, it is necessary to learn about their distribution, by area and by depth, at what time of the year they occur, and when and where they occur in usable concentrations. In fisheries such as salmon and cod, historic knowledge created an information bank that could be built upon for studying these species further, particularly when larger fishing vessels and changes in fishing gear led to more concentrated exploitation. However, as markets have opened up for previously unused or underutilized species, systematic information has had to be gathered on their distribution and movement.

The Pacific salmon, in particular, has been an interesting study in distribution and movement, because it is an anadramous fish with a life cycle that takes it from fresh water to salt water, and back again. It is also made up of many different stocks. Until the 1920s, the Pacific salmon was considered to be a coastal species that rarely left the region of its spawning rivers. But after 1926, this view became discredited, on account of tagging work carried out by H.C. Williams in 1925 and 1926. This work was entitled *Pacific Salmon Migration: Report on the Tagging Operations in 1925* (*Contributions to Canadian Biology and Fisheries*, New Series, v. 3, no. 9, 1926–27. pp. 265–306); and *Pacific Salmon Migration: Report on the Tagging Operations in 1916, with Additional Returns from the Operations of 1925* (*Contributions to Canadian Biology and Fisheries,* New Series v. 4, no. 29, 1929. pp. 453–70). These tagging experiments showed that Chinook salmon, bound for the Columbia and Sacramento rivers, occurred in large numbers all along the British Columbia coast.

The movement of Pacific salmon also involves the additional factor of the stock concept. This stock concept recognized that stocks of the different rivers of British Columbia were discrete, and varied in abundance and migration time. Besides that, there was also a relationship between spawning and recruits, and a recognition that rivers differed in the numbers of salmon they could produce. The latter meant that discrete stocks had to be managed through the development of scientific information about them. Various methods have been used to distinguish stocks, including scale characteristics, parasites of the fish, and the tagging and marking of adults or juveniles, with their subsequent recovery in the catch or on the spawning grounds. This work continued over four decades, and, by 1967, K.V. Arco and M.P. Shepard had been able to describe, for each species or stock, its distribution and abundance levels of spawning populations in British Columbia. Their data covered the period 1952–62, and was published by the International North Pacific Fisheries Commission and entitled *Pacific Salmon in Canada* (*International North Pacific Fisheries Commission Bulletin*, v. 23, 1967. pp. 225–327). Overall, there are more than 1,200 streams and rivers in British Columbia in which one or more salmon species spawn. The number of streams utilized varies, according to the species, but ranges from about 250 for chinook salmon to nearly 1,000 for coho. Spawning populations ranged in size from less than 100 salmon to more than 1 million.

Since this major work, the study and refinement of the stock concept for Pacific salmon has continued. For example, W.E. Ricker and H.D. Smith, in their paper entitled *A Revised Interpretation of the History of the Skeena River Sockeye Salmon*

*(*Onchorynchus nerka*)* (*Journal of the Fisheries Research Board of Canada*, v. 32, 1975. pp. 1369–81), examined the historic exploitation of this stock, and the interaction of 'year-class' survival rates, to explain variations over time of this stock. By the mid-1980s, D.B. Beacham and R.E. Withler used electrophoresis (a chemical procedure that can be used for the analysis of mixtures of proteins for DNA sequencing) for the identification of salmon stocks. Their first paper was entitled *Biochemical Genetic Stock Identification of Chum Salmon (*Onchorynchus keta*) in Southern British Columbia* (*Canadian Journal of Fisheries and Aquatic Sciences*, v. 42, 1985. pp. 437–48). With this procedure, the authors examined the genetic variability of 33 chum salmon stocks in southern British Columbia. Their second paper was entitled *Biochemical Genetic Stock Identification of Pink Salmon (*Onchorynchus gorbuscha*) in Southern British Columbia and Puget Sound* (*Canadian Journal of Fisheries and Aquatic Sciences*, v. 42, 1985. pp. 1474–83). In this paper, the authors again used electrophoresis to examine genetic variability of bloodline stocks of pink salmon at 22 river and creek locations in southern British Columbia. In later papers, these two researchers used genetic variations and other characteristics to examine salmon stocks. One paper was entitled *Age, Morphology, Developmental Biology, and Biochemical Genetic Variation of Yukon River Fall Chum Salmon (*Onchorynchus keta*) and Comparisons with British Columbia (Canada) Populations* (*US National Marine Fisheries Service Fishery Bulletin*, v. 86, 1988. pp. 663–74). The other paper was entitled *Genetic Variation in Body Weight and Flesh Colour of the Coho Salmon (*Onchorynchus kisutch*) in British Columbia* (*Aquaculture*, v. 119, 1994. pp. 135–48). Finally, K.M. Miller and R.E. Withler used the electrophoresis procedure again to examine by DNA sequencing the geographic variations in 14 populations of coho salmon and 15 populations of chinook salmon inhabiting rivers of British Columbia. Their paper was entitled *Mhc Diversity in Pacific Salmon: Population Structure and Trans-Species Allelism* (*Hereditas*, v. 127, 1997. pp. 83–95).

Of all the salmon stocks on the Pacific coast of Canada, perhaps the most important commercially is the sockeye salmon, which received intensive study at an international symposium entitled *Sockeye Salmon (*Onchorynchus nerka*) Population Biology and Future Management: Proceedings of the International Sockeye Salmon Symposium – Sockeye '85, Held in Nanaimo, British Columbia, November 19–22, 1985*. Edited by H.D. Smith and L. Margolis (Ottawa: Department of Fisheries and Oceans, 1987. 486 p.). This three-day symposium was sponsored by the Department of Fisheries and Oceans and attended by 73 contributors to the 39 papers presented. The symposium was organized under two main subject areas. The first dealt primarily with population biology and the second with sockeye salmon management. Separate from this symposium, all Pacific salmon stocks – the sockeye salmon (*Onchorynchus nerka*), the pink salmon (*Onchorynchus* *gorbuscha*), the chum salmon (*Onchorynchus keta*), the chinook salmon (*Onchorynchus tshawytscha*), the coho salmon (*Onchorynchus kisutch*), and the Asian Pacific only masee and amago salmon (*Onchorynchus masee* and *Onchorynchus rhodurus*) – have been discussed in detail in the monograph entitled *Pacific Salmon Life Histories*. Edited by C. Groot and L. Margolis (Vancouver: UBC Press, 1991. 564 p.). This volume was published in cooperation with the Department of Fisheries and Oceans.

In the Atlantic fisheries, cod, a groundfish, has been the historic catch, although herring (an open ocean–pelagic fish) has been extensively documented, and more recently shellfish, such as sea scallops, have been documented. The Portuguese, Spanish Basques, and the French began fishing in Newfoundland waters in the early 1500s, followed by the English at a later date. Historic knowledge provided fisherman with information about the stock and its migration on the northwest Atlantic.

The earliest systematic exploration for Atlantic fish occurred at the beginning of the World War I. The Canadian Fisheries Expedition of 1914–15, manned jointly by Norwegian and Canadian scientists, worked mainly in the Gulf of St Lawrence, mapping stocks of herring, and making extensive oceanographic observations and plankton collections. Although this work was interrupted by the war, the work completed was published as a series of study supplements by the Biological Board. These studies were entitled *Canadian Fisheries Expedition Investigations in the Gulf of St. Lawrence and Arctic Waters of Canada, 1914–1915* (Ottawa: Department of Naval Services, 1919. 495 pp. Studies Supplements 177–86). Atlantic fish can be divided into pelagic – fish that range the open sea, such as herring, Atlantic salmon, mackerel, swordfish, bluefin tuna, eels, smelt, gaspereau, and capelin – and groundfish, which feed and dwell near the bottom, such as codfish, flatfish such as plaice and flounder, haddock, pollack, halibut, turbot, and hake. Until recently, the warm waters of the northerly flowing Gulf Stream, mingling with the frigid Labrador Current, created conditions conducive to sustaining large populations of groundfish, particularly cod (*Gadus morhua*). Cod occur throughout the Canadian Atlantic area that extends from the Bay of Fundy and northern Georges Bank to northern Labrador. In each of the different regions, there are one or more identifiable cod stocks, each with its own set of characteristics. Overall, there are at least 12–14 recognized stocks, of which the most important is the southern Labrador–east Newfoundland stock, but others include the northern Labrador stock, the northern and southern Gulf of St Lawrence stocks, the southern Grand Bank stock, and the Banquereau–Sable Island stock.

The cod in some of these stocks undergo extensive migrations, and knowledge of the stocks, as well as their migration patterns, has been determined from tagging cod which have been later caught and the tags returned by fishermen.

Much of this tagging was carried out by researchers from the then Fisheries Research Board's Biological Station at St Andrews, New Brunswick. The results of tagging were reported by researchers, such as F.D. McCracken, who published the papers entitled *Cod and Haddock Tagging off Lockeport, N.S.* (Fisheries Research Board of Canada, Atlantic Progress Report No. 64, 1956. pp. 10–15); and *Cod Tagging off Northern New Brunswick in 1955 and 1956* (Fisheries Research Board of Canada Atlantic Progress Report No. 72, 1959. pp. 8–19); and R.A. McKenzie, who published *Atlantic Cod Populations along the Southern Canadian Mainland As Shown by Vertebral Count Studies* (*Journal of the Fisheries Research Board of Canada*, v. 12, 1955. pp. 698–705) and *Atlantic Cod Tagging off the Southern Canadian Mainland* (Ottawa: Queen's Printer, 1956. 93 p. Fisheries Research Board of Canada Bulletin No. 105). One of the major studies, however, based on research vessel surveys conducted from 1957 to 1962, was published by Y. Jean and entitled *Seasonal Distribution of Cod (*Gadus morhua *L.) along the Canadian Atlantic Coast in Relation to Water Temperature* (*Journal of the Fisheries Research Board of Canada*, v. 27, 1964. pp. 429–60). Migration and stocks were also summarized later by W.H. Lear and J.M. Green in their paper entitled *Migration of the 'Northern' Atlantic Cod and the Mechanisms Involved.* In: *Mechanisms of Migration in Fishes*. Edited by J.D. McCleave et al. (New York: Planum Press, 1984. pp. 109–315).

It is also interesting to note that in 1949, after the formation of the International Commission of the North Atlantic Fisheries, cod stocks were subdivided (e.g., Labrador–East Newfoundland Stock Complex: ICNAF Subarea 2, Division 3K and 3L) for detailed study and monitoring. Much of this monitoring and assessment was, again, carried out from the Biological Station at St Andrews. In the early 1970s, A.T. Pinkhorn documented cod assessments for the international committee. These assessments were entitled *Vertical Population Assessment of St. Pierre Bank–Newfoundland South Coast Cod Stock ICNAF Sub-division 3PS* (*International Commission of the Northwest Atlantic Fisheries Research Bulletin*, v. 9, 1972. pp. 75–80), and *Combined Virtual Population Assessment for ICNAF Divisions 2J, 3K and 3L Cod* (*International Commission of the Northwest Atlantic Fisheries Red Book*, Pt. III, 1972. pp. 35–38). In the mid-1970s, the Department of Fisheries and Oceans undertook its own groundfish studies and began issuing systematic four-year reports. The first report by T.D. Beacham and B.A. Fowler was entitled *1980 Analysis of Inshore Cod Stock in Subdivision 4Vn (May–Dec)* (Dartmouth, NS: Department of Fisheries and Oceans. Canadian Atlantic Fisheries Scientific Advisory Committee. Research Document, 1980. CRD80/16. 24 p.).

An example of other interesting work on distribution and movement in the Atlantic fisheries has been undertaken with sea scallops (*Placopecten magellanicus*), one of the better-known shellfishes. Sea scallops are found in the northwest Atlantic, from the north shore of the Gulf of St Lawrence to Cape Hatteras, North Carolina. In the northern part of their habitat range they occur in water shallower than 18 metres, and may occur in dense populations called beds. The major Canadian scallop beds were found to be off Georges Bank, in the Bay of Fundy, particularly off Digby, Nova Scotia, on the southwestern Scotian Shelf, and in the Gulf of St Lawrence, particularly in Northumberland Strait, around the Magdalen Islands, and in Port-au-Port Bay, Newfoundland. These bed assessments were done in the late 1970s and early 1980s by G.S. Jamieson and others. They were entitled *Status and Assessment of Northumberland Strait Scallop Stocks* (Ottawa: Department of Fisheries and Oceans. Technical Report of Fisheries and Aquatic Sciences No. 904, 1979. 12 p.); *Bay of Fundy Scallop Stock Assessment – 1978* (Ottawa: Department of Fisheries and Oceans. Technical Report of Fisheries and Aquatic Sciences No. 915, 1979. 14 p.); *Assessment of Northumberland Strait Scallop Stocks and Review, 1978–81* (Ottawa: Department of Fisheries and Oceans. Technical Report of Fisheries and Aquatic Sciences No. 1150, 1982. 50 p.); *Assessment of Northumberland Scallop Stocks – 1979* (Ottawa: Department of Fisheries and Oceans. Technical Report of Fisheries and Aquatic Sciences No. 1013, 1981. 31 p.); *Assessment of Scallop Stocks on Browns and German Banks – 1979* (Ottawa: Department of Fisheries and Oceans. Technical Report of Fisheries and Aquatic Sciences No. 1014, 1981. 17 p.); and *Assessment of Northumberland Strait Scallop Stocks – 1980* (Ottawa: Department of Fisheries and Oceans. Technical Report of Fisheries and Aquatic Sciences No. 1017, 1981. 44 p.).

Finally, a major overview of the distribution and movement of all northwest Atlantic stocks, including groundfish, pelagic fish, anodramous-catadromous fish, invertebrates, including crustaceans, echinoderms, molluscs, marine mammals, and marine plants, was published in 1976. The publication, edited by A.T. Pinhorn, was entitled *Living Marine Resources of Newfoundland–Labrador. Status and Potential* (Fisheries Research Board of Canada Bulletin No. 194, 1976. 64 p.).

C Dynamics of Fish Stocks

A major aspect of Canada's fisheries research activities has been the ongoing investigations to support, or improve, the management of fishery stocks under Canadian jurisdiction, or to provide input into decision-making international bodies. All this involves the study of stock dynamics, which includes the estimate of each stock size, the fraction of the stock taken by Canada and by other nations, its rate of replenishment from growth and from new recruits (surviving new hatches), the effects of utilization on the age and size of fish caught, variations in recruitment from year to year, forecast of abundance of stock, and the interaction between stocks. Some of this work has been discussed in the previous section in relation to stock descriptions.

Two major factors affecting fish stocks are over-exploitation, and pollution. As well, climatic conditions may also affect stocks. Pollution studies, which began in earnest in the 1960s, are discussed elsewhere. However, the mechanics of over-exploitation began to be studied much earlier. In 1918, A.G. Huntsman demonstrated quantitatively how fishing reduces the average age and size of fish in a population. And, if the recruitment remains the same, the abundance of the stock declines too. This effect, called fishing-up, was described by Huntsman in a research paper entitled *The Growth of Fishes* (*American Fisheries Society. Transactions*, v. 49, 1919. pp. 19–23). Fishing-up means that large initial catches persist while the accumulated stock of old fish is being fished up. When it is gone, the fishery must depend on current additions, or recruitments, to the stock. In the case of Pacific salmon stocks, the capture of spawning fish in the last year of their life also affects stock abundance. Historically, the Pacific salmon fisheries have not suffered from the extreme over-exploitation that has occurred on the Atlantic coast, although there has been concern about overfishing by Japanese and Russian fleets because of the migratory range of salmon species. Only more recently have fluctuations in the Pacific salmon fishery caused interest and concern. It is the Atlantic fisheries have been plagued, over time, by over-exploitation in the salmon and the cod fisheries. As well, there has been research concerning the dynamics of the lobster fishery.

In 1977, the federal government placed a ban on commercial fishing of Atlantic salmon in the principal mainland salmon-producing rivers, because of concern over low levels of adult Atlantic salmon spawning. This was not a new problem, for in 1949 the federal government, and the governments of the Atlantic provinces, set up a coordinating committee to investigate the factors limiting the production and utilization of Atlantic salmon (*Salmo salar*) and to improve the management of the fishery. The stimulus was a steady decline, since the early 1930s, in the commercial catches in tidal waters of the Canadian Atlantic coast, and concern expressed by anglers over the future of the sports fishery in freshwater sections of the salmon rivers. The report of the coordinating committee by W.J. Menzies was entitled *A Report on the Present Position of the Atlantic Salmon Fisheries of Canada, with Recommendations for the Regulation and Improvement* (St Andrews, NB: Atlantic Salmon Association, 1951. 40 p. Doc. 17). The report encouraged further research and documentation of the Atlantic salmon.

Further studies of marked, naturally produced Atlantic smolts (the stage in the life cycle of the salmon when the sub-adults acquire a silvery colour and migrate down rivers to begin their adult life in the open sea), from 1950 to 1961, on tributaries of the Miramichi River, indicated that the commercial fisheries accounted for 60% of utilization, with anglers at 7%, and 20% accounting for spawning escapement. This long-term work was described by C.J. Kerswill in the paper entitled *Relative Roles of Utilization by Commercial and Sports Fisheries of Atlantic Salmon (*Salmo salar*) from the Miramichi River, New Brunswick* (*Journal of the Fisheries Research Board of Canada*, v. 28, 1971. pp. 351–63). Another paper by C.P. Ruggles and G.E. Turner, based on data collected from 1954 to 1971, showed that the decline in the abundance of Miramichi River Atlantic salmon had been accompanied by changes in the ratio of grilse (mature salmon returning to rivers after only one winter at sea) to large salmon (two sea-years or older entering the river). An increase in commercial fishing mortality of large salmon was believed responsible for shifting the river escapement in favour of grilse. This paper was entitled *Recent Changes in Stock Composition of Atlantic Salmon (*Salmo salar*) in the Miramichi River, New Brunswick* (*Journal of the Fisheries Research Board of Canada*, v. 30, 1973. pp. 779–86). Studies covering the period 1969 to 1982 also indicated that the Miramichi River had received inadequate egg depositions during those years. This was reported by E.M. Chadwick and R.G. Randall in their paper entitled *A Stock-Recruitment Relationship for Atlantic Salmon (*Salmo salar*) in the Miramichi River, New Brunswick (Canada)* (*North American Journal of Fisheries Management*, v. 6, 1986. pp. 200–203).

The ban on commercial fishing remained in effect until 1982, when the Miramichi was opened again for commercial fishing. During that time, 490 fisherman were compensated and they were given the choice of resuming fishing activities, or returning their licences for an amount not less than $2,100. Since the reopening of the fishery, studies have continued on factors affecting the productive capacity of the Atlantic salmon. J.R. Gibson and R.A. Myers, in their paper entitled *Influence of Seasonal River Discharge on Survival of Juvenile Atlantic Salmon,* Salmo salar (*Canadian Journal of Fisheries and Aquatic Sciences*, v. 45, 1988. pp. 344–48), studied the survival of eggs and under-yearlings of Atlantic salmon in six rivers in Newfoundland and New Brunswick and noted that there was evidence that summer discharge enhanced survival in the Miramichi and Northwest Miramichi rivers of New Brunswick. R.G. Randall, in the paper entitled *Effects of Sea-Age on the Reproductive Potential of Atlantic Salmon (*Salmo salar*) in Eastern Canada* (*Canadian Journal of Fisheries and Aquatic Sciences*, v. 46, 1989. pp. 2210–18), observed that reproductive potential of Atlantic salmon (defined as average eggs per fish) varied according to the sea-age of the spawners. E.M. Chadwick and R.G. Randall, in their paper entitled *Production of Atlantic Salmon (*Salmo salar*) in Freshwater and at Sea at High and Low Densities* (*Ecology of Freshwater Fish*, v. 2, 1993. pp. 67–72), noted that for all life stages, freshwater and marine production was related to the year-class of the stock and adequate egg deposition. Finally, S.M. Komadina-Douthwright and D. Caissie, in their paper entitled *Winter Movement of Radio-Tagged Atlantic Salmon (*Salmo salar*) Kelts in Relation to Frazil Ice in Pools of the Miramichi River* (Ottawa: Department of Fisheries and Oceans. Canadian Technical Report of Fisheries and Aquatic Sciences No. 2161, 1997. 66 p.), noted that kelps (post-spawning salmon that survive) which overwintered in frazil laden pools (ice-reaching toward the bottom of a stream) stayed in areas with less frazil deposition and lower water velocities, or in areas close to spring ice breakings for protection.

The major exploited fish in the Atlantic fisheries has been the Atlantic cod. Cod has been the object of international exploitation for several centuries, but only with the establishment of the International Commission for the Northwest Atlantic Fisheries (ICNAF), in 1949, did the stock come under any conservation regulations. This commission brought together the statistics and research data of all the nations which fish the region, which extends from Greenland to the eastern United States. Although groundfish research by Canada began much earlier, major work on the stock and its movements was carried out in the 1950s and early 1960s, as noted in the previous section. Cod stocks began to decline at the end of the 1960s. During the 1950s, the cod harvest averaged about 900,000 tonnes and increased sharply, in the 1960s, to almost 2 million tonnes. Then it declined dramatically during the 1970s, to below 500,000 tonnes by 1977. In 1973, the major cod stocks in the northwest Atlantic, and in particular in the Canadian area, were placed under quota regulation. The total allowable catch (TAC), based upon scientific advice, was set by ICNAF, and later by the North Atlantic Fishing Organization (NAFCO). This quota, however, has not been effective outside Canada's 200-mile limit, declared in 1977. And, in the mid-1980s, the federal Department of Fisheries and Oceans began to issue the annual *Atlantic Groundfish Management Plan* (Ottawa: Atlantic Fisheries Service, 1983–). These management plans indicated the total allowable catch acceptable in ICNAF subdivisions, as well as areas of the fishery to be shut down. By the early 1990s, it appeared that the cod stocks were in a permanent decline.

Although it is politically tempting to ascribe the major role in the decline of the northwest Atlantic cod stocks to the intense pressure exerted by distant water-travelling fleets, just outside Canada's 200-mile protection limit, some researchers do question this reason for the cod decline. They suggest that the rapid collapse of the stock at the end of the 1960s was primarily caused by large-scale meteorological, physical, and biological changes (sea surface temperature, atmospheric pressure systems, and freshwater outflows) at that time. Over-fishing only aggravated the problem. These researchers based their findings on the fact that the northwest Atlantic cod fisheries (where there were favourable environmental conditions at that time for recruitment) saw stocks increase markedly despite intense fishing. This important paper by J.A. Kowslow and C.R. Thompson was entitled *Recruit-*

*ment to Northwest Atlantic Cod (*Gadus morhua*) and Haddock (*Melanogrammus aelgrefinus*) Stocks: Influence of Stock Size and Climate* (*Canadian Journal of Fisheries and Aquatic Sciences*, v. 44, 1987. pp. 26–39). It is interesting that this research was published in 1987, and at the same time, the Department of Fisheries and Oceans formed a task group to review reports, documents, and other publications, dealing with catch statistics, population trends, and environmental and biological factors influencing the distribution and abundance of the Atlantic cod. The group's review was entitled *A Study of Trends of Cod Stocks off Newfoundland and Factors Influencing Their Abundance and Availability to the Inshore Fishery* (St John's: Task Force, 1987. 101 p.) and suggested that over-estimation of the stocks' growth, the reduction in the size/age of fish taken, climatic factors, and prey-predator relations had affected the abundance of the Atlantic cod. Then in 1989, the department established an independent panel to review its scientific advice regarding the northern cod stock off southern Labrador and eastern Newfoundland. The panel, which was chaired by L. Harris, the President and Vice-Chancellor of Newfoundland's Memorial University, and included fisheries experts from Canada and abroad, issued its report in March 1990. The report was entitled *Independent Review of the State of the Northern Cod Stock: Final Report. Prepared for the Honourable Thomas Siddon. Submitted by Leslie Harris, Northern Cod Review Panel* (Ottawa: Communications Directorate, Department of Fisheries and Oceans, 1990. 154 p.), and recommended the 'immediate reduction of fishing pressure, examination of the gear types in order to eliminate catches of two-to-five-year-old cod and reduce by-catches of cod in fisheries directed at other species; a more focused, controlled management approach to the stock; more study of the biology and environment of the stock; reduced foreign overfishing and illegal fishing; recognition of the importance of predator/prey relationships and conducting a harp and hooded seal census; and a peer review of departmental scientific advice by independent scientists and inclusions of the broader scientific community in the overall research program.'

Just before the review was released, the federal government announced two specially funded fisheries research programs. One was called the Northern Cod Science Program (NCSP), and the other, a government/industry-funded program, was called the Ocean Production Enhancement Network (OPEN). The NCSP was designed as a five-year (1990–95) program to improve knowledge of the basic biology and environment of the northern Atlantic cod stock. The program consisted of 25 study initiatives, involving about 50 fisheries and oceanographic researchers in Atlantic Canada, and it provided funding for studies on cod reproduction, growth, migration, distribution, and diet, as well as new studies about seal abundance, movements, and diets. Also included in the program were physical oceanography initiatives to clarify the features of the physical environment off Newfoundland and Labrador. Finally, research was also undertaken on hydroacoustic and survey gear technology to enhance the development of systems or tools available to fisheries science. This program was funded to a total amount of $28 million. The OPEN program was established to investigate the processes that control survival, growth, reproduction, and distribution of fish and shellfish, principally cod and scallops. In particular, the OPEN program focused on recruitment, distribution, physiology, marine genetics, and coastal ocean dynamics of the Atlantic cod, and ran from 1990 to 1994 with funding at about $20 million.

The result of these research programs were most visible during a symposium held in St John's, Newfoundland in October, 1994, to review the recent research on the Atlantic cod that had been completed. The symposium was attended by over 180 scientists and fisheries specialists from the Department of Fisheries and Oceans science branches, and from universities in Atlantic Canada. It was organized into eight sessions, where 80 papers were presented orally, and 30 papers displayed. The symposium was entitled *Selected Proceedings of the Symposium on the Biology and Ecology of Northwest Atlantic Cod, held at St. John's Newfoundland, October 24–28, 1994*. Edited by J.S. Campbell and P. Schwinghamer (*Canadian Journal of Fisheries and Aquatic Sciences*, v. 54, 1997. 235 p. Supplement 1). Some of the papers presented were by P. Pepin and D.C. Orr, entitled *Time to Hatch and Larval Size in Relation to Temperature and Egg Size in Atlantic Cod (*Gadus morhua*)* (*Canadian Journal of Fisheries and Aquatic Sciences*, v. 54, 1997. pp. 2–10. Supplement 1); by S.J. Prinsenberg and I.K. Peterson, entitled *Interaction between Atmosphere, Ice Cover, and Ocean off Labrador and Newfoundland from 1962 to 1992* (*Canadian Journal of Fisheries and Aquatic Sciences*, v. 54, 1997. pp. 30–39. Supplement 1); by J.T. Anderson and E.L. Dalley, entitled *Spawning and Year-Class Strength of Northern Cod (*Gadus morhua*) as Measured by Pelagic Juvenile Cod Surveys, 1991–1994* (Canadian Journal of Fisheries and Aquatic Sciences, v. 54, 1997. pp. 158–67. Supplement 1); and by G.B. Stenson and R.A. Myers, entitled *Pup Production and Population Growth of Hooded Seals (*Cysophora crisata*) near Newfoundland, Canada* (*Canadian Journal of Fisheries and Aquatic Sciences*, v. 54, 1997. pp. 209–19. Supplement 1).

Another important Atlantic fishery, which has developed haphazardly over 90 years, is the lobster fishery. The American lobster (*Homarus americanus*) is an invertebrate crustacean which lives along the east coast of North America, from North Carolina to Labrador. In particular, it is abundant off Maine, southwest Nova Scotia, and the Gulf of St Lawrence coast of the Maritimes. Canada's commercial lobster fishing began before 1870, but by the end of the century the stocks above 220 grams had been exhausted and catches declined, from a peak of 45,000 metric tonnes, to 12,000 tonnes in 1918. Although productivity has increased since 1920, the overall landings have declined, mainly because most female lobsters are caught before they are able to produce eggs. This problem is aggravated because the management of the stock is divided into just over 20 separate areas, which prevents proper coordination of conservation efforts.

Researchers also say that the current imprecision associated with stock definition, the extent to which other factors (temperature, ecosystem changes), other than spawning stock size, seem to cause variability in recruitment (the number surviving from a spawning population into subsequent life stage), and the lack of understanding of larval recruitment processes makes it particularly difficult to determine the stock recruitment relationship for effectively managing the lobster fishing. In order to narrow down these problems, there have been two major workshops held to fill gaps in lobster research. The first one was held in 1977, in Perth, Western Australia, and resulted in the publication of a two-volume review entitled *Biology and Management of Lobsters*. Edited by J.S. Cobb and B.F. Phillips (New York: Academic Press, 1980. 2 v.). The Canadian contribution to this review, by D.E. Aiken and S.L. Waddy, was entitled *Reproductive Biology*. In: *Biology and Management of Lobsters*. Edited by J.S. Cobb and B.F. Phillips (New York: Academic Press, v. 1, 1980. pp. 215–76). This review described male and female reproduction in lobsters, including environmental factors and hormone regulation affecting reproduction. The second major workshop was the International Workshop on Lobster Recruitment, held at St. Andrews, New Brunswick, in July, 1985. J.E. Stewart, in the *Introductory Address* (*Canadian Journal of Fisheries and Aquatic Sciences*, v. 43, 1986. pp. 2068–69), said that 'since lobsters form the basis of economically valuable fisheries around the world and are characterized often by declining stocks, it behoves us to move quickly to define that information required to provide adequate management regimes, and to institute the necessary research programs to collect and interpret this information.'

The workshop focused upon seven objectives: (1) lobster fisheries and recruitment variability, (2) larval distribution dynamics, (3) behaviour and movement, (4) recruitment enhancement, (5) ecology and population structure, (6) factors affecting growth and reproduction, and (7) models of stock-recruitment relationships. Seventy-nine scientists from lobster-producing countries attended the workshop, and 88 papers were presented. Early on, one of the major papers given at the workshop by G.P. Ennis was entitled *Stock Definition, Recruitment Variability, and Larval Recruitment Processes in the American Lobster,* Homarus americanus*: A Review (Canadian Journal of Fisheries and Aquatic Sciences*, v. 43, 1986. pp. 2072–84). In this paper, the author said that 'knowledge of the relationship between stock size and subsequent recruitment to the stock is important in order to effectively manage a fishery.' This required '(1) a population that is more or less discrete both geographically and biologically, (2) a measure of stock size over a period when abundance ranged widely, (3) a measure of subsequent recruitment to the stock which coincides with the same time period, (4) an understanding of the effect of variation in factors other than stock size on recruitment variability, and (5) an understanding of recruitment processes for the stock. The foregoing has not been achieved for any population of the American lobster, *Homarus americanus*.'

Since that workshop was held, there have been additional studies carried out both on stock assessments and movements. G. Robichaud and F. Savoie, in their data reports entitled *Lobster (*Homarus americanus*) Fishery Sea Sampling Data from 1989 to 1994, for Fishery Area 23 in the Southern Gulf of St. Lawrence* (Ottawa: Department of Fisheries and Oceans, 1996. 40 p. Canadian Data Report of Fisheries and Aquatic Sciences No. 988); *Lobster (*Homarus americanus*) Fishery Sea Sampling Data from 1989 to 1994, for Fishing Area 24 in the Southern Gulf of St. Lawrence* (Ottawa: Department of Fisheries and Oceans, 1996. 33 p. Canadian Data Report of Fisheries and Aquatic Sciences No. 991); and *Lobster (*Homarus americanus*) Fishery Sea Sampling Data from 1989 to 1994, for Fishery Area 26B in the Southern Gulf of St. Lawrence* (Ottawa: Department of Fisheries and Oceans, 1996. 27 p. Canadian Data Report of Fisheries and Aquatic Sciences No. 992), documented lobster populations in those areas. Previous to this data collecting, the major study of lobster stocks had involved the tagging of approximately 1,000 lobsters per year for the years 1944 to 1968, and 1978 to 1980, in order to study stock movements. This work was described by A. Campbell in the report entitled *Movements of Tagged Lobsters Released off Port Maitland, Nova Scotia (Canada), 1944–1980* (Ottawa: Department of Fisheries and Oceans, 1992. 41 p. Canadian Technical Report of Fisheries and Aquatic Sciences No. 1136). One other study of lobster movement has been undertaken in the southern Gulf of St Lawrence by M. Comeau and M. Lanteigne and described in their paper entitled *Lobster (*Homarus americanus*) Movement in the Southern Gulf of St. Lawrence – Summary Sheets of Tagging Projects Conducted between 1980 and 1997* (Ottawa: Department of Fisheries and Oceans. Canadian Industry Report of Fisheries and Aquatic Sciences No. 249, 1999. 111 p.).

Most recently the environmental factors affecting the life cycle of the American lobster have been discussed by G.C. Harding in a paper entitled *American Lobster (*Homarus americanus *Milne Edwards): A Discussion Paper on Their Environmental Requirements and the Known Anthropogenic Effects on Their Populations* (Ottawa: Department of Fisheries and Oceans, 1992. 16 p. Canadian Technical Report of Fisheries and Aquatic Sciences No. 1887). This paper provided a general evaluation of the life history and habitat requirements of the American lobster, including environmental requirements such as water temperature, salinity, and oxygen, along with various anthropogenic (man-made) effects such as oil spills, chlorination, and dredging activities. The influence of temperature alone on the American lobster has been discussed by K.F. Drinkwater and G.C. Harding in their paper entitled *Temperature as Possible Factor in the Increased Abundance of American Lobster (*Homarus americanus*) during the 1980s and Early 1990s* (*Fisheries Oceanography*, v. 5, 1996. pp. 176–93). Here, the authors pointed out that through most of the geographical range of the American lobster, catches reached historical heights in the late 1980s and early 1990s, perhaps due to recruitment (survival rates) involving increased ocean temperatures. However, when catches were correlated with temperature fluctuations, no link was found to changes in the ocean temperature.

Parallel to the problems of the Atlantic fishery have been the most recent concerns about the Pacific coast salmon fishery, which has experienced stock fluctuations. However, the level of research to date has not been as intensive as that pertaining to the Atlantic fisheries. This lack of research was noted by P.H. Pearce, head and commissioner of the Commission on Pacific Fisheries Policy. The report, entitled *Turning the Tide: A New Policy for Canada's Pacific Fisheries: The Commission on Pacific Fisheries Policy Final Report* (Vancouver: The Commission, 1983. 292 p.), said that the present knowledge of fish habitats was 'seriously deficient.' As well, the report noted the need for resource conservation and better federal fisheries management of the commercial and recreational fishery. Into the late 1980s, the Department of Fisheries and Oceans was still analysing the data from salmon surveys, as noted in technical reports by B.J. Waddell and M.C. Healey, entitled *Data Analysis of 1986 and 1987 Hecate Strait (British Columbia, Canada) Juvenile Salmon Surveys* (Ottawa: Department of Fisheries and Oceans, 1989. 65 p. Canadian Technical Report of Fisheries and Aquatic Sciences No. 1719); and by J.F. Morris and M.C. Healey, entitled *The Distribution, Abundance, and Feeding Habits of Chinook and Coho Salmon on the Fishing Banks off Southwest Vancouver Island (British Columbia, Canada) May 23–30, 1988* (Ottawa: Department of Fisheries and Oceans, 1990. 75 p. Canadian Technical Report of Fisheries and Aquatic Sciences No. 1759).

However, by the 1990s, papers began to appear concerning the impact of environmental practices on the abundance of salmon stocks in the freshwater phase of their life cycle as well as the effect of climate on salmon stocks in the ocean phase of their life cycle. For example, G.F. Hartman and J.C. Scrivener, in their bulletin entitled *Impacts of Forestry Practices on a Coastal Stream Ecosystem, Carnation Creek, British Columbia (Canada)* (Ottawa: Department of Fisheries and Oceans, 1990. 148 p. Canadian Bulletin of Fisheries and Aquatic Sciences No. 223), reviewed the results from the first 17 years of a multi-disciplinary study about the effects of logging activities on a small-stream ecosystem in the coastal rainforest of British Columbia. They noted that, in the short term, forest practices increased stream isolation, water temperatures, and nutrient levels, which in turn increased the numbers, growth period, and size of coho salmon fry (young salmon in the second year), but reduced the marine survival of chum salmon. Over the long term, however, stream instability due to logging, coupled with changes in the composition of spawning gravel, reduced fish survival and numbers. Another important paper by M. Waldichuk entitled *Fish Habitat and the Impact of Human Activity with Particular Reference to Pacific Salmon* (Canadian Bulletin of Fisheries and Aquatic Sciences No. 226, 1993. pp. 295–337) reviewed various types of human impact (logging, mining, dams, water diversion, pulpmill effluent, and ocean dumping), and said that 'the scientific basis for developing universally applicable models for habitat management is still inadequate.'

Papers on the importance of climate, and its effect on salmon abundance, began to appear in the early 1990s. All these papers emphasized the fact that the climate patterns of the North Pacific Ocean, which is dominated in the winter and spring by the Aleutian Low Pressure System, corresponded to the long-term trends in the salmon catch. R.J. Beamish and D.R. Bouillon, in their paper entitled *Pacific Salmon Production Trends in Relation to Climate* (*Canadian Journal of Fisheries and Aquatic Sciences*, v. 50, 1993. pp. 1007–16), noticed this pattern by studying the all-nation (Canada, Japan, the United States, and Russia) catches of the three species of salmon (pink – *Onchorynchus gorbuscha*; chum – *Onchorynchus keta*; and sockeye – *Onchorynchus nerka*), which represent 90% of the commercial catch of Pacific salmon, taken each year from 1925 to 1989. They were the first to notice catch patterns indicating that the connection of climate and the marine environment probably played an important role in salmon production. This pattern was noted again by the authors in another paper entitled *Marine Fish Production Trends off the Pacific Coast of Canada and the United States* (Canadian Special Publication of Fisheries and Aquatic Sciences No. 121, 1995. pp. 585–91). Here, the authors reviewed trends in production for salmon and non-salmon species in the principal commercial fisheries off the Pacific coast of Canada and the United States. Again, catches followed a pattern similar to the Aleutian Low Pressure Index.

More specifically, this parallel pattern of salmon production and climate trends (the warmer the climate and hence the sea water, the slower the stock growth) was noted in Fraser River sockeye salmon by R.J. Beamish and C.E. Neville in their paper entitled *Production of Fraser River Sockeye Salmon (*Onchorynchus nerka*) in Relation to Decadal – Scale Changes in the Climate and the Ocean* (*Canadian Journal of Fisheries and Aquatic Sciences*, v. 54, 1997. pp. 543–54); by S.P. Cox and S.G. Hinch, in their paper entitled *Changes in Size at Maturity of Fraser River Sockeye Salmon (*Onchorynchus nerka*) (1952–1993) and Associations with Temperature* (*Canadian Journal of Fisheries and Aquatic Sciences*, v. 54, 1997. pp. 1159–65); and, by R.J. Beamish and G.A. McFarlane in their paper entitled

*Recent Declines in the Recreational Catch of Coho Salmon (*Onchorynchus kisutch*) in the Strait of Georgia Are Related to Climate* (*Canadian Journal of Fisheries and Aquatic Sciences*, v. 56, 1999. pp. 506–15).

D Artificial Propagation of Fish, Including Aquaculture

With the heavy commercial exploitation of some fisheries, plus the effect of environmental stress, artificial propagation or stock enhancement, particularly of Pacific and Atlantic salmon, has been followed since the end of the 19th century. Most recently, aquaculture and the introduction of new species into aquatic environments have been encouraged.

In the early part of the century, the artificial hatching and release of the larvae, or fry, of salmon and other species was a standard propagation technique, but a 12-year experiment by R.E. Foerster, at Cultus Lake in British Columbia, showed that this procedure was questionable. In the paper entitled *An Investigation of the Relative Efficiencies of Natural and Artificial Propagation of Sockeye Salmon (*Onchorynchus nerka*) at Cultus Lake, British Columbia* (*Journal of the Fisheries Research Board of Canada*, v. 4, 1938. pp. 151–61), Foerster showed that hatcheries were contributing very little to the total sockeye production. What was important was the availability of freshwater habitat life, and the necessity for eggs to be deposited in gravel, rather than troughs, for the development of a hardy fish. These lessons were later applied to the improvement of natural spawning beds, or the construction of special spawning channels, for salmon in situations where existing gravel areas were inadequate. The most important example of this type of work began, in the mid-1950s, on Babine Lake in northern British Columbia.

Babine Lake produces 90% of the sockeye salmon (*Onchorynchus nerka*) from the Skeena River system. However, studies by W.E. Johnson, entitled *On the Distribution of Young Sockeye Salmon (*Onchorynchus nerka*) in Babine and Nilkitkwa Lakes, B.C.* (*Journal of the Fisheries Research Board of Canada*, v. 13, 1956. pp. 695–708), indicated that the lake might be capable of supporting a larger number of young sockeye. Although Babine Lake, and the connected Nikitkwa Lake, totalled 500 square kilometres, catch estimates indicated that at least 67%, and possibly 88%, of young sockeye salmon were concentrated in Nilkitkwa Lake and the North Arm of Babine Lake only, which represented about 11% of the total lake area. This unequal distribution of the young sockeye was apparently a result of the distribution of the spawning parent population and a limited dispersal of sockeye fry from their points of hatching into the lake. Accordingly, in 1965, an artificial spawning channel was completed in the Fulton River (one of the major spawning streams tributary to the lake) to augment the number of sockeye fry entering the lake. An additional channel on the Pinkut Creek, down-lake, was also opened. In 1969, a second spawning channel adjacent to the Fulton River was begun, and completed in 1972. At that time, it was estimated that a record total of 102.5 million sockeye fry (newly hatched to two years old) were produced in the channels from the 1970 brood. In 1973, it was estimated that 88 million sockeye smolts (those heading downstream to sea) from the 1972 brood migrated seaward from Babine Lake. This work has been described by C.J. West in the report entitled *A Review of the Babine Lake Development Project, 1961–1977* (Vancouver: Department of Fisheries and Environment, 1978. 105 p. Technical Report No. 812).

One concern about this second channel was that temperature data collected during the first winter of operation indicated that the channel was warmer than the river. Therefore, sockeye eggs incubated in the warmer channel might hatch earlier than river sockeye and reach the lake at a time when food was scarce and starvation could cause mortality. The temperature studies were carried out by J.R. Paine, and entitled *Fulton River Fry Quality and Ecology Program. Report of 1970 Studies* (Ottawa: Department of Forestry and Fisheries, 1971. 38 p. Technical Report No. 1). Laboratory studies on starvation of channel fry were carried out by H.J. Bilton and G.L. Robins and reported in the article entitled *The Effects of Starvation and Subsequent Feeding on Survival and Growth of Fulton Channel Sockeye Salmon Fry (*Onchorynchus nerka*)* (*Journal of the Fisheries Research Board of Canada*, v. 30, 1973. pp. 1–35). Under laboratory conditions, Fulton River sockeye fry were capable of withstanding up to three or four weeks of starvation with less than 10% mortality. However, many were incapable of recovery later when presented with food. Beyond four weeks of starvation, mortality increased sharply to 90%. Paine's studies indicated that for starved channel and river fry, mortality occurred at 25 to 32 days, indicating an association with the water temperature.

One might ask how successful this enhancement program has been. J.M. McDonald and J.M. Hume, in their paper entitled *Babine Lake Sockeye Salmon (*Onchorynchus nerka*) Enhancement Program: Testing Some Major Assumptions (Canadian Journal of Fisheries and Aquatic Sciences*, v. 41, 1984. pp. 70–92), said that the enhancement program had limited success. Their conclusions were based on the fact that the project proceeded on four basic assumptions: '(1) the artificial spawning channels would provide an effective means of producing sockeye fry, (2) the fry produced would be as viable as those produced from natural spawning beds, (3) the lake nursery area had the capacity to support large juvenile populations, and (4) increased smolt outputs would result in increased adult returns.' By checking data from before and after the beginning of the enhancement program, the authors found that 'the assumptions regarding juvenile production were found to be generally valid,' but 'adult returns did not meet expectations.' This was 'due largely to the lack of response to increased smolt outputs from even-numbered brood years.'

Salmon enhancement programs have also been carried out in Newfoundland since the late 1950s on the Exploits River, but, in the late 1960s, a major salmon collection facility on that river was completed, providing access to 3,700 square kilometres of spawning area above Grand Falls. This spawning area was to be stocked in part with fry from the spawning channel at Noel Paul's Brook. The total number of fry stocked ranged from 139,880 in 1968 to 369,689 in 1973. From 1976 to 1980, stocking was expanded to include several other tributaries, in addition to Noel Paul's Brook, and the total number of fry stocked ranged from 696,897 in 1979 to 1,827,374 in 1980. Adult returns, corresponding to the first three years of high-level stocking, ranged from 2,388 in 1982 to 4,022 in 1981. This also seemed to result in a dramatic increase in commercial catches in the Bay of Exploits, but the effect on recreational fishing was not as pronounced. The enhancement work was documented by M.F. O'Connell and J.P. Davis in their technical report entitled *An Assessment of the Stocking of Atlantic Salmon (*Salmo salar *L.) Fry in the Tributaries of the Middle Exploits River, Newfoundland* (Technical Report No. 1225, 1984. 142 p.). Three years later, M.F. O'Connell and C.E. Bourgeois, in their paper entitled *Atlantic Salmon Enhancement in the Exploits River, Newfoundland (Canada), 1957–1984* (*North American Journal of Fisheries Management*, v. 7, 1987. pp. 207–14), noted that stocking with adults as well as fry had been used in the enhancement program and they evaluated both methods in terms of adult returns. Results indicated that fry stocking was the most effective approach. 'Fry stocking resulted in the establishment of runs in Exploits River tributaries in roughly, one-half the time required for adult stocking and with less than one-half the brood-stock requirement.'

Another approach to stock enhancement has seen the use of inorganic nutrients in specific lakes where salmon spend their growing years. The hypothesis behind this method of enhancement is the hope that nutrient additions to an oligotrophic lake (a lake that is characterized by a deficiency of plant nutrients) would increase the production of zooplankton in these lakes and enhance the production of juvenile sockeye salmon.

Long-term nutrient fertilization was carried out in Great Central Lake, British Columbia from 1969 to 1976. To this end, 100 tons of fertilizer were added each year to increase the primary production of zooplankton. The growth rates of young sockeye in the lake increased such that the average weight of the fish at the onset of winter was 30%–92% greater than previously documented. However, when all factors were taken into account, researchers did find that the net effect on the population was such that production was enhanced more by increased numbers of fish surviving to migrate seaward than it was by the increase in average weight per migrant. The results of this work were reported by D.G. Robertson and W.E. Barraclough and entitled *Population Estimates of Sockeye Salmon (*Onchorynchus nerka*) in a Fertilized Oligotrophic Lake* (*Journal of the Fisheries Research Board of Canada*, v. 35, 1978. pp. 851–66).

In addition to individual enhancement research studies, the federal government, and the provincial government of British Columbia, created the Salmonid Enhancement Program (SEP), a multimillion dollar federal program that aimed to double the salmonid populations on the BC coast by 1996. By 1980, hatching production was at 226 million juvenile salmonids at hatcheries located on the Puntledge, Chilliwack, and Nitimat rivers, Loon Creek on the Upper Fraser, and Tlupana and Robertson creeks on Vancouver Island's west coast. And, by 1984, when SEP was in a transition phase to long-term planning, the federal government budgeted $40.7 million for the program. That year, 432 million juvenile salmonids were released from SEP facilities, and returns were estimated at 6.2 million adults. Then, in June 1987, the federal government announced a $208 million package for SEP, or approximately $40 million a year, to carry the program into the 1990s and assure ongoing salmonid research on the Pacific coast.

The types of research that have emerged from the Salmonid Enhancement Program have been related to the artificial rearing and growth of salmon in hatcheries or streams that are part of the program. For example, W.E. Mclean and P.G. Lim, in their report entitled *Some Effects of Water Flow Rate and Alevin Density in Vertical Incubation Trays on Early Development of Chinook Salmon (*Onchorynchus tshawytscha*)* (Vancouver: Department of Fisheries and Oceans. Salmonid Enhancement Program, 1985. 46 p.), observed that low water flows and high density of alevins (the newly hatched yolk-bearing larval stage) in incubation trays tended to inhibit weight gains in the later-stage fry in the trays. Another factor that has been examined in the rearing of salmon fry in hatcheries is aeration of water, as noted by W.E. Mclean and P.B. Wright, in their report entitled *The Effects of Aeration on the Rearing of Summer Chinook Fry (*Onchorynchus tshawytscha*) at the Puntledge Hatchery (Vancouver Island, British Columbia, Canada)* (Vancouver: Department of Fisheries and Oceans. Salmonid Enhancement Program, 1985. 60 p. Canadian Technical Report of Fisheries and Aquatic Sciences No. 1390). Other factors affecting rearing of salmon that have been studied pertain to stream conditions. B.G. Shepherd and G.F. Hartman, in their paper entitled *Relationships between Stream and Intra-gravel Temperatures in Coastal Drainages, and Some Implications for Fisheries Workers* (*Canadian Journal of Fisheries and Aquatic Sciences*, v. 43, 1986. pp. 1818–22), noted that intra-gravel temperatures were warmer in the winter than in the summer, with the temperature cross-overs occurring in March and October. These temperature variations could affect salmonid egg incubation and fry emergence. D.E. Marshall and E.W.

Britton, in their report entitled *Carrying Capacity of Coho Salmon Streams* (Vancouver: Department of Fisheries and Oceans, 1990. 32 p. Canadian Manuscript Report of Fisheries and Aquatic Sciences No. 2058), noted that the rearing space available in streams, namely stream width, was involved in determining the carrying capacity of coho salmon streams.

Although the federal government boasted about the success of the Salmonid Enhancement Program, some researchers were not as enthusiastic. R. Hilborn and J. Winton, in their paper entitled *Learning to Enhance Salmon Production: Lessons from the Salmonid Enhancement Program* (*Canadian Journal of Fisheries and Aquatic Sciences*, v. 50, 1993. pp. 2043–56) said that the original goal of SEP was to produce 86,000 metric tons of salmon per year, and it relied primarily on three major technologies to achieve this goal – hatcheries, spawning channels, and lake enrichment. The authors estimated SEP's annual production to be about 12,400 metric tons per year. 'Thus, the program has fallen short of its objectives.' The authors also said that 'SEP devoted considerable effort in evaluating individual facilities and sites. However, SEP has been unable or unwilling to evaluate overall program success and direction. Due to natural variation in survival rates, it is difficult to determine, even after 15 years, which technologies are likely to be successful. Furthermore, there is a paucity of information on wild stocks and the extent to which these enhancement activities have an effect on their numbers.' However, J. Gough, in a paper entitled *A Historical Sketch of Fisheries Management in Canada* (Canadian Bulletin of Fisheries and Aquatic Sciences No. 226, 1993. pp. 5–53), was not as negative. This researcher said that the Canadian fishing industry and its management had developed differently in different regions. The Pacific management was more far-reaching than that on the Atlantic, but the latter had become more interventionist since the 1970s. The author thought that initiatives like the Pacific Salmonid Enhancement Program had produced positive results, but 'Canada's fisheries are still subject to major fluctuations.'

Perhaps the most visibly encouraged form of artificial propagation is aquaculture, or fish farming, because of its more immediate economic impact for expanding Canada's fishing industry. Through experiments on artificial rearing of finfish, molluscs, and crustaceans, Canada has developed a mosaic of small family operations, publicly owned companies, and subsidiaries of a large corporations, in the international seafood markets. In fact, in 1976, Canada hosted the *13th Pacific Science Congress: Aquaculture Symposium* (*Journal of the Fisheries Research Board of Canada*, v. 33, 1976. pp. 875–1111). Although Canadian papers were not published as part of this Congress (most papers were from Japan, the USA, Scotland, Norway, and India), Canadian researchers chaired the proceedings, which covered the topics of economic and legal considerations in aquaculture; the current status and future prospects for the culture of economically important species; endocrinology and reproductive physiology in aquaculture; the current status and future prospects for seaweed culture; health, nutrition, and growth in aquaculture; and genetics in aquaculture. In 1983, the Department of Fisheries and Oceans sponsored a conference in which the planners 'ensured that these papers were of Canadian origin in keeping with a desire to obtain an accurate reflection of Canadian industry and its opportunities.' In the *Proceedings of the National Aquaculture Conference: Strategies for Aquaculture Development in Canada.* Edited by G.I. Pritchard (Ottawa: Department of Fisheries and Oceans, 1984. 131 p.), a keynote address and 13 papers were given. The topics covered were an overview of aquaculture practices in Canada; risk reduction in enterprise planning; planning for aquaculture; the marketplace; and socioeconomic dimensions.

A major assessment of Canadian aquaculture was completed in 1989, by the Department of Fisheries and Oceans, and entitled *Long Term Production Outlook for the Canadian Aquaculture Industry* (Ottawa: Department of Fisheries and Oceans, 1990. 100 p.). The report confirmed that the industry had expanded rapidly in the last 15 years and would likely continue to grow at even higher rates. The industry has grown in value from $7 million in 1984 to $62 million in 1987, and $109 million in 1988. The latter represented 6% of the landed value of all Canadian fish that year.

Five years later, D.E. Aiken, in the paper entitled *Aquaculture in Canada* (Canadian Bulletin of Fisheries and Aquatic Sciences No. 226, 1993. pp. 411–46), noted that by the early 1990s, the production of east and west coast salmon had expanded to 22,500 tonnes per year, the blue mussel farms of Prince Edward Island were producing 3,000 tonnes per year, 2,300 tonnes of Atlantic oysters and 3,900 tonnes of Pacific oysters were being produced, and freshwater trout had expanded to just over 3,000 tonnes per year. Finally, in 1995, the Department of Fisheries and Oceans published the *Federal Aquaculture Development Strategy* (Ottawa: Department of Fisheries and Oceans, 1995. 18 p.), which defined the federal role in aquaculture development to include research, technology, transfer, environmental sustainability, and interaction, as well as other components such as resource allocation and access, product safety and inspection, and performance measurement and improvement.

Because aquaculture is a productive and expanding industry, the Department of Fisheries and Oceans has produced more publicity-type papers than it would for other fisheries. Two such publications are *Developing Aquaculture for Canada: A Discussion Paper* (Ottawa: Department of Fisheries and Oceans, 1986. 14 p.), and *Aquaculture Development in Canada: A Guide to Federal Government Programs* (Ottawa: Department of Fisheries and Oceans, 1988. 83 p.). This publication was reissued in 1993 with the title *Aquaculture: A Guide to Federal Government Programs. Revised Edition* (Ottawa: Department of Fisheries and Oceans, 1993.

40 p.). Other publications that have appeared are *Cultivating the Future: Aquaculture Strategies for the 90s* (Ottawa: Department of Fisheries and Oceans, 1990. 40 p.); and *Fisheries and Oceans Action Plan: Science and Technology for the New Century* (Ottawa: Department of Fisheries and Oceans, 1996. 17 p.).

Rearing of fish to maturity in artificial environments brings concerns about fish health and mortality in the stock being reared. Accordingly, fisheries scientists study the nutritional requirements and the control of viral and bacterial diseases in fish stocks. As well, fishery scientists also search for new fish species adaptable to artificial rearing for aquaculture.

Nutritional requirements have involved the study of not only Atlantic and Pacific salmon, but also crustaceans such as lobsters and bivalves such as scallops. For salmon, commercial feed mixes are available, and these have been tested against experimental mixes in order to compare their effect on the growth rates. D.A. Higgs and J.R. Markert, in their paper entitled *Development of Nutritional and Environmental Strategies for Maximizing the Growth and Survival of Juvenile Pink Salmon (*Onchorynchus gorbuscha*)* (*Aquaculture*, v. 47, 1985. pp. 113–30), noted that steam-dried fish meal, coupled with a high level of freeze-dried zooplankton, provided a better survival and growth rate for pink salmon than did conventional hatchery food. This was particularly noticeable at the critical time of transfer of the fish from the freshwater environment into the saline environment. The superiority of an experimental dry diet for steelhead, in particular, was noted by V.H. Fargerhend and I. McCallum in their paper entitled *Diet Composition as a Factor in the Anabolic Efficacy of 3,5,3-Triiodo-L-Thyronine Administered Orally to Steelhead Trout (*Salmo gairdneri*)* (*Aquaculture*, v. 36, 1984. pp. 49–60). The authors said that when this amino acid was added to experimental diet T-3, it stimulated growth of steelhead trout through increased carbohydrate utilization.

Unlike diet formulations for salmon, those for crustacean aquaculture have a much more recent history, and reasons have been given for the need of a standard formulation to serve as a basis for crustacean nutritional needs. J.D. Castell and J.C. Kean, in their papers entitled *A Standard Reference Diet for Crustacean Nutrition Research: I. Evaluation of Two Formulations* (*Journal of the World Aquaculture Society*, v. 20, 1989. pp. 93–99); and *A Standard Reference Diet for Crustacean Nutrition Research: II. Selection of a Purification Procedure for Production of the Rock Crab* Cancer irroratus *Protein Ingredient* (*Journal of the World Aquaculture Society*, v. 20, 1989. pp. 100–106), said that the development and use of a standard reference diet, which was mentioned at the 1984 World Mariculture Society, would 'provide a more accurate assessment of non-nutritional growth factors and permit comparisons among laboratories, species and experiments.' For bivalves, scallops and mussels have been the major source of study for nutritional needs. J.N. Whyte, in the papers entitled *Biochemical Composition and Energy Content of*

Six Species of Phytoplankton Used in Mariculture of Bivalves (*Aquaculture*, v. 60, 1987. pp. 231–42); and *Nutritional Condition of Rock Scallop,* Crassodoma gigantea *(Gray), Larvae Fed Mixed Algal Diets* (*Aquaculture*, v. 86, 1990. pp. 25–40), noted that the accumulation of fatty acids by the larvae suggested that they needed these acids during larval growth and development. Whyte also added that the importance of carbohydrates in providing a balanced diet for the effective conversion of micro-nutrients to tissue and energy reserves had been overlooked in larval nutrition.

In the area of fish diseases, a major concern has been the ongoing study of bacterial kidney disease in salmon species. For example, research has been carried out by J.C. Sanders and K.S. Pilcher on the relationship of water temperature to the disease in a paper entitled *Relation of Water Temperature to Bacterial Kidney Disease in Coho Salmon (*O. nerka*) and Steelhead Trout (*Salmon gairdneri*)* (*Journal of the Fisheries Research Board of Canada*, v. 35, 1978. pp. 8–11). Research has also been carried out by G.R. Bell and D.A. Higgs on the effect of minerals on experimentally induced kidney disease. Their paper was entitled *The Effect of Dietary Ascorbate, Zinc and Manganese on the Development of Experimentally Induced Bacterial Kidney Disease in Sockeye Salmon (*Oncorhyncus nerka*)* (*Aquaculture*, v. 36, 1984. pp. 293–311). Methods of controlling bacterial disease have focused upon immunization or drug treatment of fish. The ability of teleost fish (e.g. salmon) to develop active immunity to bacterial pathogens has been well established in laboratory experiments involving oral and perenteral vaccination, but this has disadvantages for large-scale immunization of an aquaculture stock. Accordingly, R. Antipa and D.F. Amend found that vaccination by immersion (hyperosmotic infiltration) was an improved method of immunizing cultured salmon. Their paper was entitled *Immunization of Pacific Salmon: Comparison of Intraperetonial Injection and Hyperosmotic Infiltration of* Vibrio anguillarum *and* Aeromonas salmonicida *Bacterines* (*Journal of the Fisheries Research Board of Canada*, v. 34, 1977. pp. 203–8). By the early 1990s, oral treatment (via treated food) of coho salmon infected with kidney disease was attempted using the drug fumagillin. M.J. Higgins and M.L. Kent, in their paper entitled *Field Trials with Fumagillin for the Control of Proliferative Kidney Disease in Coho Salmon* (*Progressive Fish Culturalist*, v. 58, 1996. pp. 268–72), found the oral treatment reduced bacterial infection by 25% to 50%.

Another aspect of aquaculture is the adaptability to artificially rear marine species that have market potential. One recent fish in this category has been the Arctic char, and the factors which seem to affect its successful rearing in the aquaculture environment have been examined. J.A. Tabachek, in the paper entitled *The Effect of Feed Particle Size on the Growth and Feed Efficiency of Arctic Charr (*Salvelinus alpinus *(L.)* (*Aquaculture*, v. 71, 1988. pp. 319–30), noted that feed particle size did have a significant effect on specific growth rate, including length and weight, and that specific feed particle sizes were correlated to specific mouth-sizes. As well, M.H. Papst and T.A. Dick, in their paper entitled *Effect of Rearing Density on the Early Growth and Variation in Growth of Juvenile Arctic Charr,* Salvelinus alpinus *(L.)* (*Aquaculture and Fisheries Management,* v. 23, 1992. pp. 41–47), noted that increased density of fish in a tank enclosure caused an 'overall higher mean specific growth rate.' And the study by B.G. DeMarch entitled *Social Genetic Determinants of Size Variation in Tanks of Navyuk, Norwegian, and Hybrid Arctic Charr,* Salvelinus alpinus *(L.)* (*Aquatic Research,* v. 28, 1997. pp. 305–15), concluded that 'social factors are the most important in determining the growth of individual charr, and may have confounded previous experiments examining the genetic basis of growth.'

With the development of the aquaculture industry in Canada over the last two decades with primarily salmonid species, federal, provincial, and regional regulations have been written to protect the health of salmon and trout. T.G. Carey and G.I. Pritchard, in their paper entitled *Fish Health Protection: A Strategic Role in Canadian Fisheries Management (North American Journal of Fisheries Management,* v. 15, 1995. pp. 1–13), said that the creation of regulations which relate to the control of potential hazards pertaining to the use of live fish for research, the manufacture of feeds, and the development of drugs and therapeutics, coupled with a strong research capability and training program, have created a 'favourable climate for major investments in salmonid farming and its allied industries.'

E The Study and Management of Marine Mammals

Marine mammal management is a critical area for research and assessment by marine biologists because of the slow rate of replacement of the stock at only 10% a year. In fact, without quotas on hunting, or the 'harvest,' stocks would rapidly become extinct. In Canada, marine mammal management focuses on seals (primarily harp, ringed, and bearded), walruses, beluga whales, and narwhals, ranging from the region of the Gulf of St Lawrence, the North Atlantic, and the High Arctic, to the Pacific Ocean, and involves the monitoring of populations and their distribution.

Research on harp seals (with thin white-coat pups) has been carried out in order to develop a knowledge of migration patterns and population structures, which can provide advice to the federal government and the International Commission for the Northwest Atlantic Fisheries (ICNAF) on the management of the stock. Major work in this area has been done by D.E. Sergeant in research papers entitled *Migration of Harp Seals,* Pagophilus groenlandicus *(Erxleben)* (*Journal of the Fisheries Research Board of Canada,* v. 22, 1965. pp. 433–64); *Reproductive Rates of the*

Harp Seals, Pagophilus groenlandicus *(Erxleben)* (*Journal of the Fisheries Research Board of Canada*, v. 23, 1966. pp. 757–66); *History and Present Status of Populations of Harp Seal and Hooded Seals* (*Biological Conservation*, v. 10, 1976. pp. 96–118); and *Results of Tagging and Branding of Hooded Seals, 1972–1978* (Canadian Atlantic Fisheries Scientific Advisory Committee CRD Research Document 78/36, 1978. 5p.).

In the early and mid-1970s, quotas were placed on these two species, and research was undertaken to develop a statistical model that could be used to estimate the seal populations. This work was carried out by W.D. Bowen and D.E. Sergeant and described in the research article entitled *Mark-Recapture Estimates of the Harp Seal Pup (*Phoca groenlandica*) Production in the Northwest Atlantic* (*Canadian Journal of Fisheries and Aquatic Sciences*, v. 40, 1983. pp. 728–42). This research indicated that the population size of harp seals had increased since the imposition of catch quotas in 1971. An overview of harp seal studies was written by D.E. Sergeant, in 1991, and entitled *Harp Seals, Man and Ice* (Ottawa: Department of Fisheries and Oceans, 1991, 153 p. Canadian Special Publication of Fisheries and Aquatic Sciences No. 114). In this publication, the author, who has studied harp seals for 35 years, said that although a great amount of research has been carried out on harp seals since the 1950s, 'research has opened as many problems as it has solved. Thus practically nothing is known of the behaviour of harp seals, how they navigate, the nature of groups of animals, and the bearing of these features on their migration....' Some of the topics discussed in this publication are physiology and behaviour; reproduction and growth; distribution and migration; exploitation and management; and pollution and harp seals.

In 1983, the European Community banned the import of white-coat pup pelts after a major campaign against the seal harvest. This outcry prompted the federal government to appoint a Royal Commission to look into all aspects of seals and sealing in Canada. It reported its findings, in 1986, and entitled the report *Seals and Sealing in Canada: Report of the Royal Commission* (Ottawa: Supplies and Services Canada, 1986. 3 v.). In response to the recommendations of the Royal Commission, the government announced a new seal policy in December, 1987. Under this, the large-vessel off-shore seal hunt was no longer permitted, and all commercial hunting of white-coat harp seals and blue-black hooded seals ended. In addition, the netting of seals was to be phased out, except for traditional aboriginal hunts north of 53 degrees latitude.

The harp seal migrates north during the summer months, but it is only the ringed and bearded seal and the walrus that are native to the High Arctic. In the 1970s, concern was expressed about the environmental impact upon marine mammal life of oil tanker passages based upon proposals for large-scale resource extraction in the High Arctic, Melville Island, and the Beaufort Sea. Site-specific environmental studies had been undertaken, but large-scale studies to evaluate the distribution and abundance of seals in the High Arctic in the major inter-island channels, where activities were proposed, needed to be done. These three species were studied by means of an aerial, strip, north-south intersect survey.

The major findings indicated that the ringed seal's preferred habitat was annual ice, moving and cracking, and not deep water. Bearded seals, which were fewer in number, preferred floe, or rotten ice, and shallow waters. These survey results were reported by M.C. Kingsley and I. Stirling and in their research paper entitled *The Distribution and Abundance of Seals in the Canadian High Arctic, 1980–1982 (Canadian Journal of Fisheries and Aquatic Sciences*, v. 42, 1985, pp. 1189–1210). Additional studies on ringed and bearded seals by these authors confirmed the pattern of ice behaviour of these two species. In their paper entitled *Haul-Out Behaviour of Ringed and Bearded Seals in Relation to Defence against Surface Predators* (*Canadian Journal of Zoology*, v. 69, 1991. pp. 1857–61), the authors observed that the ice habitat patterns of the two species reduced their vulnerability to capture by polar bears. M.C. Kingsley's knowledge of the ringed seal has also been summarized in the paper entitled *Status of the Ringed Seal, Phoca hispida, in Canada* (*Canadian Field Naturalist*, v. 104, 1990. pp. 138–45), which said that 'although it has been hunted for its meat, blubber, and hide by the native people of the north for millennia, and is also the mainstay of the polar bear, its populations remain stable, varying only in response to variations in the weather. It is adaptable to human disturbance of many kinds.' Further observations of the bearded seal's habitat have been undertaken by H.J. Cleator and I. Stirling in their paper entitled *Winter Distribution of Bearded Seals (*Erignathus barbatus*) in the Penny Strait Area, Northwest Territories (Canada), As Determined by Underwater Vocalizations* (*Canadian Journal of Fisheries and Aquatic Sciences*, v. 47, 1990. pp. 1071–76). Here, the authors said that vocalization surveys could not be used to determine the absolute number of bearded seals at sites with this technique, but it was possible to measure the relative abundance of seals. Later, H.J. Cleator, alone, in the paper entitled *The Status of the Bearded Seal,* Erignathus barbatus, *in Canada* (*Canadian Field Naturalist*, v. 110, 1996. pp. 501–10) said that population estimates of the bearded seal were still not available, 'although summing different indices obtained over the past 35 years for several different regions suggest that a minimum of 190,000 bearded seals may inhabit Canadian waters.'

The Atlantic walrus ranges in the Canadian Arctic waters from Baffin Bay and Davis Strait, and westward, to include most of the Hudson Bay. However, interest in this species only began in the 1950s with the publications by A.W. Mansfield entitled *The Biology of the Atlantic Walrus* Odobenus, rosmarus rosmarus *(Linnaeus) in the Eastern Canadian Arctic* (Ottawa: Fisheries Research Board of Canada, 1958. 146 p. Manuscript Report Series (Biological) No. 653); and *The Walrus in the Canadian Arctic* (Montreal: Fisheries Research Board of Canada, 1959. 13 p. Fisheries Research Board Circular No. 2). It was at this time that the first estimates of walrus populations were undertaken. Aerial surveys were carried out in the Hudson Bay and Foxe Basin in the summer and fall months of 1954, the seasons when walruses are concentrated on land. This permitted a reasonably accurate count to be made. These estimates were reported by A.G. Loughrey in the publication entitled *Preliminary Investigation of the Atlantic Walrus* (Odobenus rosmarus rosmarus *(Linnaeus)* (Ottawa: Canadian Wildlife Service, 1959. 123 p. Wildlife Management Bulletin. Series 1, No. 14). At that time, maximum counts were estimated at 2,500. Further aerial surveys to estimate walrus populations in the northern Hudson Bay region were not undertaken again until summer of 1976, using photographic equipment. At this time, A.W. Mansfield and D.J. St. Aubin, in their paper entitled *Distribution and Abundance of the Atlantic Walrus,* Odobenus rosmarus, rosmarus, *in the Southampton Island–Coats Island Region of Northern Hudson Bay* (*Canadian Field Naturalist,* v. 105, 1991. pp. 95–100), noted that these newer estimates of 1,500 were somewhat lower than Loughrey's, perhaps because of the distribution of walruses at the various haul-out sites (land or ice-floes) where the surveys were conducted. The distribution was greatly affected by weather conditions and human and animal disturbances.

As with distribution surveys, the breeding behaviour of walruses has only recently been studied. B. Sjare and I. Stirling, in their paper entitled *The Breeding Behaviour of Atlantic Walruses,* Odobenus rosmarus rosmarus, *in the Canadian High Arctic* (*Canadian Journal of Zoology,* v. 74, 1996. pp. 897–911), studied the breeding behaviour of Atlantic walruses over-wintering at the Dundas Island polynya, from 1981 to 1990. They described the mating system of walruses breeding in the area as being 'female-defense polygny' (one male with two or more mates), where a mature male had exclusive access to a herd of walruses containing several females for one to five days at a time. The authors concluded that the 'inherent gregariousness of female walruses [appears to be an important factor] influencing the evolution of social behaviour in the mating system of walruses.'

Like seal or walrus populations, whaling stocks are monitored and studied, not only for management advice given to the federal government, but also for advice given to the International Whaling Convention from 1946 to 1982 (at which time Canada withdrew its membership). The two most abundant cetacean species in Canada are belugas (white whales) and narwhals, which are closely related and confined to the northern extremes of the Northern Hemisphere. The beluga occur in many discrete stocks throughout the Arctic and sub-Arctic regions (e.g. the Mackenzie Delta stock, the Lancaster Sound stock, the Cumberland Sound stock, the Hudson Bay/Hudson Strait/Ungava Bay stock, and the St Lawrence stock). On the other hand, the narwhal's range and habitat requirements are more restricted, and the largest population in the world inhabits the deep inlets, sounds, and channels of the eastern Canadian Arctic and northwest Greenland. It is interesting to note that, prior to 1975, information on the summer and winter distribution, movements, and population estimates of beluga whales and narwhal was based on limited field studies and data reported by D.E. Sergeant and P.F. Brodie and entitled *Identity, Abundance and Present Status of Populations of White Whales,* Delphinapterus leucas, *in North America* (*Journal of the Fisheries Research Board of Canada*, v. 32, 1975, pp. 1047–54); and A.W. Mansfield and J.G. Smith, entitled *The Narwhal (*Monodon monoceros*) in Eastern Canadian Waters* (*Journal of the Fisheries Research Board of Canada*, v. 32, 1975. pp. 1041–46).

However, the advent of exploration for oil and gas reserves in the High Arctic raised concerns about the potential effects of commercial development on marine mammals, even though Canada had banned all commercial whaling in 1972, except by aboriginal peoples. Beginning in 1975, aerial surveys began to provide much new information on seasonal distribution, movements, and population estimates of belugas and narwhals. One such survey by J.G. Smith and M.O. Hammill provided population data on belugas and narwhals occupying the High Arctic during the summer months. This study, based upon reconnaissance surveys since 1974, and on two strip-transect surveys of Prince Inlet, Barrow Strait, and Lancaster Sound, carried out in 1981, was entitled *Distribution and Abundance of Belugas,* Delphinapterus leucas, *and Narwhals,* Monodon monoceros, *in the Canadian High Arctic* (*Canadian Journal of Fisheries and Aquatic Sciences*, v. 42, 1985. pp. 676–84). It was estimated from these surveys that a total of 6,300–18,600 belugas and approximately 13,200–18,000 narwhals summer in Lancaster Sound and adjoining waters. Additional aerial surveys of narwhals only were also conducted in August, 1981, and July of 1982, 1983, and 1984, and in March, 1983 in the Repulse Bay area between Roes Welcome Sound and Lyon Inlet, north of Southampton Island. This survey was reported by P.R. Richard in the paper entitled *Abundance and Distribution of Narwhals (*Monodon monocerus*) in Northern Hudson Bay (Canada)* (*Canadian Journal of Fisheries and Aquatic Sciences*, v. 48, 1991. pp. 276–83). About 1,300 narwhals were estimated to occupy waters between northern Roes Welcome Sound and Lyon Inlet, but they were concentrated in three areas – the northern Roes Welcome Sound, southern Frozen Strait, and Lyon Inlet.

One other major aerial survey in the 1980s, of the beluga whale only, was undertaken in the coastal waters of northern Quebec – namely James Bay, most of the eastern Hudson Bay, offshore to the Belcher Islands, and the whole of Ungava Bay. This had been an area of high commercial harvests during the period 1854 to 1868, where the original population had been estimated at 6,000–7,000 whales. Nearshore surveys in the late 1970s indicated the populations were very low, and aerial coastal reconnaissance surveys and systematic strip-transect surveys in 1985 confirmed these earlier indications.

In Ungava Bay, the number of belugas was so low that no population estimates could be derived. In James Bay, 1,213 whales were estimated, and in eastern Hudson Bay, the aerial survey estimated 968 whales. Because belugas continued to be hunted in Ungava Bay, and eastern Hudson Bay, by aboriginal peoples, the authors of the survey, T.G. Smith and M.O. Hammill, recommended a conservation management strategy. Their paper was entitled *Population Estimates of White Whale,* Delphinapterus leucas, *in James Bay, Eastern Hudson Bay, and Ungava Bay* (*Canadian Journal of Fisheries and Aquatic Sciences*, v. 43, 1986. pp. 1982–87).

In the 1990s, surveys of beluga whales only have been carried out in the Mackenzie River estuary in the Arctic and in the St Lawrence River. In July, 1992, a systematic survey of the southeast Beaufort Sea, Mackenzie estuary, and west Amundsen Gulf was carried out using three Twin Otter aircraft. L.A. Harwood and S. Innes, in their paper entitled *Distribution and Abundance of Beluga Whales in the Mackenzie Estuary, Southwest Beaufort Sea and West Amundsen Gulf during Late July 1992* (*Canadian Journal of Fisheries and Aquatic Sciences*, v. 53, 1996. pp. 2262–73), said that beluga whales were found throughout the estuary and offshore, with the highest density in Kugmallit Bay of the Mackenzie estuary. After adjustments to the primary counts for missed-at-surface and about-to-surface beluga, the stock size of visible beluga whales was estimated at 19,629.

In August, 1995, a survey flight to estimate the St Lawrence population of belugas, using two aircraft each equipped with a metric mapping camera, was undertaken by flying a systematic grid of transects, aligned from northwest to southeast, that stretched from the Ile Aux Coudres to Ile du Bic. M.C. Kingsley, in the report entitled *Population Index Estimate for the Belugas of the St. Lawrence in 1995* (Ottawa: Department of Fisheries and Oceans, 1996. 38 p. Canadian Technical Report of Fisheries and Aquatic Sciences No. 2117), said the final total estimated population index was 705.2, which was 34% higher than a survey undertaken in 1992 using the same methods. This survey was part of a series of nine aerial surveys of the region going back to 1973. The same author re-examined all the data and reported the results in an article entitled *Population Index Estimates for the St. Lawrence Belugas, 1973–1995* (*Marine Mammal Science*, v. 14, 1998. pp. 508–30). Here, Kingsley said that 'factors were developed to correct for differences in survey coverage and in treatment of data and for the effect of hunting on population trajectory. Revised assessments of the population trajectory and of the present smoothed index value changed little as a result.' The most recent of these surveys was undertaken in 1997, and reported, again, by M.C. Kingsley in the technical report entitled *Population Indices and Estimates for the Belugas of the St. Lawrence Estuary* (Ottawa: Department of Fisheries and

Oceans, 1999. 6 p. Canadian Technical Report of Fisheries and Aquatic Sciences No. 2266). Here, the author said, 'from linear smoothing of the most recent five surveys, the current estimate of the standard index was 700.'

F Environmental Pollution Studies

Environmental pollution studies began in the early 1970s and have ranged geographically from the Gulf of St Lawrence to the Canadian Arctic. These studies range from organochlorines in marine mammals, the potential effect of oil pipeline construction on streams and fisheries and of oil spills on the Atlantic coast, to the effect of pulp and paper emissions in waters of the Great Lakes. Unique in these pollution studies has been the biological pollution of the Great Lakes by the sea lamprey, and attempts to control its production in the lakes.

Organochlorines, which include PCBs (polychlorinated biphenyls), and pesticides such as DDT residues, have been documented in Arctic marine mammals such as seals, whales, polar bears, walruses, narwhals, and Arctic char fish. R.F. Addison and S.R. Kerr studied blubber samples from 18 harp seals taken near the Saguenay River in the Gulf of St Lawrence for PCB and DDT residues, as well as dieldrin, in January, 1971. In their paper, entitled *Variations of Organochlorine Residue Levels with Age in Gulf of St. Lawrence Harp Seals (*Pagophilus groenlandicus*)* (*Journal of the Fisheries Research Board of Canada*, v. 30, 1973. pp. 595–600), the authors noted that PCB and DDT residues increased with age of the seal, but dieldrin concentrations didn't correlate with age. The next study, by R.F. Addison and P.F. Brodie was undertaken in the Mackenzie Delta with beluga whales and reported in their paper entitled *Occurrence of DDT Residues in Beluga Whales (*Delphinapterus leucas*) from the Mackenzie Delta, N.W.T.* (*Journal of the Fisheries Research Board of Canada*, v. 30, 1973. pp. 1733–36). In this study, the authors said 'there is little information about the distribution of organochlorine pesticide residues in Canadian Arctic biota,' and the beluga whales in the Mackenzie Delta were of special interest since their migratory movements were believed to be limited to the Beaufort and Chukchi Seas. Hence, residue concentrations observed might reflect the level of pesticide contamination of the marine environment in this area. Samples of blubber, muscle, and liver tissue were obtained from 14 whales in July, 1972, and they contained DDT pesticide residues. Because of the migratory routes of the whales, the authors concluded that the residues were accumulated from feeding sources such as shrimp, worms, squid, and small fishes in the area, and suggested that the routes of entry of the pesticides into these food sources could have been from DDT used for spraying of biting flies around armed forces bases in that area in the 1950s and 1960s. Another route of residue transport into the Canadian Arctic was suggested by G.W. Bowes and C.J. Jonkel in their paper entitled *Presence and Distribution of Polychlorinated Biphenyls (PCB) in Arctic and Subarctic Marine Food Chains* (*Journal of the Fisheries Research Board of Canada*, v. 32, 1975. pp. 2111–23). The authors examined tissues of polar bears, ringed and bearded seals, and Arctic char, collected from 1968 to 1972, from sampling sites that extended west to east from Inuvik, NWT, to Baffin Island, and north to south, from Ellesmere Island to North Twin Island in James Bay. The pattern of residues of industrial chemicals found suggested further study was needed for a better understanding of the routes of pollutant transport into these high latitudes. In another study by R.F. Addison and T.G. Smith the correlation of residue levels in seals by age and sex were studied. In their paper entitled *Organochlorine Residue Levels in Arctic Ringed Seals: Variation with Age and Sex* (*Oikos*, v. 25, 1974. pp. 335–37), the authors analysed blubber samples for pesticide residues from 13 female seals and compared the results with their previous study of the correlation of residue quantity and age in St Lawrence seals. Here, the authors said that 'in the females, the clear relationship between residue concentration and age found in the males was absent. We believe that this is the first explicit demonstration that the rates, or mechanisms, of residue accumulation in marine mammals may differ between the sexes.' The reason for this differentiation was suggested by R.F. Addison and P.F. Brodie in their paper entitled *Organochlorine Residues in Maternal Blubber, Milk and Pup Blubber from Grey Seals (*Halichoerus grypus*) from Sable Island, Nova Scotia* (*Journal of the Fisheries Research Board of Canada*, v. 34, 1977. pp. 937–41). After measuring DDT group insecticides, and PCB residues, in samples of maternal blubber, milk, and pup blubber of 11 pairs of mother and pup grey seals, in 1974 and 1976, the authors noted that 60% of the DDT concentrations and 30% of the PCB concentrations of residues were found in milk lipid. Concentrations in pup blubber were approximately the same as in the milk lipid. The authors estimated that female grey seals lost 30% of their DDT residues, and 15% of their PCB residues, through lactation, and this balanced the estimated annual intake of residues from food. This observation explained why female seals showed no increase in organochlorine residues with age (in contrast to males). These authors did further work on residue transfers, in 1984 and 1985, including not only measurements in blubber and milk but also in blood serum. In their paper, entitled *Transfer of Organochlorine Residues from Blubber through the Circulatory System to Milk in the Lactating Grey Seal (*Helichoerus gryphus*)* (*Canadian Journal of Fisheries and Aquatic Sciences*, v. 44, 1987. pp. 782–86), the authors found that concentrations of DDT and PCB residues in 1985 blubber samples from females and pups were slightly higher than in the 1984 samples for unknown reasons. However, an examination of the residues based on lipid weight indicated that the differences were due to a 'non-selective barrier that existed to quantitative transfers of residues from blubber lipid to circulatory lipid,' particularly for PCB residues as compared with DDT residues.

R.F. Addison, alone, continued research on organochlorines in marine mammals. In the paper entitled *Organochlorines and Marine Mammals Reproduction* (*Canadian Journal of Fisheries and Aquatic Sciences*, v. 46, 1989. pp. 360–68), this author examined earlier data concerned with reproductive failures in California sea lions, Bothnian Bay ringed seals, and Dutch Waddensee harbour seals, where reproductive failures had been attributed to organochloride pollution. The author questioned the results, because '(a) crucial biological information in the light of which OC residue data can be interpreted is often missing, (b) the nature of the supposed reproduction effect is variable, and (c) the OC residues to which these effects are attributed vary qualitatively and quantitatively. Correlations between observed residue concentrations and apparent reproductive changes do not provide any firm evidence for a cause-effect relationship.' The author concluded by saying that 'although there appears to be no firm correlation evidence linking residue concentrations in marine mammals with reproductive effects, there is enough concern about possible links to justify more experimental studies, with the aim of identifying the possible biochemical mechanisms involved.'

Although research described above documented the presence of organochlorine contaminants in the Arctic marine food chains for seals, beluga whales, polar bears, and Arctic fish, there was still insufficient or no documentation overall of organochlorines for cetaceans, such as the beluga whales, narwhals, and walruses. The tissue of beluga whales taken by Inuit hunters at four Arctic locations (Jones Sound, Cumberland Sound, Beaufort Sea, and Hudson Bay), and the St. Lawrence River, between 1983 and 1987, was frozen and later analysed, and the results reported in the paper by D.C. Muir and C.A. Ford entitled *Organochlorine Contaminants in Belugas,* Delphinapterus leucas *from Canadian Waters.* In: *Advances in Research on the Beluga Whale.* Delphinapterus leucas. Edited by T.G. Smith and D.J. St Aubin (Ottawa: Department of Fisheries and Oceans, 1990. pp. 165–90. Canadian Bulletin of Fisheries and Aquatic Sciences No. 224). The highest concentrations of organochlorines (PCB and DDT groups) were found in animals from the St Lawrence estuary, which indicated that these animals were clearly exposed to higher levels of pollutants than the Arctic animals. Among the Arctic animals, there were few major differences in organochlorine concentrations. This lack of geographical variation was consistent with the diffuse atmospheric source, and with the wide geographic range of some stocks such as the Hudson Bay animals. Finally, as noted in previous research, there was a higher level of PCBs and organochlorine pesticides in males than in females, which was attributed to residues losses through lactation. Further work on organochlorines in beluga whales of the St Lawrence estuary was undertaken by D.C. Muir in the early 1990s, and reported in two papers entitled *Persistent Organochlorines in Beluga Whales (*Delphinapterus leucas*) from the St. Lawrence River Estuary – I: Concentrations and Patterns of Specific PCBs, Chlorinated Pesticides and Polychlorinated Dibenzo-p-Dioxins and Dibenzofurans* (*Environmental Pollution*, v. 93, 1996. pp. 219–34); and *Persistent Organochlorines in Beluga Whales (*Delphinapterus leucas*) from the St. Lawrence River Estuary – II. Temporal Trends, 1982–1994* (*Environmental Pollution*, v. 93, 1996. pp. 235–45). Blubber samples taken from 16 beluga adults in 1993 and 1994 were analysed for persistent organochlorines as well as chlorinated dibenzo-dioxins/furans and approximated residue levels found in Arctic animals. The author suggested that the levels of organochlorine residues in the St Lawrence whales were due to high levels of organochlorines in the food chain of the whales. However, reviews of previous documentation of residues in whale blubber from 1982 to 1985 and from 1986 to 1990 indicated that concentrations and relative proportions of major individual organochlorine residues were within the same range as these earlier samples.

D.C. Muir, with C.A. Ford, also undertook studies of organochlorine contaminants in narwhal, based upon blubber and liver samples collected during 1982 and 1983 from Pond Inlet on northern Baffin Island. The narwhal habitat is Hudson Bay, Baffin Bay, Lancaster Sound, and adjacent waters in the eastern Canadian Arctic. Until then, there had been no documentation on levels of organochlorine contaminants in narwhal in their Canadian habitat. In their paper entitled *Organochlorine Contaminants in Narwhal (*Monodon monoceros*) from the Canadian Arctic* (*Environmental Pollution*, v. 75, 1992. pp. 307–10), the authors said that polychlorinated camphenes (PCCs) were the predominant organochlorines in narwhal blubber, followed by PCB residues. This pattern suggested that narwhal are exposed to proportionately more volatile compounds and may have less capacity to metabolize some of them. This pollutant residue pattern also indicated the importance of long-range atmospheric-oceanic transport as a pathway for introducing more volatile organochlorine compounds into the Arctic marine food chain.

Finally, D.C. Muir and C.A. Ford examined organochlorines in walruses in the paper entitled *Can Seal Eating Explain Elevated Levels of PCBs and Organochlorine Pesticides in Walrus Blubber from Eastern Hudson Bay (Canada)?* (*Environmental Pollution*, v. 90, 1995. pp. 335–48). The authors said that although this species was important in the traditional hunt of the Inuit of the eastern Canadian Arctic, little was known about levels of persistent organochlorines in walruses in Canadian waters. Samples of walrus blubber were collected in 1989 from two communities on the east coast of Hudson Bay, as well as from Foxe Basin and Loksland (East Baffin Island). A total of 53 samples were taken, and those from the east coast of Hudson Bay showed elevated levels of PCBs, DDT-related compounds, and toxophene. This may have been due to the fact that these animals have a predominantly seal diet that may be the vector for higher contaminant levels.

Since these individual studies, D.C. Muir and B. Braune *et al.* have published a review of Arctic marine contamination entitled *Spatial and Temporal Trends and Effects of Contaminants in the Canadian Arctic Marine Ecosystem: A Review* (*Science of the Total Environment*, v. 230, 1999. pp. 83–144). Here, the authors said that 'many of the temporal trend data for contaminants in Canadian Arctic biota are confounded by changes in analytical methodology' but that 'MFO enzyme data for polar bear, beluga and seals suggest that even in Arctic animals [ecosystem contaminants] may not be without biological effects.'

Unlike marine mammal-related pollution studies, other environmental pollution studies in the Arctic, which were triggered by proposed hydrocarbon production and transportation in the Beaufort Sea, Mackenzie Delta, and Mackenzie Valley regions, encompassed the strands of political decision-making on northern development by successive federal governments. It all began with the Diefenbaker government's 'vision of the north,' and ended with the Trudeau government's National Energy Program in the late 1970s.

During the 1960s, there was a laissez-faire approach to northern policy and development, but during the 1970s, there was a shift to environmental concerns and an awareness of aboriginal concerns for self-government. These factors saw the initiation of environmental impact studies by the federal Cabinet for the northern regions in 1977. These environmental studies involved the Mackenzie Environmental Monitoring Project, the Beaufort Sea Environmental Assessment Panel, the Norman Wells Oilfield Development, and the abandoned Arctic Islands Pipeline Project. All involved the federal departments of Indian and Northern Affairs, Environment, Fisheries and Oceans, as well as the Department of Renewable Resources of the Northwest Territories, and the Yukon Territorial government. Potential problems that concerned fisheries researchers, besides the effects on Arctic marine mammals, were: (1) that improved access to fishing areas and increased fishing would decrease fish populations and change their distribution, especially Arctic char and lake trout, (2) the clearing of corridors by seismic lines or pipeline right-of-ways could affect the number and distribution of fish due to sediment increases in rivers and streams and the changes in the location, size, and shape of river channels, and (3) waste discharge and accidental oil or chemical spills would lead to decreased acceptability of fish as a food resource.

Fisheries research related to these problems was done through the regional Freshwater Institute of the then-Fisheries Research Board of Canada located in Winnipeg and was written up in a number of technical reports. For example, G.J. Brunskill and S.E. Campbell presented one entitled *The Chemistry, Mineralogy and Rates of Transport of Sediments in the Mackenzie and Porcupine River Water-*

sheds, N.W.T. and Yukon 1971–1973 (Winnipeg: Freshwater Institute, Fisheries Research Board of Canada, 1975. 69 p. Fisheries Research Board of Canada Technical Report No. 546). D.M. Rosenberg and N.B. Snow wrote a report entitled *Ecological Studies of Aquatic Organisms in the Mackenzie and Porcupine River Drainages in Relation to Sedimentation* (Winnipeg: Freshwater Institute, Fisheries Research Board of Canada, 1975. 86 p. Fisheries Research Board of Canada Technical Report No. 547). And finally, P.J. Campbell and G.W. Dentry wrote a report entitled *General Physical and Chemical Data for Water and Sediment of the Mackenzie and Porcupine Watersheds and Rates of Transport of Dissolved and Suspended Elements at Selected Stations in the Mackenzie and Porcupine Watersheds, 1971–1974* (Winnipeg: Freshwater Institute, Fisheries Research Board of Canada, 1975. 2 Parts. 396 p. Fisheries Research Board of Canada Technical Report No. 556).

More immediate than the potential impact of hydrocarbon production and transportation in the Arctic has been the fate and effects of petroleum hydrocarbons from oil spills. The environmental damage from an oil spill is immediate and devastating and presents a long-term problem of research interest concerning environmental recovery. Studies have shown that environmental recovery following a spill can take several years, but neither the pattern of recovery nor the process of recovery is understood. In order to obtain a greater understanding, J.H. Vandermeullen, of the Bedford Institute of Oceanography, in Halifax, chaired a symposium in 1978 at which 160 scientists from nine countries discussed all known spill studies. From the discussions, he hoped that the group would develop some common principles of environmental impact and recovery. Canadian research used at the symposium was based upon the 1970 tanker *Arrow*'s oil spill in Chedabucto Bay, and included papers by J.H. Vandermeullen and D.C. Gordon, Jr entitled *Re-entry of Five-Year-Old Stranded Bunker C Fuel Oil from a Low Energy Beach into the Water, Sediments, and Biota of Chedabucto Bay, Nova Scotia* (*Journal of the Fisheries Research Board of Canada*, v. 33, 1976. pp. 2002–10); by J.E. Stewart and L.J. Marks, entitled *Distribution and Abundance of Hydrocarbon-Utilizing Bacteria in Sediments of Chedabucto Bay, Nova Scotia in 1976* (*Journal of the Fisheries Research Board of Canada*, v. 35, 1978. pp. 581–84); by another by E.S. Gilfillan and J.H. Vandermeullen, entitled *Alterations in Growth and Physiology and Soft-Shell Clams* Mya arenaria *Chronically Oiled with Bunker C from Chedabucto Bay, Nova Scotia, 1970–76* (*Journal of the Fisheries Research Board of Canada*, v. 35, 1978. pp. 630–36). The conference concluded that oiled marine environments eventually return to a stable state in from five to 15 years because nature has a battery of defence mechanisms, including self-cleaning by wave action and beach erosion, biological conversion, and microbial degradation. However, little remains known of the process of contamination and the manner in which foreign compounds interfere with the normal functioning of living systems. Yet, the final question remained, whether or not the idea of long-term recovery was acceptable.

Other important pollution studies carried out by fisheries scientists in Canada have been the effect of pulp and paper mills on the aquatic environment.

The effect of pulp and paper discharges was first studied in the Upper Great Lakes system, at the Red Rock Mill, in Nipigon Bay, on the north shore of Lake Superior. Research was carried out at the request of the International Joint Commission, began in 1974, and focused upon effluent dispersion and plume processes as well as ecosystem responses to those processes. Pulping involves the use of sodium hydroxide and sodium sulphate, which dissolve lignin (a complex polymer that provides rigidity to plant cells) from the wood so that cellulose fibres can be pulped. Residual lignin is removed by chlorination, which also increases paper brightness (at this mill site 80% of wood was pulped by this process, and the remaining 20% for newsprint was pulped by physical groundwood pulping with friction against stones). The chemical by-products of these processes (along with carbon dioxide and water) are chlorinated and unchlorinated aromatic compounds, resin acids, fatty acids, and terpenes, all of which are toxic to marine plant and animal life. Although the bay once supported a wide variety of fish species, including lake trout, walleye, white fish, and lake herring, by the mid-1970s, these stocks were rare, or in severe decline.

The result of the Nipigon Bay research appeared in 1977. Plume studies, which gave an understanding of the effluent flows and areas affected, were done by C. Minns and entitled *Analysis of a Pulp and Paper Mill Effluent Plume* (*Journal of the Fisheries Research Board of Canada*, v. 34, 1977. pp. 776–87); and by M.E. Fox and entitled *Persistence of Dissolved Organic Compounds in Kraft Pulp and Paper Mill Effluent Plumes* (*Journal of the Fisheries Research Board of Canada*, v. 34, 1977. pp. 798–804). Aquatic ecosystem disruption, which was ascertained to be localized, with bacteria being most affected and fish the least sensitive, was described by D.A. Rokosh in the paper entitled *Extent of Effluent Influence on Lake Water Determined by Bacterial Population Distributions* (*Journal of the Fisheries Research Board of Canada*, v. 34, 1977. pp. 844–49); and by L. Kaiser in the paper entitled *Organic Contaminant Residues in Fishes from Nipigon Bay, Lake Superior* (*Journal of the Fisheries Research Board of Canada*, v. 34, 1977. pp. 850–55).

Finally, a unique form of aquatic pollution studies in the Great Lakes Basin has involved biological pollution by the sea lamprey, which has greatly affected fish populations in the Great Lakes. The sea lamprey (*Patromyzon marinas*) entered the upper three Great Lakes in the late 1930s, and began making sharp inroads into commercial and sports fish stocks by the mid-1940s in lakes Huron and Michigan, and in the mid-1950s in Lake Superior.

The first Canadian studies began in 1946, in streams that fed into the North Channel of Lake Huron, an area considered severely affected by lamprey predation on fish stocks. In 1964, the Great Lakes Fishery Commission approved the installation of several electrical barriers on Canadian tributaries to the North Channel. Some of these ran continuously from 1967 to 1975. There was an indication that sea lamprey spawning peaked there in 1968, but the inventory of sea lamprey spawning streams feeding into Lake Michigan indicated that every major watershed was infested with spawning runs. In Lake Superior, on the American side of the lake, the US Fish and Wildlife Service surveys indicated 87 infested streams. The Fisheries Research Board surveys of the Ontario side also indicated that 87 streams contained sea lampreys. In the lower Great Lakes, Lake Erie, and Lake St Clair have been excluded from control because of the low lamprey populations and lack of predation in fish stocks. However, the systematic surveying of Lake Ontario began in 1963, and by 1978, sea lamprey had been found in 51 of 620 tributaries of Lake Ontario and the St Lawrence River that had been surveyed.

Although electrical barriers had been installed, by 1960, in 132 tributaries of the Great Lakes, effective control measures only began to occur after 1958, when a selective toxicant, the lamprecide TFM, was used to destroy larval lamprey in streams. From 1958 to 1978, 1,223 treatments of 334 tributaries of the upper three lakes were completed. And by 1978, 105 lamprecide treatments had been carried out on 46 tributaries to Lake Ontario.

Evidence of effectiveness of the control program was obvious through the reduced sea lamprey spawning runs, as measured by the number of adults taken at electrical barriers. As well, there were significant decreases in the incidence of sea lamprey wounds on lake trout. Finally, there were excellent responses of major fish stocks to sea lamprey control. All three of the upper lakes increased the number of lake trout, coho salmon, and chinook salmon available to the sport fishery, and in some areas for the commercial fishery. But, because of historic overfishing, researchers have found it difficult to clearly assess the true effectives of lamprey predation. In 1988, R.A. Bergstedt and C.P. Schneider, in their paper entitled *Assessment of Sea Lamprey (*Petromyzon marinas*) Predation by Recovery of Dead Lake Trout (*Salvelinus mamaycush*) from Lake Ontario, 1982–1985* (*Canadian Journal of Fisheries and Aquatic Sciences*, v. 45, 1988. pp. 1406–10), estimated that half the trout deaths, after restocking, have been caused by lamprey predation. However, the control programs and surveys have continued in the upper and lower Great Lakes, and by 1985, lamprey populations were estimated to have sunk below 1962 levels.

The sea lamprey research and control program has been well documented by fisheries biologists, such as B.R. Smith and J.J. Tibbles, in their research paper entitled *Sea Lamprey (*Petromyzon marinas*) in Lakes Huron, Michigan and Superior: History of Invasion and Control, 1936–78* (*Canadian Journal of Fisheries and Aquatic Sciences*, v. 37, 1980. pp. 1780–1801; by W.A. Pearce and R.A. Braem, in their paper entitled *Sea Lamprey (*Petromyzon marinas*) in the Lower Great Lakes* (*Canadian Journal of Fisheries and Aquatic Sciences*, v. 37, 1980. pp. 1811–26); and, by P.J. Manion and L.H. Hanson, in their paper entitled *Spawning Behaviour and Fecundity of Lampreys from the Upper Three Great Lakes* (*Canadian Journal of Fisheries and Aquatic Sciences*, v. 37, 1980. pp. 1635–40).

G Fishing Gear for the Effective Harvesting of the Resource

The commercial fishing on Canada's seas and inland waters is conducted by thousands of fishermen operating a variety of fishing craft according to their needs, and using different types of gear or devices for locating and harvesting fish. Another aspect of fisheries scientists work is the examination of different types of gear in the Canadian environment, and through demonstration, the encouragement of fishermen to shift to gear more appropriate to particular types of fish harvesting. This applies also to the development and use of locating devices on board ships. When studying fishing gear, a considerable amount of information must be gathered from harvest surveys using different types of gear. These surveys usually appear as technical reports in the literature. Two major areas of fishing-gear studies – midwater trawling and pair seining – occurred in the 1970s and 1980s in Canadian waters.

Midwater trawls are used mostly for catching herring, mackerel, red-fish, pollack, capelin, and shrimp. The trawls are cone-shaped nets constructed of webbing, and are attractive because they use fewer weights and can be adjusted for towing at various depths. This adjustment is made by increasing the vessel's speed, or by increasing the length of the cable between the vessel and the net. Conservation regulations limit the size of the mesh used for individual species. It was in the early 1970s that Canadian fisheries scientists encouraged a shift to midwater trawling, rather than after-trawling (this trawl uses a similar type of net but heavily weighed for trawling the ocean bottom for groundfish), because of the former's ability to function at various levels for successful catches. Midwater trawling first took hold on the Atlantic coast and was also successfully introduced to the Pacific coast.

Two examples of Atlantic coast midwater trawling surveys were by L.S. Parsons and D.G. Parsons, entitled *Exploratory Surveys for Redfish off Eastern Newfoundland-Labrador, Cape Farewell, June–August, 1973* (St John's: Department of the Environment, Fisheries and Marine Service, 1979. 73 p. Technical Report No. 433); and by A. Tobey, entitled *Shrimp Midwater Trawl Development, 1976, Gulf of St. Lawrence* (Halifax: Department of Fisheries and the Environment, Fisheries and Marine Service, 1977. 98 p. Technical Report No. 701). An example of a later west-coast survey by L.W. Barner and F.H. Taylor was entitled *Midwater and Bottom Trawl Tours and Catches Made on M.V. Blue Waters BW 78–3, October 3–23, 1978 in Queen Charlotte Sound* (Nanaimo, BC: Department of Fisheries and Oceans, Pacific Biological Station, 1979. 73 p. Data Report No. 138).

Pair seining can be carried out using two methods, one called Danish or Scottish seining, and the other Canadian pair seining. Seining methods are used to catch species of groundfish such as flounder and cod. Both methods use similar nets and a series of ropes spread out in a pear-shaped form along the ocean floor. The action of the ropes stirs up a mud cloud and herds the fish into the path of the net. Danish or Scottish seining uses one vessel, whereas Canadian pair seining utilizes two vessels to coordinate the sweep of the ropes and haul the trawl net. The attractiveness of this type of seining is the fact that less towing power is required and it can be used by smaller inshore vessels. Pair seining surveys began to appear in the mid-1970s. One survey by C.A. Bishop was entitled *Some Results of a Spanish Pair Trawling Experiment in ICNAF Sub-Area 3 with Observations on Pair Trawling of the Main Spanish Fleet* (St John's: Department of Fisheries and the Environment, Fisheries and Marine Service, 1976. 16 p. Technical Report No. 596).

Another form of seining used is purse seining. It features a wall of encircling webbing rather than towing, or trawling, for its catching power. The purse sein has floats on the top and weights on the bottom, to keep it vertical in the water, and wire ropes that pass through rings on the bottom of the net, which enables the net to be drawn together to entrap the fish. Accordingly, this is most effective when used to capture schools of fish near the ocean bottom. One example of purse sein studies, written by F.R. Scarsbrook and J. Mcdonald, was entitled *Purse Sein Catches of Sockeye salmon (*Onchorhynchus nerka*) and Other Species of Fish at Babine Lake, British Columbia, 1973* (Nanaimo, BC: Department of the Environment, Fisheries and Marine Service, 1975. 43 p. Technical Report No. 515).

The use of electronic acoustic devices to evaluate fish abundance or distribution, for catch or conservation assessments, began to be evaluated in the 1960s. Until then, traditional methods such as catch assessments, which required substantial effort, expenditure, and evaluation with historical data were used. In particular, the Digital Echo Counter and the Analog Echo Integrator were tested and compared. There were two major problems encountered with the testing. One was the need to obtain reliable counts of the number of fish per unit volume detected by the echo sounder. The other was to determine the effective volume of water sampled uniquely by each sonic pulse. The testing and its problems were described by R.G. Dowd and E. Balken, in their paper entitled *A Comparison between Two Sonic Measuring Systems for Demersal Fish* (*Journal of the Fisheries Research Board of Canada*, v. 27, 1970. pp. 737–42).

Further investigations studied the quantifying of return echo signals in order to estimate the number of fish within the echo. Echograms were compared with net trawl results so that an equation for the credible estimation of the fish stock could be developed. This work was carried out by E. Thorne and the findings reported in the paper entitled *Investigations into the Relation between Integrated Echo Voltage and Fish Density* (*Journal of the Fisheries Research Board of Canada*, v. 28, 1971. pp. 1269–73).

By the mid-1970s, digital echo counters were being applied to various fish environments to obtain rapid assessments of fish abundance and its relationship to distribution. For example, the system was applied to examining smaller-scale, near-shore pelagic (open sea) fish populations in lakes Ontario and Erie. Comparisons of acoustic estimated abundance showed the system was adequate for rapidly assessing abundance of dominant stocks. This work was reported by J.R.Kelso and E.E. Picket in their paper entitled *A Digital Echo-Counting System Used in Determining Abundance of Freshwater Pelagic Fish in Relation to Depth* (*Journal of the Fisheries Research Board of Canada*, v. 31, 1974. pp. 1101–4).

H Handling and Storage of Fish

The goal of Canada's fishing industry is constantly to improve fish quality, and lengthen shelf life of fish products for both the domestic and export market. This means that fishermen must use the best possible conditions for handling and storage, which have been studied and developed by fisheries researchers in Canada, or other countries, and introduced into Canadian fishing fleets, usually through financial assistance.

There were two major improvements in fish storage which came out of research at Fisheries Research Board Technical Stations. One improvement was the use of aluminium rather than wood for storage bins on vessels. It reduced both labour costs and bacterial contamination. This improvement was developed on the Atlantic coast and described in a Progress Report, written by W.A. MacCallum, entitled *Cost of Metal Unit Pen Fish Holds* (Halifax: Fisheries Research Board of Canada. Progress Reports of FRB Atlantic Coast Stations No. 61, 1955. pp. 13–21). This information was reported to fishermen in a circular by C.H. Castell entitled *A Note to the Fishing Skippers Regarding the Factors That Hasten Deterioration of Fish in Boats at Sea* (Halifax: Fisheries Research Board of Canada. Circular [New Series] of the Technical Station, Halifax, No. 1, 1953. 6 p.).

The other major advance in fish storage was the concept of holding fish in refrigerated sea water (RSW) at temperatures slightly below zero degrees Centigrade, or 31 degrees Fahrenheit. Because the actual freezing point of sea water is below the actual freezing temperature of fresh water, this method looked promising for the short-term storage of fish on vessels at sea. This method was first referred to by S.W. Roach and J.S. Harrison in their bulletin entitled *Storage and Transport of Fish in Refrigerated Sea Water* (Ottawa: Queen's Printer, 1961. 61 p. Fisheries Research Board of Canada Bulletin No. 126). Refinements of this technique, and its effectiveness as a storage medium for different types of fish, did receive considerable study over the next 20 years. One of the early refinements involving refrigerated sea water was the use of ultraviolet radiation (UVR) to control bacterial contamination. J.W. Boyd and B.A. Southcott, in their paper entitled *Ultraviolet Irradiation of Circulating Refrigerated Fish Storage Brines* (*Journal of the Fisheries Research Board of Canada*, v. 21, 1964. pp. 37–43), found that irradiation of refrigerated sea water was effective in controlling bacterial developments in the brine for storage of lingcod, and sockeye salmon. However, discolouration did develop in the irradiated brine on exposed surfaces of the stored fish. As well, intermittent or continuous irradiation of brines produced off odours in the water and the stored fish.

Another refinement in the use of refrigerated sea water for fish storage at sea was the further reduction of the water temperature to 25 to 29 degrees Fahrenheit (–4 to –1 degrees C) for Pacific (sockeye) salmon storage. The problem encountered with salmon storage was 'belly-burn,' a condition that occurs in uneviscerated fish that were feeding when caught. The technical term for 'belly-burn' is proteolysis, and it is caused by the diffusion of pigment from the kidney, liver, and spleen into flesh adjacent to the visceral cavity. This causes darkening of the flesh and bacterial spoilage. N. Tomlinson and S.E. Geiger, in their paper entitled *Partial Freezing as a Means of Preserving Pacific Salmon Intended for Canning* (*Journal of the Fisheries Research Board of Canada*, v. 22, 1965. pp. 955–68), found that partial freezing was effective in preventing proteolysis in canned fish. However, if the fish were initially in good condition, the quality of unfrozen or partially frozen fish was adequate for canned fish.

Another observation noted with canned fish that had been stored in refrigerated sea water, or partially frozen sea water, was that it had a lower free oil content. This was noticed by E. Bilenski and G. Lang and reported in their paper entitled *Evidence for the Binding of Oil in Salmon Canned after Storage in Ice or in a Partially Frozen Condition* (*Journal of the Fisheries Research Board of Canada*, v. 23, 1966. pp. 917–19). However, there was no corresponding decrease in the total lipid (fat) content of the canned product. Whether the binding of the oil occurred during storage, or during canning, or because of some other factor, was unknown. That the use of partially frozen sea water for storage led to a loss of free oil in canned fish was noted again by E. Bilinski and G.A. Gibbard, 15 years later, in their report entitled *Evaluation of the Quality of Sockeye Salmon,* Oncorhyncus nerka*, Stored in Ice or Partially Frozen in Salt Fortified Refrigerated Sea Water* (Vancouver: Department of Fisheries and Oceans, 1981. 19 p. Canadian Technical Report of Fisheries and Aquatic Sciences, No. 1055). By the 1980s, sockeye salmon were also being transported by air in ice, and the storage time was only one to two days before the canning process, compared with sea transport of up to two weeks in partially frozen refrigerated sea water. The authors wanted to evaluate the quality of the salmon transported by these two methods, before and after canning. Before canning, there were some differences such as some fading of skin and gills in the partially frozen fish, but 'belly-burn' was similar in both iced or partially frozen fish. However, the quality of the fish transported by air, or sea, was judged to be good after canning, with the flesh remaining firm, with a decrease of free oil in the cans after canning. The authors concluded that the sea transport of salmon in partially frozen sea water was still a good method for transport storage, provided that good chilling equipment and proper operating procedures were used.

Besides studying the effect of using refrigerated sea water for transport of salmon, other species of fish, such as ocean perch, silver hake, and capelin, have been studied for their ability to cope with transport by refrigerated sea water. A.A. Longard and L.W. Regrer, in their paper entitled *Color and Some Composition Changes in Ocean Perch (*Sebastes marinus*) Held in Refrigerated Sea Water with and without Carbon Dioxide* (*Journal of the Fisheries Research Board of Canada*, v. 31, 1974. pp. 456–60), observed that ocean perch stored up to 12 days in refrigerated sea water with carbon dioxide were kept in better condition (as measured by trimethylamine production in the fillet tissue) than in ice or refrigerated sea water alone. The skin colour of the perch, however, was equally good when held in either refrigerated sea water or ice. Silver hake also responded better when held in refrigerated sea water saturated with carbon dioxide than in RSW alone, as the onset of bacterial spoilage was retarded. This was noted by D.F. Hiltz and B.S. Lall in their paper entitled *Deteriorative Changes during Frozen Storage in Fillets and Minced Flesh of Silver Hake (*Merluccius bilinearis*) Processed from Round Fish Held in Ice and Refrigerated Sea Water* (*Journal of the Fisheries Research Board of Canada*, v. 33, 1976. pp. 2560–67). As well, holding round fish, such as hake, in refrigerated sea water for up to six days before processing extended the frozen storage life of fillets at –10 degrees Celsius by two to three weeks, and minced flesh by one week, compared to round fish held on ice. Capelin, on the other hand, responded better to storage on ice rather than in refrigerated sea water. D.H. Shaw and J.R. Botta, in their papers entitled *Preservation of Inshore Male Capelin (*Mallotus villosus*) Stored in Ice* (*Journal of the Fisheries Research Board of Canada*, v. 32, 1975. pp. 2039–46); *and Preservation of Inshore Male Capelin (*Mallotus villosus*) Stored in Refrigerated Sea Water* (*Journal of the Fisheries Research Board of Canada*,

v. 32, 1975. pp. 2047–53), noted that, contrary to results found with other small pelagic fish, such as herring, mackerel, and sardines, major taste and texture changes were obvious by the second day with capelin stored in refrigerated sea water, as compared with capelin stored on ice, where deterioration was not noticeable until the eighth day.

Apart from general problems with fish handling and storage, some individual species offer unique problems which can affect the internal or export market for that fish. For example, herring in Canada were traditionally processed for oil or animal feed only, but there was a shift in the fishing industry to use it as a human food source. As well, herring roe was marketable in Asia as a specialty product. However, herring presents a preservation problem during its storage period. The preservation of herring before freezing or processing as food is complicated, because of the rapid spoilage and physical fragility of the muscle tissue. Besides freezing, herring has been traditionally stored through icing, but there was a trend toward using refrigerated sea water for initial storage. Scientists knew that during frozen storage, the deterioration of the quality of the herring was primarily due to the development of oxidative rancidity in the muscle. However, there had been little research done on the effects of icing or refrigerated sea water on these oxidative changes. Researchers did know that rancidity was accelerated during frozen storage, if the initial chill storage had been with refrigerated sea water rather than ice, and if oxygen was present during frozen storage (rather than using a cryovac-vacuum packed freezing method). This aspect of the storage problem was described by E. Bilinski and R.E. Jones in their paper entitled *Chill Stowage and Development of Rancidity in Frozen Pacific Herring (*Clupea harengus pallase*)* (*Journal of the Fisheries Research Board of Canada*, v. 35, 1978. pp. 473–77).

This preliminary study of herring rancidity led to further work on the problem. Again, E. Bilinski and R.E. Jones discovered that a 1% treatment with sodium erythorbate (NaER) was effective in retarding the development of rancidity in frozen herring fillets, and that protection was best when the fillets were glazed, or frozen in ice blocks. However, treatment with NaER gave less protection than vacuum packaging. The paper describing this work was entitled *Control of Rancidity in Frozen Pacific Herring,* Clupea harengus pallase*: Use of Sodium Erythorbate* (*Journal of the Fisheries Research Board of Canada*, v. 36, 1979. pp. 219–22).

Research on the processing and preservation of roe herring appeared in the technical literature. One report by N. Tomlinson and G.A. Gibbard was entitled *Storage of Roe Herring at Sea* (Vancouver: Department of the Environment, Fisheries and Marine Service, 1975. 16 p. Technical Report No. 536). Another report by J.W. Boyd and J. Cheng was entitled *Roe Herring Processing: Preservation and Factors Affecting Firming of Roe* (Vancouver: Department of Fisheries and Oceans, 1980. 26 p. Technical Report No. 921).

I Popularization of Canada's Marine Resources

Besides the fisheries research, which is documented in the journal literature, technical reports, and books, the Department of Fisheries and Oceans has been publishing a series of popular but academic publications called Underwater World since the 1980s. To date, just over 60 have been published, and they describe specific fish and fish groups, shellfish, specific marine mammals, and marine mammal groups, and the harvesting of some marine groups. Those publications describing individual species provide a general taxonomic description, distribution and migration, life history, economics and exploitation, and research and management. The range of information covered can be seen in some of the following titles:

Smith, K.E. *The Atlantic Salmon* (Ottawa: Department of Fisheries and Oceans, 1983. 8 p. Underwater World);
Bailey, R. *The Atlantic Snow Crab* (Ottawa: Department of Fisheries and Oceans, 1981. 8 p. Underwater World);
Reeves, R. *Cetaceans of Canada* (Ottawa: Department of Fisheries and Oceans, 1987. 27 p. Underwater World);
Comeau, P.A. *Sealing – A Canadian Perspective* (Ottawa: Department of Fisheries and Oceans, 1989. 8 p. Underwater World);
Atlantic Fishing Methods (Ottawa: Department of Fisheries and Oceans, 1989. 12 p. Underwater World).

IV Departmental Sources of the Publications

The huge output of fisheries research in Canada has been thoroughly documented in printed indexes issued by the Fisheries Research Board or its successor federal departments over the years. There are two cumulated indexes that have been issued. The first, by Neil M. Carter, was entitled *Index and List of Titles, Fisheries Research Board and Associated Publications, 1900–1964* (Ottawa: Fisheries Research Board of Canada, 1968. 649 p. Fisheries Research Board of Canada Bulletin No. 164). The second, also by Neil M. Carter, was entitled *Index and List of Titles: Fisheries Research Board of Canada and Associated Publications, 1965–1972* (Ottawa: Fisheries Research Board of Canada, 1973. 588 p. Fisheries Research Board of Canada Miscellaneous Special Publication No. 18).

After 1973, there are annual indexes, which are located in the last issue of the *Canadian Journal of Fisheries and Aquatic Sciences*, with the format being the same as the cumulative indexes.

Chapter 6

Forestry Research

I Historical Background

Before the first European settlements, at the beginning of the 17th century, Canada's indigenous peoples made use of the Canadian forests for shelter, transportation, and food. But this pre-European situation, in which the forests provided sustenance, also left the forests undiminished. In 1608, Samuel de Champlain created the first European settlement in Canada with the creation of his One Hundred Associates.

Although the French reserved oak and white pine for naval use, only a modest trade in ship masts developed. This same modesty applied to the shipbuilding industry, and to sawmills for building materials. After 1763, however, with the advent of British rule, the forests started to be more actively exploited. This was aided partly by the American Revolution as England shifted its need for mast timbers to Canadian sources, and by the influx into New Brunswick, Nova Scotia, the Eastern Townships of Quebec, and Southern Ontario of loyalists who used the forests, not primarily for their wood, but for cleared and burnt land where agriculture could flourish. After the French Revolution, and Napoleon's enforcement of the Continental system, the British need for timbers from Canada increased, and it continued to do so during the 19th century. This need was both for timber masts and for the domestic requirements of square-cut timbers. It was in the middle of the 19th century that an active timber trade also developed with the United States, and by the beginning of the 20th century this trade was as important as the exports to Great Britain. After World War I, Canada's eastern timber industry declined as the easy access to marketable timber was exhausted. However, the British Columbia market continued to grow, and today remains the major source of Canadian building timbers.

As the Canadian timber trade continued to develop so did the growing demand for wood fibre for papermaking, as the demand for paper exhausted the supply of rags from which most paper was then made. The first wood pulp mills began in Quebec, in 1866, but it was not until the turn of the century that wood-pulping operations began in British Columbia, and they have continued to expand ever since. According to the most recent (1999/2000) annual publication entitled *The State of Canada's Forests* (Ottawa: Forestry Canada, 1991–), Canada's current forest industry in timber, wood products, and pulp employs over 352,000 people, and generates over $69 billion in economic activity, with just under two-thirds of that being in exports. This forest industry derives approximately 240.9 million cubic metres of industrial round wood from commercial forests out of a total of 418 million hectares of available forest land in Canada. Of this forest land, 71% is provincially owned, with each province having its own forest legislation, regulations, and standards, through which it confers harvesting rights, and assigns forest management responsibilities to the private sector. The other 23% is federally owned, most of it in two Arctic territories. Only 6% of Canada's forests is held by more than 425,000 private owners. Although forest management, excluding federal lands, is a provincial jurisdiction, the federal government is actively involved in its scientific research, trade, international affairs, and economic development.

Additional data about Canada's forests can be found at the National Forestry Database Program (NFDP) website <<http://www.nrcan.gc.ca/cfs/proj/iepb/nfdp>>. The website contains detailed information gathered between 1990 and 1997 by the program, as well as important historical information for data gathered previously by other surveys. The data contained is: (1) forest inventory, 1991, (2) allowable annual cut, 1970–99, (3) forest fires, 1970–99, (4) forest insects, 1975–99, (5) forest products, 1940–99, (6) silviculture, 1975–99, (7) forest management expenditures, 1977–99, (8) revenue and economic profit, 1970–, and (9) pest control product use, 1988–99. This data was previously a printed compilation entitled *Compendium of Canadian Forestry Statistics* (Ottawa: Canadian Council of Forest Ministers, 1992–97. 6 v.).

Despite the importance of Canada's forest resources, and the wealth those forests generate, its importance at the federal level has been varied. Most recently, in 1989, when Parliament passed the Department of Forestry Act, Forestry Canada (the Department of Forestry) again became an independent department. This was not the first time, however, that forestry had a departmental status. The first autonomous Department of Forestry (and Rural Development) operated from 1960 to 1966. Then in 1966, it was reorganized as the Forestry Branch of the Department of Forestry and Rural Development. In 1967, it became the Canadian Forestry Service, at first within the Department of Fisheries and Forestry, and later within the Department of the Environment. The federal government elevated the status of the Canadian Forestry Service (now the Canadian Forest Service) in 1984, moving it to the Department of Agriculture and establishing a cabinet post of Minister of State (Forestry). However, in 1994, with the reduction of Cabinet, the Canadian Forestry Service was again absorbed, this time into the newly created Ministry of Natural Resources (Natural Resources Canada).

Whether as a separate department, or as part of another federal department in the form of the Canadian Forest Service, forestry research at the federal level is highly decentralized. It has five regional research centres. They are the Pacific Forestry Centre, Victoria, British Columbia; the Northern Forestry Centre, Edmonton, Alberta; the Great Lakes Forestry Centre, Sault Ste Marie, Ontario; the Laurentian Forestry Centre, Sainte-Foy, Quebec; and the Atlantic Forestry Centre, Fredericton, New Brunswick. These five regional forestry centres work to solve forestry problems specific to their areas of operation.

Overall, research in all areas has focused upon forest insects and diseases (forest entomology and forest biology), forest fireresearch, harvesting (logging) research, forest regeneration or reforestation research, including site preparation, tree breeding and tree growth, pollution research, and overall forest management research. From time to time, however, there has been special initiatives research such as the ENFOR (Energy from the Forest) biomass program, which came out of the energy crisis of the 1970s. This program, which was established in 1978, sought to assess the quantity of forest wood available as an energy source. Included in the research program have been studies of fast-growing tree species that could be used as energy plantations, studies of harvesting methods and their effects on forest soil, and studies for improving on methods used for converting the forest biomass into liquid or gaseous fuels. The most recent overview of the ENFOR program was entitled *Strategic Plan for Bioenergy Research, 1998–2003* (Ottawa: Natural Resources Canada, Canadian Forest Service, 1998. 20 p.). This plan stated that the latest objectives of the ENFOR program were to: (1) generate sufficient knowledge and technology on forest biomass production to help increase the contribution of forest biomass energy to Canada's overall energy supply, and (2) improve the understanding of the role of biomass production in the global carbon cycle to help make a marked contribution to Canada's commitment to stabilize its greenhouse gas emissions.

As well, there are ongoing federal forestry responsibilities, which include the Forest Insect and Disease Survey (FIDS) under which the forestry centres carry out regional surveillance programs that detect, monitor, and report on the distribution and impact of forest insects and diseases. Also, since the late 1970s, with the Forest Inventory and Management Statistics Program, the former Petawawa National Forestry Institute (now part of the Great Lakes Forestry Centre) has compiled and maintained a computerized inventory of Canada's forest resources.

The coordination of forest inventories, which are a provincial responsibility, has been described by B.D. Haddon in the publication entitled *Canadian Forest Inventory Committee: The First Ten Years* (Ottawa: Canadian Forestry Service, 1987. 27 p.); and by M.D. Gillis and D.G. Leckie in the publication entitled *Forest Inventory Mapping Procedures across Canada* (Ottawa: Forestry Canada, 1993. 79 p. Petawawa National Forestry Institute, Information Report PI-X-114).

Forest inventories are just one example of the ongoing, and complex, federal-provincial arrangements undertaken for the management of Canada's forest resources. The Forestry Canada Annual Report, 1989–90, indicated that since 1982, the Department of Forestry, or the Canadian Forestry Service, has negotiated 17 forest resource development agreements and other development programs with the provinces that totalled $1.4 billion, with the federal share at these agreements having a value of $772 million, or 56% of the total. Most of this money has gone into boosting provincial silvicultural programs, which have also helped private landowners to improve the management of their lands. For example, since 1982, in Atlantic Canada, more than 370 million seedlings have been planted on 170,000 hectares of land; 112 hectares have been prepared for regeneration; and 287,000 hectares have undergone stand tending. In Quebec, since 1984, about 820 million seedlings have been planted on 366,000 hectares of prepared sites, and an additional 620,000 hectares have undergone fertilization. In Ontario, 160,000 hectares have been prepared for planting and natural regeneration; 224 million seedlings have been planted on 104,000 hectares; and an additional 98,000 hectares of forest stands were tended to increase their productivity. The prairie provinces have planted 66 million trees on 30,000 hectares of land, and an additional 23,000 hectares of stand tending have been completed. Finally, in British Columbia, these federal-provincial agreements have concentrated upon clearing the reforestation backlog. About 267 million seedlings have been planted on more than 142,000 hectares of land that did not properly regenerate naturally, or artificially, after harvesting. Also, an additional 176,000 hectares have been prepared for future forests.

This intensity of federal-provincial cooperation on the management of Canada's forests was preceded by efforts to develop a national forestry policy for Canada. These efforts have resulted in a series of policy initiatives, the first one appearing in the late 1970s. It was entitled *Forestry Imperatives for Canada: A Proposal for Forest Policy in Canada. Prepared for the Annual Meeting of the Canadian Council of Resource and Environment Ministers, Kelowna, British Columbia, June 6 and 7, 1979* (Kelowna, BC: Canadian Council of Resource and Environment Ministers, 1979. 29 p.). More recently, these policy initiatives have been issued by the Canadian Council of Forest Ministers. They are entitled *A National Forest Sector Strategy for Canada* (Ottawa: Canadian Council of Forest Ministers, 1987. 22 p.);

and *Sustainable Forests: A Canadian Commitment* (Ottawa: Canadian Council of Forest Ministers, 1992. 51 p.). This latest strategy emphasized an approach to forest management that included both 'timber and non-timber values (such as recreation, wilderness, biodiversity and wildlife) while protecting the integrity, health and diversity of our forest eco-systems.' As well, the Canadian Forest Service has issued its own strategic plans. The first one was entitled *Toward a New Era in Sustainable Forestry: The Canadian Forestry Service Strategic Plan for Science and Technology, 1995–2000* (Ottawa: Canadian Forest Service, 1994. 11 p.). The second was entitled *Canadian Forest Service Strategic Plan: Beyond the Millennium, 1998–2003* (Ottawa: Canadian Forestry Service, 1998. 18 p.). In this plan, the mission of the CFS was 'to promote the sustainable development of Canada's forests and the competitiveness of the Canadian forest sector for the well-being of present and future generations of Canadians.' The most recent strategic plan has been entitled *Setting Priorities, Delivering Results: Canadian Forest Service Science and Technology Business Plan 2000–2003* (Ottawa: Canadian Forest Service, 2000. 32 p.). Finally, the Forest Insect and Disease Survey has issued one strategic plan, written by J.P. Hall and entitled *Forest Insect and Disease Survey: Strategic Plan, 1993–1998* (Hull, Quebec: Canadian Forestry Service, 1994. 8 p.).

All the forestry initiatives and research by the federal government have generated varying types of publications over the years, ranging from annual reports, or research reports in series, to specific research reports, as well as research articles appearing in the open literature. However, the tracking of forestry information over time, other than in the open literature, is not straightforward because of the various locations of what became the Canadian Forest Service in one federal department or another, as well as the decentralization of the service into regions.

II Publications

The publications of the Canadian Forest Service, or its predecessors, are considerable in their range, and consist of annual reports, program reviews, research reports, miscellaneous series, regional information reports, and the serial literature.

A Annual Reports

The information found in annual federal reports can be traced as follows:

Annual Report of the Department of the Interior (Ottawa: Department of the Interior, 1873/74–1934/35. 62 v.). There was a Forestry Service within the department's Forestry Branch and the Director of Forestry issued *Reports of the Director of Forestry* (Ottawa: Department of the Interior, 1926/27–1935/36. 7 v.) for those years. In 1936, the then-named Dominion Forest Service became part of the Lands, Parks and Forestry Branch of the Department of Mines and Resources [*Annual Report*] (Ottawa: J.O. Patenaude, Printer to the King's Most Excellent Majesty, 1936/37–1948/49. 13 v.). From 1950 to 1960, the Forestry Branch was shifted to the briefly named Department of Resources and Development, and information can be found in the *Annual Report of the Department of Northern Affairs and National Resources, 1950–1960* (11 v.). From 1960 to 1966, a separate Department of Forestry was created and forestry information can be found in the *Annual Report, Department of Forestry* (Ottawa: Queen's Printer, 1960/61–1965/66. 6 v.). At this time, two of the departmental branches also published annual reports. They are: *Annual Report, Forest Entomology and Pathology Branch* (Ottawa: Department of Forestry, 1961/62–1964/65. 4 v.); and *Annual Report of the Forest Products Research Branch* (Ottawa: Department of Forestry, 1961/62–1964/65. 4 v.). From 1966 to 1968, the Department of Forestry and Rural Development was created and forestry information can be found in the *Annual Report, Department of Forestry and Rural Development* (Ottawa: Queen's Printer, 1966/67 and 1967/68. 2 v.). Another shift occurred from 1968 to 1970, with the creation of the Department of Fisheries and Forestry, with forestry information found in the *Annual Report, Department of Fisheries and Forestry* (Ottawa: Department of Fisheries and Forestry, 1968/69–1970/71. 4 v.). Next, the years 1971 to 1976 found the Canadian Forestry Service as part of Environment Canada and forestry information can be found in the *Annual Report, Environment Canada* (Ottawa: Environment Canada, 1971/72–1975/76. 5 v.). And, from 1976 to 1979, the service became part of *Fisheries and Environment Canada* (Ottawa: Fisheries and Environment Canada, 1976/77–1978/79. 3 v.). During most of the 1980s, the Canadian Forestry Service was part of a Environment Canada, again, with information found in the *Annual Report, Environment Canada* (Ottawa: Environment Canada, 1979/80–1983/84), and then as part of the Department of Agriculture, with information found in the *Annual Report, Agriculture Canada* (Ottawa: Agriculture Canada, 1984/85–1987/88. 4 v.). The Canadian Forest Service came into its own again in 1988, with the re-creation of the Department of Forestry, or Forestry Canada [*Annual Report*] (Ottawa: Forestry Canada, 1988/89–). Annual reports, however, were not released after 1990/91, and in 1994, the department became part of a new department called Natural Resources Canada. The source of overall forestry information is now found in their website at <<http://www.nrcan.gc.ca/cfs/scf/>>.

Other related annual reports are the *Annual Reports Forestry Research Advisory Council of Canada* (Ottawa: Canadian Forestry Service, 1987–). The latter was also issued, for one year, as the *Annual Report to the Associate Deputy Minister, Canadian Forestry Service* (Hull, Quebec: Research and Technical Service 1986. 1 v.). This advisory council was established by the Canadian Forestry Service in 1983 'to address forest research priorities and policies at a national level and to advise CFS.' It has thirteen members – five from industry, five from the provinces, and three from forestry schools. Finally, there are two other annual reviews of interest. One is the *ENFOR Review: Energy from the Forest* (Ottawa: ENFOR Secretariat, Forestry Service, v.1, 1980–). The other is the *Model Forest Program Year in Review* (Hull, Quebec: Model Forest Program, Natural Resources Canada, 1993/94–). The latter program focuses on sustained development in forestry management.

Besides the annual reports mentioned above, the Canadian Forest Service (under its earlier names) also issued progress, or research notes, from 1945 until 1984. The breakdown of these reports is as follows: Bi-monthly Progress Report (Ottawa: Department of Agriculture, Forestry Biology Division, v. 1, 1945–v. 16, 1960. 16 v.); *Bi-monthly Progress Report* (Ottawa: Department of Forestry, v. 17, 1961–v. 22, 1966. 5 v.); *Bi-monthly Research Notes* (Ottawa: Department of Forestry and Rural Development, v. 23, 1966-v.24, 1968. 2 v.); *Bi-monthly Research Notes – Forestry Service* (Ottawa: Canadian Forestry Service, v. 25, 1969–v. 36, 1980. 12 v.); and *Canadian Forestry Service Research Notes* (Ottawa: Canadian Forestry Service, v. 1, 1981–v. 4, 1984. 4 v.).

B Program Reviews

A more detailed source of forestry activities comes from the annual program reviews of the regional centres. These program reviews describe specific research projects, including those related to forest economics. The reviews that have been issued are:

Program Review, Newfoundland Forest Research Centre (St John's: The Centre, 1981–85. 4 v.); *Program Review, Newfoundland Forestry Centre* (St John's: The Centre, 1985/86 and 1987/88. 3 v.); and most recently, *Program Review, Forestry Canada, Newfoundland and Labrador Region* (St John's: The Region, 1989–); *Annual Review, Maritimes Forest Research Centre* (Fredericton: Canadian Forestry Service, 1981–82. 1 v.); *Maritimes Forest Research Centre Program Review* (Fredericton: Maritimes Forest Research Centre, 1982/83–1983/84. 2 v.); and *Program Review, Canadian Forestry Service – Maritimes* (Fredericton: CFS – Maritimes, 1984/85–1987/88. 4 v.). This review ceased with the 1987/88 issue; *Program, Laurentian Forest Research Centre* (Sainte-Foy, Quebec: Canadian Forestry Service, 1980/81. 1 v.); *Program Review, Laurentian Forest Research Centre* (Sainte-Foy, Quebec: Laurentian Forest Research Centre, 1981/83–1983/84. 2 v.); and *Program Review, Laurentian Forestry Centre* (Sainte-Foy, Quebec: The Centre, 1984/86–1986/87. 2 v.). This review ceased with the 1986/87 issue; *Program, Petawawa National Forestry Institute* (Chalk River, Ontario: Petawawa National Forestry Institute, 1979/81. 1 v.); *Program Review, Petawawa National*

Forestry Institute (Chalk River, Ontario: Petawawa National Forestry Institute, 1981/82–1988/89. 8 v.). This review ceased with the 1988/89 issue; *Program Review, Great Lakes Forest Research Centre* (Sault Ste Marie, Ontario: Canadian Forestry Service, 1981/82–1984/85. 4 v.); *Program Review, Great Lakes Forestry Centre* (Sault Ste Marie, Ontario: Great Lakes Forestry Centre, 1985/86–). This review may have ceased with the 1987/88 issue; *Program Review, Forest Pest Management Institute* (Sault Ste Marie, Ontario: Forest Pest Management Institute, 1981/82–1989/90. 8 v.); *Program Review, Northern Forest Research Centre* (Edmonton: The Centre, 1980/81–1984/85. 5 v.); *Program Review, Northern Forestry Centre* (Edmonton: The Centre, 1985/86–1987/88. 3 v.). This review may have ceased with the 1987/88 issue; *Program Review, Pacific Forest Research Centre* (Victoria: The Centre, 1981/82–1984/85. 4 v.); *Program Review, Pacific Forestry Centre* (Victoria: Canadian Forestry Service, 1985/86-1988/89. 4 v.); *Annual Program Review, Forestry Canada in the Pacific and Yukon Region* (Victoria: Forestry Canada, Pacific and Yukon Region, 1992/93–). This review may have ceased with the 1992/93 issue.

Since the mid-1990s with the development of the Internet and websites, information about the CFS research centres (Pacific Forestry Centre, Northern Forestry Centre, Great Lakes Forestry Centre, Laurentian Forestry Centre, and the Atlantic Forestry Centre) is available through their website address at <<http://www.nrcan.gc.ca/cfs/cfs-/index_e.html>>. The general heading Science Programs and Activities leads into information about each centre under broad research subject categories.

C Research Reports

Like the annual reports pertaining to the Canadian Forest Service, in whatever federal department the Service has resided over time, the research reports have also had a varied series history and format, with publications first issued by the central headquarters, and later by the regional centres also. Major report series have been:

Forestry Branch Bulletins, No. 1, 1928–No. 127, 1962. These bulletins were successively issued by the Forestry Branch of the Department of the Interior from 1928 to 1935, then from 1936 to 1949 by the Forestry Branch of the Department of Mines and Resources, and from 1950 to 1960 by the Forestry Branch of the Department of Northern Affairs and National Resources. After 1960, they were published by the Department of Forestry. An example from this series by I.C. Place was entitled *The Influence of Seed-Bed Conditions on the Regeneration of Spruce and Balsam Fir* (Ottawa: Department of Northern Affairs and National Resources, Forestry Research Division, 1955. 87 p.

Forestry Branch Bulletin No. 117). These bulletins were replaced in the early 1960s by the following publications:

Canadian Forestry Service Publications. These publications possibly retained the numbering of the previous bulletins while the Service was part of the Department of Forestry (1960 to 1966), and successively the Department of Forestry and Rural Development (1966 to 1968), the Department of Fisheries and Forestry (1968 to 1970), Environment Canada (1971 to 1976), Fisheries and Environment Canada (1977 to 1979), and Environment Canada again (1980 to 1984). This series disappeared in 1981 with the publication by W.M. Stiell and C.R. Stanton entitled *An Introduction to Christmas Tree Growing in Canada* (Ottawa: Canadian Forestry Service, 1981. 32 p. Canadian Forestry Service Publication No. 1330). Other major series have been:

Silviculture Leaflets, No. 3, 1941–No. 100, 1954. These documented findings, or the results, of individual research projects pertaining to silviculture management and research, forest inventory research, and forest fire protection research, that may have been incorporated into later Forestry Branch Bulletins. As example of a publication from this series by J.S. Rowe was entitled *Squirrel Damage to White Spruce* (Ottawa: Department of Northern Affairs and National Resources. Forestry Branch, 1952. 2 p. Silviculture Leaflet No. 61). This series was continued by:

Technical Notes, No. 1, 1955–No. 123, 1962. An example of a publication from this series by G.L. Baskerville was entitled *Response of Young Fir and Spruce to Release from Shrub Competition* (Ottawa: Department of Forestry, 1961. 14 p. Technical Note No. 93). This series appears to have been continued, or revived by:

Forestry Technical Reports, No. 1, 1974–. These reports have been issued by the headquarters of the Canadian Forest Service and, since 1981, by the regional centres and cover topics similar to the Notes. An example from this series by A.H. Rose and O.H. Lindquist was entitled *Insects of Eastern Larch, Cedar and Juniper*. Revised Edition (Ottawa: Forestry Canada, 1992. 100 p. Forestry Technical Report No. 28).

D Miscellaneous Series

Other minor series have included Fire Research Leaflets, such as one by D.G. Fraser, entitled *Polarizing Screens and Range-Finders as Aids in Forest Fire Detection* (Ottawa: Forest Research Division, 1952. 4 p. Fire Research Leaflet No. 7); Miscellaneous Publications, such as one by W.M. Stiell, entitled *Christmas Tree Growing in Canada* (Ottawa: Forest Research and Forestry Operations Division,

1957. 21 p. Miscellaneous Publication No. 7); Forest Air Survey Publications, such as one by H.E. Seely, entitled *The Forestry Tri-Camera Method for Air Photography* (Ottawa: Dominion Forest Service and Forest Research Division, 1948. 12 p. Forestry Air Survey Publication No. 3); Forest Fire Research Notes, such as one by R.W. Chorlton, entitled *Preparation of Visible Area Maps by Field Sketching* (Ottawa: Dominion Forest Service and Forest Research Division, 1951. 16 p. Forest Fire Research Note No. 16); Aerial Forest Survey Research Notes, such as one by H.E. Seely, entitled *Determination of Tree Heights from Shadows in Air Photographs* (Ottawa: Dominion Forest Service, 1942. 10 p. Aerial Forest Survey Research Note No. 1); and Tree Pamphlets, such as one entitled *Western Cedar* (Ottawa: Department of the Interior, Forest Service, 1926. 6 p. Tree Pamphlet No. 11).

E Regional Information Reports

Beginning in 1981, with the reorganization of the Canadian Forest Service, with its headquarters in Ottawa and most research carried out in the regional forest research centres, research reports began to be issued as Information Reports with numbering unique to the particular regional centre. For example, a report with unique numbering from the Pacific Forestry Centre by M. Penner and K. Power is entitled *Canada's Forest Biomass Resources Deriving Estimates from Canada's Forest Inventory* (Victoria: Pacific Forestry Centre, 1997. 33 p. Information Report BC-X-370). These information reports now make up the majority of the research reports issued by any of the regional centres. However, some centres have also issued technical notes, such as one by A.W. McKay, entitled *Forest Insects and Diseases in Kejamkujik National Park in 1985* (Fredericton: Maritimes Forest Research Centre, 1986. n.p. Technical Note 154). Miscellaneous reports have also been issued as noted, such as one by H. Brodersen and H.J. Evans, entitled *Results of Forest Insect and Disease Surveys in the Southwestern Region of Ontario, 1985* (Sault Ste Marie, Ontario: Great Lakes Forestry Centre, 1986. 25 p. Miscellaneous Report No. 25). Finally, some centres issue different types of leaflet publications series. For example, the Laurentian Forestry Centre issues information leaflets, such as one by A. Lavallée and G. Laflamme, entitled *Cherry Tree Black Knot* (Sainte-Foy, Quebec: Laurentian Forestry Centre, 1995. 7 p. Information Leaflet LFC 16). The Northern Forestry Centre issues Forest Leaflets, such as one by P.J. Maruyama, entitled *Silverleaf* (Edmonton: Northern Forestry Centre, 1991. 1 leaf. Forestry Leaflet No. 1); and the Pacific Forestry Centre issues pest leaflets, such as one by N. Humphreys, entitled *Douglas-Fir Beetles in British Columbia* (Victoria: Pacific Forestry Centre, 1995. 4 p. Forest Pest Leaflet No. 14).

F Serial Publications

The Canadian Forest Service or Forestry Canada has some serial publications of general, regional, or research interest. Examples of these are *Timberlines* (Northern Forestry Research Centre, 1986–); *Solutions* (Ottawa: Canadian Forest Service, 1998-), and electronically at <<http://www.nrcan.gc.ca/cfs/solutions/latest.html>>; and *Forest Management Note* (Northern Forest Research Centre, 1980–). The major research journal, the *Canadian Journal of Forest Research* (Ottawa: NRC Press, v. 1, 1971–), is published by the National Research Council of Canada. As well, the journal entitled *Forestry Chronicle* (Toronto: Canadian Society of Forest Engineers, v.1, 1925–) is unique in its research and information coverage of Canadian forestry.

III Forestry Research

A Forest Insects and Diseases

Along with forest fires and windstorms, forest insects and diseases have historically been an integral part of the slow but natural pattern of forest succession. In the early days of settlement, when the demands on forests were slight, forest destruction, or degradation by insects or diseases, was of minor consequence. However, from the early 20th century on, national and international demands for wood and wood products have forced the federal and provincial governments, through their forestry departments, to interfere in the natural succession of the forest to ensure the wood supply for current and future harvesting. All of Canada's forest lands are subjected to a continuous surveillance, with respect to forest insects and diseases. The latter is the responsibility of the Forest and Insect Disease Survey of the Canadian Forestry Service, but in close collaboration with provincial forest agencies, forest industries, and private organizations. The Forest Insect and Disease Survey (FIDS) is a nationally coordinated program consisting of six regional FIDS units, and its activities are managed by the regional units. The individual units prepare annual estimates of pest occurrences by species, location, area of forest affected, and the intensity of damage. These data are used to provide continuous monitoring of pest activity during and between outbreaks, to support research, and to design detailed surveys of specific and operational interest. The most recent review of FIDS was in the publication by J.P. Hall entitled *Forest Insect and Disease Survey Strategic Plan, 1993–1998* (Ottawa: Natural Resources Canada, Canadian Forest Service, 1994. 8 p.).

Most of the data that the survey collects is retained in a computerized information system called FIDSINFOBASE. The information system has been described by J.M. Power in a report entitled *FIDSINFOBASE. The Forest Insect and Disease Survey Information System* (Ottawa: Canadian Forestry Service, 1986. 30 p. Petawawa National Forestry Institute. Information Report PI-X-65). Overall, the system, which provides nationwide access from a centralized computer, contains: (1) data on the ground survey of insects and diseases and the associated specimen analysis collected by FIDS, and its predecessors, across the country over the past four decades, (2) ground plot data from the Acid Rain National Early Warning System, and (3) insect and disease catalogues.

More recently, FIDS has looked at the potential application of a National Geographic Information System (GIS) to the Survey, which could handle geographic locations and associated pests. Specifically, these data consist of point locations of plots and outbreaks, areas of infestations, and areas or grids delineating forest stands, or summaries of stand inventory data. These additional automation requirements to the Forest Insect and Disease Survey have been described by J.M. Power and P.H. Williams in the report entitled *A National Geographic Information System for the Forest Insect and Disease Survey – Requirements and Selection* (Ottawa: Canadian Forestry Service, 1987. 14 p. Petawawa National Forestry Institute. Information Report PI-X-75), and by J.M. Power and S.P. D'Eon in the report entitled *Quantifying Pest-Caused Forest Depletion Using Geographic Information Systems and Database Technologies* (Chalk River, Ontario: Petawawa National Forestry Institute, Forestry Canada, 1991. 17 p.).

The structure of the reports generated by the Forest and Insect Disease Survey (FIDS), which was merged with the Long Range Transport of Air Pollutants Program (LRTAP) in 1996 to form the Forest Health Network (FHN), includes annual national surveys and reviews, as well as regional surveys and reviews. The annual national surveys are: *Annual Report of the Forest Insect Survey* (Ottawa: Forest Biology Division, Science Service, 1936/37/38–1950. 13 v.); *Annual Report of the Forest Insect and Disease Survey* (Ottawa: Forestry Service, 1951–79. 29 v.); and *Forest Insect and Disease Conditions in Canada* (Ottawa: Forest Insect Disease Survey, Canadian Forestry Service, 1981–). The recent publications are broken down into six topics. They are: (1) Major Insects and Diseases in Canadian Forests, (2) Abiotic [non-biological] damage, (3) Acid Rain Early Warning System (ARNEWS), (4) North American Maple [a network to access the condition of sugar maple], (5) Other Insects; Diseases, and Damage, and (6) Select Bibliography.

The national reviews, which summarily document forest depletion caused by harvesting, wildfires, and storm damage, as well as forest insects and diseases, are organized to assist forest managers and planners who develop pest control policies and research programs. An example of this type of national review is entitled *Forest Depletions Caused by Insects and Diseases in Canada, 1982–1987* (Ottawa: Natural Resources Canada, 1994. 14 p. Forest Insect and Disease Survey. Information

Report ST-X-8). In this report, depletions are listed by insect diseases, abiotic losses, and nursery losses.

The annual regional surveys are carried out by the regional forestry centres. Those published are: (1) *Forest Insect and Disease Conditions in Newfoundland and Labrador* (St John's: Newfoundland Forest Research Centre, 1983–1994/95); *Forest Insects and Diseases in the Quebec Region 1936–1988*, [by] D. Lachance and J. Thibault (Ottawa: Forestry Canada, Quebec Region, 1991. Information Report (Laurentian Forestry Centre), LAU-X-97E); *Forest Insect and Disease Conditions in Ontario* (Sault Ste Marie, Ontario: Great Lakes Forest Research Centre, 1969–95. 27 v.). The Ontario region is further broken down into northwest, northeast, and southwest sectors, which published *Results of Forest Insect and Disease Surveys in the Northwestern Region of Ontario* (Sault Ste Marie, Ontario: Great Lakes Forest Research Centre, 1979–95. 17 v.); *Results of Forest Insect and Disease Surveys in the Northeastern Region of Ontario* (Sault Ste Marie, Ontario: Great Lakes Forest Research Centre, 1979–95. 17 v.); and *Results of Forest Insect and Disease Surveys in the Southwestern Region of Ontario* (Sault Ste Marie, Ontario: Great Lakes Forest Research Centre, 1979–95. 17 v.). As noted, all the above ongoing publications ceased in 1995; *Forest Insect and Disease Conditions in Alberta, Saskatchewan, Manitoba, and the Northwest Territories in [year] and Predictions for [year]* (Edmonton: Northern Forest Research Centre, 1979–91. 13 v.); *Forest Insect and Disease Conditions in Alberta, Saskatchewan, Manitoba, and the Northwest Territories in 1992*, [by] H.F. Cerezke and J.P. Brandt (Edmonton: Northern Forestry Centre, 1993. 24 p. Information Report NOR-X-332); *Forest Insect and Pest Conditions in West-Central Canada in 1993 and Predictions for 1994*, [by] J.P. Brandt (Edmonton: Northern Forestry Centre, 1994. 39 p. Information Report NOR-X-335); *Forest Insect and Disease Conditions for 1995*, [by] J.P. Brandt (Edmonton: Northern Forestry Service, 1995. 49 p. Information Report NOR-X-340); and *Forest Insect and Disease Conditions in West-Central Canada in 1995 and Predictions for 1996*, [by] J.P. Brandt (Edmonton: Canadian Forestry Service, 1996. n.p. Information Report NOR-X-340); *Forest Insect and Disease Conditions, British Columbia and Yukon* (Victoria: Pacific Forest Research Centre, 1979–95).

The regional reviews are broad summaries of significant forest insects and diseases encountered in particular districts over a 30–year period. An example of this type of publication, compiled by M.J. Thomson and L.S. Macleod, is entitled *A Review of Important Forest Insects and Disease Problems in the IGNACE DISTRICT of Ontario, 1950–1980* (Ottawa: Canadian Forestry Service, 1988. 68 p. Appendices. Great Lakes Forestry Centre. Miscellaneous Report No. 64). The contents of this report cover: (1) Major Insects or Diseases – those capable of causing serious injury to or death of living trees or shrubs, (2) Minor Insects or Diseases – those capable of causing sporadic or localized injury but not usually a serious threat to living trees or shrubs, and (3) Abiotic Damage – damage caused by non-living factors such as frost, wind, hail, or winter drying. The contents of these reviews are abstracted from the annual FIDS report and the annual district or regional reports.

From the reports, reviews, and research, there have been two major publications written about forest insect pests in Canada. The most recent and research-oriented publication is entitled *Forest Insect Pests, in Canada*. Edited by J.A. Armstrong and W.G.H. Ives (Ottawa: Natural Resources Canada, Canadian Forest Service, 1995. 732 p.). It contains 70 authored chapters covering individual insect pest research as well as pest management control and technology by region. The second, which is more of an overview, is by R. Martineau, and entitled *Insects Harmful to Forest Trees* (Montreal: Multiscience Publications, 1984. 261 p.). This publication, which was sponsored by the Canadian Forestry Service, lists and describes 75 insects that are a major or minor menace to the various trees that make up Canada's forests. One other unique publication by R.A. Simpson which describes forest pest infestations is an atlas, entitled *An Ecological Atlas of Forest Insect Defoliation in Canada, 1980–1996* (Fredericton: Atlantic Forestry Centre, 1999. 15 [58] p. Information Report M-X-206E). It describes in maps, figures, and tables, forest defoliation for five major forest pests in Canada – the spruce budworm, the jack pine budworm, the mountain pine beetle, the hemlock looper, and the forest tent caterpillar – for the period 1980–1996. According to the *Forest Insect and Disease Conditions in Canada, 1995* (Ottawa: Forest Insect and Disease Survey, Natural Resources Canada, Canadian Forest Service, 1998. 72 p.), some of the current major forest pests are the spruce budworm, the jack pine budworm, the mountain pine beetle, the hemlock looper, the forest tent caterpillar, and the gypsy moth. Also, the major forest diseases are dutch elm disease, and sclerodermis canker.

The spruce budworm, *Choristoneura fumiferana* (Clem), is a native to North America and is considered the most destructive pest of fir and spruce forests due to its cyclic epidemics throughout the forests of eastern Canada. Through the methodological study of the growth rings of very old firs and spruces, J.R. Blais, in the papers entitled *Spruce Budworm Outbreaks in the Past Three Centuries in the Laurentides Park, Quebec* (*Forestry Science*, v. 11, 1965. pp. 130–38), and *Regional Variation in Susceptibility of Eastern North America Forests to Budworm Attack Based on History of Outbreaks* (*Forestry Chronicle*, v. 44, 1968. pp. 1–6), determined that seven outbreaks had occurred in the northeastern part of the continent between 1704 and 1877. There was an average interval of 29 years between outbreaks. Since those outbreaks, there have been three others in eastern North America. The first of these outbreaks occurred from 1909 to 1920, and ended with no indication of why the population declined. The second outbreak began in

1937, and ended in 1958, because of natural factors and the use of large-scale spraying with chemical insecticides after 1954. Then, in 1967, populations increased again in all regions of eastern Canada. This latest outbreak began to subside in western Quebec in 1976, but has persisted in other areas of eastern Canada in spite of massive programs of aerial spraying with chemical insecticide during the mid-1970s. At that time, approximately 15 million acres of forest in New Brunswick, Quebec, and Ontario were sprayed each year over a four-year period. These more recent outbreaks have been described by J.R. Blais in the paper entitled *Trends in Frequency, Extent and Severity of Spruce Budworm Outbreaks in Eastern Canada* (*Canadian Journal of Forest Research*, v. 13, 1983. pp. 539–547), and by H. Morin and D. Laprise in the paper entitled *Recent History of* Choristoneura fumiferana *Epidemics North of Lac St.-Jean (Quebec) (Canada): A Dendochronological Analysis* (*Canadian Journal of Forest Research*, v. 20, 1990. pp. 1–8).

The life-cycle of the spruce budworm is simple in eastern Canada with one generation produced per year. However, in western Canada, the generation is over a two-year period. In eastern Canada, the adult moth emerges in July, and the females lay eggs on needles exposed to sunlight. Upon hatching after 10 days, the larvae overwinter toward the interior of the crown of the tree, or among bushes, or in the cupules of staminate flowers. In the spring, the larvae emerge from their shelter to feed until July, and then pupate for 10 days before emerging as adults. Their life cycle has been described by F.C. Craighead in the publication entitled *Studies on the Spruce Budworm (Cacoeai* fumiferana Clem.*)* (Ottawa: Department of Agriculture, 1924. Technical Bulletin New Series No. 37).

The Canadian Forest Service itself is not involved in control programs, which are carried out by the provinces, but has been responsible for the development of spray strategies and guidelines to prevent adverse effects from the sprays used for control of the spruce budworm. Studies with spray strategies have been described by I.W. Varty and M.E. Godin in their report entitled *Identification of Some of the Factors Controlling Aerial Spray Efficacy* (Fredericton: Maritimes Forest Research Centre, 1983. 29 p. Information Report M-X-142), and by I.W. Varty and S.E. Holmes in their report entitled *Heterogeneity of Spray Deposit and Efficacy within a Single Swath Applied by Aircraft over Forest Infested with Spruce Budworm* (Fredericton: Canadian Forestry Service – Maritimes, 1988. 60 p. Information Report M-X-168). In Canada, the chemical insecticides commonly used for spruce budworm control are aminocarb and fenitrothion. Studies on the latter chemical have been undertaken by K.M. Sundaram in the paper entitled *Cuticular Penetration and In Vivo Metabolism of Fenetrothion in Spruce Budworm* (*Journal of Environmental Science and Health. Part B: Pesticides, Food Contaminants and Agricultural Wastes*, v. 23, 1988. pp. 643–59). However, in more recent years, the biological insecticide *Bacillus thuringiensis* has also been used. Studies on the efficiency of *B. thuringiensis* have been carried out by K. Van Frankenhuyzen and N.J. Payne in the paper entitled *Theoretical Optimization of* Bacillus thuringiensis *Berliner for Control of the Eastern Spruce Budworm,* Choristoneura fumiferana *Clem. (Lepidoptera: Tortrecidae): Estimates of Lethal and Sublethal Dose Requirements, Product Potency, and Effective Droplet Sizes* (*Canadian Entomologist*, v. 125, 1993. pp. 473–78), and by K. Van Frankenhuyzen and L. Gringorten in their paper entitled *Susceptibility of Different Instars of the Spruce Budworm (Lepidoptera: Tortrecidae) to* Bacillus thuringiensis *var. Kurstaki Estimated with a Droplet-Feeding Method* (*Journal of Economic Entomology*, v. 90, 1997. pp. 560–65).

The jack pine budworm, *Choristoneura pinus pinus* (Free), is a close relation of the spruce budworm and was recognized as a distinctive species only in 1953. The work on identifying the jack pine budworm was carried out by I.M. Campbell and T.N. Freeman and described in their papers entitled *Morphological Differences Between the Pupae and Egg Cluster of* Choristoneura fumiferana *(Clem.) and C.* Pinus *Free. (Lepidoptera: Tortricidae)* (*Canadian Entomologist*, v. 85, 1953. pp. 134–35); and *The Spruce Budworm,* Choristoneura fumiferama *(Clem.), and an Allied New Species on Pine (Lepidoptera: Tortricidae)* (*Canadian Entomologist*, v. 85, 1953. pp. 121–27). Further refinements on defining the species have been more recently undertaken by G.T. Harvey and reported in the papers entitled *Genetic Relationships among Choristoneura Species (Lepidoptera: Tortricidae) in North America as Revealed by Isozyme* [distinct forms of an enzyme with identical activities] *Studies* (*Canadian Entomologist*, v. 128, 1996. pp. 245–62), and *Interspecific Crosses and Fertile Hybrids among the Coniferophagous Choristoneura (Lepidoptera: Tortricidae)* (*Canadian Entomologist*, v. 129, 1997. pp. 519–36). The jack pine budworm is native to North America, and its range coincides with that of its preferred host, the jack pine. In Canada, the jack pine budworm is found from New Brunswick to British Columbia, and to date has caused the most damage in Ontario. Outbursts last two to three years, on a six- to eight-year cycle in eastern Canada, and 10-year cycles in western Canada. The life cycle of the jack pine budworm follows that of the spruce budworm with which it has been closely identified. This budworm is of concern to foresters, because of the damage it causes to natural stands of jack pine, and the damage caused to plantations of scots pine in central Canada. Although climatic conditions can be important in the rise and decline of outbreaks of jack pine budworm, infestations have been linked closely with the abundance of staminate flowers in jack pine stands. This relationship has been discussed by V.G. Nealis in the paper entitled *Jack Pine Budworm Populations and Staminate Flowers (Canadian Journal of Forest Research*, v. 20, 1990. pp. 1253–55). In this paper, the author noted that 'although there was some indication that early-instar budworm may be more successful at establishing feeding sites when flowers are abundant ... the assumption that staminate flowers greatly influence jack pine budworm population dynamics is questioned.'

Several species of parasites provide some form of natural control against the jack pine budworm, but research in Canada has been aimed at improving control methods by experimental tests with chemical insecticides applied both from the ground and from the air. Work on insecticides applied from the ground has been described by P.F. Deboo and V. Haldahl in their paper entitled *Chemical Control of the Jack Pine and Spruce Budworms with Ground Application Equipment in Manitoba* (*Manitoba Entomologists*, v. 5, 1971. pp. 57–67). Aerial applications of chemical insecticides in the laboratory setting have been described by P.C. Nigam in the paper entitled *Toxicity of Insecticides to Six Instar Jack Pine Budworm Larvae under Laboratory Conditions* (*Canada: Department of Fisheries and Forestry Bi-monthly Research Notes*, v. 26, 1970. pp. 2–3). Biological control has also been examined by B.L. Cadogen and B.F. Zylstra in their paper entitled *Evaluation of a New Futura Formulation of Bacillus Thuringiensis on Populations of Jack Pine Budworm,* Choristoneura pinus pinus *(Lepidoptera: Tortricidae)* (*Proceedings of the Entomological Society of Ontario*, v. 117, 1986. pp. 59–64).

The mountain pine beetle, *Dendroctonus ponderosae* Hopk, is a native to North America also, and is the most damaging forest insect in British Columbia. Outbreaks have been recorded at varying geographic locations within British Columbia and Alberta, at irregular intervals since at least 1910. The current outbreaks started during the late 1960s and early 1970s, with rapid increases in the early 1980s.

The lodgepole pine is the principal host of the mountain pine beetle, but several western pine species are susceptible. This relationship has been studied by D.W. Langer and J.R. Spence in their paper entitled *Host Effect on Fertility and Reproductive Success of* Dendroctonus ponderosae *Hopkins (Coleoptera: Scolytidae)* (*Evolution*, v. 44, 1990. pp. 609–18), and by H.F. Cerezke in the paper entitled *Egg Galley, Brood Production, and Adult Characteristics of Mountain Pine Beetle,* Dendroctonus ponderosae *Hopkins (Coleoptera: Scolytidae), in Three Pine Hosts* (*Canadian Entomologist*, v. 127, 1995. pp. 955–65). Infected trees are usually killed by the combined action of blue stain fungus transmitted by the beetle to the tree, and by the beetle larvae. The foliage of the trees usually turns red-brown in the year following an attack, and the tree mortality can then be detected by aerial surveys. In order to predict outbreaks, the pattern of infestation has been studied by T.L. Shore in the paper entitled *The Development of Systems for Rating the Risk and Hazard to Lodgepole Pine Stands of Mountain Pine Beetle Epidemics* (*Northwest Environment Journal*, v. 61, 1990. pp. 386–87).

According to *Forest Insect and Disease Conditions in Canada, 1994* (Ottawa: Natural Resources Canada, Canadian Forest Service, Forest Insect and Disease

Survey, 1996. 112 p.), over 230 million mature pine trees have been killed since the current outbreak began in the late 1960s, and declines in areas infested are due primarily to host depletion, including the harvesting of dead trees, which is successful if done quickly. The size of these trees makes them very suitable for saw logs. As well, cut and burn programs are carried out for management of this beetle.

The hemlock looper, *Lambdina fiscellaria fiscellaria* (Guen.), is an insect defoliator native to North America that occurs from the Atlantic coast westward to Alberta. In the east, its main host is balsam fir, whereas in the west it is hemlock. The hemlock looper has been a particular menace in Newfoundland. The first outbreak was noted in 1912, and five other outbreaks, each lasting from three to six years, have occurred every 10 to 15 years since that initial occurrence. These outbreaks have been described by W.J. Carroll in the paper entitled *History of the Hemlock Looper* Lambdina fiscellaria fiscellaria *(Guen) (Lepidoptera: Geometridae) in Newfoundland and Notes on Its Biology* (*Canadian Entomologist*, v. 88, 1956. pp. 587–99). In the Maritime provinces, outbreaks have been less frequent, but in Quebec there have been six waves of epidemics, between 1928 and into the 1970s, occurring at 10-year intervals and lasting about three years each time. These epidemics have been described by P. Benoit and R. Desmaulniers in their paper entitled *Epidémies passées et présentes de l'arpentense de la pruche au Québec* (*Environment Canada Bimonthly Research Review*, v. 28, 1972. pp. 11–12). In Ontario, outbreaks have been local.

The hemlock looper produces one generation per year. The eggs of the adult overwinter and hatch as larvae the following June. The larvae feed, initially, on new needles of the balsam, and then on mature needles, until the pupal stage in August. The latter lasts about 20 days. Then, the adult emerges to mate and lay eggs. What is interesting about the life cycle is the vulnerability of the larval stage to fungal infections, parasites, and excessively humid weather. These natural controls have been described by I.S. Otvos in the paper entitled *Biological Control Agents and Their Role in the Population Fluctuations of the Eastern Hemlock Looper in Newfoundland* (Ottawa: Canadian Forest Service, 1973. 34 p. Newfoundland Forestry Research Centre. Information Report N-X-102).

Artificial control of the hemlock looper by aerial spraying with the biological insecticide *Bacillus thuringiensis* and two other chemical insecticides was undertaken near Hawkes Bay in northwestern Newfoundland in 1981. The result of spraying of this experimental area indicated that the biological insecticide and the chemical insecticide fenetrothion reduced larval populations by up to 98%, and pupal populations by 100%. This spraying program has been described by A.G. Raske and R.J. West in their publication entitled *The Effectiveness of* Bacillus thuringiensis, *Diflubenzurons and Fenitrothion against the Hemlock Looper,* Lambdina fiscellaria, *in Newfoundland in 1987* (Ottawa: Forestry Canada, 1991. 74 p. Newfoundland and Labrador Region Information Report N-X-279). Additional studies on the effectiveness of *B. thuringiensis* against the hemlock looper and other forest lepidopterans have been undertaken by K. van Frankehuyzen and R. Milne and reported in the papers entitled *Comparative Toxicity of the HD-1 and NRD-12 Strains of* Bacillus thuringiensis *ssp. Kurstabi to Defoliating Forest Lepidoptera* (*Journal of Invertebrate Pathology*, v. 59, 1992. pp. 149–54); and *Toxicity of Activated Cryi Proteins from* Bacillus thuringiensis *to Six Forest Lepidoptera and* Bombyx mori (*Journal of Invertebrate Pathology*, v. 62, 1993. pp. 295–301).

The forest tent caterpillar, *Malacosoma disstria* Bhn., unlike the hemlock looper, is widely distributed from coast to coast. It is native to North America and the most common leaf eater in North American hardwood forests. Its preferred host is the trembling aspen, but will attack all hardwood trees. Damage caused by the forest tent caterpillar has been identified back to the year 1791. Since then, records indicate outbreaks at 10- to 12-year intervals. Ontario, itself, has had 12 outbreaks since 1867, with the last one beginning in 1972. Quebec outbreaks parallel the Ontario ones, but outbreaks in the Maritime provinces have occurred at different times, and for shorter intervals. These outbreaks have been described by W.L. Sippell in the paper entitled *Outbreaks of the Forest Tent Caterpillar* Malacosoma disstria *Hbn., a Periodic Defoliator of Broad-Leafed Trees in Ontario* (*Canadian Entomologist*, v. 94, 1962. pp. 408–10). Only one generation of forest tent caterpillar occurs per year. The adult moth is active in the summer, but egg masses laid by the female over-winter, with the larvae emerging the following spring. The larvae, which are voracious eaters, go through five instars and then retreat into a spun net, or tents, until they are ready to pupate in individual cocoons. The life cycle has been described by V. Hildahl and A.E. Campbell in their publication entitled *Forest Tent Caterpillar in the Prairie Provinces* (Ottawa: Canadian Forest Service, 1975. 11 p. Northern Forest Research Centre Information Report NOR-X-35). Because the larvae are voracious eaters, damage to hardwood trees may include buds and flowers as well as leaves. Nevertheless, severely defoliated trees seldom die because typical invasions in one location seldom last more than three years. This damage pattern has been observed by V. Hildahl and W.A. Reeks in their paper entitled *Outbreaks of the Forest Tent Caterpillar,* Malacosoma disstria *Hbn., and Their Effects on Stands of Trembling Aspen in Manitoba and Saskatchewan* (*Canadian Entomologist*, v. 92, 1960. pp. 199–209). Because of the relatively short time-period of forest tent caterpillar infestations, which acts against tree mortality, natural control rather than chemical spraying is the usual control pattern followed with this insect. Natural control is facilitated by the fact that larvae are very sensitive to frost. This has been observed by B.W. Wetzell and H.M. Kulman in their paper entitled *Effects of Cold*

Temperatures on Hatching of the Forest Tent Caterpillar, Malacosoma disstria *(Lepidoptera: Lasiocampidae)* (*Canadian Entomologist,* v. 105, 1973. pp. 1145–49). As well, parasites and predators such as the techinid fly, or birds, are important in the reduction of larval populations. The only time that chemical control is undertaken is in recreation and camping areas with the use of chemical insecticides. The latter has been discussed by A. Retnakaran and L. Smith in their paper entitled *Application of Dimilin Effectively Controls Forest Tent Caterpillar Populations and Affords Foliage Protection* (*Environment Canada Bimonthly Research Notes,* v. 32, 1976. pp. 26–27). There has also been some success with biological control using a protozoan parasite. This work has been described by G.G. Wilson in the papers entitled *Effects of* Nosema disstriae *(Microsporida) on the Forest Tent Caterpillar,* Malacosoma disstria *(Lepidoptera: Lasiocampidae)* (*Proceedings of the Entomological Society of Ontario,* v. 110, 1979. pp. 97–100); and *Pathogenicity of* Nosema disstriae, Pleistophora schubergi *and* Vairimorpha necatrix *(Microsporidia) to Larvae of the Forest Tent Caterpillar,* Malacosoma disstria (*Zeitschrift fuer Parasitenkunde,* v. 70, 1984. pp. 763–68).

Unlike the previously mentioned insects, the gypsy moth, *Lymantria dispar* (L.), is a common pest on all continents except for the islands of the Pacific. Its European form was introduced accidentally into North America near Boston in 1869, and, despite the flightless nature of the adult female in this founding population, the gypsy moth has steadily extended its range in North America. By the early 1990s, this moth inhabited most hardwood forests in eastern North America. However, it was not until 1969 that the first defoliation was noted in southeastern Ontario. By 1992, the gypsy moth was established throughout all of Ontario to the south and east of Lake Huron. Established populations of the moth also exist throughout the St Lawrence River and Ottawa River valleys in Quebec and Ontario, and there is evidence of new infestations in the Maritimes which go back as far as 1981.

The overall pattern of the dispersal of this moth has been described by J.S. Elkinton and A.M. Liebolt in their review paper entitled *Population Dynamics of the Gypsy Moth in North America* (*Annual Review of Entomology,* v. 35, 1990. pp. 571–96). The female adult lays egg masses in the late summer on trunks or branches of tree hosts, which include not only hardwood species but also pine and spruce trees. The egg masses develop and overwinter until the larval stage emerges the following spring to begin the defoliation of the trees. During severe outbreaks, all trees and shrubs in an infested area will be stripped. Hardwoods may survive two or three years of repeated defoliation, but pines succumb within the first year. After feeding, the larvae disperse to pupate on trunks, branches, or forest debris. The Canadian pattern of the gypsy moth has been described by P. Benoit and D. Lachance in their publication entitled *Gypsy Moth in Canada, Behaviour and*

Control (Ottawa: Forestry Canada, 1990. 21 p. Quebec Region Information Report DPC-X-32).

The introduction of the gypsy moth from Europe meant that only factors such as climate and temperature could check outbreaks, as young larvae will succumb during the winter or during spring frosts. Although insect-eating predators adapted to the new prey, the American government had to import European parasites to establish some natural control in infested areas. These historic American developments have been described in the publication entitled *The Gypsy Moth: Research towards Integrated Pest Management.* Edited by C.C. Doane and M.L. McManus (Washington, DC: USDA Forest Service, 1981. 757 p. Technical Bulletin 1584).

In Canada, artificial control of the gypsy moth, by spraying with chemical insecticides, has been attempted. Aerial spraying has been reviewed by G.S. Brown in the paper entitled *Gypsy Moth in Aerial Control of Forest Insects in Canada.* Edited by M.L. Prebble (Ottawa: Department of the Environment, 1975. pp. 208–12). Overall, work on the gypsy moth in Canada has been described by V.G. Nealis and S. Erb in their publication entitled *A Sourcebook for Management of the Gypsy Moth* (Sault Ste Marie, Ontario: Great Lakes Forestry Centre, 1993. 48 p.). However, since this publication was issued, experiments in biological control with viruses have had some success. This work has been carried out by J.C. Cunningham and N.J. Payne and reported in their papers entitled *Aerial Spray Trials with Nuclear Polyhedrosis Virus and* Bacillus thuringiensis *on Gypsy Moth (Lepidoptera: Lymantriidae) in 1994: I. Impact in the Year of Application* (*Proceedings of the Entomological Society of Ontario*, v. 127, 1996. pp. 21–35); *Aerial Spray Trials with Nuclear Polyhedrosis Virus and* Bacillus thuringiensis *of Gypsy Moth (Lepidoptera: Lymantriidae) in 1994: II. Impact One Year after Application* (*Proceedings of the Entomological Society of Ontario*, v. 127, 1996. pp. 37–43); and *Aerial Spray Trials in 1992 and 1993 against Gypsy Moth,* Lymantria dispar *(Lepidoptera: Lymantriidae). Using Nuclear Polyhedrosis Virus with and without an Optical Brightener Compared to* Bacillus thuringiensis (*Crop Protection,* v. 16, 1997. pp. 15–23).

The Dutch Elm disease also involves a forest insect, but as a vector for a fungal disease, which has devastated elm trees in Ontario and Quebec, and is a major problem in the Maritimes. The forest insect involved is the elm bark beetle *Hylurgopinus rufipes* (Eichh.), which is native to North America, and has infected Canadian trees. In the United States, however, it is the smaller European elm bark beetle *Scolytus multistriatus* (Marsh.), which was accidentally introduced in 1909, that has infected trees. This latter beetle is more aggressive than the native bark beetle but has had difficulty establishing itself in Canada because of harsher climatic conditions. Nevertheless, both beetles serve as the vector of the fungus *Ceratocystis ulmi* (Buism) C. Moreau.

In Canada, the first infected trees were found near Sorrel, Quebec, in 1944, and by 1959 the disease had killed over 600,000 elms in that province. The history of the disease in Quebec has been documented by R. Pomerleau in the paper entitled *History of the Dutch Elm Disease in the Province of Quebec* (*Forestry Chronicle*, v. 37, 1961. pp. 356–67). The first infected trees in Ontario were found in 1946 at St Isidore, Prescott County. By the mid-1960s, the disease had spread as far south as the Niagara Peninsula, and as far north as the Georgian Bay area, and Mattawa on the Ottawa River. This movement of the beetle has been documented by R.J. Finnigan in the paper entitled *Elm Bark Beetles in Southwestern Ontario* (*Canadian Entomologist*, v. 89, 1957. pp. 275–80). In the meantime, the first diseased trees in New Brunswick were found at Woodstock, on the St John River, in 1957. By 1963, the disease had spread through half of New Brunswick, as reported in the *Annual Report of the Forest Insect and Disease Survey, 1963* (Ottawa: Forestry Service, 1964. 1 v.). By 1979, the disease had spread into Prince Edward Island, and currently, the disease has spread into Manitoba and Saskatchewan, according to *Forest Insect and Disease Conditions in Canada, 1994* (Ottawa: Forest Insect and Disease Survey, Natural Resources Canada, Canadian Forest Service, 1996. pp. 18–19).

The life cycle of the native elm bark beetle is fairly complex and varies with the region, climatic conditions, and the year. The insect overwinters in all stages, except the egg, but it is the adult beetles that appear in the late spring, and mate, feed, and bore egg galleries in the bark of healthy trees. The emerging larvae also feed and create additional galleries under the bark until they pupate and emerge again as new adults that can infect new healthy trees. In Canada, the biology of the elm bark beetle was intensely studied by A. Robert during the late 1940s, and the results reported in the paper entitled *Recherches sur les insectes de l'orme* (Quebec: Ministre Ternes Forêts, Rapport pour l'année 1947–48. pp. 131–37).

Although the damage caused by the boring of the beetle does not necessarily kill the tree, the beetle does carry the fungus *Ceratocystis ulmi*, which is transmitted into healthy trees by the feeding of the adult elm bark beetles. The feeding wounds can reach the xylem of the vascular system of the tree, and the fungus spreads as it invades the xylem cells. The fungus can also spread by spores, which are distributed by the sap to other parts of the tree. The pathology of the disease itself is still not clearly understood, and there have been toxin theories about the disease, questions about the role of enzymes, as well as about the role of the plugging of the vessels in the tree by the spores and mycelium of the fungus itself. The fungus has been intensely studied by G.B. Ouellette in papers such as *Studies on the Infection Process of* Ceratocystis ulmi *(Buism.) C. Moreau in American Elm Trees* (*Canadian Journal of Botany*, v. 40, 1962. pp. 1567–75). There has been little research done on the natural control of the native elm bark beetle, but it appears that climate, particularly low temperatures in winter, may be responsible for the killing of large numbers of hibernating individuals. Natural control has been studied by A.G. Kais and E.B. Smalley in their paper entitled *Environment and Development of Dutch Elm Disease* (*Phytopathology*, v. 52, 1962. pp. 1191–96). Artificial control by means of quarantine regulations, sanitation, the use of chemical insecticides and chemotherapy have all been tried, but most methods tested have given only partial or temporary results. These methods have been described by A.G. Davidson in the publication entitled *Dutch Elm Disease* (Ottawa: Department of Forestry and Rural Development, 1967. 23 p.).

Another major fungal disease which is present in Canada is the Scleroderris canker, *Gremmeniella abietina* (Lagerb.) Morelet, of which two races (North American and European) have been determined. This canker has been detected throughout Canada, except in Prince Edward Island, Manitoba, and Saskatchewan. This disease of conifers is caused by a fungus, of which at least two species as mentioned above have been determined. The North American race of *G. abietina* is widely distributed in Canada and has been a serious problem in nurseries and young plantations for years. In eastern Canada, the disease has been studied by G. LaFlamme and L. Archambault in their paper entitled *Evaluation of Microclimatic Factors Affecting Ascospore Release of* Gremmeniella abietina var. *balsamea* (*Canadian Journal of Plant Pathology*, v. 12, 1990. pp. 190–94).

Finally, an emerging forest pest of concern is the pinewood nematode *Bursaphelenchus xylophilus* (Steiner and Buhrer) Nickle. Although surveys during the 1980s had shown that the pinewood nematode existed at only extremely low levels in the Maritimes, and has never been known to kill any trees, its presence has created an economic problem in the forest product export industry because of plant quarantine regulations in potential importing countries in Europe. In 1990, surveys carried out in the Maritimes and in British Columbia showed no pinewood nematodes in hemlock and less than 0.7% in other logs. The purpose of the survey was to propose hemlock as an exception from mandatory kiln-drying requirements which were to come into effect at the end of 1992 in Europe. These surveys were reported in *Forest Insect and Disease Conditions in Canada, 1990* (Ottawa: Forestry Canada, Forest and Insect Disease Survey, 1993. pp. 57–58). Since these surveys, further studies on conifers as hosts to these nematodes have been undertaken as described, in the papers by J.R. Sutherland and F.M. Reed, entitled *Canadian Conifers as Hosts of the Pinewood Nematode (*Bursophelenchus xylophilus*): Results of Seedling Inoculations (Scandanavian Journal of Forest Research*, v. 6, 1991. pp. 209–16); and by T.A. Forge and J.R. Sutherland, Bursaphelenchus xylophilus entitled *Distribution and Reproduction of Populations in Wood and Bark of the Western North American Conifers* (*Fundamental and Applied Nematology*, v. 19, 1996. pp. 341–47).

B *Forest Fire Research*

Fires are an important part of the natural cycle of change and renewal in the forest, but fires have both positive and negative effects – positive for forest regeneration, and negative for the costly damage to the resource. According to C.E. Van Wagner, in the publication entitled *Forest Fire Research in the Canadian Forestry Service* (Ottawa: Canadian Forestry Service, 1982. 39 p. Petawawa National Forestry Institute. Information Report PI-X-48), the average number of forest fires per year in Canada is about 9,200. About 33% of these are caused by lightning. The area burned each year is about 2,160,000 hectares of forest and represents about 0.6% of the whole Canadian forest. The annual cost of fire control is about $250 million. Although over 70% of all fires are caused by humans, lightning-caused fires account for about 89% of the national burned area. Finally, it is interesting to note that the largest 3% of fires account for 90% of the total burned area. Individual fires of over 100,000 hectares occur almost every year, yet half of all fires reported are held to less than 0.1 hectares in size. However, C.E. Van Wagner also noted in the paper entitled *The Historical Pattern of Burned Area in Canada* (*Forestry Chronicle*, v. 64, 1988. pp. 182–85) that since the 1970s there has been a sharp rise in forest fires compared to the decades after 1918.

Because of the tenure of forest land, the provincial governments are responsible for forest fire control over most of Canada's productive forests. The federal government, however, is directly concerned with forest fire protection in the Yukon and Northwest Territories, the forests of the national parks, the national forest research stations, and other federally owned land. As well, the federal government, through the Canadian Forest Service, is also responsible for forest fire research in Canada. This forest fire research focuses upon the study of fire behaviour, specialized fire prediction studies, the study of prescribed fires, using the basic information gained from the fire behaviour studies, fire ecology, and the creation of fire management systems. Fire suppression and fire economies are also a component of forest fire research, but more directly a concern of the forest industry.

The development of systems for evaluating daily forest fire danger has a relatively long history, starting in the 1970s. What evolved was something called the Forest Fire Danger Rating System. The present Canadian Forest Fire Danger Rating System began to take shape, in 1968, as described by S.J. Muraro in the paper entitled *A Modular Approach to a Revised National Fire Danger Rating System.* In: *Contributions on the Development of a National Fire Danger Rating System* (Ottawa: Canadian Forest Service, 1969. 18 p. Forest Research Laboratory, Victoria, BC. Information Report BC-X-37). By 1987, all the parts of the system had been developed, or revised, and issued as the *Canadian Forest Fire Danger*

Rating System – Users' Guide (Ottawa: Canadian Forest Service, Fire Danger Group. Three-ring binder). This present system has been described by B.J. Stocks and B.D. Lawson in their paper entitled *Canadian Forest Fire Danger Rating System: An Overview* (Forestry Chronicle, v. 65, 1989. pp. 258–65). It is actively made up of two components, the Canadian Forest Fire Behaviour Prediction System, and the Canadian Forest Fire Weather Index System.

The Forest Fire Behaviour Prediction System reflects the Forestry Canada approach to fire behaviour research. This involves the use of 'field observation and documentation of readily measured variables on experimental fires in forest stands and clearcut logging slash, followed by analysis of the data using sample mathematical models and correlation techniques. Well documented prescribed fires have been used as well, the latter being particularly useful at the extreme end of the fire behavior scale, where experimental fires are difficult to schedule and manage. Laboratory-based fire research in moisture physics and heat transfer theory provides the models and framework by which field data are analyzed and explained.' Structurally the FBP System consists of inputs and outputs affecting fire behaviour. Generally, the inputs that affect fire behaviour are fuels (fuel type), weather (wind speed and direction), and topography. Other inputs such as latitude and longitude, which affect foliar moisture content, the season, and elapsed time from ignition are included to estimate fuel condition or spread distance. The outputs of the system consist of primary and secondary components. The primary are the rate of spread, fuel consumption, head fire intensity, and fire description (surface or crown burning of the tree). The secondary output components consist of lead fire spread distance, elliptical fire area and perimeter, and flank and back rates of spread, and fire intensities. The inputs and outputs are expressed by mathematical equations, which are part of a fire behaviour database that can be used to manipulate the data. The database itself consists of observations from more than 300 fires from Canadian sources, and some selected fires in the United States, near the Canadian border. This forest fire prediction system has been described in detail in the publication entitled *Development and Structure of the Canadian Forest Fire Behavior Prediction System* (Ottawa: Forestry Canada. Fire Danger Group, 1992. 63 p. Forestry Technical Report ST-X-3), and in the user's guide to the system written by K.G. Hirsch and entitled *Canadian Forest Behavior Prediction (FBP) System: User's Guide* (Edmonton: Northern Forestry Centre, 1996. 121 p. Special Report 7). A field guide by S.W. Taylor and R.G. Pike entitled *Field Guide to the Canadian Forest Fire Behavior Prediction (FBP) System* (Edmondon: Northern Forestry Centre, 1997. 60 p. Special Report No. 11) has also been written. And a training manual in CD-ROM format by K.G. Hirsch has been produced. It is entitled *The Canadian Forest Fire Behavior Prediction (FBP) System: Interactive Training and Reference CD-ROM* (Edmonton: Northern Forestry Centre, 1998. 1 CD-ROM).

As mentioned above, the other component of the Canadian Forest Fire Danger Rating System is called the Canadian Forest Fire Weather Index System. The first system was issued in 1970, and had its fourth edition in 1984. This latest edition was published with the title *Tables for the Canadian Forest Fire Weather Index System.* Fourth Edition (Ottawa: Canadian Forest Service, 1984. 48 p. Forestry Technical Report No. 35).

The system comprises six standard components: 'three primary sub-indexes representing fuel moisture, two intermediate sub-indexes representing rate of spread and fuel consumption, and a final index representing fire intensity as energy output rate per unit length of fire front.' The system depends solely on weather readings taken each day at noon, local standard time. These readings consist of temperature, relative humidity, wind speed, and rain (if any) during the previous 24 hours. The current month is also specified. The different components of the system are used in equations that can be manipulated by computer to provide index tables to indicate the potential for fire. The equations were worked out by C.E. Van Wagner and T.L. Pickett and described in their report entitled *Equations and FORTRAN Program for the Canadian Forest Fire Weather Index System* (Ottawa: Canadian Forest Service, 1985. 18 p. Technical Report No. 33). The development of the whole system has been described by C.E. Van Wagner in the publication entitled *Development and Structure of the Canadian Forest Fire Weather Index System* (Ottawa: Canadian Forest Service, 1987. 37 p. Forestry Technical Report No. 35).

The evolving nature of forest fire research and systems has been described most recently at the Canadian Forest Service website <<http://www.nrcan.gc.ca/cfs/-scf>>. By clicking on 'Research,' information about the Fire Research Network will be found.

Related to forest fire danger systems is the more specific work carried out which attempts to predict the daily occurrence of people-caused and lightning-caused forest fires. As previously mentioned, people-caused fires are responsible for about 70% of all forest fire occurrences in Canada. This means that human activity and machines are responsible for igniting about two out of three forest fires that occur during an average fire season. This fact has been confirmed, over the years, from national statistics gathered and reported by the Canadian Forest Service, or its equivalent, beginning in 1908 with the publication by H.R. Macmillan entitled *Forest Fires in Canada during 1908* (Ottawa: Department of the Interior, Forestry Branch, 1909. 8 p. Forestry Branch Bulletin No. 7). However, from 1908 until 1945, data were not consistently reported in separate publications (data appeared in the branch's annual reports and were confined to federally protected lands only) until 1949, when separate statistical publications reappeared as *Forest Fire Losses in Canada* (Ottawa: Queen's Printer, 1949–78/79. 31 v.); *Canadian Forest Fire Statistics* (Ottawa: Canadian Forestry Service, 1979–87. 9 v.), and most recently as reported by D.G. Higgins and G.S. Ramsay in their publication entitled *Canadian Forest Fire Statistics: 1988–1990* (Ottawa: Forestry Canada, 1992. 76 p. Petawawa National Forestry Institute. Information Report PI-X-107E/F).

The proportion of people-caused fires varies from province to province, ranging from 46% to 82%, and the area burned by these fires ranges from 13% to 83%, but averaging out at about 17%. To effectively suppress these fires while they are small, one must be able to predict their numbers and location one day in advance. This advance prediction is needed in order to plan and direct aerial detection patrols and to position fire-fighting resources in anticipation of the occurrences of these fires. According to Forestry Canada, these fires do not occur at purely random times, or in random locations. Instead, they are started under rather specific conditions and their numbers are predictable, at least in mathematical probability terms. Various prediction methods have been developed and tested over the last 50 years by the Canadian Forest Service, initially using primitive correlation methods that related fuel moisture indexes to average fire occurrences. More recent approaches to the fire forecasting problem have been more sophisticated correlation methods that take into account historical and current patterns with respect to weather, fuel moisture, and geographic location. This approach, which is currently used in the Canadian Forest Service, has been described by P.H. Kourtz in the paper entitled *Decision-Making for Centralized Forest Fire Management* (*Forestry Chronicle*, v. 60, 1984. pp. 320–27). Further refinements to the approach have involved looking at the specific causes of people-caused fires. One approach uses logistic regression to correlate occurrences of specific causes within a small forest region. This approach has been described by D.L. Martell and S. Otukol in their paper entitled *A Logistic Model for Predicting Daily People-Caused Fire Occurrence in Ontario* (*Canadian Journal of Forest Research*, v. 17, 1987. pp. 394-401). The other approach uses expert system technology to consider various people-created risk situations and the likelihood of resulting fires. The overall development of predicting people-caused forest fires has been described by B. Todd and P.H. Kourtz in their publication entitled *Predicting the Daily Occurrence of People-Caused Forest Fires* (Ottawa: Forestry Canada, 1991. 16 p. Petawawa National Forestry Institute Information Report PI-X-103).

Lightning-caused fires represent a major concern to Canadian forest fire control agencies for the following reasons. Although one out of three forest fires is caused by lightning (about 34%), these types of forest fires destroy over 87% of the total area burned. It is also estimated by the Canadian Forest Service that those lightning-caused fires use up three-quarters of the Canadian fire suppression budget of about $109 million.

These high suppression costs and losses associated with lightning-caused fires are related to their remote locations and multiple occurrences pattern, as localized thunderstorm cells drift across remote forest locations igniting a variable number of fires, depending upon fuel type and fuel moisture conditions. The failure in either detection, or initial attack, can lead to the development of large forest fires under optimal burning conditions. Hence the positioning of sufficient detection and initial attack resources in anticipation of expected lightning-caused fires is vital. And, again, as with human-caused fires, a good lightning-caused fire prediction system, coupled with a strong air-attack capability, can potentially eliminate these large lightning-caused forest fires. The prediction system developed by the Canadian Forest Service, over the last 20 years, combines real-time lightning occurrence information gathered from provincial lightning sensor networks with forest fuel type, fire weather, and fire behaviour information. To accomplish this, a large region of up to 100,000 square kilometres is partitioned into rectangular geographic units, or cells, approximately 50 square kilometres in size. These cells constitute the basic structure for fire, weather, and fuel information. Data on lightning flashes are detected by sensors, fuel types are estimated from satellite images, and precipitation rates are derived from the closest weather station. A lightning-caused fire prediction is made for each cell, and then, in turn, these are accumulated to make regional fire predictions. The lightning-caused fire prediction process has four components. The first involves gathering information on the number of cloud-to-ground lightning flashes, their occurrence times, and types, for each cell. The second component predicts the number of lightning-caused firebrands that can be expected. Because relatively few lightning flashes cause forest fires, lightning occurrence information must be combined with fuel type, fuel moisture, fire weather, and fire behaviour information to estimate the number of firebrands that can be expected from lightning activity. The third component predicts both the chance of a firebrand becoming a smouldering fire as well as the chance that this smouldering fire can survive, if necessary, from one day's burning to the next. The fourth component in the process estimates the number of smouldering fires that will become visually detectable during the forecast burning period.

The prediction system developed to date has relied heavily upon published knowledge of lightning physics, as well as information that makes up the Forest Fire Danger Rating System. The current system has been described by P.H. Kourtz and B. Todd in the publication entitled *Predicting the Daily Occurrence Lightning-Caused Forest Fires* (Ottawa: Forestry Canada, 1991. 18 p. Petawawa National Forestry Institute. Information Report PI-X-112).

As explained above, these prediction models are designed to allow the positioning of fire detection resources into the forest areas where fires may occur. Hence, aircraft have played an increasingly prominent role in detecting forest fires in Canada since they were first used in the 1970s. Besides visual scanning by pilots, these aircraft may be equipped with infrared scanners that can plot fire boundaries, even through dense smoke. Within the detection models developed has been the component for the scheduling, and routing, of airborne detection patrols. Early work on this specific aspect of fire prediction was the development of a computer simulation model that was constructed to help determine the worth of using a high-altitude infrared system for supplementary forest fire detection. The model made use of a scheme for forecasting detection demand that predicted, on an hourly basis, the expected number of fires burning in 25-mile-square cells. The study area was located in northwestern Ontario. A programming algorithm was devised that determined the patrol route to be taken, in order to maximize the expected number of detected fires, given the limited aircraft range, the cloud situation, and the forecast of detection demand. Trials of the model, using actual fire and weather data, indicated that the infrared detection system had a useful role to play when added to existing systems. This early work was reported by P. Kourtz in the publication entitled *A Forest Fire Detection Demand Model – For Scheduling and Routing of Airborne Detection Patrols* (Ottawa: Department of the Environment, 1973. 34 p. Canadian Forest Service Publication No. 1322).

Because the study of forest fire behaviour in Canada relies on field observation and documentation, the use of prescribed, or experimental, fires is vital to this research. Prescribed fires are also used as a method of rebalancing vegetation, and for forest renewal. The latter type of fires depends on the knowledge gained from fires used for the study of fire behaviour. An example of the first type of prescribed fire was an experimental burning that was carried out at Porter Lake, which is situated 290 kilometres northeast of Fort Smith, NWT, in July, 1982. The reason given for carrying out this experimental fire was based on the fact that, until then, there had been no serious attempt to investigate fire behaviour over a range of burning conditions and fuel types in the Northwest Territories, except for some small-scale test fires, conducted in a boreal mixed-wood within the vicinity of Fort Smith during the 1961 fire season.

A peninsula at the northeast end of Porter Lake was selected for this prescribed fire on the basis of the control possibilities that the site offered, including the availability of water. As well, the black spruce and lichen woodland fuel type is widespread in the Northwest Territories. The objective of the study 'was to develop a practical quantitative scheme of fire behaviour prediction in the open upland black-spruce-lichen woodland fuel type for use by northern fire managers in the western sub-Arctic region of Canada.' For this study, plots for five line-ignition fires, three point-ignition fires, and two test fires were all located within a 28–hectare study area. The particular plot sites were selected on the peninsula, because the flat terrain would eliminate topography as an influence on fire behaviour, the uniform and homogeneous nature of the black spruce tree cover would ensure all plots were relatively similar in term of fuel consumption, and as previously mentioned, the lowland site and water surrounding this point of land would increase control capabilities. Before the actual fires could be carried out, preliminary studies involving a vegetation inventory, preburn fuel assessments, fire weather observations, and fire danger rating computations, using the Forest Weather Index System components previously described, had to be undertaken. As well, the actual burning procedures, and controls, had to be organized. Then, the different types of experimental fires could be undertaken. What was interesting about this prescribed fire experiment was the unexpected development of a wildfire after the ignition of the last, and largest, plot on the site. After the fire escaped the control plot, possibly due to unexpected flame size, it advanced in a southeasterly direction for 11 days, until it was brought under control by fire suppression crews. This wildfire exhibited the narrow elliptical shape characteristic of a wind-driven crown fire and the wildfire's escape did provide a unique unplanned opportunity to study actual, as opposed to simulated, wildfire behaviour of the black spruce and lichen woodland fuel type. In fact, the spread rate for this wildfire was very similar to that on one of the experimental plots under nearly identical burning conditions. The final analysis of fire behaviour in the black spruce and lichen woodland fuel complex indicated a fire deception. The continuous layer of lichen in this particular forest type, combined with the scattered black spruce tree clumps, produces high-intensity fires, but the lichen by itself doesn't. This major experimental fire was described by M.E. Alexander and B.J. Stocks in their publication entitled *Fire Behaviour in Black Spruce–Lichen Woodland: The Porter Lake Project* (Ottawa: Forestry Canada, 1991. 44 p. Northwest Forestry Centre, Northwest Region Information Report NOR-X-310). An overview of the use of prescribed forest fires in Canada has been written by M.G. Weber and S.W. Taylor in their paper entitled *The Use of Prescribed Burning in the Management of Canada's Forested Lands* (*Forestry Chronicle*, v. 68, 1992. pp. 324–34).

The use of prescribed fires for ecological purposes is a relatively recent concept, because one has to accept the fact that most of the Canadian forest is, in its natural state, dependent on periodical fire for its cyclic renewal and long-term stable existence. The tree species in the Canadian forest include primarily all the pines, spruces, Douglas fir, aspens, and white birch. They have evolved over long biological ages to depend, in varying degree, on fire for their existence in large numbers. The Canadian forest we know may be less than 10,000 years old, but it has clearly developed to its present state under the continuous influence of random periodic fire. The periodicity of fire in the natural forest is in rough equilibrium with the fire climate, and species longevity. The so-called fire cycle may vary, from 50 years for natural jack pine forests to 300 years in natural Douglas fir forests, but whatever the forest composition, each type is dependent on this periodic renewal. Based on these observations, the relatively new discipline of forest ecology has accepted that part of the ecological side of forest management is the process of the fire-cycled system, and using fire as the agent in forest renewal. Overall, there has even been some modelling attempted, based on historic fire episodes, stand composition, and vegetation changes over time, to better understand the natural diversity of forests. This modelling approach has been described by S. Gauthier and A. Leduc in their paper entitled *Forest Dynamics Modelling under Natural Fire Cycles: A Tool to Define Natural Mosaic Diversity for Forest Management* (*Environmental Monitoring and Assessment*, v. 39, 1996. pp. 417–34).

More specifically, in Canada, the natural forest succession has been studied in the Kananaskis Watershed on the eastern slopes of the Rocky Mountains, in southwestern Alberta. As well, prescribed fires have been used in the western national parks as a method of vegetation management. The 130,000-hectare forest of the Kananaskis Watershed shows a variable fire cycle over the last 400 years, along with an indication of a warm, dry climate in the 16th century, and from then on, a cooler, moister climate. The fire cycles have ranged, from 60 years to 130 years in the main range, and from 30 to 95 years in the front range, with the cycles related to the climatic change over those time periods. These fire cycles have been studied by G.I. Fryer and E.A. Johnson in their paper entitled *Reconstructing Fire Behavior and Effects in a Subalpine Forest* (*Journal of Applied Ecology*, v. 25, 1988. pp. 1063–72). This fire pattern indicates that the lower subalpine forests of lodgepole pine and Engelmann spruce require fires within the life-span of the post-fire generation of trees to effectively reproduce. The high-intensity nature of the wildfires in this watershed kills the canopy trees, and the debris provides a preferred seedbed for pine and spruce. These latter observations have been described by E.A. Johnson and G.I. Fryer in their paper entitled *Population Dynamics in Lodgepole Pine–Engelmann Spruce Forests* (*Ecology*, v. 70, 1989. pp. 1335–45).

The use of prescribed fires as a means to simulate controlled wildfires in the national parks has evolved since 1979, when the Canadian Parks Service formally recognized fire as a natural process, and allowed for the development of fire and vegetation management programs that incorporate prescribed fires. For example, in the Elk Island National Park, prescribed fires are used for the purpose of restoring aspen parkland vegetation that existed prior to the park's establishment and the era of traditional fire suppression. As well, Banff National Park has conducted numerous prescribed fires in their montane subalpine forests, and collected and analysed the data, in order to implement the use of fire for resource management purposes. This work in the national parks has been described by G.D. Fenton and A.L. Westhaver in their paper entitled *Prescribed Fires in*

National Parks of Western Canada: Past Efforts and Future Initiatives. In: *The Art and Science of Fire Management.* Edited by M.E. Alexander and G.F. Bisgrove (Ottawa: Forestry Canada, 1990. p. 95. Northern Forestry Centre, Northwest Region. Information Report NOR-X-309).

C Forest Harvesting or Logging Research

Perhaps the best-known and most controversial aspect of forestry in Canada is logging, or – under its more current name – forest harvesting. Historically, logging has been the domain of the lumberjack, but in modern times, it involves power chain saws, particularly in certain west coast areas, as well as multi-armed harvesting machines controlled by a cab-operator that fell and remove limbs, or use sky cranes that can lift logs from stump areas, up or down a slope, to a loading or landing. The latter may also be the job of rubber-tired skidders, which can haul timbers to landing areas. Currently, there are also hydraulic tree shears, which operate like a giant pair of scissors capable of snipping off trees close to the ground, or machines that can convert whole trees into chips. The transport of logs may be by river, truck, rail, or ship, including self-loading and self-dumping transport ships, on the west coast. The equipment used in forest harvesting has been described by F. Humar in the publication entitled *Compendium of Canadian Manufactured Forest Machinery and Tools* (Ottawa: Forestry Canada, 1989. 225 p.).

In Canada, as previously mentioned, there are just over 450 million hectares of forest land, and 245 million hectares of that total amount is considered to be productive forest land. Although the federal government controls only about 11% of that productive land, the harvesting of timber on the federal lands has been documented since the formation of the Dominion Forest Service, in 1940. From 1940 to 1971/72, these were unpublished internal records. After that, they became published on a regular basis as *Timber Harvest on Federal Lands* (Ottawa: Canadian Forestry Service, 1971/72–1987/88. 16 v.). Since 1985/86, data have also been collected and published for the stumpage revenues received from wood harvested on federal lands; and since 1987/88, information regarding silvicultural activities on federal lands has also been included. Starting in 1988, all the timber harvest data have been issued every two years. These publications, by R.P. Hirvonen and R.A. Woods to date, have been entitled *Timber Harvest on Federal Lands, 1987–1988* (Ottawa: Forestry Canada, 1989. 29 p. Information Report DPC-X-27); and *Timber Harvest on Federal Lands, 1988–1990* (Ottawa: Forestry Canada, 1991. 23 p. Information Report DPC-X-36). This harvesting also creates political-environmental concerns, which seem to be at the forefront of forest harvesting in Canada. However, there is a solid amount of research carried out on the effects of harvesting on forest productivity, including nutrient balances, sedimentation effects, and the effect of vegetation on forest productivity of post-harvested lands. As well, there has been some work done on developing a model for documenting the environmental sensitivity of land to forest harvesting.

A major concern about intensive logging, or harvesting, is the effect on the post-productivity of the forest lands affected. The consequences on soil fertility of the loss of organic matter, resulting from whole-tree, or intensive, harvesting, have to be approached from the perspective of how forest stands develop. In temperate and northern forests, the process of forest floor development, as the forest stand ages, has been studied by H.G. Miller in the paper entitled *Forest Fertilization: Some Guiding Concepts* (*Forestry*, v. 54, 1981. pp. 157–67). Miller has stated that the forest floor cumulation of organic litter increases rapidly in young stands, during the growth period between canopy closure (the forest cover of branches and foliage) and maximum foliage accumulation, and then slows down to a steady-state balance, where the rate of litter cumulation balances the rate of forest floor decomposition. This, in turn, is in balance with the rate of organic matter build-up in the soil. Any major disruption of the stand canopy, such as thinning or clear-cutting, disturbs the balance. The changes in the forest floor, and soil organic matter resulting from stand harvesting, are primarily the result of increased solar radiation, and moisture reaching the forest floor after canopy removal. This has the effect of accelerating the decomposition processes. In conventional harvesting, the crown components left on the harvest site act as a mulch protecting the forest floor and soil organic matter from the drastic impact of canopy opening. In contrast, on intensively harvested sites, the crown components are removed, exposing the forest floor to increased water and heat input. The net effect is a higher decomposition rate of the forest floor and soil organic matter. However, the magnitude of this increase also depends on other factors, such as litter quality, including nutrient and lignan content, pH (acid-base balance), and the predominant decomposing material.

Aside from increasing the rate of forest floor breakdown, the high biomass removal, caused by intensive harvesting, also affects the organic matter status of the site by reducing the amount of logging slash added to the forest floor. That, coupled with the increased floor decomposition due to intensive harvesting, will, in the long run, reduce the amount of soil organic matter build-up, since soil organic matter decomposition will also be increased. Because organic matter, in turn, influences water and mineral retention properties for a given area, both of which are important attributes of site productivity, the reduction of soil organic matter may lead to lower forest site productivity. Finally, intensive harvesting can affect soil acidification in forest stands. In closed forest stands, atmospheric acid input is partly neutralized by the stand canopy. Both conventional and intensive harvesting result in the opening of the canopy, allowing more acid to reach the forest floor. However, the base-rich crown components left on the site in conventional harvesting help to buffer the soil from the effects of atmospheric acid input. Intensely harvested sites lack this protection and may be more acidified. As well, the base-rich crown components left on the site during conventional harvesting may add more base cations (components of organic compounds) to soil upon their decomposition. Both the reduced buffering against atmospheric acid inputs and base depletion, as a result of the removal of crown components in intensive harvesting, may lead to lower future site productivity. This potential for acidification of forest stand soils has been described by E.C. Krug and C.F. Frink in their paper entitled *Acid Rain on Acid Soils: A New Perspective* (*Science*, v. 221, 1983. pp. 520–25); and by R.D. Fuller and C.T. Driscoll in their paper entitled *Processes Regulating Sulphate Flux after Whole-Tree Harvesting* (*Nature*, v. 325, 1987. pp. 707–10).

The impact of different harvesting methods on nutrient losses has been evaluated by C.T. Smith, Jr, and M.C. McCormack in their paper entitled *Nutrient and Biomass Removal from a Red Spruce–Balsam Fir Whole Tree Harvest* (*Canadian Journal of Forest Research*, v. 16, 1986. pp. 381–88). They described three basic approaches to the problem. The first involves static comparisons of site nutrient capital, with nutrients removed through different harvesting regimes. The second approach, added on to the first, couples input-output estimates (inputs include precipitation and nutrients and outputs include leaching and erosion losses). The third approach involves data collected from the first two approaches and developed into simulation models which facilitate the examination of different scenarios, including various harvesting levels, to predict possible changes in site productivity. These simulation models have been discussed by J.P. Kimmins in the paper entitled *Future Shock in Forest Yield Forecasting: Implications for Forest Management* (*Forestry Chronicle*, v. 50, 1985. pp. 27–31). Another study on the effect of whole-tree harvesting of natural forest stands on future productivity has been carried out in the Maritime provinces, where intensive harvesting is widespread. This study was done by S.M. Maliondo, and entitled *Possible Effects of Intensive Harvesting on Continuous Productivity of Forest Lands* (Ottawa: Forestry Canada, 1988. 26 p. Information Report M-X-171). The study concluded that the effects of whole-tree harvesting on site productivity will vary, with species and site quality, but that long-term site studies, including nutrient cycling studies, will be needed to qualify such effects. One other Canadian study, which has discussed whole-tree harvesting and nutrient balances, has been written by I.K. Morrison and N.W. Foster and entitled *Carbon Reserves, Carbon Cycling, and Harvesting Effects in Three Mature Forest Types in Canada* (*New Zealand Journal of Forestry Science*, v. 23, 1993. pp. 403–12).

Related to the nutrient balances disturbed by forest harvesting is the 'by-product' problem of soil disturbance and sedimentation of watersheds through erosion.

Soil disturbance can be caused on clearcuts logged by ground skidding methods (removal of felled trees by rubber-tired skidders or crawler tractors), or by high-lead cable-yarding systems (occurs uphill via a high steel spar so that the leading end of the log is kept clear of the ground). There has been an assumption that ground-skidding on steep high elevation sites generated more soil disturbance than cable-yarding. This assumption was studied by R. Krag and K. Higginbottom in the Nelson Forest Region of British Columbia. In their paper, entitled *Logging and Soil Disturbance in Southeast British Columbia (Canada)* (*Canadian Journal of Forest Research*, v. 16, 1986. pp. 1345–54), the authors verified that soil disturbance was significantly greater on ground-skidded than on cable-yarded cutovers, averaging 40%–45% vs 22%–30% respectively, regardless of season. Analysis of soil disturbance by source revealed skidroads as the major cause of disturbance on ground-skidded cut-overs, regardless of season. The primary source of disturbance in cable-yarded areas was yarding, in the summer, and haul roads, in the winter. Ground-skidding also caused more deep to very deep disturbance.

Soil disturbance, by erosion and sedimentation into watersheds, is also a problem caused by clearcut, or intensive, logging. The problem of sedimentation in watersheds was monitored in four small drainage basins, in the Queen Charlotte Islands, over a period of a decade by R.G. Roberts and M. Church and reported in their paper entitled *The Sediment Budget in Severely Disturbed Watersheds, Queen Charlotte Ranges, British Columbia (Canada)* (*Canadian Journal of Forest Research*, v. 16, 1986. pp. 1092–1106). In this study, the data gathered indicated that landslides, riparian (riverbank) erosion, and soil creep were important sediment sources before logging. And sediment production on hillslopes, and delivery to stream channels, may increase following logging, but not invariably. However, these drainage basins showed significant stream bank erosion following logging, and resulted in the accumulation of a substantial wedge of coarse sediment in the stream channel. In fact, sediment transport through the channels increased, by up to 10 times.

Besides nutrient losses and soil disturbances, vegetation growth can also be a problem in logged areas, as it can affect natural regeneration of the forest. In particular, the rhizomatous grass *Calamagrostis canadensis* can seriously inhibit the growth of white pine seedlings in the boreal forests of North America. V.J. Lieffers and S.E. MacDonald, in their paper entitled *Ecology of and Control Strategies for* Calamagrostis canadensis *in Boreal Forest Sites* (*Canadian Journal of Forest Research*, v. 23, 1993. pp. 2070–77), noted, after a review of this grass during four successive stages, that if the grass is found in nearly every square metre in the understorey of the forest before logging, there will be a rapid spread when the stand is clearcut, unless clones of the grass are killed using herbicides, or by deep burning of the exposed area. For effective regeneration, large spruce seedlings planted on large prepared soil mounds, coupled with herbicide use, are necessary for the spruce seedlings to establish themselves under conditions where there is encroachment by the grass.

With the increased concerns about the effects of logging on the environment, studies have been undertaken to attempt to map environmental considerations into logging operations. This approach requires forest managers to think of forestry harvesting as a specific land use problem, and to develop evaluation studies using recognized and accepted ecological land classification methods. These methods make frequent use of air photo interpretation techniques, and use defined levels of mapping between living and non-living features of the areas concerned. Such a study, based upon relative land sensitivity ratings, using the three parameters of relative soil erodability, slope, and groundwater conditions, was carried out in two areas of central and western Newfoundland in the mid-1960s by A.W. Damman and published as a monograph and thesis with the titles *Some Forest Types of Central Newfoundland and Their Relation to Environmental Factors* (Bethesda, Maryland: Society of American Foresters, 1964. 62 p.); and *The Forest Vegetation of Western Newfoundland and Site Degradation Associated with Vegetation Change*. Ph.D. Thesis (Ann Arbor, Michigan: University of Michigan, 1967. 319 p.). To these three parameters, a range of values was assigned so that the relative land sensitivity could be clearly distinguished through air photo interpretation. This work was carried out by A.R. Van Kesteren and published as a report entitled *A Method for Assessing the Environmental Sensitivity of Land to Forest Harvesting in Central and Western Newfoundland* (Ottawa: Canadian Forest Service, 1986. 24 p. Newfoundland Forestry Centre Information Report N-X-250). In the study, the author concluded that the land-sensitivity-mapping methodology developed could be applied using conventional air photo interpretation to determine relative land sensitivity, in terms of readily identifiable parameters. And the approach and methods developed should provide a practical framework for addressing environmental concerns inherent in the planning phases of a logging operation. As well, adherence to the recommended guidelines would promote considerable reductions in the level of on-site disturbance associated with logging operations. In the late 1980s, this methodology was applied by A.R. Van Kesteren in the Corner Brook Lake Watershed of Western Newfoundland and the results were reported in the publication entitled *An Application of Ecosite Mapping to Assess Land Sensitivity to Forest Harvesting in the Corner Brook Lake Watershed, Western Newfoundland* (St John's: Forestry Canada, Newfoundland and Labrador Region, 1992. 55 p. Information Report N-X-280). From this work, the author said that land sensitivity assessment was successful, based on the methodology previously developed, but 'quantitative data on the intensity and extent of timber harvesting impacts cannot be presently derived from this methodology. However, the classificatory approach achieved in this method provides a planning data base as well as a stratification for future modelling and empirical impact prediction.'

D Forest Regeneration Research

Until the second half of the 19th century, there was little thought, or understanding, given to the long-term productivity of the forest. The emphasis had been on the utilization of the forest and the removal of the best trees. However, with the diminished supply of high-quality white pine in eastern Canada, and the awareness that forest shortages could occur, the concept of sustainable yield and reforestation, as a grudging principle in forest management, came into being. Only today is that concept being questioned because of current concepts about integrated forest management, which include biological diversity, global ecological cycles, and societal responsibilities. These concepts have been elaborated upon in the publication entitled *Criteria and Indicators of Sustainable Forest Management in Canada, 1997: Progress to Date* (Ottawa: Canadian Council of Forest Ministers, 1997. 47 p.).

Reforestation can be achieved by natural regeneration, or by artificial restoration. The latter, which involves human interference in the forest cycle, requires high-quality site preparation, planting stock, and planting. The former depends on the scale, and type, of disturbance that initiates the regeneration cycle, the reproductive potential of the trees involved, the non-crop vegetation within the reforestation area, and, as with artificial regeneration, the site preparation of the area to be reforested. Site preparation can be accomplished by prescribed burning, mechanical means, chemical methods, or a combination of all three methods. Each has advantages and disadvantages, as discussed by J.R. Morawski in the provincial publication entitled *Site Preparation Methods and Recommendations for Equipment Usage* (Toronto: Department of Lands and Forests, Timber Branch, Silviculture Section, 1966. 25 p. Silviculture Notes No. 3).

Site preparation by prescribed fire, or slash-burning, is used to remove organic materials from an area designated for regeneration. The positive reasons given for using slash-burning are to reduce the fire hazard, facilitate planting, or provide an environment favourable to seedling establishment and growth, and reduce brush competition to eliminate disease or insect problems. Slash-burning has been discussed by M.C. Feller in the provincial report entitled *The Ecological Effects of Slash-Burning with Particular Reference to British Columbia: A Literature Review* (Victoria: Ministry of Forestry, 1982. [60 p.], BC Land Management Report No. 13). There are negative effects, however, to slash-burning. Slash-burning has a major effect on surface soil erosion, which in turn is affected by climatic characteristics, the inherent soil erodability at the site, the severity of the fire, and the rate of revegetation. As well, slash-burning affects the soil moisture status of the site. This is caused by the loss of vegetation, which may increase moisture going into the soil, or may cause a loss of moisture, because vegetation has moisture-holding capacities. Soil nutrients are also affected, because of chemical changes in the soil caused by the fire. This has been discussed by L.F. Debano and L.D. Mann in their paper entitled *The Transfer of Heat and Hydrophobic Substances during Burning* (*Soil Science Society of America Journal*, v. 40, 1976. pp. 779–82). Finally, certain insect and disease organisms may be encouraged, or retarded, by slash-burning. The latter has been specifically discussed by T.F. Maher in the paper entitled *The Effects of Black Army Cutworms on Backlog Regeneration Efforts in North Thompson Valley* (Ottawa: Canadian Forest Service, and Victoria, BC, Ministry of Forests, 1988. 21 pp. Forest Resource Development Agreement Report M022).

Mechanical and manual site preparation for regeneration can produce favourable planting spots, which can assist seedling trees to take hold, or ease the ability to plant seedlings on the site. What actually happens is that mechanical site preparation by different implements manipulates the soil surface and creates a variety of microsites suitable for planting. In these microsites, soil temperature, soil moisture, soil aeration, soil physical conditions, soil nutrients, competing vegetation, light, frost susceptibility, and insect and disease susceptibility can be modified. The creation of these microsites has been discussed in the workshop entitled *Learning from the Past, Looking to the Future: Proceedings of the Northern Silviculture Committee's 1988 Winter Workshop, February 2–3, 1988, Prince George, B.C.* Edited by B.A. Scrivener and J.A. Mackinnon (Ottawa: Forestry Canada, and Victoria: Ministry of Forests, 1988. 140 p. Forest Resource Development Report 030). However, M.G. Weber and R.S. McAlpine, in their paper entitled *Prescribed Burning and Disk Trenching Effects on Early Plantation Performance in Eastern Ontario, Canada* (*Forest Ecology and Management*, v. 78, 1995. pp. 159–71), noted that both site preparation techniques (prescribed fire or mechanical) improved environmental conditions for seedling tree regeneration.

In mechanical, or manual, site preparation, there are a number and variety of mechanical site preparation implements that can be used. Examples of implements are brush blades and rakes, front mounted V-blades, rear-mounted ploughs and rippers, drag scanifiers, shear blades, downing chains and roller drum choppers, patch scarifiers, maunders, disc trenchers and cone scarifiers, and mixing equipment. Brush blades and rakes are used primarily to rearrange slash material for hazard abatements, and improve planter access. Front mounted V-blades are used to improve planter access, by aligning slash, and removing competing vegetation in appropriate sites. Rear-mounted ploughs and rippers are used to remove vegetation and organic layers, where deep or extensive disturbance is required. Drag scarifiers are used primarily on lodgepole pine sites to create seedbeds and distribute cone material. Shear blades, downing chains, and roller drum choppers are used primarily to flatten vegetation, including dense lodgepole pine and aspen. Patch scarifiers create scalped depressed prepared spots which create a microsite that promotes seedling survival. Maunders create inverted humus and mineral mounds, which provide raised prepared spots. Disc trenchers and cone scarifiers create pairs of continuous trenches, which offer a variety of planting positions. Finally, mixing equipment is designed to incorporate surface organic matter into the mineral soil, to improve nutrient status, to increase soil temperature and to improve soil physical properties. Overall, the type of equipment used in forest regeneration will depend on the site itself, and the type of planting environment required. The range of site preparation machinery and its uses has been discussed at the *Proceedings of the 1984 Mechanized Silviculture Workshop*. Compiled by J.R. Gorman (Ottawa: Canadian Forest Service, 1985. 47 p. Northern Forestry Centre Information Report NOR-X-272).

Effective site preparation with chemicals, using herbicides alone, can be achieved when the weed vegetation is susceptible, when the slash density is low, and when the amount of forest floor litter is light enough to permit seeding, or the brush is sparse enough to permit planting to be done at a reasonable cost. Herbicides are also useful in carrying out site preparation in unstocked or understocked portions of plantations, or natural stands, where the use of fire is impractical, and mechanical equipment would damage crop trees. A disadvantage to using chemical herbicides, alone, is that the treatment leaves a residue of standing dead, or dying, vegetation, which may impede planting or seeding. A further disadvantage also occurs when a minor component of the weed population is resistant to a herbicide treatment and can then develop into a major problem. The use of chemicals in site preparation has been discussed by R.F. Sutton in the publication entitled *Vegetation Management in Canadian Forestry* (Ottawa: Canadian Forest Service, 1985. 34 p. Great Lakes Forest Research Centre. Information Report O-X-369). As well, specific chemicals are assessed for their suitability in site preparation. For example, J.E. Wood and F.W. Von Althen, in their paper entitled *Black Spruce Outplant Performance: Effect of Winter Application of Hexazinone on Shear-Bladed Sites in Boreal Ontario (Canada) (Canadian Journal of Forestry Research*, v. 20, 1990. pp. 1541–48), experimented with the suitability of this herbicide and found that it had little or no effect on the survival of black spruce seedlings. And D.G. Maynard, in the paper entitled *Soil Nutrient Dynamics in a Boreal Mixedwood Cutover Following the Application of Hexazinone* (*Ecological Applications*, v. 7, 1997. pp. 416–30), said that a seven-year experiment with hexazinone in site preparation indicated 'that there was little change to the total nutrient pools and that the potential loss of nutrients at this site was minimal.'

In addition to appropriate site selection, successful reforestation by planting depends upon planting stock that contains genes that allow for adaptation to the environment in which the trees are to grow (climate, soils, pests, etc.). The first step in developing appropriate planting stock is the use of provenance tests. These are experiments in which seeds are collected from a number of natural stands distributed over a species range, or part of the range, and raised in one or more common environments. The objective of the tests is to determine the best available source of seed to use in reforestation programs as well as to elucidate the overall patterns of genetic variation of a species in order to better understand the species' evolutionary development, and relate this to environmental factors. Also, the tests allow for a genetically variable collection of seed material to assist in future selection and breeding efforts. The development of provenance testing in Canada has been described by W.J. Libby in the paper entitled *Domestication Strategies for Forest Trees* (*Canadian Journal of Forest Research*, v. 3, 1973. pp. 265–76).

When provenance information is available, natural stands of trees can be designated as seed production areas, because seed still represents nearly the only form in which genetic improvement can be carried out. In Canada, natural stands remain the common source of tree seed. However, the designation of seed production areas is not uniform across Canada. Collection of seed is usually done on standing trees, by picking cones by hand, or cones can be picked from felled trees. Seed production and collection areas are usually developed provincially, with some advice from the Canadian Forest Service. The state of seed production in Canada has been described by E.K. Morgenstern in the paper entitled *Tree Seed Improvement in Canada – The National Scene*. In: *Tree Seed Production and Tree Improvement in Canada – Research and Development Needs, 1977–1987.* Edited by E.K. Morgenstern and L.W. Carlson (Ottawa: Canadian Forest Service, 1979. pp. 3–10. Petawawa Forest Experimental Station Information Report PS-X-74).

More recently, biotechnology has been incorporated into tree improvement in order to help reduce the time constraints, which are part of conservation, selection, breeding, and propagation in traditional tree improvement programs. The biotechnological approaches that can assist in tree improvement are the use of tissue cultures, molecular genetics, and genetic engineering. These new techniques in forestry research have been discussed by D.R. DeYoe in the publication entitled *The Role of Biotechnology in Regeneration of Canadian Forests* (Ottawa: Biotechnology Unit, Ministry of State for Science and Technology, 1988. 94 p.), and by W.M. Cheliak and D.L. Rogers in their paper entitled *Integrating Biotechnology into Tree Improvement Programs* (*Canadian Journal of Forestry Research*, v. 20, 1990. pp. 452–63).

The establishment of seed orchards is the principal way in which genetically improved seed can be produced. Seed orchards are collections of individual trees selected for specific tracts. The trees are brought together in one location by the grafting, or rooting, of many shoots from the selected parent to produce a clone. Trees in seed orchards are arranged randomly to produce a thorough mixing of genes in the orchard seed. Between 30 and 100 clones are included, and progeny tests are used to evaluate the genetic worth of each parental clone. Clones with poor genetic value are removed from orchards in order to increase the genetic quality. Clonal propagation offers several advantages over seed, the major one being that the parental genotype is duplicated in the clonal progeny. As well, it allows improved genotypes to be used more quickly, and it also eliminates inbreeding in planting stock. Unfortunately, routine use of clonal propagation is not yet possible for more than a few species such as yellow-cedar in British Columbia. Seed orchards, however, do have some disadvantages. Sometimes useful gene combinations are lost as genes recombine, and interact, in the sexual process of pollination. As a consequence, the progeny, or successor, trees may be somewhat genetically different from the parents. Nevertheless, until seed orchards, which have a relatively recent history, reach their full production potential, too little improved seed will be available to cover all areas needing reforestation. The development of tree orchards from clones, or seed, has been discussed by J.I. Klein in the paper entitled *Selection Programs and Genetic Parameters*. In: *Tree Seed Production and Tree Improvement Needs, 1977–1987*. Edited by E.K. Morgenstern and L.W. Carlson (Ottawa: Canadian Forest Service, 1979. pp. 73–82. Petawawa Forest Experimental Station Information Report PS-X-74).

The seed that is used in reforestation programs is also used by forest nurseries for the production of bareroot seedlings, or container seedlings. For bareroot stock, a two-year production cycle is necessary, but for containerized stocks (a development since the 1970s), a one-year cycle is possible. With bareroot stock, seed is sown in early spring, cultured for two growing seasons, and harvested in the fall of the second year. Bareroot seedling production has been described by T.G. Daniels and D.G. Simpson in their paper entitled *Seedling Production and Processing: Bareroot*. In: *Regenerating British Columbia's Forests*. Edited by D.P. Lavender and R. Parish (Vancouver: University of British Columbia Press, 1990. pp. 206–25). Containerized seedling production has been described by E. Van Earden and J.W. Gates in their paper entitled *Seedling Production and Processing: Container*. In: *Regenerating British Columbia's Forests*. Edited by D.P. Lavender and R. Parish (Vancouver: University of British Columbia Press, 1990, pp. 226–34).

Nearly all artificial regeneration and reforestation programs involve the planting of seedlings, rather than direct seeding. Manual planting is still the primary planting method in Canada, using planting spades, which can be used for bareroot planting or container stock. Mechanized or machine planting, however, has been found to be effective on abandoned farmland, particularly in Ontario, where the area to be planted is fairly large and regular in shape, not overly steep or rocky, and reasonably clear of stumps and heavy brush. In British Columbia, however, where the terrain is more variable, mechanical planters have had limited use. Manual and mechanical planting have been discussed in the publication by J.H. Cayford and A. Bickerstaff entitled *Man Made Forests in Canada* (Ottawa: Department of Fisheries and Forestry, 1968. 68 p. Publication No. 1240).

Direct seeding is a method of artificial regeneration in which seeds, rather than seedlings, are planted on a prepared site. Direct seeding of jack pine and black spruce on large-scale sites (up to 30,000 hectares) has been carried out using the so-called broadcast seed method via fixed-wing aircraft. This method has been described by L.F. Riley in the publication entitled *The Effect of Seeding Rate and Seedbed Availability on Jack Pine Stocking and Density in Northeastern Ontario* (Ottawa: Canadian Forest Service, 1980. 36 p. Great Lakes Forestry Centre. Information Report O-X-318). In Alberta, helicopters have been used for broadcast seeding. This method has been described by C.R. Smith in the paper entitled *Status of Mechanization of Silviculture in Canada.* In: *Mechanization of Silviculture: Increasing Quality and Productivity – Proceedings of CPPA/CFS Seminar on Mechanization of Silviculture, Thunder Bay, Ontario* (Ottawa: Canadian Forest Service, 1984. pp. 41–49). Other methods of direct seeding are furrow and spot seeding, using mechanical equipment, and shelter seed cones (plastic protective cones). The former method has been described by R.L. Fleming and D.S. Mossa in their publication entitled *Scarification Trials for Direct Seeding on Upland Black Spruce Sites in Northwest Ontario* (Ottawa: Canadian Forest Service, 1987. 47 p. Canadian Forestry Research Centre, Information Report O-X-385). The latter method has been described by S.W. Dominy and J.E. Worst in the paper entitled *Shelter Spot Seed Trials with Jack Pine, Black Spruce, and White Spruce in Northern Ontario* (*Forestry Chronicle*, v. 62, 1986. pp. 446–50).

E Pollution Research

Forest pollution studies by the Canadian Forest Service involve national and regional forest monitoring systems, the documentation of decline in specific tree species, and research aimed at understanding the process of damage by acid rain to forest trees. Other studies involve the impact of carbon dioxide, fluorides, and heavy metals on forest health.

A national monitoring system was first established, in 1984, by the Forest Insect and Disease Survey, and called the Acid Rain National Early Warning System (ARNEWS). It was designed to detect early signs of damage to Canadian forests by acid rain, which encompasses all forms of air pollution – wet and dry deposition of sulphates, nitrates, ozone, and airborne particles – as well as to undertake long-term monitoring of vegetation and soils in order to detect future changes that could be attributed to acid deposition. In order to carry out the objectives of ARNEWS, permanent plots were designated in all regions of Canada to monitor forest conditions and changes. The methodology for the establishment and monitoring of the plots for the network was first described by L.P. Magasi in the publication entitled *Acid Rain National Early Warning System: Manual on Plot Establishment and Monitoring* (Ottawa: Canadian Forestry Service, 1985. 51 p.). The initial results from ARNEWS were first reported in *Forest Insect and Disease Conditions in Canada, 1984* (Ottawa: Forest Insect and Disease Survey, Canadian Forestry Service, 1984. pp. 23–25) and have since been reported yearly in summary in that annual publication. As well, the complete information is reported annually in the *ARNEWS Annual Report* (Ottawa: Forestry Canada, Science and Sustainable Development Directorate, 1991–). In 1993, ARNEWS was expanded to 150 plots and the methods for assessing crown condition in conifers and hardwoods were revised. A revised manual was rewritten by S.P. D'Eon and L.P. Magasi and entitled *Canada's National Forest Health Monitoring Plot Networks: Manual on Plot Establishment and Monitoring (Revised)* (Chalk River, Ontario: Petawawa National Forestry Institute, 1994. 32 p. Information Report PI-X-117). The monitoring data includes tree data such as light and density, regeneration and sapling survey information, a vegetation survey, softwood shoot data, a pest condition survey, soil description and sampling, foliage sampling and analysis, and growth ring analyses.

The regional monitoring system which has been developed is called the North American Maple Project (NAMP). It was established in 1988, in cooperation with the United States, to monitor the condition of sugar maple (*Acer saccharum* Marsh) across northeastern North America. The network was first described by I. Millers in the publication entitled *North American Sugar Maple Decline Project, Cooperative Work Plan* (Ottawa: Canadian Forest Service, and Washington, DC: US Department of Agriculture, Forestry Service, 1988. 8 p.)

The background to this project was the fact that maple dieback in sugar maple stands had been recorded throughout southern Quebec since 1982, first in study plots, and then by aerial surveys. One conclusion has been that dieback is the major pathological problem in hardwood forests in Quebec. From aerial surveys, flown in 1987, and covering just under 1 million hectares, 49 per cent of the hardwood forest was healthy, or little affected, and 1% was moderately to heavily affected. Comparisons made between 1983 and 1986, on study plots in other areas, showed an increase in both the extent and degree of dieback. Other plots were established throughout Quebec, during 1984 and 1985, as part of the Acid

Rain National Early Warning System. These findings were reported by C. Bordeleau and D. Guerin in their paper entitled *Dieback in Maple Stands, in Quebec 1987.* In: *Forest Insects and Diseases of Trees* (*Forest Conservation*, v. 54, 1987. Supplement 10).

NAMP, which consists of 62 plots in both unmanaged and managed stands for maple syrup production, uses the same monitoring criteria as the ARNEWS network. The reporting of NAMP data began, in 1992, with publication of the research paper by D.C. Allen and C.J. Barnett entitled *Temporal Change (1988–1990) in Sugar Maple Health, and Factors Associated with Crown Condition* (*Canadian Journal of Forest Research*, v. 22, 1992. pp. 1776–84). This information has been added to by D. Lachance and A. Hopkins in their publication entitled *Health of Sugar Maple in Canada* (Ottawa: Canadian Forest Service, 1995. 27 p.). As with the ARNEWS network, summary NAMP information also appears in the annual publication *Forest Insect and Disease Conditions in Canada.* The results, to date, from the ARNEWS and NAMP monitoring systems are positive. W.W. Bowers and A. Hopkins, in their publication entitled *ARNEWS and North American Maple Project (NAMP), 1995* (Ottawa: Canadian Forest Service, 1997. 35 p.), said that 'Canadian forests remain generally healthy. The assessment of ARNEWS plots in 1995 indicated there is no large-scale decline in the health of Canadian forests that can be directly attributed to atmospheric pollution.'

The documentation of decline in specific tree species is reported in the annual *Forest Insect and Disease Conditions in Canada* under the heading 'Declines.' For example, birch dieback in New Brunswick, which causes early leaf browning and premature leaf drop, is seen in birch along the New Brunswick and sometimes Nova Scotia coastal regions, adjacent to the Bay of Fundy. The problem was first observed in 1979, and varies in intensity from year to year. It does not appear to be caused by conventional stresses, and the causes are being studied at the Maritimes Forestry Centre in Fredericton, New Brunswick. The birch dieback was first described in *Forest Insect and Disease Conditions in Canada, 1984* (Ottawa: Canadian Forest Service, 1984. 76 p.). Other species indicating decline, and reported in *Forest Insect and Disease Conditions in Canada, 1994* (Ottawa: Canadian Forest Service, 1996. pp. 22–23), are the eastern white pine (the distal portion of the current needles turned brown and died on single branches), ash (loss of foliage on healthy trees), aspen (fungus-caused injury to the root-collar area), and, again, birch.

The process of air pollution damage to forest trees involves specific acid rain research, as well as other studies such as the impact of carbon dioxide, fluorides, and heavy metals on forest stands.

Acid rain research has ranged from precipitation studies through to germination and soil studies, and acid rain's effects on leaves. Interestingly, some of the early work on precipitation in forests was carried out by J. Baker and D. Hocking in Alberta. In their paper, entitled *Acidity of Open and Intercepted Precipitation in Forests and Effects on Forest Soils in Alberta, Canada* (*Water, Air and Soil Pollution*, v. 7, 1977. pp. 449–60), the authors noted that emissions of sulphur dioxide appeared to have an acidifying effect on open rainfall (grossfall), throughfall, stemflow, and soil pollution, near major sources of SO_2 emissions. In the Turkey Lakes Watershed, near Sault Ste Marie, an area of long-term study of the effect of air pollutants (this study has been discussed in Chapter 9 covering water research; see p. 560), N.W. Foster, in the paper entitled *Neutralization of Acid Precipitation within a Maple-Birch Ecosystem* (*Tappi Journal*, v. 68, 1985. pp. 104–6), also studied the influence of acid precipitation on the chemical composition of rain throughfall, and the mineral composition of soil, over a 12-month period. The trees involved in the watershed were a forest of sugar maple and yellow birch. The author found that the acidity of the precipitation was reduced 50% by initial contact with the forest canopy, and acids entering the mineral soil were largely neutralized within the base-rich surface soils and did not pass through the acid soil into ground or drainage water. However, base minerals and strong acid anions, from mineral compounds in the soil, did leach into drainage water. The effects of these nutrient losses, nevertheless, was unknown. Additional studies on acid precipitation by M.K. Mahendrappa entitled *Tree Species and Urea Treatment Effects on Sulfur and Metals in Throughfall and Stem Flow of Some Eastern Canadian Forest Stands* (*Canadian Journal of Forest Research*, v. 17, 1987. pp. 1035–42) indicated that the quantities of sulphur and metals cycled in forest ecosystems are functions of tree species, and of the nutritional status of the tree species.

Besides acidic rain precipitation, snowfall, and snowmelt in snowpacks, have also been found to be a source of acidic precipitation. M.C. English and D.S. Jeffries studied acidic snowmelt in their paper entitled *A Preliminary Assessment of the Chemical and Hydrological Interaction of Acidic Snowmelt Water with the Terrestrial Portion of a Canadian Shield Catchment* (*Water, Air and Soil Pollution*, v. 30, 1986. pp. 27–34). In this paper, the authors examined certain aspects of the hydrological and chemical interaction of springmelt runoff with the terrestrial portion of a Canadian Shield catchment. What was interesting to note here was the fact that acidic sulphide breakdowns increased in runoff, at the end of the winter season, and became the dominant chemical formation in the snowpack meltwater runoff. The effect of runoff on forest soils themselves has been studied by J.A. Hern and G.K. Rutherford in their paper entitled *Chemical and Paedogenetic Effects of Simulated Acid Precipitation on Two Eastern Canada Forest Soils: Non Metals* (*Canadian Journal of Forest Research*, v. 15, 1985. pp. 839–47). In this study, soils were subjected to treatments of acidic ranges of water containing nitric and sulphuric acids in a 1:2 ratio in laboratory columns over a two-year period.

The results of this study showed that there was considerable base buffering by the soils, with the soil pH unchanged even under the most acidic treatment.

Simulated acid rain has also been used in order to try to understand the effect of acid rain on the very beginning of the plant cycle – on both pollination and generation. R.M. Cox, in the paper entitled *Sensitivity of Forest Plant Reproduction:* In: Vitro *and* in Vivo *Sensitivity of* Oenothera parviflora *L. Pollen to Simulated Acid Rain* (*New Phytologist*, v. 97, 1984. pp. 63–70), found that the response of pollen *in vitro* indicated significant inhibiting effects, both to germination and tube growth. Cox also indicated, in the paper entitled *The Sensitivity of Pollen from Various Coniferous and Broad-Leaved Trees to Combinations of Acidity and Trace Metals* (*New Phytologist*, v. 109, 1988. pp. 193–202), that pollen sensitivity to pH *in vitro* was related to the species' position in the forest canopy, and was different for conifers and broad-leaved trees.

The effect of carbon dioxide levels on forests has not been as well researched as the effects of acidic precipitation, and the effects of increased carbon dioxide budgets have been more surmised than documented. This was emphasized by D.F. Pollard in the paper entitled *A Forestry Perspective on the Carbon Dioxide Issue* (*Forestry Chronicle*, v. 61, 1985. pp. 312–18). In this paper, the author stated that the global atmosphere was being enriched with carbon dioxide through the combustion of fossil fuels, the reduction of the forest biomass, and the reduction of soil organic matter. The estimated pre-industrial concentration of 260 parts per million, by volume, is expected to be doubled by the year 2065, with consequential disturbances in the global and regional climates. This enrichment will certainly have direct impact on the forest sector, probably favouring fast-growing species, in particular certain hardwoods and weed species. Impacts of climatic disturbance are much more difficult to predict, largely because of uncertainty in current climate response theory. The best available information indicates the development of a serious mismatch between Canadian forests and the climatic regions they will occupy. And past climatic variations suggest that forest pest problems will intensify.

Unlike acidic precipitation or carbon dioxide studies, the study of fluoride emissions has concentrated on uptake by lichen in the forest environment. These studies have focused upon fluoride emissions from a phosphorus plant in Newfoundland. The first study was reported, in 1979, by B.A. Roberts and L.K. Thompson in their paper entitled *Terrestrial Bryophytes as Indicators of Fluoride Emission from a Phosphorus Plant, Long Harbour, Newfoundland, Canada* (*Canadian Journal of Botany*, v. 57, 1979. pp. 1583–90). The authors reported that certain species of bryophytes in the vicinity of an industrial plant producing elemental phosphorus had damage symptoms, ranging from mild phyllid chlorosis (yellowing) to severe browning of up to 95% in the sample sites used in the survey. Further observations, by the same authors, were reported a year later in another paper entitled *Lichens as Indicators of Fluoride Emission from a Phosphorus Plant, Long Harbour, Newfoundland, Canada* (*Canadian Journal of Botany*, v. 58, 1980. pp. 2218–28). In this paper, the authors noted that fluoride concentrations in the terrestrial lichens *Cladina rangiferina* (L.) Harm. and *Cladina stellaris* (Opiz.) Brodo were correlated inversely with the distance from the industrial plant producing phosphorus, in the direction of the prevailing wind. The lichen tissue had, on average, twice as much fluoride as occurred in soil humus from the same site. Six years later, S.S. Sidhu and R.J. Staniforth discussed the effects of fluorides on forest trees in their paper entitled *Effects of Atmospheric Fluorides on Foliage and Cone and Seed Production in Balsam Fir, Black Spruce and Larch* (*Canadian Journal of Botany*, v. 64, 1986. pp. 923–31). The authors noted that various degrees of foliar damage (chlorosis, necrosis, needle damage, and depletion) occurred where fluoride accumulations in the foliage exceeded 20 ppm (parts per million).

Probably, the most wide-ranging studies of air pollutants involve the heavy metals such as zinc, copper, lead, nickel, and aluminum. These studies have noted the pollution effects on hardwoods, conifers, and conifer seedlings. F.C. Lozano and J.K. Morrison studied hardwoods in their paper entitled *Disruption of Hardwood Nutrition by Sulphur Dioxide, Nickel and Copper Air Pollution near Sudbury, Canada* (*Journal of Environmental Quality*, v. 10, 1981. pp. 198–204). The authors sampled and analysed various chemical constituents from soil beneath the trees of four hardwood species – white birch, red oak, red maple, and trembling aspen – that were severely or moderately damaged by fumes and dustfall on sites near Sudbury. Results indicated that on sites damaged by fumes and dustfall, soils were impoverished with respect to organic matter content, cadmium, magnesium, and trace elements (manganese and zinc). As well, levels of sulphur, iron, copper, and nickel were also increased. The analysis of tree foliage indicated that the soil contained near-toxic to toxic levels of nickel and copper. In another paper, this one by G.D. Hogan and D.L. Wotton, entitled *Pollutant Distribution and Effects in Forests Adjacent to Smelters* (*Journal of Environmental Quality*, v. 13, 1984. pp. 377–81), pollutant levels were examined in a boreal forest system that had been exposed to smelter effluents over a 50–year period. High levels of zinc, copper, and lead were found in the surface soils close to the source, and significant deposition of metals was probably occurring at sites up to 35 kilometres from the source. However, the data didn't indicate that metal levels in the soil had affected the foliar nutrient levels of the major conifer tree species examined.

Studies of aluminium toxicity have been undertaken with seedlings. T.C. Hutchinson and L. Boyer, in their paper entitled *Responses of Fives Species of Conifer Seedlings to Aluminium Stress* (*Water, Air and Soil Pollution*, v. 31, 1986.

pp. 283–94), studied the range of aluminium concentration in seedlings of red, white, and black spruce, and white and jack pine, grown for 12 to 14 weeks in sand culture in the greenhouse environment. Aluminium was added to the regular plant nutrients in varying doses and the results indicated that red, white, and black spruce seedlings were inhibited in growth within a range of increasing concentrations, whereas white and jack pine showed stimulated growth.

F Forest Management Research

The framework for forest management is the concept of site classification, which can provide basic estimates of forest productivity for use by forest managers. Site or ecosystem classifications take the components of climate, physiography (land forms), soil, and vegetation (in combination), on a regional basis, in order to develop a classification scheme that can assist in making forest management decisions.

Of all the components, climate is the principal ecosystem driving force, as the forest structure, composition, productivity, and the genetic differentiation of plant populations are controlled primarily by photoperiodicity, temperature, and precipitation. However, for reasons noted below, climate is not used as a primary component in site classification. At the local level, the above- and below-ground climate strongly control plant productivity and the rate and nature of the biological succession in the forest. There can even be local genetic differentiation, where sharp climatic differences occur over short distances. The latter has been pointed out by the American researcher R.K. Campbell, in the paper entitled *Genecology of Douglas Fir in a Watershed in the Oregon Cascades* (*Ecology*, v. 60, 1979. pp. 1036–50). Because physiography may influence the climatic regime significantly at both the regional and local levels (e.g. mountainous areas), these two components are treated together. However, because of the relatively limited availability of reliable, long-term regional weather data, and the absence of detailed macroclimatic data, climate is rarely the primary component in forest land classification. Yet, there are some Canadian studies of climatic classification that have been of direct value in ecosystem classification. A.W. Damman discussed plant distribution, in relation to summer temperatures, in the paper entitled *Plant Distribution in Newfoundland Especially in Relation to Summer Temperatures Measured with the Sucrose Inversion Method* (*Canadian Journal of Botany*, v. 54, 1976. pp. 1561–85). In British Columbia, R.M. Newham undertook climate classification in the paper entitled *A Classification of Climate by Principal Component Analysis and Its Relationship to Tree Species Distribution* (*Forest Science*, v. 14, 1968. pp. 254–64). And in New Brunswick, H. Van Groenewoud used climate data in the paper entitled *The Climatic Regions of New Brunswick: A Multivariate Analysis of Meteorological Data* (*Canadian Journal of Forest Research*, v. 14, 1984. pp. 389–94).

Physiography (geomorphology and landform terrain) pertains to the surface form of the land. G.A. Hills, in his pioneering work on the concept of forest ecosystem regions in Ontario entitled *Regional Site Research* (*Forestry Chronicle*, v. 36, 1960. pp. 401–23), chose physiography as the stable framework of the classification within which vegetation and surface soil characteristics could be used. Another pioneer in classification, J.S. Rowe, also emphasized physiography as the most stable ecosystem component in site classification, as it strongly controls regional and local climate, soil moisture, and the related nutrient status. He discussed this in his paper entitled *Soil, Site, and Land Classification* (*Forestry Chronicle*, v. 38, 1962. pp. 420–32). Because of these inputs, physiography also determines the patterns of soil and vegetation. Rowe's early work culminated in the major publication entitled *Forest Regions of Canada* (Ottawa: Department of the Environment, 1972. 172 p. map. Canadian Forest Service Publication No. 1300).

The next major feature of site classification is soil itself, because within an area of homogeneous climate, soil factors have a strong influence upon the establishment, composition, structure, growth, and productivity of vegetation. More directly, soil factors also affect road location, road construction, and the accessibility of a forest area to management operations. In the United States, soil has become the most standard and widely applied taxonomic structure of all ecosystem components. And some classification approaches use soil as the predominant component in site classification. In other cases, where vegetation is the predominant component, correlations may be made with soil survey information. However, B.V. Barnes, in the paper entitled *Varieties of Experience in Classifying and Mapping Forest Land Ecosystems in Site Classification in Relation to Forest Management: Proceedings of a Symposium Sponsored by the Ontario Ministry of Natural Resources and the Great Lake Forestry Centre under the Auspices of the Canada-Ontario Joint Forestry Research Committee, Sault Ste. Marie, Ontario, August 27–29, 1985.* Edited by G.M. Wickware and W.C. Stevens (Ottawa: Canadian Forest Service, 1986. pp. 5–23. COJFRC Symposium Proceedings O-P-14), stated that the interpretation of soil productivity for tree growth may not be reliable, and soil survey reports may not adequately communicate landform differences and their effect on soil distribution and use.

Vegetation is used in most multiple-component methods for site classification, as well as single-component methods. The reason for this is the fact that vegetation is a measurement of the integrated effect of climate, soil, and land forms. And there is a large amount of vegetation data available for computer processing at both regional and local levels of classification, where soil and climatic factors can be integrated. This has been discussed by A.W. Damman in the paper entitled *The Role of Vegetation Analysis in Land Classification* (*Forestry Climate*, v. 55, 1979. pp. 175–82). Although vegetation is recognized as a component of major importance in site classification, vegetation does present problems when used as a classification factor in disturbed ecosystems, because vegetation is complex, highly sensitive, and varies in occurrence. Vegetation, therefore, can be used much more effectively in relatively undisturbed old growth forests, where it is easier to identify vegetation units. These problems have been discussed by R.F. Daubenmire in the paper entitled *The Use of Vegetation in Assessing the Productivity of Forest Lands* (*Botanical Review*, v. 42, 1976. pp. 115–43).

More recently, the features and refinements of site classification have been applied in Ontario and Newfoundland. In Ontario, R.A. Sims and B.G. Mackey, in their paper entitled *Stand and Landscape Level Applications of a Forest Ecosystem Classification for Northwestern Ontario, Canada* (*Annales des Sciences Forestières* [Paris], v. 52, 1995. pp. 573–88), added landform topology, wildlife preferences, and moisture to the classification system. The authors said that in the future, 'classification systems ... will be used for advanced simulation modelling problems at various spatial scales.' In Newfoundland, B.A. Roberts and E.F. Woodrow, in their paper entitled *A Cooperative, Integrated Project to Classify Forest Sites in Newfoundland (Environmental Monitoring and Assessment*, v. 39, 1996. pp. 353–64), said that site classification is more relevant when a cooperative approach among groups with differing expertise is used. The resulting mix of expertise allowed a 'focus of resources that would not be possible within one agency.'

The outcome of site classification is an estimation of forest site quality, or the ability of the forest land to grow trees. Considerable research on forest site quality has been carried out in the North Central Region of the Ontario Forest Region, at Lakehead University in Thunder Bay, and described by W.H. Carmean in the paper entitled *Forest Site Quality Research in the North Central Region*. In: *Site Classification in Relation to Site Management* (Ottawa: Canadian Forestry Service, 1986. pp. 62–67). Here, the direct measurement of forest site quality has been a site index based upon height growth in a forest stand. Height and age measurements from free-growing uninjured trees in fully stocked, even-aged stands are used with height-age curves to estimate total tree height at a specified index age. Fifty years is the normal index age. An alternative method for determining site quality, which is useful in forest stands, or plantations of trees younger than 50 years, is the growth intercept method, which is a period of early height growth, rather than the long-term 50-year height growth. This method was developed for conifers that have distinct internodes marking annual height growth. Additional work on forest site quality has been done, again by W.H. Carmean, and described in the paper entitled *Forest Site-Quality Estimation Using Forest Ecosystem Classification in Northwestern Ontario* (*Environmental Monitoring and Assessment*, v. 39, 1996. pp. 493–508). Here, soil types used in site classification were re-examined as to their effect on tree rooting systems and site quality.

In the application of site classification and site quality indexes to forest lands, specific evaluation criteria are taken into account. These evaluations may address about 12 concerns. They are the season of the harvest, site preparation intensity, the soil compaction hazard, the soil puddling hazard, the soil erosion hazard, the reforestation method, species selection, the seedling frost heave hazard, potential productivity (using the site index), vegetation competition, windthrow hazard, and snowshoe hare damage hazard. The evaluation criteria have been described in detail by I.G. Gorns and R.M. Annas in their paper entitled *Ecological Classification of Alberta Forests and Its Application in Forest Management.* In: *Site Classification in Relation to Forest Management* (Ottawa: Canadian Forest Service, 1986. pp. 53–61).

Foresters can carry out management evaluations, as described above, through knowledge and subject analysis of sites. However, with the advent of computerization, forest management has also been subjected to computerized models that may also be of assistance to forest managers. In 1985, a Forest Management Systems Program was established at the Petawawa National Forestry Institute in order to develop a comprehensive decision-support system for forest managers. It would take into account the recent advances in information technology, such as database management systems, geographic information systems, and artificial intelligence programs, in order to provide the practising forester with an easy-to-use tool for testing difficult management strategies. In the general area of forest management models, there are two types of models – optimization models and simulation models. Optimization models use mathematical methods, such as linear, integer, and dynamic programming, for projecting maximized or minimized forest conditions, or restraints, as a solution to a problem. Simulation models, on the other hand, are not designed to provide an optional solution to a planning problem, but are designed to answer 'what if' questions. Ideally, the components of a forest management program would consist of modules that take into account stand development, forest fire prediction, pest management, forest management itself, silviculture (the theory and practice of controlling the establishment composition, growth, and quality of forest stands), harvesting schedules, road networks and locations, and wildlife management.

In practice the forest management model called FIRFOR, developed at the Petawawa National Forestry Institute, so far has some shortcomings. The model used one forest type (black spruce), and three rigidly assigned site classes. According to its developers, 'in theory, the model could be readily modified to permit many forest types by adding another "dimension" to the basic forest. Site index, instead of site class, could be used in the growth function for each forest type.' And although the model allowed for a number of forest management options, these all related to the method of determining the annual harvest only. It did not take into account future regeneration projections. The FIRFOR model has been described by R.M. Newham in the publication entitled *FIRFOR – A Simple Forest Management Model* (Ottawa: Canadian Forest Service, 1987. 94 p. Petawawa National Forestry Institute. Information Report PI-X-72). Other work on forest management databases has suggested that models developed in Ontario could be developed into national databases. The concept of a national database has been discussed by D.W. McKenney and B.G. Mackey in their paper entitled *Primary Databases for Forest Ecosystem Management – Examples from Ontario and Possibilities for Canada: NatGRID* (*Environmental Monitoring and Assessment*, v. 39, 1996. pp. 399–415).

Among forest managers, more recent discussions have focused upon the practicability of forest management in so-called non-classic forest stands (i.e., mixed-wood forests of coniferous and deciduous trees) of the prairie provinces, including parts of British Columbia. In particular, these discussions have been about the management of aspen in the mixed forest, and the management of second-growth forest stands. Until recently, mixed-wood forest stands have been, by and large, ignored. However, changes in the market, as well as technology changes, have made these coniferous and deciduous trees more attractive, especially the aspen, which was once considered a 'weed' tree. The tendency has been to eradicate the aspen tree, and replace it with ever-marketable conifers after clearcut harvesting, rather than using more costly but selective harvesting techniques, which may be more beneficial for environmental reasons, including watershed management. With research on the uses of woods by groups such as FORINTEK, which is Canada's wood products research institute, funded by the federal government (50%), some provincial governments (25%), and industry (25%), the use of hardwood as pulp, as well as in other wood products for the market, has shown that these mixed-wood forests need to be exploited more effectively through better management from the harvest, right through to end use. A major discussion of the mixed-wood forests was covered in a series of symposium papers, edited by J.K. Samoil, and entitled *Management and Utilization of Northern Mixedwoods: Proceedings of a Symposium Held April 11–14, 1988, in Edmonton, Alberta* (Ottawa: Canadian Forest Service, 1988. 163 p. Northern Forestry Service Information Report NOR-X-296).

As mentioned above, a major component of these mixed-wood forests is the trembling aspen, a major deciduous tree of the boreal zone, which cuts across the central and northern portions of the prairie provinces. Because of the renewed interest in the hardwood component of these boreal forests, there has been more interest in the trembling aspen over the last 15 years. The most recent example was a major symposium on the aspen, held in Edmonton, Alberta. The symposium papers, edited by S. Navratil and P.B. Chapman, were entitled *Aspen Management for the 21st Century: Proceedings of a Symposium Held November 20–21, 1990, in Edmonton, Alberta in Conjunction with the 12th Annual Meeting of the Poplar Council of Canada* (Edmonton: Forestry Canada, Northern Forestry Centre, 1991. 172 pp.).

Unlike the trembling aspen, which is just finding a market, the Douglas fir, on the west coast, has been heavily exploited, and in 50 to 70 years it will be second-growth trees that will be harvested. This so-called juvenile wood will have physical, mechanical, and chemical properties substantially different from the old-growth resource harvested today. For example, the longitudinal shrinkage of juvenile wood is more than twice that of mature old stand wood. And, in terms of chemical properties, the lignin content of the juvenile wood is higher than that of mature wood. As well, log-size, an important factor in the grading of wood, will be smaller. Overall, these factors will have an effect on the potential grade and value reduction, caused by reduced strength and stiffness of the juvenile wood. Because of reduced value and potential end-product quality, which are problems resulting from this increase of juvenile wood, forest managers may have to expand the range of optimum planting densities and delay harvesting until the second growth trees are 90 years old. These problems with second growth Douglas fir, based upon a report of the Douglas Fir Task Force, were compiled and edited by R.M. Kellogg, and entitled *Second Growth Douglas-Fir: Its Management and Conversion for Value* (Vancouver: Forintek Canada Corp., 1989. 173 p. Special Publication No. SP-32).

IV Departmental Sources of the Publications

A Federal Cumulations

The first cumulation of federal forestry publications was issued by the Department of Public Printing and Stationery in 1963 and entitled *Department of Forestry [Publications]* (Ottawa: Queen's Printer, 1963. 137 p. Canadian Government Publications Sectional Catalogue No. 13). This cumulation lists monographs, conference proceedings (the British Empire Forestry Conferences), and series publications of the Forest Research Division and the Forest Products Laboratories of Canada. Annuals and periodicals are also listed. There have also been annual listings of the publications of the Canadian Forest Service. These are the *Publications [Canadian Forestry Service], 1981–1985* (Ottawa: Canadian Forestry Service, 1982–86. 5 v.), which includes scientific, technical, and interpretive publications, issued by the regional forestry centres and institutes, as well as the headquarters of the Canadian Forestry Service. Next has been a two-year listing by D. Rochon entitled *Forestry Canada Publications, 1986–1988* (Ottawa: Forestry Canada, 1989. 63 p. Information Report DPC-X-29). Parallel to these listings are the *Information Reports, Digest, 1980–1989* (Ottawa: Canadian Forestry Service, 1980–89. 4 v.), and the *Publications Digest: Canadian Forest Service* (Natural Resources Canada, 1991–). The latter lists, by subject, the publications (including journal articles appearing in the international literature) of both the regional research establishments and the headquarters. Some listings provide abstracts. The *Publications Digest* is also available on-line from 1995 to date at the website: <<http://www.nrcan.gc.ca/cfs/pub/digest/common/>>.

B Regional Cumulations

Some of the regional centres also produce their own publication lists. The Pacific Forest Research Centre has published annual listings of its research, since 1971, under two titles. They are *Reports and Publications 1970/71–1974/75* (Victoria: Canadian Forestry Service, 1971–76. 6 v.); *Reports and Publications, 1984–1993* (Victoria: Pacific Forestry Centre, 1985–94. 10 v.); and *Recent Publications, 1994–* (Victoria: Pacific Forestry Centre, 1994–). Some publications are also listed at their website <<http://bookstore.pfc.cfs.nrcan.gc.ca/>>.

The most complete of these separate listings has been issued by the Northern Forestry Centre. The first was compiled by V. Hildahl and R. Prott and entitled *Bibliography of Forest Research* (Winnipeg: Department of Forestry and Rural Development, 1968. n.p. Liaison Note MS-L-2). This covered research in the Manitoba-Saskatchewan region prior to 1968. The second listing was compiled by F.M. Dendwick *et al.* and entitled *Forest Research Bibliography, 1968–1975, with 1976–1979 Supplement* (Edmonton: Northern Forestry Research Centre, 1980. 76 p. Information Report NOR-X-152). The third listing was compiled by J.K. Samoil and G.B. Turtle and entitled *Northern Forestry Centre Publications, 1980–1986* (Edmonton: Northern Forestry Centre, 1988. 36 p. Information Report NOR-X-297). The fourth listing was compiled by D.A. Leroy and entitled *Northern Forestry Centre Publications, 1987–1990* (Edmonton: Northern Forestry Centre, 1991. 30 p. Information Report NOR-X-321). The fifth listing, also by D.A. Leroy, was entitled *Northern Forestry Centre Publications, 1991–1993* (Edmonton: Northern Forestry Centre, 1994. 27 p. Information Report NOR-X-336E). The sixth listing by D.A. Leroy was entitled *Northern Forestry Centre Publications, 1994–1996* (Edmonton: Northern Forestry Centre, 1998. 30 p. Information Report NOR-X-354E); and the most recent listing by D.A. Leroy was entitled *Northern Forestry Centre Publications, 1997–1999* (Edmonton: Northern Forestry Centre, 2000. 21 p. Information Report NOR-X-372E). The arrangement of these publication lists is by author and subject. As well, there is an appendix, listing reports prepared under the federal-provincial forestry development agreements for Alberta, Saskatchewan, and Manitoba.

The Great Lakes Forestry Centre has issued a series of compilations by C.A. Plexman of its publications. The first one was entitled *Bibliography of Great Lakes Forest Research Centre Publications, 1965–1977* (Sault Ste Marie, Ontario: Canadian Forestry Service, 1978. 108 p. Report 0704-7797 279). The second one was entitled *Bibliography of Great Lakes Forest Research Centre Publications, 1978–1982* (Sault Ste Marie, Ontario: Great Lakes Forest Research Centre, 1982. 81 p.). The third one was also entitled *Bibliography of Great Lakes Forestry Centre Publications, 1983–1987* (Sault Ste Marie, Ontario: Great Lakes Forestry Centre, 1987. n.p. Information Report O-X-388). These provide an author listing of information reports, joint reports, miscellaneous reports, contributions to symposium proceedings, leaflets, survey bulletins, and newsletter articles. A keyword subject index is also provided. A supplement to the latter publication was compiled by C.A. Plexman and entitled *1988 Supplement to Bibliography of Great Lakes Forestry Centre Publications* (Sault Ste Marie, Ontario: Great Lakes Forestry Centre, 1988. 20 p. Appendix). For the years 1990 and 1991, annual bibliographies were also issued. They were entitled *1990 Bibliography: Forestry Canada, Ontario Region, Great Lakes Forestry Centre* (Sault Ste Marie, Ontario: The Centre, 1991. 12 p. Information Report O-X-416); and *1991 Bibliography: Forestry Canada, Ontario Region, Great Lakes Forestry Centre* (Sault Ste Marie, Ontario: The Centre, 1992. 14 p. Information Report O-X-424). These listings are alphabetical, by author, with a separate listing for information reports, miscellaneous reports, joint reports, technical notes, and COFRDA [Canada-Ontario Forest Resource Development Agreement] Reports. As well, they include a subject index. For publications issued since 1990, the centre has also issued a rolling author index by year for the centre's publications from their website: <<http://www.glfc.forestry.ca/index-en/index-en.html>>. Click on Publications & Products.

Publication lists have also been produced by the Laurentian Forestry Centre. The first was compiled by J.M. Murphy and M. Poulin and entitled *Centre de foresterie des Laurentides / Laurentian Forestry Centre Publications, 1980–88* (Sainte-Foy, Quebec: Laurentian Forestry Centre, 1989. 44p. Information Report LAU-X-86). A second compilation has been done by C. Bilodeau and M. Poulin and entitled *Centre de foresterie des Laurentides/Laurentian Forestry Centre – Publications, 1989–1991* (Sainte-Foy, Quebec: Laurentian Forestry Centre, 1992. 30 p. Information Report LAU-X-100B). A third listing, entitled *Centre de foresterie des Laurentide, Publications 1992–94* (Sainte-Foy, Quebec: Canadian Forest Service, 1995. 44 p. Information Report LAU-X-111B), has also been published. These three publication lists are alphabetical, by author, with a short subject index provided. Publications issued since 1989 are also listed alphabetically, by author and by year through their website: <<http://www.cfl.forestry.ca/CFL-LFC/home.html>>. Click on Publications et produits.

The Atlantic Forestry Centre also has its publications issued since 1992 listed at its website: <<http://atl.cfs.nrcan.gc.ca/cfs/publications_e.html>>. The only other Maritime publication list was an early one by E.C. Salter entitled *Bibliography of Forest Research: Newfoundland Region* (Ottawa: Canadian Forest Service, 1971. 43 leaves. Information Report N-X-64).

C Research Institute Cumulations

The two forest research institutes (now part of the Great Lakes Forestry Research Centre) have also issued publication lists. The Petawawa National Forestry Institute issued four bibliographies. The first one was compiled by A.C. Yapa and M.H. Mitchell and entitled *Bibliography/Liste des publications pour les années 1979–1985* (Chalk River, Ontario: Petawawa National Forestry Institute, 1987. 150 p. Information Report PI-X-70 E/F). A second was done, covering the year 1986–87, and entitled again *Bibliography/Liste des publications pour les années 1986–1987* (Chalk River, Ontario: Petawawa National Forest Institute, 1988. 1 v.). A third one, compiled by P.A. Boross and M.H. Mitchell, was entitled *Bibliography: 1988–1990* (Chalk River, Ontario: Petawawa National Forestry Institute, 1991. 125 p. Information Report PI-X-106E/F). A fourth bibliography, also by P.A. Boross and M.H. Mitchell, was entitled *Annotated Bibliography of the Petawawa National Forestry Institute: 1990–1992* (Chalk River, Ontario: The Institute, 1994. 77 p. Information Report PI-X-116E/F). These are annotated bibliographies, listed alphabetically, by author, but including a separate author, title, information report number, technical report number, species and keyword index.

The Forest Pest Management Institute at Sault Ste Marie also issued bibliographies. The first, compiled by K.B. Griffiths, was entitled *Forest Pest Management Institute Bibliography (1977–1982)* (Sault Ste Marie, Ontario: Forest Pest Management Institute, 1983. 60 p. Information Report FPM-X-60). The second one was also compiled by K.B. Griffiths and entitled *Forest Pest Management Institute Bibliography Supplement (1982–1985)* (Sault Ste Marie, Ontario: Forest Pest Management Institute, 1985. 21 p. Information Report FPM-X-60, rev. ed.). The third bibliography, compiled by K.B. Jamieson, was entitled *F.P.M.I. Bibliography 1985–1990* (Sault Ste Marie, Ontario: Forest Pest Management Institute, 1991. 76 p. Information Report FPM-X-86). These are annotated bibliographies, the first one by subject, with an author index, the second one an author listing; and the third one, by author, with an author, species, and subject index, as well as a listing by information report number.

D Subject Bibliographies

Besides the regional or research institute bibliographies, or lists of publications, the Canadian Forest Service from time to time has also published some special subject bibliographies. For example, I.S. Alemdag and J.R. Richardson have compiled an *Annotated Bibliography of ENFOR Biomass Reports, 1979–1990* (Ottawa: Natural Resources Canada, 1993. 288 p. Information Report ST-X-6). This bibliography is a compilation of all published and unpublished reports produced during that time span under the ENFOR (Energy from the Forest) program. The reports are organized by broad subject category, and include an author, title, province and region, tree species, keyword, productivity, project, and project-by-year index. Another specialized bibliography by R.S. McAlpine and A. Mellors has been entitled *Annotated Bibliography of Fire Behavior and Ecology Research at the Petawawa National Forestry Institute 1979–1994* (Chalk River, Ontario: Petawawa National Forestry Institute, 1995. 27 p. Information Report PI-X-118E/F). Other examples of specialized bibliographies are, first, by R.M. Smith, entitled *Bibliography of Forest Pathology Research Publications, Western and Northern Region* (Edmonton: Northern Forest Centre, 1976. 37 leaves. Information Report NOR-X-163); and *Bibliography of Forest Entomology Research, 1927–77* (Canadian Forest Research Centre, 1978. 34 p. Information Report NOR-X-212); second, by J.P. Kimmins, entitled *Whole-Tree Harvest – Nutrient Relationships: A Bibliography* (Chalk River, Ontario: Petawawa National Forestry Institute, 1985. 377 p.); a third, by K. Nietmann, was entitled *Catalogue of Canadian Forest Inventory Publications and Manuals* (Chalk River, Ontario: Petawawa National Forestry Institute, 1987. 24 p. Information Report PI-X-76E/E); and a fourth, by P.A. Addison, was entitled *The Canadian Forestry Service Air Pollution Program and Bibliography* (Ottawa: Canadian Forestry Service, 1988. 133 p. Information Report DPC-X-26).

Chapter 7

Earth Sciences Research

I Historical Background

The historic studies of earth sciences research in Canada trace back at least to the 19th century. It was in September, 1841, that the Legislature of the Province of Canada (formed in 1840 by the union of Upper and Lower Canada) passed the resolution 'that a sum not exceeding one-thousand five hundred pounds sterling be granted to Her Majesty to defray the probable expense in causing a Geological Survey of the province to be made.' The motivation, beyond the intense Victorian curiosity in the natural sciences, was the fact that if Canada were to expand beyond an agriculture-based colony, an industrial economy was needed for the country. And that industrial economy would depend, to a considerable extent, on a viable mineral industry, particularly coal.

The provision of funds to carry out a survey made possible the appointment of an officer (based upon English recommendations) by the name of William Logan, who had been born in Montreal, and educated in Edinburgh. To assist him, Alexander Murray, a naval person who undertook training in land surveying and geology, was appointed. Their initial survey began in 1843. From its beginnings, until 1877, the Geological Survey acted as a Crown agency. At that time, it was incorporated into the Department of Interior, which had been created in 1873. Then, in 1890, its role was more clearly defined and expanded by an Act of Parliament. This expanded role for the Geological Survey included mines and mining resources. However, in 1907, a separate Department of Mines was established, and these two areas did not come together again until 1936, when the Department of Mines and Resources was created. In 1950, the latter became the Department of Mines and Technical Surveys, which also included geophysical surveys, such as the emerging airborne geodetic surveys, aeromagnetic surveys, and seismology and gravity programs. Then, in 1966, the name changed to the Department of Energy, Mines and Resources, until 1994, when the Department of Forestry was incorporated, and the new departmental name became the Department of Natural Resources, or Natural Resources Canada. In 1995, the reorganization was stabilized into four sectors. Earth sciences research is centred in the Earth Sciences Sector, which consists of the Geological Survey of Canada, Geomatics Canada, containing the Geodetic Survey and the Canada Centre for Remote Sensing (the latter mentioned below), and the Minerals and Metals Sector, which also contains CANMET (mentioned below). Along with these changes has been the growth of specialized research centres, or divisions.

In 1971, the Canada Centre for Remote Sensing was organized 'to collect, process, disseminate and develop applications applicable to resource management and environmental control ... obtainable from specialized remote sensing.' In 1974–75, the Mines Branch was reorganized to become the Canada Centre for Mineral and Energy Technology (CANMET) 'to highlight research in the energy field and to reflect its wide range of mining and mineral interests.' In 1980, the Atlantic Geoscience Division (later Centre), located in the Bedford Institute of Oceanography in Dartmouth, Nova Scotia, was created to be 'responsible for the Atlantic and Arctic offshore regions of Canada, including the Beaufort Sea, and for the sedimentary basins of the Appalachian region.' At the same time, the Institute of Sedimentary and Petroleum Geology, located in Calgary, Alberta, became 'responsible for the sedimentary basins of western and northern mainland Canada and the Arctic.' The Cordilleran Geology Division, located in Vancouver, BC, became responsible for studies of the 'geology of the Cordilleran region and the adjacent Pacific continental shelf and margin'; and the Marine Geology Section and the Pacific Geophysics Division, based at the Pacific Geoscience Centre, located at the Institute of Ocean Sciences, near Sidney, BC, were made responsible for bedrock, surficial, and geomorphic process studies of the Pacific and continental shelf and adjacent areas as well as for geophysical studies of the Canadian Cordillera and the adjacent offshore regions. In 1985, with the exception of the Canada Centre for Remote Sensing, there was a realignment of these research centres and divisions under the Geological Survey of Canada. And, in 1989, a Quebec Geoscience Centre, 'which focused initially on structural and sedimentological investigations,' was formed.

Today, earth sciences research in Canada encompasses, besides the geological surveys, the study of geochronology in aid of geological exploration, which covers surveys and mapping over the territory of Canada, and the use of geochemistry in the search for mineral deposits, their mining, extraction, and processing, including mining safety. Besides these large areas of exploration and research, earth sciences research in Canada also encompasses geophysical studies of the physical forces that affect the whole earth, or that part which is Canadian territory. These studies involve the earth's magnetic forces, including palaeomagnetism, its gravity and gravity anomalies, its shape, or geoid, and seismology. All this research is documented in a variety of publications and maps, mainly issued by the Department of Energy, Mines and Resources, or its successor Natural Resources Canada, or in the international research literature, or issued in the form of communications at conference proceedings.

II Publications

Historically, the Department of Energy, Mines and Resources (or its successor, the Earth Sciences Sector of Natural Resources Canada) has issued and continues to issue a large number of publications through the Geological Survey of Canada, the Mines Branch and the Mineral Resources Division (both now CANMET – the Canada Centre for Mining and Energy Technology) and the Earth Physics Branch. In the late 1980s and early 1990s, the Earth Physics Branch became the Geophysics and Terrain Sciences Branch under the Geological Survey of Canada Sector. Then, after the 1993 reorganization of the federal government, both these areas came under the Earth Sciences Sector as Geomatics Canada and their publications show a range of imprints. Only the publications of the Geological Survey of Canada have shown a continuous consistency in publication, and numbering, throughout these changes.

A Geological Survey of Canada

The publications of the Geological Survey of Canada are issued in the form of bulletins, memoirs, papers, the economic geology series, and maps.

(a) Bulletins

The Bulletins began in 1913, and were issued by the National Museum, when it was part of the Geological Survey of Canada. They covered not only the geological sciences, but also biology and anthropology. With the separation of the National Museum from the Geological Survey in 1945, the latter continued to issue bulletins, with their numbering beginning from 1945. The bulletins are considered to be the proper vehicle for material 'not properly the subject of a memoir.' An example from the bulletin series by B. Dubé and K. Lauziere is entitled *Gold Metallogeny of the Cape Ray Fault Zone, Southwestern Newfoundland* (Ottawa: Geological Survey of Canada, 1997. 90 p. Geological Survey of Canada Bulletin No. 508).

(b) Memoirs

Unlike the bulletins, the memoir series, which commenced in 1910, embodied final reports of major undertakings of the Geological Survey. From 1910 to 1922, the series was divided into geological, topographical, biological, and anthropological series. Only the geological series was continued after 1922. An example from the memoir series by B.D. Ricketts is entitled *Basin Analysis, Eureka Sound Group, Axel Heiberg and Ellesmere Islands, Canadian Arctic Archipelago* (Ottawa: Geological Survey of Canada, 1994. 119 p. Geological Survey of Canada Memoir No. 439).

(c) Papers

The Papers began in 1935, and were designed for speed of production in order to supply provisional information until final, more complete reports with maps were printed. An example from the paper series by D.R. Grant is entitled *Quaternary Geology of Port Saunders Map Area, Newfoundland* (Ottawa: Geological Survey of Canada, 1994. 59 p. Geological Survey of Canada Paper No. 91–20).

(d) Economic Geology Reports

The Economic Geology Reports consist of monographs on specific economic minerals and natural mineral products. This series began in 1909. An example from the economic geology series by D.C. Harris is entitled *The Mineralogy and Geochemistry of the Hemlo Gold Deposit, Ontario* (Ottawa: Energy, Mines and Resources Canada, 1989. 88 p. Geological Survey of Canada Economic Geology Report No. 38).

(e) Maps

These are coloured maps, which go back as far as 1843, and the first Geological Survey by William Logan. These may be mineral maps, geology maps, and geophysical maps, such as gravity maps, gravity anomaly maps, magnetic anomaly maps, and seismology maps. These latter maps are most visible as part of the *Canadian Geophysical Atlas* (Ottawa: Geological Survey of Canada, 1987–). To date, 15 maps have been completed. An example of an individually issued map from this series by K.E. Wallace-Dudley and B.H. Ortman is entitled *Gas Pools of Western Canada* (Ottawa: Geological Survey of Canada, 1991. 1 map on 2 sheets. Geological Survey of Canada. 'A' Series Map). Not unrelated to the map series are the data files and data services from which geophysical data is obtained. These are the Seismological Data Services, the Geomagnetic Data Services, the Gravity Data

Services, the Geodynamics Data Services, and the Geothermal Data Services, all under Geomatics Canada.

B Mines Branch and CANMET Publications

Like the geological literature, the related mining literature from the Mines Branch (now CANMET or the Canada Centre for Mineral and Energy Technology) is well documented. There are, in effect, three series. The earliest Mines Branch publications cover the period 1902–59. This is the 'old series.' The 'new series' began in 1958, along with the Mineral Resources Division publications, and continued until 1974–75, when CANMET was created, and developed its own publication series.

(a) Mines Branch

The old series of the Mines Branch publications consisted of reports and maps, bulletins, the memorandum series, technical papers, and information circulars.

i. Reports and Maps The Reports and Maps were numbered No. 1, 1902 to No. 865, 1959. This series consisted primarily of minerals of economic interest, and sometimes included maps. Also covered were reports on mining and smelting techniques, and some industrial water reports. An example of a publication from this series by J.P. Phillips is entitled *Improving the Properties of Clays and Shales* (Ottawa: Department of Mines and Resources, 1938. 39 p. Mines Branch Report No. 793). This series was continued as the monograph series.

ii. Bulletins The bulletins were numbered No. 1, 1909 to No, 33, 1921. This series was also issued and listed within the report and map series. An example of a publication in this series by H. Fréchette is entitled *Western Portion of Torbrook Iron Ore Deposits, Annapolis County, N.S.* (Ottawa: Department of Mines, 1912. 20 p. Report No. 110. Bulletin No. 7).

iii. Memorandum Series The Memorandum Series were numbered No. 1, 1921 to No. 137, 1958. This series consisted of reports of general interest available to the public. An example of a publication in this series by A.A. Swinnerton is entitled *Fifty Years of Fuel Testing and Research* (Ottawa: Department of Mines and Technical Surveys, 1957. 15 p. Mines Branch Memorandum No. 136). This series was superseded by Mineral Surveys, issued by the Mineral Resources Division.

iv. Technical Papers The Technical Papers were numbered No. 1, 1953 to No. 17, 1957. This series was restricted to reports on methods, and results of research, carried out in Mines Branch Laboratories. An example of a publication from this series by J.R. Zimmerman is entitled *The Determination of Aluminum by the Fluorophotometric Method* (Ottawa: Department of Mines and Technical Surveys, 1953. 12 p. Mines Branch Technical Paper No. 4). This series was continued by the Mines Branch as Research Reports.

v. Information Circulars The Information Circulars were numbered No. MR1, 1954 to No. MR26, 1957. This series was issued by the Mineral Resources Division and consisted of brief papers about rare or less common metals, in Canada. Also included were surveys of particular mining industries. An example of a publication in this series by T.H. Janes is entitled *Survey of the Iron Ore Industry in Canada during 1956* (Ottawa: Department of Mines and Technical Surveys Mineral Resources Division Information Circular No. MR22). This series was continued under the name Mineral Information Bulletins by the Mineral Resources Division of the Mines Branch.

(b) Mines Branch – New Series

The new series of the Mines Branch consisted of monographs, information circulars, research reports, and technical bulletins.

i. Monographs The Monographs were numbered No. 866, 1959 to 1976. This series was a continuation of reports of the old series, and continued its numbering. An example of a publication in this series by J.S. Ross is entitled *Bentonite in Canada* (Ottawa: Department of Mines and Technical Surveys, 1964. 61 p. Mines Branch Monograph No. 873.

ii. Information Circulars The Information Circulars were numbered No. IC102, 1958 to 1986. This series consisted of modest descriptions of milling techniques, specific mineral industries, and bibliographies of high-temperature research. An example of a publication in this new series by R.K. Collings in entitled *Canadian Gypsum Industry* (Ottawa: Department of Mines and Technical Surveys, 1959. 41 p. Mines Branch Information Circular No. IC 114).

iii. Research Reports The Research Reports were numbered No. R1, 1958 to 1976. This series consisted of 'original contributions to science and technology based primarily on laboratory research conducted in the Mines Branch laboratories or under Mines Branch auspices.' An example of a publication in this series by E.H. Nickel and J.F. Rowland is entitled *Composition and Crystallography of*

Niocalite (Ottawa: Department of Mines and Technical Surveys, 1958. 9 p. Mines Branch Research Report No. R26).

iv. Technical Bulletins The Technical Bulletins were numbered No. TB1, 1959–76? The technical bulletins consisted of papers on separation techniques and metallurgical techniques. An example of a bulletin from this series by R.J. Guest and J.C. Ingles is entitled *Analytical Procedures for a Vanadium Recovering Process* (Ottawa: Department of Mines and Technical Surveys, 1966. 62 p. Mines Branch Technical Bulletin No. TB-79).

(c) Mineral Resources Division Publications

The publications of the Mineral Resources Division of the Mines Branch consisted of Mineral Information Bulletins and Mineral Maps.

i. Mineral Information Bulletins The Mineral Information Bulletins were numbered No. MR27, 1958–76. These bulletins consisted of Canadian mineral industry surveys. An example of a publication in this series by D.B. Fraser and J.W. Patterson is entitled *Lead and Zinc in Canada, 1961* (Ottawa: Department of Mines and Technical Surveys, 1963. 90 p. Mines Branch. Mineral Resources Division Information Bulletin No. MR60).

ii. Mineral Maps The Mineral Maps were numbered MR[no.], 1958–76. An example of a map from this series is entitled *Iron Ore Trade, Canada and the World* (Ottawa: Department of Mines and Technical Surveys, 1963. 1 Map. Mines Branch Mineral Resources Division Map MR2).

(d) CANMET Publications

In 1974–75, the Mines Branch became the Canada Centre for Mineral and Energy Technology (CANMET), and the former Mines Branch publications became CANMET publications. The CANMET publications consist of CANMET reports, CANMET special reports or publications, contractual research reports, and divisional reports. Some are numbered by type and year, and others have accession numbers only.

i. CANMET Reports The CANMET Reports consist of mineral reports and technical studies on mineral ores. An example of a publication in this series by L. Cabri and J. Laflamme is entitled *Mineralogical Study of the Platinum-Group*

Element Distribution and Associated Minerals from Three Stratigraphic Layers, Birds River Sill, Manitoba (Ottawa: CANMET, 1988. 60 p. CANMET Report No. 88-01E).

ii. CANMET Special Reports or Publications CANMET Special Reports or Publications consist of proceedings of CANMET-sponsored conferences, abstracts of relevant international meetings, and mining technology reports. An example of a publication in this series has the title *Mining Automation: Proceedings of the Second Workshop Sponsored by CANMET/Mining Research Laboratories, Sudbury, Ontario, October 17, 1986* (Ottawa: CANMET, 1987. 93 p. CANMET Special Report SP87-1).

iii. Contract Research Reports The Contract Research Reports consist of research contracted out by CANMET on topics ranging from mineral or fuel research to energy performance studies. An example of a publication in this series by B. Clark is entitled *Tidal Power and Canada* (Ottawa: CANMET, 1986. 158 p. CANMET Contract Research Report EMR-FPS-86-0003).

iv. Divisional Reports The Divisional Reports consist of reports generated in the Coal Research Laboratories, the Energy Research Laboratories, the Metals Technology Laboratories, the Mineral Sciences Laboratories, and the Mining Research Laboratories. An example of a publication in this series by Y.C. Lezotte is entitled *Empirical Procedures for Prediction of Rock Fragmentation by Blasting* (Ottawa: CANMET, 1990. 64 p. Mining Research Laboratories Technical Report MRL90-006TR).

C *Earth Physics Branch Publications (now under the Continental Geoscience Division of the Geological Survey of Canada)*

These publications, which incorporated the publications of the Dominion Observatory in 1970, consist of contributions, publications, gravity map series, seismological series, geomagnetic series, geothermal series, aerodynamic series, and maps.

(a) Contributions

The Contributions (dating from No. 1, 1947–) are scientific research papers that have appeared in the international journal literature, and have been reprinted for distribution. An example from this series by A. Hicken and E. Irving is entitled *Tectonic Rotation in Western Canada* (*Nature*, v. 268, 1977. pp. 219–20. Earth Physics Branch Contribution No. 673).

(b) Publications

The Publications, dating from No. 1, 1913–, consist of reports and papers published directly by the Earth Physics Branch. An example of a publication in this series by K. Whitham and H.S. Hasegawa is entitled *The Estimation of Seismic Risk in Canada – A Review* (Ottawa: Energy, Mines and Resources Canada, 1975. pp. 137–62. Physics Branch Publication, v. 45, No. 2).

(c) Gravity Maps

The Gravity Map Series, dating from No. 1, 1961–, consist of quarterly seismological bulletins (to 1963), annual seismological bulletins (since 1963), and descriptions of seismological operations. An example of a publication in this series by W.G. Milne and A.G. Davenport is entitled *Earthquake Probability* (Ottawa: Department of Energy, Mines and Resources, 1969. 19 p. Observations Branch Seismological Series No. 54).

(d) Geomagnetic Series

The Geomagnetic Series, dating from No. 1, 1975–, consist of annual reports for magnetic observations, aeromagnetic surveys, and catalogues of paleomagnetic directions and poles. An example of a publication in this series by G.V. Haines and W. Hannaford is entitled *A Three Component Aeromagnetic Survey of Central Canada* (Ottawa: Energy, Mines and Resources Canada, 1978. 22 p. Earth Physics Branch Geomagnetic Series No. 13).

(e) Geothermal Series

The Geothermal Series, dating from No. 1, 1974–, consists of geothermal data and geothermal studies. A publication in this series by A.M.Tessop is entitled *Energy from Sedimentary Basins* (Ottawa: Energy, Mines and Resources Canada, 1976. 10 p. Earth Physics Branch Geothermal Series No. 8).

(f) Geodynamic Series

The Geodynamic Series, dating from No. 1, 1961–, consists of daily PZT (piezoelectric transducer) observations and time signals. Numbers 1–63 were issued as monthly bulletins, and cover the period from January, 1961, to December, 1971. An example of a publication in this series by J. Popelar and S.B. Sin is entitled

PZT Observations of Time and Latitude Ottawa and Calgary (Otawa: Energy, Mines and Resources Canada, 1984. 31 p. Earth Physics Branch Geodynamic Series Bulletin No. 74).

(g) Geomagnetic Map Series

The Geomagnetic Map Series consists of bouguer gravity anomaly (a value that corrects the observed gravity for latitude and elevation variations) maps. An example of a map published in this series is entitled *Bouguer Gravity Anomaly Map of Canada* (Ottawa: Department of Energy, Mines and Resources, 1969. 1 map. Survey and Mapping Branch. G-MS-1).

D Other Publications

Besides the specific publication series, there are also numerous annual geological and mining commentaries that have been published, or are still being published, by sections or branches within Energy, Mines and Resources Canada, or its successor Natural Resources Canada.

(a) Annual Publications of the Geological Survey of Canada

Report of Progress for the Year [1843–84]: Geological Survey of Canada (Montreal: Geological Survey of Canada, 1845–85. 27 v.). None published for the years 1859–62; 1869–70;

Annual Report: Geological and Natural History Survey of Canada, v. 1, 1885–v. 16, 1904 (Montreal: Dawson Brothers, 1886–1907. 16 v.);

Summary Report: Geological Survey of Canada, 1904–1933 (Ottawa: King's Printer, 1904–33. 30 v.);

Report of Activities: Geological Survey of Canada, 1934–1977 (Ottawa: Geological Survey of Canada, 1934–77. 44 v.);

Current Research: Geological Survey of Canada, 1978–1993 (Ottawa: Geological Survey of Canada, 1978–93. 16 v.). This publication was divided into six parts in 1988. They were: *Part A: Abstracts*; *Part B: Eastern and Atlantic Canada; Part C: Canadian Shield*; *Part D: Interior Plains and Arctic Canada*; *Part E: Cordillera and Pacific Margin*; and *Part F: National and General Programs.* In 1989, Part G: Frontier Geoscience Program, Arctic Canada; and Part H: Frontier Geoscience Program, Queen Charlotte Islands, British Columbia were added;

Current Research: Geological Survey of Canada, 1994–1999 (Ottawa: Geological Survey of Canada, 1994–99. 6 v.). This series was divided into four parts.

They were: Part A: *Cordillera and Pacific Margin*; Part B: *Interior Plains and Arctic Canada*; Part C: *Canadian Shield*; and Part D: *Eastern Canada and National and General Programs*; and

Current Research: Geological Survey of Canada [computer file], *2000–* (Ottawa: Geological Survey of Canada, 2000–). This is available in CD-ROM format.

The Geological Survey of Canada also publishes an annual review. Its publishing history is:

Annual Review: Geological Survey of Canada (Ottawa: Energy, Mines and Resources Canada, 198?-1985/86);

Geological Survey of Canada: [Review] (Ottawa: Energy, Mines and Resources Canada, 1986/87–1992/93. 7 v.); and

Geological Survey of Canada (Ottawa: Natural Resources Canada, 1993/94–).

(b) Annual Publications of the Earth Sciences Sector

The Earth Sciences Sector of the Department of Energy, Mines and Resources also issued an annual publication entitled *Earth Sciences Sector* (Ottawa: Energy, Mines and Resources Canada, 1981/82 to 1985/86). This highlighted the work of the Geological Survey of Canada, the Earth Physics Branch, the Surveys and Mapping Branch, and the Polar Continental Shelf Project, all of which were part of that sector during that period of time. There have also been, or are, individual, or combined annual reports, issued by components of the Earth Sciences Sector. They are: *Canada Centre for Remote Sensing: Annual Review* (Ottawa: Energy, Mines and Resources Canada, Surveys, Mapping and Remote Sensing Sector. 1988/89–1992/93. 5 v.). This was superseded in part by *Geomatics Canada: Annual Review* (Ottawa: Geomatics Canada, 1993/94–1994/95. 2 v.) and *Geomatics Canada* (National Resources Canada, 1996–).

Finally, the Canada Centre for Mineral and Energy Technology (CANMET), which was organized in 1974/75 from the old Mines Branch, also issued an annual report of activities. It was called the *CANMET Annual Report* (Ottawa: Canada Centre for Mineral and Energy Technology, 1987/88–94/95. 8 v.).

Highlights of current research in the Earth Sciences Sector are available from the Natural Resources Canada website at <<http://www.nrccan.gc.ca>>.

(c) Annual Publications of the Canadian Geoscience Council

The Canadian Geoscience Council, which provides a central cooperative forum for earth sciences societies in Canada, and provides advice to the federal government on geoscience policy, also issues an annual report entitled *The Geosciences in Canada* (Ottawa: Geological Survey of Canada, 1975–). These reports give a broad outline of the yearly activities of member societies, such as the Mineralogical Association of Canada, or the Canadian Society of Petroleum Geologists. The Canadian Geoscience Council has also issued two review reports on the state of the geosciences in Canada. They are, first, *The Geosciences in Canada, 1974. A Status Report.* Edited by E.R. Neale and A.C. Clague (Ottawa: Energy, Mines and Resources Canada, 1975. 51 p. Geological Survey of Canada. Paper No. 75-6), and secondly, by A.J. Naldrelt, entitled *The Geosciences in Canada, 1985. Part 2: Report of the Canadian Geoscience Council Advisory Committee on Mineral Deposits Research at the Geological Survey of Canada* (Ottawa: Geological Survey of Canada, 1986. 43 p. Geological Survey of Canada Paper No. 85–6, Part 2).

The Geoscience Council also publishes an annual volume entitled *Current Research in the Geological Sciences in Canada* (Ottawa: Geological Survey of Canada, 1951–). This series indexes, by geological subject, all current research done in Canada. Federal, provincial, and university research, including aerial mapping and mining geology, is included. Until 1965, this publication was a section of the *Annual Report – National Advisory Committee on Research in the Geological Sciences* (Ottawa: Geological Survey of Canada, 1953/54–1964/65).

(d) Annual Publications of the Department of Mines and Its Successors

Perhaps the broadest source of ongoing earth sciences research in Canada comes from the historical and more recent relevant federal departmental annual reports. Of these, the Department of Mines was the earliest body, and, interestingly, it did not prepare an annual report distinct from the annual summary reports issued by the Geological Survey of Canada and the Mines Branch from 1907 to 1920. After that there were annual reports as follows:

Report of the Department of Mines, 1921–1936 (Ottawa: Department of Mines, 1921–36. 16 v.). This report included statements from the Geological Survey of Canada and the Mines Branch;
Report of the Department of Mines and Resources, 1937–1949 (Ottawa: Department of Mines and Resources, 1937–49. 13 v.);
Report of the Department of Mines and Technical Surveys, 1950–1965 (Ottawa: Department of Mines and Technical Surveys, 1950–65. 16 v.); and
Annual Report – Energy, Mines and Resources Canada, 1965–1992 (Ottawa: Department of Energy, Mines and Resources, 1965–92. 28 v.).

Besides the annual reports, the Department of Energy, Mines and Resources also issued a public interest journal entitled *GEOS* (Ottawa: Energy, Mines and

Resources Canada, v. 1, 1972–v. 22, no. 1, 1992). This journal had articles which interpreted the research activities of the department for the layperson.

III Earth Sciences Research

A Geochronology Techniques in the Service of Geology

In order to set the chronological framework for the history of the earth and of life, geologists, over time, have evolved two methods for estimating the geological time-scale. The first, which provides an estimate of the relative length of geological time, uses the principle of superimposition that was first enunciated over 300 years ago. This principle is based upon the fact that sedimentary rocks cover about 75% of the surface area of the continents, and the historical sequence of the laying of sedimentary rocks is easily defined because their deposition, including organic fossil remains, proceeds from bottom to top. In practice, however, deformations may have caused tilting, or overturning, and erosion may have blurred the record because of the removal of a portion of the sedimentary rock. Thrusting and faulting of the rocks can also create additional problems in reading the geological record. Hence, what may have been the original top, or bottom, causes problems in identifying what is the bottom and top in the sequences of deposition.

Relative age can also be deduced in metamorphic rock (an aggregate of minerals formed by the recrystallization of pre-existing rocks), or as new minerals form at the expense of older ones in response to those changing temperatures and pressures, or with igneous rocks formed from magma. Here, a second and more recent method, called radiometric dating, is used to determine the relative age of rocks. In fact, radiometric dating is required for studying the Precambrian Record, which began 4,000 million years ago and continued to up to 570 million years ago, where there are no organic fossil remains to confirm relative age. Radiometric dating is based upon the fact that the time which has elapsed since certain minerals were formed can be determined because of the presence of small amounts of natural radioactive atoms in their structures. Although radiometric dating was discovered about 100 years ago, it has only been since 1950 that extensive work has been carried out using this technique to come up with absolute dates of rocks formed during Precambrian time.

In order to understand radiometric dating, one must be aware of the structure of chemical elements. All chemical elements are composed of atoms, held together by electrical forces, and each atom consists of a relatively heavy nucleus, made up of positively charged protons, and neutral particles, called neutrons. Almost 100% of the mass of an atom resides in the nucleus, but orbiting the latter are one or more negatively charged electrons, which balance the positive charge of the nucleus. The sum of the protons and neutrons in the atomic nucleus is called the element's mass number, whereas the total tally of protons alone provides the atomic number. Accordingly, those atoms having the same atomic number belong to the same chemical element, but those having the same atomic number, but a different tally of neutrons in the nucleus, are called isotopes of the element.

All elements have at least two isotopes, and some have more. Some elements have naturally occurring isotopes that are radioactive, and may decay spontaneously to lighter isotopes of other chemical elements. This disintegration takes place at a fixed rate, for particular isotopes, to so-called daughter elements, and is the basis for the methodology of radiometric dating used by geologists to date certain minerals that commonly occur in crushed rocks. Taking uranium for example, one of its isotopes is uranium-238, which decays to the non-radioactive isotope of lead, called lead-206. The ratio of decay is expressed by the isotope's half-life, which refers to the time it takes for one-half of the original isotope to decay into its daughter element. The rate of decay for uranium-238 is 4,500 million years to its half-life. Uranium may be the best-known element with radioactive isotopes, but other more common elements such as thorium, potassium, argon, and rubidium also have radioactive isotopes. The isotope thorium-232 decays to a non-radioactive isotope of lead called lead-208. Other parent-daughter isotopes are uranium-235, with a daughter, lead-207, and a half-life of 710 million years, rubidium-87, with a strontium-87 daughter, and a half-life of 4,700 million years, and potassium-40, with a daughter, argon-40, and a half-life of 1,300 million years. In order to find out the time ratios of radiogenic leads, strontiums, or argons, to the parental radioactive isotopes, mass spectroscopy is used. The latter is a chemical technique that allows the measurement of atomic and molecular masses.

In the mid-1960s, a major review of geochronology activities in Canada was published, and entitled *Geochronology in Canada.* Edited by F. Fitz Osborne (Toronto: University of Toronto Press, 1964. 156 p. Royal Society of Canada Special Publication No. 8). This monograph reviewed dating correlation problems in various geological regions of Canada, as well as the chronological value of fossils and the limitations of radiometric dating. Radiometric dating at the Geological Survey of Canada has been described in the publication by W. Dyck entitled *The Geological Survey of Canada Radiocarbon Dating Laboratory* (Ottawa: Department of Energy, Mines and Resources, 1968. 45 p. Geological Survey of Canada Paper No. 66-45); and by R.W. Wanless in the publication entitled *Rubidium-Strontium Isotopic Age Studies* (Ottawa: Department of Energy, Mines and Resources, 1972, v. 1. Geological Survey of Canada Paper No. 72–73). The former publication was updated by J.A. Lowdon in a paper with the same title – *The Geological Survey of Canada Radiocarbon Dating Laboratory* (Ottawa: Geological Survey of Canada, 1985. 19 p. Geological Survey of Canada Paper No. 85–24), and the latter was updated by the same author (R.W. Wanless), in the paper entitled *Rubidium-Strontium Isotopic Age Studies, Report 2 (Canadian Shield)* (Ottawa: Energy, Mines and Resources Canada, 1978. 70 p. Geological Survey of Canada Paper No. 77–14).

Specific geochronology studies have a regional geology focus, with papers such as that by R.W. Klassen entitled *Quaternary Stratigraphy and Radiocarbon Chronology in Southwestern Manitoba* (Ottawa: Department of Energy, Mines and Resources, 1969. 19 p.); or by R.J. Fulton entitled *Radiocarbon Geochronology of Southern British Columbia* (Ottawa: Department of Energy, Mines and Resources, 1971. 28 p. Geological Survey of Canada Paper No. 71-37); or by L.E. Jackson entitled *Alberta Radiocarbon Dates with Regional Significance to the Late Quaternary Geochronology and Geomorphology of Alberta* (Ottawa: Energy, Mines and Resources Canada, 1984. 27 p. Geological Survey of Canada Paper No. 83-25). Other recent examples of geochronology research are by R.W. Macdonald and E.C. Carmack entitled *Age of Canada Basin Deep Waters: A Way to Estimate Primary Production for the Arctic Ocean* (*Science*, v. 254, 1991. pp. 1348–50); by P.E. Smith and R.M. Farquhar, entitled *Direct Radiometric Age Determination of Carbonate Diagenesis Using U-P6 in Secondary Calcite* (*Earth and Planetary Space Letters*, v. 105, 1991. pp. 474–91); by R.J. Scammal and K.M. Bethune, entitled *Preliminary U-Pb Geochronology of Greenstone Belts, Granitic Plutons and Metamorphic Rocks, Ege Bay Region, Baffin Islands, Northwest Territories.* In: *Exploration Overview, 1994. Northwest Territories Mining Exploration and Geological Investigations.* Compiled by R. Kusick and S.P. Goff (Yellowknife: Northern Affairs Program, 1994. pp. 54–55); and, by W.J. Davis entitled *U-Pb Zircom and Rutile Ages from Granulite Xenoliths in the Slave Province: Evidence for Mafic Magmatism in the Lower Crust Coincident with Proterozon Dike Swarms* (*Geology*, v. 25, 1997. pp. 343–46).

B Geological Research in Canada

Because of Canada's geographical size and position on the earth, the country is a geologist's delight and challenge. Its physical and geological area encompasses two major features. There is a core of ancient Precambrian rocks, 2,600 million or more years old, making up the Canadian Shield, and a crescent of younger rocks, forming what is called the Borderlands. The Borderlands consist of an inner ring of lowlands, plains, and plateaux of sedimentary rocks, and an outer ring consisting of areas of mountains and plateaux. For geological purposes both the shield and the Borderlands are divided into regions, and provinces within regions, based on geological research. The regional divisions are the Canadian Appalachia and St Lawrence Lowlands; the Canadian Shield, which is divided into the Slave, Grenville, and Churchill geological provinces; the Arctic Archipelago, consisting of the Innuitian Orogen, the Arctic Platform, or Lowlands, and the Arctic Coastal Plains; the sedimentary Basins of the Interior Plains; and the Western Cordillera. Geological studies in Canada, which are described below, have also culminated in the ongoing publication of the *Geological Atlas of Canada* (Ottawa: Geological Survey of Canada, 1982–). This atlas edition is a series of 100 maps, which will cover all the geological regions of Canada. The first map, issued by K.P. Card and B.V. Sanford, is entitled *Geology: Timmins, Ontario-Quebec* (Ottawa: Geological Survey of Canada, 1982. Geological Atlas, Map NM-17–G. Sheet 1 of 5, Scale 1:1000 000).

(a) The St Lawrence Lowlands, Appalachia, and the Atlantic Continental Shelf

Starting in the east, the Borderlands consist of the St Lawrence Lowlands and the Appalachian Region. The latter was formed during the Early Palaeozoic (over 500 million years ago), when sediment-filled basins along the continental margin of ancient North America became deformed to create a system of orogenic (mountain building) belts called the Appalachian Orogen.

The St Lawrence Lowlands border the Canadian Shield on the southeast, extending from the Lake Huron, Lake Erie region, northeasterly to the Strait of Belle Isle, and consist of three units – the West, Central, and East St Lawrence Lowlands. The Lowlands are made up of unfolded Paleozoic strata. The St Lawrence Lowlands are described in the volume edited by H. Williams, entitled *Geology of the Appalachian-Caledonian Orogen in Canada and Greenland* (Ottawa: Geological Survey of Canada, 1995. 994 p. Geology of North America, Volume F-1. Geology of Canada No. b).

Longitudinally, the Canadian Appalachia now extends from the international border, north and east, to the Gaspé, and on to Newfoundland. The horizontal traverse of this is belt after belt of deformed rocks from Rivière du Loup on the south shore of the St Lawrence through parts of Maine, New Brunswick, and Nova Scotia, and on to Newfoundland. Physiographically, the Appalachia Region of Canada is divided into a number of distinctive units. Overall, the area is dominated by a peneplain (a featureless, undulating plain created by erosion), probably Cretaceous in age (100 million years old). Its elevation is highest in the northwest, and slopes southeastward to the ocean. The units of the peneplain have been formed by the varying erosion of soft and hard rocks, to form lowlands, highlands, and uplands. Seawards, the whole region is bounded by the Atlantic Continental Shelf.

Overall, the geological evolution of the Canadian Appalachia has been described by W.H. Poole in the paper entitled *Tectonic Evolution of the Appalachian*

Region of Canada. In: *Collected Papers on Geology of the Atlantic Region – Hugh Lilly Memorial Volume* (Toronto: Geological Association of Canada, 1967. pp. 9–51. Special Paper No. 4); and in the volume, edited by H. Clark, entitled *Appalachian Tectonics* (Toronto: University of Toronto Press, 1967, 99 p. Royal Society of Canada Special Publication No. 10). As well, the Canadian Appalachia has been more recently described in the volume, edited by P. St-Julian and J. Béland, entitled *Major Structural Zones and Faults of the Northern Appalachians* (Waterloo, Ontario: Geological Association of Canada, 1982. 280 p. Geological Association of Canada Special Paper No. 24), and in the paper by H. Williams entitled *Appalachian Orogen in Canada* (*Canadian Journal of Earth Sciences*, v. 16, 1979. pp. 792–807).

Within the Appalachia, Newfoundland itself comprises three highlands, two uplands, and four lowlands, varying in height from 2,600 feet down to 200 feet. Geologically, these areas range from little metamorphosed rock, in the lowland areas, to crystalline Palaeozoic and Precambrian rock, aged from 500 to over 2,500 million years. The evolution of the Newfoundland portion of the Appalachia has been described by S. Hada in the paper entitled *Pre-Carboniferous Evolution of the Newfoundland Appalachians* (Ottawa: Ministry of Supply and Services Canada, 1980. 21 p. Geological Survey of Canada Paper No. 80–8). Furthermore, recent work on the historic metamorphosis of the Newfoundland Appalachia has been undertaken by B. Dube and G.R. Dunning in their paper entitled *New Insights into the Appalacian Orogen Geology and Geochronology along the Cape Ray Fault Zone, Southwest Newfoundland* (*Geological Society of America Bulletin*, v. 108, 1996. pp. 101–16).

Nova Scotia is divisible into three highland areas, three uplands, and several small lowlands, ranging in height from 1,500 feet in the Cape Breton Highlands, down to 600 feet on the southeast shore of the mainland. Geologically, Nova Scotia consists of hard Triassic basalt, about 200 million years old from the Mesozoic era, as well as late Palaeozoic sediments over 200 million years old. A good review of the geology of Nova Scotia has been written by A.E. Roland entitled *Geological Background and Physiography of Nova Scotia* (Halifax: Nova Scotia Institute of Science, 1982. 311 p.). Some of the earlier works upon which this publication relied were area studies, such as W.A. Bell's paper, entitled *Mississippian Horton Group of Type Windsor-Horton District, Nova Scotia* (Ottawa: Department of Mines, 1960. 112 p. Geological Survey of Canada Memoir No. 314); R.H. Maehl's paper, entitled *The Older Palaeozoic of Pictou County, Nova Scotia* (Halifax: Department of Mines, 1961, 10 p. Nova Scotia Department of Mines Memoir No. 4); and D.G. Kelley's paper, entitled *Basal Mississippian Volcanic Rocks in Cape Breton Island* (Ottawa: Department of Mines and Technical Surveys, 1965. 10 p. Geological Survey of Canada Paper No. 64-34). More recent work on the Nova Scotia portion of the Appalachia has been undertaken by S.M. Bair and R.P. Raeside in their papers entitled *Tectono-Stratigraphic Terranes in Cape Breton Island, Nova Scotia: Implications for the Configuration of the Northern Appalachian Orogen* (*Geology*, v. 17, 1989. pp. 822–25); and *Terrane Evolution and Accretion in Cape Breton Island, Nova Scotia*. In: *Current Perspectives in the Appalachian-Caledonian Orogen*. Edited by J.P. Hibbard and C.R. Van Staal (Toronto: Geological Association of Canada, 1995. pp. 391–407).

Geologically, New Brunswick falls into a single New Brunswick Highlands, in the west and central part of the province, reaching heights of 2,000 feet, the Chaleur uplands, astride the New Brunswick–Quebec border, rising to about 1000 feet, and the Maritime Plain, or lowland area, which includes Prince Edward Island and the Magdalen Islands. Some of the work on the geology of New Brunswick has covered specific areas, such as R.G. Skinner's paper, entitled *Geology of Tetagouche Lakes, Bathurst and Nepisiquit Falls Map-Area, New Brunswick with Emphasis on the Tetagouche Group (21 O/9, 21 P/12, 21 P/5)* (Ottawa: Department of Energy, Mines and Resources, 1974. 133 p. Geological Survey of Canada Memoir No. 371). But other work is broader in scope, such as the paper by M. Lamoth, entitled *Pleistocene Stratigraphy and Till Geochemistry of the Miramichi Zone, New Brunswick* (Ottawa: Supply and Services Canada, 1992. 58 p. Geological Survey of Canada Bulletin No. 433); the paper by K.L. Currie entitled *Repeated Basement Reactivation in the Northeastern Appalachians* (*Geological Journal*, v. 18, 1983. pp. 223–39); and the paper by C.R. Van Staal and J.A. DeRoo entitled *Mid Paleozoic Tectonic Evolution of the Appalachian Central Mobile Belt in Northern New Brunswick, Canada: Collision, Extensional Collapse and Dextral Transpression*. In: *Current Perspectives in the Appalachian Caledonian Orogen*. Edited by J.P. Hibbard and C.R. Van Staal (Toronto: Geological Association of Canada, 1995. pp. 367–89).

The Quebec portion of the Appalachia is primarily mountainous, where the tops feature the flat and undulating surface of the Appalachia peneplain. The eastern section consists of the Notre Dame Mountains, which run from Thetford Mines to the Gaspé Bay, and reach heights of 4,000 feet. The southwestern part consists of two other highland areas, one, the Sutton Mountains, which are a continuation of the Green Mountains of Vermont and reach 3,000 feet, and the Megantic Hills, which are a continuation of the White Mountains of the New England states. The rocks of this area are Devonian, Silurian, and Ordovician, of the Palaeozoic era, formed 300 million to 500 million years ago. Some examples of the work done on this region are by L.M. Cumming, entitled *Silurian and Lower Devonian Formation in the Eastern Part of the Gaspé Peninsula, Quebec* (Ottawa: Queen's Printer, 1959, 45 p.); by D. Lavoie, entitled *Peritidal Origin of the Lower Ordovician Upton Group, Southern Quebec Appalachians* (*Canadian Journal of*

Earth Sciences, v. 29, 1992. pp. 1106–18); and by S. Castonguay and A. Tremblay, entitled *Ordovician and Silurian Metamorphic Cooling Ages along the Laurentian Margin of the Quebec Appalachians: Bridging the Gap between New England and Newfoundland* (*Geology*, v. 25, 1997. pp. 583–86).

The Canadian Appalachia is bounded to the north by the St Lawrence River and the St Lawrence Lowlands, and to the east by the Gulf of St Lawrence and the Atlantic Continental Shelf. The Atlantic Continental Shelf is less than 200 metres in depth, until it slopes steeply at the Atlantic Continental Slope. Two major areas of study of this continental shelf are the Scotian Shelf and the Labrador Shelf, the former in particular, because of its petroleum potential. The Scotian Shelf is about 125 miles wide with some depressions or basins 200 metres deep, until it reaches the Slope.

The geology of the Scotian and Labrador Shelves was described by government researchers in the 1970s. L.H. King wrote a paper entitled the *Geology of the Scotian Shelf* (Ottawa: Environment Canada. Fisheries and Marine Service, 1976. 30 p. Geological Survey of Canada Paper No. 74-31); and D.C. Umpleby wrote a paper entitled *Geology of the Labrador Shelf* (Ottawa: Energy, Mines and Resources Canada, 1979. 32 p. Geological Survey of Canada Paper No. 79-13).

Other studies of these shelves have included G. Vilks's study, entitled *Postglacial Basin, Sedimentation of Labrador Shelf* (Ottawa: Energy, Mines and Resources Canada, 1980, 28p.); and J.A. Wade's paper, entitled *Petroleum Resources of the Scotian Shelf* (Ottawa: Energy, Mines and Resources Canada, 1989. 28 p. Geological Survey of Canada Paper No. 88–19). As well, B.C. Maclean and J.A. Wade have undertaken additional studies, entitled *Petroleum Geology of the Continental Margin South of the Islands of St. Pierre and Miquelon Offshore Eastern Canada* (*Bulletin of Canadian Petroleum Geology*, v. 40, 1992. pp. 222–53); and *Mezozoic and Cenozoic Stratigraphy, Eastern Scotian Shelf: New Interpretations* (*Canadian Journal of Earth Sciences*, v. 32, 1995. pp. 1462–73). Other research on the Scotian Shelf has been contract work, such as J.J. Judge's paper, entitled *Sea-Floor Stability Study, Inner Scotian Shelf* (Halifax: Mortec, 1987. 80 p.); and D.O. Hodgens's study, entitled *Field Measurements of Sediment Transport on the Scotian Shelf* (Vancouver: Seaconsult Marine Research, 1986. 2 v.).

(b) The Canadian Shield

The oldest part of North America is a triangular block of crust called the North American Craton. It is made up of ancient sedimentary strata that perhaps accumulated in subsidiary basins, formed in response to fracturing and down-warping of the ancient crust of the earth. The oldest strata, which date from Precambrian times, are located in western Ontario. During Precambrian times, areas of this strata were deformed by at least four episodes of crustal upheaval – one about 3,000 million years ago, another around 2,400 million years ago, a third about 1,600 million years ago, and a fourth somewhere between 1,000 million and 700 million years ago. This is the Canadian Shield. These upheavals gave a highly metamorphosed structure to the rocks of the Shield. Typical rocks are granitic, and include some highly metamorphosed types called gneisses (coarse grain bonded rocks), along with much-deformed sedimentary and volcanic rock found in Archian (early Precambrian) greenstone belts called greywackes. Since these upheavals, many long periods of erosion have influenced the levelling of broad expanses of the Canadian Shield, and have reduced it to a near-featureless surface. However, the whole area retains its geologically distinct ancient boundaries, with some areas physio-geologically distinct, but separated, from each other.

Today, there are four types of terrain in the Shield that can be distinguished. First are the plains, which are areas of unmetamorphosed rocks that lie more or less flat. Examples are the Thelon and Athabasca Plains (the former lies in the Arctic, west of Hudson Bay, and the latter in northern Saskatchewan), and the Nipigon and Cobalt Plains, which lie north of the Great Lakes in Ontario. These plains can range in height from 300 to 2,000 feet. Second are the hills, ranging in elevation from 1,000 to 2,000 feet, and composed of low-grade metamorphosed rocks of tilled or folded sediments. Examples of these hills are the Bathurst and East Arm Hills of the western Arctic, and the Penobean and Labrador Hills of southern Ontario and northwestern Quebec. Third are the mountains composed of massive anorthosites (feldspar), which stand out above the surrounding terrain. The only mountains in the Shield are the Mealy Mountains, which rise above the south side of the Melville Plain in southeast Labrador. They rise to a height of 3,700 feet. The final terrain type of the Canadian Shield is the highlands, which are uplifted areas made of massive crystalline rocks. Examples of the highlands are the Labrador Highlands, which extend along the northeastern flank of Baffin and Bylot Islands, and then north over eastern Devon Island and Ellesmere Island. These highlands reach a height of 7,000 feet.

The Shield has been studied intensively, and written about in overviews, or more narrowly about its geological provinces (for geological purposes the Shield is divided into 34 provinces and sub-provinces). The most recent overview study by A.S. Dyke and J.S. Vincent is entitled *Quaternary Geology of the Canadian Shield.* In: *Quaternary Geology of Canada and Greenland.* Edited by J.R. Fulton (Ottawa: Geological Survey of Canada, 1989. pp. 175–318. The Geology of North America, Volume K-1. Geology of Canada, No. 1). Another broad study, edited by J.A. Fraser and W.W. Heywood, was a 1977 symposium entitled *Metamorphism in the Canadian Shield: Proceedings of a Symposium Held in Ottawa, Canada, May 5–6,*

1977 (Ottawa: Energy, Mines and Resources Canada, 1978, 367 p. Geological Survey of Canada Paper No. 78-10). An earlier overview, edited by J.S. Stevenson, was entitled *The Tectonics of the Canadian Shield* (Toronto: University of Toronto Press, 1962, 180 p. Royal Society of Canada Special Publication No. 4). This overview was actually a series of papers, presented as a symposium, which formed part of the program of the annual meeting of Section IV (Geological Sciences) of the Royal Society of Canada held at McGill University, Montreal, in 1961. More recent work is also represented in the publication edited by J.M. Moore and A. Davidson entitled *The Grenville Province* (St John's: Geological Association of Canada 1986. 358 p. Geological Society of Canada. Special Publication No. 31). This, too, is a series of papers presented at a symposium, held in 1984 in London, Ontario. Finally, another publication at the 'province' level by J. King and H. Helmstaedt is entitled *The Slave Province, Northwest Territories, Canada.* In: *Greenstone Belts.* Edited by J. Dewit Maarten and L.D. Ashwal (Oxford: Oxford University Press, 1997. pp. 459–79).

Besides overview volumes about the Canadian Shield, or its provinces, specific areas are constantly researched. For example, some recent work on the Thelon Plain, or Zone, has appeared in the journal literature. One paper by N. Culshaw is entitled *Post-Collisional Oblique Convergence along the Thelon Tectonic Zone, North of the Bathurst Fault, NWT, Canada* (*Journal of Structural Geology*, v. 13, 1991. pp. 501–16). Another paper by J.B. Henderson and P.H. McGrath is entitled *Intracratonic Indentation of the Archean Slave Province into the Early Proterozoic Thelon Tectonic Zone of the Churchill Province, Northwestern Canadian Shield* (*Canadian Journal of Earth Sciences*, v. 27, 1990. pp. 1699–1713).

Recent work on specific hill areas of the Shield has been done by K.D. Card and entitled *Geology of the North Shore of Lake Huron: Penokean Fold Belt and Sudbury Structure.* In: *Mineral Deposits of North America* (Washington, DC: American Geophysical Union, 1989. v. 2, pp. 3–19.); by B.J. Johnson, in a paper entitled *Stratigraphy and Structure of the Early Proterozoic Wilson Island Group, East Arm Thrust Fold Belt, NWT* (*Canadian Journal of Earth Sciences*, v. 27, 1990. pp. 552–69); and, by D.E. Boerner and R.D. Kurtz, entitled *Electrical Conductivity and Paleo-Proterozoic Foredeeps* (*Journal of Geophysical Research, B., Solid Earth and Planets*, v. 101, 1996. pp. 13775–91).

Work on the only mountain area of the Shield has been done by R.J. Wardle and U. Schaur in a paper entitled *Inter-relationships between Thrusting, Magnetization and Generalization of a Mongonite-Anorthosite Suite in the Mealy Mountains Terrane, Grenville Orogen, Canada* (*Geological Association of Canada, Annual Technical Meeting Abstracts*, v. 25, 1989. p. 217). Other work on the mountains has been done by J.K. Park and E.I. Tanczyk in a paper entitled *Magnetic Fabric and Its Significance in the 1400 Ma Mealy Diabase Dykes of Labrador, Canada* (*Journal *of Geophysical Research, B. Solid Earth and Planets*, v. 93, 1988. pp. 13689–704); and by R.F. Emslie and E. Hegner, entitled *Reconnaissance Isotopic Geochemistry of Anorthosite-Mangerite-Charnockite-Granite (AMCG) Complexes, Grenville Province, Canada* (*Chemical Geology*, v. 106, 1993. pp. 279–98).

(c) The Arctic Archipelago

The Arctic Archipelago consists of the Innuitian Orogen, in the High Arctic, and includes the northwestern tip of Greenland, and a middle band called the Arctic Platform, consisting of some southern islands and the Arctic Coastal Plain. To the south, it is bounded by the Canadian Shield, which includes most of Baffin Island.

The Innuitian Orogen, which is the northern margin of the ancient North American Craton, is comprised of belts of folded sedimentary rocks, and older fault-bounded metamorphosed rocks. The area indicates four major stratigraphic successions, or deformations, in its geological history. The first is a concealed Upper-Middle and Upper Palaeozoic succession, over 1,000 million years old as indicated around the northernmost limits of the Canadian Shield. The second is an Upper Proterozoic to Upper Devonian (Mid-Proterozoic) Succession, ranging from 350 million to 600 million years old, which covers the Franklinian Mobile Belt over much of Melville Island and northernmost Ellesmere Island. The third succession, or deformation, occurred during the Carboniferous, down to Pleocene times, and ranges from 300 million to 500 million years ago. This succession makes up the Sverdrup Basin, located west and south of the Ellesmere and Axel Heiberg Islands. The last succession of the Archipelago occurred during the Cretaceous down to Holocene period, which ranged from 100 million down to one million years ago. This succession makes up the Arctic Coastal Plain, which skirts the western edge of Banks Island and the Queen Elizabeth Islands.

Because of these successions of geological upheavals, the Innuitian Orogen includes mountains, uplands, and lowlands, divided into four regions. These regions are the North Greenland Region and Axel Heiberg–Ellesmere Region of mountains and uplands, and Parry Upland and Sverdrup Lowland. The belt of uplands and mountains is rather continuous, from northern Greenland in the east to Melville Island in the west, and is only broken by deep channels such as Nares Strait. These uplands and mountains range in height from 3,000 feet up to just over 8,000 feet. On the other hand, the Sverdrup Lowland is less than 500 feet above sea level, and half the total area is covered by water with the seafloor at depths of 200 metres–500 metres.

The Arctic Platform borders the Canadian Shield at about 110 degrees longitude in the western Arctic, and touches the northernmost reaches of the Shield along the Arctic coastline to the east. Northward, the Platform makes a belt across most of Banks Island and Victoria Island, to the northwestern corner of Baffin Island. Geologically, the Arctic Platform is a series of Palaeozoic, Mesozoic, and Cenozoic sedimentary successions, ranging from 570 million years down to 55 million years in age.

These sedimentary rocks are mainly shallow-marine in origin, and their boundary stretches as far as northwestern Greenland. These rocks actually rest on the Precambrian basement of the Canadian-Greenland Shield. Cutting through the area are the Minto Arch of Victoria Island, an area of uplift, which developed during the Devonian to Jurassic geological period, 400 million down to 150 million years ago, and the Boothian Uplift, which cuts through the centre of the Arctic Archipelago, in the area of the Boothia Peninsula. This latter uplift occurred during the Silurian and Devonian period, over a 50-million-year period, starting about 400 million years ago.

For descriptive purposes, the Arctic Platform is divided into three large regions. The first is the Kane Basin–Independence Fjord Region, which stretches across northwestern Greenland and east-central Ellesmere Island. The second is the Lancaster Region, to the east of the Boothia Peninsula, and the third is Banks-Victoria Region, to the west of Boothia Peninsula. The southernmost part of the Platform, which touches the Canadian Shield, is called the Foxe Plain, and lies to the west of Baffin Island. Overall, the Platform consists of a broad belt of plains and plateaux, with an undulating surface in the 'interior,' but dips toward the continental margins at the northwest.

The Arctic Coastal Plain includes the coastal terrain along the shores of the Arctic Ocean, from Meighen Island to Alaska. The plains border, and lie seaward, of the Innuitian Orogen, the Arctic Platform, the Interior Plains, and the Cordilleran Region. The plains are underlain by Tertiary (65 down to 25 million years old) and Quaternary (25 million down to 2 million years old) deposits. Hence, its features are relatively young.

The most recent overviews of the geology of the Innuitian Orogen and the Arctic Platform, including the Arctic Coastal Plain, are in volumes edited by H.P. Trettin, entitled *Geology of the Innuitian Orogen and Arctic Platform of Canada and Greenland* (Ottawa: Geological Survey of Canada, 1991. 2 v. Geology of North America, Volume E. Geology of Canada No. 3); and in volumes edited by D.F. Stott and J.P. Aitken, entitled *Sedimentary Cover of the Craton in Canada* (Ottawa: Geological Survey of Canada, 1993. 2 v. Geology of North America, Volume D-1. Geology of Canada No. 5). These publications are part of the Geology of North America series produced as part of the Decade of North America Geology project.

Another earlier review was written by Y.O. Fortier and L.W. Morley in their paper entitled *Geological Unity of the Arctic Islands* (*Transactions of the Royal Society of Canada*. Series 3, v. 50, 1956. pp. 3–12).

In the mid-1950s and 1960s, the Arctic Archipelago was examined primarily by geological reconnaissance. One example of this type of geological work was by E.T. Tozer, and entitled *Geological Reconnaissance: Prince Patrick, Eglinton and Western Melville Islands, Arctic Archipelago, Northwest Territories* (Ottawa: Department of Mines and Technical Surveys, 1956. 320 p.). Another paper of this type was by R.G. Blackadar and entitled Geological Reconnaissance of the Precambrian of Northwestern Baffin Island, Northwest Territories (Ottawa: Department of Mines and Technical Surveys, 1965. 25 p. Geological Survey of Canada Paper 64–42).

Other work has focused upon specific geological features or geological time-periods. One example here is a paper by M.R. Gibling and G.M. Narbonne entitled *Siluro-Devonian Sedimentation of Somerset and Cornwallis Islands, Arctic Canada* (*Bulletin of Canadian Petroleum Geology*, v. 25, 1977. pp. 1145–56). Another example by F.G. Fox is entitled *Structural Geology of the Parry Islands Fold Belt* (*Bulletin of Canadian Petroleum Geology*, v. 33, 1985. pp. 306–40). And other examples are found in papers by N.D. Stewart entitled *Late Proterozoic to Early Tertiary Stratigraphy of Somerset Island and Northern Boothia Peninsula, District of Franklin, N.W.T.* (Ottawa: Geological Survey of Canada, 1987. 73 p. Geological Survey of Canada Paper No. 83-26); and by L.S. Lane and J.R. Dietrich entitled *Tertiary Structural Evolution of the Beaufort Sea–Mackenzie Delta Region, Arctic Canada* (*Bulletin of Canadian Petroleum Biology*, v. 43, 1995. pp. 293–314).

By the 1980s, there was much more exploration being carried out on the energy potential and environmental risks for the Arctic. For example, one paper by B.D. Ricketts and A.F. Embry was entitled *Summary of Geology and Resource Potential of Coal Deposits in the Canadian Arctic Archipelago (Bulletin of Canadian Petroleum Geology*, v. 32, 1984. pp. 339–71). Another paper by J. Utting was entitled *Thermal Maturity of Carboniferous and Permian Rocks of the Sverdrup Basin, Canadian Arctic Archipelago* (Ottawa: Energy, Mines and Resources Canada, 1989. 20 p. Geological Survey of Canada Paper 89-19). And yet another by C.D. Anglin was entitled *Preliminary Mineral and Energy Resource Assessment, Northern Bathhurst Island.* In: *Explorations Overview, 1994. Northwest Territories: Mining, Exploration and Geological Investigations.* Compiled by R. Kusick and P.S. Goff (Yellowknife: Northern Affairs Program, NWT Geology Division, 1994. pp. 21–22). Finally, environmental concerns are shown in the paper by P. McLaren entitled *The Coastal Morphology and Sedimentology of Eastern Lancaster Sound and Northeast Baffin Island: A Study of Shoreline Sensitivity to a Potential Oil Spill*

(Ottawa: Geological Survey of Canada, 1985. 32 p. Geological Survey of Canada Paper No. 83-24).

(d) Sedimentary Basins

The Interior Plains are part of the ancient North American Craton, which also contains the Canadian Shield and is bordered by the Appalachia in the east and the Cordillera in the west. The area was covered, from Early Palaeozoic times (over 500 million years ago), with marine sediments from a shallow sea. During Early Cambrian times, 500 million years ago, the craton warped, causing a separation of the shallow sea, and the creation of different forms and sizes of basins. As the marine sea subsided during the Permean period, about 250 million years ago, a rhythmic sedimentation of shales, sandstone, and limestone, along with fossil remains of plants and animals that settled in organic muds, occurred and gave rise to coal and hydrocarbon fluids that migrated into these sandstones and porous limestones.

Today, the Interior Plains of Canada are underlain by flat-lying Late Proterozoic, Paleozoic, Mesozoic, and Cenozoic strata, laid down during the sedimentation period, and occupying the region between the Shield on the east, and the mountains of the Cordillera to the west. The southern part is semi-arid grassland, the central part, tree-covered, and the northern part is tundra. The whole region is divided into several units of plains, hills, and plateaux, with the smaller and more varied units located in the north and the larger, less varied units in the south.

At the Arctic Slope, where the drainage is directed to the Arctic Ocean, are the Horton and Anderson Plains. The Horton Plain is underlain by flat-lying Paleozoic and late Proterozoic sediments, and its surfaces reach elevations of 1,200 to 2,000 feet in the south. The western part is rocky, and the eastern part is rolling areas of glacial drift. To the north, the Late Proterozoic strata are slightly folded and faulted, producing a rolling surface of low scarps (cliffs) and mesas (flat-topped hills). The Anderson Plain is undulating, and rises inland in rocky escarpments, 600 to 1,000 feet in elevation. Moving southward is the Peel Plain, which lies southwest of the Mackenzie River. Its southwestern part is a broad, shallow hollow, about 400 feet deep, and made up of Mesozoic sediments. To the northeast, the Paleozoic strata form hills that reach 1,000 feet in elevation. Between the Peel Plain and the Mackenzie Mountains is the Peel Plateau, which rises in a series of steps. The first step is made up of eroded Palaeozoic and Mesozoic strata, followed by a second and third rise of horizontal strata. These rising plateaux are separated from the Mackenzie Mountains by a broad, shallow valley. South of the Peel Plain are the Coville Hills, made up of ridges of Paleozoic strata, that rise above the surrounding plains. The hills and ridges enclose hollows and some large lakes, and the elevation ranges from 800 to 2,200 feet. Next follows the Great Bear Plain, which is largely composed of Mesozoic strata in a rolling surface, just under 1,000 feet. Between the Great Bear Plain and the Great Slave Plain are the Cartridge Mountains, which overlook the Great Slave Plain made up of Palaeozoic strata. South of the Great Slave Plain is the Alberta Plateau, which is made up of Cretaceous sediments. It is actually a ring of plateaux, separated by wide valleys, and includes the Cameron Hills and the Caribou Mountains, which reach a height of 3,200 feet toward the Great Slave Plain. To the east is the Saskatchewan Plain. The southern border of the plateau consists of the Cheecham Hills to the southeast, and the northern border consists of the Athabasca River. To the west are the foothills of the Rocky Mountains.

South of the Athabasca River is the Alberta Plain, largely composed of Mesozoic sediments, but with some Tertiary (Cenozoic) sediments in the south, along the Canada–USA border. Much of the Plain is about 2,500 feet in elevation, but the Cypress Hills in the southeast rise to 4,700 feet. The eastern edge of the Alberta Plain, called the Missouri Coteau, 'steps' down to the Saskatchewan Plain. The Saskatchewan Plain rises to about 2,600 feet, and is formed of Mesozoic sediments. Through it flows the Saskatchewan River. Finally, on the east of the Saskatchewan Plain is the Manitoba Escarpment, overlooking the Manitoba Plain, which is made of undulating to flat Palaeozoic sediments. The surface of the Plain has an elevation of about 800 feet, and is largely covered by lakes.

A major overview of the geological structure of the Interior Plain has been the volume, edited by C. Beaumont and A.J. Tankard, entitled *Sedimentary Basins and Basin-Forming Mechanisms* (Halifax: Atlantic Geoscience Society, 1987. 527 p. Special Publication No. 5). The basins are also studied regionally. Some recent publications pertaining to the Arctic Interior Plains and basins include work by J.B. Anderson entitled *Geology of the Yellowknife–Hearne Lake Area, District of Mackenzie: A Segment across an Archean Basin* (Ottawa: Ministry of Supply and Services Canada, 1985. 135 p. Geological Survey of Canada Memoir 414); and by C.J. Yorath entitled *Mesozoic and Cenozoic Depositional History of the Northern Interior Plains of Canada.* In: *The Mesozoic of Middle North America: A Selection of Papers from the Symposium on the Mesozoic of Middle North America, Calgary, Alberta, Canada, May 1983* (Calgary: Canadian Society of Petroleum Geologists, 1984. pp. 69–83). Other studies have been written by J.W. Porter and entitled *Early Surface and Subsurface Investigations of the Western Canada Sedimentary Basin* (Calgary: Alberta Association of Petroleum Geologists Memoir 55, 1992. pp. 125–57); and by T.G. Kolzer and T.K. Kyser entitled *Palaeomagnetism and the Evaluation of Fluids in the Proterozoic Athabasca Basin, Northern Saskatchewan, Canada* (*Canadian Journal of Earth Sciences*, v. 29, 1992. pp. 1474–91). In particular, D.G. Cooke and B.C. MacLean have studied the Anderson Plain, in a series of papers entitled *The Influence of Proterozoic Structures on the Development of Laramide Structures, Northern Interior Plains, Northwest Territories, Canada* (*Bulletin of Canadian Petroleum Geology*, v. 40, 1992. pp. 207–21); *The Intracratonic Paleoproterozoic Forward Orogeny, and Implications for Regional Correlations, Northwest Territories, Canada* (*Canadian Journal of Earth Sciences*, v. 32, 1995. pp. 1991–2008); and *Mid-Continent Tectonic Inversions, Northwest Territories, Canada* (*Journal of Structural Geology*, v. 18, 1996. pp. 791–802).

(e) The Western Cordillera

The Cordillera of North America forms an elevated mountain chain that runs from Alaska to Mexico. At its broadest, the North American Cordillera is about 1,600 kilometres wide at the latitude of the states of Colorado and California, but moving northward, the Canadian section of the Cordillera draws together into a narrower bundle of ranges between the Pacific coast and the Interior Plains, as far north as the Beaufort Sea. Geologists now agree that the growth of the North American Cordillera was caused by plate movements involving the Pacific, Juan de Fuca, Cocos, and North American Plates. These movements, which may have began prior to Tertiary times, over 65 million years ago, began a phase of mountain building, about 30 million years later, as the Juan de Fuca Plate became almost totally subducted under the continental crust of North America. The descent and burial of perhaps a 7,000-kilometre-wide strip of the Juan de Fuca Plate may have caused the heating and expansion at the shallower end of the buried plate, creating volcanic areas. This, and the further lateral contact movements of the Pacific and North American Plates, created further distortions of the continental crust. Overall, these crustal displacements were gradual, but continuous, over millions of years.

Geological research on the Canadian Cordillera indicates there were four major phases encompassing a wide variety of processes. First, there is an indication of a succession of carbonate, shale, and sandstone sediments, along the western passive margin of ancient North America, beginning in the middle Proterozoic era, over 1,500 million years ago. This phase created the foreland basins that follow the eastern edges of the Cordillera. The second phase, which began during the late Palaeozoic era around 400 million years ago, was an orogenic (mountain building) event that resulted in rifting volcanism and plutonism (the creation of volcanic rocks and granites). The third phase involved the building up and amalgamation of volcanic areas, caused by the subduction of the oceanic plates and by volcanic deformation and metamorphism, beginning during the Mesozoic era, over 200 million years ago. A fourth phase, beginning during the Cretaceous period,

100 million years ago, and lasting down through the Cenozoic era as late as five million years ago, caused the major displacements that are associated with the North American and Canadian Cordillera today.

Today, the Cordillera in Canada is divided longitudinally into three great belts called the Eastern System, the Interior System, and the Western System. The Eastern System is composed almost entirely of folded sedimentary strata. The Interior System is made up of folded sedimentary and volcanic strata and igneous rocks, along with massive metamorphic rocks. The Western System, which is the largest belt (the Coastal mountains), is composed mainly of volcanic rocks and volcanic and sedimentary strata. Within these belts are particular mountain ranges, plateaux, hills, valleys, trenches, basins, and plains. The Cordillera is also divided transversely into a number of east-west belts of relatively low terrain, or plateaux, which separate the mountain ranges. The most southerly is the Interior Plateau and in the north are the Liard Plateau, the Liard Plain and Yukon Plateau, and finally the Porcupine and Peel Plateau. Research on the Canadian Cordillera includes overviews based upon cumulative research to date, regional research of areas of the Cordillera, and specific studies.

There have been two major overviews written, covering the knowledge to date about the Canadian Cordillera, as well as individual articles. The first major overview by H.S. Bostock was entitled *Physiography of the Canadian Cordillera, with Special Reference to the Area North of the Fifty-Fifth Parallel* (Ottawa: Department of Mines and Technical Surveys, 1948. 106 p. Geological Survey of Canada Memoir No. 247). The most recent and far-reaching publication, which is part of the Decade of North American Geology Project, is a two-volume work, edited by H. Gabrielse and C.J. Yorath, entitled *The Geology of the Cordilleran Orogen in Canada* (Ottawa: Geological Survey of Canada, 1992, 2 v. Geology of North America, Volume G-2. Geology of Canada No. 4). Examples of specific research articles on the Canadian Cordillera are one by B.E. Nesbitt and K. Muehlenbachs entitled *Origins and Movement of Fluids during Deformation and Metamorphism in the Canadian Cordillera* (*Science*, v. 245, 1989. pp. 733–46); and another by M.P. Cecile and D.W. Morrow entitled *Early Paleozoic (Canadian to Early Devonian) Tectonic Framework, Canadian Cordillera* (*Bulletin of Canadian Petroleum Geology*, v. 45, 1997. pp. 54–74).

The more regional research covers the Eastern, Interior, and Western Systems of the Cordillera. From the Western System, one example of research by C.J. Yorath and R.D. Hyndman is entitled *Subsidence and Thermal History of Queen Charlotte Basin* (*Canadian Journal of Earth Sciences*, v. 20, 1983, pp. 135–59). Another example of research on the Western Systems is by G.J. Woodsworth and M.J. Orchard and entitled *Upper Palaeozoic to Lower Mesozoic Strata and Their Conodonts, Western Coast Plutonic Complex, British Columbia* (*Canadian Journal of*

Earth Sciences, v. 22, 1985. pp. 1329–44). From the Interior System, one example of research by K.D. Tempelman and D. Parkinson is entitled *Extension across the Eocene Okanagan Crustal Shear in Southern British Columbia* (*Geology* (Boulder), v. 14, 1986. pp. 318–21). Another paper by C.A. Evenchick is entitled *Geometry, Evolution, and Tectonic Framework of the Skeena Fold Belt, North Central British Columbia* (*Tectonics*, v. 10, 1991, pp. 527–46). Folding in the Interior System has also been studied by M.E. McMechan and R.I. Thompson in their paper entitled *The Canadian Cordilleran Fold and Thrust Belt South of 66 Degrees N and Its Influence on the Western Interior Basin.* In: *Evolution of the Western Interior Basin.* Edited by W.G. Caldwell and E.G. Kauffman (Toronto: Geological Association of Canada, 1993. pp. 73–90. Geological Association of Canada Special Paper No. 39). Covering the Eastern Systems, some of the regional research has been done by D.W. Morrow and H.H. Geldsetzer and entitled *Devonian of the Eastern Canadian Cordillera* (*Proceedings of the Second International Symposium on the Devonian System*, v. 1, 1988. pp. 85–121), and by L.C. Struik and entitled *Eastern Canadian Cordillera: Implications of a Balanced Basement* (*Proceedings of the Seventh International Conference on Basement Tectonics, Kingston, Ontario, 1987*, v. 1, pp. 305–11). As well, the Eastern System has been studied by R.J. Enkin and K.G. Osadetz in their paper entitled *Paleomagnetic Constraints on the Tectonic History of the Foreland Belt, Southern Canadian Cordillera: Preliminary Results* (*Canadian Journal of Earth Sciences*, v. 34, 1997. pp. 260–70).

Specific work on the Canadian Cordillera has been done by C.F. Roots and R.P. Parrish and entitled *Age of the Mount Harper Volcanic Complex, Southern Ogilvie Mountains, Yukon.* In: *Radiogenic Age and Isotopic Studies, Report 2.* Edited by J.K. Mortensen (Ottawa: Geological Survey of Canada, 1988. pp. 29–35. Paper No. 88–2). Other examples of specific work are papers by J.D. Aitken entitled *Uppermost Proterozoic Formations in Central Mackenzie Mountains, Northwest Territories* (Ottawa: Energy, Mines and Resources Canada, 1989. 26 p. Geological Survey of Canada Bulletin No. 368); and by P.B. Jones entitled *Structural Geology of the Modern Frank Slide and Ancient Bluff Mountain Slide, Crowsnest, Alberta* (*Bulletin of Canadian Petroleum Geology*, v. 41, 1993. pp. 232–43). This latter paper is an example of research carried out by a consultant group based in Calgary, Alberta, and called International Tectonic Consultants.

C Geochemical Techniques in the Service of Geology and Mining

In its broadest sense, geochemistry is the study of the chemistry of the earth, but more narrowly, it involves the study of the origin of ore deposits, chemical reactions involved in the weathering of rocks, the chemical changes that occur in the compaction and cementation processes in the formation of sedimentary rocks, and the changes that take place as rocks undergo metamorphism. Another aspect of geochemistry is the behaviour of biological materials and their subsequent disposition. Major problems of organic biogeochemistry include the incorporation of organic materials into rocks, including sedimentary rocks. The latter study involves the nature and chemical transformation of organic materials present in the deposits of coal, petroleum, and natural gas. Finally, organic geochemistry involves the study of geochemical processes involved in the weathering of rocks and the production of soil (pedogeochemistry), and the solution, precipitation, and secretion of dissolved materials, such as calcium carbonates (hydrogeochemistry).

The practical applications of geochemistry are in its use in geochemical prospecting, and the basis for this prospecting is a phenomenon called the geochemical halo, or dispersion train. A primary halo was formed in surrounding rocks at the time ore bodies were deposited in geological time. A secondary halo, or train, results from later weathering and transport processes. These halos, and trains, reflect the presence of mineralization in rocks, soils, water, stream sediments, rock gases, atmospheric gases, and biological materials. It is these halos, and trains, that provide the anomalies for which geochemists search when prospecting. Some of the methods used for the chemical analysis of rocks are X-ray fluorescent spectroscopy, in which the intensity of the characteristic X-radiation is in proportion to the concentration of the element in the rock, atomic absorption spectroscopy for routine geochemical analysis, and gas chromatography, the latter particularly useful for traces of hydrocarbons in soils, rocks, and water.

Geochemical prospecting research and applications began in the former USSR, in the 1930s, but were not developed in Canada until after the World War II, particularly by H.V. Warren and C.H. Howatson, when their paper entitled *Biogeochemical Prospecting for Copper and Zinc* (*Geological Society of America Bulletin*, v. 38, 1947. pp. 803–20) was published. Studies on geochemical halos, or the dispersion of ore metals in glacial materials, soil, and water, began in 1954, in the Keno Hill area of the Yukon, with the publication by R.W. Boyle of the report entitled *The Geology and the Geochemistry of the Silver-Lead-Zinc Reports of Keno Hill and Sourdough Hill, Yukon Territory: Preliminary Report* (Ottawa: Department of Energy, Mines and Resources, 1956. 78 p. Geological Survey of Canada Paper No. 55–30). In addition to the large amount of basic information gathered on the geochemistry of lead-zinc-silver deposits of the area, this early work focused upon the problems of applying geochemical prospecting techniques in regions underlain by permafrost. Geochemical work in this area has continued since the 1950s, with the publication of papers by C.F. Gleeson entitled *The Hydrogeochemistry of the Keno Hill Area, Yukon Territory* (Ottawa: Energy, Mines and Resources Canada, 1976. 22 p. Geological Survey of Canada Paper No. 75-14); and the *Litho-*

geochemistry of the Keno Hill District, Yukon Territory (Ottawa: Energy, Mines and Resources, 1980. 19 p. Geological Survey of Canada Paper No. 77-31). Additional research has also been undertaken by K.W. Watson in the paper entitled *Silver-Lead-Zinc Deposits in the Keno Hill–Galena Hill Area, Central Yukon* (*Yukon Geology*, v. 1, 1986. pp. 83–88); and by J.P. Franzen in the paper entitled *Metal-Ratio Zonation in the Keno Hill District, Central Yukon* (*Yukon Geology*, v. 1, 1986. pp. 98–108).

In the 1950s, also, reconnaissance geochemical surveys of waters and stream sediments were begun, using portable geochemical laboratories. These studies commenced in Nova Scotia and New Brunswick, with the publication of papers by R.W. Boyle entitled *Geochemistry of Pb, Zn, Cu, As,Sb, Mo, Sn, W, Ag, Ni, Co, Cr, Ba, and Mn in the Waters and Streams Sediments of the Bathurst-Jacquit River District, New Brunswick* (Ottawa: Department of Mines and Technical Surveys, 1966. 50 p. Geological Survey of Canada Paper No. 65-42); and E.W. Present, entitled *Geochemistry of Iron, Manganese, Lead, Copper, Zinc, Arsenic, Antimony, Silver, Tin, and Cadmium in the Soils of the Bathurst Area, New Brunswick* (Ottawa: Department of Energy, Mines and Resources, 1971. 93 p.). Geological Survey of Canada Bulletin No. 174).

In Ontario, major geochemical studies have been carried out on minerals in northwestern Ontario by N.J. Theis in the paper entitled *Uranium-Bearing and Associated Minerals in Their Geochemical and Sedimentological Context, Elliot Lake, Ontario* (Ottawa: Energy, Mines and Resources Canada, 1979. 50 p. Geological Survey of Canada Bulletin No. 304); and by D.C. Harris in the paper entitled *The Mineralogy and Geochemistry of the Hemlo Gold Deposit, Ontario* (Ottawa: Energy, Mines and Resources Canada, 1989. 88 p. Geological Survey of Canada Economic Geology Report No. 38). The mineralogy and geochemistry of gold deposits have also been studied in other Canadian regions, using a variety of techniques. For example, J.R. Belanger, in the paper entitled *Remote Sensing of Geochemical Anomalies Related to Ultrabasic Rocks and Gold and Uranium Mineralization* (*Canadian Journal of Remote Sensing*, v. 17, 1991. pp. 112–21), used Landsat imagery to study uranium ores in Saskatchewan and gold ores in southeastern Quebec. And C.E. Dunn and G.E. Ray have studied gold ores in British Columbia, in their paper entitled *A Comparison of Lithogeochemical and Biogeochemical Patterns Associated with Gold Mineralization in Mountainous Terrain of Southern British Columbia* (*Economic Geology and the Bulletin of the Society of Economic Geologists*, v. 90, 1995. pp. 2232–43).

Besides the use of geochemistry in the study of mineral and ore deposits, organic geochemistry has been particularly important for the study and understanding of the processes that control the formation, migration, and accumulation of sedimentary deposits, and the evaluation of their petroleum potential. Early

Canadian work in this area was by C.E.B. Conybeare, and entitled *Hydrocarbon Generation Potential and Hydrocarbon Yield Capacity of Sedimentary Basins* (*Canadian Petroleum Geologist, Bulletin*, v. 13, 1965. pp. 509–28). Other major work has been the study by G. Deroo entitled *The Origin and Migration of Petroleum in the Western Canadian Sedimentary Basin, Alberta: A Geological and Thermal Maturation Study* (Ottawa: Energy, Mines and Resources Canada, 1977. 136 p. Geological Survey of Canada Memoir No. 262). And, by the early 1980s, major work such as the paper by L.R. Snowdon entitled *Organic Geochemistry of the Upper Cretaceous-Territory Delta Complexes of the Beaufort, Mackenzie, Sedimentary Basins, Northwest Territories* (Ottawa: Energy, Mines and Resources Canada, 1981. 46 p. Geological Survey of Canada Bulletin No. 29) began to appear. And since then, particular studies on petroleum migration have been noted in the research paper by D.L. Stasiuk and L.R. Snowdon entitled *Fluorescence Micro-Spectrometry of Synthetic and Natural Hydrocarbon Fluid Inclusions: Crude Oil Chemistry, Density and Application to Petroleum Migration* (*Applied Geochemistry*, v. 12, 1997. pp. 229–41).

D The Search for Economic Minerals in Canada

Mineral exploration, which is a part of geological studies, has a long history that goes back to the 16th and 17th centuries, with Canada's early explorers such as Cartier, Frobisher, and Champlain. However, the systematic search for minerals began with the founding of the Geological Survey of Canada, as previously described. Although the coal and iron-and-steel industries, in the Maritime provinces, were under way by the 1850s, and asbestos was found in the Eastern Townships of the Canadian Appalachia, the search for minerals in Canada has been extricably linked to the Canadian Shield.

Nickel was found at Sudbury, Ontario, in 1883. And in the early part of the 20th century, cobalt was found, and the Kirkland Lake and Porcupine gold developments opened up in northern Ontario. As well, there were copper discoveries at Noranda, in Quebec. Toward the half-way mark of 1900s, the Quebec-Labrador iron ore opened up, uranium areas in Beaverlodge, Saskatchewan, and Elliot Lake, Ontario were discovered, and copper deposits were opened up in the Gaspé and Chibougamau areas of Quebec. Major copper-zinc deposits were opened up in Manitouwadge, Ontario, and lead-zinc deposits at Pine Point of Great Slave Lake were explored. In the Cordillera, gold, silver, and base metal discoveries were made in the 1880s and 1890s, in southern British Columbia, including the lead-zinc-silver deposits at Kimberley. And, in the north, Yukon gold was discovered in 1869 and 1897–98. On the eastern slopes of the Cordillera, the mid-20th century was notable for the hydrocarbon discoveries in Alberta.

Beginning with nickel, Canada has been a world leader in the production of that mineral since the exploitation of the Sudbury deposits began in 1887. Still today, more than 70% of Canada's total nickel production is derived from the Sudbury district. The Sudbury ore-bodies, including others in the Canadian Shield, are sulphide in nature, with the principal ore mineral being pentlandite. These nickel deposits are located in, or near, magmatic rocks. At Sudbury, these nickel sulphide ore-bodies occur at, or near, the outer margins of a body of igneous rock known as the Sudbury Irruptive. These Sudbury ores have been described in detail by J.E. Hawley and R.L. Stanton in three related and consecutive papers entitled *The Sudbury Ores, Their Mineralogy and Origin – Pt. 1, The Geological Setting* (*Canadian Mineralogist*, v. 7, 1962. pp. 1–29); *The Sudbury Ores, Their Mineralogy and Origin – Pt. 2, The Facts – The Ores, Their Minerals, Metals and Distribution* (*Canadian Mineralogist*, v. 7, 1962. pp. 30–145); and *The Sudbury Ores, Their Mineralogy and Origin – Pt. 3. Interpretations – The History and Origin of the Sudbury Ores* (*Canadian Mineralogist*, v. 7, 1962. pp. 146–207). The most recent and most comprehensive treatise on the Sudbury ore deposits, however, has been a special volume, edited by E.G. Pye and A.J. Naldrelt, entitled *The Geology and Ore Deposits of the Sudbury Structure* (Toronto: Ontario Geological Survey, 1984. 603 p. Ontario Geological Society Special Volume 1). The chapters in this volume were written by provincial, federal, and university geologists, as part of Ontario's bicentennial year.

Another major nickel reserve was discovered, in the mid-1950s, at Thompson, Manitoba. These deposits were first described by H.D. Wilson and W.B. Brisbin in their paper entitled *Regional Structures of the Thompson–Moak Lake Nickel Belt, Manitoba* (*Canadian Mining and Metallurgy Bulletin*, v. 53, 1960. p. 195). They have also been described by J.F. Davies in the paper entitled *Mineral Deposits Related to Major Structures in the Precambrian of Manitoba* (*Canadian Mining and Metallurgical Bulletin*, v. 57, 1964. pp. 661–65); by A.D. Paktene, in the paper entitled *Petrogenesis of Ultramafic and Mafic Rocks of the Thompson Nickel Belt, Manitoba* (*Contributions to Mineralogy and Petrology*, v. 88, 1984. pp. 348–53); and by F. Fueten and P.Y. Robin in the paper entitled *Deformation in the Thompson Belt, Central Manitoba: A Progress Report* (Ottawa: Geological Survey of Canada. Current Research, Part B-, 1986. pp. 797–809). Finally, an overview of nickel, its production and export, has been written by B.W. Mackenzie and entitled *Nickel – Canada and the World* (Ottawa: Department of Energy, Mines and Resources, 1968. 176 p.).

Unlike nickel, cobalt is found only in the Canadian Shield, and the greatest part as a by-product of nickel–copper ores at Sudbury, Ontario, and Lynne Lake, Manitoba. Again, the mineral ore pentlandite is the main source of cobalt. J.W. Hawley and R.L. Stanton, in their papers *The Sudbury Ores, Their Mineralogy and*

Origin, referred to above, estimated the cobalt content in pentlandite to be at about 1%. More recently, cobalt has been linked with the silver production from Cobalt and Gowganda areas of Ontario. Shipping of by-product cobalt from nickel mines began from these areas in 1954. An overview of this mineral was written at that time by R.J. Jones and entitled *Cobalt in Canada* (Ottawa: Department of Mines and Technical Surveys, 1956. 96 p.). Much more recently, however, the area was re-examined by D.R. Sharpe, of the Geological Survey of Canada, in the paper entitled *The Gowganda Formation: Revisited to Ponder Resedimentation* (*Geoscience Canada*, v. 12, 1985. pp. 109–11). It is interesting to note that research on cobalt was initiated, after the turn of the last century, by the first Superintendent of Mines, Eugene Haanel. The impetus for this research was the fact that cobalt could be retrieved from mixed silver-cobalt ores mined at Cobalt. By 1970, most of Canada's cobalt was being produced from silver-cobalt ores, as well as nickel-cobalt ores.

In Canada, silver is produced, either as a mining metal, or as a by-product, from mixed ores such as the silver-cobalt ores found at the Cobalt and Gowganda areas of Ontario. These Cobalt area ores were producing as much as 33 million ounces a year by 1920, and today this region produces over 26% of the silver obtained in the Canadian Shield. These deposits have been most recently described by J.A. Andrews and L.Owsiacki in two papers entitled *The Silver Deposits at Cobalt and Gowganda, Ontario: I, Geology, Petrology, and Whole-Rock Geochemistry* (*Canadian Journal of Earth Sciences*, v. 23, 1986. pp. 1480–1506); and *The Silver Deposits at Cobalt and Gowganda, Ontario: II, An Experiment in Age Determinations Employing Radiometric and Palaeomagnetic Measurements* (*Canadian Journal of Earth Sciences*, v. 23, 1986. pp. 1507–18).

However, the largest silver production in the Shield comes from massive copper-zinc-sulphide deposits, called the Flin Flon type, mined essentially for their copper and zinc contents, with silver as the by-product. About 61% of the silver obtained from the Shield comes from these deposits. They are generally large, irregular ore-bodies, consisting of pyrite, pyrotite, chalcopyrite, sphalerite, and galena. One of the largest of these types of sulphide bodies is the Kidd Creek Deposit near Timmins, Ontario. These deposits have been described by W. Petruk and D.R. Owens in two papers entitled *Part 1: Mineralogical Investigation of Zinc and Copper Ore from the North and South Zones of the Texas Gulf Sulphur Deposit, Kidd Township, Ontario* (Ottawa: Department of Energy, Mines and Resources, 1975. 1 v. Mines Branch: Investigation Report No. IR65–89); and *Part 2: Mineralogical Investigation of a Zinc-Copper-Silver-Lead Ore from the North Zone of the Texas Gulf Sulphur Deposit in Kidd Township near Timmins, Ontario* (Ottawa: Department of Energy, Mines and Resources, 1975. 1 v. Mines Branch Investigation Report No. IR65-105). However, most recently, seismic imaging has been used to study these large sulphide deposits. These studies were undertaken by M.H. Salisbury and D. Eaton, and reported in three papers entitled *Seismic Imaging of Massive Sulphide Deposits: Part I, Rock Properties* (*Economic Geology and the Bulletin of the Society of Economic Geologists*, v. 91, 1996. pp. 821–28); *Seismic Imaging of Massive Sulphide Deposits: Part 2, Reflection Seismic Profiling* (*Economic Geology and the Bulletin of the Society of Economic Geologists*, v. 91, 1996. pp. 829–34); and *Seismic Imaging of Massive Sulphide Deposits: Part III, Borehole Seismic Imaging of Near – Vertical Structures* (*Economic Geology and the Bulletin of the Society of Economic Geologists*, v. 91, 1996. pp. 835–40).

Copper and zinc ores come primarily from deposits located in the Canadian Shield and account for over 70% of Canadian production of copper alone. And zinc itself became Canada's leading base metal by the 1970s. About half of the copper mined comes from nickel-copper mines, and the rest from copper-zinc mines. The ores, themselves, come from meta-volcanic and meta-sedimentary rocks which are mixtures of pyrite, pyrotite, sphalerite, chalcopyrite, and galena. Most production has been from the Abitibi Belt in the Superior Province of the Shield, which extends from Timmins, Ontario, to Matagami and Chibougamau districts in Quebec. The copper deposits of the Abitibi Belt have been discussed in various papers. One such by M. Latulippe was entitled *The Relationship of Mineralization to Precambrian Stratigraphy in the Matagami Lake and Val d'Or Districts of Quebec* (Geological Association of Canada Special Publication No. 3. Precambrian Symposium, 1966. pp. 21–42). Another paper was by J. Kutina and A. Fabbri and entitled *Relationship of Structural Lineaments and Mineral Occurrences in the Abitibi Area of the Canadian Shield* (Ottawa: Department of Energy, Mines and Resources, 1972. 32 p. Geological Survey of Canada Paper No. 79-9). Examples of papers discussing specific areas of the Abitibi Belt are by W.A. Maclean, entitled *Geology and Ore Deposits of the Matagami District* (Canadian Institute of Mining and Metallurgy Special Volume 34, 1984. pp. 483–95); and by R. Kanwar, entitled *Geology and Ore Potential at Chibougamau* (*Western Miner*, v. 57, 1984. pp. 18–21).

Other important deposits are found in the Manitouwadge area of Ontario and the southern part of Churchill Province at Flin Flon, Manitoba. Further north, in the Bear Province of the Shield, much testing has been done on the copper occurrences in the basalts of the Coppermine River region. However, the origin of the deposits remains to be established. These deposits have been described by E.P. Kindle in the paper entitled *Preliminary Report on the Copper Deposits: Coppermine River Area, District of Mackenzie* (Ottawa: Department of Energy, Mines and Resources, 1970. 13 p. Geological Survey of Canada Paper No. 70–49). And some work on the origin of these deposits has been done by F.H. Campbell, and reported in the paper entitled *Stratigraphy of the Rae Group, Coronation Gulf Area,*

Districts of Mackenzie and Franklin (Ottawa: Geological Survey of Canada. Current Research, Part A, 1983. pp. 43–52).

Surprisingly, lead, which may occur in conjunction with zinc and copper ores, has had little production in the Canadian Shield, compared to that geological region's output of other metals already mentioned. Only about 1.5% of lead production is derived from the Shield. However, there has been more interest in the Appalachian areas of New Brunswick in the form of copper-lead-zinc sulphides usually in the form of pyrites. These deposits have been described by S. Roy in the paper entitled *Mineralogy and Paragenesis of Lead-Zinc-Copper Ores of the Bathurst-Newcastle District, New Brunswick* (Ottawa: Department of Mines and Technical Surveys, 1961. 19 p. Geological Survey of Canada Bulletin No. 72-30); and by R.W. Boyle in the paper entitled *The Geology, Geochemistry, and the Origin of the Barite Manganese, and Lead-Zinc-Copper-Silver Deposits of the Walton-Cheverie Area, Nova Scotia* (Ottawa: Department of Energy, Mines and Resources, 1971. 181 p. Geological Survey of Canada Bulletin No. 166). The only recent work on these deposits, however, has been carried out by D.F. Sangster and M.D. Burtt in the paper entitled *Geology of the B Baseline Zone, Walton CU-Pb-Zn-Ag-Ba Deposit, Nova Scotia, Canada* (*Economic Geology and the Bulletin of the Society of Economic Geologists*, v. 93, 1998. pp. 869–82).

Also linked to the Appalachia of Canada has been the historic production of iron ore since the mid-19th century. This early mining was from relatively small deposits of various types of iron ore in Nova Scotia, and New Brunswick, but after World War I production shifted to Wabana, Newfoundland. Iron deposits were discovered, in 1820 and 1825, near Annapolis, Nova Scotia, and mined until 1916. These deposits are hematite-greenalite-magnetite ores that occur in limestone. The New Brunswick ores are magnetite concentrations, found near Bathurst, and were mined between 1910 and 1915. These deposits have been described by B.W. Boyd in the paper entitled *Iron Ore in Canada, 1886–1986* (Ottawa: Energy, Mines and Resources Canada, 1988. 21p. Mineral Policy Sector Internal Report MRI 88/2). The Wabana iron deposits were discovered on Bell Island, in Conception Bay, Newfoundland, in 1892. These are hematite ores in limestone, and were worked until 1966, at which time they were closed because of more cost-efficient production elsewhere. These deposits have been described by W.K. Coughlan in the paper entitled *Geology of the Wabana Deposit (Canadian Mining and Metallurgy Bulletin*, v. 59, 1966. pp. 171–75).

The major source of iron ore in Canada has been derived from the Precambrian rocks of the Canadian Shield. In fact, about 95% of Canadian production comes from the Shield. This production began in 1939, with the opening of the Helen Mine in the Michipicoten district of Ontario. This deposit was described by Algoma Ore Properties Limited, in their paper entitled *Helen Mine Geology, Pt. 2 of Algoma Ore Properties Limited (Ontario)* (*Canadian Mining Journal*, v. 77, 1956. pp. 80–87). Then, during World War II, the Steep Rock iron range, north of Lake Superior, was developed. The Steep Rock deposits have been described by H.S. Hicks in the paper entitled *Geology of the Iron Deposits of Steep Rock Iron Mines Ltd. (Ontario)* (*Precambrian*, v. 23, 1950. pp. 8–10, 13); by A.W. Joliffe, in the paper entitled *Stratigraphy of the Steep Rock Group, Steep Rock Lake, Ontario* (Geology Association of Canada Special Paper No. 3. Precambrian Symposium 1966. pp. 75–98); and by M.E. Wilks and E.G. Nisbet in the paper entitled *Stratigraphy of the Steep Rock Group, Northwest Ontario; A Major Archaean Unconformity and Archaean Stromatolites* (*Canadian Journal of Earch Sciences*, v. 25, 1988. pp. 370–91). And, after World War II, the first ores from the Labrador-Quebec ranges were developed.

There are many types of iron deposits in the Precambrian, but they predominantly fall into the Algoma or Superior type. The Algoma type of iron formation was deposited with volcanic and sedimentary rocks, and composed of grey or red-jasper chert (hard crystalline rock), interlayered with magnetite and hematite rich layers. These formations tend to be located in Superior and Churchill geological provinces of the Shield. These have been described by A.M. Goodwin in the paper entitled *Mineralized Volcanic Complexes in the Porcupine–Kirkland Lake–Noranda Region* (*Economic Geology*, v. 60, 1965. pp. 955–71). Iron formations of the Superior type are thin-banded cherty (silica type) rocks, with granular oolite (limestone) textures. These formations contain large, naturally enriched, hematite-goethite iron deposits. Superior type iron formations in the Labrador Geosyncline extend more than 700 miles west of Ungava Bay, then southeast, beyond Schefferville, Quebec, and Wabush Lake, Labrador, and finally southward for 200 miles within Grenville Province of the Shield. These deposits have been described by G.A. Gross in the monograph entitled *Iron Ranges of the Labrador Geosyncline* (Ottawa: Department of Energy, Mines and Resources, 1968. 179 p. Economic Geology Report No. 22). This is volume III of the author's work, entitled *Geology of Iron Deposits in Canada.* Volume I of the series was entitled *General Geology and Evaluation of Iron Deposits* (Ottawa: Department of Energy, Mines and Resources, 1967. 181 p. Economic Geology Report No. 22); and Volume II was entitled *Iron Deposit: in the Appalachian and Grenville Regions of Canada* (Ottawa: Department of Energy, Mines and Resources, 1968. 111 p. Economic Geology Report No. 22).

Like iron deposits, gold is found throughout the Shield, but has also been mined extensively in every province and territory in Canada, particularly in the Cordillera of British Columbia, in the Greenwood and Rossland areas, and in the Yukon River, in the form of placer gold, during the 1869 gold rush. In the Shield, considerable gold is recovered as a by-product in the mining of copper, zinc, and nickel deposits. The greater part of the gold production has come from six principal gold camps – Porcupine, Kirkland Lake/Larder Lake, Northwestern Quebec, Red Lake, Yellowknife, and Little Long Lac. The first three camps account for about three-quarters of the total production, and are grouped in a relatively small area that extends a hundred miles east and west of the Ontario-Quebec boundary. The other camps are in northwestern Ontario and in the Northwest Territories. Geologically, most mines are located in the Superior Province of the Shield, with the exception of the Yellowknife deposit, which is located in the Slave Province. Gold deposits in the Churchill and Grenville provinces have little production. In the Canadian Shield, gold deposits show a close relationship to the belts of Archean volcanic rocks (formed 2,400 million to 3,000 million years ago), with most of the deposits occurring in basic lavas, tuffs, and sediments associated with those rocks. The most important of these belts, as with iron deposits, are in the Abitibi Belt.

An overview of Canadian gold deposits has been written by R.W. Boyle in the monograph entitled *The Geochemistry of Gold and Its Deposits: Together with a Chapter on Geochemical Prospecting for the Element* (Ottawa: Energy, Mines and Resources Canada, 1979. 584 p. Geological Survey of Canada Bulletin No. 280). Cordilleran deposits have been described by L.C. Strurk in the paper entitled *Structural Geology of the Caribou Gold Mining District, East-Central British Columbia* (Ottawa: Energy, Mines and Resources Canada 1988. 100 p. Geological Survey of Canada Memoir No. 421). Research pertaining to the gold deposits of the Canadian Shield includes papers by W.R. Baragan entitled *Geochemistry of the Yellowknife Volcanic Rocks* (*Canadian Journal of Earth Sciences*, v. 3, 1966. pp. 9–30), by S.A. Ferguson entitled *The Relationship of Mineralization to Stratigraphy in the Porcupine and Red Lake Areas, Ontario* (Geological Association of Canada Special Paper No. 3, 1966. pp. 99–120); and by A.M. Goodwin, entitled *Mineralized Volcanic Complexes in the Porcupine–Kirkland Lake–Noranda Region* (*Economic Geology*, v. 60, 1965. pp. 955–71). Finally, one of the more recent gold deposits in the Shield, described in the early 1980s, is the Hemlo gold deposit near the northeast shore of Lake Superior. This has been described by P.C. Harris in the paper entitled *The Mineralogy and Geochemistry of the Hemlo Gold Deposit, Ontario* (Ottawa: Energy, Mines and Resources Canada, 1989. 88 p. Geological Survey of Canada Economic Geology Report No. 38).

Like the gold deposits of the Shield, uranium ores of the Shield have had a high profile. In fact, Canada is one of the leading producers of uranium, and has the largest known reserves of ore, all located in the Shield. Production from the Shield began, in 1942, at the Eldorado Mine on Great Bear Lake. These deposits were first described by R. Murphy in the paper entitled *Geology and Mineralogy at Eldorado Mine, Northwest Territories* (*Canadian Institute of Mining and Metallurgy*

Transactions v. 49, 1946. pp. 426–35), and were noted again much later, in passing, by H.R. Steacy and R.W. Boyle in the paper entitled *Mineralogical Notes on the Uranium Occurrences at South March and Eldorado, Ontario.* In: *Report of Activities, Part B, November, 1972 to March, 1973* (Ottawa: Geological Survey of Canada Paper 73-1B, 1973. pp. 103–5). Larger deposits, however, were found in northern Saskatchewan, at Beaverlodge, in 1946, and in Ontario, at Blind River, in 1949. These deposits have been described by V. Koeppel in the paper entitled *Age and History of the Uranium Mineralization of the Beaverlodge Area, Saskatchewan* (Ottawa: Department of Energy, Mines and Resources, 1968. 111 p.), and by J.A. Robertson in the paper entitled *The Blind River Uranium Deposits: The Ores and Their Setting* (Ottawa: Ministry of Natural Resources, Division of Mines, 1976. 45 p.).

Although uranium has been actively sought, and many occurrences found since the late 1940s, it is a mining industry that is extremely dependent upon demand, and exploitation has varied with that demand. Most recent reserves have been concentrated at the Elliot Lake area in Ontario, and in the Athabasca region of Saskatchewan. The Elliot Lake reserves have been described by N.J. Theis in the paper entitled *Uranium-Bearing and Associated Minerals in Their Geochemical and Sedimentological Context, Elliot Lake, Ontario* (Ottawa: Energy, Mines and Resources Canada, 1979. 50 p.). The Athabasca area reserves have been described by E.M. Cameron in an edited volume entitled *Uranium Exploration in Athabasca Basin, Saskatchewan, Canada* (Ottawa: Energy, Mines and Resources Canada, 1983. 310 p.). Finally another area of uranium exploration has been in the Keewatin area of the Northwest Territories. These explorations have been described by A.R. Miller in the paper entitled *Uranium Geology of the Eastern Baker Lake Basin, District of Keewatin, Northwest Territories* (Ottawa: Energy, Mines and Resources Canada, 1980. 63 p. Geological Survey of Canada Bulletin No. 330).

Uranium has a strong tendency to form oxides, rather than sulphides, and to be dispersed, rather than concentrated in high-grade deposits. It occurs as uranium oxide (U_3O_8). Canadian uranium ores also occur as quartz-pebble conglomerates containing brannerite, uraninite, and uriniferous monazite in stringer systems, or veins, of pitcheblende, and in dykes of syenite and granite. The nature of these occurrences has been described by H. Baadsgaard and G.C. Cumming in the paper entitled *U-Pb Geochemistry of Minerals from the Midwest Uranium Deposit, North Saskatchewan* (*Canadian Journal of Earth Sciences*, v. 21, 1984. pp. 642–48); by J.A. Robertson, in the paper entitled *Uranium Deposits of Ontario. Their Distribution and Classification* (Toronto: Ontario Ministry of Natural Resources, 1981. 37 p.); by S.M. Roscoe, in the paper entitled *Huronian Rocks and Uraniferous Conglomerates* (Ottawa: Department of Energy, Mines and Resources, 1969. 203 p. Geological Survey of Canada Paper No. 68-40); and by the same author in another paper entitled *Uranium in Granites* (Ottawa: Energy, Mines and Resources Canada, 1982. 173 p. Geological Survey of Canada Paper No. 81-23).

Like the specificity of uranium deposits, which are restricted to the Canadian Shield, important deposits of coal, a historic energy source, occur only in the Canadian Appalachia and the Canadian Cordillera. The first coal mined in Canada occurred in the 17th century in New Brunswick and Cape Breton. Geologically, these deposits occur in the Appalachian Mineral Province, which has a complex tectonic and geological history involving several orogenic (mountain building) disturbances. These disturbances occurred mainly in southwestern New Brunswick, western Newfoundland, and northeastern Nova Scotia. It is interesting to note that with the last two orogenic disturbances, the faulting and folding of the Carboniferous coal-bearing beds of the Pennsylvanian type occurred partly at the same time as their deposition.

Although coal has long been one of the important mineral products of the Canadian Appalachia, production peaked in 1940, and has declined since that time. The major deposits mined have been at Sydney on Cape Breton Island. The Sydney coalfield underlies a narrow fringe of lowland, extending just over 30 miles along the coast of the Atlantic Ocean, and covers a land area of about 200 square miles, as well as a large area beneath the sea. The reserves of this deposit represent about 90% of the total remaining reserves of the entire Canadian Appalachia. This field has been described by T.B. Haites in the paper entitled *Some Geological Aspects of the Sydney Coalfield with Reference to Their Influence on Mining Operations* (*Canadian Institute of Mining and Metallurgy Transactions*, v. 54, 1951. pp. 215–25); and by P.A. Hacquebard in the paper entitled *Geological Development and Economic Evaluation of the Sydney Coal Basin, Nova Scotia* (Dartmouth, NS: Atlantic Geoscience Centre, 1983. Geological Survey of Canada Paper No. 84-1A. pp. 17–28). The Sydney coalfield is part of the Pictou Group of deposits that were geologically the youngest laid down in the Canadian Appalachia. Besides the Sydney coalfield, this group includes the Inverness and Matou mines, and the Pictou coalfield itself in Nova Scotia, and the Minto coalfield in New Brunswick. The Pictou Group has been described by M.S. Barss and P.A. Hacquebard in the paper entitled *Age and Stratigraphy of the Pictou Group in the Maritime Provinces as Revealed by Fossil Spores* (Geological Society of Canada Special Paper No. 4, 1967. pp. 267–82). There are two other distinctive coal deposits in the Canadian Appalachia. They are the Riverdale Group, which includes the St George's coalfield of southeast Newfoundland, the St Rose–Chimney Corner and the Port Hood coalfields of Nova Scotia, and the Cumberland Group, which are in the northern part of the Cumberland Basin of Nova Scotia. The latter includes the Joggins, River Hebert, and Springfield coalfields. These groups have been described by P.A.

Hacquebard in the paper entitled *The Correlation by Petrographic Analyses of No. 5 Seam in the St. Rose and Chimney Corner Coalfields, Nova Scotia* (Ottawa: Department of Mines and Technical Surveys, 1951. 33 p. Geological Survey of Canada Bulletin No. 19); and in the paper by M.J. Copeland entitled *Coalfields, West-Half Cumberland County, Nova Scotia* (Ottawa: Department of Mines and Technical Surveys, 1959. 89 p. Geological Survey of Canada Memoir No. 298). Overall, the most recent description of the coalfields of the Canadian Appalachia has been written by R.A. MacLean in the chapter entitled *Maritime Coal Fields: Geological Setting.* In: *Coal in Canada.* Edited by T.H. Patching (Montreal: Canadian Institute of Mining and Metallurgy, 1985. pp. 39–45. Canadian Institute of Mining and Metallurgy Special Volume 31).

In the Canadian Cordillera, and the forelands, coal deposits are widely distributed, as far west as Vancouver Island, and as far east as Saskatchewan, but the largest deposits are in the southern Rocky Mountains and foothills. Coal deposits occur mainly in the Kootenay Formation, which was formed during the Cretaceous period, 65 million years ago. The coal is mostly bituminous. In southeastern British Columbia, these coal deposits are found in the Crowsnest area where the strata have been folded and faulted, and the coal squeezed into pockets on the crests and troughs of the folds. The coal is medium bituminous in quality and has excellent coking qualities. To the north, along the Bow Valley in the vicinity of Canmore, in Palaeozoic strata, coal has been mined since 1888. Again, this coal is bituminous. The Kootenay Formation has been described by D.W. Gibson in the paper entitled *Stratigraphy, Sedimentology, and Depositional Environments of the Coal-Bearing Jurassic-Cretaceous Kootenay Group, Alberta and British Columbia* (Ottawa: Ministry of Supply and Services, 1985. 108 p. Geological Survey of Canada Bulletin No. 357).

In the forelands of the Interior Plains, coal has been mined as far east as Saskatchewan, from the Tertiary Ravenscrag Formation formed about 65 million years ago. This coal, which has been used primarily for electrical generation, has been described by S.H. Whitaker in the monograph entitled *Coal Resources of Southern Saskatchewan: A Model for Evaluation Methodology* (Ottawa: Energy, Mines and Resources Canada, 1978, 2 v. Geological Survey of Canada Economic Geology Report No. 30). The coal deposits of the Alberta Plains occur mostly in Upper Cretaceous formations (65 million years old), particularly the Foremost and Oldman formations of southwestern Alberta. These deposits and others have been described by G.D. Williams in the paper entitled *Deep Coal Resources of the Interior Plains, Estimated from Petroleum Borehole Data* (Ottawa: Energy, Mines and Resources Canada, 1981. 15 p. Geological Survey of Canada Paper No. 81-13).

The coal deposits of Vancouver Island have been described by J.E. Muller in the paper entitled *Geology, History and Potential of Vancouver Island Coal Deposits* (Ottawa: Department of Energy, Mines and Resources, 1971. 50 p. Geological Survey of Canada Paper No. 70–53). Finally, two overall surveys of Canada's coal resources have been written, the first an edited volume by T.H. Patching in a special volume entitled *Coal in Canada* (Ottawa: Canadian Institute of Mining and Metallurgy, 1985. 327 p. Special Volume No. 31), mentioned above, and by G.G. Smith, in the monograph entitled *Coal Resources of Canada* (Ottawa: Energy, Mines and Resources Canada, 1989. 146 p. Geological Survey of Canada Paper No. 89-4).

The last major 'mineral' resources of Canada are its hydrocarbon resources. Hydrocarbons occur in the form of oil and natural gas. In southwestern Ontario, in the Lake Erie–Lake Huron region, oil and gas are found in primarily Cambrian, Ordovician, Silurian, and Devonian formations, which are 345 million to 570 million years old. In New Brunswick, oil and gas deposits are found in Mississippian strata, which are just under 345 million years old. And, in the western foothills of the Interior Plains, Canada's major hydrocarbon production, including the far ends of northeastern Alberta, is from Ordovician and Devonian fields. There is also the offshore petroleum potential of the Beaufort Sea, in the Arctic, and the Hibernian field of Eastern Canada.

The fields in southwestern Ontario occur along the Algonquin and Findley Arches, and the flanks of the Michigan and Allegheny Basins. The first wells in Ontario were drilled into bedrock formations at Oil Springs, and with the discovery at Petrolia, in 1861, and at Bothwell, in 1862, Ontario's oil output increased, and peaked, in 1894. Natural gas was discovered in the Kingsville field in Essex County in 1889, and since that time numerous gas fields have been discovered and occur over a wide area extending from Lake Huron to the Niagara River. These oil and gas fields have been described in the map by W.E. Koepler and B.V. Sanford entitled *Oil and Gas Fields of Southwestern Ontario* (Ottawa: Department of Mines and Technical Surveys, 1965. Geological Survey of Canada Map 6–1965), and in their paper entitled *The Silurian Oil and Gas Fields of Southwestern Ontario* (Ottawa: Department of Energy, Mines and Resources, 1965. 138 p. Geological Survey of Canada Paper 45-30).

In New Brunswick, the Stony Creek oil and gas field, near Moncton, has been the only commercially productive field, to date, in the Canadian Appalachia. This field was discovered in 1909, and has been described by R.D. Howie in the paper entitled *Structure and Stratigraphy of the Stony Creek Field, New Brunswick* (Ottawa: Department of Mines and Technical Surveys 1965. Geological Survey of Canada Paper No. 65-2 Part II, p. 71).

Since the first Canadian oil and gas discoveries in the 19th century, the major producing oil and gas fields and reserves, to date, have been in western Canada.

These fields were formed during the Middle and Upper Devonian period about 400 million years ago, mostly in the province of Alberta. Middle Devonian fields have been located in northwestern Alberta, and include the Granite-Wash field, the Keg River Formation, the Slave Point, the Sulphur Point Formations, the Gilwood Formation, and the Kee Scarp Formation. Some of these fields have been described by M.H. Portugal and R.M. Creed in the paper entitled *Oil and Gas Developments in Western Canada in 1987* (*Bulletin of Canadian Petroleum Geology*, v. 37, 1989. pp. 334–45); and by P. Wong and M. Gilhooly in a conference paper entitled *Sequence Stratigraphy of the Middle to Late Devonian Carbonate Strata of Alberta, Canada* (Abstracts of the American Association of Petroleum Geologists and Society of Economic Paleontologists and Mineralogists Annual Meeting, 1992. p. 145). These Upper Devonian fields include the Beaverhill Lake Formation, south of the Peace River Arch in the Swan Hills area northwest of Edmonton, as well as the famous Leduc Formation discovered in 1947.

Besides the Devonian fields, the Mississippian fields are the second most productive system in western Canada. Mississippian oil and gas have come primarily from the Turner Valley Formation in west-central Alberta, which also takes in the Banff Formation, and the Pekisko Formation. The Mississippian fields have been described by J.A. Majorowicz and M. Rahman in the paper entitled *The Paleogeothermal and Present Thermal Resources of the Alberta Basin and Their Significance for Petroleum Occurrences* (*Bulletin of Canadian Petroleum Geology*, v. 33, 1985. pp. 12–21). The Turner Valley Field itself has been described by W.B. Gallup in the paper entitled *Geology of the Turner Valley Oil and Gas Field, Alberta* (*American Association of Petroleum Geologists Bulletin*, v. 35, 1951. pp. 797–871).

Other oil- and-gas producing sediments are in Permian rocks (225–280 million years old) in the north-central border areas of British Columbia and Alberta, in Triassic sediments (190–225 million years old) in north-eastern British Columbia that contain the greater part of the oil and gas reserves of British Columbia, in Jurassic fields (135–190 million years old) that have oil and gas reserves, occurring mostly in southern Alberta and Saskatchewan, and in Upper and Lower Cretaceous fields (65–135 million years old) which include, among others, the Bow Island and Medicine Hat Formations, which contain large oil and gas reserves. The most complex of the Alberta oil and gas fields are the Foothills fields, because they are part of a late Cretaceous deformation of folds and thrusts that are a mix of Mississippian and Devonian strata. These fields are found in the Waterton and Pincher Creek areas of southwestern Alberta. These various formations have been described by A.W. Bally and P.L. Gordy in the paper entitled *Structure, Seismic Data, and Organic Evolution of Southern Canadian Rocky Mountains* (*Canadian Petroleum Geologists Bulletin*, v. 14, 1966. pp. 337–81); by D.C. Barss and N.

Meyers, in the chapter entitled *Triassic.* In: *Geological History of Western Canada* (Calgary: Society of Petroleum Geologists, 1965. pp. 113–36); and by R.J.W. Douglas and N.K. Norris in the paper entitled *Geology and Petroleum Potentialities of Northern Canada* (Ottawa: Department of Mines and Technical Surveys, 1963. 28 p. Geological Survey of Canada Paper 63–31). Overall, the conventional oil potential of western Canada has been described in the paper entitled *Conventional Oil Resources of Western Canada [Light and Medium]* (Ottawa: Geological Survey of Canada, 1987. 149 p. Geological Survey of Canada Paper No. 87-26).

Perhaps one of the largest, if not the largest, known oil reservoirs in the world are the Athabasca Oil Sands of northeastern Alberta. These deposits were first reported by explorers as early as 1778. J.R. Pow and G.H. Fairbanks, in their paper entitled *Descriptions and Reserve Estimates of the Oil Sands Alberta* (Edmonton: Research Council of Alberta, 1963. 14 p.), estimated that these sands contain more than 626 billion barrels of heavy oil in place, 369 billion barrels of recoverable raw oil-sands oil, and 266 billion barrels of recoverable upgraded synthetic crude oil. The sands, which became economically feasible for oil recovery after the international oil crises of the early 1970s, extend over an area of five and three quarter million acres. Since then, more technical papers, such as one by A. Settari and Y. Ito, entitled *Geotechnical Aspects of Recovery Processes in Oil Sands* (*Canadian Geotechnical Journal*, v. 30, 1993. pp. 22–33) have been written. Finally, the most recent overview of all western Canadian hydrocarbons, including the oil sands, has been the atlas, compiled by N.L. Anderson and L.V. Hills, entitled *The CSEG/CSPG Geophysical Atlas of Western Canadian Hydrocarbon Pools* (Calgary: Canadian Society of Exploration Geophysicists/Canadian Society of Petroleum Geologists, 1989. 344 p.).

Although the western Canadian hydrocarbons are still Canada's largest producing sources, the 1980s saw a shift of focus to Canada's offshore hydrocarbon potential in the Beaufort Sea of the Arctic, and the Hibernia field of off-shore eastern Canada. The former has been written about by F.G. Young and V.B. Lyatsky of Home Oil Company, in their paper entitled *Beaufort Mackenzie Oil Exploration and Evolution of Geological Concepts* (*Oil and Gas Journal*, v. 84, 1986. pp. 65–67, 70, 72); and the latter by A.C. Grant and K.D. McAlpine in the paper entitled *Off-Shore Geology and Petroleum Potential of Eastern Canada* (*Energy Exploration and Exploitation*, v. 4, 1986. pp. 5–52).

E Applied Mineral Research

Besides the techniques used, and the search for economic minerals in Canada, research is also carried on in the fields of mineral processing and extraction, metallurgical research, fuel research, and mining research.

(a) Mineral Processing

In recent times, the three major areas of mineral processing and extraction have been the recovery of uranium, the handling of sulphide ores, and improved iron ore extraction.

It was in 1930 that a high-grade pitchblende was discovered in the Northwest Territories. These ores from the Port Radium Mine were a mix of pitchblende, silver, and lead, with a uranium oxide content of 30%–50%. Milling of these ores began in 1933 with the recovery of a gravity concentrate of pitchblende and silver, which was transported to Port Hope, Ontario, for refining. The refining process, however, presented a persistent problem of low overall recovery of uranium at the Port Radium site. But, after 1945, the Radioactivity Division of the then Department of Mines began to search for a solution to this problem, as well as improving physical and chemical analytical methods for accurate determination of uranium in low-grade ores and concentrates that the Geological Survey was locating in other areas of Canada. After a study of the ores, researchers decided that physical concentration would not achieve an improved recovery rate, but that chemical leaching of the ore would improve recovery.

The development of an acid leaching process began in 1948, and was perfected by the mid-1950s. This acid leaching process involved the recovery of uranium from solution, with the use of aluminium powder that caused the chemical precipitation of insoluble uranous compounds that were then converted to metal-grade uranium trioxide. This procedure increased the recovery of U_3O_8 to about 30% for Port Radium Eldorado ores. However, when lower-grade ores were discovered at the Beaverlodge deposits in northern Saskatchewan, the recovery of uranium by leaching from these more pitchblende carbonate minerals was facilitated with the use of a sodium carbonate leach to reduce the chemical acidification of the process. The development of this process was described, by the staff of the Radioactivity Division, in a paper entitled *Development of the Port Radium Leaching Process for the Recovery of Uranium* (Ottawa: Department of Mines and Technical Surveys, Mines Branch, 1955. 22 p.). Since the mid-1950s, however, other leaching processes have continued to be studied, as noted in papers by A. Himsley, entitled *Uranium Extraction from Turbid Liquids by Continuous Ion Exchange Process* (*Transactions of the Canadian Institute of Mining and Metallurgy and of the Mining Society of Nova Scotia*, v. 80, 1977. pp. 182–88; by K.E. Haque, entitled *Aqueous-Chlorine Leaching of Typical Canadian Uranium Ores* (*Uranium*, v. 1, 1982. pp. 65–78); and by C.R. Edwards entitled *Uranium Extraction Process Alternatives* (*CIM Bulletin*, v. 85, 1992. pp. 112–36).

Another major research development in mineral recovery was the sulphide program that began in 1964. A major reason for undertaking this was the realization that a substantial portion of Canada's mineral wealth was in base metals contained in sulphides, which were often complex and lean. The Mineral Physics and Mineralogy sections of the Mineral Sciences Division undertook this job. The proposal for this program was described by A.T. Prince in the paper entitled *A Proposal for a Sulphide Research Programme in the Mineral Sciences Division of the Mines Branch* (Ottawa: Department of Mines and Technical Surveys, 1964. 1 v. Internal Report MS64–43). The purpose was to study the particular mineral substance to determine inter-atomic bonding, electron energies, and other fundamental properties of its crystal lattice. Hence, chemical, crystallographic, optical, electrical, and magnetic measurements, as well as surface properties of minerals, would be required. The surface properties were particularly important, as surface energies play an important part in mineral processes such as dissolution and flotation. The specific mineral initially selected for study was sphalerite, or zinc sulphide, but the major focus of the program was the study of iron atom bonding in pyrites, which occurs to varying degrees in all mixed sulphides, by using the Mossbauer effect technique. This method was described by R.H. Goodman in the paper entitled *The Mossbauer Effect* (*Chemistry in Canada*, v. 18, 1966. pp. 31–36). (The Mossbauer effect is the recoil movement of an atomic nucleus alone, or of the crystal lattice structure if the nucleus is tightly bound in the lattice, when the nucleus emits a gamma ray photon from its structure.) The results of the research were reported, again by R.H. Goodman, in the paper entitled *Iron Impurity States in Cubic Zinc Sulphide* (Ottawa: Department of Energy, Mines and Resources, 1966. 1 v. Internal Report MS 66-22). The general conclusion of this work was that synthetic mineral crystals of known composites needed to be studied before going on to the study of the natural mineral compound. Work continued, however, on the study of surface properties, using the absorption of oleic acid on hematite and magnetite as the basis of this work. This research was reported by S.M. Ahmed in the paper entitled *Studies of the Dissociation of Oxide Surfaces at the Liquid-Solid Interface* (*Canadian Journal of Chemistry*, v. 44, 1966. pp. 1663–70). Other work carried out in the sulphide research program was undertaken by L.G. Ripley, and described in two papers entitled *Crystal Growth: Part I, Background to Crystal Growth* (Ottawa: Department of Energy, Mines and Resources, 1971. 36 p. Sulphide Research Contribution No. 39), and *Crystal Growth: Part IV. The Arsenides and Sulpharsenides of Cobalt, Iron, and Nickel* (Ottawa: Department of Energy, Mines and Resources, 1973. 112 p. Sulphide Research Contribution No. 56).

(b) Metallurgical Research

Perhaps a more immediate success in metallurgical research at the federal level was the influence of the Ungava Bay iron ore deposits upon research carried out on the beneficiation (ore dressing) of iron ores. Although the Schefferville ores were high-grade, other large deposits further south of this region were not only of a lower grade, but they also contained other minerals that would require evaluation for mineral processing. Major beneficiation research has been carried out in the Mineral Dressing and Process Metallurgy Division, since the 1950s, on the processing of iron ores. These have involved flotation separation, filtration studies, and grinding of ores. Initial studies were carried out with low-grade ores with high silica, phosphorus, or titanium contents, that needed to be enriched in concentrate or pellet form to qualify as feed for blast furnaces, direct iron processing, powder metallurgy, and the manufacture of segments.

Early flotation studies were undertaken by L.L. Sirois and T. Takamori, and described in papers entitled *Experimental Flotation Cell* (Ottawa: Energy, Mines and Resources, 1964. 6 p. Mines Branch. Research Report R135); *Simple Low Rate Feeder for Water – Insoluble Flotation Reagents* (Ottawa: Energy, Mines and Resources, 1966. 41 p. Mines Branch. Technical Bulletin TB81); and *Study of Dodecyclammonium Chloride on Labradorite* (*Nippon Koggo Kaisshi*, v. 83, 1967. pp. 19–24). In this latter study, this chemical was found to be a good depressant for iron oxide in the flotation process. Another aspect of flotation studies was to compare the process of recovering finely divided minerals in the one- to 20-millimicron range. Studies were made on bubble formation, and behaviour, on surface electrical characteristics of minerals. This work was reported by L.L. Sirois and G.E. Alexander in the paper entitled *Development of an Automated Potentiometric Titrator for the Study of Surface Electrical Phenomena* (*Canadian Institute of Mining Bulletin*, v. 67, 1969. pp. 410–14). Finally, this work on flotation separation was summarized by K.S. Moon and L.L. Sirois in the paper entitled *Theory and Application of Column Flotation* (Ottawa: CANMET, 1988. 28 p. CANMET Report 0705–5196).

Filtration studies began in 1966. At that time, an experimental project was started wherein water was continuously passed through a porous bed. A number of factors that controlled liquid-solid separation, including surface electrical properties and ultrasonic vibration, were investigated. Following laboratory tests, synthetic fibre filter cloths were selected for testing at the Iron Ore Company of Canada, and at Halton Mines, over a period of time. This project was described by W. Nemith in the paper entitled *Filtration Basic Mechanisms and the Medium* (*Canadian Mining Journal*, v. 92, 1970. pp. 71–76). This work was discontinued in 1972.

Grinding of ores and minerals, however, has been an ongoing area of research for innovations, as it is the first and most costly step in treatment after coarse ore crushing occurs in the mining operation. An experimental grinding project was started, in 1966, by the Metallic Minerals Research Laboratory in the Mines Branch. A 20-by-30-inch ball mill was used with a feed, first of limestone, and then of quartz. The study was designed with variables measured in a continuous open circuit grinding. These studies were reported by T. Nagahana and L.L. Sirois in a paper entitled *Study of Grinding Parameters by Statistical Analysis* (*Commonwealth Mining and Metallurgical Congress, 9th Proceedings*, v. 3, 1968. pp. 603–30). Other studies with metal ores, such as titanium ores, have been carried out by G.E. Agar and W.B. Kipkie and reported in the paper entitled *Ilmenite Concentration from the Laurentian Titanium Deposit* (*CIM Bulletin*, v. 73, 1980. pp. 140–47). And an overview of these technologies is discussed in the volume entitled *Minerals, Metals and Mining Technologies*. Compiled and edited by L.L. Sirois and J.R. MacDonald (Ottawa: Canada Centre for Mineral and Energy Technology, 1984. 18 p.).

Another important area of research was in physical metallurgy, which involves metal physics, the study of the fundamental properties and structure of metals; metal processing, which involves the melting and solidification sequence of metals; and metal applications, which is the applied side of metal physics such as metal strength and fatigue. Overall, it involves the research and development of alloys. Some of the early work carried out in metal physics involved the study of determining the orientation of single crystals of metals, or minerals, by ion bombardment that ejected atoms in crystallographic directions. This work was done by R.L. Cunningham and J. Ng-Yelem and reported in a series of papers entitled *Ion Bombardment of Single Crystals of Aluminum* (*Journal of Applied Physics*, v. 34, 1963. pp. 984–89); *Orientation Determinations of Crystals Using Ejection Patterns Resulting from Ion Bombardment* (*Journal of Applied Physics*, v. 35, 1964. pp. 2185–88); and *Ejection of Atoms from Metallic Single Crystals* (*American Journal of Physics*, v. 33, 1965. pp. 1064–69). The technical aspect of this work was described in their paper entitled *Description and Use of an Ion Bombardment Camera and Ancillary Equipment* (Ottawa: R. Duhamel, 1968. 18 p.).

The early research concerned with metal processing was carried out in the Metal Forming Laboratory, which was set up in 1947. Some of the early work included basic alloy research. An example of this type of research by J. Perry was entitled *An ABC of Cold Extrusion of Steel* (*Canadian Institute of Mining. Transactions*, v. 57, 1954. pp. 64–67). Other examples of research are by D.W. White entitled *Surface Tension of Molten Zinc and Some Zinc Alloys* (Ottawa: Energy, Mines and Resources, 1965. 62 p. Mines Branch Research Report R160); and by A. Couture and J.O. Edwards entitled *Hot-Tearing of Copper Alloys* (Ottawa: Energy, Mines and Resources, 1965. 54 p. Mines Branch Research Report R164). Later research has included the examination of metal processing in different geographical and climate settings, as well as the use of computer modelling in metal fabrication. One collection of papers was entitled *Metals and Alloys for Arctic Use*. Edited by R.C. Thurston (Ottawa: Energy, Mines and Resources Canada,

1976. 234 p. CANMET Report 76-1). Another paper by D.E. King was entitled *Alternative Techniques for Metal Processing: Implications for the Yukon* (Ottawa: Energy, Mines and Resources Canada, 1982. 29 p. Mineral Policies Sector Internal Report 0713-7044. MRI82/3). Computer modelling of metal processes was the focus of an Ottawa conference entitled *Computer Modelling of Fabrication Processes and Constitute Behaviour of Metals: Proceedings of an International Conference Held in Ottawa, Ontario, May 15–16, 1986.* Edited by J.J. Too (Ottawa: Energy, Mines and Resources Canada, 1986. 655 p.).

Another aspect of metal processing is welding, which is a metal-joining process. Early on, there was considerable work done on the factors influencing heat-affected zone cracking, which is one of the difficulties in the welding process. One paper in this field by K. Winterton was entitled *Weldability Prediction from Steel Composition to Avoid Heat-Affected Zone Cracking* (Weld Journal, v. 40, 1961. Supplement. pp. 2535–85. The same author also studied the weldability of other metals, as noted in the title *Weldability of Titanium and Titanium Alloys* (Ottawa: Energy, Mines and Resources, 1965. 19 p. Mines Branch Technical Bulletin TB71). Finally, K. Winterton also wrote the report entitled *Thirty Years of Welding Research at PMRL* (Ottawa: Energy, Mines and Resources Canada, 1976. 18 p. Physical Metallurgy Research Laboratories Report PMRL-76-13). Since then, welding research has also examined potential welding problems in adverse cold environments. For example, J.E. Braid and J.A. Gianetto examined marine environments in their papers entitled *Toughness of Weld Metals for Fabrication of Cold Marine Structures and Vessels. Part 1, Shielded Metal Arc Welding* (Ottawa: Metals Technology Laboratories, 1988. 69 p. Metals Technology Laboratories Report MTL 880-21); and *Toughness of Weld Metals for Fabrication of Cold Marine Structures and Vessels. Part 2, Submerged Arc Welding* (Ottawa: Metals Technology Laboratories, 1988. 50 p. Metals Technology Laboratories Report MTL 88-25). And J.A. Gianetto, alone, wrote the report *Effect of High Heat, Input on the Microstructure and Mechanical Properties of Submerged-Arc Welds in Steels for Arctic and Offshore Structures* (Ottawa: Metal Technology Laboratories, 1988. 46 p. Metals Technology Laboratories Report MTL 88-29).

Other work has involved the study of the effect of pollutants on the corrosion of metals, such as the paper by G.K. Lee, entitled *Control of Pollutant Emission and Sulphuric Acid Corrosion from Combustion of Residential Fuel* (Ottawa: Department of Energy, Mines and Resources, 1968. 1 v. Mines Branch Research Report R195); and the departmental paper entitled *Underground Corrosion of Water Pipes in Canadian Cities: Case, the City of Calgary* (Ottawa: Energy, Mines and Resources Canada, 1985. 94 p.). It is interesting to note that Atomic Energy of Canada has also undertaken corrosion studies in terms of nuclear waste management. P.M. Mathew discussed this problem in two papers. One was entitled

Metal-Matrix Corrosion Studies in the Canadian Shield Granitic Groundwaters (*Materials Research Society Symposium Proceedings*, v. 84, 1987. pp. 261–70). The other was *Metal Matrices for Nuclear Fuel Waste Disposal.* In: *The Geological Disposal of High Level Radioactive Wastes.* Edited by D.G. Brookins (Athens, Greece: Theophrastus Publ., 1987. pp. 315–47).

The applied side of metal physics, which is called engineering physics, has been researched in the Engineering Physics Section of the Mines Branch. This section has devoted considerable effort to the study of metal fatigue, fatigue cracking, and failure phenomena, as well as the solution to problems in this area. Some pioneering work in this field was carried out in the late 1960s by researchers such as S. Nonomura and R.C. Thurston, who discussed fatigue cracking in the paper entitled *Fatigue Crack Propagating Rate in Metals* (Proceedings of the International Conference on the Strength of Metals and Alloys, Tokyo, 1967. pp. 1015–20).

As well, E.G. Eeles and R.C. Thurston have undertaken research on the environmental influences relating to fatigue properties of light metals. Their work was published in the paper entitled *Atmospheric and Surface Effects in the Fatigue of Aluminium Alloys* (*Journal of the American Institute of Aeronautics and Astronautics*, v. 8, 1970. pp. 224–28).

(c) Fuel Research

Fuel research in Canada has concentrated upon the bulk use of coal in combustion and in metallurgical applications. This research, mostly carried out in the 1960s and early 1970s, was necessitated by the fact that Canadian coal production had peaked, in 1949, with the highest proportion of production in Alberta, although coal was also produced in British Columbia, Saskatchewan, and Nova Scotia. The outlook for coal use was uncertain, particularly with the development of hydrocarbon industry in Alberta. The problem with Canadian coal was a high sulphur content of maritime coals, and the difficulty with it in combustion furnace grates, due to a low ash fusion temperature. Better-quality coals could be had for Ontario, and Quebec, from the United States. On the other hand, Alberta and British Columbia coals produced a low-rank high-moisture coal, and high-ash coals. These types of coals break up in storage unless they are treated. By the mid-1950s, the Canadian coal industry could survive only by the bulk use of coal in combustion and metallurgical applications. This required the Minerals Dressing and Process Metallurgy Division to initiate a series of technical studies which did result in the reversal of the erosion of markets for Canadian coal, both domestically and for export to Japan, during the early 1970s.

Two areas of study concentrated upon the cleaning of western coals to produce a more efficient burning fuel, and a more effective carbonization of the coals. As mentioned above, one of the problems with western coals was their fragility, which degraded during the cleaning process. At the western Regional Laboratory in Edmonton, J. Visman evaluated the use of dry separation techniques, but turned to water separation techniques using a modified Dutch technique. Bulk testing of Alberta coals, and preliminary results, were described by J. Visman in the paper entitled *Cleaning Fine Coal in a 3-Stage Water Cyclone Circuit* (*Canadian Institute of Mining Transactions*, v. 64, 1961. pp. 273–77). Further work led to the design of a different type of cyclone, called the Compound Water Cyclone, which was patented in 1962. A paper on the final design, and use, involving this cyclone process was described, again, by J. Visman in the paper entitled *Bulk Processing of Fine Materials by Compound Water Cyclones* (*Canadian Institute of Mining Transactions*, v. 69, 1966. pp. 85–98).

The carbonization of coal into coke for domestic use in the production of steel, or for the export of coking coal from Alberta and southeastern British Columbia to Japan, was the other success in coal fuel research in Canada. Work on western coals involved both new furnace designs for the heat treatment of coal feed, and the study of maximum temperature processing for the treatment of coal for coking. An interesting aspect of this work was studies done on the loss of coking coal stockpiled, while awaiting export. An effective chemical method for determining the loss of coking capacity, by heating the coal to 330 degrees centigrade and analysing the carbon monoxide and carbon dioxide gases given off, was developed by B.S. Ignasiak and B.N. Nandi, who described this work in the paper entitled *Comparison of Dilatation and the Chromatographic Analysis of the Pyrolysis Gases as Methods for Studying Coal Heating* (*Canadian Institute of Mining. Transactions*, v. 73, 1970. pp. 70–74).

Besides coal research, there has been a long-term interest into hydrocarbon research of the Alberta oil sands. Early work was carried out after World War II, in attempts to separate the sand from the bitumen contained within, and to evaluate the technical and economic feasibility of the commercial production of oil products from the bitumen. However, with the rapid development of conventional hydrocarbon production in the 1950s, and the high capital costs required for the development of an integrated recovery and refining plant, research was not pursued vigorously until the early 1970s, during the first international oil crisis and the increase in hydrocarbon prices, which made recovery of hydrocarbons from the oil sands more attractive. In fact, during the period 1971 to 1979, the Fuels Research Centre, near Ottawa, undertook research on the hydrocracking of bitumen, to produce oils and residues as an alternative energy source. During this period, a series of papers by E.C. McColgan and R.G. Draper were written and entitled *The Hydrocracking of Residual Oils and Tars* (Ottawa: Department of Energy, Mines and Resources, Mines Branch, Fuels Research Centre, 1971–74.

6 parts). Other papers written during this period were by W.H. Merrill and R.B. Logie entitled *A Pilot Scale Investigation of Thermal Hydrocracking of Athabasca Bitumen* (Ottawa: Department of Energy, Mines and Resources, 1973. 25 p. Mines Branch Research Report No. 281).

With the increased development of the oil sands since the 1970s, there has been a growth of studies into the environmental impact of the development of the sands area. Some of these were joint federal and provincial endeavours for the Alberta Oil Sands Environmental Research Program. One study by D.B. Tripp and P.T. Tsui was entitled *Fisheries and Habitat Investigations of Tributary Streams in the Southern Portion of the AOSERP Study Area* (Edmonton: The Program, 1980. 2 v. AOSERP Report No. 92). Another study by K. Machniak was entitled *Fisheries and Aquatic Habitat Investigations in the MacKay River Watershed of Northeastern Alberta* (Edmonton: The Program, 1980. 273 p. AOSERP Report No. 93). And another study by G.R. Ash and L.R. Noton was entitled *A Fisheries and Water Quality Survey of Ten Lakes in the Richardson Tower Area, Northeastern Alberta* (Edmonton: The Program, 1980. 2 v. AOSERP Report No. 94). Finally, one Environment Canada contract study, by CH2M Hill Engineering Ltd, was entitled *Evaluation of Options to Reduce Sulphur Dioxide Emissions from the Natural Gas Processing and Tar Sands Industries* (Ottawa: Environment Canada, 1993. 64 p.).

(d) Mining Research

The mining of recoverable mineral deposits brings into play mining safety and the study of rock mechanics and stress, for both underground and open-pit mining. Since the early 1960s, the Mines Branch has pursued notable research in the study of rock mechanics, and stress measurement in rocks.

The study of rock mechanics in Canada was given its first major impetus at the Wabana submarine iron mines at Belle Island, Newfoundland, in the mid-1950s. At that time, the Dominion Ore Company wanted to increase ore recovery from their submarine mines, and the Mines Branch was able to transfer some of its knowledge of stress problems in coal mines to that environment. The Wabana project was conducted in the same manner as a coal-mining study, and started with a geological investigation by D.K. Norris entitled *Structural Conditions at the Wabana Iron Mines, Newfoundland with Reference to Increased Ore Recovery* (Canadian Mining and Metallurgical Bulletin No. 543, 1957. pp. 539–49). To study stress, a new model was first developed and described by K. Barron and G.E. Larocque in the paper entitled *Development of a Model for a Mine Structure* (Rock Mechanics Symposium. Proceedings 1st, McGill University, 1963. pp. 145–62).

An important feature of underground mining may be the use of rock pillars that may also contain ore, but are used for the stability of underground mining structures. In 1963, a cooperative research project was established with Algoma Steel Corporation to conduct rock mechanic studies at its MacLeod Mine in Ontario. This study was concerned with the stability of pillars, with the aim of assisting in solving problems of mining at depths greater than 1,000 feet in an ore-body inclined steeply. Both a mathematical model and actual stress measurements were studied. The results were described by D.G. Hedley and G. Zahary in the paper entitled *Underground Measurement in a Steeply Dipping Ore Body* (Canadian Rock Symposium. Proceedings 5th, Toronto, 1968. pp. 105–25). At the same symposium, M.J. Royea presented a paper entitled *Rock Stress Measurements at the Sullivan Mine* (Canadian Rock Symposium. Proceedings 5th, Toronto, 1968. pp. 59–74). This study was important because Cominco, the mining company, wanted to assess the stress changes in a pillar extraction area around a large pillar containing about 200,000 tons of ore, which it was planning to break in a single blast in that mine. Later, a major analytical study based on case histories was undertaken by K. Barron of the Coal Mining Research Centre for CANMET. The study was entitled *An Analytical Approach to the Design of Pillars in Coal. Part I: The Analytical Method: Final Report, Phase I; Part 2: Application of the Model to Case Histories: Final Report, Phase 2* (Edmonton: Coal Mining Research Centre, 1980. 516 p.). A decade later, K. Barron and D. Kullman did model actual outbursts of coal seams at a Glace Bay, Nova Scotia Colliery. The papers were entitled *Modelling of Outbursts at '26 Colliery Glace Bay, Nova Scotia; I, Outburst History and Field Data* (*Mining Science & Technology*, v. 11, 1990. pp. 253–60); *Modelling of Outbursts at '26 Colliery Glace Bay, Nova Scotia; II, Proposed Outburst Mechanism and Model (Mining Science & Technology*, v. 11, 1990. pp. 261–68); and *Modelling of Outbursts at '26 Colliery Glace Bay, Nova Scotia; III, Comparison of Model Results and Field Data* (*Mining Science & Technology*, v. 11, 1990. pp. 269–80). Since then, CANMET has established its own CANMET Experimental Mine at Val-d'Or, Quebec (the former Beacon gold mine), to provide in-site testing and research in a true underground mining environment. This mine is described at the website <<http://www.nrcan.gc.ca/mms>>.

Unlike underground mining, open-pit mining requires knowledge of the stability of rock slopes. Important studies on this topic were written by D.F. Coates and Y.S. Yu entitled *Analysis of Rock Slopes Using the Finite Element Method* (Ottawa: Department of Energy, Mines and Resources, 1971. 74 p. Mines Branch Research Report No. 229); *Rock Anchor Design Mechanics* (Ottawa: Department of Energy, Mines and Resources, 1971. 13 p. Mines Branch Research Report No. 233); and *Block Flow Slope Instability* (Ottawa: Canada Centre for Mineral and Energy Technology, 1978. 12 p. CANMET Report 78-10). When rock slopes are deemed to be stable, mining productivity can be increased by increasing the angle of the slope, without loss of stability, by using deep cable anchors and mesh support. A comprehensive engineering feasibility study assessing the maximum slope angle was carried out by K. Barron and D.F. Coates and described in the paper entitled *Artificial Support of Rock Slopes* (Ottawa: Department of Energy, Mines and Resources, 1979 (rev.). 145 p. Mines Branch Research Report No. 228). One other overall contribution to the study of rock mechanics by D.F. Coates was the monograph entitled *Rock Mechanic Principles*. 4th Edition Revised (Ottawa: CANMET, 1981. 460 p.).

F Geophysical Research and Geophysical Surveys

In Canada, the geophysics aspect of earth sciences research, and related surveys, involves (1) the ongoing estimation, and correction, of the size and shape of the earth through geodetic surveys, (2) the measurement and the relationship of gravity variations to the internal structure of the earth, (3) the study of the magnetic forces which surround the earth, along with related paleomagnetic studies and magnetic anomalies, which aid in filling the gap about the evolution of the earth, (4) the study of seismology, and the monitoring of earthquake activity, and (5) the use of remote sensing techniques for observing surface changes in the earth's structure.

(a) Geodetic Surveys

Geodesy is concerned with not only the study of the rotation of the earth but also the size and shape of the earth, which is not a uniform ellipsoid in shape, but a flattened sphere called a geoid. These studies are complicated, because all theoretical measurements of the earth are based upon a calculation of a perfect sphere taken at sea level, and the latter is affected by gravity anomalies, or variations, over the surface of the earth. To most people, however, the practical side of geodesy is seen in the mapping of the earth and national territories, as well as specific surveys for boundaries.

In order to carry out these measurements, systems of coordinates must be established for horizontal positioning (longitudes and latitudes), and for vertical positioning, or levelling, according to theoretical spherical trigonometry. The geodetic surveying of horizontal and vertical controls at selected points within Canada, or globally, constantly extends and intensifies the national global network for calculating the geoid, and more locally for mapping by government or private mappers. The traditional methods of positioning have been well described in annual reports of the department. For example, in the 1970/71 Annual Report of the Department of Energy, Mines and Resources, there is a description of the year's survey work. It said, 'the extension and densification of the first-order horizontal-control framework was carried out in Yukon and Northwest Territories and four provinces. In the Territories, the aerodist network was extended westward from the Yukon-Northwest Territories boundary. The project covered 91,200 square miles and provided horizontal mapping control for three-hundred eighteen 1:50,000 map sheets. Three astronomic parties established 11 Laplace azimuth stations, two sets of inter-visible Laplace azimuth stations, and 12 plumb-line-deflection stations at various locations throughout the country. The precise astronomic position of the satellite-triangulation station at Cambridge Bay, NWT, was also determined.' These traditional methods of positioning have also been documented in early papers, for example by C.K. McElroy entitled *Instructions for Building Triangulation Towers* (Ottawa: Department of Mines and Technical Surveys, 1952. 32 p. Canada Geodetic Survey Publication No. 10); and by L.P. Robertson in the paper entitled *Precise Levelling in British Columbia South of Dawson Creek* (Ottawa: Department of Mines and Technical Surveys, 1961. 314 p. Geodetic Survey of Canada Publication No. 24A).

More and more, satellites are used to assist in geodetic surveying. The first satellite designed for geodetic purposes (using the Doppler frequency shift described below) was launched in the early 1970s in the United States, and called a Laser Geodynamic Satellite. It was placed in a nearby circular orbit, at a height of approximately 6,000 kilometres above the earth, and carried over 400 reflectors suitable for reflecting laser beams back along their path. The elevation was chosen to minimize both the effects of atmospheric drag and local gravity anomalies. Coupled with the use of computers, the technical feasibility of satellite geodesy has not only allowed the creation of a global framework of coordinates and networks over land and sea, but has also generated the data that can be applied to the refining of geodetic theory.

In Canada, the use of the Doppler frequency shift of satellites to assist in geodetic surveying began in 1973, when 21 Doppler stations from the Atlantic region to Manitoba were established. The Doppler method is based on the principle that electromagnetic waves coming from a moving object appear to undergo a frequency shift in relation to a stationary observer. The position of a Doppler ground station can be calculated by recording this frequency shift, because the position of the satellite is known in terms of latitude and longitude at all times. With this technology, Canadian geologists found there were distortions in the national network of horizontal control, because measurements made by the new high-accuracy instruments did not conform with data from older surveys. Accordingly, 52 additional Doppler stations in the Arctic, Quebec, and Greenland were established to assist in a Doppler Survey of Canada, carried out from 1974 to 1976. Reports on Doppler positioning used for geodetic surveys appeared at the

9th Annual Meeting of the Geological Society of Canada, North-Central Section, held in Waterloo, Ontario in 1975. At that meeting, both J. Kouba and D.E. Wells gave papers. The former's was entitled *Doppler Satellite Positioning in Establishing a Geodetic Control Network* (*Geological Society of America. Abstract of Programs*, v. 7, 1975. p. 799). The latter's was entitled *Semi-Dynamic Doppler Satellite Positioning* (*Geological Society of America. Abstract of Programs*, v. 7, 1975. pp. 878–79).

By the 1980s, the United States began to develop a Global Positioning System (GPS) for use by the military, with availability for civilian use in the late 1980s. A global positioning system would mean even more advances being made in the methods used for geodetic surveying.

As well, in Canada, the Geodetic Survey Division of Energy, Mines and Resources began to work toward a Global Positioning System, called the Active Control system (ACS), that would initially consist of 10–15 stations throughout Canada, each housing a GPS satellite receiver, and ensure that surveys done in any part of the country had the same frame of reference and would define Canada's territorial limits in a global reference system. As well, the division participated in the North American Datum 83 project, a comprehensive readjustment and redefinition of North American geodetic survey networks, and a minor redesign of the National Geodetic Network. This work was described in the National Report to the International Association of Geodesy, and entitled *Geodesy and Gravity in Canada, January 1979–December 31, 1982* (Ottawa: Energy, Mines and Resources Canada, 1983. 85 p.). The Canadian territorial geoid has been described in detail by P. Vanicek and A. Kleusberg in the report entitled *The Canadian Geoid* (Ottawa: Department of Energy, Mines and Resources, 1986. 117 p. Geodetic Survey Division Contract Report 86–001).

The advantage of using the satellite Global Positioning System for height determination, as compared with conventional levelling using spirit level instruments, has been demonstrated by A. Mainville and R. Forsberg in their paper entitled *Global Positioning System Testing of Geoids Computed from Geopotential Models and Local Gravity Data: A Case Study* (*Journal of Geophysical Research*, v. 97, 1992. pp. 11137–47). Here, the authors used GPS stations along a 900-kilometre. first-order levelling line around the Great Slave Lake area as a 'control for inter-comparison of geoid predictions,' which used global spherical harmonic models that had been tested. The models fit the geoid with standard errors of 30 cm to 50 cm. The authors concluded that 'in large regions like Canada ... it would be, practically, far too expensive and time-consuming to cover all parts of the country by conventional levelling. The computation of gravimetric geoids is especially important for replacing levelling with satellite methods in these cases.' Also, as the amount, complexity, and diversity of data required for geoid modelling continue to increase, data may have to be handled differently. S.D. Pagiatakis and C. Armenaksis, in their paper entitled *Gravimetric Geoid Modelling with GIS* (*International Geoid Service Bulletin*, v. 8, 1998. pp. 105–12), suggested that the use of Geographic Information Systems (GIS) technology Aoffers an effective and efficient approach to higher level data processing, analysis, systhesis, retrieval, representation and mapping.

As already mentioned, geodesy also includes the study of the changes in the ratio of the earth's rotation, and in the orientation of the rotation axis itself relative to celestial bodies. Since the mid-1970s, precise monitoring of the earth's rotation and polar motions has been done by astronomical and satellite Doppler methods from observations near Ottawa and Calgary. This has been described by J. Kouba in the paper entitled *Satellite Doppler Positioning for Geodynamics* (*Annales de Géophysique*, v. 37, 1981. pp. 205–12). An excellent review of this field of research was also written by M.G. Rochester and entitled *Global Geodynamics in Canada, 1971–79* (*Canadian Journal of Earth Sciences*, v. 16, 1979. pp. 621–46).

(b) Gravity Surveys

Another form of earth surveying is the gravity survey. Everyone knows that an object released from above the earth's surface accelerates toward the earth. This is an example of universal gravitation, where all mass within the universe attracts all other mass, and the acceleration is due to gravity or 'g.' As noted previously on the discussion of geodetic surveys, gravity varies over the earth's surface, and this variation is related to the sea-level surface or geoid. There is, however, another use of gravity, and its documentation, for earth sciences research. It is the measurement and the relationship of gravity variations that occur inside the earth to the internal structure of the earth. This concept is called isostasy, in which differences in elevation, as indicated by gravity measurements, are compensated by either low-density roots or lower-density surface. (For example, below the visible portion of a mountain chain is the root zone of low-density rocks which is different from the mantle material of high density.) These measurements are achieved with the use of an instrument called a gravimeter. It is about the size of a wine bottle, and consists of a weight on a spring that stretches or contracts in response to either an increase, or decrease, in gravity from place to place. These instruments can detect variations as small as one-hundred-millionth of the earth's gravity. The standard unit of measurement is the milligal, which is equal to a gravitational acceleration of 0.001 centimetres per second.

Many of the gravity measurement techniques involving the earth's interior have come from the oil industry, with the realization that oil-bearing areas often produced variations from the normal gravity field. These anomalies are caused by a change in the sub-surface mass, due to the presence of structures such as salt domes. Because any anomalous mass will have an effect on the gravitational field, gravitational surveys (after taking into account adjustments made for the distance from the equator, because the earth is a flattened sphere at the poles and the earth's rotation causes the force of gravity to be less at the equator than at the poles, as well as the gravitational contribution of near-surface rock masses) are an important tool for clarifying the earth's sub-surface geology. Although sub-surface variations are an important factor in gravity anomalies, the whole area is not fully understood, because gravity anomalies may also be associated with deep mantle convections, plate tectonics, and temperature variations.

In Canada, gravity surveys are ongoing for both the national and regional collections of data. This has been facilitated by the development of dynamic gravimeters for aircraft and ships, in the mid-1970s. These have allowed a much greater area of coverage over a relatively short period of time. For example, in 1978/79, 'the personnel of the Gravity Service and the contract personnel occupied about 5,100 gravity stations in the Yukon, Northwest Territories... and logged more than 36,000 line km of dynamic gravimeter profiling in Hudson Bay, off the Coasts of the Queen Charlotte Islands and in Dixon Entrance.' Currently, approximately 87% of the landmass and 75% of the off-shore continental margins have been surveyed by gravity measurements, using LaCoste, Romberg, and Scintrex gravimeters, and a network of 5,737 control stations that make up the Canadian Gravity Standardization Net. The Net is managed by the National Gravity Database (NGDB), developed over the last 30 years. Today, the NGDB contains information pertaining to gravity standardization networks in Canada and abroad, gravity mapping in Canada, instrument parameters, digital terrain, and crustal motion. For example, the database 'contains information on more than 700,000 gravity anomaly stations from more than 400 projects, more than 10,000 gravity control stations, 244 instruments and more than 118,000 gravity network ties.' The earlier development of this database has been described by R.K. McConnell in the publication entitled *The Management of the Canadian National Gravity Data Base* (Ottawa: Energy, Mines and Resources Canada, Earth Physics Branch Publications, v. 40, 1977. pp. 107–12). Current descriptions of the National Gravity Program are found at the website <<http://geod.nrcan.gc.ca>>. This continuous surveying has resulted in national maps, the most recent being part of the Canada Geophysical Atlas, described below, as well as separate regional map series. For example, a new gravity-Bouguer anomalies series of 100 maps to the scale of 1:1,000,000 commenced in 1991. The first one was entitled *Gravity-Bouguer Anomalies, Shelburne, Nova Scotia* (Geological Survey of Canada, 1991. Map NK-20–GR (BA), Scale 1:1,000,000).

Besides the generation of maps, the Earth Physics Branch (now part of the

Geodetic Survey Division of Geomatics Canada) also carries out considerable research and study of crustal formations using the gravity anomaly surveys in order to obtain a greater understanding of geological history and features. In the Canadian Shield, for example, researchers have suggested that some form of plate tectonics process played a major role in the formation of the Archean crust (the earliest of the two great divisions of the Precambrian period). Geologists have proposed that there are Proterozoic sutures between the older and younger geological provinces that make up the Shield. Based upon a mix of gravity, magnetic, seismic, and palaeomagnetic data related to these sutures, this view of Shield plate tectonics tends to support the theory that the Shield comprises a mosaic of once-separated, but now joined, cratonic blocks. This concept of Shield tectonics has been put forward by R.A. Gibb and M.D. Thomas in papers entitled *Proterozoic Sutures in Canada* (*Geoscience Canada*, v. 7, 1980. pp. 149–54); and *Gravity Anomalies at Structural Province Boundaries of the Canadian Shield: Products of Proterozoic Plate Interaction* (*Geo Skifter*, v. 10, 1978. pp. 21–57).

Finally, very specific gravity anomaly surveys are carried out in delineated zones in order to estimate the potential of mineral or petroleum deposits. For example, in 1992, in the Mackenzie District of the Northwest Territories, a gravity survey was carried out over the Sue-Dianne precia zone (an area of angular volcanic rocks possibly 1.8 million years of age). The incentive for this stemmed from the recognition that this area contained large polymetallic deposits similar to an important deposit in South Australia. A total of 70 gravimeter readings were taken during a five-day survey. The gravity readings indicated massive magnetite concentrations (iron oxides), formed geologically during the creation of the Great Bear Magnetic zone of the Early Proterozoic period, about 1,800 million years ago. This survey was carried out by S.S. Gandhi and D. Holliday and entitled *Gravity Survey of the Sue-Dianne Deposit, Northwest Territories* (In Current Research, Part E, Geological Survey of Canada Paper No. 91-1E, 1993. pp. 231–238).

(c) Magnetic Surveys

The earth behaves like a huge magnet, and the generally accepted theory of the earth's magnetic field hinges primarily on the probability that the outer part of the earth's core is composed of liquid metal (iron and nickel) that can conduct electrical currents, and also generate, or interact with, a magnetic field rather like a dynamo. The earth's magnetic field takes the form of a magnetic dipole, behaving in the same way as the field produced by a bar magnet. This dipole can be pictured as a huge bar magnet running through the centre of the earth, with the axis of the magnet almost paralleled to the geographic north-south line of the spin axis, but actually off about 18 degrees. The north magnetic pole is located in the islands of the Canadian Archipelago, and the south magnetic pole is south of Tasmania. However, the position of the poles, which is a separate geodynamic study, is not constant and is probably caused by shifts in the liquid core of the earth.

The magnetic lines of force act as a shield around the planet, protecting the earth from harmful radiations such as cosmic rays. In fact, after the era of satellites began in the late 1960s, investigations showed that there are zones of intense radiation surrounding the earth, now called Van Allan radiation belts. These belts are contained in the magnetosphere, which is actually an extension of the magnetic force field in a doughnut-shaped region surrounding the earth. On the sun-facing side, the magnetosphere extends outward about 10 earth radii, but on the side away from the sun it streams out into a tail for several millions of kilometres before losing itself in the overall magnetic field that encompasses the solar system. This disparity is caused by the solar wind, a stream of ionized particles or plasma that is discharged from the sun in all directions. Upon striking the earth's force field, the particles decelerate and a bow shock is created, inside of which is a turbulent region called the magnetosheath where the pressures from the solar wind and the earth's field are equal. Within the magnetosheath is the magnetopause, which encompasses the magnetosphere on the sun-facing side and the side away from the sun.

As mentioned above, the earth's magnetic field is like a dipole, but it is not in line with the geometric north-south axis of the earth and it shifts. This variation with the geographic axis is called the declination. It is this declination, along with six other trigonometrically related elements, plus the fact that the field varies from the equator to the north or south magnetic poles, that creates variation in the magnetic force field around the earth. The field strength at any point on the earth's surface is measured by a delicate instrument called a magnetometer. The unit of field strength is the gauss, named after the German mathematician, J.K. Gauss. At the equator, the field is about three gauss, and at the poles it is about double that strength. Both the declination and the field strength are surveyed, and charted for changes, on an ongoing basis. Severe changes caused by solar storms (there was a major one in 1972) affect the earth's magnetic field to the extreme, and can cause power interruptions in electric currents, as well as affect navigation instruments.

Currently, in Canada, magnetic observations are part of the National Geomagnetism Program of the Geological Survey of Canada. Previously, they were part of the Geophysics Division of the Earth Physics Branch of Energy, Mines and Resources Canada. The automatic digital recording of magnetic data began in 1969. The system is called the Canadian Geomagnetic Observatory Network, and originally consisted of 12 standard observatories incorporating an Automatic

Magnetic Observatory System (AMOS). The first network was called AMOS 1, which was later replaced by the AMOS MK III System. The latter has been described by D.F. Trigg and A. Nandi in their paper entitled *The Automatic Magnetic Observatory System AMOS III* (Ottawa: Energy, Mines and Resources Canada, Earth Physics Branch, 1984. 25 p. Geomagnetic Series No. 27). The system now consists of a network of 14 magnetic observatories across Canada. Twelve of these use the updated CANMOS (CANadian Magnetic Observatory System) instrumentation, and the two others use the older AMOS MK III instrumentation. The CANMOS automated system is described in detail in the National Geomagnetism website at <<http://www.geolab.nrcan.gc.ca>>. This network of standard magnetic observations, operating on a permanent basis, and with its careful measurement of absolute field values, is an important control for repeat station surveys and for ground and airborne surveys elaborated on below. This was discussed in detail by E.R. Niblett and R.L. Coles in a paper entitled *Scientific Requirements for Geomagnetic Observatory Data.* In: *Proceedings of the International Workshop on Magnetic Observatory Instruments: Ottawa, Canada.* 30 July–9 August, 1986. Edited by R.L. Coles (Ottawa: Energy, Mines and Resources Canada, 1988. 94 p. Geological Survey of Canada Paper 88-11. Geomagnetic Series No. 32).

Besides the standard observatories, there is a Canadian Magnetic Repeat Station Network consisting of 59 stations. Four of these are designated Class A and are revisited every two years, 37 are Class B stations and revisited every four years, and the remaining 18 Class C stations are visited approximately once a decade. At each repeat station, a temporary magnetic observatory is set up for a period of approximately 72 hours to record absolute and variations in the magnetic field. This system of repeat stations has been described by L.R. Newitt and R.L. Coles in their paper entitled *Canadian Magnetic Repeat Stations, 1962–1983: Instruments, Procedures and Observations* (Ottawa: Energy, Mines and Resources Canada, 1985. 31 p. Geomagnetic Series No. 29). This paper is also a good source for historical development of magnetic observations. Since 1985, the data from the magnetic observatories and the repeat stations has been collected and published in the *Annual Report for Magnetic Observations and Repeat Stations* (Ottawa: Energy, Mines and Resources Canada, 1985–). Besides the magnetic data calculations for each station, the annual report discusses the instrumentation used, the derivation of absolute observations and baseline calculations, and the quality of the data. From 1970 to 1984, the *Annual Report for Magnetic Observatories* (Ottawa: Energy, Mines and Resources Canada, 1970–84. 14 v.) reported the magnetic data for the 17 standard magnetic observatories only. And, before 1970, data was published individually for the standard magnetic observatories as they were established. For example, the Victoria Magnetic Observatory was established in 1957, and from that date until 1964, the *Record of Observations at Victoria Magnetic Observatory* (Ottawa: Department of Energy, Mines and Resources, 1957–64. 8 v.) was published.

Besides the geomagnetic network across the country, aeromagnetic surveys covering particular sections of the country have also been taken. For example, in the late 1970s and early 1980s, G.V. Haines and W. Hannaford documented aeromagnetic and data in three papers, entitled *A Three-Component Aeromagnetic Survey of Saskatchewan, Alberta, Yukon, and the District of Columbia* (Ottawa: Energy, Mines and Resources Canada, 1976. 28 p. Geomagnetic Series No. 8); *A Three-Component Aeromagnetic Survey of Central Canada* (Ottawa: Energy, Mines and Resources Canada, 1978. 22 p. Geomagnetic Series No. 3); and *A Three-Component Aeromagnetic Survey of Eastern Canada* (Ottawa: Energy, Mines and Resources Canada, 1980. 18 p. Geomagnetic Series No. 16). The aeromagnetic program is ongoing. For example, in 1993, high resolution geomagnetic surveys, totalling 147,000 line kilometres, were flown in British Columbia, Saskatchewan, and Manitoba. These surveys, which are described by R. Dumont and P.E. Stone in the paper entitled *Aeromagnetic Survey Program of the Geological Survey of Canada, 1993–94 in Current Research, 1994, D – Eastern Canada and National Programs* (Ottawa: Energy, Mines and Resources Canada, 1994, pp. 165–68), were expected to contribute to the understanding of deep structures beneath the Cordillera as well as the Precambrian basement beneath the Phanerozoic of the Canadian Shield.

The recurring magnetic data obtained over the years is incorporated into maps and research papers. The most recent cumulation of data appears in the Canadian Geophysical Atlas, described below. The techniques used, and the analysis of data obtained from magnetic observations, have appeared in papers; for example, by J.M. DeLaurier and E.R. Niblett entitled *Geomagnetic Depth Sounding over the Central Arctic Islands, Canada* (*Canadian Journal of Earth Sciences*, v. 17, 1980, pp. 1642–52); by E.L. Loomer and E.R. Niblett, entitled *Deviation of K Indices Using Magnetograms Constructed from Digital Data* (*Geophysical Surveys*, v. 6, 1984. pp. 431–37); by E.R. Niblett and R.D. Kurtz, entitled *Magnetotelluric Measurements over the Alpha-Ridge* (*Physics of the Earth and Planetary Interiors*, v. 45, 1987. pp. 101–18); and by R.D. Kurtz and J.A. Craven entitled *The Conductivity of the Crust and Mantle beneath the Kapuskasing Uplift: Electrical Anisotropy in the Upper Mantle* (*Geophysical Journal International*, v. 113, 1993. pp. 483–98). Data is also provided to world data centres, located at the National Geophysical Data Center, Boulder, Colorado; the Goddard Space Flight Center, Greenbelt, Maryland; the Russian Geophysical Committee of the Academy of Sciences of Russia, Moscow; the British Geological Survey, Edinburgh; and, the Faculty of Science, Kyoto, Japan. The data is for the International Geomagnetic Reference

Field (IGRF), which is a series of mathematical models of the earth's main magnetic field, and its secular (cumulative) variation. The Canadian part of the program is called the Canadian Geomagnetic Reference Field, and is described in the National Geomagnetism Program website at <<http://www.geolab.nrcan.gc.ca>>. The most recent revision of IGRF is the seventh generation, adopted by the International Association of Geomagnetism and Aeronomy. The Mathematical Models are based on five-year periods, or epochs, that go back to 1900. The seventh generation revisions were announced by C.E. Barton and R.T. Baldwin, et al., in the letter entitled *International Geomagnetic Reference Field, 1995 Revision Presented by IAGA Division V, Working Group 8* (*Physics of the Earth and Planetary Interiors*, v. 97, 1996. pp. 23–26). Details about the derivation of the 1995 revision of the IGRF were described by C.E. Barton, et al., in the paper entitled *International Geomagnetic Reference Field – The 7th-Generation* (*Journal of Geomagnetism and Geoelectricity*, v. 49, 1997. pp. 123–48). The eight generation five-year period revisions, which cover the years 2000–2005, are also under way.

Besides the ongoing measurement of magnetic intensities and dipole shifts, historic magnetic intensities are also monitored for clues to the geological formation of the earth's crust. This study of the fossil magnetization of rocks is called palaeomagnetism. Almost all rocks contain very small quantities of iron compounds capable of acquiring a magnetization. They may be igneous or sedimentary rocks. Igneous rocks acquire their magnetism, called thermoremanent magnetism (TRM), as they gradually cool (i.e., the minute basic magnets within their magnetic grains become aligned in the ambient field direction at the place and historic time of cooling). Sedimentary rocks, by contrast, may acquire a detrital remanent magnetism (DRM), as the small magnetic particles settled through water and oriented themselves, according to the influence at the magnetic field at that time. What is particularly interesting is not so much that the rocks acquired their magnetism, but that the original direction of the magnetism has been retained over millions of years through all vicissitudes of geological history. This palaeomagnetism also makes possible the study of the geomagnetic pole, and its historic shifts called pole-wander. The latter has helped to verify the concept of continental drift. Besides field strengths in ancient rocks, examples of field reversals have also been documented and confirmed, in the mid-1960s, to indicate that the north and south magnetic poles have been reversed from time to time throughout our geological history.

In Canada, palaeomagnetic studies have concentrated upon the perimeter of extreme geologic regions, such as the Arctic Archipelago, the Canadian Cordillera, and the Canadian Appalachia. The purpose of these studies is to aid in filling the gaps concerning the evolution of North America. Examples of Arctic palaeomagnetic studies are by J. Vincent and W.A. Morris, entitled *Glacial and Nonglacial Sedi-*

ments of Matuyama Palaeomagnetic Age on Banks Island, Canadian Arctic Archipelago (*Geology*, v. 12, 1984. pp. 139–42); by P.J. Wynne and E. Irving, entitled *Paleomagnetic Results from the Sverdrup Basin Theoleiites and a Mid-Cretaceous Tectonic Reconstruction of the Arctic Region* (Program with Abstracts of the Joint Annual Meeting of the Geological Association of Canada, Mineralogical Association of Canada and the Canadian Geophysical Union, 1986. 11 p.); by R.D. Knight and G.D. Jackson, entitled *Sedimentology and Stratigraphy of the Mesoproterozoic Elwin Subgroup (Aqiqlik and Sinasiuvik Formations), Uppermost Bylot Supergroup, Borden Rift Basin, Northern Baffin Island* (Ottawa: Geological Survey of Canada, 1994. 43 p. Geological Survey of Canada Bulletin No. 455); and by S.J. Pehrsson and K.L. Buchan, entitled *Borden Dykes of Baffin Island, Northwest Territories: A Franklin U-Pb Baddeleyite Age and a Paleomagnetic Reinterpretation* (*Canadian Journal of Earth Sciences*, v. 36, 1999. pp. 65–73). Papers on palaeomagnetic studies of the Cordillera have been written by E. Irving and G.J. Woodsworth, entitled *Palaeomagnetic Evidence of Northward Motion of the Western Cordillera in Latest to Early Tertiary Times* (*Abstracts with Programs. Geological Society of America*, v. 17, 1985. p. 636); and again by E. Irving and J.W. Monger in a paper entitled *New Palaeomagnetic Evidence for Displaced Terranes in British Columbia* (*Geological Association of Canada. Special Paper*, v. 20, 1980. pp. 444–56). Finally, a major paper published in the mid-1970s by E. Irving and F.K. North used palaeomagnetic data to show the relationship between oil occurrences and climate, nutrients and tectonics. The paper was entitled *Oil, Climate, and Tectonics* (*Canadian Journal of Earth Sciences*, v. 11, 1974. pp. 1–17).

E. Irving has also compiled five worldwide palaeomagnetic data catalogues. For example, the fifth was entitled *Catalogue of Palaeomagnetic Directions and Poles: Fifth Issue Cenozoic Results, 1927–1975* (Ottawa: Energy, Mines and Resources Canada, 1976. 87 p. Geomagnetic Series No. 10). Examples of palaeomagnetic studies of the Canadian Appalachia are by E. Irving and D.F. Strong, entitled *Evidence against Large-Scale Carboniferous Strike-Slip Faulting in the Appalachian-Caledonian Orogen* (*Nature*, v. 310, 1984. pp. 762–64); by F.L. Jansa and G. Pe-Piper, entitled *Appalachian Basement and Its Intrusion by Cretaceous Dykes, Offshore Southeast Nova Scotia, Canada* (*Canadian Journal of Earth Sciences*, v. 30, 1993. pp. 2495–2509); and by K.L. Buchan and F.W. Chandler, entitled *Palaeomagnetism of the Distal Member of the New Glasgow Formation, Nova Scotia, Canada: New Constraints on Magnetic Polarity Stratigraphy near the Base of the Kiaman Superchron* (*Journal of Geology*, v. 107, 1999. pp. 271–86).

Lastly, much of the detailed magnetic surveying in Canada pertains to magnetic anomalies, because of their importance in mineral exploration work. For example, P.J. Hood and J. Irvine, in their paper entitled *The Application of the Aeromagnetic Gradeometer Survey Technique to Gold Exploration in the Val d'Or*

Mining Camp (Ottawa: Geological Survey of Canada, 1983. 5 p. Geological Survey of Canada Paper No. 83-8), discussed new techniques for mineral exploration. In the paper by P.H. McGrath and R.S. Hildebrand, entitled *An Estimate, Based on Magnetic Interpretation of the Minimum Thickness of the Hornby Bay Group, Leith Peninsula, District of Mackenzie* (Ottawa: Geological Survey of Canada, 1984. Geological Survey of Canada Paper No. 84–1A. pp. 223–28), the authors were interpreting data to assist in the unravelling of the tectonic history of the Canadian Archipelago.

Perhaps the most visible signs of magnetic anomaly studies, within Canada, are the series of magnetic anomaly maps that are issued from time to time by the Geological Survey of Canada. The most recent series of 102 magnetic anomaly maps began to be issued in 1991. The first map was entitled *Magnetic Anomaly Map, Labrador Sea South* (Ottawa: Geological Survey of Canada, 1991. Geological Survey of Canada Map NN-22–M, Scale 1:1,000,000). As well, national magnetic anomaly maps are part of the *Canadian Geophysical Atlas*, described below.

(d) Seismology Operations and Research

Seismology often concerns the study and research of earthquakes, with their variable damage to buildings, and loss of life in inhabited areas. But seismology more positively also concerns the observation and study of the elastic seismic waves generated by earthquakes and man-made explosions in order to generate information about the earth's interior.

These elastic waves, which involve a sudden release of energy, are caused by the fact that the earth is continuously undergoing deformation, due to stresses often caused by the tectonic plates on which the earth's crust rests. There are three types of elastic or seismic waves involved. There are compressional or primary waves, called P-waves, which move in a push-pull manner; shear waves, or secondary waves, called S-waves, which vibrate back and forth in a direction perpendicular to their travel path; and longitudinal, or L-waves, which travel along the surface of the earth. Both P- and S-waves can transmit freely through solid rocks, but only S-waves can pass through liquids. These seismic waves are similar to light waves, because they can be reflected from, transmitted through, or refracted within layers of rock inside the earth. With these waves, the denser the material, the greater the velocity with which the waves travel, and when they encounter a change in density of the medium through which they pass, they will be refracted. Geologists can build up a picture of the interior of the earth by analysing the effects that internal rocks have upon seismic waves. Seismic waves, either man-made or from natural causes, are measured using a device called a seismometer, which measures any change between itself and the earth beneath it. This is an approximate measurement and it is achieved by mounting a small mass on either a pendulum or the end of a spring, so that when the ground moves, the mass remains still because of its inertia, and the earth is displaced relative to it. It is this displacement that is recorded by the seismometer to give a seismograph.

In order for geologists to study the sub-surface strata, explosive devices are used to create man-made mini-earthquakes. To record the reflected waves created by these man-made earthquakes, small recorders called geophones, for land use, and hydrophones, for marine use, are used. Computers process the signals produced by these devices, apply corrections, and generate a seismic profile. However, there is one other method used by the oil and gas industry for creating seismic profiles. This method uses a logging tool called a sonde, which is lowered on a steel cable into a bore hole, and makes a variety of measurements aimed at establishing the porosity and permeability of the strata encountered. In Canada, seismology studies, and documentation, are currently carried out by the Seismology and Electromagnetism Section of the Continental Geoscience Division of the Geological Survey of Canada, where there has been considerable seismic research carried out on the earth's upper mantle and crust, studies on interpreting P-waves, and analytical studies on potential seismic situations. As well, major work has been carried out on the detection of underground nuclear explosions.

Crustal and mantle research has been carried out to enlarge basic knowledge, as well as assist in the Arctic exploration for hydrocarbon resources. Examples of the range of research consist of one paper by A.G. Green and D.H. Hall entitled *A Sub-Critical Seismic Crustal Reflection Survey over the Aulneau Batholith, Kenora Region, Ontario* (Canadian Journal of Earth Sciences, v. 15, 1978. pp. 301–15); and, other papers by A.G. Green and B. Milkereit entitled *Crustal Structure of the Grenville Front and Adjacent Terranes* (*Geology*, v. 16, 1988. pp. 788–92); *Deep Structure of an Archean Greenstone Terrane* (*Nature*, v. 344, 1990. pp. 327–30); and *Origin of Deep Crustal Reflections: Terranes in Canada* (*Tectonophysics*, v. 173, 1990. pp. 627–38). Work concentring on the crustal structure of the Arctic region has been carried out by D.A. Forsyth and J.A. Mair in their research papers entitled *Crustal Structure of the Central Sverdrup Basin* (*Canadian Journal of Earth Sciences*, v. 16, 1979. pp. 1581–98); and *Crustal Structure of the Lomonosov Ridge and the Fram and Makarov Basins near the North Pole* (*Journal of Geophysical Research. B*, v. 89, 1984. pp. 473–81). Other papers by I. Asudeh and D.A. Forsyth have been entitled *Delineation of Sedimentary Basins Using Seismic Techniques on Canada's Arctic Continental Margin*. In: *The Potential of Deep Seismic Profiling for Hydrocarbon Exploration*. Edited by B. Pinot and C. Bois (Paris: Editions Technip, 1990. pp. 225–36); and *Sedimentary Basins and Basement Highs beneath the Polar Continental Shelf North of Axel Heiberg and Meighen Islands* (*Bulletin of Canadian

Petroleum Geology, v. 46, 1998. pp. 12–29). As well, the Beaufort Sea area has been studied by H.S. Hasagawa and C.W. Chou in their paper entitled *Seismotectonics of the Beaufort Sea* (*Canadian Journal of Earth Sciences*, v. 16, 1979. pp. 816–30). Finally, between 1984 and 1992, geophysical data, derived from 6,000 kilometres of seismic profiling of the Appalachian mobile belt between the Canadian Maritimes and the continental margin of the Atlantic Ocean, was interpreted in a major paper by S. Dehler and J. Hall entitled *Geophysical Studies of the Structure of the Appalachian Orogen in the Atlantic Borderlands of Canada* (*Canadian Journal of Earth Sciences*, v. 35, 1998. pp. 1205–21).

The study of P-waves focused upon the Malbaie area of the north shore of the St Lawrence River, in 1974–75, and produced important data to help outline the seismic patterns between Quebec City and the Malbaie and provided a greater understanding of the physics of earthquakes. This area is the site of a large meteorite crater, and is the location of the highest seismicity and largest earthquakes recorded in Canada. Papers generated from this study by G.G. Buchbinder were entitled *P-wave Residuals in Canada* (*Canadian Journal of Earth Sciences*, v. 14, 1977. pp. 1292–1304), and *Stability of P-Travel Times in the Region of La Malbaie, Quebec* (*Bulletin of the Seismology Society of America*, v. 69, 1979. pp. 463–81).

It was in the late 1960s that considerable progress was made in research into the problems of detecting underground nuclear explosions. To that end, the Yellowknife seismological array was established by the Department of Energy, Mines and Resources, in 1962, with assistance from the scientists of the Atomic Weapons Research Establishment of the United Kingdom Atomic Energy Authority, to collect seismological data for research only. The array was not intended to be an operational monitoring station. Positive identification levels were achieved with these installations, using the relative excitation of long-period surface waves by an underground explosion, and by an earthquake in the same area. This research was summarized by P.W. Basham and K. Whitlam, in their publication entitled *Seismological Detection and Identification of Underground Nuclear Explosions* (Ottawa: Department of Energy, Mines, and Resources. Earth Physics Branch Publications, v. 41, no. 9, 1971. pp. 145–82).

Besides the Yellowknife array, Energy, Mines and Resources Canada operates 14 standard seismological stations, 48 regional stations, which are used in seismically active areas, to supplement the standard station network; two regional telemetered networks, based in Ottawa, and near Victoria; two local telemetered networks; two strong-motion seismograph networks on the east and west coasts; and several special or temporary operations. These networks transmit data by UHF radio, and/or telecommunication lines, to a central processing site, such as Ottawa, or to a control centre, such as the Yellowknife array. Overall, seismograph operations in

Canada are described in the annual publication entitled *Canadian Seismograph Operations* (Ottawa: Energy, Mines and Resources Canada, 1974–). This is the successor to the *Seismological Bulletin of the Earth Physics Branch*, which was published from 1954 to 1973. The data collected is used both for the ongoing documentation of earthquakes occurring in Canada, and for research and analysis purposes. The ongoing documentation is published in the annual series entitled *Canadian Earthquakes* (Ottawa: Energy, Mines and Resources Canada, 1960–). This annual publication not only describes the Canadian seismograph network, but also summarizes seismic activity according to region. Until 1984, the publication also included an Appendix of all earlier publications, which documents Canadian earthquakes as far back as 1534. An example of this latter type of publication was one by S.S. Meidler, entitled *Seismic Activity in the Canadian Arctic, 1899–1955* (Ottawa: Department of Mines and Technical Surveys, 1961. 9 p. Seismological Series of the Dominion Observatory, 1961–3). This publication was a compilation, from available sources, of earthquakes which occurred during that time period in the region north of latitude 60°.

The study of existing seismic data has resulted in a range of analytical publications. For example, after a major earthquake in Quebec in 1973, R.B. Home and R.J. Wetmiller published a paper entitled *The St. Donat, Quebec Earthquake Sequence of February 18–23, 1978* (*Canadian Journal of Earth Sciences*, v. 16, 1979. pp. 1892–98). This tremor was felt over an area of 70,000 square kilometres in western Quebec, eastern Ontario, and northern New York State. An analysis of the data suggested that there was a systematic change in focal depths across the western Quebec zone, from shallow (upper crustal) depths, north of Montreal, to deeper (mid-crustal) depths, north of Ottawa. Another paper which focused upon a later earthquake event in Quebec has been written by D.A. Forsyth and entitled *The Mont Laurier, Quebec, Earthquake of 19 October, 1990 and Its Seismotectonic Environment* (*Bulletin of the Seismological Society of America*, v. 84, 1994. pp. 1506–22). One very broad paper by K. Whitham and H.S. Hasegawa was entitled *The Estimation of Seismic Risk in Canada – A Review* (Ottawa: Energy, Mines and Resources Canada. Earth Physics Branch Publications, v. 45, 1975. pp. 137–62). This paper discussed the problem of estimating seismic risk and expressing it in national building codes. The prediction of future ground motions or their causative earthquakes could only be estimated on the reliability and probability associated with projecting past data into the future. In fact, the 2001 National Building Code, for the first time, includes seismic hazard maps, as noted in the survey publication Geological Survey of Canada 1997–98 (Ottawa: Natural Resources Canada, 1998. p. 24). Finally, the overall documentations of seismicity data from seismograph operations are to be found in the *Canadian Geophysical Atlas* described below.

(e) Remote Sensing in Canada

Remote sensing is the term usually applied to observations made of the earth's surface from aircraft, or satellites. Aerial photography was first employed in the mid-19th century for military intelligence, using kites and balloons, and every subsequent development in remote sensing has been triggered directly from military applications. Initially, infrared photography was devised to distinguish between live plants, which reflect infrared radiation, and infrared-absorbing camouflage and dead vegetation. Thermal imagery, which measures emitted infrared, came from night-time bombing reconnaissance. Airborne radar scanning was developed for the monitoring of missile sites. Today, remote sensing is an important tool for geologists, as well as geographers, meteorologists, and environmental scientists, because of its high and wide field for viewing of the earth's features.

These methods of remote sensing depend upon the way in which matter absorbs, reflects, refracts, or radiates electromagnetic radiation, as it is measured by airborne instruments or by scanners mounted on board space satellites. Conventional photography can detect only that part of the electromagnetic spectrum between long-wavelength ultraviolet and short-wavelength infrared. For conventional photography, visible ordinary panchromatic, black and white film, or natural colour film, is used in conjunction with a filter to eliminate the blue end of the spectrum that results in the scattering effect of atmospheric produced haze. Infrared, or short-wavelength, however, can be detected on ordinary film, using a red filter, or infrared-sensitive black-and-white film. Colour photography is also possible, using specially prepared colour infrared film.

At present, there are two methods that are used by spacecraft, or orbiting satellites, for gathering images of the earth. One uses a multispectral scanner (MSS), which is a remotely operated system employing a mirror to scan a swath of ground beneath the satellite or spacecraft. The radiation collected is split into discrete wavebands by series of prisms, and passed onto one of a number of photo cells, and then recorded on multi-track magnetic tape. This stored information is periodically transmitted back to receiving stations on the earth's surface. Another method, called SLAR (side-looking airborne radar), actually uses an energy source generated on board an orbiting satellite. The energy source is a transmitted radar pulse. The information reflected back from the target area is collected aboard the satellite, and then recorded onto magnetic tape. This data can subsequently be processed into visual images. Despite these advanced techniques for monitoring the earth's surface, much geological mapping still involves the study of overlapping photographs, photographed vertically downward, which allows for stereoscopic views of the earth.

In Canada, aerial surveys and photography have long played an important part in mapping the Canadian territory, and in helping to assess various resources, such as forest cover, agricultural crops, and wildlife, as well as applications to geology. However, with the development of sophisticated cameras, films, filters, and other types of scanners, mentioned above, a new era for mapping and re-evaluating the Canadian environment has opened up. It was in 1971 that remote sensing in Canada became an established fact with the creation of the Canada Centre for Remote Sensing as a branch of the Department of Energy, Mines and Resources. (The Centre is now part of Geomatics Canada of the Earth Sciences Sector of Natural Resources Canada.) In May of that year, an agreement was signed between EMR and NASA (the US National Aeronautics and Space Administration), to receive remote sensing imagery from an orbiting satellite, to be launched by NASA in 1972. That satellite, which went up on 23 July 1972, was the first of NASA's Earth Resources Technology Satellites (ERTS-1 or LANDSAT-1). To read out Canadian data, an existing 85-foot parabolic disk antenna system at Prince Albert, Saskatchewan, was converted and equipped with appropriate tracking and recording facilities. The ERTS imagery was produced as black and white and colour composite images, each encompassing an area 183 by 183 kilometres. As well that year, the Canada Centre for Remote Sensing expanded its airborne remote sensing activities to a fleet of four aircraft. That year, 35,346 sensor line-miles were flown, involved in 116 projects. Besides these flights, work was also carried on in the development of sensors. One system, called HISS, for measuring ice thickness, was contracted to the University of Toronto, and a laser-ranging device, called LIDAR, was contracted to York University, in Toronto, for measuring atmospheric constituents near that city. By 1975, NASA had also launched NOAA-3, a weather satellite, and in the early 1980s, participated in the European Space Agency Earth Resources Satellite (ERS-1) program, which was developed to carry a range of microwave sensors for ice, oceans, and weather applications research. As well, the Centre for Remote Sensing planned a Canadian remote sensing satellite, called RADARSAT, which was launched in the early 1990s. RADARSAT is equipped with a synthetic aperture radar (SAR) as its main sensor, and operates in the 10-centimetre wavelength, with a ground resolution of 25 metres, and a swath width of 200 kilometres in all weather conditions. Its ability to see through clouds, storms, and darkness makes it particularly useful for supporting Arctic and off-shore operations, as well as being invaluable for applications in geology, forest management, agricultural production, water resources, and for oil, gas, and mineral explorations. Today, the RADARSAT-1 program is operated by the Canadian Space Agency, and its activities are documented in the publication entitled *RADARSAT Annual Review* (Saint-Hubert, Quebec: Canadian Space Agency, 1997–).

As remote sensing developed after 1972, with the additional use of satellites and new sensors, research scientists at the Canada Centre for Remote Sensing began to issue research reports and manuals on the applications of remote sensing in Canada. For example, A.K. McQuillan wrote one report, in 1975, entitled *The Value of Remote Sensing in Canadian Frontier Petroleum Operations* (Ottawa: Canadian Centre for Remote Sensing, 1975. 60 p. Research Report 75–4 AJ). Another report by J.M. Harvie and J. Cihlar was entitled *Surface Cover Mapping in the Arctic through Satellite Remote Sensing* (Ottawa: Canada Centre for Remote Sensing, 1983. 61 p. Users' Manual 82-1). Again, another report by F.J. Aherin and W.J. Bennet was entitled *Landsat MSS Enhancements for Forestry Applications* (Ottawa: Canada Centre for Remote Sensing, 1985. 25 p. Users' Manual 85-1). An overview of the importance of remote sensing to geography, edited by M.D. Thompson, was entitled *Landsat for Monitoring the Changing Geography of Canada* (Ottawa: Canada Centre for Remote Sensing 1982. 84 p. Special Report for COSPAR by the Geography Working Group, Canada Advisory Committee on Remote Sensing). This paper, which reviewed the use of remote sensing for monitoring hydrologic and coastal changes, agricultural and land use intensification, urbanization, and resource development, suggested that after 10 years of remote sensing, the technique was well suited for monitoring environmental changes in Canada, particularly in view of the large area of the country to be monitored. Papers pertaining to agricultural applications of remote sensing have been written by B. Brisco and R.J. Brown and entitled *Temporal Ground-Based Scatterometer Observations of Crops in Western Canada* (*Canadian Journal of Remote Sensing*, v. 18, 1992. pp. 14–22); *Potential Applications of RADARSAT Data to Agriculture and Hydrology* (*Canadian Journal of Remote Sensing*, v. 19, 1993. pp. 317–29); and *Precision Agriculture and the Role of Remote Sensing: A Review* (*Canadian Journal of Remote Sensing*, v. 24, 1998. pp. 315–327).

G The Canadian Geophysical Atlas

Perhaps an appropriate conclusion to this survey of geophysical research, and data gathering, in Canada is a look at the most visible result of this ongoing work undertaken at Energy, Mines and Resources Canada (now absorbed into Natural Resources Canada). This is the *Canadian Geophysical Atlas* (Ottawa: Geological Survey of Canada. Geophysical Data Centre, 1987. scale 1:10,000,000). To date, this Atlas consists of a series of 15 coloured maps. Each map contains an excellent introduction, or survey, of the topic covered, as well as a bibliography of relevant references. These maps are listed below, with the authors responsible for their publication.

Observed Gravity Values of Canada by A.K. Goodacre, R.A.F. Grieve, and J.F. Halpenny (Ottawa: Geological Survey of Canada. Canadian Geophysical Atlas, 1987. Map 1, Scale 1:10,000,000).
Free Air Gravity Anomaly Map of Canada by A.K. Goodacre, R.A.F. Grieve, and J.F. Halpenny (Ottawa: Geological Survey of Canada, Canadian Geophysical Atlas, 1987. Map 2, Scale 1:10,000,000).
Bouger Gravity Anomaly Map of Canada by A.K. Goodacre, R.A.F. Grieve, and J.F. Halpenny (Ottawa: Geological Survey of Canada, Canadian Geophysical Atlas, 1987. Map 3, Scale 1:10,000,000).
Isostatic Gravity Anomaly Map of Canada by A.K. Goodacre, R.A.F. Grieve, and J.F. Halpenny (Ottawa: Geological Survey of Canada, Canadian Geophysical Atlas, 1987. Map 4, Scale 1:10,000,000).
Horizontal Gradient of the Bouger Gravity Anomaly Map of Canada by A.K. Goodacre, R.A.F. Grieve, J.F. Halpenny, and V.L. Sharpton (Ottawa: Geological Survey of Canada, Canadian Geophysical Atlas, 1987. Map 5, Scale 1:10,000,000).
Vertical Gradient of the Bouger Gravity Anomaly Map of Canada by A.K. Goodacre, R.A.F. Grieve, and J.F. Halpenny (Ottawa: Geological Survey of Canada, Canadian Geophysical Atlas, 1989. Map 6, Scale 1:10,000,000).
Gravimetric Geoid Map of Canada by D. Nagy (Ottawa: Geological Survey of Canada, 1989. Canadian Geophysical Atlas, Map 7, Scale 1:10,000,000).
Total Intensity Chart of Canada, 1990.0 by L.R. Newitt and G.V. Haines (Ottawa: Geological Survey of Canada, 1990. Canadian Geophysical Atlas, Map 8, Scale 1:10,000,000).
Magnetic Inclination Chart of Canada 1990.0 by L.R. Newitt and G.V. Haines (Ottawa: Geological Survey of Canada, 1990. Canadian Geophysical Atlas, Map 9, Scale 1:10,000,000).
Magnetic Declination Chart of Canada 1990.0 by L. R. Newitt and G.V. Haines (Ottawa: Geological Survey of Canada, 1990. Canadian Geophysical Atlas, Map. 10, Scale 1:10,000,000).
Magnetic Anomaly Map of Canada by S.D. Dods, D.J. Teskey, and P.J. Hood (Ottawa: Geological Survey of Canada, 1989. Canadian Geophysical Atlas, Map 11, Scale 1:10,000,000).
Vertical Gradient of the Magnetic Anomaly Map of Canada by D.J. Teskey, P.J. Hood, and S.D. Dods (Ottawa: Geological Survey of Canada, 1989. Canadian Geophysical Atlas, Map 12, Scale 1:10,000,000).
Magnetic Anomaly Map of Canada – Upward Continued to 40 km. by D.J. Teskey, P.J. Hood, and S.D. Dods (Ottawa: Geological Survey of Canada, 1989. Canadian Geophysical Atlas, Map 13, Scale 1:10,000,000).
Shaded Relief Presentation of the Magnetic Anomaly Map of Canada by D.J.

Teskey, P.J. Hood, and S.D. Dods (Ottawa: Geological Survey of Canada, 1989. Canadian Geophysical Atlas, Map 14, Scale 1:10,000,000).

Seismicity Map of Canada by F.M. Anglin, R.J. Wetmiller, R.B. Horner, G.C. Rogers, and J.A. Drupdale (Ottawa: Geological Survey of Canada, 1989. Canadian Geophysical Atlas, Map 15, Scale 1:10,000,000).

All the above maps are available from the Geological Survey of Canada, 601 Booth Street, Ottawa, Ontario K1A 0E8.

IV Departmental Sources of the Publications

The two earliest major catalogues covering the publications of the Geological Survey of Canada were by W.F. Ferrier, entitled *Annotated Catalogue of and Guide to the Publications of the Geological Survey of Canada, 1845–1917* (Ottawa: J. de Labroquerie, 1920. 544 p.); and by L.B. Leafloor entitled *Publications of the Geological Survey of Canada (1917–1952)* (Ottawa: Department of Mines and Technical Surveys, 1952. 82 p.). More recently, the the historic documentation of geology and mining has been well served in two cumulative publications. The first was compiled by A.G. Johnston, and entitled *Index of Publications of the Geological Survey of Canada (1845–1958)* (Ottawa: Department of Mines and Technical Surveys. Geological Survey of Canada, 1965. 403 p.). Part I of this catalogue lists all publications of the Geological Survey of Canada, from its inception to the end of 1958. Part II of the catalogue lists all reports and maps classified by area, according to the National Topographic System. Since then, this Index has been updated with cumulations covering the years 1959 to 1974, 1975–79, 1980–84, and 1985–87. Since 1975, the indexes have been produced from the GEOSCAN database, which coordinates input from the 10 provinces, several federal agencies, and other geoscience agencies. These later indexes list by publication title, author, and National Topographic System area.

The second cumulative publication, for mines publications, is the sectional catalogue by the Department of Public Printing and Stationery entitled *Canadian Government Publications, Mines Branch and Mineral Resources Division, Energy, Mines and Resources*. 3rd Edition (Ottawa: Queen's Printer, 1967. 401 p. Sectional Catalogue No. 12). This catalogue, in five parts, covers all Mines Branch publications from 1902, to the end of 1960. It is divided into the early Mines Branch publications and Mineral Resources Division publications, to 1959, and the new publication series of the Mines Branch and Mineral Resources Division, since 1958. This publication was partially and separately updated, in 1980, by Energy, Mines and Resources Canada, with the publication of the *Index of Selected Publications, 1979* (Ottawa: Energy, Mines and Resources Canada, 1980. 254 p.).

The latter is a title and subject index of general publications on energy, minerals, and earth sciences, published primarily in 1976–79. More consistent have been the separate publication lists of the Canada Centre for Mineral and Energy Technology. They are: *Catalogue and Index of CANMET Scientific and Technical Papers Published in or Submitted to Periodicals and Presentations, 1967–1973* (Ottawa: Canada Centre for Mineral and Energy Technology, 1976. 110 p.); *Catalogue of CANMET Publications, 1976–1981/82* (Ottawa: Canada Centre for Mineral and Energy Technology, 1976–81/82. 7 v.); *Catalogue of Publications, 1982/83* (Ottawa: Canada Centre for Mineral and Energy Technology, 1982/83. 1 v.); and *CANMET Catalogue of Publications, 1983/84–* (Ottawa: CANMET Library and Documentation Services Division, 1988–).

Other sectional lists that have been published from time to time are: (1) the Gravity Division's *Index of Publications, 1948–1973* (Ottawa: Department of Energy, Mines and Resources, Earth Physics Branch, 1974. 22 p.), which is a list of publications, contributions, and map series of that division, along with a listing of their information services (e.g. Earth Tide Data Service). This publication was supplemented, in 1979, with the *Gravity and Geodynamics Divisions Index of Publications, 1973–1978 Supplement* (Ottawa: Department of Energy, Mines and Resources, Earth Physics Branch, 1979. 2 p.), and (2) Earth Physics Branch's *Index of Geophysical Publications, Series and Contributions of the Dominion Observatory and Its Successor the Earth Physics Branch of the Department of Energy, Mines and Resources, Canada, through September, 1978.* 2nd Edition (Ottawa: Energy, Mines and Resources Canada, Earth Physics Branch, 1978. 53 p.).

Finally, with the creation of the Natural Resources Canada website at <<http://www.nrcan.gc.ca>>, the Earth Sciences Sector and the Minerals and Metals Sector web-pages do list some publications by sector or division. However, these lists are not comprehensive, and the sites vary in quality. The two most comprehensive sites are the Geological Survey of Canada and the Canada Centre for Remote Sensing. For example, the Geological Survey of Canada provides links to its library site for its major series (papers, etc.) but not for its research papers published in the international journal literature. On the other hand, the Canada Centre for Remote Sensing has a CCR Bibliographic Data Base which is searchable by author, keyword, date range (back to 1982), and classification (subject) for their research appearing in the international literature.

Chapter 8

Climate Research and Services

I Historical Background

Meteorology studies and climate research in Canada began with the arrival, in 1839, of a detachment of the Royal Artillery, commanded by C.J.B. Riddell, to establish a magnetic and meteorology observatory for the British government. Toronto was chosen rather than Montreal as the site, because of the interference caused to magnetic observations by the underlying geology of the St Lawrence River valley. Accordingly, in 1840, the Toronto Magnetic and Meteorological Observatory was erected on property owned by King's College. The stimulus for creating the observatory was provided by the early-19th-century European interest in global magnetism, and the desire to create fixed observatories around the world in order to document and study magnetism. By the late 1840s, however, the British government decided to withdraw support for the observatory, and the latter passed to the government of the United Province of Canada, but to be administered by the University of Toronto, with funds for support coming from the government.

It was during the 1850s, under the guidance and encouragement of G.T. Kingston, who became professor of meteorology at the University of Toronto and Director of the Magnetic and Meteorological Observatory from 1855 until he retired in 1880, that the first meteorological observing stations were established at senior grammar schools in Upper Canada. At the time of his retirement, there were 176 active stations, and an informal network that stretched from Nova Scotia to British Columbia. By the end of the century, there were over 300 climate observing stations, including the west and northwest. By the 1920s, the developing commercial aviation industry required weather services, and after the establishment of the Department of Transport in 1936, the Meteorological Service was transferred to it. During that period, the observing network had expanded to 775 stations, and by the 1960s there were over 2,000 stations. Finally, in 1971, one hundred years after the creation of the Meteorological Service of Canada, weather service activities, including meteorological and climate research, were transferred to the new Department of the Environment as the Atmospheric Environment Service (AES). It is interesting to note that in 1999, the AES reverted back to its original name as the Meteorological Service of Canada.

Today, the mandate of the Meteorological Service of Canada is to 'ensure the safety and well-being of all Canadians by providing high quality environmental information and advice ...' In order to carry out that broad mandate, the Service consists of a network of directorates, branches, and regional structures. For example, the Atmospheric Monitoring and Water Division (formerly the Weather Service Directorate), located at Downsview, Ontario, in suburban Toronto, is responsible for the overall management of present and predicted weather data. But, for geographical reasons, these services are broken down into five regional directorates. They are the Pacific and Yukon Region, the Prairie and Northern Region, the Central Region, the Ontario Region, the Quebec Region, and the Atlantic Provinces Region. The prediction activities themselves are located at nine weather forecast centres within the regions. And these forecast centres are supported by the Canadian Meteorological Centre, and its computer system located in Montreal (this is part of the Atmospheric Environment Prediction Division of the MSC), where all data is verified for use. The archived weather data itself is stored in the National Climatological Archive at Downsview. Finally, there are another 62 smaller weather offices, located across Canada, which serve as distribution and consultation points for the forecasts and warnings issued by the Weather Forecast Centres. Besides being responsible for the management of weather services, and the generation of weather information, the Meteorological Service of Canada also carries out studies on climate processes, and atmospheric research. In particular, this consists of national, regional, mesoclimatology, and Canadian city studies, weather phenomena, weather forecasting, boundary layer, and air quality research. The latter is carried out under the Long Range Transport of Air Pollutants Program (LRTAP) and covers topics ranging from surface ozone and its relationship to other pollutants, such as nitrogen oxide, and volatile organic compounds, to acid rain monitoring and research, and the phenomenon of global warming and its impact on climate. This research is carried out in the Atmospheric and Climate Science Division, which consists of five branches. They are the Air Quality Research Branch, the Climate Research Branch, the Meteorological Research Branch, the Adaptation and Impacts Research Group, and the Science Assessment Branch.

II Publications

Other than M.K. Thomas's work, entitled *The Beginnings of Meteorology in Canada* (Toronto: ECW Press, 1991. 308 p.), an overview and ongoing source of the development of weather services and meteorology in Canada are the annual reports of the federal departments to which the Meteorological Service of Canada has been attached. As previously noted, the Meteorological Service of Canada became part of the Department of Transport in 1937, because of the former's importance to aviation developments. The Meteorological Service's activities were reported, yearly, in the *Department of Transport Annual Report*, from 1937 to 1970 (Ottawa: Department of Transport, 1937–70. 43 v.). It is interesting to note in the 1937 report that 'members of the staff published 11 papers in various journals during the year, generally of a technical nature.' From 1971 on, with the formation of the Atmospheric Environment Service, the latter's activities were reported in the *Environment Canada Annual Report*, 1970–71 to 1975–76 (Ottawa: Environment Canada, 1971–76. 6 v.); the *Fisheries and Environment Canada Annual Report*, 1976–77 to 1978–79 (Ottawa: Fisheries and Environment Canada, 1977–79. 3 v.); and then again, the *Environment Canada Annual Report*, from 1980 onward (Ottawa: Environment Canada, 1980–). It is also in these annual reports that the international activities of the Service are mentioned. For example, in the early 1970s, the AES participated in the International Field Year for the Great Lakes (IFGL). As well, at that time, the Service participated in the Global Atmospheric Research Program (GARP), including the latter's (Global) Atlantic Tropical Experiment (GATE). And AES participated in the 100-year celebration of the World Meteorological Organization, in 1973, and continues under its new name to sit on its Technical Commissions, and participates in the data gathering and publication of Ozone Data for the World. Finally, the one other source that is useful for current information on programs, activities, and administrative structures is the Meteorological Service of Canada website at <<http://www.msc.gc.ga.ca>>.

The climate data and research activities carried out within the Meteorological Service of Canada have been documented in a range of current and historical publications, either produced by the MSC (previously the Atmospheric Environment Service), or appearing in the international scientific literature. The whole range of publications, issued over time, has been documented to a great extent in the publication *Handbook on Climate Data Sources of the Atmospheric Environment Service*. Compiled by D.W. Phillips and D.W. Gullet (Downsview, Ontario: Atmospheric Environment Service, Canadian Climate Centre, 1988. 6 Sections). Pertinent publications described below are Current Climate Data Publications; Climatological Normals Data; the Climate Atlas Climatique-Canada; Current Climatology Publication Series; Technical Report Series; Unpublished Reports; Solar Radiation Data; Ozone Data; and, Miscellaneous Publications. Most currently, both print and electronic publications and services are listed at Environment Canada's website <<http://www.cmc.ec.gc/climate/>>.

A Current Climate Data Publications

Historically, some form of meteorological data collecting and publication has been going on in Canada since 1840, with the establishment of the Magnetic and Meteorological Observatory in Toronto, in late 1839. A listing of stations, with climatological records dating back to 1840, can be found in the publication entitled *Canadian Stations with Climatological Records Dating Back to 1881* (Downsview, Ontario: Atmospheric Environment Service, 1961. 2 p. Climate Data Summary CDS 61-1). There are 38 general climatological elements (temperature etc.) that may be collected at weather stations, and there are over 300 data elements (e.g. maximum and minimum daily temperature) that may be documented by individual stations. The data that is collected at particular stations across Canada is listed in a multi-volume catalogue, entitled *Climatological Station Data Catalogue: British Columbia, the North Y.T and N.W.T., Prairie Provinces, Ontario, Quebec, Atlantic Provinces* (Downsview, Ontario: Atmospheric Environment Service, 1976–81, 6 v.). This publication refers to all climatological stations, active or inactive, and to the type and availability of data, to the end of 1980. Information listed includes the station number, station name, province, latitude and longitude, elevation, time periods with no change in program, location or name, and the type of observing or data-collecting program. Although all regional offices produce climatological summaries such as *Prairie Provinces Weekly Weather Summary* (Winnipeg: Atmospheric Environment Service, Central Region, v. 1, 1936–), and reports for specialized users, these are generally restricted to a limited number of stations and climatological elements, allowing for quick and economic publication of valuable data. The most familiar of the climatological summaries produced by the Weather Services Directorate, however, are the *Monthly Record of Meteorological Observations in Canada*; the Monthly and Annual Meteorological Summaries by Individual Stations (produced by principal stations); and the *Monthly Review of Canadian Climate.* As well, the Climate and Water Products Division produces an updated CD-ROM product entitled the *Canadian Daily Climate Data on CD-ROM.*

(a) *Monthly Record of Meteorological Observations in Canada*

The *Monthly Record of Meteorological Observations in Canada* (Downsview, Ontario: Atmospheric Environment Service, 1983–) provides basic meteorological data from about 2,700 stations operated by the AES. Although monthly summaries of meteorological observations were first published, from 1871 to 1915, as a Supplement to the Annual Report of the Department of Marine and Fisheries, and in the Monthly Weather Review from 1877 to 1915, the current record began as *The*

Monthly Record of Meteorological Observations in the Dominion of Canada, and the Colonies of Bermuda and Newfoundland, in 1916, and except for minor changes in title and data, remained unaltered until 1977. From 1977 to 1982, the *Monthly Record* was published in three parts (with all data appearing in metric units): Part 1, Western Canada (British Columbia, Alberta, Saskatchewan, and Manitoba); Part 2, Northern Canada (Yukon and Northwest Territories); and Part 3, Eastern Canada (Ontario, Quebec, New Brunswick, Nova Scotia, Prince Edward Island, and Newfoundland). In 1983, significant changes in the *Monthly Record* were introduced. It was published again in one part, but only in microfiche, with some changes in format. Data for each province and territory start on a new fiche, and the monthly summary of 'days with' has been added for all climate stations, not just for the principal stations as previously published.

The *Monthly Record* includes: daily values of precipitation (rainfall, snowfall, and total precipitation; temperature (maximum, minimum, and mean); sunshine and degree-days less than (<) 18°C; a monthly summary of precipitation, temperature, sunshine, and degree days less than (<)18°C, giving the total percentage of normal or departure and the extreme(s) for each; a monthly summary of the 'number of days with ...' for a selection of meteorological elements (temperature below or equal to (<) 0°C, rain or drizzle, measurable precipitation, snow, thunderstorms, freezing precipitation, hail, fog, smoke or haze, dust or blowing sand, and blowing snow); mean synoptic (forecasting) observations; a wind summary; recording rain gauge data; and class 'A' pan evaporation data.

A Station Data Index, on each microfiche, lists all the stations located within the corresponding province or territory. This serves as a guide for finding data on that fiche, or on subsequent fiches, for stations in the corresponding province/territory. Only stations that have data published in that monthly issue of the *Monthly Record* are listed in the Index, which also indicates the basic elements (temperature, precipitation, sunshine, synoptic report, wind, rain, and evaporation) that are published that month for each station. Climatological data received too late for inclusion in the appropriate issue of the *Monthly Record* appear in a supplement at the end of the year.

(b) Monthly and Annual Meteorological Summaries by Individual Stations

Besides the *Monthly Record of Meteorological Observations in Canada*, monthly and annual summaries are prepared by some individual stations. These stations are indicated in the publication entitled *A List of Stations Preparing Monthly and Annual Meteorological Summaries* (Downsview, Ontario: Atmospheric Environment Service, 1987. 4 p. Climate Data Summary CDS 87–12). This publication indicates that the first *Monthly Meteorological Summary* was produced in January,

1941, for Toronto, and since then monthly and annual summaries have been issued for an increasing number of stations. At the end of 1987, there were 62 stations issuing monthly summaries, of which 46 also issued an annual summary.

Typical of the monthly and annual meteorological summaries for an individual station are those for the Toronto Pearson International Airport. The *Monthly Meteorological Summary at Toronto Pearson International Airport* (Toronto: Atmospheric Environment Service, 1987–) is four pages in length. The first two pages display standard data tables of daily temperature, degree days, relative humidity, thunderstorms, precipitation, snow on the ground, wind, and bright sunshine. Means, extremes, and normals for the month, as well as 10-year summaries, are also included. In these monthly summaries, some tables are optional. Some may include hourly averages for dry-bulb temperature, precipitation occurrences, prevailing wind direction and wind speed, and wet-bulb temperature. Others may include a brief textual summary of the month's weather, or any other relevant information, if there are local or regional requirements for this.

The *Annual Meteorological Summary, Toronto Pearson International Airport* (Downsview, Ontario: 1984–) is 10 pages in length, and provides the following tables: monthly and annual temperature and degree-days; monthly and annual precipitation, including rain and snow, monthly and annual sunshine and rain; daily maximum, minimum, and mean temperatures; daily rainfall, snowfall, and total precipitation; 'number of days with' for 15 parameters (e.g., measurable rain); and monthly and annual extremes of record for temperature and precipitation.

(c) *Canadian Climate Summary: Monthly Review of Canadian Climate and Water*

The *Canadian Climate Summary: Monthly Review of Canadian Climate and Water* (Downsview, Ontario: Atmospheric Environment Service, v. 1, 1996–v. 3, 1998. 3 v.), which was based upon reports from approximately 225 Canadian synoptic (forecasting) weather stations, gives a list of weather extremes for the month, by location, as well as temperature and precipitation data for selected stations, by province across Canada. The national maps included show the departure from normal mean temperature, mean temperature; percent of normal precipitation; total precipitation; heating degree-days; total snowfall; percent of normal snowfall; extreme minimum temperature for the month; extreme maximum temperature for the month; and atmospheric circulation. The *Canadian Climate Summary* was preceded by *Climatic Perspectives: A Weekly Review of Canadian Climate and Water* (Downsview, Ontario: Atmospheric Environment Service, 1979–95. 17 v.) and the *Canadian Weather Review* (Toronto: Atmospheric Environment Service, 1963–83. 21 v.).

(d) *Canadian Daily Climate Data on CD-ROM*

The *Canadian Daily Climate Data on CD-ROM* (*CDCD*) contains daily temperature, precipitation, and snow-on-the-ground information for 7,276 Canadian climate observing sites. The volume comprises two discs, one for Eastern Canada and the North, and one for Western Canada and the North. 'Data is available for the entire period of record for each station, with the most recent information being the end of 1998. The accompanying software, which provides access to the data, may be run directly from the CD-ROM and allows for various viewing and output options.'

B Climatological Normals Data

The climatological normals data described below consist of: (1) *Canadian Climate Normals, 1961–90*, (2) *Canadian Climate Normals, 1951–80*, and (3) other climate normals publications.

According to the *Meteorological Glossary*, compiled by D.H. McIntosh (London, Meteorological Office, 1972, 319 p.), a normal is the name given to the average value over a period of years of any meteorological element such as pressure, temperature, rainfall, or duration of sunshine. Normal meteorological values are subjected to uncertainty, owing to year-to-year variability of the observations (the computed standard error of the mean decreasing with the increasing length of period over which averages are taken is a measure of this uncertainty). In the selection of a suitable length of period, a compromise must be struck. A period of about 30 years is thought to be a reasonable compromise, and is generally used in deriving normals. In the 1930s, the 30-year period, from 1901 to 1930, was selected as the first international standard period for climate normals. In 1960, further international agreements confirmed the use of 30-year periods, but it was decided that normals should be updated every decade rather than every 30 years, beginning with the period 1931 to 1960. In Canada at the beginning of the century, there was a limited number of stations with observations over the 1901 to 1930 period. As a result, averages were computed based on all available data, rather than the normals period. The first set of 30-year normals was computed for the 1921–50 period using adding machines, and sets were then produced after the end of each subsequent decade. Computerization of normals production began in 1965.

(a) *Canadian Climate Normals, 1961–90*

The most recent series of normals, for which 20 years of data are available from over 1,000 locations, is entitled *Canadian Climate Normals, 1961–90* (Downsview,

Ontario: Atmospheric Environment Service, 1993. 6 v.). Volume 1 covers British Columbia; Volume 2 covers the Prairie Provinces; Volume 3 covers the Yukon and Northwest Territories; Volume 4 covers Ontario; Volume 5 covers Quebec; and Volume 6 covers the Atlantic Provinces. Maps indicating stations of most observation sites are located at the back of each volume. For the normals, data for nine climate elements are listed, and except where the element is labelled 'extreme,' 'date' or 'direction,' the values in the tables are averages for the 1961–90 period, or for a portion of that period, no shorter than 20 years. Extreme values are the highest or lowest occurrence, for all years for which data are available. Starting and ending dates for station observing programs are given after station names, and stations are categorized as either 'ordinary' or 'principal.' The former are typically stations which record daily temperature and precipitation amounts. On the other hand, principal stations take hourly observations for all or parts of the day. As mentioned above, there are nine climate elements in the tables. For temperature, the highest maximum temperature is the highest recorded in a 24-hour period ending in the morning of the next day. The minimum values are for a period of the same length, beginning in the evening of the previous day. Mean temperature is the average of the two.

Degree-days for a given day represent the number of Celsius degrees that the mean temperature is above, or below, a given base. Values above, or below, the base of 18° C are used primarily to estimate the heating and cooling requirements of buildings. Values above 5° C are frequently called growing degree-days and are used in agriculture as an index to crop growth.

Rain, drizzle, freezing rain, freezing drizzle, and hail are usually measured using the standard Canadian rain gauge. The precipitation is funnelled into a plastic graduate which serves as the measuring device. Snowfall is the measured depth of newly fallen snow, measured using a snow ruler. In the tables, 'precipitation' is the water equivalent of all types of precipitation.

Snow cover is the depth of accumulated snow on the ground, measured at several points which appear representative of the immediate area, and then averaged.

The 'number of days with' specified parameters give the average number of days per month, or year, on which a specific meteorological event (rainfall, freezing precipitation, fog, thunderstorms) occurs.

Sunshine observations are made using the Campbell-Stokes sunshine recorder, first developed in 1863. Sunshine instruments were first installed in the 1880s at six stations in eastern Canada, and since then the network has grown steadily. The recorder allows the observer to determine, to the nearest tenth of an hour, the amount of sunshine that occurs on a given day.

The station pressure is the atmospheric pressure, or force, exerted on the earth's surface by a column of air of a cross-sectional area extending from the surface of the earth to the outer limit of the atmosphere. The standard instrument for the measurement of atmospheric pressure is the mercury barometer, in which the air pressure is balanced against the weight of a column of mercury in a glass tube that contains a vacuum. The measurement is given in millibars; a millibar is 0.2953 inches of mercury, or 0.145 pounds per square inch.

Moisture readings consist of vapour pressure and relative humidity. Vapour pressure is the moisture exerted by the air. It increases with both atmospheric temperature and moisture content. It is related to relative humidity in that the latter is the ratio of the actual value of vapour pressure to its value if the air were saturated at the same temperature. Vapour pressure is measured in kilopascals (one kilopascal is equal to 10 millibars, 0.2953 inches of mercury, or 0.145 pounds per square inch). Relative humidity is listed as a percentage of the moisture content of the volume of air, at a given temperature, to the saturation level of that column of air.

The majority of wind measurements are made by anemometers (an instrument designed to measure wind speed), installed at 10 metres above the ground. Winds are normally measured at level, open sites, removed as much as possible from obstacles to wind flow such as trees, buildings, or hills. At a majority of principal stations, wind is measured by taking a one- or two-minute mean at each observation. Averaging periods may vary from one minute to an hour. The extreme or maximum gust speed is the instantaneous peak wind observed from the anemometer dials, or abstracted from a continuous chart recording.

All this data for the normals is derived from the National Climatological Archive, located at the Downsview headquarters. Except for extremes, most of the values in the tables have been derived from monthly averages, or counts for each year in the 1961–90 period. In the case of daily temperatures (maximum, minimum, and mean values), the 'three-five' rule has been used, that is, a month was counted as missing if more than five daily observations were missing (more than three if consecutive). In the case of extreme values, no data have been excluded from the selection process.

(b) *Canadian Climate Normals, 1951–80*

Unlike the 1961–90 Normals, which are consolidated into six volumes by region, the 1951–80 Normals are in a total of seven volumes, by climate element, for stations across Canada. As well, there is a separate six-volume set that combines temperature and precipitation by region. The first volume of the 1951–80 seven-volume set is entitled *Solar Radiation, 1951–1980* (Ottawa: Atmospheric Environment Service, 1982. 57 p. *Canadian Climate Normals*, v. 1).

This publication contains mean monthly and hourly radiation data, falling on a horizontal surface, for four radiation fields from 58 stations in Canada. Data has been computed from the observations taken at each station, over the period 1956 to 1980, although the period of record at the different stations varies from five to 25 years. The table of daily values includes means, extremes, and standard deviations. The radiation fields are: RF1 – Global Solar Radiation; RF2 – Sky Diffuse Radiation; RF3 – Reflected Solar Radiation; and RF4 – Net Radiation. Finally, a code, from one to eight, is given for the length of record (e.g., code 1: complete 30 years).

The second volume of the 1951–80 Normals is entitled *Temperature, 1951–1980* (Ottawa: Atmospheric Environment Service, 1982. 306 p. *Canadian Climate Normals*, v. 2). This volume contains tables of monthly and annual normals of temperature, in degrees Celsius, for 2,250 stations, in alphabetical order by province and territory that have an appreciable length of record. Data are included for mean daily temperature; mean daily maximum temperature; mean daily minimum temperature; and standard deviation of mean daily temperature. Again, a code, from one to eight, is given for length of record (e.g., code 1: complete 30 years).

The third volume of the 1951–80 Normals is entitled *Precipitation, 1951–1980* (Ottawa: Atmospheric Environment Service, 1982. 602 p. *Canadian Climate Normals*, v. 3). This volume contains tables of monthly and annual normals of precipitation for 2,700 stations that have an appreciable length of record. Ten tables of monthly and annual precipitation are presented for those stations, and arranged in alphabetical order by province or territory. The tables are as follows: mean rainfall; mean snowfall; total precipitation; standard deviation of total precipitation; number of days with measurable rainfall; number of days with measurable snowfall; number of days with measurable precipitation; greatest rainfall in 24 hours; greatest snowfall in 24 hours; and greatest precipitation in 24 hours. These precipitation values are monthly and annual totals of the daily amounts, and have been averaged over the 30-year period of record for each month. Standard deviations (extremes observed) of monthly and annual precipitations are calculated from the same data used to calculate total precipitation. The number of days with measurable amount of rain, snow, and precipitation recorded in each month over the 30 years have been added and the monthly totals have been averaged to obtain the normals. Again, a code, from one to eight, is given for length of record (e.g., code 1: complete 30 years).

The fourth volume of the 1951–80 Normals is entitled *Degree-Days, 1951–1980* (Ottawa: Atmospheric Environment Service, 1982. 280 p. *Canadian Climate Normals*, v. 4). Here, degree days listed are based on seven reference temperatures: above 24°C, 18°C, 10°C, 50°C, 0°C, and below 0°C, and 18°C. An additional table of monthly and annual standard deviations, for degree days below 18°C, is also included. To obtain degree days for a single day, first, the mean temperature is calculated by adding together the maximum and minimum temperature and dividing the total by two. The mean is then subtracted from the selected reference temperature. Then again, a code, from one to eight, is assigned to each entry indicating the average number of years used in the calculation (e.g., code 1: complete 30 years).

The fifth volume of the *Canadian Climate Normals* is entitled *Wind, 1951–1980* (Ottawa: Atmospheric Environment Service, 1982. 283 p. *Canadian Climate Normals*, v. 5). This volume contains tables of monthly and annual average wind data for 485 stations across Canada, for the period from 1955 to 1980. For each station, the principal table contains a listing of the mean direction frequency for either eight or 16 points of the compass (e.g., N, NNE, NE, etc.), depending on wind instrumentation, by month and year, along with the mean speed for each direction, and for all directions combined. Two other tables are also included in this volume. One is the mean wind speed and prevailing direction, and the other is mean vector speed and direction. Finally, reporting stations are also listed in a separate index, along with the number of reporting years for each station.

The sixth volume of *Canadian Climate Normals* is entitled *Frost, 1951–1980* (Ottawa: Atmospheric Environment Service, 1982. 276 p. *Canadian Climate Normals*, v. 6). This volume presents, in a concise form, the means, extremes, and probabilities of frost occurrence recorded at 2,250 stations across Canada. Normals are from the 30-year period 1951–80, but extremes are for the full period of station record, with this period in years indicated in the tables. Probability values are based on a minimum 10 years, and a maximum 50 years of data from 1931 to 1980. Three significant frost events are listed; the last occurrence of the frost in spring; the first occurrence of the frost in fall; and the mean number of days between these occurrences (the frost-free period).

The seventh volume of the *Canadian Climate Normals* is entitled *Bright Sunshine* (Ottawa: Atmospheric Environment Service, 1982. 7 p. *Canadian Climate Normals*, v. 7). This publication contains tables of mean monthly and annual totals of bright sunshine for 320 stations. Although the means are calculated for the 30-year period, from 1951 to 1980, the actual period of record used to compute each average is indicated by the code numbers one through six (e.g., code 1=30 years).

The eighth volume of this series is entitled *Atmospheric Pressure, Temperature, and Humidity, 1951–1980* (Ottawa: Atmospheric Environment Service, 1984. 334 p. *Canadian Climate Normals*, v. 8). This volume contains monthly means of atmospheric pressure, temperature, and humidity, at selected hours, for principal climatological stations across Canada. Since no hourly data exist in the archives before 1953, the averages appearing in this volume have been derived from all available hourly observations, at the selected hours, for the period 1953 to 1980 inclusive. The actual number of years of record at each station is given in the Index of Stations at the beginning of the data tables, along with a map of principal stations and time zones.

The ninth and final volume of the 1951–80 Normals is entitled *Soil Temperature, Lake Evaporation, Days with ... Blowing Snow, Hail, Fog, Smoke or Haze, Frost, 1951–1980* (Ottawa: Atmospheric Environment Service, 1984. 109 p. *Canadian Climate Normals*, v. 9). This volume contains normals for the seven elements, in seven tables that have been grouped together, more for convenience of publication than for compatibility of data. Each station in this volume has been assigned the usual code, from one to eight, to indicate the length of record used to obtain the average.

Table 1 contains monthly and annual soil temperature normals for standard depths for 67 stations, mostly in agricultural areas across Canada. Table 2 contains monthly lake evaporation normals, and standard deviations, calculated for 145 stations in the MSC evaporation network. Tables 3 to 6 (blowing snow, hail, fog, and smoke or haze) contain monthly and annual normals of the number of days with the occurrence of specific meteorological elements observed at about 350 principal climate stations. Tables 3 to 6 are identical in format. Table 7 contains monthly data on days with frost for approximately 2,250 ordinary and principal climate stations across Canada. All tables are arranged alphabetically by station within each province or territory.

Besides the nine volumes of Normals for the period 1951 to 1980, the Atmospheric Environment Service (now the Meteorological Service of Canada) has also published, in six volumes, the climate normals for temperature and precipitation for 1951–86, from 2,700 sites by region. This is the same data as covered in Volume 2 (*Temperature, 1951–1980*), and Volume 3 (*Precipitation, 1951–1980*), of the normal series described above. This set is entitled *Canadian Climate Normals: Temperature and Precipitation, 1951–1980* (Downsview, Ontario: Atmospheric Environment Service, 1982. 6 v.). The regions covered by these volumes are British Columbia, the Prairie provinces, Ontario, Quebec, the Yukon and Northwest Territories. Data presented for the 1951–80 period are the mean daily maximum and minimum temperatures; mean daily temperature; standard deviation of daily temperature; mean rainfall; mean snowfall; total precipitation; standard deviation of total precipitation; and the number of days with measurable rain, snow, and precipitation. Extremes for all available data, up to the end of 1980, are listed in this series for maximum and minimum temperatures, and for greatest rainfall, snowfall, and precipitation occurring in 24 hours. The period for which records exist (a minimum of five years) is listed with a one- to eight-digit code for each station.

(c) Other Climate Normals Publications

Besides the above series of Canadian Climate Normals for 1951–80, the Meteorological Service of Canada, or its predecessors, has also published other series, and specific data publications, covering normals. For example, the series entitled *Climate Normals* (Toronto: Meteorological Office, 1968. 6 v.), covers the period 1931–60. Included in this series are normals for temperature; precipitation; sunshine; cloud cover; pressure and thunderstorms; humidity; wind; and frost data. As well, a six-volume set entitled *Temperature and Precipitation* (Toronto: Meteorological Office, 1967. 6 v.) was published to give climate normals for those two climate elements. Examples of individual publications are: *Temperature and Precipitation, 1941–1970: Mean for the Period 1941 to 1970 and Extremes up to 1970: Ontario* (Downsview, Ontario: Atmospheric Environment Service, 1971. 90 p.); *Mean Monthly and Annual Days with Fog, 1941–1970.* By G.M. Hemmerick (Downsview, Ontario: Atmospheric Environment Service, 1971. 5 p.); *Daily Bright Sunshine, 1941–1970.* By B.J. Yorke and G.R. Kendall (Downsview, Ont.: Atmospheric Environment Service, 1972. 9 p. Maps); *Frost Data 1941–1970.* By G.M. Hemmerick (Ottawa: Atmospheric Environment Service, 1972. 19 p. Maps); *Station, Sea Level and Vapour Pressure Normals Based on the Period 1953–1972* (Downsview, Ontario: Atmospheric Environment Service, 1976. 76 p.); and *Soil Temperature Averages, 1958–1978* (Downsview, Ontario: Atmospheric Environment Service, 1979. 17 p.). A complete list of other historical normals is given in the *Handbook on Climate Data Sources of the Atmospheric Environment Service.* 2nd Edition, compiled by D.W. Phillips and D.W. Gullet (Downsview, Ontario: Atmospheric Environment Service, Canadian Climate Centre, 1988. 6 Sections).

C 'Climatic Atlas Canada'

The atlas equivalent of the *Canadian Climate Normals, 1951–1980* is entitled the *Climatic Atlas Canada: A Series of Maps Portraying Canada's Climate* (Ottawa: Supply and Services Canada, 1984. 5 Series) and was prepared by the Canadian Climate Program of the Atmospheric Environment Service. Unlike the massive data format of the climate normals, climatic maps are a graphic means of summarizing climate data in a readily comprehensible visual way. The first *Climate Atlas of Canada* was published in 1953 as a joint project of the Division of Building Research of the National Research Council and the Meteorological Division of the Department of Transport. In 1967, the Meteorological Branch began publishing a climatic map series in which the different climatic elements were mapped on separate sheets that were distributed as they became available. In all, 10 series of monthly and annual maps based primarily on the 1931–60 climate normals were issued and later grouped together into an *Atlas of Climatic Maps.*

The present atlas is an update of that 1967 series with metric units used rather than English units.

Overall, the Atlas is divided into five series: Series 1, Temperature and Degree Days; Series 2, Precipitation; Series 3, Pressure, Humidity, Cloud, Visibility, and Days with Thunderstorms, Hail, Smoke/Haze, Fog, Freezing Precipitation, Blowing Snow, Frost, Snow on the Ground; Series 4, Bright Sunshine and Solar Radiation; and Series 5, Wind. Although mentioned in the introduction, Soil Temperature and Evaporation were not included in the Atlas. The map scales themselves range from 1:12,500,000 to 1:42,000,000 according to the number of maps per page. Because the map scales are small, the isoline pattern (representing isotherms of equal temperature) had to be smoothed, particularly so for mountainous terrains.

Map Series 1 – Temperature and Degree Days – contains: (1) monthly and annual maps of mean daily temperatures, (2) mean daily maximum and minimum temperatures, (3) all-time extreme temperatures for 100 selected stations, (4) mean annual number of days with minimum temperatures greater than 30°C, (5) mean annual days with minimum temperatures less than or equal to 0°C, (6) mean annual days with temperatures below –20°C, (7) annual total cooling degree days above 18°C, (8) annual total heating degree days below 18°C, (9) annual total growing degree-days above 5°C, and (10) annual total freezing degree-days below 0°C. The tabular data for this series are published in the *Canadian Climate Normals 1951–1980*, vol. 2 – *Temperature*; and vol. 4 – *Degree Days*.

Map Series 2 – Precipitation – contains monthly and annual maps of mean rainfall, snowfall, and total precipitation. Maps in this series are grouped by element (rainfall, snowfall, and total precipitation). The sheet for each month contains maps for the mean rainfall; the mean number of days with measurable rainfall; the mean number of days with rainfall exceeding 10 millimetres; and the greatest rainfall in one day. Corresponding sets of maps for snowfall, and total precipitation, follow in succession. The final maps in the set, for each element grouping, are the annual map for mean rainfall, snowfall, or total precipitation, and for the mean number of days with measurable rainfall, snowfall, or total precipitation. The series is completed with maps of the mean annual extreme rainfall for durations of 24 hours, 12 hours, six hours, one hour, 15 minutes, and five minutes. Again, the monthly and annual means, or normals, in this series were calculated by summing and averaging the monthly and annual totals over the 30-year period, 1951–80. The tabular data for this series are published in the *Canadian Climate Normals, 1951–1980*, v. 3, *Precipitation*.

Map Series 3 comprises four major climate element groupings, namely atmospheric pressure, atmospheric humidity, cloud cover and visibility, and cumulative days with, for eight separate weather elements. The first group consists of maps of the monthly and annual means and extremes of atmospheric sea-level pressure measured in kilopascals (1 kilopascal – kPa – is equal to 10 millibars – mb). The second group of maps depicts atmospheric humidity (that is, vapour pressure and relative humidity). Displayed are the monthly and annual means of vapour pressure as well as the monthly means for early morning and afternoon observations of vapour pressure and relative humidity. An annual map of extremes of vapour pressure at selected stations follows. The third group of maps shows morning to afternoon variations in cloud cover and visibility in each month. Here, the archived data, in absolute numbers of observations, were first converted to mean percentage frequencies to show the monthly mean percentages of observations with clear and cloudy skies, and with good and poor visibilities at 7 a.m. and 1 p.m., local standard time. The fourth and final group of maps in this series shows the monthly and annual means of the number of days with eight separate weather elements, including thunderstorms; hail; smog/haze; fog; freezing precipitation; blowing snow; frost and snow on the ground. Ideally, the averaging period is 1951 to 1980, but the hourly reported elements have been archived, digitally, only since 1953. As well, the 'days with' data have been archived, digitally, only since 1955. These two data deficiencies, however, are considered insignificant for mapping purposes.

Map Series 4 – Bright Sunshine and Solar Radiation – provides maps representing these two meteorological elements. The first set of maps represents: (1) the mean monthly percentage of possible numbers of hours of bright sunshine (assuming cloudless skies), (2) the monthly mean number of days with some bright sunshine, (3) the monthly extremes (maxima and minima) of bright sunshine for selected stations, (4) the annual mean number of hours of bright sunshine, and (5) the annual mean number of days with some bright sunshine. The data used are based upon the *Canadian Climate Normals, 1951–1980*, volume 7 – *Bright Sunshine.*

The second set of maps from Series 4 consists of the following charts: (1) the monthly mean amounts of daily global, diffuse, and direct solar radiation, (2) the monthly extremes of daily global solar radiation for selected stations, (3) the annual mean amounts of daily global, diffuse, and direct solar radiation, and (4) the annual extremes of daily global radiation. For these data, the *Canadian Climate Normals, 1951–1980*, Volume 1 – *Solar Radiation* (Measurements of the RF1 – total Solar Radiation Incident on an Upward-Facing Horizontal Surface and RF2 – Diffuse Sky Radiation) were used, along with the *Solar Radiation Data Analyses for Canada, 1967–1976* (Downsview, Ontario: Atmospheric Environment Service, 1985. 6 v.), which used simulation models to fill in gaps in Canadian radiation data.

Map Series 5 – Wind – contains: (1) monthly and annual maps of mean percentage frequencies of wind directions, and accompanying mean wind speeds (kilometres per hour), for all hours, (2) prevailing wind speeds (kilometres per hour) for all hours, (3) hourly wind speed extremes, and gust extremes (kilometres per hour), with accompanying directions, and (4) monthly maps of prevailing wind directions, and accompanying mean speeds, to the nearest 5 kilometres per hour at 7 a.m. and 1 p.m. (LST). Again, many of the data for Series 5 were published in the *Canadian Climate Normals, 1951–1980*, Volume 5: *Wind*. However, the hourly wind speed extremes were derived from the National Climatological Digital Archive of the MSC.

D Current Climatology Publication Series

These series consist of: (1) the Climate Change Digest – CCD, (2) Climate Data Summaries – CDS, (3) Documentation Sheets – DS, and (4) Climatological Circulars.

(a) Climate Change Digest

One of the components of the Canadian Climate Program, established in 1978, has been studies to assess and identify the potential social and economic impacts of climate warming expected under a doubled carbon dioxide scenario. Most of these studies have been conducted, under contract, by Canadian universities. The first publication of the Climate Change Digest appeared, in 1987, with the title *Canadian Climate Impacts Program* (Ottawa: Environment Canada, 1987, 11 p. These publications are described in detail in Section V.

(b) Climate Data Summaries (CDS)

Climate Data Summaries include various tabular summaries of means, extremes, frequencies, and new records, of climatological elements or their derivations. They have been prepared as aids in answering correspondence from the general public, but none have been issued since 1987. An example from this series by M.K. Thomas is entitled *Mean Monthly Sea-Level Pressure Maps for North and South America* (Downsview, Ontario: Atmospheric Environment Service, Meteorological Applications Branch, 1977. 13 p. Climate Data Summary 77–4).

(c) Documentation Sheets (DS)

These sheets contain documentary information on a variety of subjects, including brief explanations of techniques used for processing certain types of data, lists of publications, microform inventories, price lists, statistical tables, conversion tables, and maps. Documentation sheets have been published since the 1970s. An example of a documentation sheet by K. Johnstone is entitled *Water Balance Tabulations for Canadian Climate Stations* (Downsview, Ontario: Atmospheric Environment Service, 1983. 16 p. Documentation Summary 88–3).

(d) Climatological Circulars (CLI)

The Climatological Circulars expand and interpret the information presented in the Climate Data Summaries and the Documentation Sheets, and address specific topics, or contain bibliographies. An example from this series by L.D. Mortsch and I. Kalnins is entitled *A Bibliography on Storm Surges and Seiches* (Dowsview, Ontario: Atmospheric Environment Service, 1984. 42 p. CLI No. 84–3).

(e) Climatological Studies

This series of publications is designed to provide climate information to scientific community engineers, planners, and other users, and was first issued in 1965. These are described in detail in Section III.

E Technical Report Series

These series consist of: (1) Technical Circulars, (2) Technical Memoranda, (3) Canadian Meteorological Research Reports, (4) Canadian Climate Centre Reports, and (5) Climate and Atmospheric Research Reports – CARD.

(a) Technical Circulars

These were issued by the Meteorological Branch of the Department of Transport as TEC 1, 1964 to TEC 666, 1967. This series also had a double numbering as CIR 1002 to CIR 4545. An example of one of these publications by A.H. Mason and M.K. Thomas is entitled *The October 15–16, 1955 Storm 'Hurricane Hazel' in Ontario* (Toronto: Department of Transport, Meteorological Branch, 1967. 9 p. CIR-2606 TEC-210). This series was continued by the Technical Memoranda.

(b) Technical Memoranda

This successor series to the Technical Circulars was published by the Meteorological Branch, with the numbering TEC 667, 1968 to TEC 750, 1971, and by the

Atmospheric Environment Service, with the numbering TEC 751, 1971 to TEC 882, 1982. An example from this later series by G. Vigeant is entitled *Summer Severe Weather: An Update* (Downsview, Ontario: Atmospheric Environment Service, 1980. 61 p. TEC 873).

(c) Canadian Meteorological Research Reports

These published research reports were issued by various branches, or divisions, of the Atmospheric Environment Service, or its predecessor, the Meteorological Branch of the Department of Transport, from No. 1, 1968 to No. 11, 1979. In 1979, they merged with the Canadian Climate Centre Reports (CCC Reports). An example from this series by A.K. Lo is entitled *On the Modelling of the Atmospheric Boundary Layer over a Sudden Change of Surface Roughness* (Downsview, Ontario: Atmospheric Environment Service, 1975. 23 p. Canadian Meteorological Research Report (MRR 4-75).

(d) Canadian Climate Centre Reports

Canadian Climate Centre Reports No. 1, 1978–94. The CCC Reports have the designation 'unpublished manuscript,' but are listed for public sale in the publication entitled *List of Canadian Climate Centre Reports – CCC Reports* (Downsview, Ontario, Atmospheric Environment Service, 1987. 11 p. Addendum, 1988–94. 5 p.). An example from this series by T.A. Agnew and A. Silis is entitled *Spring Season Climate Variability at Resolute Bay, NWT* (Downsview, Ontario: Atmospheric Environment Service, 1994. 26 p. CCC Report No. 94–1). With the merging of the Canadian Climate Centre and the Atmospheric Research Directorate in 1994, this series was also merged into a new CARD Report Series.

(e) CARD (Climate and Atmospheric Research Directorate) Reports

CARD (Climate and Atmospheric Research Directorate) Reports 1994–. An example from this new series by M.E. Shokr and N.K. Sinha is entitled *Physical, Electrical and Structural Properties of Arctic Sea Ice Observed during SIMMS'92 Experiment* (Downsview, Ontario: Atmospheric Environment Service, 1995. 148 p. Research Report CARD No. 95-005).

F The Unpublished Reports

Within the Atmospheric Environment Service (now the Meteorological Service of Canada), there have been, or still are, numerous series of unpublished internal project, or research, reports. One such cluster has been issued over time, with some overlaps, by the Atmospheric Research Directorate itself, or the Atmospheric Processes Research Branch, or the Meteorological Services Research Branch.

The cluster contains:

Internal Report CARD, No. 1, 1975–94 (nos. 2 and 3 were issued by the Atmospheric Processes Research Branch). An example from this series by D.E. Worthy is entitled *An Inter-Comparison of Methane Standard Gas Scales between AES and NOAA/CMDL* (Downsview, Ontario: Atmospheric Research Directorate, 1994. 17 p. Internal Report CARD 94-008). This series was merged with the CARD Reports series mentioned above.

Internal Report MSRB, No. 1, 1974–93. An example from this series from the Meteorological Services Research Branch, written by W.R. Burrows, is entitled *Climatology of Daily Total Ozone and Ultraviolet – B Radiation Levels* (Downsview, Ontario: Atmospheric Research Directorate. Meteorological Services Research Branch, 1992. 204 p.). In 1994, with the amalgamation of the Canadian Climate Centre and the Atmospheric Research Directorate to form the Climate and Atmospheric Research Directorate, this series was merged to form the CARD Report series mentioned above.

Internal Report APRB, No. 1, 1975 to No. 117, 1985. An example from this series of the Atmospheric Processes Research Branch by G.M. Shah and D.I. Wardle is entitled *One Year's Data on UVB Spectral Irradiance at Toronto* (Downsview, Ontario: Atmospheric Research Directorate, Atmospheric Processes Research Branch, No. 115. n.d. n.p).

Another smaller group of internal reports has been issued by the Science Services Division and the Atmospheric Issues Division of the Ontario Region. They are the Internal Report SSD, 1978–93, and the Internal Report AIP, No. 1, 1994–.

The other large cluster of internal reports is the project reports, which were first issued by the Meteorological Branch of the Climatology Division. These reports focused on climatology studies. They are:

Project Report, Department of Transport, Meteorological Branch, No. 1, 1969–No. 10, 1973. An example from this series by R.J. Smith is entitled *A Climatology Study of Toronto* (Toronto: Atmospheric Environment Service, 1972. 85 leaves. Project Report No. 8).

Project Report, Atmospheric Environment Service, Meteorological Applications branch, No. 11, 1973–No. 33, 1977.

G Solar Radiation Data

Solar radiation is the single most important factor affecting climate, as air temperatures have their origin in the absorption, by the earth, of radiant energy from the sun. Because of the slight ellipticity of the earth's orbit around the sun, the amount of solar energy intercepted by the earth steadily rises and falls by about 3.5% throughout the year. Although about 37% of this energy is not used, because it is scattered back into space, the remaining amount is sufficient to power the movement of atmospheric winds and oceanic currents and to sustain nearly all the biospheric activity of the earth. Of the incident radiant energy reaching the top of the atmosphere, about 46% is absorbed by the earth's surface, on average (depending on cloudiness, surface type, and elevation). An additional 23%, or 10% of this incident radiation, is absorbed on average in the atmosphere, especially by water vapour and clouds at lower altitudes, and by ozone (O_3) in the stratosphere. Absorption of solar radiation by ozone shields the terrestrial surface from harmful ultraviolet light and warms the stratosphere. The constant gain of solar energy by the earth is systematically returned to space in the form of thermally emitted radiation, in wavelengths in the infrared portion of the solar spectrum. The difference between the solar radiation absorbed and the thermal radiation emitted to space determines the radiation budget of the earth, which in turn affects the rate of temperature change in any region of the earth, and again is the single most important factor for our climate.

There have been two major publications for the recording of solar radiation data issued by the Meteorological Service of Canada. The first was entitled the *Monthly Radiation Summary* (Ottawa and Downsview: Atmospheric Environment Service, 1960–89. 30 v.). Beginning with the January, 1985, issue, this summary was published in microfiche only. Nothing has been issued since 1989.

The *Monthly Radiation Summary* provided research scientists, engineers, and others with monthly radiation data from Canadian stations between four and eight months after the end of each observation month. In January, 1978, the publication became bilingual, and the unit of radiation measurement became recorded in megajoules per square metre, rather than langleys, and more summary data was made available for each hour. In the Summary, the data are published for five radiation fields: RF1 – global (total) solar radiation on a horizontal surface. This direct beam radiation is the solar radiation coming from within 2.5 degrees of the centre of the geometric disc of the sun; RF2 – sky radiation (diffuse), which is all the solar radiation from the sky, except that from the direct beam; RF3 – reflected solar radiation; RF4 – net radiation; and, RF7 – natural illumination.

The second major publication of solar radiation is the compilation by D.C. McKay and R.J. Morris entitled *Solar Radiation Data Analyses for Canada, 1967–*

1976 (Downsview, Ont.: Atmospheric Environment Service, 1985. 6 v.). The six volumes are by region – Volume 1: the Atlantic Provinces; Volume 2: Quebec; Volume 3: Ontario; Volume 4: the Prairie Provinces; Volume 5: British Columbia; and Volume 6: the Yukon and Northwest Territories.

In the introduction, the authors state the 'burgeoning awareness of renewable energy and activity in solar energy applications have resulted in the need for more comprehensive information on solar radiation climatology.' Most of the data archived for 50 sites in Canada is solar radiation incident on a horizontal surface (RF1), and few climatological data have been published on solar radiation on tilted surfaces, or on the direct beam sky diffuse, or the ground-reflected components of solar radiation. However, these solar radiation quantities are of 'great interest and use to planners, designers and retailers of modern solar energy collection systems.'

In order to meet these emerging needs, the Atmospheric Environment Service implemented computer simulation models to estimate solar radiation and its components on horizontal and various tilted surfaces, at sites where hourly meteorological observations are taken and archived. The models were executed for 129 sites in Canada, for the period 1967 to 1976. The data were calculated and the results presented in maps and tables.

H Ozone Data

As mentioned in the previous section, stratospheric ozone absorbs solar radiation and shields the terrestrial surface from harmful ultraviolet light, besides warming the stratosphere. Under the sponsorship of the World Meteorological Organization, and in cooperation with the Meteorological Service of Canada, ozone data has been collected, published, and archived at the Downsview headquarters since 1960.

Annual publication of data began, in 1960, and continued until 1965, when a bi-monthly publication was introduced to disseminate the data among researchers as quickly as possible. This publication is entitled *Ozone Data for the World* (Downsview, Ontario: Atmospheric Environment Service in Co-operation with the World Meteorological Organization, 1960–97). Since 1997, the data has been available electronically through the website <<http://www.msl.ec.gc.ca/woodc>>, for data from 1994 onward produced by the World Ozone and Ultraviolet Radiation Data Centre. In these summaries, six data tables are included in each issue. They are: (1) daily surface ozone measurements from selected stations, including maximum, minimum, and mean values, (2) total ozone amounts for each day from selected stations, including the monthly mean, (3) a monthly mean vertical distribution of ozone from Umkehr observations (an effect which is used to infer the vertical distribution of ozone from surface measurements) for nine layer-mean ozone partial pressures, (4) vertical distribution of ozone from ozonesonde observations for seven-day periods, (5) observation of the Umkehr effect for 14 zenith angles, and (6) Lidar profiles, obtained from layer measurements.

A separate publication, based on the data collected above, has been issued and entitled *Ozone Data for the World: Second Umkehr Measurements at Toronto, June 1968–September, 1969* (Toronto: Department of Transport Meteorological Branch in Co-operation with the World Meteorological Organization, 1971. 77 p.). The second Umkehr values, like the first, are a stationary value of the ratio of intensities at the two wavelengths in the 3,000–3,300A region, in light from the zenith sky (extensive stretch of sky immediately overhead). The first Umkehr, which occurs when the sun is above the horizon, has been extensively measured. The second occurs during twilight, at the solar zenith distance of about 0=93–94°, and since the intensities are much lower, measurements are more difficult.

I Miscellaneous Publications

From time to time, the Meteorological Service of Canada staff write small publications to assist the layperson in understanding weather phenomena, the archiving of climate data, and their overall climate program. One of the most basic of their printed publications has been entitled *What You Can Do about Weather* (Downsview, Ontario: Atmospheric Environment Service, 1977. 7 p.). This small publication attempts to show basic weather behaviour, by explaining how the layperson can keep simple meteorological records for water evaporation, humidity, snowfall, rainfall, daily minimum temperature, wind direction, visibility, and sunshine. A second publication, entitled *Knowing Weather: Facts and Myths* (Downsview, Ontario: Atmospheric Environment Service, 1980. 27 p.), not only describes how to construct simple instruments for measuring wind speed and direction, rainfall, and air pressure, but also provides easy to understand descriptions of air pollution, radiosondes and the upper atmospheric, weather satellites, the use of radar in weather observations, the signals, or indicators, of weather change, unusual weather phenomena, such as hurricanes and tornados, the moon and weather, weather lore and legends, and animals and weather. The most recent publications, written to explain meteorological topics to the general public, are located through the Meteorological Service of Canada website <<http://www.msc.ec.gc.ca>>. These on-line publications include Fact Sheets (Acid Rain; Climate Change; The Ozone Layer; Smog; Weatheradio; and Windchill), Brochures on Air Pollution (Smog; Understanding the Ozone Layer; UV and You), and Weather (Blizzards and Winter Hazards; El Nino; Humidity) and Weather Fun (Weather Quiz 2000).

Climate data archiving has been described in the printed publication entitled *Accessing Digital Data in the National Climatological Archive* (Downsview, Ontario: Atmospheric Environment Service, 1980. 30 p. Documentation Sheet DS 80-14). This publication provides a brief survey of the climatological stations and networks across Canada, including maps, as well as a description of the national archive of digital climate data, and the formats used for entering data. Information is also provided on how to select and obtain digital climate data. Currently, this information is also available at the Canadian Meteorological Centre website <<http://www.cmc.ec.gc.ca/climate/climate_archive.htm>>.

Finally, an interesting publication by H.L. Ferguson and D.W. Phillips entitled *Canadian Climate Program: Developing a National Climate Program – A Decade of Canadian Experience* (Ottawa: Environment Canada, 1986. 30 p.), explains the development of the climate program at Downsview in 1978 and after.

III Climatological Studies

As mentioned in the introduction, the climatological studies of the Meteorological Service of Canada are a unique series, first issued in 1965. The majority of these studies fall into various groupings – national and regional, or provincial studies, including the Arctic, plus one overview study, mesoclimatological studies, which are small area studies, and climatological studies of Canada's major cities. However, there are also some specialized climatological studies, including weather phenomena analyses.

A National and Regional Studies

(a) National Studies

For the general reader, the most recent overview of all the regional climates of Canada has been written by D. Phillips and entitled *The Climates of Canada* (Toronto: Atmospheric Environment Service, 1990. 176 p). This book is 'dedicated to Morley Thomas who has contributed enormously to the history, science and service of Canadian meteorology for nearly half a century.' The book discusses climate controls, including the sun, the shape of the land, the oceans and lakes, and the city and the country. As well, the climate elements (temperature, precipitation, evaporation, etc.) are clarified and discussed. Geographically, the climate regions of Canada are discussed province by province, and the work concludes with a discussion of past and future climates, a climate glossary, a bibliography, and climate data tables. This book is geared for the general reader and provides an excellent summary of Canadian climate.

The most recent detailed overview of Canadian climatology by region, with an emphasis on climate impacts of global warming, has been the multi-volume study available in electronic format, entitled *The Canada Country Study: Climate Impacts and Adaptation [Computer File]* (Downsview, Ontario: Environment Canada, 1998. <<http://www.ec.gc.ca/climate/ccs/ccs_e.htm>>. The purpose of the study is to provide a 'national assessment of potential impacts of climate change and variability including consideration of existing and potential adaptive responses by region.' Besides providing a review of climates by region, specific topics discussed pertaining to climate change and impact are sea-levels, river flows, glaciers, geological formations, plant communities, forests, agriculture, fish and wildlife, and air quality. As well, the possible human response to climate change has been discussed. The results of these assessments have been published in eight electronic volumes. Six volumes are by region, and include a News Release, Background, and Executive Summary, and two organized by sector (agriculture, etc.), and by cross-cutting topics (domestic trade and commerce, etc.). They include an Executive Summary, and a list of chapters. Also included with this electronic publication is *The Canada Country Study: Climate Impacts and Adaptation National Symposium, November 24–26, 1997 in Toronto, Ontario*. Edited by B. Maxwell *et al.* (328 KB).

The regional volumes are:

Volume I: *Responding to Global Climate Change in British Columbia and Yukon.* Edited by E. Taylor (6.3 MB);

Volume II: Responding to *Global Climate Change in Canada's Arctic.* Edited by B. Maxwell (20 MB);

Volume III: *Responding to Global Climate Change in the Prairies.* Edited by R. Herrington and B. Johnson (1.5 MB);

Volume IV: *Responding to Climate Variability and Change in Ontario.* Edited by J. Smith *et al.* (850 MB);

Tome V: *Impacts et adaptation à la variabilité et au changement du climat au Québec.* Edited by L. Bergeron and G. Vigeant (2.4 MB);

Volume VI: *Responding to Global Climate Change in Atlantic Canada.* Edited by J. Abraham and T. Canavan (1.2 MB);

Volume VII: *National Sectoral Volume.* Edited by G. Koshida and W. Avis (3.2 MB); and

Volume VIII: *National Cross-Cutting Issues Volume.* Edited by N. Mayer and W. Avis (1.3 MB).

Some of these above-listed electronic publications are available in print or CD-ROM format. For example, the Executive Summaries are available under the titles *National Summary for Policy Makers*; *Highlights for Canadians*; *National Sympo-*

sium Proceedings; *British Columbia and Yukon Executive Summary*; *Canadian Arctic Summary*; *Canadian Prairies Summary*; *Ontario Region Executive Summary*; and *Atlantic Canada Summary*. All but Volume III of the Reports are also available in print form. However, Volumes I to VI of the Regional Reports, and Volumes VII and VIII of the National Reports are available in CD-ROM format. Queries about print or CD-ROM formats can be made by E-mail to: enviroinfo@ec.gc.ca.

(b) The Provinces

The first of these climatological studies by D.M. Brown and G.A. McKay was entitled *The Climate of Southern Ontario* (Toronto: Atmospheric Environment Service, 1980. 67 p. Climatological Studies No. 5). The reason for selecting this region for the first of these studies is the fact that southern Ontario is one of the most densely populated and highly industrialized areas in Canada. Its manufacturing industry produces about one-half of the Canadian output, and employs about one-third of the Canadian labour force. Although the region contains only 11% of the total Canadian improved farmland, the area produces about 30% of the total value of farm products in Canada. For example, nearly half the fruit, two-thirds of the vegetables, and over 40% of Canadian dairy products come from this area. This achievement is, to a large extent, the result of the region's favourable climate. However, not all areas are equally productive, because the climate is quite varied, as marked differences occur owing to the effects of the Great Lakes and variations in topography.

The Great Lakes regulate the climate near their shores, modifying extremes of heat and cold, and extending the growing season in the autumn. The Lakes are also an important source of atmospheric moisture that helps to ensure a fairly uniform seasonal distribution of precipitation in amounts which are usually adequate for agricultural and other water requirements. All told, 17 regions of distinctive and fairly uniform climate have been defined. In the publication, climate controls (weather systems), climate regions, and weather records (temperature, the growing season, precipitation, sunshine, humidity, and wind) are discussed. Climate tables and figures are also included.

The second volume of these climatological studies by L.J. Chapman and M.K. Thomas is entitled *The Climate of Northern Ontario* (Toronto: Atmospheric Environment Service, 1968. 58 p. Climatological Studies No. 6) and serves as a companion volume to the above-mentioned publication. Northern Ontario is that part of the province between the upper Great Lakes and Hudson Bay. Most the land is uninhabited and forested, except for a few well-defined farming areas, south of latitude 50°. The contrast between the two regions is the result of geological differences, caused by the Precambrian cover of the Laurentian Shield and other younger areas of sedimentary rock. The climate is described as modified continental, the modification being mainly due to the pressure of the Great Lakes on the south, and, to a minor extent, to Hudson Bay on the north. A pattern of relatively low winter and high summer precipitation prevails. In the summer, a succession of cyclonic storms sweeps over the area, and warm, humid air masses from the south alternate with cooler and drier air from the north, to create a repeated pattern of weather sequences. Following the climate description, temperature, moisture, humidity, sunshine, and wind data, climate maps are provided.

Like the climatology study of Ontario, the study of Quebec is also in two parts. This study by C.V. Wilson is entitled *The Climate of Quebec* (Ottawa: Canadian Meteorological Service, 1971. 2 Parts). Part I is a *Climate Atlas of the Province of Quebec*, and Part II is entitled *The Application of Climate Information.* Part I consists of an introduction which discusses the geography and economics of Quebec, both of which are affected by the climate. As well, the cold season, the warm season, and local climates are described. The climate data are presented in 44 maps, and there are also diagrams, ranging from topographical maps and meteorological stations to climate normals (1931–60) and human bioclimate. At the end of Part I, there is an extensive bibliography. Part II is designed to meet both the general requirements for planning and the public need for information. There is a detailed discussion of climate controls, such as atmospheric characteristics and the nature of the land surface, the vertical distribution of temperature, cloud visibility and winds, temperature near the ground, precipitation, evaporation and run-off, and again, human bioclimate. Each chapter indicates the appropriate companion section in Part I. Included in Part II are 20 tables based on climate normals.

Besides this two-part study, an additional volume was prepared by C.V. Wilson for later publication and entitled *The Climate of Quebec, Energy Considerations* (Ottawa: Information Canada, 1975. 120 p. Climatological Studies No. 23). This discusses radiation energy and the energy budget, as it affects Quebec, for the purposes of regional and resource planning. The discussion covers incoming solar radiation, reflected solar radiation, net radiation, and the heat balance. Included with the discussion are tables of radiation observation data.

To the west, the climatology of the prairie provinces has been studied by R.W. Longely and entitled *The Climate of the Prairie Provinces* (Toronto: Atmospheric Environment Service, 1972. 79 p. Climatological Studies No. 13). It followed an earlier, more specific study by G.A. McKay and J. Maybank entitled *The Agricultural Climate of Saskatchewan* (Toronto: Department of Transport Meteorological Branch, 1967. 18 p. Climatological Studies No. 10). This later study includes a discussion of the basic elements of climate and the physiography of the prairie region, which is made up of mountains and foothills, the prairie, and part of the Laurentian Shield. The core of the study is a discussion of spring, summer, autumn, and winter weather, and the overall climate during the year. A bibliography is also included.

The most recent of the regional climatological studies, edited by A. Robertson and S. Porter, is entitled *Climate and Weather of Newfoundland* (St John's: Creative Publishers, 1993. 152 p.). This study was the result of 14 papers presented at a workshop in St John's by the Newfoundland and Labrador Climate Advisory Committee of the Canadian Climate Program, and the Canadian Meteorological and Oceanic Society. The only previous study of this region was entitled *The Climate of Newfoundland* (Toronto: Department of Transport, Meteorological Branch, 1964. 11 p. CLI 30 CIR 4019). The workshop papers covered past and present climate, including local climate effects, the effect of climate and weather on the Canadian Coast Guard, construction of roads and buildings, utility operations, energy conservation and resources, forestry, and recreation. The workshop concluded that the Atmospheric Environment Service 'must increase its efforts to promote intelligent use of both climatic knowledge and weather services, and involve experts in practical applications in those efforts.'

(c) The Arctic

The first climatological study of the Canadian Arctic was written by R.W. Rae and entitled *Climate of the Canadian Arctic Archipelago* (Toronto: Department of Transport, Meteorological Branch, 1951. 90 p.). Then, in 1967, H.A. Hammond wrote a study for the Canada Year Book, 1967, which was enlarged and published separately by the Dominion Bureau of Statistics, with the title *The Climate of the Canadian Arctic* (Ottawa: The Bureau, 1967. 32 p.). This latter survey discusses the climate controls that work on the Canadian Arctic, the climate in general, the climate from December to April, May and June, July and August, and September to November. Climate tables for temperature and precipitation from climate stations in the Arctic area are also included.

More recently, a climatological study of the Canadian Arctic was written by J.B. Maxwell and entitled *The Climate of the Canadian Arctic Islands and Adjacent Waters* (Ottawa: Atmospheric Environment Service, 1980. 2 v. Climatological Studies No. 30). Again, the purpose of this study was to provide a comprehensive handbook on all aspects of the climate of Canada's Arctic Islands, because of accelerated resource exploitation and transportation, and its effects on the environment. The systematic collecting of meteorological data from a planned network of observing stations on the Canadian Arctic Islands was not begun until 1947, but by 1980, there were over 30 years of data, and Maxwell's study complements the earlier Burns's study of the Mackenzie Valley, mentioned below, in that its geographic range included all the islands north of the Canadian mainland, plus the Boothia Peninsula and all the waters adjacent to those areas, except for the Beaufort Sea. Again, the solar energy impact, and latitude, provide the basic character of the Arctic Islands climatology, coupled with the overriding consideration that the snow and ice cover affects both land and water for approximately 10 months of the year. Volume I discusses the climate controls, and all other factors, that affect the climate of the islands such as temperature, wind, precipitation, and snow cover. Volume II continues with a discussion of the sea, and sea ice, aviation-related aspects of the climate, the Arctic temperature inversion, pollution potential, atmospheric phenomena, synoptic systems (summary weather patterns), and climate regions. Throughout, the emphasis is on probability estimates of extreme temperature, wind, and precipitation. The study is backed up by extensive climate tables and climate diagrams.

A third Arctic study by H.E. Wahl and D.B. Fraser has been entitled *Climate of Yukon* (Ottawa: Canadian Government Publishing Centre, 1987. 343 p. Climatological Studies No. 40), and was initiated by the Atmospheric Environment Service as a support study for planners, engineers, environmentalists, and others involved in weather-sensitive activities in the Yukon. The first six sections discuss the history of Yukon development, geography, sources of data, climate controls, and a survey of the climate regime by region, which is the most variable in North America. The last eight sections contain detailed information, and analyses, for the climate elements of wind, temperature, precipitation, humidity, cloud cover, thunderstorms, visibility, and obstructions to vision. The synoptic weather patterns for four common elements (average weather, wet unsettled weather, summer hot spells, winter cold spells) are also given, as are summaries for air pollution potential, wind energy, and climatic change. All summaries have detailed climate maps and tables.

More specifically, another Arctic climatology study, edited by L. Mortsch, has the title *Eastern Canadian Boreal and Sub-Arctic Wetlands: A Resource Document* (Ottawa: Atmospheric Environment Service, 1990. 169 p. Climatological Studies No. 42). This is the outcome of the Canadian Northern Wetlands Study, a comprehensive research program whose objective is to assess the role of Canadian wetlands as sources of biogenic gases. For a study of this scope, a resource document that summarized, not only the role of the wetlands and their distribution, but also the climate, air chemistry, physiography, geology, hydrology, soils, vegetation, and socio-economic characteristics (population, agriculture, forestry, mining, transportation), was necessary. The climate data includes a section on the solar energy budget, temperature, growing degree days, while wind and precipitation climate maps are also included.

Probably, the most detailed of the Arctic climatology studies was a six-year collaborative research project (federal government, Northwest Territories government, BC Hydro, the University of Victoria, and others), pertaining to the Mackenzie Basin. 'The purpose of the study was to look at the effect that a change in climate might have on the Mackenzie Basin, its lands, waters, and the communities that depend on them.' In particular, this basin was chosen for a case study 'in order to look at a high latitude or northern region which had sensitive ecosystems and a large number of aboriginal people who were still following the traditional ways or lifestyle.' However, this was not the first time that this region had been studied.

The first of the detailed climate studies of the Mackenzie Basin was the two-volume work by B.M. Burns entitled the *Climate of the Mackenzie Valley – Beaufort Sea* (Toronto: Atmospheric Environment Service, 1973–74. 2 v. Climatological Studies No. 24). The object of this study was to draw together climatological information pertinent to all types of transportation, communications, engineering design and construction, defence, industry, agriculture, fishing, and foresting as well as hydrological, environmental, and ecological problems. Before the 1960s, climatological information was primarily of importance to the military, but since then the growing demand for oil, gas, and minerals provided the impetus for more exploration, and the need for more climatological data of this region. There were, however, restrictions imposed on this study, because prior to the mid-1950s, usable weather data were available from only 12 locations in the western district of Mackenzie and the coastal areas of the Beaufort Sea. The observation network was expanded, after the mid-1950s to include DEW (distant early warning) sites, plus various additional meteorological stations, for a total of 22 (at the time of this study). The overlying feature of Arctic climatology is the solar energy input into a region that depends basically upon its latitude. Keeping that in mind, this study provides a historic introduction to the area, climate controls, the Arctic inversion (an increase in temperature with height, caused by the negative radiation balance over the snow and in surface, which are present during the greater part of the year), temperature, wind, precipitation, snow cover, ocean areas, ice, synoptic systems, and cloud cover, visibility, and icing. As well, there are maps and tables in both volumes that document the descriptive climatology presented.

The most current Mackenzie Basin study began in 1990, and along the way, background studies, journal articles, and newsletters were published, some by researchers at the Atmospheric Environment Service. For example, D. Etkin published the report entitled *Mackenzie Valley Climate Stations* (Downsview, Ontario: Atmospheric Environment Service, 1991. 125 p. Canadian Climate Centre Report No. 91-8); and, T. Staple and G. Wall wrote the publication entitled *Climate Change and Recreation in Nahanni National Park Reserve* (Ottawa: Minister of Supply and Services, 1996. 11 p. Climate Change Digest No. 96-02). (This publication was later issued in a longer version with the same title in the *Canadian Geographer*, v. 40, 1996. pp. 109–20). As well, two interim reports were written by S.J. Cohen. The summary of the Mackenzie Basin Impact Study Interim Report #1 appeared as Appendix 4 of the Final Report, described below, but the second report was published with the title *Mackenzie Basin Impact Study: Summary of Interim Report #2* (Ottawa: Minister of Supply and Services, 1995. 9 p. Climate Change Digest No. 95–01). Journal articles published by S.J. Cohen and Y. Yin were entitled *Identifying Regional Goals and Policy Concerns Associated with Global Climate Change* (*Global Environment Change*, v. 4, 1994. pp. 245–60); and *An Integrated Data Base for Mackenzie River Basin Climate Change Impact Assessment* (*Canadian Journal of Remote Sensing*, v. 20, 1994. pp. 426–34). The newsletter issued was entitled *Mackenzie Basin Impact Study Newsletter* (Downsview, Ontario: Environment Canada, 1992–95. 5 issues). The final report, edited by S.J. Cohen, was entitled *Mackenzie Basin Impact Study (MBIS): Final Report* (Ottawa: Environment Canada, 1997. 372 p.). This report contained the proceedings of the MBIS Final Workshop held 5–8 May 1996 in Yellowknife, Northwest Territories and consisted of (1) summaries of climate impact assessments of land, waterways, forests, wildlife, and people and their communities and economic activities, (2) condensed versions of round-table discussions, (3) contributions from studies conducted in other countries, and (4) government representatives' statements.

B Mesoclimatology Studies

Besides the regional climatology studies, the Atmospheric Environment Service, now the Meteorological Service of Canada, has also published studies involving small regions. They are called mesoclimate studies. Two important mesoclimate studies have involved the Great Lakes Basin as a whole, and the north shore, or Toronto region, of Lake Ontario. A third study has looked at the climate of the Maritime provinces from the point of view of recreational needs.

The first study by D.W. Phillips and J.A. McCullough is entitled *The Climate of the Great Lakes Basin* (Toronto: Atmospheric Environment Service, 1972. 42 p. Climatological Studies No. 20). The reason given for studying this region as a whole is the fact that it remains a rapidly growing industrial and urban complex, or perhaps a Great Lakes megalopolis. Although the Lakes themselves occupy about 95,000 square miles, the total region for the drainage area is 295,000 square miles. The unique features of the climate of the basin are four distinct seasons, and a variety of precipitation types and sources, with almost month-to-month variation in precipitation amounts. But there are marked temperature contrasts over only 750 miles of latitude, as well as a variety of weather events from one location to another. Perhaps the most important feature, however, is the influence of the Great Lakes in modifying the continental air. The detailed topics covered in the publication include a discussion of climate controls, weather records, temperature, precipitation, humidity, evaporation, radiation, sunshine, sky-cover, fog, ice, and wind. There are 27 tables and figures explaining the climate, as well as 57 charts of the Great Lakes Basin, which indicate everything from the location of weather stations used in the study to October wind patterns for selected airport stations.

Another study relating to the Great Lakes by G.J. Irbe is entitled *Great Lakes Surface Water Temperature Climatology* (Ottawa: Atmospheric Environment Service, 1992. 215 p. Climatological Studies No. 43). The study is important, because surface water temperature is an essential parameter in physical and biological studies of the Great Lakes, triggered by the rapid growth in population, industry, and economic activity in the Great Lakes Basin, and the environmental degradation of the Lakes. Physical and biological studies require two kinds of lake-wide water temperature data. Real-time surface water temperature data are required in investigations of ongoing processes and currently prevailing conditions (e.g., the interaction of the atmosphere with the lake surfaces and circulation patterns within the water masses). Other studies require a definition of the long-term average of surface water temperature field in order to establish the common, or normal, state of various environmental parameters and processes, as well as to detect trends and anomalous behaviour in these parameters and processes. These information needs cannot be fully met by the conventional method of surveying the Great Lakes temperature from ships. Accordingly, the Hydrometeorology and Marine Division of the Atmospheric Environment Service turned to remote sensing technology, and developed a technique for retrieving surface water temperatures using airborne infrared radiometers.

The Airborne Radiation Thermometer (ART) program was carried out from 1966 to 1979, and made the regular collecting of lake-wide temperature data economically feasible. However, there was still the problem of spatial coverage of the lake surface, because ART data collecting was along transects spaced 50 kilometres apart. In the late 1970s, after the installation of a satellite receiving station, using a TIROS-N polar orbiting environmental satellite of the United States National Oceanographic and Atmospheric Administration (NOAA), water temperatures began to be retrieved on a year-round basis by satellite radiation thermometer (SRT). The ART and SRT programs have met at least some of the needs for real-time information on Great Lakes water temperatures, as well as assisting in the study of the surface water climatology of the Great Lakes. The data documented in this publication first describes the overall water temperature regime, including the annual surface water temperature cycle, surface water temperature patterns, and the seasonal (winter, spring, summer, and fall) thermal structure of the Lakes. Then, the water temperature regime for each Great Lake (Erie, Ontario, Huron-Georgian Bay, Superior) is given in parallel detail.

Another mesoclimatology study of note is by D. Greenland, and entitled *Aspects of the Mesoclimatology of the Toronto Area* (Downsview, Ontario: Atmospheric Environment Service, 1977. 77 p.). This study deals with some aspects of the mesoclimatology of an area of land on the northwest shore of Lake Ontario, where the City of Toronto is the most important urban centre. The author hoped that the climate information presented would be useful in the decision-making processes pertaining to land use and human activities. The study begins with an estimate of the mean annual energy budget of the area. Then, a series of mesoclimatic analyses of the same area, but descending through shorter time periods, are provided. Finally, there is a discussion of the potential use of the results for planning (e.g., location of airports), as well as areas needing further investigation (e.g., the effect of the Niagara escarpment on climatological parameters).

Another example of a mesoclimatological study, as applied to a specific planning area, has been written by A.D. Gates and entitled *The Tourism and Outdoor Recreation Climate of the Maritime Provinces* (Toronto: Atmospheric Environment Service, 1975. 133 p. Publications in Applied Meteorology REC-3-73). This study was prepared to promote the use of meteorological information in recreation and tourism facility planning, and used the two-season approach, where the year is divided into a winter, or snow-cover season, and a summer season. The first part of the study examines the seasons in detail, and includes dates and durations for these seasons. The second part of the study examines those elements of climate which affect tourism and outdoor recreation. These include temperature, wind, sunshine, precipitation, fog, water temperatures, and humidity, with a special emphasis of their effect on human comfort and convenience. Heating requirements are also discussed. Finally, the climatological information presented has been combined in the concept of suitable days. This method indicates the percentage of time throughout the year when weather conditions are suitable for each of several general categories of outdoor activities (e.g., landscape touring, skiing and snowmobiling).

C Climate of Canadian Cities

The sub-series on the climate of Canadian cities began in 1984. This was initiated in response to a need to distribute to the public general and applied climatic information for Canadian urban areas. The first one by R.B. Crowe is entitled *The*

Climate of Ottawa-Hull (Ottawa: Atmospheric Environment Service, 1984. 42 p. Climatological Studies No. 35. The Climate of Canadian Cities No. 1). In this publication, climate data is broken down by weather spells and extremes, applied climate (wind and heat in recreation, tourism and gardening), and similar climates around the world. In particular, the author has described the climate as 'stimulating and variable,' with a cold and snowing winter and a warm summer, and overall, no pronounced dry season. There are two main seasons – summer and winter, and two transition periods – spring and autumn. Finally, the climate description includes a series of climate data sources and tables.

Related to this study is one by J.M. Masterton and D.W. McNichol entitled *A Recreational Climatology of the National Capital Region* (Downsview, Ontario: Atmospheric Environment Service, 1981. 120 p. Climatological Studies No. 34). This was undertaken to help major groups of users of the National Capital Region, such as recreational facility managers, facility developers and planners, and governments, with long-term land-use planning. The National Capital Region does enjoy a stimulating variable climate characterized by cold snowing winters and warm sunny summers, which encourages participation in a variety of outdoor recreational activities, including skiing in winter and beach activities in the summer. Covered in the study are the relationships between weather and recreation, primary climate controls, average climate conditions in Ottawa, winter and summer weather, weather throughout the National Capital Region, the seasons for outdoor recreation, and the quality of winter and summer weather for outdoor recreation. These topics are rounded out with climate tables. Unique to the study is a chapter about incorporating weather into the decision-making process by planners.

The second publication of the sub-series was by R. Olson and entitled the *Climate of Edmonton* (Ottawa: Atmospheric Environment Service, 1985. 54 p. Climatological Study No. 37. The Climate of Canadian Cities No. 2). In this publication, climate data is broken down by climate controls, the seasons – broken down by winter, spring, summer, fall – and applied climate, as related to comfort, winter heating, summer cooling, recreation, and gardening. There is also a comparison of similar climates around the world. In particular, the author describes Edmonton as having a northern, mid-latitude continental climate, where the summers are warm, the winters are cold, and most of the precipitation falls during the summer months. In general, the climate is described in terms of two main seasons, summer and winter, and two transition periods, spring and fall. Along with the climate description, there are appendices of climate data sources and tables of climate data.

The third publication of the sub-series by P. Klivokiotis and R.B. Thompson is entitled *The Climate of Calgary* (Ottawa: Atmospheric Environment Service,

1986, 52 p. Climatological Studies No. 38. The Climate of Canadian Cities No. 3). This publication follows the format of its predecessor with the climate data broken down by climate controls, the seasons, and applied elements. The Calgary climate is described as being a northern, mid-latitude continental climate, where the summers are warm, the winters cold, and most precipitation falling during the summer months. However, the proximity of Calgary to the Rocky Mountains allows the ameliorating effect of Chinook winds to break winter cold spells. Along with climate description are the climate data sources and list of tables.

The fourth publication in this sub-series is entitled *The Climate of Montreal* (Ottawa: Atmospheric Environment Service, 1987. 62 p. Climatological Studies No. 39. The Climate of Canadian Cities No. 4). This publication follows a slightly different format from its predecessors. It is divided into four sections, which incorporate discussions, tables, and figures, together. Section One discusses meteorological and topographical factors that influence the climate of Montreal. Section Two discusses the climate elements of the metropolitan area, including temperature, winds, precipitation, humidity and fog, sunshine and cloud, and spells of extreme heat and cold. Section Three discusses the climate and human activities, such as agriculture, heating and air conditioning, and lawn watering. Section Four discusses the Montreal climate in relation to the climate of other cities of the world. Meteorologists view Montreal's climate as being under the influence of two very different regimes – one continental, and the other maritime. Overall, the Montreal region is classified as a moderate temperature zone that is rather wet, and continental in type, because of the large differences in temperature between summer and winter. Finally, four appendices list the climatological stations in the Montreal region, climate normals at Dorval, climate normals in the region, and post-meteorological events.

The most recent publication of this sub-series is entitled *The Climate of Metropolitan Toronto* (Ottawa: Atmospheric Environment Service, 1989. 73 p. Climatological Studies No. 41. The Climate of Canadian Cities No. 5). This title follows the format of the previous publication, but has a larger introduction, and five sections. Section One discusses climate influences, including climate-observing stations, and factors that influence climate. Section Two discusses the climate elements – temperature (including winter cold and mild spells), precipitation (including prolonged heavy precipitation and dry spells), wind (including wind chills), humidity (including fog), sunshine and cloud cover, and severe weather (including tornados, severe thunderstorms, and hurricanes). Section Three discusses the Toronto region climate, and human activities (including gardening, heating, air-conditioning, and recreation). Section Four discusses climate variability and change, including anomalies, and Section Five discusses Toronto and the world.

In the last section, Toronto is described as the most southerly in latitude of Canada's large cities, with some of the warmest summers. On a global scale, meteorologists describe the climate of Toronto as 'moderate in temperature, moderate for precipitation totals, and intermediate in its continentality, which is a measure of the annual variation of temperature.' Finally, the Appendices include the climate normals of Toronto (city), the Pearson International Airport, and the surrounding areas, as well as meteorological events of the past.

Besides the above-mentioned current series of city climates, there are still two earlier publications available for Vancouver and Winnipeg, neither of which has been updated. One, by K.F. Harry, is entitled *The Climate of Vancouver* (Ottawa: Department of Transport, Meteorological Branch, 1967. 53 p. Technical Memoranda No. 258). The other, by J.J. Labelle and R.J. Brown, is entitled *Climate of Winnipeg* (Toronto: Department of Transport, Meteorological Branch, 1966. 57 p. Circular 4437. CLI 35).

D Specialized Climatological Studies

Besides the above regional, mesoclimatological, and city studies, the climatological series also covers specialized focus studies. One such study by M.K. Thomas is entitled *Recent Climatic Fluctuations in Canada* (Toronto: Atmospheric Environment Service, 1975. 92 p. Climatological Studies No. 28). The purpose of this study was 'to provide information on those climatic fluctuations that have occurred in Canada over the past third of a century.' What triggered the study was a general concern about unusual weather conditions during the early 1970s, and the possibility of a permanently deteriorating climate and its effects on agriculture, food production, energy, water requirements, and transportation. For practical purposes, the day-to-day climate can be considered to be stable, with its seasonal rhythms. However, when long-range planning is considered, climatic fluctuations (not major climate change) must be taken into account. The author thought that there was no reason why minor fluctuations, or climate anomalies, would not continue to occur as they have in the past, but it was important to increase one's knowledge about these anomalies. The climate information provided in the publication, to further the knowledge of anomalies, was gleaned from the archival records of 37 surface synoptic observing stations chosen to provide a good geographic coverage of the country. The period of record selected was 1940 through 1974. The climate information provided in text, and figures, covered temperature and precipitation (station trends, district trends and national indexes), bright sunshine (station trends), cloudiness (station and district trends), snowfall (station and northern and southern Canada trends). As well, the departures from normal temperatures for 1973, and temperature trends in the Northern

Hemisphere and southern Canada, have been charted for the period 1890 to 1970.

Another example of a specialized study is by V.R. Swail and L.D. Mortsch, entitled *A Review of the State-of-the-Art in Marine Climatology on the East Coast* (Ottawa: Atmospheric Environment Service, 1984. 82 p. Climatological Studies No. 36). The purpose of this two-part report was to provide an up-to-date survey of marine climatology off the east coast of Canada for the Royal Commission on the Ocean Ranger Marine Disaster. Part I describes the hazards to human safety posed by the atmosphere, a review of research activities, the agencies involved (the federal government, the oil industry), and knowledge gaps identified. Part II describes meteorological observing and analysis techniques for identified climate parameters (wind, icing, visibility, sky condition, dew points temperature, precipitation, and station pressure).

E Weather Phenomena Analysis

The study of particular weather phenomena, with storm rainfall, tornadoes, blocking, climate variability, and winter storms as examples, is also undertaken by the Meteorological Service of Canada, or its predecessor, the Atmospheric Environment Service. One early study by D.M. Pollock is entitled *An Index to Storm Rainfall in Canada* (Downsview, Ontario: Atmospheric Environment Service, 1975. 37 p. CL1 75-1). Storm rainfall had been both documented and published since 1961, as reference material for evaluating the risk of large rainstorms causing flooding in various Canadian watersheds. This documentation covered only storms in Quebec and the prairies. Storms in Quebec were selected if they produced more than two inches of precipitation in one day, three inches in two days, or four inches in three days. As the selection of storms in other areas began, criteria similar to those used in Quebec were applied elsewhere. Storms for the prairies were selected for analysis if they produced over four inches of rain in two days, or exceeded the 10–year return period point rainfall for the month. However, over time, the selection criteria were changed to eliminate all but the very largest storms. For Pollock's study, 448 storm analyses were presented, although, by 1975, over 900 rainstorms had been analysed (none going back before 1906, and few before 1920, because insufficient observing stations were operating). Included with the Index is a list of errors, with corrections, for storm analyses previously published.

A similar type of study by M.J. Newark, undertaken for the Atomic Energy Control Board, was entitled *Tornadoes in Canada for the Period 1950 to 1979* (Ottawa: Atomic Energy Control Board, 1981. 88 p. CL1 83-2). This presented a set of statistics concerning the physical characteristics of tornadoes in Canada during that 30-year period. Unlike rainstorm data gathered from the National Climatological Archive, at Downsview, tornado data had to be obtained from primary sources, such as individuals involved with tornadoes, newspapers, local histories, and literature held by the AES library. The reason for this type of search was the paucity of any other records. Once the data, as such, had been collected, it was processed into a form suitable for data storage and the creation of a master file. The author accepted that such data would have three sources of error. The first derived from the fact that most information had been recorded by reporters with no special knowledge of tornadoes. The second source of error stemmed from incomplete observations and recording. For example, tornadoes in unpopulated areas are seldom observed at all, and it is common for only part of each tornado in populated areas to be observed and recorded. The third source is random statistical errors. This type of error includes local residents' estimates of where they saw a tornado touch down. Included in the study were incidence and risk maps, and an estimate of recurrence for particular regions.

Another type of probability study, pertaining to 'blockings,' used data for a 33-year period. In climatological terms, blocking is generally understood to be an obstruction, on a large scale, of the normal west-to-east motion of mid-latitude pressure systems. If blocking lasts from one to several weeks, the resulting prolonged weather pattern (either warm or cold) may have serious economic and social consequences. Statistical studies of blocking have traditionally used purely subjective means of identifying the blocking anticyclone (an area of high central pressure relative to the surrounding where the air circulation is clockwise about the centre in North America). However, J.L. Knox, in the study entitled *A Blocking Signature Sequence Catalogue for the 500 MB Level Northern Hemisphere 1946–1978 Inclusive* (Downsview, Ontario: Atmospheric Environment Service, 1982. 39 p. CL1 82–2) objectified the means of identifying blocking episodes, by studying atmospheric pressure data over five-day periods for 500 millibar pressure anomaly centres.

Long data periods have also been used to examine the weather phenomena of climate variability in the Central Canadian Arctic and the Canadian prairies. The Arctic study by T.A. Agnew and A. Silis was entitled *Spring-Season Climate Variability in the Central Canadian Arctic Islands* (*Annals of Glaciology*, v. 21, 1995. pp. 330–36). The historic record of temperature at Resolute Bay, Northwest Territories, over the 1948 to 1993 period, was examined for the pattern of spring temperature variables, particularly the anomalies of extreme cold or warm springs, and the atmospheric circulation patterns which caused them. The extremes occurred in conjunction with a change in the circulation pattern called the Baffin Island/West Atlantic Oscillation. On the other hand, there was no significant correlation between temperature and precipitation, or between precipitation and atmospheric circulation in the region. Attempts to apply general circulation models to these variables, in order to look for global climate warming patterns, were not practical, because of the relatively local scale of the climate area studied.

A longer period of data was used to examine precipitation variability and atmospheric circulation on the Canadian prairies. B.R. Bonsal and X. Zhang, in their paper entitled *Canadian Prairie Growing Season Precipitation Variability and Associated Atmospheric Circulation* (*Climate Research*, v. 11, 1999. pp. 191–208), examined five decadal scale periods (1905 to 1996) to document the phenomena of growing season precipitation variability. Although no century scale precipitation trends were discernible, there were inter-decadal variabilities, with some increase in precipitation over the last 40 years. In the early 1970s, this increase occurred in the month of May, and then shifted to late July and August in the 1980s. Although these variations were accounted for by low atmospheric pressure in the south-central United States, higher atmospheric pressure over the Canadian Arctic, and low pressure over eastern Canada, the authors said that 'further analysis both regionally within the Prairies and temporally within the growing season may provide more insight ... in identifying and understanding the causes of growing season precipitation variability on the Canadian Prairies.'

Perhaps the most major program undertaken by the Atmospheric Environment Service (now the Meteorological Service of Canada) to better understand weather phenomena has been the Canadian Atlantic Storms Program (CASP), whose first phase was carried out from January to March, 1986. The purpose of the meteorological component of the program was to 'improve the understanding and prediction of mesoscale features within East Coast storms as well as the storms themselves.' These storms are considered unique (in comparison with the storms on the west coast of North America and Europe), because they develop on the east coast, are affected by land, and are often characterized by sub-freezing conditions at the surface. Over the three-month period of the first phase, 16 storms were documented, and this included storm fronts, precipitation bands, heavy snow, and freezing precipitation. This preliminary investigation was described by R.E. Stewart and R.W. Shaw in their paper entitled *Canadian Atlantic Storms Program: The Meteorological Field Project* (*Bulletin of the American Meteorological Society*, v. 68, 1987. pp. 338–45). Prior to Phase I of the CASP, there had been a few numerical modelling studies of rain-snow boundaries, and freezing precipitation episodes, concerning east coast storms, as noted in the paper by R.E. Stewart entitled *Precipitation Types in Winter Storms* (*Pure and Applied Geophysics*, v. 123, 1985. pp. 597–609); and in the paper by R.E. Stewart and P. King entitled *Freezing Precipitation in Winter Storms* (*Monthly Weather Review*, v. 115, 1987. pp. 1270–79). However, the CASP information on rain-snow boundaries, and freezing precipitation, from 14 storms, allowed this data to be compared with the modelling studies, and R.E. Stewart and L.M. Patinaude, in their paper entitled *Rain-Snow Boundaries and Freezing Precipitation in Canadian East Coast Winter Storms* (*Atmosphere-Ocean*, v. 26, 1988. pp. 377–98), said that 'theoretical predictions of a mesoscale circulation ... are consistent with the observed characteristics of the rain-snow boundaries and the freezing precipitation episodes.' In conclusion, the authors said that 'all winter storms, regardless of location, may be similarly dependent on these processes.' Accordingly, R.E. Stewart, in the paper entitled *Canadian Atlantic Storms Program: Progress and Plans of the Meteorological Component* (*Bulletin of the American Meteorological Society*, v. 72, 1991. pp. 364–71), said that 'although CASP was very successful, it has also become apparent that another field project is needed. This time the focus will be on the weather affecting the Avalon Peninsula/Hibernia region. Consequently, the CASP II field experiment will be conducted over this region from 15 January to 15 March 1992.' The reason for selecting this region was because, in the late winter and spring, the southern advance of sea ice moves into the area and both pack ice and icebergs can occur there. And the east coast storms can play a major role in determining the behaviour of the ice, which is a major concern to offshore activities. As well, the jagged Avalon Peninsula coastline would probably have a complicating effect on these storms. The data collected from this field project was organized and published by R.W. Crawford and M. Medej with the title *CASP II Meteorological Field and Data Summary* (Downsview, Ontario: Atmospheric Environment Service, 1992. 500 p.).

The data collected from the CASP projects allowed scientists to study, in detail, the meteorological characteristics of the storms, and the precipitation types (snow, rain, ice pellets), particularly the snow-rain boundary, and the separations between precipitation types during these storms, as noted in the paper by J.M. Hanesiak and R.E. Stewart entitled *The Mesoscale and Microscale Structure of a Severe Ice Pellet Storm* (*Monthly Weather Review*, v. 123, 1995. pp. 3144–62); in papers by R.W. Crawford and R.E. Stewart entitled *Precipitation Type Characteristics at the Surface in Winter Storms* (*Cold Regions Science and Technology*, v. 23, 1995. pp. 215–29); and *Some Characteristics of the Precipitation Formed within Winter Storms over Eastern Newfoundland* (*Atmospheric Research*, v. 36, 1995. pp. 17–37); and in the paper by K.K. Szeto and A. Tremblay, entitled *The Mesoscale Dynamics of Freezing Rain Storms over Eastern Canada* (*Journal of the Atmospheric Sciences*, v. 56, 1999. pp. 1261–81).

The interest in east coast winter storms has also generated interest in storm phenomena in other regions of Canada, as noted in the paper by R.E. Stewart and P. King, entitled *Precipitation Type Transition Regions in Winter Storms over Southern Ontario* (*Journal of Geophysical Research*, v. 95, 1990. pp. 22335–68). As well, R.E. Stewart *et al.* have published a review article entitled *Winter Storms over*

Canada (*Atmosphere-Ocean*, v. 33, 1995. pp. 223–47). This review paper focused on Canadian storms which produce precipitation 'near or below a surface temperature of 0° C. The ensuing precipitation is in the form of snow, ice pellets or freezing precipitation although many of the storms also contain regions where rain reaches the surface. Such storms batter all regions of Canada.'

IV Climate Research

Major topics in climate research are weather forecasting, including techniques and evaluations, and boundary-layer research.

A Weather Forecasting Techniques and Evaluations

Historically, weather forecasting has been, and continues to be, carried out by making observations and predicting changes. Generally, the type of forecasting we understand is synoptic, that is, characterizing the weather over large regions, at exactly the same time, in order to organize and plot the information on a map for a visual and meteorological view of a region. Since the mid-20th century, computers have made this task much easier. It is now possible to calculate changes in atmospheric conditions, and to adopt numerical weather prediction models, both global and mesoscale (middle-sized), to study particular features (thunderstorms, tornadoes, etc.), or for short-range or long-range forecasting with the aid of station networks, radar, and earth-orbiting satellites. At the Atmospheric Environment Service (Meteorological Services of Canada), work has been carried out over the years on the use and evaluation of forecasting techniques, or system and prediction models, for both mesoscale forecasting and national forecasting.

Mesoscale studies have looked at precipitation predictions and wind forecasting. One precipitation study by R.V. Tyner is entitled *The Prediction of Liquid or Frozen Precipitation in the Atlantic Provinces* (Downsview, Ontario: Atmospheric Environment Service, 1972. 36 p. Technical Memoranda TEC 775). In this study, an analysis of the 1,000–700 millibar, and 1,000–500 millibar, air pressure levels associated with the boundaries between liquid and frozen precipitation, and between snow and freezing precipitation, was made over a three-year period, from 1968 to 1970, for the four Atlantic provinces. The study concluded that for continental air masses, which affect the Atlantic provinces in winter, the 1,000–500 millibar air pressure thickness, at about 5,400 metres in height, appeared to be a useful indicator on the rain-snow boundary. However, the positions of the rain-freezing precipitation, and the snow-freezing precipitation boundaries, appeared to be more closely linked with the presence of shallow layers of cold air, and not the 1,000–500 millibar air pressure thickness.

Another precipitation study, this one by W.L. Gutzman, is entitled *Heavy Snowfall in Eastern Canada* (Downsview, Ontario: Atmospheric Environment Service, 1972. 19 p. Technical Memoranda TEC 766). In this, the winter of 1970–71 was used to conduct an examination of all snowfalls in eastern Canada, which amounted to over five inches in 24 hours, in order to attempt to forecast snowfall amounts for that region. The problem faced in this study was the fact that the effects of a large body of open water throughout eastern Canada, and orography (the forced uplift of moist air associated with counter-clockwise air movements around a central low pressure depression), not only produce heavy snowfalls by themselves but also distort the major snowfall areas. By relating snowfall patterns, air pressures, and thicknesses the author was able to determine and chart that Great Lakes snowfalls would occur in the 528–540 millibar band of the 1,000–500 millibar air pressure range.

Besides precipitation studies, there have also been attempts to forecast winds, particularly unique winds. For example, M.D. Drews studied the forecasting of Chinook winds in the paper entitled *An Aid to Forecasting Chinooks over Southern Alberta* (Downsview, Ontario: Atmospheric Environment Service, 1982. 20 p. Technical Memoranda TEC 880). Chinook winds can occur almost anywhere east of, and close to, the Rocky Mountains, from Colorado to the Mackenzie Valley. But this study focused on the southern Alberta Chinook belt area, centred on Lethbridge. Chinook winds, which cause a rapid artificial warming trend during winter or early spring periods, can be useful for urban street maintenance and snow removal cost, or they can be destructive in their force and 'eat up' the snow pack that is important for soil moisture content. Hence, there is a need to be able to forecast this warming trend with its strong winds. These strong westerly winds originate as Pacific warm fronts that cross the Rockies and are forced down slope on the eastern side of the Rockies. The air that manages to reach the eastern surface should have the characteristics of mid-level temperatures at 2,400–37,000 metres, and temperatures should resemble the 700 millibar air pressure winds. Accordingly, the paper concluded that analyses of 700 millibar patterns of winds, in particular, should allow for the forecasting of Chinook winds and temperatures at surface levels.

Moie national studies began to appear in the mid-1980s, with the unpublished manuscript by A. Shabbar entitled *AES Experimental Monthly and Seasonal Forecasts* (Downsview, Ontario: Atmospheric Environment Service, 1985. 11 p. Canada Climate Centre Report No. 85-10). At that time, monthly and seasonal forecasts were identified as the most important forecasting periods. The forecasts themselves consisted of a monthly and a seasonal outlook of temperature and precipitation anomalies, based upon the 1951–80 Climate Normals, previously described. The analyses and predictions for these outlooks were for the 50 kPa pressure surface, but included analyses for the range of 50 kPa to 100 kPa for the height and thickness of the circulation field described. The monthly mean forecast itself was based on an average sequence of six five-day mean predictions, and shown on two sets of contour maps, one for temperature, and one for precipitation. However, because of the scale of seasonal forecasting, the 50 kPa height anomalies and numerical weather prediction were not used. Nevertheless, these predictions were guided by 'statistical relationships at individual grid points over the Northern Hemisphere ... established relationships between ocean temperature anomalies and atmospheric circulation [were] utilized ... and consideration [was] given to the climatology of snow cover, sea ice state and soil moisture' in the preparation of the seasonal forecasts.

After two years of experimentation with the monthly and seasonal forecasting techniques described above, A. Shabbar, in the unpublished manuscript entitled *AES Monthly Techniques and Verification* (Downsview, Ontario: Atmospheric Environment Service, 1991. 52 p. Canadian Climate Centre Report No. 91-3), said 'it was found that the seasonal forecasts had negligible skill and the monthly precipitation forecasts also showed no skill compared with chance for a two-year sample. Only monthly temperature forecasts showed significant skill.' As a result, the AES decided that the Canadian Climate Centre would begin to issue monthly and mid-monthly temperature forecasts for Canada. By 1990, the AES said that these forecasts showed 'significant skill for winter and spring but not for summer or autumn. Regionally, the forecast skills were highest for Western Canada and for the East Coast in the spring' and 'scores were high over the Canadian Arctic, especially in summer and autumn.' The author concluded that 'we have demonstrated that monthly prediction skills varies [*sic*] with location and situation' but that 'elimination of the systematic errors in large-scale model climatology during the next few years along with the introduction of new techniques to post-correct NWP [numerical weather prediction] output ... offer the best hope for deriving skillful forecasts from dynamical extended-range integrations.'

Precipitation predictability was discussed in further detail by A. Shabbar in the unpublished manuscript entitled *Potential Predictability of Precipitation in Canada* (Downsview, Ontario: Atmospheric Environment Service, 1993. 26 p. Canadian Climate Centre Report No. 93–8). The author said that 'estimates of the potential predictability of monthly precipitation were based on an idealized view of interannual variability.' One component of this variability was time-averaged precipitation, which resulted from daily weather fluctuations. The other component arose from changes in external conditions, such as the El Nino–Southern Oscillation (ENSO) phenomena. This latter component was discussed in detail by A. Shabbar and M. Khandekar in the paper entitled *The Impact of El-Nino-Southern Oscillation on the Temperature Field over Canada* (*Atmosphere-Ocean*, v. 34, 1996.

pp. 401–16). Here, the authors examined surface temperature for 91 years (1900–90), and 500–1,000 kPa thickness data for 49 years (1946–94), in relation to the two-phase El Nino and La Nina oscillation and normal years. What they found out was that the western Canadian surface temperatures were influenced during both phases of ENSO (warm and then colder), but eastern Canadian surface temperatures were affected during the El Nino phase only (lower temperatures). However, the largest negative anomalies were found to exist over the Yukon and just west of Hudson Bay. The authors concluded by suggesting 'the possibility of developing a long-range forecasting technique for Canada using ENSO related indexes.'

In another paper by A. Shabbar and A. Barnston entitled *Skill of Seasonal Climate Forecasts in Canada Using Canonical Correlation Analysis* (*Monthly Water Review*, v. 124, 1996. pp. 2370–85), the authors suggested that the long-term trend in global SST (sea surface temperature) should also be taken into account as 'another important skill source for Canadian winter forecasts.' This suggestion was based upon the examination by an empirical system, called the canonical correlation analysis (CCA), for forecasting three-month mean surface temperature and total precipitation for Canada for the 1956–90 data period.

Finally, R. Laprise *et al.*, in the paper entitled Climate and *Climate Change in Western Canada As Simulated by the Canadian Regional Climate Model* (*Atmosphere-Ocean*, v. 36, 1998. pp. 119–67), for the first time used a climate simulation performed with the new Canadian Regional Climate Model, partly developed by the Atmospheric Environment Service, to assess longer-term regional climate change. Here, the authors said that the increased resolution afforded by fast computers, with their improved model initiation and data assimilation methods, has led to better weather analysis and short-term forecasts. It is at the Informatics Branch of the Canadian Meteorological Centre in Montreal that weather forecasts, model simulations, and analyses are carried out using NEC supercomputers. The CMC Super-Computation Infrastructure is described at their website <<http://www.cmc.ec.gc.ca>>.

B Boundary-Layer Research

The earth's surface is a boundary for the atmosphere, which touches it, and creates atmospheric transport processes because the earth's surface warms and cools in response to solar radiation. These processes occur from 100 metres to one kilometre above the earth's surface in a band called the boundary layer that can be divided into the surface boundary layer, with a depth of 100 metres, and the planetary boundary layer, which reaches a height of 600 metres or more. The boundary layer is part of the troposphere that extends from the earth's surface to a height of 11 kilometres.

One of the most frequently occurring processes, and one which characterizes the boundary layer, is turbulence, which consists of irregular swirls of motions called eddies. (Beyond the boundary layer, turbulence is primarily found only near the jet stream, where strong wind shears can create clear-air turbulence.) Some examples of boundary layer turbulence are heat thermals (rising warm air due to solar heating), wind shears caused by a frictional drag on the air flowing over the ground, and turbulent wakes caused by obstacles such as trees or buildings that deflect air flow.

As mentioned by R.W. Stewart, in the paper entitled *Atmospheric Boundary Layer*. In: *Meteorological Challenges: A History*. Edited by D.P. McIntyre (Ottawa: Information Canada, 1972. pp. 267–82), which was one of a series of papers presented at the opening of the new Atmospheric Environment Service facilities at Downsview, Ontario, the concept of the atmospheric boundary layer was derived from fluid mechanics studies in the early part of the twentieth century. This concept initially showed up in a paper by G.I. Taylor entitled *Eddy Motion in the Atmosphere* (*Philosophical Transactions of the Royal Society. Series A*, v. 215, 1915. pp. 1–26), which discussed heat transfer by turbulent motions. There was also interest shown by the military during both World Wars in the dispersal of gases and aerosols in the atmosphere. However, it was not until the late 1960s that more and continuing research was undertaken on the boundary layer.

For example, until the 1980s, little work had been done on wind flows over low hills or flat terrains, or over water, where unique features could affect air flows. Wind flow over low hills was first given serious discussion by P.S. Jackson and J.C. Hunt in their paper entitled *Turbulent Wind Flow over a Low Hill* (*Quarterly Journal of the Royal Meteorological Society*, v. 101, 1975. pp. 929–55). Further studies were encouraged by the International Energy Agency Programme of Research and Development on Wind Energy Conversion Systems (WECS), because of the search for alternate energy sources triggered by the energy crises of the 1970s. In 1980, P.A. Taylor and J.L. Walmsey of the Atmospheric Environment Service described wind flow studies at the Kettles-Hill, in Alberta, in their papers entitled *Boundary-Layer Flow over Kettles-Hill, Alberta* (*Bulletin of the American Meteorological Society*, v. 61, 1980. p. 1510); and *Estimates of Wind-Speed Perturbation in Boundary Flow over Isolated Low Hills – A Challenge* (*Boundary-Layer Meteorology*, v. 20, 1981. pp. 253–57). The authors then attempted to apply a computer model of these near-surface winds, based on two-dimensional topographic features (ridges, valleys, and escarpments). This study was described in their paper entitled *On the Application of a Model of Boundary-Layer Flow over Low Hills to Real Terrain* (*Boundary-Layer Meteorology*, v. 23, 1982. pp. 17–46). The results of this work led the authors to then look at the three-dimensional models in order to improve air flow predictions.

Further work in the 1980s was carried out in the relatively well-known study called 'The Askervein Hill Project.' This hill is located on the west coast of one of the islands of the Outer Hebrides, Scotland. The reasons given for this study location were that the hill was a potentially good WECS site, that it was suitable for numerical modelling, and that representative measurements could be made at moderate heights on the hill and over surrounding terrain. And the hill was readily accessible. The physical advantages of the hill were that it was reasonably isolated and had a 'uniform upward fetch' for the prevailing wind, and it had uniform ground cover. It was also hoped that these further studies would contribute to large-scale weather prediction models, and models for local wind loading of buildings on hills and the dispersal of pollutants around those buildings. The project was described by P.A. Taylor and H.W. Teunessen in their paper entitled *The Askervein Hill Project: Overview and Background Data* (*Boundary-Layer Meteorology*, v. 39, 1987. pp. 15–39); and, again, by P.A. Taylor and P.J. Mason in the review paper entitled *Boundary-Layer Flow over Low Hills* (*Boundary-Layer Meteorology*, v. 39, 1987. pp. 107–32).

A particular hope of the project was to use the data from field experiments to evaluate wind-tunnel experiments and mathematical modelling techniques. Wind-tunnel simulations were carried out after the field data was gathered in order to assess the reliability with which changes in mean wind and turbulence structure induced by the Askervein hill on boundary-layer flow could be reproduced in the wind tunnel. As well, researchers hoped to determine the relative impact of certain modelling approaches (surface roughness, model scale, measurement techniques, etc.) on the quality of the simulations. This work was undertaken by H.W. Teunessen and M.E. Shokr and described in their paper *The Askervein Hill Project: Wind-Tunnel Simulations at Three-Length Scales* (*Boundary-Layer Meteorology*, v. 40, 1987. pp. 1–29). In their paper, these researchers concluded that wind-tunnel simulations provided an excellent means for modelling boundary-layer flow over an isolated low hill. Changes in mean air flow could be reproduced very well, and changes in the longitudinal component turbulence level also modelled well.

Another practical application of boundary-layer meteorology can be seen in wind flow studies over Arctic waters involving pack ice and polynyas (isolated open water areas in Arctic waters). In 1979, W. Feldman and P.J. Howarth wrote a paper entitled *Estimating the Surface Wind Speed over Drifting Pack Ice from Surface Weather Charts* (*Boundary-Layer Meteorology*, v. 16, 1979. pp. 421–29). However, this paper received comments from M.L. Khandekar and V.R. Neralla of the Atmospheric Environment Service, because they had done considerable operational work on ice dynamics which were not taken into account in the paper. The ice dynamics can affect wind surface directions and speeds. These comments were given in the paper entitled *On Estimating the Surface Wind Speed over Drifting Pack*

Ice from Surface Weather Charts (*Boundary-Layer Meteorology*, v. 23, 1982. pp. 247–49). The other interesting paper concerning the boundary-layer and Arctic features, in particular polynyas, was written by A.K. Lo and entitled *On the Boundary-Layer Flow over a Canadian Archipelago Polynya* (*Boundary-Layer Meteorology*, v. 35, 1986. pp. 53–71). This AES researcher developed a numerical model for predicting boundary-layer parameters for air flow over a polynya. The model took into account the fact that as the air flow encounters a polynya, it experiences a sudden change from a cold, dry, rough, snow-covered ice surface to a warmer, wetter, and smoother open water surface. The sudden change is a buoyancy force for the air flow, which is caused by the underlying warmer surface.

On a much wider scale, boundary-layer processes have been incorporated into the creation of two versions of the Canadian Climate Centre General Circulation Model, keeping in mind that boundary-layer features, such as large temperature and humidity fluctuations, are difficult to fully represent in general circulation models, according to K. McDuffie and A. Henderson-Sellers in their monograph entitled *A Climate Modelling Primer*, 2nd Edition (New York: Wiley, 1997. pp. 150–52). G.J. Boer and N.A. McFarlane, in their paper entitled *The Climatology of the Canadian Climate Centre General Circulation Model As Obtained from a Five-year Simulation* (*Atmosphere-Ocean*, v. 22, 1984. pp. 430–73), said that 'the purpose of a general circulation model is to simulate the climate of the atmosphere from first principles.' The principles are mathematical equations of motions that represent physical processes of the atmosphere, which 'may then be used to investigate various aspects of the physical and dynamical processes occurring in the real climate system.' The Canadian General Circulation Model itself was described in a paper by G.J. Boer and N.A. McFarlane entitled *The Canadian Climate Centre Spectral Atmospheric General Circulation Model* (*Atmosphere-Ocean*, v. 22, 1984. pp. 397–429). This model was considered successful in producing realistic climate simulations, but was restrictive in the way it incorporated 'cloud amounts, radioactive properties, specific sea surface temperatures and sea-ice mass and extent.' However, a second generation of the model described by N.A. McFarlane and G.J. Boer in the paper entitled *The Canadian Climate Centre Second-Generation General Circulation Model and Its Equilibrium Climate* (*Journal of Climate*, v. 5, 1992. pp. 1013–44) did not have these limitations but still had some deficiencies, particularly in the 'treatment of surface processes, clouds, and the representation of the vertical mixing processes ...' These shortcomings affected calculations involving both the surface boundary layer and the planetary boundary layer.

Nevertheless, these concerns did lead to a number of revisions of the surface boundary-layer treatment (the formulation of heat and moisture transfer) in the second-generation General Circulation Model (GCMII), including the development of a new surface flux parameterization scheme. These changes were described and discussed by K. Abdella and N.A. McFarlane in the paper entitled *Parameterization of the Surface-Layer Exchange Coefficients for Atmospheric Models* (*Boundary-Layer Meteorology*, v. 80, 1996. pp. 223–48). The authors said that 'the parameterization is based on the Monin-Obukhov similarity theory using well established flux-profile relationships for the unstable conditions ... it also includes a parameterization for the free-convective boundary layer, which often occurs over warm surfaces with light winds.' When these modifications were tested with the current operational GCMII, there was confirmation that the new parameterization had significant impacts on surface fluxes and the lowest level air temperatures in the surface boundary layer, and would be incorporated into a third-generation general circulation model. Shortcomings of the GCMII concerning turbulence in the planetary boundary layer were addressed, again, by K. Abdella and N.A. McFarlane in the paper entitled *A New Second-Order Turbulence Closure Scheme for the Planetary Boundary Layer* (*Journal of Atmospheric Sciences*, v. 54, 1997. pp. 1850–67). Here, the authors said that 'parameterizations do not adequately represent the complex mixing processes including counter-gradient and entrainment effects that affect the profiles of mean quantities.' The authors also added that 'part of the motivation for the present work is to develop a more physically realistic ABL parameterization for use in the atmospheric general circulation model of the Canadian Centre for Climate Modelling and Analysis.'

V Air Quality Research and Global Warming

Air quality deterioration, or air pollution, is not just a by-product of industrialization, but has been documented historically to the use of wood, or coal, for heating in England in 13th century, or the problem of city air in London by the 1660s. However, intensive industrialization, coupled with the use of hydrocarbons in transportation, has aggravated the problem in the 20th century.

Air pollutants are made of particulate matter represented by solid particles and liquid droplets that are small enough to remain suspended in the air. Collectively they are known as aerosols, and include solid particles, such as carbon, dust, smoke, and pollen, that are irritants but not usually poisonous. Some of the more dangerous substances, however, are carbon monoxide, a poisonous gas that forms during the incomplete combustion of carbon-containing fuels; sulphur dioxide (SO_2), which is a colourless gas that comes primarily from the burning of sulphur-containing fossil fuels, such as coal and oil, and is a contributor to acid rain; volatile organic compounds (VOCs), which are made up of hydrogen and carbon, and include benzene, formaldehyde, and chlorofluorocarbons, which are carcinogenic; oxides of nitrogen (nitrogen dioxide and nitric acid), derived from motor vehicles, power plants, and waste disposal systems. Because these compounds are highly reactant, they play an important role in producing surface-level ozone, and other ingredients of photochemical smog.

In Canada, the federal government monitors air pollution and examines the methodologies used. Also, research is undertaken on the broader problems of acid rain, Arctic air pollution, stratospheric and surface ozone levels, and the overall concern about global warming.

A Air Quality Monitoring

The documentation of air pollution in Canada by the Air Pollution Control Directorate of Environment Canada began, in 1970, with the publication of the ongoing annual inventory entitled:

National Air Pollution Surveillance: Annual Summary (Ottawa: Air Pollution Directorate, 1970–92, 23 v.); and
National Air Pollution Surveillance (NAPS) Network: Annual Summary for [year] (Ottawa: Environmental Protection Service, Environment Canada, 1993–).

Another irregular series has been:

A Nationwide Inventory of Air Pollutant Emissions, 1970 (Ottawa: Air Pollution Control Directorate, Environmental Protection Service, Department of the Environment, 1973. 123 p. Appendices. Technical Appraisal Report EPS 3-AP-73-2).
A Nationwide Inventory of Air Pollutant Emissions: Summary of Emissions for 1972 (Ottawa: Environment Canada, Environmental Protection Service, Air Pollution Control Directorate, 1976. 13 p. Economic and Technical Review Report EPS 3-AP-75-5);
A Nationwide Inventory of Emissions of Air Contaminants (1974) (Ottawa: Pollution Data Analysis Division, Air Pollution Programs Branch, Air Pollution Control Directorate, 1978. 145 p. Economic and Technical Review Report EPS 3-AP-78-2);
A Nationwide Inventory of Emissions of Air Contaminants (1976) (Ottawa: Environment Canada, Environmental Protection Service, Air Pollution Control Directorate, 1981. 183 p. Economic and Technical Review Report EPS 3-AP-80-1);
A Nationwide Inventory of Emissions of Air Contaminants (1978) (Ottawa: Program Management Branch, Environmental Protection Programs Directorate,

Environmental Protection Service, Environment Canada, 1983. 181 p. Economic and Technical Review Report EPS 3–EP-83-10); and Kosteltz, A.M. *Canadian Emissions Inventory of Common Air Contaminants (1985)* (Ottawa: Conservation and Protection, Environment Canada, 1990. 116 p.).

As well, one decadal summary has been published as:

Emissions and Trends of Common Air Contaminants in Canada 1970–1980 (Ottawa: Program Management Branch, Environmental Protection, Environment Canada, 1986. 104 p.).

Inventories of sources and emissions of specific substances have also been published. They are:

National Inventory of Sources and Emissions of Asbestos, Beryllium, Lead, and Mercury: Summary of Emissions for 1970 (Ottawa: Data Analysis Division, Air Pollution Programs Branch, Air Pollution Directorate, 1974. 19 p. Economic and Technical Review 74-1);
National Inventory of Sources and Emissions of Copper and Nickel (1976) (Ottawa: Environment Canada, Environmental Protection Service, Air Pollution Control Directorate, 1981. 46 p. Economic and Technical Review Report EPS 3-AP-81-4);
National Inventory of Sources and Emissions of Phosphorus (1978) (Ottawa: Program Management Branch, Environmental Protection Programs Directorate, Environmental Protection Service, Environment Canada, 1983. 22 p. Economic and Technical Review Report EPS 3-EP-83-7);
National Inventory of Sources and Emissions of Chlorine (1978) (Ottawa: Program Management Branch, Environmental Protection Service, Environment Canada, 1983. 19 p. Economic and Technical Review Report EPS 3-EP-83-8);
Jan, W. *National Inventory of Sources and Emissions of Lead (1978)* (Ottawa: Environment Canada, Environmental Protection Service, Environmental Protection Programs Directorate, 1983. 28 p. Economic and Technical Review Report EPS 3-EP-83-6);
Ambient Air Particulate Lead Concentrations in Canada 1975–1983 (Ottawa: Environmental Protection Service, Environment Canada, 1985. 30 p. Report EPS/7/AP/15).

Beyond these above-listed publications, researchers have also published individual papers on air pollutants, as part of the Long-Range Transport of Air

Pollutants (LRTAP) program. For example, in 1975, E.C. Voldner and Y. Shah prepared a paper entitled *A Preliminary Canadian Emissions Inventory for Sulphur and Nitrogen Oxides* (*Atmospheric Environment*, v. 14, 1980. pp. 419–28). This inventory, which covered Canada south of 60 degrees latitude, was important for the data it provided for the development of mathematical models in, or to access and predict air quality and deposition in Canada for, the Long-Range Transport of Air Pollutants program. More narrowly, J.N. Galloway and D.M. Whelpdale estimated deposition, in their paper entitled *An Atmospheric Sulphur Budget for Eastern North America* (*Atmospheric Environment*, v. 14, 1980. pp. 409–17). These estimates were formulated for the eastern United States, eastern Canada, and eastern North America, based on natural and man-made emissions, wet and dry deposition, and atmospheric inflows and outflows.

Besides sulphuric acidic deposition, polychlorinated biphenyls (PCBs) and organo-pesticides are monitored because of their capacity for bio-accumulation through the food chain. The first data set, obtained over an annual cycle, was done by R.M. Hoff and D.C. Muir, and published in their papers entitled *Annual Cycle of Polychlorinated Biphenyls and Organohalogen Pesticides in Air in Southern Ontario: 1. Air Concentration Data*; and *2. Atmospheric Transport and Sources* (*Environmental Science and Technology*, v. 26, 1992. pp. 266–75, and pp. 276–83). Another paper by R.M. Hoff and D.C. Muir entitled *Measurement of PCCs in Air in Southern Ontario* (*Chemosphere*, v. 27, 1993. pp. 2057–62), examined the annual behaviour of the polychlorinated boranes or camphenes (PCCs) in southern Ontario in the air. The results showed the 'annual cycle of PCCs peaks in the summer months, indicating volatilization of PCCs from the earth's surface.'

B Methodologies

Besides the monitoring and documentation of air quality, the methodologies used must also be examined. As K.G. Anlauf and P. Fellin said, in their paper entitled *A Comparison of Three Methods for Measurement of Atmospheric Nitric Acid and Aerosol Nitrate and Ammonium* (*Atmospheric Environment*, v. 19, 1985. pp. 325–33), 'accurate measurements ... are of importance to the validation of atmospheric photochemical reactions and to the acid rain issue.' Their paper reported on an environmental impact study carried out in southwestern Ontario, in June, 1982, which provided an opportunity to compare filter methods for measuring atmospheric nitrogen compounds at much lower concentrations than previous studies. Differences were observed between techniques, and between daytime and nighttime sampling, which meant that further studies were necessary. These continued to be carried out, the latest by S.V. Herin and D.R. Lawson, and others, as part of an eight-day field study in the eastern portion of the Los Angeles Basin in late

1985. The study was entitled *The Nitric Acid Shootout: Field Comparison of Measurement Methods* (*Atmospheric Environment*, v. 22, 1988. pp. 1519–39).

At the same time, a study called the Nitrogen Species Methods Comparison Study, at Claremont, California, was carried out by K.G. Anlauf and D.C. MacTavish and reported in the paper entitled *Measurement of Atmospheric Nitric Acid by the Filter Method and Comparisons with the Tuneable Diode Laser and Other Methods* (*Atmospheric Environment*, v. 22, 1988. pp. 1579–86). This study used 'two tuneable diode laser spectrometer systems,' and measurements showed 'strong diurnal patterns reaching maximum values in the early afternoon and decreasing rapidly in the late afternoon and evenings.' One other study carried out at Glendora, California, and called the Carbonaceous Species Methods Comparison Study, was undertaken by K.G. Anlauf and H.A. Wiebe and entitled *Intercomparison of Atmospheric Nitric Acid Measurements at Elevated Ambient Concentrations* (*Atmospheric Environment: Part A. General Topics*, v. 25, 1991. pp. 393–400). Other methodology work carried out by these researchers has been entitled *Evaluation of a Comprehensive Eulerian Air Quality Model with Multiple Chemical Species Measurements Using Principal Component Analysis* (*Atmospheric Environment*, v. 28, 1994. pp. 3449–61); and *Emission Ratios and Photochemical Production Efficiencies of Nitrogen Oxides, Ketones, and Aldehydes in the Lower Fraser Valley during the Summer Pacific 1993 Oxidant Study* (Atmospheric Environment, v. 31, 1997. pp. 2037–48).

C Acid Rain Studies

Acidic deposition involves the measurement of wet and day dry deposition, and the study of its chemical transport mechanisms. By the early 1980s, there was a growing awareness that emissions of sulphur dioxide were contributing to acid precipitation in eastern Canada, and the deterioration of the environment would be aggravated by the increased use of high-sulphur coal, in both the United States and Canada. Consequently, models that predicted acid input from emissions, and from meteorological wind and precipitation, needed to be developed in order to assess the impact of present and future energy strategies. As part of the Long-Range Transport of Air Pollutants (LRTAP) program within Environment Canada, a model was developed to obtain an estimate of the flux of sulphur across the US-Canadian boundary, and the concentration or deposition patterns of sulphur dioxide (SO_2) and sulphates, over eastern Canada, on a seasonal and annual basis. As mentioned above, E.C. Voldner and Y. Shah had carried out preliminary measurements in Canada, in 1975, in order to gather data for the creation of a LRTP model. The model that was developed, and the comparison of model results with measurements obtained from different monitoring networks in North America, were described by E.C. Voldner and M.P. Olson in their paper entitled *Comparison between Measured and Computed Concentrates of Sulphur Compounds in Eastern North America* (*Journal of Geophysical Research*, v. 86, 1981. pp. 5339–46). By the late 1980s, this and other regional models were evaluated under the International Sulphur Deposition Model Evaluation (ISDME) program by American and Canadian researchers. In the paper by E.C. Volder entitled *The Evaluation of Long-Term Sulphur Deposition Models* (*Atmospheric Environment*, v. 23, 1989. pp. 2267–88), the author said that 'in spring, all of the models correctly predicted the location of the observed maximum amount of S[Sulphur] wet disposition' and 'except for summer, most models predicted the magnitude of the observed maximum within the data uncertainty limits.'

The components of acidic precipitation, such as sulphuric and nitric acid that are created during a complex series of chemical reactions in the air, either fall slowly to the earth by themselves, or adhere to cloud droplets and are transported and deposited as precipitation. However, studies concerning the concentration and transportation of these pollutants, including other trace methods such as mercury and cadmium, were not carried out until the mid-1980s. For example, in 1984, G.A. Isaac and P.H. Daum undertook a field project near North Bay, Ontario, in order to analyse and document the chemical and microphysical properties of the air masses, clouds, and precipitation, in that region of eastern North America during the winter season. Aircraft with newly designed cloud-water and snow collectors captured airborne pollutants, and a surface station was used to monitor pollutant concentrations on the ground. In their paper entitled *A Winter Study of Air, Cloud and Precipitation Chemistry in Ontario, Canada* (*Atmospheric Environment*, v. 21, 1987. pp. 1587–1600), the authors confirmed related studies that showed that the composition of winter precipitation in northeastern North America was significantly different from the composition of precipitation deposited during other seasons of the year. This may be caused by the seasonal variation of pollutants available for conversion to acid in a given locale; by the slower photochemical conversion of oxides to acids during the winter because of diminished solar intensity and sunlight; by the reduction of in-cloud chemical processes because the concentration of photochemically produced soluble oxidants are lower in the winter; and by the differences in 'scavenging efficiencies' between seasonal rain and winter snow.

Further studies by G.A. Isaac, and W.R. Leaitch at North Bay, Ontario, and Syracuse, NY, also showed that the vertical profiles (the earth surface to five kilometres) of aerosol particle concentration was highest in air masses from the south, and lowest in air masses from the north. And again, concentrations of pollutants were higher during the summer than the winter. These findings were reported in the paper entitled *The Vertical Distribution of Aerosols and Acid Related*

Compounds in Air and Cloud Water (*Atmospheric Environment. Part A: General Topics*, v. 24, 1990. pp. 3033–46). In addition, the authors noted that northern-originating concentrations of pollutants at North Bay, in the winter and summer, were similar to low-altitude pollutant profiles gathered near Alert, in the Northwest Territories. This similarity suggested the cause was the long-range transport of aerosols via the Arctic.

The findings of lower winter concentrations of acids in precipitation tended to be confirmed by A. Sirois and A.W. Fricke in their paper entitled *Regionally Representative Daily Air Concentrations of Acid-Related Substances in Canada, 1983–1987* (*Atmosphere Environment. Part A: General Topics*, v. 26, 1992. pp. 593–607). Beginning in 1979, the Atmospheric Environment Service had measured ambient air concentrations of particulate sulphur monoxide radicals (SO-4-2), NO-3- (nitric oxide radicals), ammonium radicals, (HH-4+) gaseous sulphur dioxide (SO_2), and nitric acid radicals (HNO-3). Between 1983 and 1987, nine sites, ranging from Saskatchewan to Newfoundland, were operational. The authors said that 'a strong statistically significant annual cycle existed for the SO_2 daily concentrations at all sites. They were highest in winter and lowest in summer. A weaker more complex annual cycle was also present in the SO-4-2, NO-3-, NH-4+ and HNO-3 daily air concentrations.' These higher levels were in the gaseous form, which would confirm the existence of lower winter precipitate forms.

Additional work on acidic deposition has been carried out under the Canadian Aerosol Measurement Program, established in 1992 'to gain a better understanding of the atmospheric behaviour of fine particle strong acidity acid aerosols and to facilitate an assessment of the potential health risks associated with acid aerosols and particles in general.' From 1992 to 1994, filter measurements were taken at 10 sites across Canada and sulphate measurements were found to be highest in southern Ontario, but 'sulphate was the most regionally homogenous species' at all sites. This work was carried out by J.R. Brook and A.H. Wiebe and reported in the paper entitled *Temporal and Spatial Relationships in Fine Particle Strong Acidity, Sulphate, PM_{10}, and $PM_{2.5}$ across Multiple Canadian Locations* (*Atmospheric Environment*, v. 31, 1997. pp. 4223–36). Later work by J.R. Brook, along with L. Zhang, showed interest in developing a model for acid deposition velocities. The first paper was entitled *Description and Evaluation of a Model of Deposition Velocities for Routine Estimates of Air Pollutant Dry Deposition over North America. Part I: Model Development* (*Atmospheric Environment*, v. 33, 1999. pp. 5037–51); and the second paper was entitled *Description and Evaluation of a Model of Deposition Velocities for Routine Estimates of Dry Deposition over North America. Part II: Review of Past Measurements and Model Results* (*Atmospheric Environment*, v. 33, 1999. pp. 5053–70).

D Air Pollution in the Arctic

The problem of air pollution in the Arctic was first noticed by aircraft pilots, who coined the phrase 'Arctic haze,' a term used to describe a situation of reduced visibility in northern regions caused by the scattering of light by suspended particulate matter or aerosols. In early 1979, the Canadian Arctic Air Pollution Program (CAAPP) was established, along with an aerosol sampling network, in response to concerns about elevated levels of pollution caused by human impact on the environment. This Canadian network is part of an international Arctic aerosol sampling network involving the United States, Norway, Denmark, Iceland, and the United Kingdom.

This network, and its initial results, were described by L.A. Barrie and R.M. Hoff in their papers entitled *The Influence of Mid-Latitudinal Pollution Sources on Haze in the Canadian Arctic* (*Atmospheric Environment*, v. 15, 1981. pp. 1407–19); and *Five Years of Chemistry Observations in the Canadian Arctic* (*Atmospheric Environment*, v. 19, 1985. pp. 1995–2010). As well, overall 20th-century trends have been described by L.A. Barrie and D. Fisher in their paper entitled *Twentieth Century Trends in Arctic Air Pollution Revealed by Conductivity and Acidity Observations in Snow and Ice in the Canadian High Arctic* (*Atmospheric Environment*, v. 19, 1985. pp. 2055–63). These last two papers were part of the Proceedings of a Third Arctic Air Symposium held at the Atmospheric Environment Service headquarters, at Downsview, Ontario, 7–9 May, 1984, and indicated that the Arctic haze is widespread, and undergoes a distinct annual cycle which peaks in the period December–April. The pollutants, including trace metals and sulphates, come from Siberia and North America, in January and February, and are acidic. During the spring months, however, European sources predominate as the origin of these aerosols.

The sulphur dioxide particulates, in particular, which make up to 30% of the Arctic haze, have been studied by L.A. Barrie and R.M. Hoff in their paper entitled *Oxidation Rate and Residence Time of Sulphur Dioxide in the Arctic Atmosphere* (*Atmospheric Environment*, v. 18, 1984. pp. 2711–22). The authors found that the 'residence time' of SO_2 was 14–20 days in the late fall, 16–32 days in midwinter, and 10–19 days in April. Further work, to confirm the source of sulphide pollution in the Arctic, used isotope measurements from samples in the archival aerosol collection, held at the AES, in Downsview. This work was carried out by J.O. Nriagu and R.D. Coker, who reported their findings in the paper entitled *Origin of Sulphur in Canadian Arctic Haze from Isotope Measurements* (*Nature*, v. 349, 1991. pp. 142–45). The authors confirmed the Eurasian and European sources, which had been noted eleven years earlier, using trajectory analyses.

Other papers written in 1990s concerning Arctic air pollution have documented additional forms of air pollution. For example, three papers by S.M. Li *et al.* have been entitled *Aerosol Silicon and Associated Elements in the Arctic High and Mid-Troposphere* (*Atmospheric Environment. Part A: General Topics*, v. 27, 1993. pp. 2907–12); *Particulate and Snow Nitrite in the Spring Arctic Troposphere* (*Atmospheric Environment. Part A: General Topics*, v. 27, 1993. pp. 2959–67); and *Seasonal and Geographic Variations of Methanesulfonic Acid in the Arctic Troposphere* (*Atmospheric Environment. Part A: General Topics*, v. 27, 1993. pp. 3011–24). Additional papers by L.A. Barrie *et al.* have been entitled *Atmospheric Toxaphene in the High Arctic* (*Chemosphere*, v. 27, 1993. pp. 2037–46); *Spatial and Temporal Variation of Polycyclic Aromatic Hydrocarbons in the Arctic Atmosphere* (*Environmental Science and Technology*, v. 31, 1997. pp. 3593–99); and *Polychlorinated Napthalenes and Coplanar Polychlorinated Biphenyls in Arctic Air* (*Environmental Science and Technology*, v. 32, 1998. pp. 3257–65).

E The Problems of Stratospheric and Surface Ozone

The majority of atmospheric ozone (about 97%) is found in the upper atmosphere in the stratosphere, where it is formed naturally when oxygen atoms (O_2) combine with oxygen molecules (O) to form stratospheric ozone (O_3). Here, the concentration of ozone averages less than 0.002%, by volume. This small quantity of stratospheric ozone is important, because it shields plants, animals, and humans from the sun's harmful ultraviolet rays. Within the boundary layer of the lower troposphere, however, low but excessive concentrations of tropospheric ozone are a secondary pollutant created by a complex series of chemical reactions in sunlight (photochemical reactions) among pollutants, ranging from nitric oxides to volatile organic compounds or VOCs (the hydrocarbons from transport emissions). In its extreme, this surface ozone can be seen as the main component of photochemical smog (Los Angeles type, or the historic London type). At the Meteorological Service of Canada, both stratospheric and surface ozone are studied, the former because of losses in the stratosphere, and the latter because it is a pollutant.

In the mid-1970s, scientists discovered that chlorofluorocarbons (CFCs) – the major propellant in spray cans – and nitrous oxide, emitted from chemical fertilizers, were drifting into the stratosphere and destroying ozone. These decreases were first noticed in the stratosphere above the Antarctic, where ozone diminutions were at their extreme in the month of September and October. This decrease of stratospheric ozone over the Antarctic is known to atmospheric scientists as the 'ozone hole.' Normally, the stratosphere over the Antarctic has a high ozone concentration, which is formed over the tropics and brought to the

Antarctic by stratospheric winds. However, during the long, cold Antarctic winters, stratospheric ice clouds facilitate chemical interactions between ozone and nitrogen, hydrogen, and chlorine atoms in the air, which contributes to the destruction of ozone. In the polar Arctic, however, where there are also high levels of ozone-destroying chlorine compounds, no real ozone hole has developed. This may be due to the fact that the Arctic circulation of air differs from the Antarctic circulation, and the stratosphere is warmer, preventing the widespread development of ice clouds that can help activate chlorine molecules. Nevertheless, a crater-like structure, or 'hole,' in the Arctic polar ozone layer was noticed in March, 1986, in images from the total ozone mapping spectrometer instrument on the NIMBUS satellite. W.F. Evans, in the paper entitled *A Hole in the Arctic Polar Ozone Layer during March, 1986* (*Canadian Journal of Physics*, v. 67, 1989. pp. 161–65), suggested that this 'hole' could have been caused by vertical air circulation upward from the floor of the crater, combined with extreme low temperatures. When the data was analysed on an annual trend basis, there was an indication that the 'hole' was developing in the Northern Hemisphere, over Spitzbergen, Norway.

Because of the uncertainty of the loss of ozone in the Arctic region, the amount of ozone research being carried out began to increase. Accordingly, the then-Atmospheric Environment Service upped the frequency of ground-based ozone measurements made from their Canadian Arctic stations at Alert and Resolute during the winter of 1987–88, and from these two sites, as well as Heiss Island in the then-USSR, during the winter of 1988–89. The increased measurement activity has been described by J.B. Kerr and C.T. McElroy in their paper entitled *Measurement of Arctic Total Ozone during the Polar Winter* (*Atmosphere-Ocean*, v. 28, 1990. pp. 383–92). However, during the winter of 1991–92, a National Aeronautics and Space Administration–National Oceanic and Atmospheric (NASA-NOAA) Airborne Arctic Stratospheric Expedition (AASE), which included H. Fast, of the AES, was undertaken to assess the possibility of an ozone hole occurring in the Northern Hemisphere during the 20th century. The results of this expedition were described by M.H. Proffitt and K. Aiken in their paper entitled *Ozone Loss inside the Northern Polar Vortex during the 1991–1992 Winter* (*Science*, v. 261, 1993. pp. 1150–54). In this article, the authors said that the polar vortex (the belt of stratospheric winds encircling the Arctic region) was ozone-rich as air entered the vortex, and ozone loss accumulated as the air descended through high concentrations of reactive chlorine. Accordingly, the depleted air that was released out of the bottom of the vortex was 'sufficient to significantly reduce column ozone at mid-latitudes.' Although the Arctic vortex contained more O_3 than normal, an ozone hole did not form.

As mentioned above, surface ozone, unlike stratospheric ozone, is a serious pollutant at ground level. Through a series of complex photochemical reactions involving nitrogen dioxide (NO_2) and volatile organic compounds (hydrocarbons), surface ozone is created, and is the main component of photochemical smog. The concentration of ozone necessary to make it a serious pollutant, with potentially damaging effects on vegetation, is very small. Yet, it was not until the 1930s that some chemical measurements of surface ozone concentrations were initiated, and measurements remained sporadic until the late 1950s during the International Geophysical Year (IGY). Before the 1930s, and widespread gasoline consumption, there was very little knowledge about concentrations of surface ozone, and this lack of background information has been a hindrance to the overall study of ozone concentrations. However, R.D. Bojokov, of the Atmospheric Environment Service, has studied 19th-century references to ozone in the scientific literature in order to interpolate 19th-century ozone measurements for 20th-century comparisons. In the paper entitled *Surface Ozone During the Second Half of the Nineteenth Century* (*Journal of Climate and Applied Meteorology*, v. 25, 1986. pp. 343–52), the author studied the measurements developed by Christian F. Schönbein, the discoverer of ozone, as well as data obtained from French and American measurements from that period. By further study of conversion factors, the author was able to approximate ozone measurements, which were about one-half of the values presently measured in the same geographic regions. There was a pattern of maximums during April–May, and minimums in October–December, similar to current measurements of 'current mid-latitude stations' generally free from local contamination.

In the 1990s, the AES (now the Meteorological Service of Canada) work on surface ozone has concentrated on assessing the accuracy of an Acid Deposition and Oxidant Model (ADoM) to project the concentrations and deposition velocity of ozone and SO_2. J. Padro and K.J. Puckett in their paper entitled *The Sensitivity of Regionally Averaged O_3 and SO_2 Concentrations to ADoM Dry Deposition Velocity Parameterizations* (*Atmospheric Environment. Part A: General Topics*, v. 27, 1993. pp. 2239–42) said that the 'ADoM considerably underpredicted the concentrations by as much as 37% for O_3 and SO_2, implying weaknesses in ADoM that are not related to dry deposition.' And S.M. Li and K.G. Anlauf in their paper entitled *Evaluation of a Comprehensive Eulerean Air Quality Model with Multiple Chemical Species Measurements Using Principle Component Analysis* (*Atmospheric Environment*, v. 28, 1994. pp. 3449–61) confirmed that the ADoM 'underpredicts ... the O_3 production but overpredicts the contributions to other species.'

F Global Warming

Although global temperature measurements can have inaccuracies, there is general agreement among scientists that the earth's surface temperature has been undergoing a slight warming for the past 100 years. Computer models, called general circulation models, that mathematically simulate the physical processes of the atmosphere and oceans, predict that, if this warming continues unabated, global climate changes could adversely affect water resources and agricultural activity, as well as cause a shift in global wind circulation, and increase precipitation. However, there are too many uncertainties in the climate models to adequately project highly complex atmospheric processes, particularly when the possible effects of clouds and oceans are added to the model equations. Many scientists believe that the main cause for the warming is the 'greenhouse' gas carbon dioxide (CO_2), whose concentration has been increasing, primarily because of the burning of fossil fuels and deforestation. However, in recent years, increasing concentrations of other greenhouse gases, such as methane (CH_4), nitrous oxide (N_2O), and chlorofluorocarbons (CFCs), have been shown to have an effect equal to carbon dioxide.

(a) Canadian Studies of Global Warming

In Canada, studies pertaining to increasing concentrations of greenhouse gases have emphasized the monitoring of carbon dioxide concentrations and the use of climate modelling as background studies for the writing of public studies on the nature and impacts of major climate warmings in the various geographic regions of Canada. The Canadian program for monitoring atmospheric CO_2 began in 1961 when time-series monitoring was undertaken at Ocean Weather Station P, in the northwest Pacific Ocean, as part of a cooperative project made up of CO_2 groups at the Scripps Institution of Oceanography in the United States. Then, in 1975, two additional Canadian sites at Alert, Northwest Territories, in the high Arctic, and at Sable Island, Nova Scotia, off the Canadian Atlantic coast, were added to the network. Evaluation of the data, to the end of 1987, was undertaken by C.S. Wong and Y.H. Chan and described in their paper entitled *Trends of Atmospheric CO_2 over Canadian WMO Background Stations at Ocean Weather Station P, Sable Island, and Alert* (*Journal of Geophysical Research*, v. 89, 1984. pp. 9527–39). The CO_2 collected, in parts per million by volume (ppmv), appeared to be consistent with global patterns. From 1975 to 1981, the increase in CO_2 concentration averaged 1.4 ppmv per year. But, in 1975–76, and in 1978, the annual growth rates of atmospheric CO_2 measurements decreased at all stations. Another dip in the growth rate of CO_2 concentrations was also observed in 1974–75, at Station P. In general, the long-term trends all showed lower growth in the beginning, and more rapid growth at the end, indicating 'a typical two-year El Niño/southern oscillation event' (referring to the cyclical atmospheric pressure–oceanic current phenomena, along the west coast of South America, where cold ocean currents move north and southerly winds promote upswelling until the end of the calendar year, when cold nutrient-rich surface water is replaced by warm nutrient-poor water). Variations in the weekly measurements of CO_2, at Alert, have been described by K. Higuchi and S.M. Daggupaty in their paper entitled *On Variability of Atmospheric CO_2 at Station Alert* (*Atmospheric Environment*, v. 19, 1985. pp. 2039–44). From the weekly data, a subset of measurements for the winter and summer season was examined, and variations in CO_2 measurements (higher in the summer than the winter) indicated that summer measurements were affected by air parcels arriving from tundra regions and the USSR, and winter measurements were affected by air parcels from northern Europe and the USSR.

The monitoring of levels of CO_2 concentrations has led to some unpublished studies that have been useful for the published documentation of the Canadian Climate Impacts Program, described below. The unpublished studies have discussed warming scenarios for southern Ontario and the Arctic. One study by R.B. Crowe was entitled *Effect of Carbon Dioxide Warming Scenarios on the Total Winter Snowfall and Length of Winter Snow Season in Southern Ontario* (Downsview, Ontario: Atmospheric Environment Service, 1985. 55 p. Canadian Climate Centre Report No. 85-14. Unpublished Manuscript). This report used the data of the 1951–80 Climate Normals for Temperature and Precipitation, from southern Ontario, in order to project theoretical warming scenarios for that region. Another study by D. Etkin and W. Macdonald was entitled *Annual and Seasonal Climate Scenarios for a Warmer Arctic* (Downsview, Ontario: Atmospheric Environment Service, 1986. 25 p. Appendices. Canadian Climate Centre Report No. 86-1. Unpublished Manuscript). In this study, a linear statistical model was used to estimate potential climate change in the Arctic, based upon Northern Hemisphere surface temperature anomalies. The results of this model indicated the Arctic warming would occur, but would vary from region to region. On the other hand, another unpublished study by D. Etkin entitled *Enhanced Greenhouse Warming: Regional Response and Believability* (Downsview, Ontario: Atmospheric Environment Service, 1991. 33 p. Canadian Climate Centre Report No. 91-6. Unpublished Manuscript) stated that 'the confidence one can have in climate predictions decreases with scale such that below a continental scale one cannot consider climate predictions from numerical models to be a reasonable guide to future trends.'

(b) The Climate Change Digest of the Canadian Climate Impact Program

One of the published results of the concern about possible climate change, due to the 'greenhouse' effect, is the Climate Change Digest series, established in 1984.

This published series attempts to assess and identify the potential social and economic impacts of climate warming expected under a doubled carbon dioxide (CO_2) scenario. Many of these studies, so far, have been conducted under contract by Canadian universities as a sub-set of the Canadian Climate Impact Program, which was organized in 1978 to integrate the efforts of various federal and provincial agencies, as well as the universities and the private sector, in the fields of climatology. The initial program identified 13 areas of study of climate change due to the 'greenhouse' effect, and has been described in the publication entitled *Climate Change Digest: Canadian Climate Impact Program* (Ottawa: Environment Canada, 1987. 11 p. Climate Change Digest, CCD 87–01). The first studies focused on climate impacts on the Great Lakes and Ontario, and covered such topics as navigation, tourism, and recreation. Similar studies were also carried out on other regions of Canada.

The very first study by B. Smit was entitled *Implications of Climate Change for Agriculture in Ontario* (Downsview, Ontario: Canadian Climate Centre, 1987. 19 p. Climate Change Digest, CCD 87-02). It was prepared in two phases with phase one assessing the impacts of climate change on Ontario's potential to support crops, and phase two identifying the extent to which climate change could enhance or diminish crop yields. The conclusion was that a warmer climate could see the regional pattern of agriculture across Ontario transformed with a warmer climate, creating crop failures in the traditional agricultural heartland, but with a growth of horticultural (fruit) crops over much of southern Ontario, and a shift of grain crops to the north. However, the advantages of warmer temperatures could be offset by moisture stress, caused by greater transpiration.

The next study by M. Sanderson was entitled *Implications of Climatic Change for Navigation and Power Generation in the Great Lakes* (Downsview, Ontario: Canadian Climate Centre, 1987. 21 p. Climate Change Digest, CCD 87–03). This study indicated lower lake levels, with increased shipping costs. However, an 11-month ice-free shipping season would ensue. The last Ontario study, to date, by G. Wall, was entitled *Implications of Climate Change for Tourism and Recreation in Ontario* (Downsview, Ontario: Canadian Climate Program, 1988. 17 p. Climate Change Digest, CCD 88-05). This report indicated that summer recreational activities were likely to have extended seasons, with an increase in the viability of summer recreational enterprises. On the other hand, the downhill Georgian Bay ski season would be eliminated and that in the Lakehead area reduced. As well, the warmer climate would mean declining lake levels with associated changes in the ecological and recreational interests associated with lakes and wetlands.

Other regional studies, similar to the Ontario studies, have focused upon unique regional topics. For example, one, by G.D. Williams, was entitled *Estimat-*

ing Effects of Climate Change on Agriculture in Saskatchewan, Canada (Downsview, Ontario: Canadian Climate Centre, 1988. 11 p. Climate Change Digest, CCD 88-06). The first publication on Quebec by the firm Lamothe and Périard Consultants was entitled *Implications of Climate Change for Downhill Skiing in Quebec* (Downsview, Ontario: Canadian Climate Centre, 1988. 12 p. Climate Change Digest, CCD 88–03). A second study on Quebec by B. Singh was entitled *The Implications of Climate Change for Natural Resources in Quebec* (Downsview, Ontario: Canadian Climate Centre, 1988. 11 p. Climate Change Digest, CCD 88–08).

The first study of the Maritime region by P. Lane and Associates was entitled *Preliminary Study of the Possible Impacts of a One Metre Rise in Sea Level at Charlottetown, Prince Edward Island* (Downsview, Ontario: Canadian Climate Centre, 1988. 7 p. Climate Change Digest, CCD 88–02). However, a slightly broader publication by N. Houtman entitled *A Regional Response to Global Climate Change: New England and Eastern Canada: Symposium Summary* (Downsview, Ontario: Canadian Climate Centre, 1994. 19 p. Climate Change Digest, CCD 94-03) has examined the impact of climate change on tourism and recreation, energy, agriculture, and wetland ecosystems. The first west coast study was written by G.A. McBean and entitled *Review of Models for Climate Change and Impacts on Hydrology, Coastal Currents and Fisheries in B.C.* (Downsview, Ontario: Canadian Climate Centre, 1992. 16 p. Climate Change Digest, CCD 92-02). A second study by G. Wall has been entitled *Implications of Climate Change for Pacific Northwest Forest Management: Symposium Summary* (Downsview, Ontario: Canadian Climate Centre, 1993. 11 p. Climate Change Digest, CCD 93-04). Here, the symposium summary suggested that 'forest managers need to take appropriate actions now to reduce the threat to sustainability of the forest resource.' And the first publication on the Canadian Arctic by D.G. McGillivray and T.A. Agnew was entitled *Impacts of Climatic Change on the Beaufort Sea-Ice Regime: Implications for the Arctic Petroleum Industry* (Downsview, Ontario: Canadian Climate Centre, 1993. 19 p. Climate Change Digest, CCD 93-01). A second publication by R.D. Brown has been entitled *Impacts of Global Climate Warming for Canadian East Coast Sea-Ice and Iceberg Regimes over the Next 50–100 Years* (Downsview, Ontario: Canadian Climate Centre, 1993. 16 p. Climate Change Digest, CCD 93-03). And a third publication by J. Brotton and G. Wall has been entitled *The Bathurst Caribou Herd in a Changing Climate* (Downsview, Ontario: Atmospheric Environment Service, 1997. 17 p. Climate Change Digest, CCD 97-01).

Besides these regional studies, some national studies have also been written. The first of these by P. Timmerman and A.P. Grima was entitled *Economic Perspectives on the Impact of Climate Variability and Change: A Summary Report* (Downsview, Ontario: Canadian Climate Centre, 1988. 7 p. Climate Change Digest, CCD 8

8-04). This emphasized that the 'issues of future water supply and demand are central to climate impact assessment and of particular concern to Canada.' Another study, by the Canadian Climate Program Board, was entitled *Climate Change and Canadian Impacts: The Scientific Perspective* (Downsview, Ontario: Canadian Climate Centre, 1991. 30 p. Climate Change Digest, CCD 91-01). This study pointed out that there was a strong consensus in the scientific community that the 'observed and continuing increases in greenhouse gas concentrations due to human activities will result in an unprecedented global warming and sea-level rise unless emissions are substantially reduced ... and Canada ... could not avoid major effects due to significant changes in world trade patterns, specially in agriculture ... and environmental refugees seeking higher ground, as sea-level rises.' Another report by C.I. Jackson entitled *Global Warming Implications for Canadian Policy* (Downsview, Ontario: Canadian Climate Centre, 1992. 17 p. Climate Change Digest, CCD 92-01) states that Canadian research and policy development on global warming should continue to devolve within the international and intergovernmental framework that has developed in the last decade. However, some of the pioneering studies of the 1980s probably needed to be reworked, because of second-generation global climate models that have been developed.

The importance of climate modelling and its overall development has been described in the publication entitled *Modelling the Global Climate System* (Downsview, Ontario: Climate and Atmospheric Research Directorate, 1994. 20 p. Climate Change Digest, CCD 94–01). However, the accuracy of using climate modelling, such as the Canadian General Climate Model, to forecast future climate changes depends 'in part, on the accuracy of its assumptions about future changes in greenhouse gases and aerosols as well as the precision with which it simulates climatic processes such as water vapour and cloud response or changes in ocean circulation. If concentrations of greenhouse gases and aerosols change at rates that are significantly different from the estimates used in the model, then, the model's projections will become increasingly unrealistic.' These comments were made by H.G. Hengeveld in a publication entitled *Projections for Canada's Climate Future: A Discussion of Recent Simulations with the Canadian Global Climate Model.* Special Edition (Downsview, Ontario: Atmospheric and Climate Science Directorate, 2000. 27 p. Climate Change Digest, CCD 00-01).

VI Departmental Sources of the Publications

Thanks to the former Atmospheric Environment Service, meteorological research and weather and climate studies are well documented in Canada. Four major bibliographies, which form a coherent series, have been published. They are:

A Bibliography of Canadian Climate 1763–1957. Compiled by M.K. Thomas (Ottawa: National Research Council, Division of Building Research, 1961. 114 p.);
A Bibliography of Canadian Climate 1958–1971. Compiled by M.K. Thomas (Ottawa: Information Canada, 1973. 170 p.);
A Bibliography of Canadian Climate, 1972–1976. Compiled by M.K. Thomas and D.W. Phillips (Ottawa: Supply and Services Canada, 1979. 135 p.);
A Bibliography of Canadian Climate, 1977–1981. Compiled by D.W. Phillips (Ottawa: Atmospheric Environment Service, 1983. 169 p.).

Each of these volumes provides an author, subject, source, and geographic index. As well, a list of periodicals with abbreviations is provided, along with a list of subjects and geographical names used in those indexes.

Another pertinent documentation source, mentioned again, is the *Handbook on Climate Data Sources of the Atmospheric Environment Service.* 2nd Edition. Compiled by D.W. Phillips and D.W. Gullit (Downsview, Ontario: Atmospheric Environment Service, Canadian Climate Program, 1988. 6 Sections). This lists the data sources available from the Atmospheric Environment Service. Information included covers the types of data and publications, the periods of record, the extent of Canada covered, the bibliographic history, cost, and where and how to order the publications.

For publications available to the mid-1980s, the Atmospheric Environment Service issued a typed bibliography, which was entitled *AES Publications Catalogue des publications du SEA* (Ottawa: Atmospheric Environment Service, 1986. 108 p.). This is particularly useful for its listings of earlier series, such as the Canadian Meteorological Research Reports, and the Technical Memoranda.

For other publications available, the Atmospheric Environment Service has a series of typed lists. They are:

Climate Change Digest (Downsview, Ontario: Canadian Information Branch, Atmospheric Environment Service, 1995. 2 p.);
Inventory of Project Reports (renamed CCC Reports in 1978) and Canadian Climate Centre Reports (CCC Reports) on Microfiche (Downsview, Ontario: Climatological Services Division, Atmospheric Environment Service, 1987. 10 p. DS 87-7);
List of Canadian Climate Centre Reports – CCC –, Reports (Downsview, Ontario: Climatological Services Division, Atmospheric Environment Service, 1987. 11 p. DS 87-11. Addendum to DS 77-4);

List of Climate Data Summaries and Documentation Sheets (Downsview, Ontario: Climatological Services Division, Atmospheric Environment Service, 1987. 10 p. DS 87-8);

List of Climatological Studies (Downsview, Ontario: Climatological Service Division, Atmospheric Environment Service, 1987. 5 p. DS 87-9);

List of CLI Circulars (Downsview, Ontario: Climatological Services Division, Atmospheric Environment Service, 1987. 3 p. DS 87-10).

Chapter 9

Water Research and Services

I Historical Background

Federal water research and services are not limited to one federal department, or service, because the British North America Act of 1867 did not mention the word water. However, the Act did allocate exclusive legislative jurisdiction over fisheries and navigation to the newly formed federal government, and jurisdiction over agriculture to both the federal and provincial governments. From these beginnings, both levels of government, over time, made arrangements concerning the nation's water resources in order to accommodate this constitutional vagueness, but some jurisdictional disputes did end up with the Privy Council, in London. For example, by 1920, the Privy Council agreed that in the realm of fisheries, the federal government had the right to regulate the inland, or non-tidal, fisheries, but the provincial governments had administrative jurisdiction. With agriculture, the Dominion Lands Act of 1872 authorized homesteading on the prairies, but did not acknowledge the need for irrigation. The National Transfer Agreement of 1930 gave the prairie provinces control over natural resources, but the federal government through its Prairie Farm Rehabilitation Administration, in 1930, had to undertake or participate in community or regional water projects to help drought-stricken areas. This became the basis for later major irrigation projects because of the harsh prairie hydrological cycle.

Since the federal government was assigned the jurisdiction over water navigation, various departments were initially involved in the surveying of river flows, depths of lakes, and harbours, in response to particular needs. By 1936, these responsibilities settled into the new Department of Mines and Resources. As well, water quality monitoring, which began in 1934, was assigned to that department. But it was not until 1945, at the Dominion Provincial Conference on Reconstruction, that the federal and provincial governments agreed on a division of water resource and research responsibilities. In the field of water resources, the federal government undertook the responsibility for basic surveys and research on a national scale. The Canada Water Act authorized Environment Canada to collect data related to water, and to collaborate with the provinces to this end. As well, the Fisheries Act authorized the Department of Fisheries and Oceans to collect water data related to fisheries. Also, the Canada Shipping Act authorized the latter to conduct hydrographic surveys.

In practice, until 1992, Environment Canada collected, analyzed, and distributed data for lake levels, river discharges, sediment, and water quality through the Water Survey of Canada of the Inland Waters Directorate. Since 1992, however, data collecting has been completely decentralized (because of government downsizing) to regional offices, or provincial counterparts. However, the National Water Research Institute (formed in 1980 and now part of the Environment Canada), and the National Hydrology Research Institute (formed in 1988, but absorbed in 1998 as the Aquatic Ecosystem Impact Branch of the NWRI), carry out research on the chemical, physical, biological, and environmental problems pertaining to water quality as well as groundwater, snow, and ice studies. Also included in this research are wetland, grassland, forest, permafrost, and urban hydrology studies. The NWRI is physically located in Burlington, Ontario, and in Saskatoon, Saskatchewan.

Separately, the Canadian Hydrographic Service, of the Department of Fisheries and Oceans, conducts hydrographic surveys, and produces charts, and tide and current tables, for both inland waters and ocean coastal areas, as well as conducting biological and chemical oceanography studies, which are a relatively recent development in Canadian water science. Other federal departments or services that carry out water-related research, as noted in previous chapters, are Agriculture and Agri-Food Canada, the Canadian Forestry Service, and the Canadian Wildlife Service. Also, the universities and private consulting firms may undertake research, the former through NSERC grants, and the latter as contracted research.

II Publications

Publications related to water research, and the hydrological cycle, consist of data gathering periodicals in print, and in CD-ROM format, as part of the Water Survey of Canada, ice thickness data, and Canadian Hydrographic Service charts and atlases. Unique is the Hydrological Atlas of Canada. As well, there are research series from the Environmental Conservation Service, and the National Water Research Institute, and technical report series from the Department of Fisheries and Oceans. And there is the annual review, or report literature, issued by the major federal departments in which these institutes and services reside. Finally, water research studies also appear in the open scientific literature.

A Data Collecting Periodicals

Data collecting involving hydrometric, sediment, and water quality surveys has been carried out historically by Environment Canada, in agreements with the provincial governments. As well, ice data is collected by the Ice Climatology Service of Environment Canada. However, the charting of Canada's inland and coastal waters is under the responsibility of the Department of Fisheries and Oceans.

(a) Water Survey of Canada

As previously mentioned, the Water Resources Branch of Environment Canada's Inland Waters Directorate was directly responsible, until 1992, for the Water Survey of Canada, which provided a systematic hydrometric survey and sediment survey program across the country. These surveys involved the measurement of the flow, level of water, and suspended sediment in rivers, streams, and lakes, as well as the computation, analysis, and publication of data. The hydrometric survey in Canada can be traced back to 1894, when some stream measurements were done in connection with irrigation surveys. However, it was not until 1908, when Parliament made its first specific appropriation for gauging streams and determining the water supply in southern Alberta and Saskatchewan, that the Hydrographic Survey was founded. In 1911, with the establishment of the Dominion Water Power Branch, there was a consolidation of the hydrometric program, already in operation for irrigation requirements, the water power investigation under way on the Winnipeg and Bow rivers, and the administration of water power on Dominion lands.

In 1913 and subsequent years, arrangements were made with the provincial governments across Canada to continue the hydrometric survey program under the Dominion Water Power Branch, while passing the administration of nearly all water resources to the provinces. Sediment survey investigations of sediment transported by streams were first undertaken by individual organizations, or agencies, in 1948, to meet the urgent needs of particular projects. While the investigations were extended to meet expanding requirements of federal and provincial agencies after 1948, it was not until 1961 that a systematic survey was begun by the Water Resources Branch. In 1975, federal-provincial agreements were established and recognized that water quantity data could be collected to meet federal or provincial needs, or both. According to those agreements, the operation of the networks would be undertaken on a shared-cost basis, according to each jurisdiction's needs.

Today, the hydrometric and sediment data survey programs still reach out across the country with hydrometric data collected at nearly 3,500 different locations and with sediment data available for approximately 500 of these sites, 100 of which are sampled on a regular basis. As mentioned earlier, since 1992, data collecting has been completely decentralized to regional offices or provincial counterparts. However, the data are still held centrally, and archived in the national water database called HYDAT, which is held now by the Climate and Water Applications section of the Meteorological Service of Canada in Toronto.

Since 1990, the Water Survey of Canada data has been published in a CD-ROM format only, entitled *HYDAT CD-ROM* (Concord, Ontario: Greenland Engineering Group, 1990–). This contains daily stream flow, water level, suspended sediment concentration, sediment particle size, and sediment load data. Also included is information describing locations for 3,500 active and about 4,200 discontinued stations, periods of operation and types of hydrometric data collected at each site. Further information about the CD-ROM version is available at the website http://www.grnland.com.

i. Surface Water Data Historically, from 1908 to 1966, surface water data were published in a variety of print formats, in imperial units, in approximately 300 publications, most of which are now out of print. However, there are still extant six regional cumulated data collections of varying date ranges. These collections are: (1) *Surface Water Data for Pacific Drainage* (Ottawa: Water Resources Branch, 1911/12–1963/64. 53 v.). This collection also has the variant title *Report on Railway Belt Hydrographic Survey to 1918*, as well as the title *Surface Water Supply of Canada, Pacific Drainage, British Columbia and Yukon Territory*, (2) *Surface Water Data for Arctic and Western Hudson Bay Drainage and Mississippi Drainage in Canada* (Ottawa: Water Resources Branch, 1919/20–1963/64. 31 v.), (3) *Surface Water Data, James Bay and Hudson Bay Drainage, 1949–1963* (Ottawa: Queen's Printer, 1967. 202 p.), (4) *Surface Water Data, St. Lawrence and Southern Hudson Bay Drainage, Ontario and Quebec* (Ottawa: Water Resources Branch, 1925/27–1963/64. 23 v.), (5) *Surface Water Data, Ungava Bay Drainage*, 1954–1963 (Ottawa: Queen's Printer, 1966. 113 p.), and (6) *Surface Water Data for Atlantic Drainage* (Ottawa: Water Resources Branch, 1919/20–1962/1964. 23 v.). Then, from 1965 to 1990, the Water Survey of Canada published four types of ongoing data issued by province or territory. They are surface water data, historical streamflow summaries, historical water levels summaries, and sediment data. These data collections are:

Surface Water Data, British Columbia (Ottawa: Water Survey of Canada, 1965–90. 26 v.);
Surface Water Data, Alberta (Ottawa: Water Survey of Canada, 1965–90. 26 v.);
Surface Water Data, Saskatchewan (Ottawa: Water Survey of Canada, 1965–90. 26 v.);
Surface Water Data, Manitoba (Ottawa: Water Survey of Canada, 1965–90. 26 v.);
Surface Water Data, Ontario (Ottawa: Water Survey of Canada, 1965–1990. 26 v.);
Données sur les eaux de surface, Québec: Surface Water Data, Quebec (Ottawa: Division des relevés hydrologiques du Canada, 1964–90. 27 v.);
Surface Water Data, Atlantic Provinces (Ottawa: Water Survey of Canada, 1965–90. 26 v.); and
Surface Water Data, Yukon and Northwest Territories (Ottawa: Water Survey of Canada, 1965–90. 26 v.).

This print data, issued in the International System of Units (SI), since 1979 provides (by river or lake): (1) tables of daily water level or discharge, (2) summaries of monthly and annual data, and (3) descriptive information, including the type of gauge used, its location, the drainage area, and whether or not the flow is natural or regulated.

The guide for using this surface water data is the publication entitled *Surface Water Data Reference Index* (Ottawa: Water Survey of Canada, 1966–91. Various editions). The first edition of the *Surface Water Data Reference Index* was published for data to 1966, by province, or territory. Subsequent editions were published annually, or biennially, for all of Canada. Commencing in 1979, however, the data were presented in the International System of Units (SI), and published biennially until the last edition, in 1991. The Index gives a tabulation of stations showing station number and name, the drainage area, gauge location, discharge records, type of gauge used, the operation schedule, and other remarks. The names of rivers and lakes for which data are available are given in a station number index and an alphabetical index. Overall, the Index is subdivided alphabetically by province, or territory, with the Quebec portion in French. Within each geographic subdivision, the stations are listed in stream order, beginning with the most upstream location, and extending to the most downstream station.

ii. Historical Streamflow Summary The first edition of the Historical Streamflow Summary was produced for data to 1970, and subsequent editions were published every three years to 1982, and from 1982 to 1990, every two years, with each new edition superseding the previous one. Since 1979, the data have been presented in the International System of Units (SI). This publication series contains a summary of historical monthly and annual mean discharges, and annual extremes, for the entire period of record for all stations where streamflow data have been collected during the history of the Water Survey of Canada. There is also an index by station location, with its station number. The last editions of this series which were issued were:

Historical Streamflow Summary: British Columbia to 1990 (Ottawa: Water Survey of Canada, 1991. 1116 p.);
Historical Streamflow Summary: Alberta, to 1990 (Ottawa: Water Survey of Canada, 1991. 629 p.);
Historical Streamflow Summary: Saskatchewan, to 1990 (Ottawa: Water Survey of Canada, 1992. 461 p.);
Historical Streamflow Summary: Manitoba, to 1990 (Ottawa: Water Survey of Canada, 1991. 301 p.);
Historical Streamflow Summary: Ontario, to 1990 (Ottawa: Water Survey of Canada, 1992. 663 p.);
Historical Streamflow Summary: Québec, to 1990 (Ottawa: Water Survey of Canada, 1992. 526 p.);
Historical Streamflow Summary: Atlantic Provinces, to 1990 (Ottawa: Water Survey of Canada, 1991. 294 p.); and
Historical Streamflow: Yukon and Northwest Territories, to 1990 (Ottawa: Water Survey of Canada, 1991. 167 p.).

iii. Historical Water Levels Summary The first edition of the Historical Water Levels Summary series was produced for data to 1976, and subsequent editions were published every three years, until 1982, and then every two years, until 1990, with each new edition superseding the previous one. Since 1979, all editions have been presented in the International System of Units (SI). This publication series contains a summary of historical monthly and annual mean water levels, and annual extremes, for the period of record for all stations on lakes, or reservoirs, where water level data have been collected during the history of the Water Survey of Canada. It also contains a summary of water level data for all stations on rivers where 'water levels only' were collected, or for certain stations on rivers where streamflow data were collected, and where the historic water level information was considered important enough to be included in the publication. The data are organized by station number. The most recent editions of this series were:

Historical Water Levels Summary: British Columbia, to 1990 (Ottawa: Water Survey of Canada, 1992. 177 p.);
Historical Water Levels Summary: Alberta, to 1990 (Ottawa: Water Survey of Canada, 1992. 71 p.);
Historical Water Levels Summary: Saskatchewan, to 1990 (Ottawa: Water Survey of Canada, 1992. 92 p.);
Historical Water Levels Summary: Manitoba, to 1990 (Ottawa: Water Survey of Canada, 1992. 142 p.);
Historical Water Levels Summary: Ontario, to 1990 (Ottawa: Water Survey of Canada, 1992. 150 p.);
Historical Water Levels Summary: Quebec, to 1990 (Ottawa: Water Survey of Canada, 1992. 231 p.);
Historical Water Levels Summary: Atlantic Provinces, to 1990 (Ottawa: Water Survey of Canada, 1992. 27 p.); and
Historical Water Levels Summary: Yukon and Northwest Territories to 1990 (Ottawa: Water Survey of Canada, 1992. 51 p.).

iv. Sediment Data For the period 1961 to 1983, sediment data were collected first for several prairie rivers to meet the needs of particular projects, and then the data collection expanded eastward, and northward, to encompass other areas of Canada. Overall, the series has the title *Sediment Data, Canadian Rivers* (Ottawa: Water Survey of Canada, 1961–83. 23 v.). Within this series, the variant titles have been *Sediment Data for Saskatchewan and Manitoba, 1961–1963* and *Sediment Data for Canadian Rivers.*

For sediment data gathered during miscellaneous surveys from 1966 to 1983 by the Water Survey of Canada and possibly published in formats different from the post-1983 series, the Water Resources Branch published the volume entitled *Miscellaneous Sediment Data: Canada: 1966–1983* (Ottawa: Ministry of Supply and Services, 1988. 248 p.).

The most recent of the printed sediment data publications issued from 1984 to 1990 by province or territory are:

Sediment Data: British Columbia (Ottawa: Water Survey of Canada, 1984–90. 7 v.);
Sediment Data: Alberta (Ottawa: Water Survey of Canada, 1984–90. 7 v.);
Sediment Data: Saskatchewan (Ottawa: Water Survey of Canada, 1984–90. 7 v.);
Sediment Data: Manitoba (Ottawa: Water Survey of Canada, 1984–90. 7 v.);
Sediment Data: Ontario (Ottawa: Water Survey of Canada, 1984–90. 7 v.);
Sediment Data: Quebec (Ottawa: Water Survey of Canada, 1984–90. 7 v.);
Sediment Data: Yukon and Northwest Territories (Ottawa: Water Survey of Canada, 1984–90. 7 v.).

This series first provides descriptive information about the sediment stations' location, latitude and longitude, drainage area, whether the flow is natural or regulated, the type of gauge and total suspended sediment load, in tonnes, for the period of record. The data provided contains information on streamflow, instantaneous suspended sediment concentration, daily suspended sediment loads, dissolved solids concentration, load summary, particle size, distribution of suspended sediment in sand, silt, and clay, bed load and bed material, water temperature when the samples were collected, the type of samples, and the single sample vertical location. Also included is a historic sediment data summary.

The guide for using the sediment data since 1972 has been the *Sediment Data Reference Index, Canada* (Ottawa: Water Survey of Canada, 1972–88. Various editions). The first edition was published for data to 1972, and contained a tabulation of locations in Canada where sediment data were collected by the Water Survey of Canada. The data were published biennially, with the latest edition superseding all previous editions, and from 1978 on the data were presented in the International System of Units (SI).

v. Water Quality Data Water quality data by province (except for Ontario, where data has been collected since 1972 by region by the provincial Ministry of Environment), drainage region, or territory is variable because it may have been collected as a special study, or survey, or as part of a long-term monitoring program. Included are physical (temperature, turbidity), chemical (calcium, magnesium, oxygen, etc. by litre), and biological (coliform and fecal counts) data, collected at the appropriate Water Survey of Canada station locality. The following printed publications have been issued over time:

Water Quality Data: British Columbia, 1961–1971 (Ottawa: Inland Waters Directorate, 1974. 1000 p.);

Water Quality Data, Alberta, 1961–1973 (Ottawa: Water Quality Branch, 1975. 500 p.);

Detailed Surface Water Quality Data, Alberta, 1974/76–1977/79 (Ottawa: Water Quality Branch, 1977–80. 2 v.);

Water Quality Data, Saskatchewan, 1961–1975 (Ottawa: Water Quality Branch, 1977. 350 p.);

Detailed Surface Water Quality Data, Saskatchewan (Regina: Saskatchewan Western and Northern Region, Water Quality Branch, 1977–80. 2 v.);

Water Quality Data, Manitoba, 1961–1976 (Ottawa: Water Quality Branch, 1978. 350 p.);

Detailed Surface Water Quality Data, Manitoba (Winnipeg, Manitoba: Western and Northern Region, Water Quality Branch, 1977–80, 2 v.);

Detailed Surface Water Quality Data, Alberta, Manitoba, Northwest Territories and Saskatchewan (Ottawa: Water Quality Branch, 1982–);

Données sur la qualité des eaux, Fleuve Saint-Laurent, Québec, 1950–1980 (Ottawa: Direction de la qualité des eaux, 1982. 159 p.);

Water Quality Data, New Brunswick, 1961–1977 (Ottawa: Environment Canada, 1979. 350 p.);

Water Quality Data, Nova Scotia, 1965–1978 (Water Quality Branch, 1980. 271 p.);

Water Quality Data, Prince Edward Island, 1961–1973 (Ottawa: Water Quality Branch, 1976. 200 p.);

Water Quality Data, Newfoundland, 1965–1980 (St John's: Department of Environment, 1982. 147 p.);

Water Quality Data, Newfoundland, 1981–1985 (Ottawa: Environment Canada, 1988. 465 p.);

Water Quality Data, Northwest Territories, 1960–1973 (Ottawa: Water Quality Branch, 1976. 292 p.);

Detailed Surface Water Quality Data, Northwest Territories, 1974/76–1980/81 (Calgary: Inland Waters Directorate, Western and Northern Region. Water Quality Branch, 1980. 3 v.); and

Water Quality Data, Mackenzie River Basin, 1960–1979 (Ottawa: Water Quality Branch, 1981. 1080 p.).

(b) Ice Thickness Data

Because of Canada's geographic position in the Northern Hemisphere, the documentation of the measurement of ice formation, thickness, and breakup in Canadian waters has been systematically carried out for over 30 years. As well, Arctic polar ice and icebergs, on the northeast seaboard, are documented and studied by conventional means and satellite imagery. Freeze-up, break-up, and ice thickness data have been collected and published annually, since 1961, and the data for the period spanning the beginning of the record, to 1974, has been compiled by W.I. Allen in the publication entitled *Freeze-Up, Break-Up and Ice Thickness in Canada* (Downsview, Ontario: Atmospheric Environment Service, 1977. 185 p. CLI-1–77).

From 1976 to 1993, the data were issued annually in two series of publications. They are:

Ice Thickness Data for Canadian Selected Stations: Freeze-Up, Break-Up (Ottawa: Atmospheric Environment Service, 1976/77–1978/79. 3 v.); and

Ice Thickness Data (Ottawa: Ice Centre, Climatology and Applications, 1979/80–1993/94. 15 v.).

In general, ice thickness is measured every Friday, throughout the winter, beginning as soon as the ice is thick enough to safely bear a person's weight, and ending in the spring when the ice becomes unsafe. In addition to ice thickness, the average depth of snow on the ice is included. Finally, where freeze-up and break-up data is available, two dates are given, marking the beginning and the end of each process.

The ice thickness stations (99 of them), and their period of record, are listed in the 1993–94 issue.

By 1992, enough ice thickness data had been documented to produce a publication covering the latest 30-year normals climatological period, from 1961 to 1990. This publication is entitled *Ice Thickness Climatology, 1961–1990* (Ottawa: Ice Climatology Services, Ice Centre, 1992. 277 p.). For this publication, ice thickness measurements for 135 locations have been summarized in both tabular and graphical formats. Of these 135 locations, only 21 sites span the complete 1961 to 1990 period.

The tabular data presents a summary of ice thickness and snow depth data for 52 weekly periods. In addition, maximum and minimum ice thickness for each period is shown.

Related to the ice thickness data are the series of ice atlases, which began publication in 1980. These, produced by M.E. Markham, are entitled *Ice Atlas: Eastern Canadian Seaboard* (Toronto: Atmospheric Environment Service, 1980. 96 p.); *Ice Atlas: Canadian Arctic Waterways* (Ottawa: Atmospheric Environment Service, 1981. 198 p.); and *Ice Atlas: Hudson Bay and Approaches* (Ottawa: Ice Centre, 1988. 123 p.). In 2000, the Canadian Ice Service began issuing an annual atlas entitled *Annual Arctic Ice Atlas* (Ottawa: Canadian Ice Service, 2000–). This series is derived from RADARSAT data. Finally, in 2001, the Canadian Ice Service began publishing a new series of sea ice atlases based on their digital database for the 30–year climatic period, 1971–2000. To date, two atlases have been published. They are: *Sea Ice Climatic Atlas: East Coast of Canada, 1971–2000* (Ottawa: Canadian Ice Service, 2001. 42 p. maps); and *Sea Ice Climatic Atlas: Northern Canadian Waters, 1971–2000* (Ottawa: Canadian Ice Service, 2002. 48 p. maps).

Ice information in electronic format is available at the Canadian Ice Service website located at: <<http://www.ice-glaces.ec.gc.ca>>.

(c) Canadian Hydrographic Service Data Collecting Periodicals and Atlases

The oldest of the water data gathering services in Canada is the Canadian Hydrographic Service. It is responsible for the surveying of Canada's inland waters (estuaries and lakes), as well as the coastal waters, and the seas of the Canadian

Arctic. This data gathering system generates a range of electronic and print publications, maps, and charts, including current and water level tables, for boats, or ships, using Canada's waterways. The service is also responsible for the training of Canada's cartographers and hydrologists working in the service.

The Canadian Hydrographic Service operates as a component of the Science Sector of the Department of Fisheries and Oceans, where planning and coordination work is done for the regional stations, or institutes, that are responsible for carrying out all hydrological work in Canada. Beginning on the east coast, these stations include the Bedford Institute of Oceanography at Dartmouth, Nova Scotia, for the Altantic region. This institute undertakes hydrographic activities on Canada's east coast, from the fishing grounds of Georges Bank, south of Nova Scotia, to the Central Canadian Arctic, including half of the Gulf of St Lawrence, and the Grand Banks of Newfoundland. The Maurice Lamontagne Institute, located in Trois Rivières, covers the Quebec Region, and includes the charting of the St Lawrence River from Montreal to the east, including the rivers feeding into the river and estuary, and the Gulf of St Lawrence itself. The Bayfield Institute of the Canada Centre for Inland Waters, located in Burlington, Ontario, covers the Central and Arctic Regions. This includes the charting of the navigable waters of Ontario and Manitoba, as well as the waters of Hudson Bay and James Bay. In addition, this region carries out bathometric (ocean mapping) and gravity surveys in the Arctic Archipelago in cooperation with the Department of Energy, Mines, and Resources. Finally, the Institute of Ocean Sciences at Sidney, British Columbia, covers the Pacific Region. This region is responsible for the surveying of Canada's west coast and fiords, the waters off Vancouver Island and the Queen Charlotte Islands, and the straits and waters and inland seas of the western Arctic.

The Canadian Hydrographic Survey formally began in 1883, when John Boulton was appointed upon a recommendation of the British Admiralty to undertake the Georgian Bay Survey. By 1894, this was completed, and the federal government decided to continue work on the remaining Canadian waters of the Great Lakes because of the increased use of deeper-draught vessels with increased speed in those waters. The first publications of the Survey began, in 1892, with the first volume of *Canadian Sailing Directions – The Georgian Bay and North Channel Pilot* (Ottawa: Ministry of Marine and Fisheries, 1892/CHIM Microfiche Series No. 57746. 3 microfiche). At the same time, Parliament placed both federal observations and hydrographic surveys under the Chief Engineer's Branch of the Department of Marine and Fisheries.

It is interesting that 86 years later, after these initial surveys, the Survey noted that only 45% of Canadian waters had been surveyed adequately for current requirements, and only 15% of Canada's Arctic waters had been surveyed. Overall, 25% of all Canadian waters remained unsurveyed, or surveyed only to a minimum 'reconnaissance level.' For the years issued, the *Activity Report* (Ottawa: Canadian Hydrographic Service, 1975–88. 14 v.) of the Canadian Hydrographic Service displayed a map of the Canadian waters surveyed under the categories 'not surveyed to modern standard,' 'spot soundings or partially surveyed,' 'modern surveys.' The annual Survey Program, which shows considerable activity, is also listed. For example, for 1987–88, the Atlantic Region noted that 'a total of 37,178 km of sounding profiles were done, 2,018 shoal examinations made, and 18 ports swept by CSS. *FCG Smith*.' Or, in the Quebec Region, 'the main 1987–1988 surveys were: continuation of a 1:20,000 scale survey of the Magdalen Islands; modern surveys of Sorel at scales of 1:20,000 and 1:10,000; surveys of Rivière-Port Neuf at a scale of 1:5,000; tide gauge surveys of the North Shore; photogrammetric survey of harbour facilities between Quebec City and Trois-Rivières at a scale of 1:2,000 and a complete resumption of complete jurisdiction over the network of permanent water level gauges at 41 stations in the region.' Or, in the Central and Arctic Region 'the St. Lawrence River survey was a continuation of a program which started in 1979 to resurvey the river from Kingston, downstream to Beauharnois, Quebec to support international charting along the Seaway.' As well, that region's Cartography Section 'produced nine new bilingual and metric charts, 17 new editions, and 64 notices to mariners ...' In the Pacific Region, 'the major survey for the season was for charts 3482 and 3494, Vancouver harbour, Western and Central portions.' And 'a network of 21 permanent gauges and four temporary tide gauges was maintained in the Western Arctic and on the Pacific coast: the data for these gauges were processed and archived on a monthly schedule for studies of mean sea level, crustal movements and storm surges, as well as hydrographic surveys, oceanic studies, and the Tide and Current Tables.'

The charts available are listed in the following publications: (1) *Catalogue of Nautical Charts and Publications. 1, Atlantic Coast* (Ottawa: Canadian Hydrographic Service, 1998–), (2) *Catalogue of Nautical Charts and Publications. 3, Ontario/Manitoba Including the Great Lakes* (Ottawa: Canadian Hydrographic Service, 2000. 1 folded sheet), and (3) *Catalogue of Nautical Charts and Related Publications* (Ottawa: Canadian Hydrographic Service, 1999. 1 folded sheet). A fourth catalogue covering the Pacific Coast has yet to be published.

The most visible outcome of all this data gathering, besides the constant preparation and revision of charts, are the Sailing Directions and Small Craft Guides, both published with varying frequencies of new editions, and the Tide and Current Tables published on a regular basis.

i. Sailing Directions The Sailing Directions are the companions to the charts, as they are used with the charts of the area, and describe in detail, using text and photographs, features shown on the charts. Each volume contains information on the history, weather, charts and publications, water levels, navigation information, regulations, and aids to navigation. Specific marine navigation information describes wharves, lights, dangers to navigation, local conditions, ice currents, anchorages, marinas, facilities, and services – all with reference to the charts of the area they cover. Sailing Directions are published currently in book and booklet formats. These publications tend to be revised about every two to four years and are kept current between new editions with the bi-weekly, monthly, and annual editions of the *National Edition of Notices to Mariners* (Ottawa: Canadian Coast Guard, v. 22, 1997–, the *Eastern Edition of Notices to Mariners*, v. 22, 1997–, and the *Western Edition of Notices to Mariners*, v. 22, 1997–). The Sailing Directions available in book form are:

Sailing Directions, Newfoundland (Ottawa: Canadian Hydrographic Service, 8th ed., 1986–). This was previously entitled the *Newfoundland Pilot* (Ottawa: Canadian Hydrographic Service, 1st ed., 1958–3rd ed., 1966);

Sailing Directions, Labrador and Hudson Bay (Ottawa: Canadian Hydrographic Service, 3rd ed., 1974–). This was previously entitled the *Labrador and Hudson Bay Pilot* (Ottawa: Canadian Hydrographic Service, 1st ed., 1955–2nd ed., 1965);

Sailing Directions, Nova Scotia (Atlantic Coast) and Bay of Fundy (Ottawa: Canadian Hydrographic Service, 1st ed., 1990–);

Sailing Directions, Gulf of Maine and Bay of Fundy (Ottawa: Canadian Hydrographic Service, 1st ed., 2001–);

Sailing Directions, Cape North to Cape Canso (including Bras d'Or Lake) (Ottawa: Canadian Hydrographic Service, 1st ed., 2001–);

Sailing Directions, Cape Canso to Cape Sable (including Sable Island) (Ottawa: Canadian Hydrographic Service, 1st ed., 2001–);

Sailing Directions, Gulf of St. Lawrence (Ottawa: Canadian Hydrographic Service, 1st ed., 1992–). This was previously entitled *Sailing Directions, Gulf and River St. Lawrence* (Ottawa: Canadian Hydrographic Service, 1st ed., 1973–6th ed., 1985);

Sailing Directions, Great Lakes. Vol. I (Ottawa: Canadian Hydrographic Service, 10th ed., 1986–). This was previously entitled the *Great Lakes Pilot. Vol. I* (Ottawa: Canadian Hydrographic Service, 1st ed., 1933–6th ed., 1967);

Sailing Directions, Great Lakes. Vol. II (Ottawa: Canadian Hydrographic Service, 7th ed., 1993–). This was previously entitled the *Great Lakes Pilot. Vol. II, Lake Huron, Georgian Bay and the Canadian Shores of Lake Superior* (Ottawa: Canadian Hydrographic Service, 1st ed., 1955–3rd ed., 1968);

Sailing Directions, Great Slave Lake and Mackenzie River (Ottawa: Canadian Hydrographic Service, 7th ed., 1989–). This was previously entitled *Great Slave Lake and Mackenzie River Pilot* (Ottawa: Canadian Hydrographic Service, 2nd. ed., 1967–);

Sailing Directions. Vol. I, British Columbia Coast (South Portion) (Sidney, BC: Canadian Hydrographic Service, 13th ed., 1984–). This was previously entitled *British Columbia Pilot. Vol. I* (Victoria: Canadian Hydrographic Service, 1st ed., 1930–7th ed., 1969);

Sailing Directions. Vol. II, British Columbia (North Portion) (Sidney, BC: Canadian Hydrographic Service, 7th ed., 1977–). This was previously entitled the *British Columbia Pilot. Vol. II* (Sidney, BC: Canadian Hydrographic Service, 1930–69);

Sailing Directions, Arctic Canada. Vol. I (Ottawa: Canadian Hydrographic Service, 4th ed., 1994–). This was previously entitled *Pilot of Arctic Canada. Vol. I* (Ottawa: Canadian Hydrographic Service, 1959–70);

Sailing Directions, Arctic Canada. Vol. II (Ottawa: Canadian Hydrographic Service, 4th ed., 1985–). This was previously entitled *Pilot of Arctic Canada. Vol. II* (Ottawa: Canadian Hydrographic Service, 1959–70); and

Sailing Directions, Arctic Canada, Vol. III (Ottawa: Canadian Hydrographic Service, 5th ed., 1994–). This was previously entitled the *Pilot of Arctic Canada. Vol. III* (Ottawa: Canadian Hydrographic Service, 1961–68).

The Sailing Directions available in booklet format are:

Sailing Directions. General Information. Atlantic Coast (Ottawa: Canadian Hydrographic Service, 1st ed., 1992–);

Sailing Directions. Newfoundland – East and South Coasts. Cape Bonavista to Ferryland Head (including Placentia Bay) (Ottawa: Canadian Hydrographic Service, 1st ed., 1995–);

Sailing Directions. Newfoundland – Southwest Coast. Ferryland Head to Port au Basques (Ottawa: Canadian Hydrographic Service, 1st ed., 1995–);

Sailing Directions. Newfoundland: Northeast and East Coast: Cape Bauld to Cape Bonavista (Ottawa: Canadian Hydrographic Service, 1st ed., 1997–);

Sailing Directions. Saint John River (Ottawa: Canadian Hydrographic Service, 1st ed., 1994–);

Sailing Directions. St. Lawrence River. Cap Whittle/Cap Gaspé to Les Escoumins (Ottawa: Canadian Hydrographic Service, 1st ed., 1992–);

Sailing Directions. St. Lawrence River. Ile Verte to Québec (Ottawa: Canadian Hydrographic Service, 1st ed., 1992–);

Sailing Directions. St. Lawrence River. Cap-Rouge to Montreal (Ottawa: Canadian Hydrographic Service, 1st ed., 1992–2000). Continued by *Sailing Directions. St. Lawrence River. Cap-Rouge to Montreal and Rivière Richelieu* (Ottawa: Fisheries and Oceans Canada, 2nd ed., 2001–);

Sailing Directions. St. Lawrence River: Montreal to Kingston (Ottawa: Canadian Hydrographic Service, 1st ed., 1996–);

Sailing Directions. Lake Ontario (Ottawa: Canadian Hydrographic Service, 1st ed., 1996–), *Sailing Directions. Welland Canal and Lake Erie* (Ottawa: Canadian Hydrographic Service, 1st ed., 1996–);

Sailing Directions. Detroit River, Lake St. Clair, St. Clair River (Ottawa: Canadian Hydrographic Service, 1996–);

Sailing Directions. Georgian Bay (Ottawa: Canadian Hydrographic Service, 1st ed., 1988–); and

Sailing Directions. General Information: Great Lakes (Ottawa: Canadian Hydrographic Service, 1st ed., 1996–).

Sailing Directions. Lake Huron, St. Marys River, Lake Superior (Ottawa: Canadian Hydrographic Service, 1st ed., 2000–).

ii. Small Craft Guides The Small Craft Guides are similar to the Sailing Directions, but are specifically designed for use by the recreational boater. The Small Craft Guides that are currently available are:

Small Craft Guide, Rideau Waterway and Ottawa River (Ottawa: Canadian Hydrographic Service, 1st ed., 1984–);

Small Craft Guide, Trent – Severn Waterway (Ottawa: Canadian Hydrographic Service, 1st ed., 1973–);

Small Craft Guide: Lake Ontario (Ottawa: Canadian Hydrographic Service, 1st ed., 1987–);

Small Craft Guide. Georgian Bay (East and North Shore Routes) (Ottawa: Canadian Hydrographic Service, 1st ed., 1983–);

Small Craft Guide, British Columbia. Vol. 1 (Ottawa: Canadian Hydrographic Service, 7th ed., 1988–);

Small Craft Guide, British Columbia. Vol. 2 (Ottawa: Canadian Hydrographic Service, 7th ed., 1988–); and

Small Craft Guide. Saint John River (Ottawa: Canadian Hydrographic Service, 1st ed., 1976–).

iii. Canadian Tide and Current Tables The annual Canadian Tide and Current Tables provide predicted times, and heights, of the high and low tides, and the predicted times for slack water, and the times and velocities of maximum current associated with the horizontal movement of the tides. They are:

Canadian Tide and Current Tables. Vol. 1, Atlantic Coast and Bay of Fundy (Ottawa: Canadian Hydrographic Service, 1967–);
Canadian Tide and Current Tables. Vol. 2, Gulf of St. Lawrence (Ottawa: Canadian Hydrographic Service, 1959–);
Canadian Tide and Current Tables. Vol. 3, St. Lawrence and Saguenay Rivers (Ottawa: Canadian Hydrographic Service, 1967–);
Canadian Tide and Current Tables. Vol. 4, Arctic and Hudson Bay, 1964–);
Canadian Tide and Current Tables. Vol. 5, Juan de Fuca Strait and Strait of Georgia (Ottawa: Canadian Hydrographic Service, 1997–);
Canadian Tide and Current Tables. Vol. 6, Discovery Passage and West Coast of Vancouver Island (Ottawa: Canadian Hydrographic Service, 2000–); This was previously issued as:
Canadian Tide and Current Tables. Vol. 6, Barkley Sound and Discovery Passage to Dixon Entrance (Ottawa: Canadian Hydrographic Service, 1959–99. 41 v.); and
Canadian Tide and Current Tables. Vol. 7, Queen Charlotte Sound to Dixon Entrance (Ottawa: Canadian Hydrographic Service, 2000–).

iv. Atlases Besides the charts, Sailing Directions, Small Craft Guides, and the Tide and Current Tables, the Canadian Hydrographic Service also publishes atlases. An interesting example of this type of publication is the *Current Atlas: Juan de Fuca Strait to Strait of Georgia* (Ottawa, Canadian Hydrographic Service, 1983. 212 p.). The Atlas consists of three parts. Part A shows 'a number of charts by means of which the user may determine the appropriate tidal stream pattern at a specified time.' Part B shows a 'number of charts by means of which the user may determine the change in tidal stream flow patterns over a 25–hour period of a typed large or small tide. The user can see the build-up and decay of eddies, strong streams and other features.' Part C shows 'the current flow patterns for three hours before low water, at low water and for three hours after in water for river discharges ...' The CHS has also produced a specialized *Atlas of Tidal Currents* (Ottawa: Canadian Hydrographic Service, 1997. 1 atlas. 108 p.), which indicates the direction and velocity of surface currents in the St Lawrence River, between Cap de Bon-Désir and Trois-Rivières, at 20–minute intervals.

As well, the Canadian Hydrographic Service has the responsibility for surveying the waters adjacent to the Canadian land mass, and producing bathymetric maps and charts, in digital and analogue formats, not for navigation but for studying the nature of the seafloor, and the material beneath it. The types of maps produced are Natural Resource Maps, which provide a geomorphological interpretation of the seafloor through the use of contours. Currently, the west coast region has been completely surveyed, and the CHS can provide base maps for regional earth science activities. This mapping project is described at the CHS website http://www.chs-shc.dfo-mpo.gc.ca>>.

Finally, for general information and news in the area of electronic charting and mapping in Canada, the Canadian Hydrographic Service produces in paper and electronically a publication entitled *Contour Magazine* (Ottawa: Canadian Hydrographic Service, 1993–); or, through their website http://www.chs-shc.dfo-mpo.gc.ca>>.

B Hydrological Atlas of Canada

The major overview source of Canadian water distribution, on and under the earth, is the *Hydrological Atlas of Canada* (Ottawa: Fisheries and Environment Canada, 1978. 7 p. 34 leaves). This was a product of the Canadian National Committee of the International Hydrological Decade (1965–74), and remains the bibliographic reference document for water resources in Canada. The data from which the maps and text were compiled covered the period 1941–70. Besides an overview of the hydrologic cycle, and the influence of man, the Atlas consists of 34 maps, each with an accompanying explanatory and interpretive text, and bibliography. The maps consist of: (1) the hydrologic cycle, (2) precipitation networks, (3) annual precipitation, (4) depth, duration, and frequency of point rainfall: 60-minute rainfall for 2-year, 5-year, 10-year, and 25-year return periods, (5) depth, duration, and frequency of point rainfall: 60-minute rainfall for 2-year, 5-year, 10-year, and 25-year return periods, (6) depth, duration, and frequency of point rainfall: 24-hour rainfall for 2-year, 5-year, 10-year, and 25-year return periods, (7) depth, duration, and frequency of point rainfall: ratios of 6-hour, 48-hour, and 72-hour rainfall, to 24-hour extremes; location of recording rain gauges used in the analyses, (8) snowcover measurement networks, (9) annual snowfall, (10) dates of formation, and loss of snowcover, (11) mean maximum depth of snow, and time of occurrence, (12) radiation, (13) air temperature measurement network, (14) mean January daily temperature, (15) mean July daily temperature, (16) winds, (17) mean annual lake evaporation, (18) large lakes, (19) freeze-up and break-up of rivers and lakes, (20) river systems of Canada, (21) streamflow gauging stations, (22) annual large river flow, (23) variation of monthly river flows, (24) annual runoff, (25) water balance-derived precipitation, and evapotranspiration, (26) suspended sediment concentration, (27) water quality networks, (28) water quality of surface waters, (29) groundwater observation wells, (30) surficial hydrogeology, (31) bedrock hydrogeology, (32) permafrost, (33) influence of glaciers on streamflow, and (34) influence of man.

C Research and Report Series Publications

(a) Environmental Conservation Service Research Series

These series originated from the Inland Waters Directorate, until 1992, when the latter became the Ecosystem Science Directorate of the Environmental Conservation Service. At that time, individual series names and numbering continued. These series, including those discontinued, are:

Scientific Series (Ottawa: Inland Waters Directorate: Environmental Conservation Science, No. 1, 1968–). This major series covers freshwater chemical and biological studies, including water quality documentation of lakes and rivers. Included in this, with joint numbering, are the papers of the former National Water Research Institute. An example of a water quality guidelines publication in this series by R.A. Kent and B.D. Pauli is entitled *Canadian Water Quality Guidelines for Dinoseb* (Ottawa: Inland Waters Directorate, Water Quality Branch, 1991. 35 p. Scientific Series No. 189)

Technical Bulletin Series (Ottawa: Inland Waters Directorate, No. 1, 1965–No. 161, 1989?). The Technical Bulletins tended to consist of papers pertaining to modelling and design but also contained pollution studies. An example of a publication in this series by J.C. Merriman is entitled *Bottom Sediment Quality of the Ottawa River* (Burlington: Inland Waters Directorate, Waters Quality Branch, 1987. 13 p. Technical Bulletin No. 153).

Report Series (Ottawa: Inland Waters Directorate, No. 1, 1968–No. 76, 1988). Work appearing in the Report Series tended to be pilot studies, or laboratory quality control studies, pertaining to hydrochemistry. An example of a publication from this series is entitled *Research Projects in Glaciology, 1974* (Ottawa: The Directorate, 1974. 130 p. Canada. Inland Waters Directorate. Report Series No. 36).

Social Science Series (Ottawa: Inland Waters Directorate: Environmental Conservation Service, No. 1, 1973–). This ongoing series covers the economic aspects of water-related developments. An example from this series by A.M. Kassem and D.M. Tate is entitled *Water Use Analysis Model (WUAM) Demonstration* (Ottawa: Environmental Conservation Service, 1994. 88 p. Social Science Series No. 28).

(b) National Water Research Institute Research Series

Until the integration into a single entity of the National Water Research Institute in Burlington, Ontario, and the National Hydrology Research Institute in Saskatoon, both institutes issued distinct numbered publications. They are:

NWRI Contributions (Burlington, Ontario: National Water Research Institute, [year – number). These publications pertain to aquatic monitoring and pollution studies. An example from this series by Y.L. Lau is entitled *Physical Processes of Pollutant Transport in the Aquatic Environment* (Burlington, Ontario: National Water Research Institute, 1990. 37 p. NWRI Contribution 90–159).

NHRI Symposia (Saskatoon: National Hydrology Research Institute, No. 1, 1988–No. 17, 1997). These publications have documented the proceedings of workshops or symposia pertaining to all aspects of hydrological studies. An example from this series has been entitled *Applications of Remote Sensing in Hydrology: Proceedings of the Second International Workshop, 18–20 October 1994, Saskatoon, Saskatchewan.* Edited by G.W. Kitt and A. Pietroniro (Saskatoon: National Hydrology Research Institute, 1995. 322 p. NHRI Symposium No. 14).

NHRI Science Reports (Saskatoon: National Hydrology Research Institute, No. 1, 1990–No. 7, 1995). These reports covered snow, ice, and glacier studies. An example from this report series has been entitled *Environmental Aspects of River Ice.* Edited by T.D. Prowse and N.C. Gridley (Saskatoon: National Hydrology Research Institute, 1993. 155 p. NHRI Science Report No. 5).

(c) Department of Fisheries and Oceans Technical Report Series

These technical report series reflect the chemical and physical oceanography program of the Department of Fisheries and Oceans (Fisheries and Oceans Canada).

Canadian Technical Report of Hydrology and Oceans (Ottawa: Department of Fisheries and Oceans, No. 1, 1982–). These reports contain scientific and technical information considered to be of sufficient importance to be preserved, but not appropriate for primary publication. An example of a publication from this series by J.A. Thompson and C. Stewart is entitled *Organotin Compounds in the Coastal Biota of British Columbia – An Overview* (Sydney, BC: Fisheries and Oceans Canada, 1994. 11 p. Canadian Technical Report of Hydrology and Ocean Sciences No. 155).

Canadian Data Report of Hydrology and Ocean Sciences (Ottawa: Department of Fisheries and Oceans, No. 1, 1992–). This series provides a medium for documentation, archiving, and dissemination of data compilations, where little or no analysis is included. These compilations, may have been prepared in support of other publications or of work related to hydrography and to the chemical and physical oceanography program of the Department of Fisheries and Oceans. An example of a publication in this series by R. Pearson is entitled *NOGAP B. 6: Physical and Chemical Data Collected in the Beaufort Sea and Mackenzie River Delta, April, May and September, 1992 and Ice Core Data Collected in 1991–1992* (Sidney, BC: Department of Fisheries and Oceans, 1993. 199 p. Canada Data Report of Hydrology and Ocean Sciences No. 129).

Canadian Contractor Report of Hydrology and Ocean Sciences (Ottawa: Department of Fisheries and Oceans, No. 1, 1982–). These reports are unedited final reports from scientific and technical projects contracted by the Department of Fisheries and Oceans. An example of a publication in this series by P.C. Chandler and S. de Margerie is entitled *Numerical Modelling of Tides in Hudson Strait and Ungava Bay* (Dartmouth, NS: Bedford Institute of Oceanography, 1985. 60 leaves. Canadian Contractor Report of Hydrology and Ocean Sciences No. 13).

D Annual Reviews or Reports

Over time, the Inland Waters Directorate, in general, or the National Water Research Institute and the National Hydrological Research Institute, in particular, have produced review or annual report literature. As well, some information can be found in the Environment Canada and the Department of Fisheries and Oceans annual reports. These publications are:

Canada Water Year Book (Ottawa: Environment Canada, 1976–86. 11 v.). This series of yearbooks, from the Inland Waters Directorate, highlighted various aspects of water resources in Canada. For example, the *Canada Water Year Book, 1979–1980: Water Research Edition* (Ottawa: Environment Canada, 1981. 132 p.) gave an overview of water resources research in Canada. Topics covered were physical processes, chemical and biological processes, water quality, waste water treatment, water quality management, and water resources planning.

National Water Research Institute, Annual Report (Burlington, Ontario: National Water Research Institute, 1980–86. 7 v.);

National Water Research Institute, Reflections, 1987–88 (Burlington, Ontario: National Water Research Institute, 1988. 1 v.);

National Water Research Institute, Digest (Burlington, Ontario: National Water

Research Institute, 1988–95. 8 v.); and via Internet access: *National Water Research Institute, Annual Report, 1998–1999* (<<http://www.cciw.ca/nwri/nwri.html>>).

National Hydrology Research Institute, Annual Report (Saskatoon: National Hydrology Research Institute, 1987–88 and 1998–99. 12 v.).

Environment Canada Annual Report (Ottawa: Environment Canada, 1980–90). Since 1990, no printed annual reports have been issued, but there is information about the department at their website <<http://www.ec.gc.ca>>.

Department of Fisheries and Oceans, Annual Report (Ottawa: Department of Fisheries and Oceans, 1980–94). Since 1993/94, no printed annual reports have been issued, but a Departmental Performance Report can be read at their website <<http://www.ncr.dfo.ca>>.

III Water Research

A Hydrological Cycle

Besides surface water documentation and studies, there are below-surface and above-surface features and factors that affect the hydrological cycle and water resources of the Canadian land mass. These factors are the groundwater beneath the land surface, the wetlands on the surface, and the influence of grasslands, forestry, permafrost, and the urban environment. These areas of research and study are carried out by the federal government, and their provincial counterparts, private research consulting firms, and the universities. Hence the research mandate is very broad.

(a) Groundwater Hydrology

Groundwater exists everywhere beneath the Canadian land surface, but the volume of fresh groundwater, as a water resource important for human consumption, is not evenly distributed across the country. In fact, few major cities in Canada use groundwater, because of the abundance of surface water. Nearly all groundwater in Canada is extracted from wells in sand or gravel deposits, or from wells in fissured bedrock. Sand and gravel aquifers (geological formations that are permeable enough for water yields to be useful are known as aquifers, and water-containing formations that are less permeable are called aquitards) vary in size and shape, but small and medium-sized ones commonly do not appear on maps of geologic deposits, or of groundwater resources, because their extent is rarely known. On the other hand, the location and scale of large regional aquifers, such as the Palaeozoic carbonate-rock aquifer, in Manitoba, are generally known.

According to J.A. Cherry, in his paper entitled *Groundwater Occurrence and Contamination in Canada.* In: *Canadian Aquatic Resources.* Edited by M.C. Healey and R.R. Wallace (Ottawa: Department of Fisheries and Oceans, 1987. pp. 387–4260. Canadian Bulletin of Fisheries and Aquatic Sciences No. 215), 'the total Canadian research effort directed at groundwater issues is extremely small ... considering that about one quarter of the population of Canada depends on water from aquifers.' Only a small amount of basic research on groundwater circulation, age and replenishing times, or natural geochemical and biochemical processes in the groundwater zone has been carried out. One such Ph.D. thesis in this area by D. Desaulniers was entitled *Groundwater Origin, Geochemistry and Solute Transport in Three Major Glacial Clay Plains of East Central North America* (Waterloo, Ontario: University of Waterloo, 1986. 455p.). However, more effort has been directed at the contamination of groundwater by municipal landfills, or by industrial waste disposal.

In Ontario, contaminated sand aquifers at three municipal landfill sites have been studied for some years. One is situated on the outskirts of North Bay, and two smaller ones are located near Alliston and Kitchener-Waterloo. The geologic and hydrologic features of all three sites have been described by J.A. Cherry, of the University of Waterloo, in a paper entitled *Occurrence and Migration of Contaminants in Groundwater at Municipal Landfills on Sand Aquifers.* In: *Environment and Solid Wastes.* Edited by C.W. Francis and S.I. Auerback. (Boston: Butterworths, 1983. pp. 127–47). Research on the North Bay landfill site has shown that seepage by inorganic and organic contaminants into the groundwater created a contaminated plume in the sand aquifer that took about four years to contaminate a series of springs that empty into Lake Nipissing. Hence, the North Bay landfill rendered the stretch of aquifer between it and the springs unsuitable for potable use.

This occurrence and distribution of the contaminants has been studied by M.N. Reinhard and N.L. Goodman in their paper entitled *Occurrence and Distribution of Organic Chemicals in Two Landfill Leachate Plumes* (*Environmental Science and Technology*, v. 18, 1984, pp. 953–61). More recently, the Gloucester landfill site near Ottawa, Ontario, has also been a source of groundwater contamination because of the disposal of chemical wastes into the site. At this landfill site, S. Lesage and R.E. Jackson reported in their paper entitled *Occurrence and Fate of Organic Solvent Residues in Anoxic Groundwater at the Gloucester Landfill, Canada (Environmental Science & Technology*, v. 24, 1990. pp. 559–66) that the disposal of organic chemicals in trenches at this waste disposal site had resulted in the contamination of the underlying aquifer by a wide variety of volatile organic compounds, particularly 1,4–dioxane, because of its toxicity and mobility.

Besides these studies of groundwater contamination by municipal landfills, subsurface water disposal and groundwater contamination have also been studied. In the mid-1970s, the Water Resources Branch of the Inland Waters Directorate carried out a reconnaissance study on the effects of the disposal of liquid industrial wastes through old oil and gas wells, in Lambton County, into the subsurface of southwestern Ontario. Government researchers suspected that a number of chemical samples, characterized by high chloride and total dissolved solids concentrations, not entirely fitting into the regional pattern of chemical water types, were indicative of contamination of the fresh water aquifer in the county by deep liquid wastes. This study, written by A. Vandenberg, was entitled *Subsurface Waste Disposal in Lambton County, Ontario: Piezometric Head in the Disposal Formation and Groundwater Chemistry of the Shallow Aquifer* (Ottawa: Inland Waters Directorate, Water Resources Branch, 1978. 64p. Technical Bulletin No. 90).

A major contamination of groundwater by industrial waste disposal occurred near the town of Mercier, Quebec, just 20 kilometres south of Montreal between the years 1968 and 1972. At that time, a company in the business of transporting used oil disposed of approximately 40,000 cubic meters of chemical and petrochemical liquid wastes into lagoons in a gravel pit area near the town. This area was located on unconfined aquifers where the water table is close to the bottom of the pits. When disposals ceased in 1972, half the original volume of waste had seeped downward into the sand and gravel aquifer, contaminating the region's source of potable well water. Because of the size of the plume of contamination, local water then had to be supplied by aqueduct.

Investigations conducted between 1973 and 1983 indicated the contaminated plume was 30 meters thick, and the lateral velocity of the groundwater flowing southwest from the lagoons moved at between 100 and 500 meters per year. By 1975, part of the residual mass of wastes in the lagoons had been buried in clay nearby. In 1983, a program of aquifer restoration began, by creating new wells for pumping contaminated water from the aquifer, and treating it and discharging it as surface water. By the mid-1980s, this disaster had cost over $10 million for changes in the local water supply and aquifer restoration. This polluted site was first studied by M. Poulin for a master's thesis entitled *Groundwater Contamination near a Liquid Waste Lagoon, Ville Mercier, Quebec* (Waterloo, Ontario: University of Waterloo M.Sc. Thesis, 1977. 158 p.). Later the findings were reported by M. Poulin and G. Simard in their paper entitled *Pollution des Eaux Souterraines par les Composés Organiques à Mercier, Québec* (*Hydrogéologie* (Orleans), v. 2, 1985. pp. 125–31).

Additional groundwater contamination sites in aquifers in Canada have been noted in both Prince Edward Island and Saskatchewan. In Prince Edward Island, shallow groundwaters of the fractured sandstone aquifer of the island were found to be contaminated with aldecarb residues, more than two years after the use of this toxic pesticide for the control of aphids and the Colorado potato beetle had ceased. The site had been subjected to groundwater monitoring from 1985 to 1989. R.E. Jackson and J.P. Mutch, in their paper entitled *Persistence of Aldecarb Residues in the Sandstone Aquifer of Prince Edward Island, Canada* (*Journal of Contaminant Hydrology*, v. 6, 1990. pp. 21–36), thought that the persistence of the residue was due to the low pH-buffering effects of ammonium fertilizer oxidation. As well, the aldecarb residues remained constant, because of their storage within the sandstone matrix and diffusion back into the fractures of this porous system. This study was written up in more detail by the authors in their publication entitled *The Fate and Simulation of Aldecarb in the Soil and Groundwater of Prince Edward Island* (Burlington, Ontario: Rivers Research Branch, National Water Research Institute, Canada Centre for Inland Waters, 1992. 96 p. Scientific Series (Canada Inland Waters Directorate) No. 194).

In Saskatchewan, a contaminant plume within the centre of the Condie Aquifer near Regina, Saskatchewan, was discovered in the early 1990s. The contamination site was a chloride plumb, 7,700 metres long and a few hundred metres wide, which was caused by the softening of water used at the Provincial Correctional Centre, starting in 1964. Water softening uses sodium chloride to regenerate an ion exchange resin, and this leads to an increased concentration of chloride in the effluent discharged into the PCC sewage reservoir, which occupies a shallow natural depression lying above the Condie Aquifer. This contaminated plume was studied by G. Van Der Kamp and L.D. Luba and described in their paper entitled *Field Study of a Long and Very Narrow Contaminant Plume* (*Ground Water*, v. 32, 1994. pp. 1008–16), which noted that the uniqueness of the plume shape was caused by the extreme steadiness of flow of the groundwater regime of the Condie Aquifer, and this resulted in very little depression-induced dilution of the contaminants.

Besides studying actual groundwater contamination, research is also now carried out at the National Water Research Institute in Burlington, Ontario, in order to try to predict groundwater contamination, particularly by pesticides. A.S. Crowe and J.P. Mutch, in their paper entitled *Assessing the Migration and Transformation of Pesticides in the Subsurface: The Role of Expert Systems* (*Water Pollution Research Journal of Canada*, v. 25, 1990. pp. 293–324), described an expert system designed as a management tool to assist regulatory personnel in their assessment of the potential for pesticides to contaminate groundwater. This expert system is called EXPRES (Expert System for Pesticide Regulatory Evaluation Simulations), and consists of existing simulation models combined with a knowledge-based system that guides the user through the choice of physical, climatic, hydrogeological, pedological, and agricultural settings, of agricultural regions across Canada, required by the pesticide model in order to come up with model predictions.

A user's manual for this system was written by J.P. Mutch and A.S. Crowe and entitled *EXPRES: An Expert System for Assessing the Fate of Pesticides in the Subsurface: User's Manual* (Burlington, Ontario: National Water Research Institute, 1993. 138 p. Scientific Series No. 201). This expert system has been further refined to provide a multi-level pesticide assessment methodology, with a regional scale that can prioritize districts having a potentially high risk for groundwater contamination from a particular pesticide application, a local scale that can identify critical areas within a district, and a soil profile scale that can allow the user to evaluate specific factors influencing pesticide leaching and persistence, including the extent and timing of leaching. The initial studies, incorporating regional scales into the EXPRES system, were described by A.S. Crowe and J.P. Mutch in the paper entitled *Regional Analysis of Pesticide Contamination of Groundwater Using an Expert System Approach* (Ground Water Management, v. 9, 1992. pp. 413–27). The additional refinements have been described by A.S. Crowe and W.G. Booty in their paper entitled *A Multi-Level Assessment Methodology for Determining the Potential for Groundwater Contamination by Pesticides (Environmental Monitoring and Assessment*, v. 35, 1995. pp. 239–62). As well as being used for overall pesticide pollution assessment in groundwater, modelling has also been used to study the breakdown of particular pesticides in aquifers. For example, J.V. Headley and J.R. Lawrence, in their paper entitled *Transformation of the Herbicide Diclofop-Methyl in a Large-Scale Physical Aquifer Model* (Water Pollution Research Journal of Canada, v. 29, 1994. pp. 557–69), studied the probable breakdown of diclofop-methyl in a meso-scale aquifer model. In general, the chemical transformation of the pesticide in the model was similar to compounds found in field studies.

In 1995, a special session of the Canadian Symposium on Water Pollution Research was held in Burlington, Ontario, in order to provide 'an opportunity to exchange information and evaluate methods and procedures related to pesticide contamination of Canada's groundwater ...' From the symposium, the editors, A. Crowe and P. Milburn, selected 10 papers that were published as *Special Issue on Contamination of Groundwater by Pesticides in Canada* (*Water Quality Research Journal of Canada*, v. 30, 1995. pp. 363–561). These papers focused upon the extent of groundwater contamination in rural areas of Nova Scotia and Ontario, current field and laboratory studies designed to assess factors influencing the migration and fate of pesticides on the subsurface, the study of pesticides in surface water, or runoff, and its relationship to groundwater contamination, and Canadian protocols for developing water quality guidelines for irrigation and livestock watering.

Apart from the specific problem of groundwater contamination into aquifers, the larger problem of the effect of climate change on groundwater quality, and quantity, particularly in the prairie regions, has also begun to be an object of study. In the paper by W. Nicholaichuk of the National Hydrology Research Institute entitled *Climate Variability and Change and Water Supply on the Canadian Prairies*. In: *Symposium on the Impacts of Climatic Change and Variability on the Great Plains, Calgary, Alberta, 1990*. Edited by G. Hall (Waterloo, Ontario: Department of Geography, University of Waterloo, 1991. pp. 173–78. Department of Geography Publication Series Occasional Paper No. 12), the author indicated that current climate change studies suggested decreases in summer soil moisture, increased irrigation demand, an increase in the likelihood of floods, and no significant reduction in the probability of droughts. Because of these probabilities, additional research would be required, including modelling for evapotranspiration and glacier behaviour, as well as improved measurements for precipitation, evaporation, soil moisture, groundwater, and runoff.

However, the measurement of groundwater discharge in particular is difficult to predict, or measure, in relation to climate change, because of the factors affecting precipitation to or from groundwater, as well as the nature and depth of aquifers containing groundwater. These problems were discussed by G. Van der Camp in the paper entitled *Groundwater Levels and Climatic Change*. In: *Using Hydrometric Data to Detect and Monitor Climate Change: Workshop on Using Hydrometric Data to Detect and Monitor Climate Change, Saskatchewan, Sask., April 8–9, 1992*. Edited by G.W. Kite and K.D. Harvey (Saskatoon: National Hydrology Research Institute, 1992. pp. 143–47. NHRI Symposium No. 8). Because of the problems noted, the author thought that 'monitoring of groundwater levels for climate change effects would be most meaningful in well-defined aquifers with constant land use over the entire extent of the aquifer, and where there is a good coverage of climatic stations,' and that 'semiconfined aquifers would probably be most suitable ... because they will respond with a characteristic time of several years.' These views were based on a research paper by G. Van der Camp and H. Maathius entitled *Annual Fluctuations of Groundwater Levels as a Result of Loading by Surface Moisture* (*Journal of Hydrology*, v. 127, 1991. pp. 137–52).

One particular study undertaken to gain insight into long-term responses of a prairie lake watershed to climate change was carried out on Lake Wabamum, near Edmonton, Alberta, and has been described by A.S. Grove. In the paper, entitled *The Application of a Coupled Water-Balance-Salinity Model to Evaluate the Sensitivity of a Lake Dominated by Groundwater to Climate Variability* (*Journal of Hydrology*, v. 141, 1993. pp. 33–73), the author used a 'dynamic hydrological model specifically designed ... to access the effects of long-term climatic variability on the lake and its surrounding watershed, specially the groundwater regime.' Comparison of simulated lake surface and salinity fluctuations with historic (1940–89) data indicated 'that very small but long-term changes to temperature and precipitation have the potential to cause a decline in lake levels, a deterioration in the quality of the lake water, and a substantial loss of groundwater resources.'

Finally, a slightly different approach to the problem of the effects of climate change on groundwater supply has been suggested by R.G. Lawford in a book chapter entitled *Regional Hydrologic Responses to Global Change in Western North America*. In: *Earth System Responses to Global Change: Contrasts between North and South America*. Edited by H.A. Mooney and E.R. Fuentes (San Diego: Academic Press, 1993. pp. 73–99). Here, the author suggested that the effects of changed land use patterns, and industrial activities, on hydrology might be greater than those produced by climate changes.

(b) Wetlands Hydrology

Unlike groundwater, wetlands, which are visible, are relatively poorly understood. However, wetlands have gained stature in recent years, because of the greater awareness of their importance to plant and animal life, as well as to the hydrological cycle. Wetlands develop because either surface drainage is poor, or surface water is in nearly constant supply. Accordingly, climate and physiography are the key factors in wetland development. In 1981, the National Wetlands Working Group of the Canada Committee on Ecological Land Classification identified five types of wetlands (acidic bogs, fens, nutrient-rich marshes, swamps, and shallow water), and 27 classes of wetland regions, covering one-seventh of Canada's land area, in their map entitled *Wetlands of Canada* (Ottawa: Environment Canada, Lands Directorate, 1981. 2 maps). The classification system itself has been most recently described by S.C. Zoltai and D.H. Vitt in their paper entitled *Canadian Wetlands: Environmental Gradients and Classification* (*Vegetatio*, v. 118, 1995. pp. 131–37). Overall, these wetlands have been described in the monograph by the National Wetlands Working Group entitled *Wetlands of Canada* (Montreal: Polyscience Publications in cooperation with Environment Canada and the Canadian Government Publishing Centre, Supply and Services Canada, 1988. 452 p.).

More specifically, the Inland Waters Directorate has also documented wetland distribution and conversion in southern Ontario. According to E.A. Snell, in the paper *Wetland Distribution and Conversion in Southern Ontario* (Ottawa: Environment Canada, Inland Waters and Lands Directorate, 1986. 53 p.), before 1800 there were about 2.3 million hectares of wetland distributed throughout southern Ontario. By 1982, however, 0.93 million hectares remained, and were more prevalent in the northern parts of the study. The original wetland area had been reduced by 61% overall, and by 68% south of the Precambrian Shield. Wetland decline has been most severe in southwestern Ontario, where over 90% of the original wetlands have been converted to other use.

According to H. Whillans, in his paper entitled *Wetlands and Aquatic Resources* In: *Canadian Aquatic Resources*. Edited by M.C. Healey and R.R. Wallace (Ottawa: Department of Fisheries and Oceans, 1987. pp. 387–426). Canadian Bulletin of Fisheries and Aquatic Sciences No. 215), the information needs or research required for a greater understanding of Canada's wetlands are theoretical information based on models of wetland systems, and greater practical information on wetland functions and characteristics. Also needed is more information on the response of wetlands to nutrient loading and contaminants. In addition, the monograph entitled *Wetlands of Canada*, noted above, suggests in its conclusions that 'significant research needs also exist in the analysis of social, economic, and legislative factors involved in the conversion of wetlands to other uses.' Nevertheless, some work has been done on the possible acidification of watersheds from wetland soils, water transportation in wetlands, and the assimilation of pollutants by wetlands, as well as studies on wetland plants, fish, and wildlife, and the beginning of research using numerical modelling.

In the Inland Water Directorate's publication by J.A. Wood and G.M. Wickware entitled *Chemistry of Wetlands and Their Effect on Receiving Waters in Kejimkujik National Park, Nova Scotia* (Regina: Inland Waters Directorate, Western and Northern Region, Water Quality Branch, 1991. 46 p. Scientific Series No. 193), a three-year field program studied wetland soil chemistry in Kejimkujik National Park in order to determine if there was any relationship to drainage water chemistry. The emphasis was upon organic acid production and metals, because researchers were trying to determine if acidification of a watershed was affected by organic acid inputs from wetland soils. On the other side of the country, drainage water chemistry has been involved in the study of water and solute (chloride) transfer between a prairie wetland and the adjacent upland. M. Hayashi and G. Van der Camp, in their papers entitled *Water and Solute Transfer between a Prairie Wetland and Adjacent Uplands, 1. Water Balance* (*Journal of Hydrology*, v. 107, 1998. pp. 42–55); and *Water and Solute Transfer between a Prairie Wetland and Adjacent Uplands, 2. Chloride Cycle* (*Journal of Hydrology*, v. 207, 1998. pp. 56–67), studied the exchange of water and solutes in order to evaluate the mechanism of the transfer. The source of the water coming into the wetland was winter snowmelt and summer precipitation, and most of this water moved rapidly to the margin of the wetland and further on to the upland. Very little went into recharge the underlying groundwater. This water movement was documented through the measurement of chloride inputs used as a natural tracer.

Other research carried out on wetland functions has been noted in the thesis by N.J. Patterson entitled *An Approach to Wetland Evaluation and Assessing Cultural Stresses on Freshwater Wetlands in Southern Ontario* (Toronto: University of Toronto, Department of Geography, M.Sc. Thesis, 1984. 192 p.). Some of the functions examined were the primary productivity, animal productivity, the effects of the hydrologic regime on wetlands, and the assimilation of pollutants by wetlands. The ability of wetlands to handle pollutants is based upon the high productivity of their plant life, and their potential for assimilating nutrients. Although most work has been carried out in Europe and the United States, some work has been done in Canada. At Listowel, in southern Ontario, artificial marshes have been created as wetland wastewater treatment facilities by the Ontario Ministry of the Environment.

In regular sewage treatment facilities, the main problems encountered are phosphorus, nitrogen, biochemical oxygen demand, and suspended solids. S.A. Black and I. Wile, in their paper entitled *Sewage Effluent Treatment in an Artificial Marshland* (Toronto: Ontario Ministry of the Environment, 1981, n.p.), state that these components were reduced by 60%–90% by treatment in an artificial wetland. Similar results were also achieved at a natural wetland near Bradford, Ontario. Nevertheless, the use of wetlands for the reduction of toxic contaminants by uptake in plants needed further examination, because of the potential transfer of toxic contaminants into herbivorous animals who occupy wetlands. This problem has been noted, not in animals, but in amphipods (beach fleas or water lice of the Phylum Crustacea), in prairie wetlands. M.T. Arts and J.V. Headley, in their paper entitled *Persistence of Herbicide Residues in Gammarus lacustris (Crustacea: Amphipoda) in Prairie Wetlands* (*Environmental Toxicology and Chemistry*, v. 15, 1996. pp. 481–88), used tandem mass spectroscopy to detect herbicide residues in wild populations of amphipods. The chief advantages of using tandem MS are 'the ability to analyze small quantities of tissue ... the ability to detect picogram levels of target analytes and user flexibility in the choice of tissue to be analyzed.'

Some wetlands contain plant communities that are among the most productive in the world, in terms of their ability to convert photosynthetically available radiation to energy in the form of dry matter. In Canada, some work has been done by R.J. Reader on sub-Arctic bog-marsh complexes, and reported in the paper entitled *Primary Production in Northern Bog Marshes*. In: *Freshwater Wetlands: Geological Processes and Management Potential*. Edited by R.E. Good and D.F. Whigham (New York: Academic Press, 1978. pp. 53–62). In particular, cattail-dominated marshes are extremely productive, as research by I. Bayly and F. MacLatchy entitled *Nutrient Behavior and Abcisic Acid Content during Fall and Winter in a Typha glauca Community*. In: *Wetlands Research in Ontario*. Edited by M.J. Bardecki (Toronto: Ryerson Polytechnical Institute, Department of Applied Geography Occasional Paper, 1982. pp. 35–52), has noted. Nutrients from organic matter settle into the sediment porewater of the wetland and are the nutrient source for wetland ponds. This feature of wetlands has been observed and noted by T. Mayer and C. Ptacek in their paper entitled *Sediments as a Source of Nutrients to Hypereutrophic Marshes of Point Pelee, Ontario, Canada* (*Water Research*, v. 33, 1999. pp. 1460–70). According to the authors, these marshes are 'one of the finest surviving wetlands on the Lower Great Lakes.'

Because of their recreational significance, perhaps the most highly visible function of wetlands is their contribution to the production of fish and wildlife, even though these populations tend to be transient and not permanent to wetlands. Waterfowl, in particular, are most closely identified with wetlands, where they pass their breeding and moulting cycles. In Canada, the most important wetlands for waterfowl breeding are in the prairie and parkland areas of the prairie provinces. D.J. Neave, in his paper entitled *Waterfowl and Wetlands: Problems and Programs in Alberta.* In: *First Western Hemisphere Waterfowl and Waterbird Symposium* (Ottawa: Canadian Wildlife Service, 1983. pp. 8–12), estimated that up to six million pond wetlands, such as potholes and sloughs, provide breeding habitats for as many as 50% of North American ducks. As well, temporary potholes and sloughs have shown that they are a potentially important source of energy in prairie food webs, because of phytoplankton production in them. R.D. Roberts and D.B. Donald, in their paper entitled *Phytoplankton Primary Production of Three Temporary Northern Prairie Wetlands* (*Canadian Journal of Fisheries and Aquatic Sciences,* v. 52, 1995. pp. 897–902), said that phytoplankton measurements in small temporary ponds were 'about three orders of magnitude greater than lakes.'

The use of wetlands by fish is far less understood than waterfowl use. However, the regular occurrence of some species such as northern pike (*Esox lucius*), largemouth bass (*Micropterus salmoides*), and bowfin (*Amia calva*) have been documented by W.B. Scott and E.J. Crossman in their monumental work *Fresh Water Fishes of Canada* (Ottawa: Fisheries Research Board of Canada. Bulletin No. 184, 1973. 966 p.). As well, some relationships between fish communities and wetlands have been documented in papers, such as that by T.H. Whillans, entitled *Historic Transformation of Fish Communities in Three Great Lakes Bays* (*Journal of Great Lakes Research*, v. 5, 1979. pp. 195–215).

Besides freshwater fowl and fish, wetlands have some fur-bearing mammals such as beaver (*Castor canadensis*), mink (*Mustela vison*), muskrat (*Ondatra zibethicus*), and others that heavily depend on wetlands. But it is reptiles and amphibians, which are either fully resident or transient users of wetlands, that are the least studied. Unlike waterfowl, fish, or fur-bearing mammals, reptiles and amphibians are not subjected to heavy exploitation, but are important for viewing. This was noted by E.L. Filion in *The Importance of Wildlife to Canadians: Highlights of the 1981 National Survey* (Ottawa: Canadian Wildlife Service, 1983. 40 p.).

Apart from the functions of wetlands on plant or animal productivity, some research on wetlands has begun to use mathematical modelling. For example, the impact in variations in water levels in wetlands has been studied via conceptual mathematical models in order to assist environmental managers to predict impacts. This modelling has been undertaken and applied to 17 shoreline marshes of various shapes in the Georgian Bay/North Channel region of the Great Lakes. R.P. Bukata and J.E. Bruton, in their paper entitled *An Evaluation of the Impact of Persistent Water Level Changes on the Areal Extent of Georgian Bay/North Channel Marshlands* (*Environmental Management*, v. 12, 1988. pp. 359–68), said that 'such a geometric model, despite its neglect of the biological complexities of marsh ecology, can serve as a valuable tool for assessing the range of impacts of both natural and man-made changes in long-term ambient water levels on shoreline marshes.'

As well, numerical modelling has been used to simulate water movement and levels in prairie sloughs, in order to study the development of salinity around them and, in particular, the development of salinity rings around a recharge slough near Hamiote, Manitoba. One reason for this type of study is the fact that soil salinization is a problem with major economic concerns on the prairies, but the mechanisms of salinization are not well known. In the paper by W.J. Stolte and S.L. Barbour entitled *A Simulation of the Mechanisms Influencing Aquatic Ecosystems in Semi-Arid Regions: Implications for Resource Management.* Edited by R.D. Roberts and M.L. Bothwell (Saskatoon: National Hydrology Research Institute, Ottawa: Rawson Academy of Aquatic Science, 1990. pp. 165–77. NHRI Symposium No. 7), the authors confirmed that the close correspondence of the salinity concentration profiles found by the modelling process to those found in the field 'is convincing evidence that the evapotranspiration regime upslope from the slough is indeed the dominant mechanism affecting the development of these salinity rings.'

(c) Grasslands Hydrology

Besides the influence of groundwater or wetlands, grasslands also influence the hydrological cycle. In Canada, the grasslands, or prairie regions, lie primarily in the Saskatchewan–Nelson River basins, and range from southwestern Alberta through to the northeastern transitional grassland area of Manitoba. Within that vast area, potholes, or sloughs, are the major feature of the prairie hydrological cycle. Sloughs are either classified as recharge (this water moves into the groundwater), discharge (groundwater moves to the surface), or transitional (water moves in both directions at different times of the year). Most of the river flow through the prairies originates in the Rocky Mountains and foothills, as the grassland area itself contributes very little river flow. Besides sloughs and rivers, the prairies do contain major man-made reservoirs with irrigation outlets, as well as permanent lakes, the latter usually being saline.

Major research on the grassland hydrologic cycle tends to focus upon the effect of cultivation on that cycle. On the prairies, precipitation, in the form of rain showers, is highest in early summer, and lowest during the winter months, but at any time there can be local highly intense storms which may cause erosion during the growing season. This pattern has been described by J.A. Toogood in the paper entitled *Water Erosion in Alberta* (*Journal of Soil and Water Conservation*, v. 18, 1963. pp. 238–40). Snow does represent about 30% of the annual grassland precipitation, but its accumulative interception depends upon surface obstructions such as grass or stubble. This pattern of precipitation has been studied by D.M. Gray and H.W. Steppuhn in their paper entitled *Estimating the Areal Snow Winter Equivalent in the Prairie Environment*. In: *Cold Climate Hydrology: Proceedings of the Canadian Hydrology Symposium, Vancouver, 1979* (Ottawa: National Research Council, 1979. pp. 302–32). Intercepted precipitation will either infiltrate into the soil, or accumulate as pond water. However, some will be lost as runoff, or through evapotranspiration.

Research on infiltration has focused upon the infiltration rate of precipitation on short or tall grasses as well as the effects on infiltration of animal grazing, or controlled burnings of field residues. Some research in this area has been done by F.T. Haupt in his paper entitled *Infiltration, Overland Flow, and Soil Movement on Frozen and Snow Covered Plots* (*Water Resources Research*, v. 3, 1967. pp. 145–61); by G.F. Gifford and R.H. Hawkins in their paper entitled *Hydrologic Impact of Grazing on Infiltration: A Critical Review* (*Water Resources Research*, v. 14, 1978. pp. 305–13); and by H.C. Reeves in the paper entitled *Use of Prescribed Fire in Land Management* (*Journal of Soil and Water Conservation*, v. 32, 1977. pp. 102–4).

It can be noted that the cultivation of grassland soils will generally reduce their ability to be infiltrated, due to an increase in bulk density and the reduction of continuous channels from the soil surface to the subsoil. This has been observed by C.S. Emmond in the paper entitled *Effect of Rotations, Tillage Treatments and Fertilizers on the Aggregation of a Clay Soil (Canadian Journal of Soil Science*, v. 51, 1971. pp. 235–41). When the rate of rainfall, or snowmelt, is greater than the infiltration rate of the soil, water builds up on the soil surface, and an overland flow or runoff will begin. According to F.J. Dragoun, in the paper entitled *Effects of Cultivation and Grass on Surface Runoff* (*Water Resources Research*, v. 5, 1969. pp. 1078–83), runoff losses are greater for cultivated than for untouched grassland fields. Particularly in the drier areas of the prairies, most of the runoff is derived from snowmelt. This feature has been observed by W. Nicholaichuk, in the paper entitled *Comparative Watershed Studies in Southern Saskatchewan* (*Transactions of the American Society of Agricultural Engineers*, v. 10, 1967. pp. 502–4). However, R. DeJong and W. Nicholaichuk, in their paper entitled *Water Retention Equations and Their Relationship to Soil Organic Matter and Particle Size Distribution for Disturbed Samples (Canadian Journal of Soil Science*, v. 63, 1983. pp. 291–302), observed that 'texture was the main soil property influencing the shape and position of the water retention curve' and that 'soil zone and cultivation history had little effect on water retention.' Nevertheless,in a later paper by B.G. McConkey and W. Nicholaichuk entitled *Effects of Fall Subsoiling and Snow Management on Water Conservation and Continuous Spring Wheat Yields in Southwestern Saskatchewan (Canada) (Canadian Agricultural Engineering*, v. 32, 1990. pp. 225–34), the authors assessed a four-year study on the subsoiling on water infiltration. Here, the authors noted that 'subsoiling in the fall to a 35 cm depth substantially increased snowmelt infiltration for the first crop year,' but after that there was not a significant increase, except for the fact that the snowmelt infiltration did reach greater depths.

Finally, the greatest change that has been noted in the studies of grassland hydrology has occurred in the area of evapotranspiration. Native grassland has been shown to use double the amount of precipitation, compared with wheat. However, with field rotations, and summer fallow, up to 85% of the precipitation is lost to evaporation. Some of the work in evapotranspiration has been described by R. De Jong and H. Steppuhn in their paper entitled *Water Conservation: Canadian Prairies*. In: *Dryland Agriculture. Agronomy 23*. Edited by R.E. Dregne and W.O. Willis (Madison, Wisconsin: American Society of Agronomy, 1983. pp. 89–104). Changes have also been noted in grassland hydrology, where there is a combination of irrigation water and boron filtration from soil and alfalfa plants. This combination infiltration, which could create possible toxicity and salinity problems, was studied at 29 irrigation projects in Saskatchewan in the early 1980s. W. Nicholaichuk and A.J. Leyshon, in their paper entitled *Boron and Salinity Survey of Irrigation Projects and the Boron Adsorption Characteristics of Some Saskatchewan (Canada) Soils* (*Canadian Journal of Soil Science*, v. 68, 1988. pp. 77–90), found that 'except for a few locations where crops sensitive to boron may experience some problems, water quality with respect to boron was acceptable for most projects,' and that salinity posed more of a threat to crop production under irrigation than did boron toxicity.

(d) Forest Hydrology

Like the grassland areas of Canada, our forest regions also influence the hydrological regime. The study of forest hydrology in Canada, however, has had a short history. This was commented on by W.W. Jeffrey in his paper entitled *Forest*

Hydrology Research in Canada. In: *Proceedings of the International Symposium on Forest Hydrology*. Edited by W.E. Sopper and H.W. Lule (Oxford: Pergamon Press, 1967, pp. 1–19). The influence of forests on the hydrologic cycle begins with the onset of precipitation, and forest canopies intercepting the amount of rain and snow reaching the ground and altering its distribution. Forests also alter snow accumulation and snowmelt patterns, by sheltering snowpacks from sun and wind. Hence, it is these forest/precipitation interactions that determine the quantity and timing of evaporation and transpiration, infiltration and water storage, and run-off processes.

Snowfall interception and accumulation studies show little research, until the 1990s, and have concentrated upon the development of formulas in order to model 'the estimate of snow interception as a function of leaf area, evergreen canopy coverage, tree species, initial canopy snow load, air temperature, wind speed, time since snowfall and snowfall amount,' as noted by N.R. Hedstrom and J.W. Pomeroy in the paper entitled *Measurements and Modelling of Snow Interception in the Boreal Forest* (*Hydrological Processes*, v. 12, 1998. pp. 1611–25). The authors said that 'the model suggested that interception efficiency is strongly controlled by leaf area and snowfall,' and that 'snowfall interception was much larger than rainfall interception.' The latter comment was based upon a study by J.W. Pomeroy and D.M. Gray entitled *An Evaluation of Snow Accumulation and Ablation Processes for Land Surface Modelling* (*Hydrological Processes*, v. 12, 1998. pp. 2339–67). This modelling work was based upon snow depth equations developed by R.S. McNay and L.D. Peterson in their paper entitled *The Influence of Forest Stand Characteristics on Snow Interception in the Coastal Forests of British Columbia* (*Canadian Journal of Forest Research*, v. 18, 1988. pp. 566–73); and work by J.W. Pomeroy and D.M. Gray entitled *Snowcover Accumulation, Relocation and Management* (Saskatoon: National Hydrology Research Institute, 1995. 144 p. Science Report No. 7).

Related to snowfall interception is the snowmelt under the forest canopy, which is a slower process than in the open environment. J.W. Pomeroy and K. Dion, in their paper entitled *Winter Radiation Extinction and Reflection in a Boreal Pine Canopy: Measurements and Modelling* (*Hydrological Processes*, v. 10, 1996. pp. 1591–1608), noted that within forests the 'daily mean solar angle exerted a strong control on light extinction in a medium density pine canopy in central Saskatchewan,' which in turn affected the energy balance for snowmelt. And J.W. Pomeroy and P.G. Granger in the paper entitled *Sustainability of the Western Canadian Boreal Forest under Changing Hydrological Conditions. I. Snow Accumulation and Ablation*. In: *Sustainability of Water Resources under Increasing Uncertainty*. Edited by D. Rosbjerg *et al.* (Wallingford, Oxfordshire: International Association of Hydrological Sciences, 1997. pp. 237–42. IAHS Publication

No. 240) measured melt energetics in stands of mature pine, mixed aspen-spruce and regenerating pine, and clear-cut, in a Saskatchewan boreal forest, and said that it was 'dependent upon species, leaf area, and removal of canopy as in clear-cutting.' As well, a threefold increase in snowmelt rate upon the removal of a mature boreal forest was inferred by the authors.

The interception of rain by forest canopies is controlled by the type and density of the trees, and by the intensity of the precipitation. In Canada, these interactions have been studied and noted by P.G. McMinn in the publication entitled *Water Relations and Forest Distribution in the Douglas-Fir Region on Vancouver Island* (Ottawa: Department of Agriculture, Forest Biology Division, 1960. 71 p. Publication No. 1091); and by B. Singh and G. Szeicz in their paper entitled *The Effect of Intercepted Rainfall on the Water Balance of a Hardwood Forest* (*Water Resources Research*, v. 15, 1979. pp. 131–38). The amount of precipitation trapped is also dependent on the moisture quality of the precipitation, and the effects of clear-cutting. Clear-cutting, in particular (but this could also apply to insect depletion or forest fires), removes the interception of precipitation by the forest canopy by reducing evaporation processes, and making more water available for runoff process. This phenomena has been described by A.P. Plamondon in the paper entitled *Ecoulement et Modification du Couvert Forestier (Nature Canada*, v. 108, 1981. pp. 289–98). In the 1990s, J.W. Pomeroy *et al.* implemented a rainfall interception model described in the report entitled *Hydrological Pathways in the Prince Albert Model Forest: Final Report to the Prince Albert Model Forest Association* (Saskatoon: National Hydrology Research Institute, 1997. 154 p.). This model was based on interception models developed in England by A.J. Rutter and A.J. Morton and described in the paper entitled *A Predictive Model of Rainfall Interception in Forests. II. Generalization of the Model and Comparison with Observations in Some Coniferous and Hardwood Stands* (*Journal of Applied Ecology*, v. 12, 1975. pp. 367–80); and by R.J. Granger and D.M. Gray in the paper entitled Evaporation from Nonsaturated Surfaces (Journal of Hydrology, v. 111, 1989. pp. 21–29).

According to T.D. Prowse, in the paper entitled *Northern Hydrology: An Overview*. In: *Northern Hydrology: Canadian Perspectives*. Edited by T.D. Prowse and C.S. Ommanney (Saskatoon: National Hydrology Research Institute, 1990. pp. 1–36), evaporation and transpiration processes in Canadian Forests are poorly understood, particularly in the northern areas of Canada. However, a major international project, called BOREAS, involving federal government scientists was carried out over the period 1993 to 1997 in order to study, in part, surface-atmosphere exchanges of water, with an emphasis on evaporation. Overall, the purpose of the project was 'to improve our understanding of basic climatic physical and dynamic processes' for input into climate modelling. Initial results of the project were reported by P.J. Sellers, *et al.*, in the paper entitled *The Boreal*

Ecosystem – Atmosphere Study (BOREAS): An Overview and Early Results from the 1994 Field Year (Bulletin of the American Meteorological Society, v. 76, 1995. pp. 1549–77). They revealed surprisingly small evaporation in forests during the growing season, because of the small root zone of trees in the Boreal Shield ecozone, and the underlying low permeability substrate that encouraged runoff rather than storage. Also, the physiological control exerted by trees led to small transpiration rates. A more complete review of the BOREAS project by P.J. Sellers, *et al.*, entitled *BOREAS in 1997: Experiment Overview, Scientific Results, and Future Directions* (*Journal of Geophysical Research*, v. 102, 1997. pp. 28731–69) also indicated that these low evaporation rates were not presented correctly in most atmosphere-climate models.

Besides interception and evaporation, water infiltration and storage are another important process in forest hydrology. An overview of water storage has been discussed by J.H. McCaughey and B.D. Amiro in the book chapter entitled *Forest Environments*. In: *The Surface Climates of Canada*. Edited by W.G. Bailey and T.R. Oke (Montreal: McGill-Queen's University Press, 1997. pp. 247–76). More specifically, recent work has focused upon water infiltration into frozen soil, the effect of flowpaths on water movement in forest soils, and the examination of soil-water balances. J.W. Pomeroy, *et al.*, in the report *Hydrological Pathways in the Prince Albert Model Forest: Final Report to the Prince Albert Model Forest Association*, mentioned above, found that water entry into frozen soils was not impeded unless there was an ice lens, or coating, on the surface. And J.M. Buttle and D.A. House, in their paper entitled *Spatial Variability of Saturated Hydraulic Conductivity in Shallow Microporous Soil in a Forested Basin* (*Journal of Hydrology*, v. 203, 1997. pp. 127–42), noted that micropores (pores larger than the overall soil matrix) allowed infiltrating water to reach specific soil depths faster, on account of bypassing the soil matrix. Finally, studies on soil-water balance, which has important implications for the forest hydrological processes of evaporation and runoff, have been undertaken by J.A. Elliot and B.M. Toth in their paper entitled *Soil Moisture Storage in Mature and Replanted Sub-Humid Boreal Forest Stands* (*Canadian Journal of Soil Science*, v. 78, 1998. pp. 17–27).

The runoff process itself is also affected by forests, because the process of water movement and storage is affected by the characteristics of the soil mantle, which is highly permeable in forests, due to the accumulations of organic matter, live roots, and decayed root channels. Hence, surface runoff is rare in forested regions, which tend to generate sub-surface into expanded channel systems. These runoff characteristics have been described by A.P. Plamondon and T.A. Block in their paper entitled *The Role of Hydrologic Properties of the Forest Floor in Watershed Hydrology*. In: *Proceedings of a Symposium on Watersheds in Transition* (Urbana, Illinois: American Water Resources Association, 1972. pp. 341–48); by D.G. Cheng and

T.A. Black, in the paper entitled *The Generation of Stormflow from Small Forested Watersheds in the Coast Mountains of Southwestern British Columbia.* In: *Proceedings of the Canadian Hydrology Symposium: 77 Proceedings* (Ottawa: National Research Council, 1977. pp. 542–51); and by J.S. Price and L.K. Hendry in their paper entitled *Water Motion in a Decidous Forest during Snowmelt* (*Journal of Hydrology*, v. 64, 1983. pp. 339–56).

From runoff through surface channels, or groundwater flow, the waterflow can be estimated for a drainage area. Because forests are consumers of water, they reduce water yields, but the latter will increase in a deforested area because of reduction in evapotranspiration by trees. Again, deforestation may be caused by logging, insect infestations, or forest fires. As well, the quality of the water yield will be affected. The interaction between forests and water yields has been studied by D.W. Schindler and R.W. Newbury in their paper entitled *Effects of a Windstorm and a Forest Fire on Chemical Losses from Forested Watersheds and on the Quality of Receiving Streams* (*Canadian Journal of Fisheries and Aquatic Sciences*, v. 37, 1980. pp. 328–34). This paper was part of a series of articles which studied water quality and pollution over the long term in the Experimental Lakes Area of northwestern Ontario. Because the area had been well documented for research studies over a four-year period before accidental windstorm and fire damage in 1973–74, these findings were particularly valid. The Experimental Lakes study is described in detail toward the end of this chapter under water pollution studies.

The most violent runoffs are stormflow, or rapid runoff, or a combination of these. Both forms of runoff are particularly affected by soil water storage capacity. In general, forested areas increase the storage opportunity through canopy interception. Research studies have indicated that stormflow runoff peaks tend to increase where there are smaller storms in small watersheds in a tree harvested area. But, if the severity and duration of the rainfall increases, the proportional increase in peak flow diminishes. These findings were reported by E.D. Heatherington in the paper entitled *Effects of Forest Harvesting on the Hydrologic Regime of Carnation Creek Experimental Watershed. A Preliminary Assessment.* In: *Proceedings of the Canadian Hydrology Symposium* (Ottawa: National Research Council, 1982. pp. 247–67). Further long-term studies of the Carnation Creek watershed have also shown that the effect of logging activities, particularly streamside logging, decreased stream channel stability, and hence organic debris, and caused a decline in fish survival and numbers. These combined hydrological-fisheries-forestry interactions were noted by G.F. Hartman and J.C. Scrivener in their paper entitled *Impacts of Forestry Practices on a Coastal Stream Ecosystem, Carnation Creek, British Columbia (Canada)* (*Canadian Bulletin of Fisheries and Aquatic Sciences*, No. 228, 1990. pp. 1–148). The complexity of the combined hydrological-fisheries-forestry interaction in forest hydrology studies was also observed by R.A. Cunjak and D. Caissie in their report entitled *The Catamaran Brook Habitat Research Project (New Brunswick, Canada): Description and General Design of the Study* (*Canadian Technical Report of Fisheries and Aquatic Sciences*, No. 1751. 1990. pp. 1–15). The problems encountered with human interactions in forest ecosystems are increasing, and also include the problems of acid deposition and climate change. These comments on the future for forest hydrology have been elaborated on by D.W. Schindler in the paper entitled *A Dim View for Boreal Waters and Landscapes* (*Bioscience*, v. 48, 1998. pp. 157–64).

With snowmelt runoff, forested areas, composed of a mosaic of openings and tree cover, generate different streamflow patterns than those arising from a complete forest cover, by creating variable snowmelt runoff. This feature was noted by R.B. Dickison and D.A. Daugharty in their paper entitled *Some Preliminary Results of the Hydrologic Effects of Clearcutting a Small Watershed in Central New Brunswick*. In: *Proceedings of the Fifth Canadian Hydrotechnical Conference* (Fredericton: Canadian Society of Civil Engineers, 1981. pp. 59–75). The mosaic pattern can also be applied to forest edges, because the latter influence the amount of precipitation trapped. This has been observed in the Northwest Territories, where wind-blown snow has accumulations eight to 12 times that of the open tundra. In this vast region, however, the intercepted precipitation accumulates sulphide and nitrate ions that show greater concentrations than in the snow on the tundra landscape. These observations were noted by J.W. Pomeroy and P. Marsh in their paper entitled *Relocation of Major Ions in Snow along the Tundra-Taiga Ecotone* (*Nordic Hydrology*, v. 24, 1993. pp. 151–68).

According to E.D. Heatherington, in the paper entitled *The Importance of Forests in the Hydrological Regime*. In: *Canadian Aquatic Resources*. Edited by M.C. Healey and R.R. Wallace (Ottawa: Department of Fisheries and Oceans, 1987. pp. 179–217. *Canadian Bulletin of Fisheries and Aquatic Sciences*, No. 215), the Canadian research done so far in forest hydrology has tended to modify strongly held views on the benefits of forests in relation to the hydrological regime. For example, forests provide the best cover for minimizing flood flows and damage, but can't protect against floods which are created by meteorological precipitation problems. Or, the notion that the forest floor acts as a sponge during heavy precipitation is inaccurate, because of the limited storage capacity of the forest floor at times of major runoff. Since E.D. Heatherington's review comments were written, another review by J.M. Buttle and L.F. Creed entitled *Advances in Canadian Forest Hydrology, 1995–1998* (*Hydrological Processes*, v. 14, 2000. pp. 1551–78) has been published. In this, the authors said that more forest hydrology research needed to be carried out on the 'hydrological processes in a greater range of forested landscapes' in Canada, the 'dynamics of storm precipitation in mountainous forested landscapes,' the 'contribution of climatic change to natural variations in forest ecosystems,' the application of large-scale data and research to 'small-basin landscape and regional scales,' and the integration of process studies into more detailed simulation models.

(e) Permafrost Hydrology

Like grasslands or the forest regions of Canada, the permafrost region is distinct, and has a unique influence on hydrologic processes. Although the permafrost region includes approximately half the land area of Canada, it supports only a small fraction of the population. As well, economic development is limited and localized, and the need for hydrological data is also limited. However, a variety of major prospects such as highway development, Arctic island pipeline proposals, or hydropower developments have focused current needs for hydrologic data.

In the permafrost region of Canada, the extreme climate may have as significant an effect on northern hydrological processes as does the presence of permafrost itself. For example, the mean annual precipitation is relatively low over much of the region, and this limits the amount of water available for either surface run-off or groundwater discharge. As well, mean annual air temperatures as low as –25° C keep rivers, lakes, and the active layer frozen for up to 10 months a year, and surface runoff and the infiltration of snowmelt, or rain, to recharge groundwater aquifers may be possible only two months per year.

According to R.O. Van Everdingen, in his paper entitled *The Importance of Permafrost in the Hydrological Regime.* In: *Canadian Aquatic Resources.* Edited by M.C. Healey and R.R. Wallace (Ottawa: Department of Fisheries and Oceans, 1987. pp. 243–76. *Canadian Bulletin of Fisheries and Aquatic Sciences,* No. 215), it is natural and man-made disturbances that affect the heat balance in permafrost, and result in the degradation, or aggradation, of permafrost. These in turn give rise to detrimental ground settlement, or local changes in surface, or sub-surface drainage patterns. These pattern changes in turn can further affect the heat balance.

Natural disturbances that can affect the heat balance at the surface of the ground include forest fires, erosion along river banks, lake shores, and coastlines, and the draining of ponds and lakes. For example, natural disturbances affecting streamflow in cold regions, which include permafrost changes, have been studied by T.D. Prowse in the paper entitled *Environmental Significance of Streamflow in Cold Regions* (*Freshwater Biology,* v. 32, 1994. pp. 241–59). Here, the author pointed out that changes in the permafrost, along with other snow and ice masses, are important to streamflow in cold regions, because they play unique roles as major moisture sources that can affect runoff pathways. Permafrost changes can also affect peatland ecosystems, another feature of colder regions, and these changes can be studied through historical time by undertaking peat macro-fossil analyses. S.C. Zoltai, in the paper entitled *Regional Variations in Peatland Ecosystems of West-Central Canada through Time* (*Gunneria*, No. 70, 1995. pp. 35–42), has indicated that south of the western boreal region the evidence of collapse scars has shown that historically several permafrost development and subsistence cycles have taken place, beginning over 3,000 years ago, with the most recent cycle beginning about 150 years ago.

Man-made disturbances of permafrost terrain are related to such activities as petroleum exploration, transportation, mining, and power generation. These types of disturbances lead to degradation of permafrost, and to local changes in drainage patterns. One specific area that has been studied has been the karst-rich area west of Great Bear Lake. (Karst is an uneven limestone topology, which is pitted with depressions caused by the collapse of underground cave-ins.) This area has the potential for future engineering developments, but has risks, because of the seasonal degradation of the permafrost. In the paper entitled *Morphology, Hydrology, and Hydrochemistry of Karst in Permafrost Terrain near Great Bear Lake, Northwest Territories* (Calgary: National Hydrology Research Institute, Inland Waters Directorate, 1981. 53 p. NHRI Paper No. 11, IWDS Scientific Series No. 14), R.O. Van Everdingen has noted that seasonal flooding of karst depressions could cause problems for engineering projects due to the collapse of karst, which is widespread in that region. The tendency of karst to collapse presents a special geotechnical hazard, in addition to the problems related to the ice-rich permafrost of the region. This karst-water system is extremely vulnerable to accidental spills and its rapid sub-surface transport features make containment and recovery difficult or impossible.

Overall, it is considered to be climate change that can be expected to have some consequences unique to the permafrost region. The effects of a general increase in mean annual ground-surface temperatures would probably cause a gradual decrease in the thickness and extent of permafrost, which in turn would cause active-layer failures, thaw slumps, thermokarst developments (sinkholes), and accelerated coastal, lakeshore, and riverbank erosion. All these processes would in turn increase sedimentation in surface waters, affect fish populations, and cause increased wear on turbines in hydropower generators. In addition, thermal erosion of highways could occur, along with the destabilization of pipeline corridors.

These types of potential changes have been studied through models, and reported at international conferences. For example, M.W. Smith and D.W. Riseborough presented a paper entitled *Permafrost Sensitivity to Climatic Change*. In: *Proceedings of the Fourth International Permafrost Conference*. (Washington, DC: National Academy Press, 1983, v. 1, pp. 1178–83) at that conference. More recently, the analyses of soil temperature data at sites near Manner's Creek, on the

Norman Wells pipeline in the Mackenzie Valley, showed that temperature data taken from the disturbed pipeline right-of-way indicated higher soil temperatures than those from the undisturbed adjacent natural vegetation. After applying numerical models to the data, H. Hayhoe and C. Tarnocai in their paper entitled *Effect of Site Disturbance on the Soil Thermal Regime near Fort Simpson, Northwest Territories, Canada* (*Arctic and Alpine Research*, v. 25, 1993. pp. 37–44) suggested that there was a continuing warming trend along the disturbed pipeline right-of-way, which could affect permafrost in the discontinuous permafrost zone. On the other hand, a comparison between permafrost occurrences surveyed along the Mackenzie Highway in 1964 by R.J. Brown and described in the publication entitled *Permafrost Investigations on the Mackenzie Highway in Alberta and Mackenzie District* (Ottawa: National Reseach Council, 1964. 1 v. Division of Building Research Technical Paper No. 175) indicated 'the distribution of permafrost ... is patchy and erratic. It occurs only in low areas. ... No permafrost occurs in high areas.' Surveys taken again in 1988 and 1989 by Y.T. Kwong and T.Y. Gan and published in their paper entitled *Northern Immigration of Permafrost along the Mackenzie Highway and Climate Warming* (*Climatic Change*, v. 26, 1994. pp. 399–419) indicated that 'the fringe of discontinuous permafrost in the region has migrated northward by about 100 km. in the past 26 years.'

At the same time, a detailed analysis of air temperature records for the area from 1949 to 1989 indicated a warming trend implying that the 'resultant increase in mean annual air temperature could readily lead to the observed northward immigration of the permafrost.' The suggestion that climate change could negatively affect permafrost had been suggested previously by M. Smith in the mid-1980s in the paper entitled *The Significance of Climate Change for the Permafrost Environment*. In: *Climate Change Impacts in the Canadian Arctic: Proceedings of a Canadian Climate Program Workshop, March 3–5, 1986, Geneva Park, Ontario*. Edited by H.M. French (Downsview, Ontario: Atmospheric Environment Service, 1986. pp. 67–81). It was discussed again by R.A. Stuart and D.A. Etkin in their publication entitled *Recent Observations of Air Temperature and Snow Depth in the Mackenzie Valley Area and Their Implications on the Stability of Permafrost Layers* (Downsview, Ontario: Atmospheric Environment Service, 1991. 178 p. Canadian Climate Centre Report No. 91-2).

In both the continuous and discontinuous permafrost regions of the Arctic, recent studies have focused upon the movement of groundwater in permafrost. Exfiltration of groundwater from below the permafrost is rare, but has been observed and studied in perennial springs found in western Axel Heiberg Island of the High Arctic by W.H. Pollard and C. Omelon, as noted in their paper entitled *Geomorphic and Hydrologic Characteristics of Perennial Springs on Axel Heiberg Island, Canadian High Arctic*. In: *Permafrost: Seventh International Conference.*

June 23–27, 1998, Proceedings, Yelloknife, Canada. Edited by A.G. Lewkowicz and M. Allard (Quebec: Centre d'Etudes Nordique. Université Laval, 1998. pp. 909–14). Movement of sub-permafrost groundwater has also been noted in discontinuous permafrost areas where taliks (perennial unfrozen zones) provide conduits for the groundwater to move to the surface where icings, or aufeis, in German (sheet-like masses of layered ice), are formed. The growth and decay of these icings in the sub-Arctic Yukon and Northwest Territories has been studied by I.D. Clark and B. Lauriol in their paper entitled *Aufeis of the Firth River Basin, Northern Yukon, Canada: Insights into Permafrost Hydrogeology and Karst* (*Arctic and Alpine Research*, v. 29, 1997. pp. 240–52). More prevalent, however, is the exfiltration of groundwater that resides above the permafrost. This feature has been studied by M.K. Woo and Z.J. Xia in gravelly terrain 'designated as polar desert soil' and in a fen, or wetland area, in the High Arctic, on Cornwallis Island, and described in their paper entitled *Effects of Hydrology on the Thermal Conditions of the Active Layer* (*Nordic Hydrology*, v. 27, 1996. pp. 129–42).

(f) Urban Hydrology

Unlike permafrost, upon which both natural and man-made disturbances may affect the hydrological regime, the urban environment, a totally man-made phenomenon, has uniquely affected the hydrological cycle. This is due to modifications to local climate, increased flooding potential on account of peak flows created by artificial drainage channels and sewer networks, and reduced water quality. As well, in Canada, urban hydrology creates challenges because of the colder climate. This has been discussed by J. Marsalek and G.A. Larkin in their paper entitled *Urban Drainage in Cold Climate: Challenges and Progress in Management Practices* (Burlington, Ontario: National Water Research Institute, 1998. 1 v. NWRI Contribution No. 98-043). The three major areas of concern about water quality changes induced by urbanization are waste water disposal, storm water runoff, and combined sewer overflows.

In Canada, most cities provide some level of treatment for their sewage, with notable exceptions, such as St John's, Halifax, and Quebec City. In newer areas of cities, storm water is collected in separate sewers, and disposed of in local water bodies. However, in older portions of most cities, storm water and domestic sewage are collected in the same sewer system and directed toward a treatment facility. But during rainfall, the capacity of the treatment facility may be overwhelmed, and some sewage discharged into a local water body directly without treatment. Wastewater disposal, and sanitary sewage disposal, are a major concern for most municipalities, because of the large volumes of water involved, and the cost of treatment. In fact, D. Tate and D. Lacelle, in their paper entitled *Municipal*

Water Use in Canada (*Canadian Water Resources Journal*, v. 3, 1978, pp. 61–78), estimated that 60% of Canada's municipal water supplies are returned as wastewater. And, unlike wastewater sewage disposal, stormwater runoff is a non-point source of pollution, and difficult to control. Non-point sources may include airborne materials washed out of the atmosphere by precipitation, metals, rubber, and asbestos fibres from automobiles, animal and bird droppings, fertilizers, and salt.

As mentioned above, one of the methods used to reduce the pollution from urban stormwater discharged into receiving bodies of water is the creation of stormwater detention ponds. Many such structures are operated in the greater Toronto area, and they were first examined by J.B. Dutka and J. Marsalek of the National Water Research Institute at Burlington, in a one-year seasonal study in the early 1990s, in order to gain a better understanding of the ecotoxicological status of sediments of these stormwaters. In their paper entitled *A Seasonal Ecotoxicological Study of Stormwater Ponds* (*Zeitschrift fuer Angewandte Zoologie*, v. 80, 1994. pp. 361–81), the authors said that they found some bacterial and toxicant load, which showed no seasonal variation, at the inlet or outlets of these ponds. As well, there was evidence of the pesticides triazine and metaclor in the suspended sediments.

Inventories of heavy metals and phosphorus in bottom-suspended sediments in stormwater ponds have also been undertaken by T. Mayer and J. Marsalek. In their paper entitled *Nutrients and Metal Contaminants Status of Urban Stormwater Ponds* (*Lake and Reservoir Management*, v. 12, 1996. pp. 348–68), the authors said that the data collected suggested land use had the 'most profound impact on the quality of suspended and deposited sediments, the ponds in an industrial/commercial area catchment having the highest metal concentrations followed by ponds located in the residential catchment.' The data also suggested that the removal of heavy metals in these reservoirs may not be adequate, with concentrations during dry summer conditions, or under winter ice cover, exceeding the levels recommended by the Canada Water Quality Guidelines of 1987 for the protection of aquatic life. One year later, J. Marsalek and W.E. Watt also measured the physical and chemical properties of sediments at the Kingston, Ontario, stormwater pond. The authors found 'gross pollution by Cr, and significant pollution by Cd, Cu, Fe, Pb, Mn, Ni and Zn.' These findings were reported in their paper entitled *Physical and Chemical Characteristics of Sediments from a Stormwater Management Pond* (*Water Quality Research Journal of Canada*, v. 32, 1997. pp. 89–100).

In Canada, considerable research was initially carried out in the 1970s, under the Research Program for the Abatement of Municipal Pollution, which was within the provisions of the Canada-Ontario Agreement on Great Lakes Water Quality. These research reports still provide a central source of information on municipal wastewater and stormwater pollution. Besides specific pollution problems examined, such as phosphorus loading, some of the reports were broad in their approach, and provided overviews of urban waste and stormwater disposal. One report by K.M. Kazan and B.I. Boyko entitled *Identification of Problem Areas in Water Pollution Control Plants* (Ottawa: Environment Canada. Canada-Ontario Agreement on Great Lakes Water Quality, 1973. 41 p. Report No. 15) studied problem areas commonly encountered in wastewater treatment facilities. The major problem identified was hydraulic overloading, which was independent of plant size. A second report by D.H. Waller and Z. Novak entitled *Municipal Pollutant Loadings to the Great Lakes from Ontario Communities* (Ottawa: Environment Canada, Canada-Ontario Agreement on Great Lakes Water Quality, 1974. 40 p. Research Report No. 94) updated and extended previous estimates of total annual pollutant loads to the Great Lakes from Ontario communities. And a third report by W.R. Hutchison and A.P. Galati entitled *Water Plant Waste Discharge in the Province of Ontario* (Ottawa: Environment Canada. Canada-Ontario Agreement on Great Lakes Water Quality, 1978, 135 p. Research Report No. 77) surveyed 105 water treatment plants in Ontario to determine the characteristics of their wastewater discharges.

Twenty years after these reports were written, the Indicators and Assessment Office of Environment Canada has written a major overview of municipal wastewater effluents, entitled *The State of Municipal Wastewater Effluents in Canada* (Ottawa: Environment Canada, 2001. 67 p.). This comprehensive report outlines the sources and the nature of contaminants entering municipal sewer systems, as well as examining how municipal wastewater is managed in Canada.

B Environmental Pollution Studies

Water research involving the federal government is most wide-ranging in studies of pollution in aquatic environments. These, which often involve the full range of federal departments, or international bodies, cover the effect of chemical and metal contaminants in the Great Lakes on freshwater plant and animal life, and long-term studies of bodies of water or whole watersheds in Ontario, Quebec, and the Canadian Arctic.

(a) Great Lakes Pollution Studies

The increased commercial production and widespread use of synthetic chemicals and metals since the 1940s have resulted in the contamination of the Great Lakes Basin. By 1985, the International Joint Commission had identified 11 of the most persistent and widespread toxic contaminants in the Basin. These critical pollutants range from PCBs (polychlorinated biphenyls) to dioxins and furans. Specific studies on the impact of contaminants became numerous by the mid-1960s, and collectively have shown that toxic substances as general metabolic poisons can affect all components of a lake ecosystem, from phytoplankton (marine plant forms), or zooplankton (marine animal forms), to fish, and include biochemical, behavioural, and genetic responses. One example of this type of research, carried out by K.E. Biesinger and G.M. Christensen, was entitled *Effects of Various Metals on Survival, Growth, Reproduction, and Metabolism of* Daphnia magna (*Journal of the Fisheries Research Board of Canada*, v. 29, 1972. pp. 1691–1700). Another example by H.L. Conway was entitled *Sorption of Arsenic and Cadmium and Their Effects on Growth, Micronutrient Utilization of Photosynthetic Pigment Composition of* Asterronella formosa (*Journal of the Fisheries Research Board of Canada*, v. 36, 1979. pp. 579–86). Examples of research involving fish species were done by W.J. Christie and entitled *Changes in the Fish Species Composition of the Great Lakes* (*Journal of the Fisheries Research Board of Canada*, v. 31, 1974. pp. 827–54); and by J.M. McKim entitled *Evaluation of Tests with Early Life Stages of Fish for Predicting Long-Term Toxicity* (*Journal of the Fisheries Research Board of Canada*, v. 32, 1975. pp. 593–98). Another example from the research literature by A. Salki and M. Turner was entitled *The Influence of Fish Zooplankton – Phytoplankton. Interactions on the Results of Selenium Toxicity Experiments within Larger Enclosures (Canadian Journal of Fisheries and Aquatic Sciences*, v. 42, 1985. pp. 1132–43). Other more recent examples of the effects of contaminants on invertebrates and vertebrates have been studies involving fresh-water mussels and the common snapping turtle. K.E. Day and J.L. Metacalfe in their paper entitled *Changes in Intracellular Free Amino Acids in Tissues of the Caged Mussel,* Elleptio complanata *Exposed to Contaminated Environments (Archives of Environmental Contamination and Toxicology*, v. 19, 1990. pp. 816–27) suggested that changed levels of amino acids were indicative of generalized stress induced by environmental factors. Deformities in Great Lakes snapping turtle eggs have been monitored by C.A. Bishop and J.R. Brooks in their paper entitled *The Case for a Cause-Effects Linkage between Environmental Contamination and Development in Eggs of the Common Snapping Turtle* (Chelydra serpentina serpentina) *from Ontario, Canada* (*Journal of Toxicology and Environmental Health*, v. 33, 1991. pp. 521–48). Here, there was a strong association between deformities and levels of organochloride contaminants in eggs. The levels of contamination in turtle eggs, from a range of Great Lakes Basin sites, had been previously documented in the paper by J. Struger and J.E. Elliot entitled *Environmental Contaminants in Eggs of the Common Snapping Turtle* (Chelydra serpentina serpentina) *from the Great Lakes–St. Lawrence River Basin of Ontario, Canada (1981, 1984)* (*Journal of Great Lakes Research*, v. 19, 1993. pp. 681–94).

Some specific sites in the Great Lakes Basin that have undergone particular examination for toxic contamination have been the Hamilton Harbour, the Toronto waterfront, and the Niagara River. Hamilton Harbour is an enclosed body of water situated at the western end of Lake Ontario, and is polluted by municipal and industrial effluents. It is connected to Lake Ontario by a ship canal, which allows for an exchange of harbour and lake water. This exchange has been observed by J. Barica and D.J. Poulton in their paper entitled *Water Exchange between Lake Ontario and Hamilton Harbour (Ontario, Canada) Water Quality Implications* (*Water Pollution Research Journal of Canada*, v. 23, 1988. pp. 213–26), who suggested that the beneficial effect of dilution by Lake Ontario far exceeds the contamination of western Lake Ontario by Hamilton Harbour water. The contamination in the Hamilton Harbour is derived from ammonia and nitrite loadings, as well as toxic hydrocarbons and phosphorus. The ammonia and nitrite loadings come from the Hamilton and Burlington sewage treatment plants. The distribution of these chemicals as un-ionized ammonia, nitrite and nitrate, and dissolved nitrogen were measured in 1987–88 by J. Barica and reported in the paper entitled *Ammonia and Nitrite Contamination of Hamilton Harbour (Ontario, Canada), Lake Ontario (Water Pollution Research Journal of Canada*, v. 25, 1990. pp. 359–86). Ammonia and nitrite concentrations did exceed chronic toxicity thresholds, particularly during the late spring and summer.

PCBs (polychlorinated biphenyls) and PAHs (polynuclear aromatic hydrocarbons) have been measured in the harbour, and the loading of these into Lake Ontario compared with other sources. M.E. Fox and R.M. Khan, in their paper entitled *Loadings of PCBs and PAHs from Hamilton Harbour to Lake Ontario* (*Water Quality Research Journal of Canada*, v. 31, 1996. pp. 593–608), noted that the Hamilton Harbour was actually not a significant source of those compounds into the lake, compared with those coming from the atmosphere or the Niagara River. Nevertheless, treatments to biodegrade these toxic hydrocarbons, including oil, have been undertaken. In 1992 and 1993, 18 tons of calcium nitrate, and five tons of nutrients, were injected into the sediments of the Dofasco Boatslip in the harbour. T. Murphy and A. Moller, in their paper entitled *In Situ Treatment of Hamilton Harbour Sediment* (*Journal of Aquatic Ecosystem Health*, v. 4, 1995. pp. 195–203), noted that the biodegradation of the organic contaminants ranged from 57% to 94%, and suggested that *in situ* treatment of contamination takes time, but with its significantly lower cost relative to dredging and confinement it is a viable alternative.

The reduction of phosphorus loading in the harbour had been previously tackled by the addition of a pickle liquor step at the Hamilton Sewage Treatment Plant. This procedure reduced phosphorus loading by 200 kilograms per day, and consequently, chlorophyl concentrations in the water. The latter reduction added to water clarity, which was beneficial for swimming and fish habitat, identified as goals in the Hamilton Harbour Remedial Action Plan. This phosphorus reduction was reported by S. Painter and L. Hampson in their paper entitled *Hamilton Harbour (Ontario, Canada) Water Clarity Response to Nutrient Abatement* (*Water Pollution Research Journal of Canada*, v. 25, 1990. pp. 351–58). Despite these efforts to improve water quality in Hamilton Harbour, sampling, covering the period 1987 to 1995, 'indicated little recent improvement in the harbour generally,' according to M.N. Charlton and R. LeSage, in their paper entitled *Water Quality Trends in Hamilton Harbour: 1987 to 1995* (*Water Quality Research Journal of Canada*, v. 31, 1996. pp. 473–84).

The Toronto waterfront is another body of water on the Great Lakes that has been documented for contaminants. This was done in the spring, summer, and fall of 1987, and concentrations of 89 toxic organic pollutants, which included 25 chlorobenzenes, and pesticides, and 64 PCB isomers were recorded. The data also indicated that Humber Bay, the inner harbour, and areas near the Toronto Main Sewage Treatment Plant were the most polluted. However, E. Halfon and D.J. Poulton, in their paper entitled *Distribution of Chlorobenzenes, Pesticides and PCB Congeners in Lake Ontario near the Toronto Waterfront* (*Water Pollution Research Journal of Canada*, v. 27, 1992. pp. 751–72), pointed out that Lake Ontario receives large amounts of pollutants from atmospheric sources and the Niagara River, and even if all local sources of pollution were removed, the Toronto waterfront area would probably remain affected by other sources, primarily the Niagara River.

The Niagara River is a source point for chlorinated hydrocarbons, particularly its surface layers. This has been documented by R.J. Maguire and R.J. Tkacz in their paper entitled *Chlorinated Hydrocarbons in the Surface Microlayer and Subsurface Water of the Niagara River (Ontario, Canada), 1985–86)* (*Water Pollution Research Journal of Canada*, v. 23, 1988. pp. 292–300). These authors found that the surface microlayer of the Niagara River, at Niagara-on-the-Lake, contained PCBs, chlorobenzenes, and chlorinated hydrocarbons at concentrations generally up to 40 times greater than concentrations in subsurface water. However, 'despite such high concentrations of chlorinated hydrocarbons in the surface microlayers at no time during this study did the microlayer contribute significantly, relative to subsurface water to the loading (i.e., amounts) of these chemicals from the Niagara River to Lake Ontario.' Another group of highly toxic substances that have also been found in the Niagara River are radionuclides, which include plutonium and its atomic derivatives, such as 238, 239, 240, and 241 Pu. S.R. Joshi, in the paper entitled *West Valley Plutonium and Americium-241 in Lake Ontario Sediments off the Mouth of Niagara River (Canada)* (*Water, Air and Soil Pollution*, v. 42, 1988. pp. 159–68), undertook sediment core radionuclide inventories of this area, which confimed the sources were from global fallout, as well as low-level releases of these radionuclides from the West Valley radioactive waste management site.

This on-going research work, to date, of contaminants in the Great Lakes Basin has been surveyed in a two-volume publication issued by Environment Canada, the Department of Fisheries and Oceans, and Health and Welfare Canada, and entitled *Toxic Chemicals in the Great Lakes and Associated Effects* (Toronto: Environment Canada, 1991. 2 v.). Volume I of this report summarized data on the concentrations of toxic chemicals in water and sediments, invertebrates and fish, wildlife, and humans. Volume II reviewed what is known about the effects of persistent toxic chemicals on fish, wildlife, and people. Both volumes contain considerable bibliographic references. This work was summarized for public reading as *Toxic Chemicals in the Great Lakes and Associated Effects: Synopsis* (Ottawa: Supply and Services Canada, 1991. 51 p.).

Since this major review, studies have continued on the contamination of the Great Lakes Basin, as noted in the paper by J.M. Azcue and F. Rosa entitled *Distribution of Major and Trace Elements in Sediments and Pore Water of Lake Erie* (*Journal of Great Lakes Research*, v. 27, 1996. pp. 389–402). Parallel to these continuing studies has been a growing interest in the use of mathematical modelling for large-scale contaminant prediction studies, as well as specific modelling of particular contaminants. Large-scale modelling of Great Lakes processes has been undertaken by C.R. Murthy and W.M. Schertzer of the National Hydrology Research Institute. Their first paper on this subject was entitled *Physical Limnology and Water Quality Modelling of North American Great Lakes: Part I. Physical Processes* (*Water Pollution Research Journal of Canada*, v. 29, 1994. pp. 129–56). This paper described the physical limnological research carried out on the coastal boundary-layer characteristics and currents, coastal circulation characteristics, and turbulence diffusion characteristics of Lake Ontario, based on large-scale experiments. Their second paper was entitled *Physical Limnology and Water Quality Modelling of North American Great Lakes: Part II. Water Quality Limnology* (*Water Pollution Research Journal of Canada*, v. 29, 1994. pp. 157–84). This paper emphasized that the incorporation of physical limnological processes (previously described) within the water quality/contaminant modelling framework of large lake systems is fundamental for the accurate simulation of the transport, distribution, and pathways of pollutants.

Small-scale modelling on specific contaminants has been undertaken on lead sedimentation in Lake Ontario. S.R. Joshi and B.S. Shukla, in their paper entitled *Lead-210 Sedimentation in Lake Ontario* (*Environmental Geology and Water Sciences*, v. 19, 1992. pp. 121–26), derived the sedimentation rates, and diffusive sediment mixing coefficients, at several Lake Ontario sites, from measurements of lead-210 profiles in sediment cores. Further work in this field of study was carried out by A.G. Bobba and R.A. Bourbonniere in their paper entitled *Numerical Simulation Model of Fatty Acids in Lake Sediments* (*Water, Air and Soil Pollution*, v. 89, 1996. pp. 77–90). These authors developed a finite element model to predict the Pb-210 (lead) and total extractable fatty acid concentrations at different depths, in two sediment cores from Lake Ontario, by using different transport parameters. These parameters were used to simulate the total extractable fatty acid concentrations at different depths. Then, the authors compared the computed results with observed data, and the results were also compared by statistical methods and 'good agreement was achieved.' They also said that by modifying this model it could be applied to contaminant transport in lake sediments, industrial wasteponds, and fish ponds, with different geological, physical, chemical, and biological parameters at different depths.

(b) Long-Term Pollution Studies

Besides the many research papers that continue to be written on contamination in the Great Lakes Basin, there have been other major large-scale research projects that have examined specific pollution problems. One was a 12-year study (1972–84) called Project Quinte, which studied the effect of phosphorus loading in the Bay of Quinte region of eastern Lake Ontario. A second study, undertaken from 1968 to 1987, involved the Experimental Lakes Area that allowed for large-scale eutrophication and acidification studies. A third was the eight-year (1980–88) Turkey Lakes Watershed (TLW) study, undertaken to obtain a better understanding of acid precipitation. And a fourth study, pertaining to Quebec lakes, was undertaken from 1981 to 1985. The most recent of these large-scale studies has been the Slave River Environmental Quality Monitoring Program (1990–94).

i. Project Quinte Study The reason for the Bay of Quinte study was the fact that by 1970, phosphorus loading in the Great Lakes was a concern to the International Joint Commission. Phosphorus loading from municipal sewage treatment plants had escalated from 1940 onward in this Lake Ontario area, and by the 1950s had caused a deterioration of water quality, which affected water supplies and sport and commercial fishing. This deterioration, or eutrophication, was caused by the loss of oxygen and nutrient budgets due to the increase and density of phytoplankton and zooplankton in the bay. By the mid-1960s, public outcry and government attention became focused upon this problem, and by 1970, a campaign to reduce phosphorus loading began with regulated decreases in phosphorus content in detergents. In 1970, the International Joint Commission recommended further reductions of the remaining phosphorus in municipal and industrial waste effluent discharging into Lake Erie, Lake Ontario, and their tributaries, and the international section of the St Lawrence River.

In 1977, both the Canadian and US governments' commitments were formalized in the Canada-US Great Lakes Water Quality Agreement, and a five-year program commenced in order to reduce industrial and municipal inputs of phosphorus, and to investigate further the significance of non-point sources. This agreement was renewed in 1978, and there was a reduction in phosphorus loading and an improvement in water quality. Nevertheless, even in the early 1970s, there had been no case studies undertaken to show that improved water quality could be achieved by phosphorus reductions. The arguments for phosphorus removal had been based on laboratory experiments only. Hence, the Bay of Quinte, which was a small water basin in eastern Lake Ontario, was ideal for case histories and studies to be undertaken by fisheries scientists and water quality managers to validate, refute, or clarify previously held assumptions about phosphorus loading in the Great Lakes Basin.

Studies were carried out over a 14-year period and, overall, 23 research papers, edited by C.K. Minns *et al.*, were published under the title *Project Quinte: Point-Source Phosphorus Control and Ecosystem Response in the Bay of Quinte, Lake Ontario* (Ottawa: Department of Fisheries and Oceans, 1986. 270 p. Special Publication No. 86). Within this collection, there were overview papers, as well as specific biological studies. Overview papers were by P.G. Sly, entitled *Review of Postglacial Environmental Changes and Cultural Impacts in the Bay of Quinte* (pp. 7–26); by C.K. Minns and G.E. Owen, entitled *Nutrient Loads and Budgets in the Bay of Quinte, Lake Ontario, 1965–1981* (pp. 59–76); and by C.K. Minns, entitled *Total Phosphorus Loading to the Bay of Quinte between the Years 1800 and 2000: A Model Study* (pp. 77–83). The individual biological studies included phytoplankton, zooplankton, and fish, and were of particular interest as they showed the complexity within previously held assumptions about phosphorus loading. K.H. Nicholls and L. Heintsch, in their paper entitled *Some Effects of Phosphorus Loading Reduction on Phytoplankton in the Bay of Quinte, Lake Ontario* (pp. 145–48), said that there was a significantly lower seasonal (May–October) biomass, as well as a shortening of the periods of growth of bloom-forming species. Also, the altered phytoplankton communities were potentially more functional in supporting the aquatic food chain, and the algal related benefits of less severe odour problems and filter clogging at municipal water treatment facilities were apparent after 1977.

The paper by M.G. Johnson and O.C. McNeil entitled *Changes in Abundance and Species Composition in Benthic Macroinvertebrate Communities of the Bay of Quinte, 1966–84* (pp. 177–200) indicated that there was a general decline in numbers and biomass of oligochaetes, sphaeriids, isopods, and chironomids between 1977 and 1984. This was probably related to the decline in productivity in the bay, in response to phosphorus input reductions. D.A. Hurley, in the paper entitled *Fish Populations of the Bay of Quinte, Lake Ontario, before and after Phosphorus Control* (pp. 201–22), said that some of the major species such as alewife (*Alosa pseudoharengers*), white perch (*Morone americana*), yellow perch (*Perca flavescens*), and walleye (*Stizostedion vitreum vitreum*) showed significant shifts in abundance that were evidently more related to a climatic anomaly, and to predator-prey interactions, than to changes in nutrient concentrations. In effect, the fish communities of the bay had changed from a percicthyid (white perch)–clapeid (alewife) dominated one of the 1960s and 1970s, to one where percids (yellow perch, walleye) were the more abundant species. The water quality studies also showed interesting contrasts. C.K. Minns, in the paper entitled *Temperature and Oxygen Conditions and Oxygen Depletion in the Bay of Quinte, Lake Ontario* (pp. 40–49), said that oxygen depletion did not respond to decreased point-source phosphorus loadings. G.W. Robinson, in the paper entitled *Water Quality of the Bay of Quinte, Lake Ontario, before and after Reductions in Phosphorus Loading* (pp. 50–58), said that water quality improvements were immediately apparent in the Bay of Quinte, following reductions in phosphorus loadings from local sewage treatment plants. However, there was no change in total phosphorus levels, but soluble reactive phosphorus and chlorophyll-a concentrations decreased.

Since this major study was published in 1986, the federal government has put out a major overview of nutrient loading in the environment, entitled *Nutrients and Their Impact on the Canadian Environment*. Edited by P. Chambers, *et al.* (Ottawa: Environment Canada, 2001. 241 p.). This documents 'the nature and extent to which nutrients derived from human activity may be impairing Canadian ecosystems and affecting the quality of life and health of Canadians.'

ii. Experimental Lakes Area Study In the late 1960s, the first long-term studies undertaken by fisheries scientists to alter whole water bodies, or enclosures, for research purposes were undertaken. In 1968, the federal government, and the government of Ontario, agreed to the establishment of the Experimental Lakes Area (ELA) in northwestern Ontario. The purpose of designating this area was to permit experimentation on a whole-lake scale to resolve questions relating to eutrophication, the effects of acidification, pollution with heavy metals or radionuclides, clear-cut logging, and netrilotriacetic acid (NTA).

The initial research studies, 14 in all, were published in 1971, and covered surveys and documentation of the physical and chemical characteristics of the lakes. Examples of this type of work were published by T.R. Cleugh and B.W. Hauser and entitled *Results of the Initial Survey of the Experimental Lakes Area, Northwestern Ontario* (*Journal of the Fisheries Research Board of Canada*, v. 28,

1971. pp. 129–37); by G.J. Brunskill and D.W. Schindler, entitled *Geography and Bathometry of Selected Lake Basins, Experimental Lakes Area, Northwestern Ontario* (*Journal of the Fisheries Research Board of Canada*, v. 28, 1971. pp. 139–55; and by F.A. Armstrong and D.W. Schindler, entitled *Preliminary Chemical Characterization of Waters in the Experimental Lakes Area, Northwestern Ontario* (*Journal of the Fisheries Research Board of Canada*, v. 28, 1971. pp. 171–87).

The first experimental results, consisting of 13 papers, appeared in 1973, and concentrated upon the dynamics of a lake's response to nutrient enrichment. Some of these were by D.W. Schindler and H. Kling and entitled *Eutrophication of Lake 227 by Addition of Phosphate and Nitrate: The Second, Third and Fourth Year Enrichment, 1970, 1971, and 1972* (Journal of the Fisheries Research Board of Canada, v. 30, 1973. pp. 1415–40).

The second group of results, consisting of 20 papers, appeared in 1980, and were considered a justification for supporting long-term interdisciplinary research. In fact, it was thought that these research papers would serve as standard reference materials in eutrophication and acid-precipitation for limnologists for years to come. One example of the titles published in this group was by D.W. Schindler and R. Wagemann and entitled *Experimental Acidification of Lake 223, Experimental Lakes Area: Background Data and the First Three Years of Acidification* (*Canadian Journal of Fisheries and Aquatic Sciences*, v. 37, 1980. pp. 342–54). Other papers from this group were by J.S. Marshall and D.L. Millenger and entitled *Dynamics of Cadmium-Stressed Plankton Communities (Canadian Journal of Fisheries and Aquatic Sciences*, v. 37, 1980. pp. 403–14); and by R.J. Flett and D.W. Schindler, entitled *Nitrogen Fixation in Canadian Shield Precambrian Shield Lakes* (*Canadian Journal of Fisheries and Aquatic Sciences*, v. 37, 1980. pp. 494–505).

The fourth series of research papers, consisting of 14 articles, was published in 1987. Again, this focused upon eutrophication and acidification studies, and emphasized the importance of long-term whole-lake experimental studies, and long-term interpretation. An example of papers from this group is by D.W. Schindler and R.H. Hesslein, and entitled *Exchange of Nutrients between Sediments and Water after 15 Years of Experimental Acidification* (*Canadian Journal of Fisheries and Aquatic Sciences*, v. 44, Suppl. 1, 1987. pp. 26–33).

iii. Turkey Lakes Watershed Study The Turkey Lakes Watershed study has involved a series of four lakes on provincial crown land, situated northeast of and between Lakes Superior and Huron. This area receives a considerable amount of atmospheric pollution that contributes to acid precipitation in the watershed. TLW was chosen in order to provide an in-depth analysis of the biogeochemical processes at work in a relatively undisturbed basin, so that system models could be developed and evaluated to give a better understanding of the aquatic and terrestrial effects of acidic atmospheric deposition. This study is an example of the descriptive, analytical, and biological studies undertaken cooperatively by the National Water Research Institute, the Atmospheric Environment Service, the Canadian Forestry Service, and the Department of Fisheries and Oceans.

Initially, the physical, chemical, and biological characteristics of the watershed were studied and documented, and sulphate and nitrate atmospheric deposition was estimated and analysed. Bulk, wet, or wet-dry precipitation was compared for acidic deposition and its chemical make-up determined. Trace metal distribution was studied. The fish community and its health were studied and documented. In all, 21 papers were written, and issued as an untitled supplement to the *Canadian Journal of Fisheries and Aquatic Sciences*, v. 45, 1988. Supplement No. 1. The first of the descriptive studies by D.S. Jeffries and J.R. Kelso, entitled *Physical, Chemical, and Biological Characteristics of the Turkey Lakes Watershed, Central Ontario, Canada* (*Canadian Journal of Fisheries and Aquatic Sciences*, v. 45, 1988. Supplement No. 1, pp. 3–13), pointed out that 'the Turkey Lakes Watershed (TLW) in central Ontario was selected for intensive research into the effects of the long-range transport of air pollutants (primarily acidic deposition) at a site on the Canadian Shield having both vulnerable terrain and an undisturbed Great Lakes forest type.' Also, the area was well removed from local sources of pollution, and receives mainly long-range transported pollutants.

Another descriptive study by R.G. Semkin and D.S. Jeffries was entitled *Chemistry of Atmospheric Deposition, the Snowpack, and Snowmelt in the Turkey Lakes Watershed* (*Canadian Journal of Fisheries and Aquatic Sciences*, v. 45, 1988. Supplement No. 1, pp. 38–46). Here, the chemistry of the snowpack, which had been previously studied and shown to store $H+$, $SO4–2–$, $NO3–$, and $NH4+$ ions, was reviewed in 1986 and described. The ion forms of these chemical elements and compounds determine the chemistry of the snowmelt and its impact on lake and stream waters. One other descriptive study by D. Craig and L.M. Johnston was entitled *Acid Precipitation and Groundwater Chemistry at the Turkey Lakes Watershed* (*Canadian Journal of Fisheries and Aquatic Sciences*, v. 45, 1988. Supplement No. 1, pp. 59–65). Here, the groundwater chemistry of the watershed was studied in order to describe 'the dominant weathering reactions and processes that govern the chemistry, particularly as they relate to the problem of acid rain.' Overall, it was found that the groundwaters of the Turkey Lakes Watershed are well buffered, and, in general, provide a reservoir of alkalinity for surface water in the basin.

Analytical papers included studies of the ion mass budgets for the lakes, water, and chemical budgets for the land, and the mechanisms of acid deposition and nutrient leaching. D.S. Jeffries and R.G. Semkin, in their paper entitled *Ion Mass*

Budgets for Lakes in the Turkey Lakes Watershed (*Canadian Journal of Fisheries and Aquatic Sciences*, v. 45, 1988. Supplement No. 1. pp. 47–58), estimated and analysed the input pathway for H+, NH4+, and NO3– ions into the four lakes of the watershed, and showed that these ions are retained by the lakes to varying degrees. However, the acid-neutralizing capacity (ANC) of the lakes showed excess output over estimated inputs. J.A. Nicolson, in the paper entitled *Water and Chemical Budgets for Terrestrial Basins at the Turkey Lakes Watershed* (*Canadian Journal of Fisheries and Aquatic Sciences*, v. 45, 1988. Supplement No. 1. pp. 88–95), monitored 20 terrestrial basins in order to measure discharge and ion loss from the terrestrial ecosystem, and to estimate terrestrial contributions to the main aquatic system of the watershed. This estimate would provide a baseline for assessing the atmosphere input and terrestrial basin influences in the watershed.

The biological studies included the zoobenthic (bottom-feeding invertebrates) and fish communities in the four lakes. D.M. McDermott, in the paper entitled *Zoobenthic Distributions and Biomass in the Turkey Lakes* (*Canadian Journal of Fisheries and Aquatic Sciences*, v. 45, 1988. Supplement No. 1. pp. 107–14), noted that the benthic fauna was dependent on the absence of fish (the uppermost lake – Batchawana – had no fish populations), the lake depths, and the oxygen concentrations in the water. Lake Batchawana had the greatest biomass, and amphipods and predatory insects dominated. However, the lake had the lowest benthic populations, because of oxygen depletion at low levels. On the other hand, Little Turkey Lake had the greatest biomass, by estimated weight, because of a large population of chenonomids (midges). The fish populations were studied by J.R. Kelso in the paper entitled *Fish Community Structure, Biomass, and Production in the Turkey Lakes Watershed, Ontario* (*Canadian Journal of Fisheries and Aquatic Sciences*, v. 45, 1988. Supplement No. 1. pp. 115–20). Again, the author noted the lack of fish stock in the headwater lake (Batchawana), but that salmonids increased with progression into the downstream lakes. Also, 'fish biomass and production in the watershed were strongly influenced by lake depth, but alkalinity and phytoplankton carbon assimilation also were relative to stock and production.'

Since this major basic study of the Turkey Lakes Watershed was completed, ongoing research has been undertaken on hydrological modelling of the watershed, as well as forest simulation studies. B.G. Ghosh and D.C. Lam, in their paper entitled *Hydrological Modelling of Acidified Canadian Watershed* (*Ecological Modelling*, v. 50, 1990. pp. 5–32), used a hydrological model called the Turkey-Mersey Watershed Acidification Model (TMWAM), and applied it to the Turkey Lakes Watershed. Results using the model were 'presented on the calibration and confirmation of the model using data from these watersheds.' Particular emphasis was placed on the hydrological factors, and their relationships to the varying degrees of acidification observed at these locations. Nevertheless, 10 years later,

B.A. Ghosh and V.P. Singh, in their paper entitled *Application of Environmental Models to Different Hydrological Systems* (*Ecological Modelling*, v. 125, 2000. pp. 15–49), said that 'there exist uncertainties in environmental models due to imperfect knowledge of processes controlling water quality parameters as well as errors in data.' Simulation studies of acid rain on maple woods at the Turkey Lakes Watershed were undertaken by C.M. Thirukkumaran and I.K. Morrison in their paper entitled *Impact of Simulated Acid Rain on Microbial Respiration, Biomass, and Metabolic Quotient in a Mature Sugar Maple* (Acer saccharum) *Forest Floor (Canadian Journal of Forest Research*, v. 26, 1996. pp. 1446–53). The authors said that 'the observed deleterious effects of H_2SO_4 application under experimental conditions suggest the possibility of adverse effects in the field over the long-term.'

iv. Quebec Lakes Study Acid precipitation on the Quebec side of the Canadian Shield north of the St Lawrence River has been monitored since the mid- to late 1980s. In Quebec, 54 lakes were studied for their concentrations of freshwater zooplankton, natural morphometry, water quality, and acidification. It was found that five factors seemed to affect the extent of acidification, which in turn affected the size of zooplankton communities. These factors were natural alkalinity, lake depth and size, colour, and sulphate and nitrate concentrations in the water. The most vulnerable lakes were located above the eastern north shore, and the lakes with the lowest acidification were located in the western region of Quebec, where there was high sulphate and high mineralization in the water. The monitoring was done over a four-year period and the results published in 1987 and entitled *Effects of Acid Precipitation on Lake and River Ecosystems in Quebec: Review of Department of Fisheries and Oceans Research Activities (1981–1985)* (Technical Report No. 1554, 1987. 65 p.). The analysis of the data was published by B. Pinel-Alloul and G. Verreault and entitled *Zooplankton Species in Quebec Lakes: Variation with Abiotic Factors, Including Natural and Anthropogenic Acidification* (*Canadian Journal of Fisheries and Aquatic Sciences*, v. 47, 1990. pp. 110–21). Thirty-eight zooplankton species were subjected to 'cluster and correlation analyses to determine characteristic zooplankton associations, relative importance of abiotic variables like typology and relationships behaviour, the integrated environmental factors derived from lake morphology, water quality, and acidification levels and the structure of zooplankton communities.' The authors found that the community structure, 'whether expressed in class sizes or in trophic groups, tends to vary among sets of lakes of different pH.'

v. Slave River Environmental Quality Monitoring Program The most recent major watershed monitoring program has been the Slave River Environmental Quality Monitoring Program, established in 1990 for a five-year period, to assess the suspended sediment quality in the Slave River Watershed, at the Alberta–Northwest Territories border. This watershed drains a large area, and is the recipient of chemical compounds from various aquatic sources, including the Athabasca and Peace Rivers. Hence, the potential for contamination from industrial and agricultural processes, upstream, is very real. Accordingly, a baseline data set, with which to determine present and future effects from upstream activities and long-range transport of contaminants, was created. Included in the baseline set for detection were polyaromatic hydrocarbons (PAHs), polychlorinated biphenyls (PCBs), and chlorinated phenolics (CPs). The first study by L.H. McCarthy and T.G. Williams was entitled *Baseline Studies in the Slave River, NWT, 1990–1994. Part I. Evaluation of the Chemical Quality of Water and Suspended Sediment from the Slave River (NWT)* (*Science of the Total Environment*, v. 197, 1997. pp. 21–53). This initial study found that there were measurable levels of metals in the water, but they occurred naturally, and not from industrial sources. Also, the chemical concentrations of contaminants in the suspended sediment were generally very low, or below analytical detection. The authors said that 'it can be included from the extensive data set collected during the five-year program that impact from upstream industrial and agricultural sources is at the present time negligible, and that the environmental quality of the Slave River at Fort Smith is pristine.'

The second part of the study was to ascertain whether the fish in the Slave River were safe to eat. This study was undertaken by L.H. McCarthy and G.R. Stephens and entitled *Baseline Studies in the Slave River, NWT, 1990–1994. Part II. Body Burden Contaminants in Whole Fish Tissue and Livers* (*Science of the Total Environment*, v. 197, 1997. pp. 55–86). Here, the authors found that the whole tissue of fish (muscle) was fit for human consumption, as toxic contaminants and heavy metals detected in tissue were all below federal fish consumption advisories. However, toxophene levels in burbot fish livers were constantly above fish consumption advisories, and 'should continue to be monitored as these are eaten extensively by the native peoples.' The third part of the study was biochemical in its scope, and studied enzyme activity in fish. This was carried out by T.G. Williams and W.L. Lockhart and entitled *Baseline Studies in the Slave River, NWT, 1990–1994. Part III. MFO Enzyme Activities in Fish (Science of the Total Environment*, v. 197, 1997. pp. 87–109). The authors undertook hepatic MFO (Mixed Function Oxidase) enzyme analyses on walleye, northern pike, lake whitefish, and burbot. They found that MFO enzyme activity indicated that some differences were evident in fish sampled, and in many instances no differences were observed. Again, this biochemical study reinforced the suggestion of a relatively pristine environment.

The last part of the study by L.H. McCarthy and K. Robertson was entitled

Baseline Studies in the Slave River, NWT, 1990–1994. Part IV. Evaluation of Benthic Invertebrate Populations and Stable Isotope Analyses (*Science of the Total Environment*, v. 197, 1997. pp. 111–25). The purpose of this was to assess benthic invertebrate populations useful for biomonitoring purposes alone, as well as to measure isotope ratios of carbon, sulphur, and nitrogen, in the Slave River fish. Ecologically, these types of measurements can be used to determine the feeding and migration pattern of fish, and to define their trophic (nutrition) position in the food chain. From a toxicological point of view, these stable isotope ratios can help explain contaminant sources and transfer through the food chain. The benthic invertebrate survey itself provided a baseline for future population analyses, and analyses of the stable isotope ratios of carbon, sulphur, and nitrogen in fish from the Slave River could help explain contaminant sources and transfer through the food chain. In particular, the stable isotope ratio of sulphur indicated at least two significant food sources – one probably the Great Slave Lake, and the other the Athabasca and Peace Rivers.

C Oceanographic Studies

Besides inland waters studies, there are three newer areas of study, which are carried out by the Department of Fisheries and Oceans. These are oceanographic studies, which break down into physical oceanography, chemical oceanography, and biological oceanography. Canada borders three oceans, the Atlantic, the Pacific, and the Arctic, and the currents and masses of these oceans have a great influence on weather and climate, and on the oceanic conditions that prevail over the continental shelves. Of these oceans, the Arctic, which is the smallest of the world's oceans, is the most poorly studied. The physical oceanography of these water masses in Canada encompasses the study of tides and tidal currents, surface waves and wave forecasting, air-sea interactions, small-scale processes such as turbulence, internal waves, thermal micro-structure, dynamics and currents of the deep sea, and the continental shelf, as well as the dynamics and currents of ocean inlets and channels.

Two reviews have been written about the state of physical oceanography in Canada. The first, by N.J. Campbell, was entitled *An Historical Sketch of Physical Oceanography in Canada* (*Journal of the Fisheries Research Board of Canada*, v. 33, 1976. pp. 2155–67). The second, by C.R. Mann, was entitled *Physical Oceanography in Canada* (*Journal of the Fisheries Research Board of Canada*, v. 33, 1976. pp. 1412–33). As well, R.E. Thomson, in the publication entitled *Oceanography of the British Columbia Coast* (Ottawa: Department of Fisheries and Oceans, 1981. 291 p. Canadian Special Publication of Fisheries and Aquatic Sciences No. 56), has written a specific overview of the physical oceanography of the whole Canadian west coast. Chemical oceanography is the most recently developed of the oceanographic sciences in Canada. It is concerned, principally, with the physical/chemical processes that occur in water columns, or sea ice. However, chemical oceanography may also include studies in the area of sedimentary geochemistry. The state of chemical oceanography in Canada has been described by C.R. Mann in the paper entitled *Chemical Oceanography in Canada (Journal of the Fisheries Research Board of Canada*, v. 33, 1976. pp. 1452–09). Finally, biological oceanography, in its broadest form, is the study of the biology of the oceans but usually applies to research of zooplankton, phytoplankton, or benthic (fauna) bottom-level research.

(a) Physical Oceanography

The tidal rise and fall of the sea in bays and harbours has been observed since early times, because it influences all activities along shorelines. More recently, it has been taken into account in the design of port facilities, the operation of pleasure craft, and fishing operations. The times of high and low water can be predicted accurately, as the major tide-generating forces – the gravitational attraction of the moon and sun – are periodic. Although the time of high and low water at different places along a coast depends of the geometry of the coast, gauges to measure the elevation of the sea surface are installed where tidal predictions are required. In Canada, these observations, as previously described, are carried out by the Canadian Hydrographic Service. The major knowledge gap in the basic knowledge of tides is the understanding of tides of the deep ocean, but there has not been a great amount of attention given to the tides on continental shelves, either. In Canada, surface studies have focused on coastal tides, and currents either in numerical modelling, or the study of inter-tidal environments. As well, tsunamis, storm surges, and waves have been examined.

i. Tidal Studies At the Atlantic Oceanographic Laboratory of the Bedford Institute of Oceanography, K.T. Lee has worked on methods of computing three-dimensional, tidally induced residual currents. His first paper, entitled *The Structure of Three Dimensional Tide-Generating Current. Part I. Oscillating Current* (*Journal of Physical Oceanography*, v. 9, 1979. pp. 930–44), examined the three-dimensional computing of first-order tidally induced residual current. His second paper, entitled *The Structure of Three-Dimensional Tide-Induced Current. Part II. Residual Currents (Journal of Physical Oceanography*, v. 10, 1980. pp. 2035–57), examined the three-dimensional computing of second-order tidally induced residual currents. These currents are of interest to environmental researchers because of the net transport of various materials in coastal waters, such as air pollutants, sediments, nutrients, and sea ice. And, in 1982, these theoretical models were verified by K.T. Lee in the paper entitled *The Structure of 3-Dimensional Tide-Generating Currents: Experimental Verification of a Theoretical Model* (*Estuarine Coastal and Shelf Science*, v. 14, 1982. pp. 27–48). To examine the validity of the theoretical model, 12 current meters were moored at three locations, at the head of the Bay of Fundy. The author said that 'by comparing the analyzed coefficients with [his] 1979 simple model and by using realistic forms of the vertical eddy viscosity, the simple theoretical model reproduced the observed vertical variation of the tidal currents in the well mixed estuary reasonably well.'

At the Centre for Inland Waters in Burlington, renewed interest has been shown in the tides of the Hudson Bay–James Bay system, because of possible estuarine changes resulting from proposed freshwater discharge modifications following the announcement, in the early 1970s, of the hydroelectric power development of the northern Quebec rivers flowing into James Bay. In order to assess the levels of tidal mixing, and water transport, it was necessary to develop a more complete understanding of tidal propagations than was presently available from the analysis of observed data.

T.H. Manning and R.W. Rae first described the tidal regime along the southern Hudson Bay and western James Bay coast in their paper entitled *Tidal Observations in Arctic Waters* (*Arctic*, v. 3, 1950. pp. 95–100). This, and other work, was largely the result of the analysis and extrapolation of shore-based gauge observations and theoretical treatment of the data. However, N. G. Freeman and T.S. Murty, in their paper *Numerical Modelling of Tides in Hudson Bay* (*Journal of the Fisheries Research Board of Canada*, v. 33, 1976. pp. 2345–61) went further, and created a two-dimensional numerical model to study the co-oscillating and independent tides in Hudson Bay. Comparison of the model with previous work and shore-based observations showed the model was in overall agreement, and could be used to interpret tidal propagation in the Hudson–James Bay system.

In the 1980s, tidal studies of Hudson and James Bay noted the effect of ice cover on the tidal regime. G. Godin and F.G. Barker, in their paper entitled *Variability of the Tide and Some Sites in the Canadian Arctic* (*Arctic*, v. 33, 1980. pp. 30–37), said that 'there can be little doubt ... that in some parts of the Canadian Arctic the tide is strongly influenced by the ice cover.' They suggested that the 'influence likely occurs through alteration by the ice cover of the physical characteristics of a water body (depth, surface area, dimensions of openings or parts to adjacent seas).' Later, G. Godin, in the paper entitled *Modification by an Ice Cover of the Tide in James Bay and Hudson Bay* (*Arctic*, v. 39, 1986. pp. 65–67), said, after a study of tidal data from gauges around Hudson and James Bay, that 'the consistency of the results and the magnitude of the changes must reflect an overall modification of the tide inside Hudson Bay by the formation of an ice cover.' In fact, 'the range of the tide was always decreased after the formation of ice in all seas where such changes were reported.' Most of Godin's observations were from the eastern shores of Hudson and James Bay. S.J. Prinsenberg, in the paper entitled *Damping and Phase Advance of the Tide in Western Hudson Bay by the Annual Ice Cover* (*Journal of Physical Oceanography*, v. 18, 1988. pp. 1744–51), said that 'data shows that the ice cover damps the tide and advances the time of arrival of high water and flood current in western Hudson Bay' and that 'during winter, the extra function due to the presence of the seasonal ice cover causes the locations of the amphidromic points [tidal action in which the tide wave progresses around a point of little or no tide] to move toward the coast along which the reflecting wave travels.'

ii. Tsunami Studies Related to tides are tsunamis (tidal waves) and storm surges. Tsunamis are created by earthquakes or volcanic eruptions on the seafloor, and they propagate across the ocean basin as a series of crests moving at high speeds. The effects of tsunamis are felt on both coasts of Canada. For example, the tsunamis from the Alaskan earthquake of 1964 swept south along the coast of British Columbia into the coastal passages and fiords. The waves crested above high water and flooded built-up areas, causing damage estimated at $3 million. M.G. Sparth and S.C. Beckman described these tsunamis in their publication entitled *The Tsunamis of March 28, 1964 as Recorded at Tide Stations* (*Coastal Geodetic Survey. Technical Bulletin*, v. 33, 1967. 86 p.). And in 1972, T.S. Murty and R.F. Henry studied the propagation of tsunamis into a complex inlet system in the Queen Charlotte Sound of the west coast of Canada. Their modelling was described in the paper entitled *Some Tsunami Studies for the West Coast of Canada* (Ottawa: Department of the Environment, Marine Science Division, Manuscript Report 28, 1977. pp. 1–46).

Additional modelling for Pacific coast tsunamis was undertaken by D. Dunbar and P. LeBlond and reported in their paper entitled *Evaluation of Tsunami Amplitudes for the Pacific Coast of Canada* (*Progress in Oceanography*, v. 26, 1991. pp. 115–77). Here, the authors described three computer models developed for maximum tsunami water levels, and currents, along the British Columbia outer coast, from waves originating from Alaska, Chile, the Aleutian Islands (Schumagin Gap), and Kamchatka. These models were verified against water level measurements made at tidal gauges, after the 1964 Alaska and 1960 Chile earthquakes, and maximum wave and current amplitudes were tabulated for each simulated tsunami, in 185 locations, along the British Columbia coast.

Further evidence of tsunamis on the west coast of British Columbia has been described by J.J. Clague and P.T. Bobrowsky in their paper entitled *Evidence for a Large Earthquake and Tsunami 100–400 Years Ago on Western Vancouver Island,*

British Columbia (*Quaternary Research*, v. 41, 1994. pp. 176–84). These authors found a peaty marsh soil, overlain by a sand sheet and intertidal mud, at tidal marshes near Tofino and Ucluelet on Vancouver Island. This buried soil was submerged quickly, and radioactive dating placed this event between 100 and 400 years ago. The sand sheet overlying the peaty soil was the record of a tsunami, triggered by a large earthquake on the northern part of the Cascadia subduction zone. Further examination of the tidal marsh at Tofino, Ucluelet, and Port Alberni also provided evidence of the tsunami, triggered by the Alaska earthquake of 1964, as well as sand deposited by a tsunami 500 to 800 years ago during another earthquake, possibly in southern Alaska. This additional evidence of tsunamis was reported, again by J.J. Clague and P.T. Bobrowsky, in their paper entitled *Tsunami Deposits beneath Tidal Marshes on Vancouver Island, British Columbia* (*Geological Society of America Bulletin*, v. 106, 1994. pp. 1293–1303). They said that this evidence indicated tsunamis could be expected in the future, and they posed a hazard to people and property in some coastal areas.

Contrary to popular belief, the eastern part of Canada is also seismic, and has produced tsunamis. In 1929, the Grand Banks earthquake generated a tsunami in Burin Inlet, on the south coast of Newfoundland, that killed 26 people. This earthquake and its resultant tsunami, were described by J.H. Johnstone in the paper entitled *The Acadian – Newfoundland Earthquake of November 18, 1929 (Proceedings and Transactions of the Nova Scotia Institute of Science, Halifax, Nova Scotia*, v. 17, 1930. pp. 223–37). Overall, the current knowledge of tsunamis has been summarized by T.S. Murty in the book *Seismic Sea Waves: Tsunamis* (Ottawa: Development of Fisheries and the Environment Fisheries and Marine Service Bulletin 198, 1977. 336 p.).

iii. Storm Surges Storm surges, unlike tsunamis, are caused by the action of severe winds and air pressure anomalies. The resulting rise in sea level along the coast can last a day or more, and be sufficient to cause severe flooding. The major Canadian studies of these storm surges have been undertaken in the southern Beaufort Sea. These were part of the coordinated government and industry investigations known as the Beaufort Sea Project, carried out in 1974 and 1975 to obtain basic biological and geophysical data, and to devise methods of predicting sea, ice, and atmospheric conditions, on the Mackenzie Bathurst Shelf during the navigable summer season. It is well known that sea level elevations well above normal are experienced at Tuktoyaktuk, on the southern Beaufort Sea coast, whenever strong northwesterly, or westerly, winds occur during the open water season. Recorded loss of life, due to surges in this sparsely populated region, has been low, but the risk was increasing as oil and gas explorations intensified. Based upon recorded data from shore-based tide gauges, the physical features of the southern Beaufort

Sea, the coastal topography, the freshwater/saline stratification from the Mackenzie River flow, ice cover, and climatological data, equations were developed for two numerical storm-surge models by R.F. Henry and N.S. Heaps. One was a large-area computerized model and the other a small-scale model.

Their consistency and accuracy were evaluated against a surge episode that had occurred in 1970, and it was found that the accurate prognosis of surface winds were important for the prediction models. The research paper, based on this work, was entitled *Storm Surges in the Southern Beaufort Sea* (*Journal of the Fisheries Research Board of Canada*, v. 33, 1976. pp. 2362–76). The storm-surge elevations in the Tuktoyaktuk region of the Beaufort Sea coast have also been estimated by J.P. Harper and R.F. Henry from a systematic survey of log debris deposited on the tundra at the surge limit along the coastline of the Tuktoyaktuk area, and Kugmallet Bay. The reason for using this technique was the fact that tidal records there only go back to 1961, and have significant gaps. In their paper, entitled *Maximum Storm Surge Elevations in the Tuktoyaktuk Region of the Canadian Beaufort* (*Arctic*, v. 41, 1988. pp. 48–52), the authors were able to document surges that occurred in 1944, as well as 1970, and compare debris and tidal gauge measurements for a surge in 1986. They also said that the 'pattern of spatial variance in maximum surge elevation is consistent with numerical model results (Henry and Heaps, 1976).'

It is also well known that storm surges can occur in internal waters, estuaries, and gulfs, and some of these surges have been the focus of individual studies. For example, C. Libicki and K.W. Bedford, in their paper entitled *Sudden, Extreme, Lake Erie Storm Surges and the Interaction of Wind Stress, Resonance, and Geometry* (*Journal of Great Lakes Research*, v. 16, 1990. pp. 380–95), modelled surge data and analyses in order to try to explain the anomalously large storm surges that occur on Lake Erie. They concluded that these unique storm surges, at the ends of Lake Erie, were the interaction of lake geometry and basin friction. Other storm surges in Canadian waters have been studied and reviewed by T.S. Murty and S. Venkatesh in their paper entitled *Storm Surges in Canadian Waters* (*Atmosphere-Ocean*, v. 33, 1995. pp. 359–87). Here, the authors examined storm surges that have occurred on the east and west coasts, the Beaufort Sea, the Gulf of St Lawrence and the St Lawrence estuary, Hudson Bay, and the Great Lakes. Included in the review was a discussion of the weather systems that cause storm surges in Canadian waters, mathematical formulations to describe the development of storm surges, and the influence of sea ice on surge development.

iv. Surface Wave Studies The one other oceanographic surface phenomenon studied in Canada is surface waves. Both the provision of wave information and research into the formation of waves are a particular focus of the physical oceanographic program in Canada. In fact, the Marine Environment Data Service (MEDS) has taken a leading role in providing climatological data on waves, which includes the average and extreme conditions in an area for use in the design of structures, and the plotting of shipping routes and schedules. The MEDS program began in 1969, with the object of establishing a set of buoys along the coast to measure wave period and height, and to obtain climatological data on the wave conditions along the seaboard. The Marine Environment Data Service has also established stations as part of a long-term wave-climate study, such as wind-waves along the east and west coasts of Canada. These developments have been described by C.R. Mann in the review entitled *Physical Oceanography in Canada* (*Journal of the Fisheries Research Board of Canada*, v. 33, 1976. pp. 1412–33), previously noted.

Wind-wave predictions depend upon the accuracy of the predicted wind field and the techniques used for forecasting. The predictions may be used in order to provide warnings for those on the water, as well as shore dwellers, or for hindcasting in order to provide an estimate in assessing some subsequent effect that may be caused by waves. Hindcasting of wind-waves does require accurate information on the wave spectrum, particularly during the important height and peak period, against which a numerical model can be developed and applied. One example of this type of wave research was carried out by M. Donelan in the paper entitled *A Method for the Automatic Measurement of Wave Frequency* (*Journal of the Fisheries Research Board of Canada*, v. 33, 1976. pp. 2318–22). More recently, wind-wave hindcast extremes have been issued for the east and west coasts of Canada by the Canadian Climate Centre, based on contracted research. They are entitled *Wind Wave Hindcast Extremes for the East Coast of Canada* (Halifax: McLaren Plansearch, 1991. 109 p.); and *Wind Wave Hindcast Extremes for the West Coast of Canada* (Halifax: McLaren Plansearch, 1992. 143 p.).

In the mid-1980s, wave forecasting was one focus of the Canadian Atlantic Storms Program (CASP), which developed and assessed model predictions for different types of storm conditions, as well as for the evaluation of winter storms that affect Atlantic Canada. This type of prediction research is important because of the amount of offshore oil exploration being carried out in that region. Field measurements were collected from a number of locations on the US east coast, the Georges Bank, the Scotian Shelf, and the Grand Banks of Newfoundland. The numerical model developed was a regional spectral model for real-time wave forecasting. Again, the model showed that accurate wind specifications were important in utilizing an operational wave model.

This work, carried out by a Nova Scotia consultant firm, has been described by B.M. Eid in the publication entitled *Operational Test of Wave-Forecasting Models during the Canadian Atlantic Storms Program (CASP)* (Halifax: MacLaren

Plansearch, 1987, 1 v.). Since the early 1990s, wave modelling in Canada has used a 'spectral energy balance equation' which incorporates second- or third-generation algorithms. The Canadian model is called the Canadian Spectral Ocean Wave Model (CSOWM), and is operated by the Canadian Meteorological Centre (CMC) in Montreal. This model has been described in the Environment Canada workshop entitled *Third International Workshop on Wave Hindcasting and Forecasting: Supplement* (Downsview, Ontario: Atmospheric Environment Service, 1992. 141 p.). Finally, the problem of storm waves in Canadian waters has been discussed by M.L. Khandekar and V.R. Swail in their paper entitled *Storm Waves in Canadian Waters: A Major Marine Hazard* (*Atmosphere-Ocean*, v. 33, 1995. pp. 329–57). This paper has provided an overview of storm waves associated with intense weather systems affecting the east and west coasts of Canada, as well as the Beaufort Sea, and Canadian Great Lakes region. The wave climatology has been presented in terms of the 100–year significant and maximum wave heights (i.e., the wave height which could be expected to occur, on average, once in 100 years). Also included is a history of the development of ocean surface wave models and their present status.

v. Ocean Channel and Inlet Dynamics Studies Below the surface of the water are the dynamics and the circulation of ocean channel and inlet waters, which may involve a scale from a few centimetres to thousands of kilometres. In Canada, it has been the smaller-scale processes that are studied. These processes are turbulence, internal waves, and thermal microstructures, all being mechanisms by which heat exchanged with the atmosphere is distributed through the upper layer of the ocean, and by which pollutants are dispersed that affect marine life, particularly plankton. Within small-scale processes, Canadian research has focused upon the dynamics and physical characteristics of inlets and channels along coasts and, in particular, the west coast including the Strait of Georgia, Johnston Strait, Hecate Strait, and the Arctic Archipelago.

The Strait of Georgia, on the west coast, is unique because it is fringed by a number of tributary inlets, bays, and sounds, with fresh water inputs from the Fraser River Estuary, and ocean water renewals from the north through Johnson Strait and from the south through the Juan de Fuca Strait. Although the physical oceanography and water properties of the Strait of Georgia had been relatively well studied, information on the circulation in the strait was more sparse, due to the fact that current measurements are more difficult to make. The surface circulation was described by M. Waldichuk in two papers, one entitled *Physical Oceanography of the Strait of Georgia, British Columbia* (*Journal of the Fisheries Research Board of Canada*, v. 14, 1957. pp. 321–486); and the other entitled *Drift Bottle Observations in the Strait of Georgia* (*Journal of the Fisheries Research Board of*

Canada, v. 15, 1958. pp. 1065–1102). Work continued throughout the 1950s and the 1960s, with measurements taken from different areas, using self-recording current meters. The later work included a study of the area west of Sturgeon Bank, which is adjacent to the Fraser River Delta. The data analysed showed low-frequency currents, which decreased linearly with depth, but tidal currents represented 40%-70% of the current energy. The examination of the Sturgeon Bank data was done by P. Chang and S. Pond and published in the paper *Subsurface Currents in the Strait of Georgia, West of Sturgeon Bank* (*Journal of the Fisheries Research Board of Canada*, v. 33, 1976. pp. 2213–41).

In the early 1980s, another review of the state of physical oceanography of the Strait of Georgia was written by P.H. LeBlond in the paper entitled *The Strait of Georgia: Functional Anatomy of a Coastal Sea* (*Canadian Journal of Fisheries and Aquatic Sciences*, v. 40, 1983. pp. 1033–63). The author said that high-frequency phenomena, particularly the tides, were the most understood aspect of the physical oceanography of the strait and nearest to the level of engineering use. However, there was 'a definite need for further development and refinement of numerical tidal models.' And, 'an enormous amount of basic as well as applied work remains to be done before upper layer modelling can be used at the operational level, i.e., in scenarios in which the Coast Guard or the Environmental Protection Service may dial it up in an emergency for information on the drift of boats in distress or of spilled oil.' Also not to be neglected were 'basic physical studies on the paramaterization of friction entrainment, and boundary conditions should go hand in hand with further numerical model development, including the problem of incorporating fronts within the model.' Finally, the author emphasized that 'certainly the most fundamental problems facing physical oceanography in the Strait of Georgia concern the nature and mechanisms of deep residual flows and slow water exchange with the ocean. This is where the mysteries lie and where fundamental research is likely to focus in the years to come.'

After this review, it is interesting to note that subsequent research undertaken on the Strait of Georgia has focused upon these suggested research areas. T. Yao and S. Pond, in their paper entitled *Profiles of Low-Frequency Subsurface Current Fluctuations in the Strait of Georgia during 1981 and 1982* (*Journal of Geophysical Research*, v. 90, 1985. pp. 7181–98), provided a statistical description of subsurface current fluctuations at subtidal frequencies in the central strait, based upon vertical profiles of currents made by cyclesondes (profiling current meters) for the these two years. (The author also noted the LeBlond article of 1983 – mentioned above – that 'emphasized the present lack of understanding of the low-frequency dynamics.') The data showed a variability of horizontal currents in the Strait of Georgia region, but 'the reasons for the interannual variation in the mean flow from spring 1981 to spring 1982 are unknown.' D.H. LeBlond with H. Ma again studied the strait, as observed in their paper entitled *Deep and Intermediate Water Replacement in the Strait of Georgia* (*Atmosphere-Ocean*, v. 29, 1991. pp. 288–312). In this paper, the authors re-examined previous hydrographic data by P.B. Crean and A.B. Ages entitled *Oceanographic Records from Twelve Cruises in the Strait of Georgia and Juan de Fuca Strait, 1968* (Ottawa: Department of Energy, Mines and Resources, 1971. 5 v.); and by M.W. Stacey and S. Pond entitled *An Analysis of the Low-Frequency Current Fluctuations in the Strait of Georgia and Influencing Factors* (*Journal of Physical Oceanography*, v. 17, 1987. pp. 326–42), which suggested that 'during the summer, warm salty water originating from the Strait of Juan de Fuca penetrates into the depths of the Strait of Georgia. In contrast, during the winter cold but fresher water spread into the strait from the south. A third, and lesser, renewal process is associated with cold water entering from Discovery Passage in the north.' However, the authors wondered 'how will our knowledge of the subsurface replacement processes progress beyond what has been deduced from the available data? Common features of the two different processes suggest an observational program that could confirm the conceptual models developed here as well as provide significant new information.' Also needed were contributions to understanding the complexity of understanding low-frequency current fluctuations in their baroclinic instability (variations in their horizontal temperature), which may contribute to the variability of low-frequency current fluctuations. This baroclinic instability was pointed out by R.H. Karsten and G.E. Swaters in their paper entitled Stability Characteristics of Deep-Water Replacement in the Strait of Georgia (Journal of Physical Oceanography, v. 25, 1995. pp. 2391–2403).

Inflows of subsurface waters into the strait create not only subsurface currents and fluctuations but also residual currents. S.G. Marinone and J. Fyfe, in their paper entitled *Residual Currents in the Central Strait of Georgia* (*Atmosphere-Ocean*, v. 30, 1992. pp. 94–119), defined residual currents as 'that part of the flow left after low-pass filtering the currents with a cut-off period of a couple of days or so. It is made up of a mean (time-averaged) component and a fluctuating component.' These residual currents had been previously documented by hydrographic data, but these authors studied the residual circulation in the central Strait of Georgia with a 'spatially nested depth-averaged numerical model,' and said that 'the modelled mean residual forms a cyclonic eddy in the middle of the Strait.' However, G. Holloway, in the paper entitled *On Residual Currents in the Central Strait of Georgia, B.C.: Research Note* (*Atmosphere-Ocean*, v. 31, 1993. pp. 167–70), suggested shortcomings in the model and the results, because 'flows in the Strait of Georgia, like other flows in channels, estuaries, lakes or oceans constitute a system of many interacting degrees of freedom. Numerical models, even on major computing facilities, realize only relatively few of these degrees of freedom' and 'conventional models omit, or grossly corrupt statistical-mechanical tendencies that should arise from interactions of the modelled flow with unresolved eddies.'

Just north of the Strait of Georgia is the narrower Johnstone Strait, about 100 kilometres long, that is part of the navigable 'inside passage' separating Vancouver Island from the North American mainland. It is important as part of the southward salmon migration runs. Despite its importance, the strait had received little attention from physical oceanographers, but R.F. Thomson, using unpublished data from the University of British Columbia and the Fisheries Research Board of Canada, along with four months of personal current meter readings taken across the strait, came to an interesting conclusion. He said that the strait, which has depths up to 300 meters, had an estuarine-type circulation, beginning with high-density salinity in the western end that gave way to lower salinity as the currents moved uni-directionally at all depths. Moreover, at lower depths, the current speed was almost twice that of the upper layer, which was affected by tidal heights. Thompson's work was entitled *Tidal Currents and Estuarine-Type Circulation in Johnstone Strait, British Columbia* (*Journal of the Fisheries Research Board of Canada*, v. 33, 1976. pp. 2242–64).

North of the Johnstone Strait and the Queen Charlotte Sound is the Hecate Strait, which separates the Queen Charlotte Islands from the mainland of British Columbia. Both winter and summer currents and transport have been studied, because 'interest in hindcasting the oceanographic conditions in Hecate Strait has grown as fisheries scientists try to understand the causes of inter-annual variability in year-class strengths of various Hecate Strait fish stocks.' Overall observations of the transport, currents, and subsurface pressure were carried out in 1983 and 1984 by W.R. Crawford and W.S. Huggett and reported in their paper entitled *Water Transport through Hecate Strait* (*Atmosphere-Ocean*, v. 26, 1988. pp. 301–20). An analysis of the data indicated 'that the mean currents, mean transport, and synoptic scale (periods of a few days to a few months) transport fluctuations are wind forced.' More detailed analysis of winter currents was undertaken by W.R. Crawford and A.V. Tyler in their paper entitled *A Possible Eddy Retention Mechanism for Ichthyoplankton in Hecate Strait, B.C.* (*Canadian Journal of Fisheries and Aquatic Sciences*, v. 47, 1990. pp. 1356–1363). As well, C.G. Hannah and W.R. Crawford investigated the 'dynamics of the subsurface pressure and transportation fluctuations during the late fall and winter' in the paper, entitled *Winter Transport and Sea Level Fluctuations in Hecate Strait, British Columbia* (*Journal of Geophysical Research*, v. 101, 1996. pp. 28365–76). Here, the analysis and modelling of data showed the 'transport through the strait is driven by differences in sea level between Queen Charlotte Sound, denoted the pressure-driven response, and by local wind stress, denoted the wind-driven response.' Besides winter circulation and transport, summer circulation and transport of both Hecate Strait and Queen

Charlotte Sound, have been documented by W.R. Crawford and W.S. Huggett in the paper entitled *Summer Circulation of the Waters in Queen Charlotte Sound* (*Atmosphere-Ocean*, v. 23, 1985. pp. 393–413); and by W.R. Crawford and M.J. Woodward in the paper entitled *Oceanographic Features of Hecate Strait and Queen Charlotte Sound in Summer* (*Atmosphere-Ocean*, v. 33, 1995. pp. 639–81). In this paper, the authors provided 'evidence of previously unresolved oceanographic features in Queen Charlotte Sound and Hecate Strait using data collected in the summer of 1990 and interpreted using a three-dimensional, finite-element diagnostic numerical model for two separate simulations.'

Unlike Canada's most southern waters, knowledge of the physical oceanography of the channels and inlets of the Canadian Arctic Archipelago tends to be poor, or lacking, because field programs in the Arctic pose problems peculiar to the area. These are caused by the fact that most channels of the archipelago, and the southern Beaufort Sea, are accessible to ships using conventional techniques only during a short summer season. As a result, winter observations are scarce, and special techniques have to be devised to deploy equipment developed to operate in Arctic conditions. Much of the data that is collected is stored and updated for future study and interpretation, as is the case of Arctic surveys carried out in the early and mid-1980s. One such compilation by J.R. Birch and P.D. Lemon for the Arctic Data Compilation and Appraisal Project was entitled *Beaufort Sea and Amundsen Gulf: Physical Oceanography – Temperature, Salinity, Currents, Water Levels and Waves, Revised and Updated to Include 1914 through 1986* (*Canadian Data Report of Hydrology and Ocean Sciences*, No. 5(12), 1990. 451 p.). Other examples are C.K. Ross's *Currents and Temperature Data from Northwestern Baffin Bay, September, 1983–September, 1984* (*Canadian Data Report of Hydrology and Ocean Sciences*, No. 79, 1990. 227 p.). Sometimes, however, the data is examined and evaluated, as was the case with field observations in 1971, during an investigation of sea ice movement in Nares Strait, the passage between Ellesmere Island and Greenland. Five years of data obtained in Nares Strait was used to provide estimates of the transport of water, heat, and salt between the polar ocean and northern Baffin Bay. These estimates were carried out by H.E. Sadler, and evaluated in the paper entitled *Water, Heat and Salt Transports through Nares Strait, Ellesmere Island* (*Journal of the Fisheries Research Board of Canada*, v. 33, 1976, pp. 2286–95).

Another but more current focus of Arctic oceanography has been the study of the relationship between currents and ice-cover. According to R.F. Marsden and R.G. Ingram, *et al.*, in their papers entitled *Currents under Land-Fast Ice in the Canadian Arctic Archipelago Part 1: Vertical Velocities* (*Journal of Marine Research*, v. 52, 1994. pp. 1017–86), and *Currents under Land-Fast Ice in the Canadian Arctic Archipelago Part 2: Vertical Mixing* (*Journal of Marine Research*, v. 52, 1994.

pp. 1037–49), 'during the spring bloom, ice algae constitute the largest photosynthetic biomass in ice-covered polar seas, forming dense mats in the bottom few centimeters of the ice and at the ice-water interface.' It has been estimated that 'approximately 25% of the annual primary production in the Arctic Ocean takes place within the sea ice and these algae are not only important ecologically as the only source of food at a critical time for many pelagic animals' but 'they also contribute to the biological pumping of carbon dioxide from the atmosphere into deep waters through the export of carbon from the surface associated with ice melt, having possible significance for the global biogeochemical flux of carbon.' It is because of this algal production that both the vertical velocities of internal waves and currents and the vertical mixing through eddy-like features, estimated by an acoustic Doppler current profiler, were important calculations, for the latter 'indicated that a minimum of 25% of the nutrient demand of the ice algae could be provided by turbulence associated with the high frequency internal wave events. Their nutrient demands are a spatial average applicable to the general Lancaster Sound, Barrow Strait region and are accurate only to an order of magnitude.'

vi. Ice Studies One other major area of physical oceanography is the study of ice movements from the Arctic Sea southward. The provision of ice predictions and the preparation of charts and ice atlases are carried out by the Ice Centre of Environment Canada. The Ice Centre prepares charts of ice distribution all the way from the Beaufort Sea to the Gulf of St Lawrence and includes the estimated dates of freeze-up. Ice patrols began in the Gulf of St Lawrence, in 1940, with the intent to determine the date that open water navigation into the St Lawrence River ports became feasible. By the late 1950s, the patrols were supplanted by the then-Meteorological Branch's ice reconnaissance and forecasting program, which operates throughout the winter months and supports shipping in all ice-encumbered Canadian waters. Besides aerial and shipboard observations, output from meteorological and remote sensing satellites has improved the range and quality of data.

Beginning in 1980, W.E. Markham produced a series of ice atlases which are entitled *Ice Atlas, Eastern Canadian Seaboard* (Toronto: Atmospheric Environment Service, 1980. 96 p.); *Ice Atlas, Canadian Arctic Waterways* (Ottawa: Atmospheric Environment Service, 1981. 198 p.); and *Ice Atlas, Hudson Bay and Approaches* (Ottawa: Ice Centre, 1988. 123 p.). The eastern seaboard atlas was updated from passive microwave remote sensing data (with passive wavelength sensors, radiation is able to penetrate the clouds) by A.E. Owens and entitled *Ice Atlas of Eastern Canada Based on Nimbus-S Passive Microwave Data (1972–1977)* (Ottawa: Atmospheric Environment Service, 1984. 145 leaves). Most recently, another update to the eastern Canadian seaboard atlases has been issued and entitled *Sea Ice Climatic Atlas: East Coast of Canada, 1971–2000* (Ottawa: Canadian Ice Service, 2001. 42 p.; 147 p., charts, maps, graphs). Nevertheless, the most comprehensive ice atlas produced to date is by E. LeDrew and D. Barker, and entitled *Canadian Sea Ice Atlas from Microwave Remote Sensed Imagery: July 1987 to June 1990* (Ottawa: Atmospheric Environment Service, 1992. 80 p. Climatological Studies No. 44). This atlas, which covers the polar region of the Northern Hemisphere, was undertaken as one of several Canadian initiatives to mark the 1992 International Space Year.

Besides collecting and publishing ice data in atlas formats (their tabular data is described in this chapter's Section II-A-b, Ice Thickness Data), the Ice Centre of the AES also carries out research on the quality of ice data collected, and, more specifically, has undertaken research on the effect of sea ice and climate change on icebergs in the Labrador-Newfoundland portion of the eastern seaboard. As mentioned, AES gathers ice data by conventional means (ship observations, reconnaissance aircraft, satellite photographs, etc.), as well as using passive microwave data (using a scanning multi-channel microwave radiometer – SMMR). But how do the two types of sources compare? D.A. Etkin and R.O. Ramseier, in their paper entitled *A Comparison of Conventional and Passive Microwave Sea-Ice Datasets for Hudson Bay* (*Atmosphere-Ocean*, v. 31, 1993. pp. 359–78), found after a three-year study of the data sets from 1979 to 1981, covering Hudson Bay, that the passive microwave data set had 'errors' caused by the computer algorithm, which interpreted ponding (melted water) on ice as open water during the seasonal ice break-up. On the other hand, the algorithm was more accurate at estimating the amount of ice during freeze-up than the conventional data.

During the break-up of sea ice on the eastern seaboard, the loss of sea ice increases the movement of icebergs from the Arctic regions of the North Atlantic Ocean off Labrador and Newfoundland. This can be hazardous to shipping and oil exploration platforms. Hence, it is necessary to not only track icebergs but to forecast their drift. S. Venkatesh and B. Sanderson, in their paper entitled *Optimum Deployment of Satellite-Tracked Drifters to Support Iceberg Drift Forecasting* (*Cold Regions Science and Technology*, v. 18, 1990. pp. 117–31), devised strategies for the optimum deployment of Tiros satellite-tracked Ocean Drifters (TOD) in order to derive measurements of ocean currents for iceberg drift forecasting off the east coast. The authors estimated that 20–100 TODS would be required for release within the Labrador current off Labrador, and off Newfoundland, to allow for variabilities in the eddy structure of the currents for iceberg drift. The severity of icebergs in that area is affected not only by sea ice variability but possibly by long-term climate change. J.R. Marko and D.B. Fissel, in work done with the Ice Centre in Ottawa, have suggested that global warming due to greenhouse gas concentrations would increase the retreat of sea ice and cause increased iceberg calving at tidal glaciers in the Arctic, and accordingly increase the iceberg numbers in the North Atlantic off Labrador and Newfoundland. These suggestions were based on general circulation models and reported in their paper entitled *Iceberg Severity off Eastern North America: Its Relationship to Sea Ice Variability and Climate Change* (*Journal of Climate*, v. 7, 1994. pp. 1335–51).

The major principal force driving the movement of pack ice in many areas is wind stress. Wind drag on ice can be separated into surface drag, on a relatively uniform terrain, and form drag, on isolated ridges on Arctic pack ice. In the mid-1970s, experiments were carried out at various locations in the Arctic Ocean on these forms of wind stress in order to develop equations that could be used to estimate actual wind stress. Surface elevations, wind profiles, and turbulence level data for uniform and ridged terrain were collected in order to come up with the equations. This work was carried out by E.G. Burke and S.D. Smith and reported in their paper entitled *Recent Measurements of Wind Stress on Arctic Sea Ice* (*Journal of the Fisheries Research Board of Canada*, v. 33, 1976. pp. 2307–17).

Another focus of study of pack ice is the small localized areas, called polynyas, that remain ice-free in the pack throughout the winter. This feature occurs in certain parts of the Arctic Archipelagos, despite the extreme temperature differences between the atmosphere and the ocean surface. One such area close to Dundas Island, between Baffin and Ellesmere Islands, and close to Devon Island, in the eastern Arctic, has been studied in detail both from the point of view of the atmospheric/oceanic heat exchange over the open water, and from its local geographic and oceanic context. A characteristic feature of polynyas is the strong modulation of the near-surface ocean temperature at tidal frequencies involving temperature differences as great as 0.2 degrees centigrade within a single tidal cycle. The surface waters themselves tend to be at 0.1 to 0.2 degree centigrade above freezing. The presence of this relatively warm surface water is thought to be due to mixing, occasioned by fast tidal flows across the shallow sill connecting Dundas and Devon Islands. This major study of polynyas was carried out in the early 1980s by D.R. Topham, *et al.*, working from the Institute of Ocean Sciences, Sidney, BC, the Bedford Institute of Oceanography, Dartmouth, Nova Scotia, and the Atmospheric Environment Service, Downsview, Ontario. This study resulted in three papers being published, the first entitled *An Investigation of a Polynya in the Canadian Archipelago. 1: Introduction and Oceanography* (*Journal of Geophysical Research*, v. 88, 1983. pp. 2888–99); the second entitled *An Investigation of a Polynya in the Canadian Archipelago. 2: Structure of Turbulence and Sensible Heat Flux* (*Journal of Geophysical Research*, v. 88, 1983. pp. 2900–10); and the third entitled *An Investigation of a Polynya in the Canadian Archipelago. 3: Surface Heat Flux* (*Journal of Geophysical Research*, v. 88, 1983. pp. 2911–16).

(b) Chemical Oceanography

Chemical oceanography is the most recently developed of the oceanographic sciences in Canada. Strictly speaking, chemical oceanography is concerned principally with the physical/chemical processes in water columns, or sea ice. The subject, however, can also include sedimentary geochemistry. The basis of the chemical studies is salinity data, which is concerned with the dissolved salts that give sea water its character, and have a fundamental influence on all physical processes in the sea. Chemical knowledge is also important for monitoring changes in estuaries, gulfs, and coastal waters, due to industrial and domestic activities. These environmental studies have become a major focus as these ocean environments become increasingly contaminated. As well, the exchange processes between the atmosphere and the sea surface are studied. The most important of these exchanges is the large-scale interaction between the ocean and atmospheric carbon dioxide, and the ability of the ocean to absorb the increased levels of atmospheric carbon dioxide caused by man's activities.

To date, chemical oceanography studies in Canada have concentrated on the Gulf of St Lawrence. The term Gulf of St Lawrence is used by chemical oceanographers to describe the entire estuarine system, from the mouths of the St Lawrence and Saguenay rivers in the west to the two connections with the open Atlantic Ocean in the east – Cabot Strait and the Strait of Belle Isle. It is a complex system because the Gulf is almost surrounded by land with the most important freshwater inputs concentrated at the western end of the Gulf, and the only connections with the Atlantic Ocean at the eastern entrances of the system. The semi-enclosed nature and the separation between inputs and outputs facilitate the study of both internal processes and exchanges with adjacent waters. As well, the tidal forces, deep channels, and freshwater input into the system, combined with its proximity to some high population and industrial concentrations, adds to the complexity of this system. Studies carried out so far have included the distribution and transport of suspended particulate matter, nutrients, organic matter in the Gulf, stable isotopes, trace metals in the water column, trace metals geochemistry, and petroleum residues in the waters of the Gulf.

Studies have shown that suspended particulate organic or inorganic matter is carried downstream in the surface layer of the river estuary, and upstream near the bottom, trapping particles that are too large to be carried from the estuary in the surface layer, but too small to be deposited. Overall, much of the St Lawrence input is deposited in the Gulf, but particulate matter in the Cabot Strait section of the deep Laurentian Channel that flows through the Gulf is produced within the Gulf, and stays there. Some of the work in this field has been carried out by B.F. D'Anglejan and E.C. Smith and described in their paper entitled *Distribution,*

Transport and Composition of Suspended Matter in the St. Lawrence Estuary (*Canadian Journal of Earth Sciences*, v. 10, 1973. pp. 1380–96); and by M.I. El-Sabh in the paper entitled *Oceanographic Features, Currents, and Transport in Cabot Strait* (*Journal of the Fisheries Research Board of Canada*, v. 34, 1976. pp. 516–28).

As suspended organic particulates fall and disintegrate in the water, their nutrients, which are primarily silicate, phosphorus, and nitrate, precipitate out. These nutrients play an important role in primary biological production in the production of plankton, and hence to the fisheries of the Gulf. Overall, the nutrient levels tend to be at higher concentrations in the freshwater surface layers of the estuary, and also at greater concentrations in the deeper levels of the Laurentian Channel of the Gulf. Some of the work on nutrient distribution was carried out by P. Greisman and G. Ingram and described in their paper entitled *Nutrient Distribution in the St. Lawrence Estuary* (*Journal of the Fisheries Research Board of Canada*, v. 34, 1977. pp. 2117–23); and by M. Sinclair and M.I. El-Sabh in their paper entitled *Seaward Nutrient Transport in the Lower St. Lawrence Estuary* (*Journal of the Fisheries Research Board of Canada*, v. 33, 1976. pp. 1271–77). A knowledge of the organic chemical composition of particulate organic matter, which is generated from phytoplankton, and water-borne vascular plants and attached algae, as well as water-borne materials of terrestrial origin, is very important for the calculation of food sources available to other organisms, including fish. The St Lawrence River and the St Lawrence Estuary are the major freshwater contributors of particulate organic matter to the gulf. However, sustained levels of primary organic production are also maintained within the gulf by sub-surface water of high nutrient content flowing in at Cabot Strait, moving along the deep Laurentian and Esquimux Channel, and mixing into the surface layer by upwelling. The study of particulate organic matter is perhaps the most complete of any of the chemical features investigated for the Gulf of St Lawrence. Major work has been done by M.I. El-Sabh in the paper entitled *Oceanographic Features, Currents, and Transport in Cabot Strait* (*Journal of the Fisheries Research Board of Canada*, v. 34, 1975. pp. 516–28); and by R. Pocklington and F.C. Tan in their paper entitled *Seasonal and Annual Variations in the Organic Matter Contributed by the St. Lawrence River to the Gulf of St. Lawrence* (*Geochimica and Cosmochimica Acta*, v. 51, 1987. pp. 2579–86).

Related to particulate organic matter is the study of the isotopes of sedimentary organic carbon in the Gulf of St Lawrence. These studies are of potential use for gaining an insight into the sources of sedimentary organic carbon, and the interaction of suspended particulate matter with it and plankton organic carbon. Two papers on this subject have been written by F.C. Tan and P.M. Strain. The one, entitled *Carbon Isotope Ratios of Particulate Organic Matter in the Gulf of St. Lawrence* (*Journal of the Fisheries Research Board of Canada*, v. 36, 1979. pp. 678–

82), described the carbon isotope ratios and concentrations of particulate organic carbon (POC) of the near-surface and intermediate depths in the central Gulf, as well as the deep Laurentian Channel. The second paper, entitled *Sources, Sinks and Distribution of Organic Carbon in the St. Lawrence Estuary Canada* (*Geochimica and Cosmochimica Acta*, v. 47, 1983. pp. 125–32), indicated that the studies of carbon isotope ratios in the St Lawrence suggested that POC in the lower and upper Estuary and Gulf were most likely produced and deposited *in situ*, and there was 'negligible transport of particulate organic carbon between these three carbon reservoirs.' The most recent data on the carbon and nutrient chemistry of the Gulf was based on water samples collected between 1992 and 1994 by C. Savenkoff and A.F. Vezina and discussed in their paper entitled *Distributions of Oxygen, Carbon, and Respiratory Activity in the Deep Layer of the Gulf of St. Lawrence and Their Implications for the Carbon Cycle* (*Canadian Journal of Fisheries and Aquatic Sciences*, v. 53, 1996. pp. 2451–65). The authors said that 'organic carbon fluxes estimated from sediment trap data, deepwater respiratory activity, and benthic respiration measurements ... suggest that approximately 10% of the local primary production reaches the deep layer and the sediments in the GSL. The vertical carbon budget is almost balanced in the eastern part of the Gulf, but approximately half of the carbon produced in the surface layer of the northwestern Gulf cannot be accounted for. The differences in hydrodynamic and biological conditions prevailing in both areas may explain our observations.'

Rivers are a main source of trace metals to the ocean. Because of this, the St Lawrence River, its Estuary, and the Gulf offer a unique area for research. The transport of river-derived metals through estuarine and coastal environments is affected by the deposition and re-suspension of metal-containing particles, precipitation or dissolution reactions, and the release of dissolved metals from sediments. The first general description of trace metals in the open Gulf itself was a study of total iron, cobalt, nickel, copper, zinc, cadmium, and lead concentrations carried out by J.M. Bewers and I.P. Macauley and entitled *Trace Metals in the Waters of the Gulf of St. Lawrence* (*Canadian Journal of Earth Sciences*, v. 11, 1974. pp. 939–50). The study was limited by the dependence on flame atomic absorption spectrophotometry for trace metal analysis but gave reasonably reliable concentrations for dissolved iron, total nickel, copper, and zinc. Cobalt, cadmium, and lead levels were below their detection limits.

The St Lawrence River itself is the single most important source of trace metals to the Gulf of St Lawrence. This has been documented by P.A. Yeats and J.M. Bewers in their paper entitled *Discharge of Metals from the St. Lawrence River* (*Canadian Journal of Earth Sciences*, v. 19, 1982. pp. 982–92). In this, the authors discussed measurements of the dissolved and particulate forms of aluminium, iron, manganese, cobalt, nickel, copper, zinc, and cadmium, made during a two-year period, at the mouth of the St Lawrence River, just upstream of the landward limit of salt intrusion. Water samples had been used earlier in their paper entitled *Oceanic Residence Times of Trace Metals* (*Nature*, v. 268, 1977. pp. 595–98), to scale Gulf of St Lawrence data to oceanic proportions in order to improve on current estimates of residence time of trace metals in the deep ocean. For their more recent paper, the authors conducted a more detailed analysis of the data, and re-examined the validity of their scaling process of the previous paper. They also defined the average concentrations of dissolved and total particulate trace metals, and determined their variability on a monthly time scale. In 1982, P.A. Yeats, alone, in the paper entitled *Trace Metals in the Water Column*. In: *Chemical Oceanography in the Gulf of St. Lawrence*. Edited by P.M. Strain (Ottawa: Department of Fisheries and Oceans, 1988. pp. 79–98. *Canadian Bulletin of Fisheries and Aquatic Sciences*, No. 220), reviewed the 1974 to 1976 data again, along with trace element data gathered by other researchers, and said 'the two main conclusions that we can draw from these results are that, in general, the levels of both dissolved and particulate metals are lower than in many other rivers and that the month-to-month variability is also quite low.'

Complications that may occur in the study of trace elements and chemical contaminants in the St Lawrence have been pointed out by R. Carignan and S. Lorrain in their paper entitled *A 50-Yr Record of Pollution by Nutrients, Trace Metals, and Organic Chemicals in the St. Lawrence River* (*Canadian Journal of Fisheries and Aquatic Sciences*, v. 51, 1994. pp. 1088–1100). These authors studied sediment cores taken from the fluvial lakes, St Francis, and St Louis, upstream from Montreal. These sites were chosen because 'sedimentation rates are high in these sites' and 'the blurring effect of biological/physical sediment mixing is relatively small, and past changes in sediment quality have left clear signals.' The cores did show that biological productivity and contamination by trace metals, and organic chemicals, were much higher in the 1960s and 1970s than now, and 'inferred chronological changes in algal pigments and nutrients generally correspond to eutrophication trends directly measured in Lake Ontario.' However, 'our data provide some indications that river discharge influences year-to-year variations in contamination levels in the St. Lawrence and that the importance of local versus upstream sources has varied with time. These hypotheses should be taken into consideration in future trend analysis of current monitoring data.'

More complicated, however, is the study of trace elements and chemical contaminants in the St Lawrence Estuary because of the bathymetry (depth), tidal currents, and turbulent vertical and horizontal mixing of fresh and salt water which complicates the understanding of the role played by suspended sediments in the partitioning between the dissolved and particulate phases of metal and organic contaminants in the system. Nevertheless, studies were undertaken in the

Estuary by P.F. Hamblin, in 1986, and reported in the paper entitled *Observations and Model Sediment Transport near the Turbidity Maximum of the Upper Saint Lawrence Estuary* (*Journal of Geophysical Research*, v. 94, 1989. pp. 14419–28). The author sought to provide more information on the processes responsible for the distribution and transport of contaminants, such as lead, cadmium, and volatile hydrocarbons in the estuary by the development of a 'simple one-dimensional transport model ... capable of accounting for the major features of the temporal and depth behaviour provided that the simultaneous flow and density fields are known.' The author concluded that 'future studies in the upper St. Lawrence estuary should be aimed at profiles of high-speed and well-resolved suspended concentrations in the region of peak turbidities where sediment floculation may be important.'

Because the Gulf of St Lawrence is the most important of Canada's marginal seas in terms of its economic and social significance, it is vital to practise sound environmental management of the region to maintain the harvest of its living resources at economically viable levels, while continuing the use of the Gulf as a transport corridor for industrial raw and manufactured materials, with their incumbent petroleum residues, and the newer explorations for hydrocarbon reserves. Petroleum residues have been monitored, since the early 1970s, by the Bedford Institute of Oceanography. From 1971 to 1979, the Institute carried out 10 surveys of petroleum residues in the Gulf of St Lawrence, to determine how their levels changed over that period of time. During the program, data from more than 2,000 samples were obtained at 448 stations. The impetus for these surveys was the need to assess the impact of the bunker oil spill from the oil tanker *Arrow*, in 1970. In spite of the 1970 readings on the Scotia Shelf that indicated discharges from the *Arrow*, the 10-year pattern indicated, by 1975, a steady state had been reached between the rate of input and the ability of the Gulf to assimilate dissolved/dispersed petroleum residues. This steady state of petroleum residues was 'attributed to a diffuse source such as fallout from the atmosphere of non-polar aromatic substances formed during the high combustion of organic materials.' After 1975, residue levels declined. This work was summarized by E.M. Levy in the paper entitled *Background Levels of Dissolved/Dispersed Petroleum Residues in the Gulf of St. Lawrence, 1970–1979* (*Canadian Journal of Fisheries and Aquatic Sciences*, v. 42, 1985. pp. 544–56).

Levy reviewed the data again in 1985, as well as passing comment on other petroleum residue data taken from studies of the Grand Banks, Baffin Bay, and the Hudson Strait/Labrador Shelf region, in the paper entitled *Petroleum Residues in the Waters of the Gulf of St. Lawrence*. In: *Chemical Oceanography in the Gulf of St. Lawrence*. Edited by P.M. Strain (Ottawa: Department of Fisheries and Oceans, 1988. pp. 139–57). Again, the author said that 'since the existing background levels in all regions are well below those known to have immediate and direct adverse affects on marine life, there is no reason at present for environmental alarm. Neither, however, is there room for complacency, since any increase in the rate of input could easily alter the present situation, possibly with serious deterioration of environmental quality.'

The above-noted collection of papers on the chemical oceanography of the Gulf of St Lawrence was entitled *Chemical Oceanography of the Gulf of St. Lawrence*. Edited by P.M. Strain (Ottawa: Department of Fisheries and Oceans, 1988. 190 p. *Canadian Bulletin of Fisheries and Aquatic Sciences*, No. 220). It presented the most current review of knowledge, to date, on the subject. The editor concluded that work on the Gulf was limited by the availability of resources, logistics, and the state of the art, but a beginning had been made. In conclusion, the editor suggested that future advances on the study of the Gulf and the Estuary would depend upon the design of field experiments rather than field surveys.

(c) Biological Oceanography

Biological oceanography, in its broadest form, is the study of the biology of the oceans, but it usually applies to research on zooplankton, phytoplankton, or benthic (bottom level) fauna, all of which are important as ocean nutrients. The most current survey of the field in Canada was carried out by T.R. Parsons in his report entitled *Biological Oceanography in Canada: A Perspective and Review* (Ottawa: Environment Canada, 1975. 53 p. Fisheries Research Board of Canada Report No. 4). At that time, Parsons indicated that the largest single biological oceanographic research program in Canada was done by the Marine Ecology Laboratory (now the Bedford Institute of Oceanography), at Dartmouth, Nova Scotia. Its program was designed to understand the biological oceanography underlying fisheries production. In the Canadian Arctic, work has concentrated upon increasing our knowledge of Arctic marine biology, and the marine food chain. On the west coast, research programs concentrated on biological processes in the Strait of Georgia.

The earliest major biological oceanographic investigation of Canadian waters was organized under the direction of the Norwegian scientist Johan Hjort, whose research was published with the title *Canadian Fisheries Expedition, 1914–1915. Investigation in the Gulf of St. Lawrence and Atlantic Waters of Canada* (Ottawa: Department of Naval Service, 1919. 495 p.). Since that time, the biological oceanography of the Estuary and Gulf of St. Lawrence still remains relatively understudied. The biological oceanography of the St Lawrence Estuary, including its contributory rivers, has been reviewed by J.C. Therriault and L. Legendre in their paper entitled *Oceanography and Ecology of Phytoplankton in the St. Lawrence*

Estuary. In: *Oceanography of a Large-Scale Estuarine System: The St. Lawrence*. Edited by M.I. El-Sabh and N. Silverberg (New York: Springer-Verlag, 1990. pp. 269–95); and by J.A. Runge and Y. Simard in the paper entitled *Zooplankton of the St. Lawrence Estuary: The Imprint of Physical Processes on Its Composition and Distribution*. In: *Oceanography of a Large-Scale Estuarine System: The St. Lawrence*. Edited by M.I. El-Sabh and N. Silverberg (New York: Springer-Verlag, 1990. pp. 296–320). In both papers, the authors pointed out that the oceanic (offshore) and coastal (estuarine and inshore) marine environments, as well as temperature, salinity, vertical mixing, and turbidity, affect phytoplankton and zooplankton biomass and production. In the second paper, the authors also noted 'there is not a broad foundation of zooplankton research in the estuary' and 'our knowledge of zooplankton in the St. Lawrence Estuary is based on relatively few studies ...'

Biological studies of the Gulf of St Lawrence can be divided, geographically, into the Gulf proper, and the lower St Lawrence Estuary. But G.L. Bugden and B.T. Hargrave in their publication entitled *Freshwater Run-Off Effects in the Marine Environment: The Gulf of St. Lawrence Example* (Ottawa: Department of Fisheries and Oceans, 1982. 89 p. Canadian Technical Report of Fisheries and Aquatic Sciences No. 1078), concluded that from a physical oceanographic point of view, the Gulf of St Lawrence can be considered a different ecosystem, within which the dynamics are controlled, to a great extent, internally rather than by large-scale oceanic processes. Going further than that, however, Y. de Lafontaine and S. Demers in their paper entitled *Pelagic Food Web Interactions and Productivity in the Gulf of St. Lawrence: A Perspective*. In: *The Gulf of St. Lawrence. Small Ocean or Big Estuary? Proceedings of a Workshop/Symposium Held at the Maurice Lamontagne Institute, Mont-Joli, 14–17 March, 1989*. Edited by J.C. Therriault (Ottawa: Department of Fisheries and Oceans, 1991. pp. 99–123. Canadian Special Publication of Fisheries and Aquatic Sciences No. 113), divided the Gulf into four regions as a geographic framework to evaluate the general hypotheses that the Gulf of St Lawrence is not homogeneous biologically within the enclosed Gulf ecosystem, as it has its own biological dynamics in the pelagic food webs. This regional view of the Gulf is further complicated by the fact that the information gathered on pelagic populations in the Gulf, to date, has essentially focused on quantitative estimations, and 'the lack of definition concerning the structure and the interaction strength between the various components of the different pelagic communities do not allow us to make any quantitative production on the nature of the biological dynamics in the Gulf of St. Lawrence.'

In the Canadian Arctic, W.G. Harrison and T. Platt, in the late 1970s, researched the eastern Canadian Arctic for their paper entitled *Primary Production and Nutrient Assimilation by Natural Phytoplankton Populations of the Eastern Canadian Arctic (Canadian Journal of Fisheries and Aquatic Sciences*, v. 39, 1982.

pp. 335–45). Their research indicated that phytoplankton readings taken in 1973, from Baffin Bay, indicated a more fertile production of phytoplankton than previously believed. During the short phytoplankton growing season (dependent on light, water temperature, and salinity), 50% of the nitrogen requirements are replenished in situ, and the rest by water column diffusion processes. To the west of Baffin Bay, D. Sameoto and A. Herman, in their paper entitled *Relations between the Thermocline Mesozooplankton and Microzooplankton, Chlorophyll-A and Primary Production Distributions in Lancaster Sound, Eastern Canadian Arctic* (*Polar Biology*, v. 6, 1986. pp. 53–61), noted that the 'westward intrusion of Baffin Bay water along the north shore of the sound which mixed with Arctic water flowing easterly resulted in a complex physical environment in the sound. The admixing of these water masses resulted in a highly variable biological structure.' In these waters, the thermocline (a region of rapidly changing temperature) was the main physical feature that affected the depth of the chlorophyll layer and the levels of primary production, with microzooplankton production concentrated in the upper levels of the thermocline and most microzooplankton found below the thermocline.

The study of ice algae in Arctic waters has also become a growing research interest because the algae are not only a major nutrient source in Arctic waters but also an example of biological metabolism under chronically low temperature and low light environments. R.E. Smith and R. Clement, in their paper entitled *Intracellular Photosynthate Allocation and the Control of Arctic Marine Ice Algal Production* (*Journal of Phycology*, v. 23, 1987. pp. 124–32), measured intracellular carbon in ice algae and found that they displayed 'marked shifts in metabolism consistent with major changes in light and inorganic nutrient supply, driven in part by the physical process of tidal mixing.' W.G. Harrison, in the paper entitled *Respiration and Its Size-Dependence in Microplankton Populations from Surface Waters in the Canadian Arctic* (*Polar Biology*, v. 6, 1986. pp. 145–52), noted that this phytoplankton biomass had high oxygen flux rates in consumption and production in near-surface waters, and although oxygen generation was greater in larger phytoplankton, over 70% of the respiration activity occurred in the larger mass of small phytoplankton. Some time later, W.G. Harrison and G.F. Cota in their paper entitled *Nitrogen Utilization in Ice Algal Communities of Barrow Strait, Northwest Territories, Canada* (*Marine Ecology Progress Series*, v. 67, 1990. pp. 275–83), studied the utilization of inorganic nitrogen (NO3– and NH4+ ions) by sea-ice algae, over two seasons. During the early stages of algal biomass accumulations under the ice, the metabolism was dominated by the NO3– ion, but when the biomass was in decline, NH4+ uptake by the algae dominated. Additional experiments also indicated that organic nitrogen, in urea and amino acid forms, was also a significant portion of algal sea-ice nutrition, but there was indirect evidence that these sea-ice communities were not nitrogen-limited in their nutrient requirements.

Phytoplankton studies have continued to be a focus of west coast near-shore biological studies, particularly the Strait of Georgia. The Strait is particularly important to biological oceanographers, because it comprises an area of 6,900 square kilometres that is a valuable fishing and rearing area for the young of several major stocks of Pacific salmon, and for large stocks of herring. The general biological oceanography of the Strait has been reviewed by P.J. Harrison and J.D. Fulton in their paper entitled *Review of the Biological Oceanography of the Strait of Georgia: Pelagic Environment* (*Canadian Journal of Fisheries and Aquatic Sciences*, v. 40, 1983. pp. 1064–94). In their paper, the authors stated that there was a basic understanding of the composition and distribution of phytoplankton in the central and southern Strait of Georgia, but the information was based upon lines of stations down the middle of the Strait. Accordingly, there was a need for cross-strait sampling, from shore to shore, and a closer examination of sub-surface concentrations. With regard to zooplankton, the authors stated that quantitative estimates of zooplankton biomass are poor, except for microzooplankton, and more effort needed to be directed toward studies on spatial and temporal distribution of microzooplankton, including life history studies. Interestingly, the link between zooplankton and fish needed much more study, with work starting on the assessment of the importance of the density and timing of zooplankton biomass in relation to larval fish survival. Finally, the authors pointed out that even less was known about the distribution of bacteria in the Strait of Georgia.

Nevertheless, one area that has definitely received particular attention is the influence of the Fraser River plume on the phytoplankton ecology of the Georgia Strait. General observations of phytoplankton production, abundance, and distribution were made from 1975 to 1977 at 16 stations in the Strait by J.G. Stockner and D.D. Cliff and reported in their paper entitled *Phytoplankton Ecology of the Strait of Georgia, British Columbia* (*Journal of the Fisheries Research Board of Canada*, v. 36, 1979. pp. 657–66). Although the purpose of the study was to describe the size and distribution of phytoplankton populations over two seasons, the authors noted that 'physical studies have demonstrated that the general northward movement of the water mass from off the Fraser delta' and 'the generalized physical transport system strongly influences the distribution and production of phytoplankton' in the southern Strait. The actual entrainment rate, or the carrying along and mixing of salt water into the freshwater Fraser River plume, was calculated from model studies carried out by R.E. Cordes and S. Pond and reported in their paper entitled *Estimates of Entrainment in the Fraser River Plume, British Columbia* (*Atmosphere-Ocean*, v. 18, 1980. pp. 15–26). Salt-water intrusions reduce the high concentration of fresh-water phytoplankton growth encouraged by nutrients coming from the Fraser River plume. As well, fluctuations in the Fraser River discharge (maximum discharge into the plume is in June and July) affect the concentrations of inorganic nitrogen that encourage phytoplankton production. These effects have been studied by P.J. Harrison and R.J. Beamish in their paper entitled *Effects of a Fluctuation in Fraser River Discharge on Primary Production in the Central Strait of Georgia, British Columbia, Canada* (*Canadian Journal of Fisheries and Aquatic Sciences*, v. 54, 1997. pp. 1015–24). Their studies showed that the entrainment of nutrients during peak Fraser River discharge increased the amount of inorganic nutrients (NO3–), and hence increased concentrations of phytoplankton production to levels comparable with spring production. The authors concluded that 'our results clearly demonstrated for the first time that entrainment of nutrients and phytoplankton production in the central Strait of Georgia are closely coupled to fluctuations in the Fraser River discharge as the estuarine plume moves seaward.'

IV Departmental Sources of the Publications

The former Inland Waters Directorate, the Water Quality Branch of Environment Canada, the National Water Research Institute, and the Fisheries Research Board of Canada and its most recent successor, the Department of Fisheries and Oceans, have produced printed publications lists.

The most recent printed list from the former Inland Waters Directorate (now the Ecosystem Science Directorate of Environment Canada) is entitled *Inland Waters Directorate Publications, 1990* (Ottawa: Supply and Services Canada, 1990. 58 p.). This publication lists documents published between 1980 and 1990. It includes general interest brochures, the scientific series, the technical bulletin series, the report series, social science series, NHRI (National Hydrology Research Institute) papers, data series, water quality reports, flood damage reduction program publications, and western and northern region reports. However, the *Inland Waters Directorate. Publications, 1986* (Ottawa: Supply and Services Canada, 1986. 1 v.) lists publications issued from 1968 to 1986.

The publication lists from the Water Quality Branch of the Inland Waters Directorate are entitled *Water Quality Branch Publications, 1987–* (Ottawa: Inland Waters Directorate, 1989–). This annual list (only one has been issued to date) provides abstracts for policy and activity reports published during the year, leaflets, interpretive reports, data reports, data management reports, water quality guidelines and objectives, laboratory and field methods, quality assurance reports, and precipitation studies. Earlier editions of this publication were titled *Bibliography of Water Quality Branch Publications, 1978* (Ottawa: Water Quality Branch,

1978. 1 v.); and *Water Quality Branch Publications, 1980–1986* (Ottawa: Water Quality Branch, 1981–87. 7 v.).

The National Water Research Institute has issued one publication list entitled *National Water Research Institute Publications, 1980–1987*. Compiled and edited by R.J. Daley and M. Hawkins (Burlington, Ontario: National Water Research Institute. Inland Waters Directorate, 1988. 53 p.). However, publication lists for 1997 and 1998 are available at their website <<http://www.cciw.ca/nwri/1997pubs.html>>, and <<http://www.cciw.ca/nwri/1998pubs.html>>.

Oceanographic studies from 1900 to 1972 have been thoroughly documented in printed indexes issued by the Fisheries Research Board of Canada. The latter issued two cumulated indexes and list of titles. The first, by Neil M. Carter, was entitled *Index and List of Titles: Fisheries Research Board and Associated Publications, 1900–1964* (Ottawa: Fisheries Research Board of Canada, 1968. 649 p. Fisheries Research Board of Canada Bulletin No. 164). The second, also by Neil M. Carter, was entitled *Index and List of Titles: Fisheries Research Board of Canada and Associated Publications, 1965–1972* (Ottawa: Fisheries Research Board of Canada, 1973. 588 p. Fisheries Research Board of Canada Miscellaneous Special Publication No. 18). After 1973, there are annual indexes, and lists of titles, which are located in the last issue of the *Canadian Journal of Fisheries and Aquatic Sciences* for any year. The format followed is the same as the earlier cumulative lists.

Appendix

Current Sources of Federal Government Publications

I Print Sources

Bishop, O.G. *Publications of the Government of the Province of Canada, 1841–1867* (Ottawa: National Library of Canada, 1963. 351 p.).

This major annotated bibliography begins with an important historical Introduction about the creation of the Province of Canada in 1841, which was based upon the recommendations of Lord Durham in 1838. For the purpose of tracing federal government publications, those documents issued by the newly created province are the precursor to federal publications issued after the Conference of 1867. From the point of view of federal science publications, Olga Bishop's monograph is useful for documenting: (1) the creation of the Fisheries Act of 1857, and the serials, and special publications, issued by the Fisheries Branch of the Department of Crown Lands, (2) the passage of the initial navigation acts, and the publication of the first reports on mines, issued by the Department of Crown Lands, (3) the act creating a Supervisor of Gullers, pertaining to the inspection and measurement of timbers, masts, etc., and the ensuing reports issued by the Crown Lands Department, or the Crown Timber Office, (4) the acts passed pertaining to navigation improvements on the St Lawrence River, and the construction of lighthouses, along with the reports of the Board of Works and the Department of Public Works pertaining to the above, and (5) the passing of an act in 1841 creating the Geological Survey of the Province of Canada, as well as the serial and Special publications issued by the Geological Survey.

Bishop, O.G. *Canadian Official Publications.* 1st Edition (Oxford: Pergamon Press, 1981. 297 p. Guides to Official Publications, v. 9).

This second publication by Professor Bishop is the standard work for students of federal government documents. The historical development and intricacies of the federal structure, federal-provincial relations, and conferences are succinctly described, but the core of the publication is the documentation of official federal publications, parliamentary and non-parliamentary, official and non-official indexes, guides to bibliographies, reference books, and statistics. There is also a description of the Public Archives of Canada. Of particular interest to students of federal science publications is a selected list of Indexes to Departmental Publications, which includes some science-oriented departments. As well, the chapter entitled *Non-Parliamentary Publications* gives examples of publications issued by science-based departments. Finally, the chapter on reference books also lists some science reference materials, as noted at the time of the publication of the monograph.

Canadian Research Index – Microlog (Toronto: Micromedia, v. 17, 1995–).

This index, in print, is a monthly indexing service, which covers government publications at the federal, provincial, and municipal levels, and includes scientific and technical research papers in the physical, natural, and social sciences. (For example, the Canadian Technical Report(s) of Fisheries and Aquatic Sciences, listed in the Weekly Checklist of Canadian Publications – described below – are available in print, or microfiche copy, only from Micromedia.) The index also includes policy papers, important statistical papers, and annual reports. Each record has a unique Microlog Accession Number that is the key to immediate retrieval of any document where institutions or libraries may subscribe to complete or partial collections of indexed records. Or, as noted above, individual copies in microfiche or paper reprints may be ordered through Micromedia's Document Delivery Service via:

Global Information Centre Canada
240 Catherine Street, Suite 305
Ottawa, ON
K2P 2G8
E-mail: gic@micromedia.on.ca

The index is organized in the following sections: Title, Index; Corporate Name Index; Personal Name Index; Subject Index; Series Titles; and Microfiche Collections (lists the contents of each fiche collection for quick reference). The full Canadian Research Index file, from 1982 to date, is available in electronic formats via the SilverPlatter SPIRS platform on CD, or it may be licensed for network use on a local server.

From 1987 to 1994, the *Canadian Research Index* was called *Microlog, Canadian Research Index* (Toronto: Micromedia, 1987–94. 7 v.). It was a monthly index, with an annual cumulation. Semi-annual cumulations were issued in CD-ROM format. Prior to that, the index was known as *Microlog Index* (Toronto: Micromedia, 1979–87. 9 v.). This index was an amalgamation of *ProFile Index* (Toronto: Micromedia, 1973–78. 6 v.), an index of Canadian provincial publications; *Publicat Index* (Toronto: Micromedia, 1977–

78. 2 v.), an index of Canadian federal publications; and *Urban Index: A Current Index to Canadian Publications in the Field of Urban and Regional Planning and Development* (Toronto: Micromedia, 1977–78. 2 v.).

Weekly Checklist of Canadian Government Publications (Ottawa: Canadian Government Publishing, Public Works and Government Services, 1978–).

The *Weekly Checklist* (known as the pink list), which began in November, 1978, superseded the *Daily Checklist of Government Publications*, issued from 1952 to 1978, and incorporates the *Special List of Canadian Government Publications* (known as the green list) that was issued from late 1969 into the early 1990s. The *Weekly Checklist* indicates those publications that may be distributed free to full depository libraries; publications that are issued free to selective depository libraries on request; publications sold to non-depository libraries by Canadian Government Publishing, Communications Canada; publications distributed to non-depository libraries by the issuing agency; publications available only from the issuing agency; and publications available on the Internet. As well, serials automatically sent to libraries that have requested them, via mailing lists through the Consolidated List of Serials, are noted. The total list is divided into two parts: Parliamentary Publications, and Departmental Publications. Since 1999, the *Weekly Checklist of Canadian Government* Publications has also produced an annual Supplement, entitled *Canadian Federal Government DATABASES*, accessible through the Internet. This list is 'published for information only.'

Government of Canada Publications Catalogue (Ottawa: Canadian Government Publishing Centre, 1953–98. 40 v.).

This publication was issued monthly, with annual cumulations from 1953 to 1978, and issued quarterly from 1979 to 1997. Its earlier publication history was: *Price List of Government Publications, [to] 1927; Catalogue of Official Publications of the Parliament and Government of Canada, 1928–1939*; and *Government Publications Annual Catalogue, 1943–1948*. In its final format, this catalogue cumulated the *Weekly Checklist*, and *Special List*, and was divided into three parts: (1) Parliamentary Publications, (2) Departmental Publications, and (3) an Index. Parts one and two were divided into two sections: an English language-section and a French-language section. The alphabetical Index was bilingual.

Canadiana (Ottawa: National Library of Canada, 1951–1991. 40 v.);

Canadiana [Microform] (Ottawa: National Library of Canada, 1973–2000); and

Canadiana [on CD-ROM]; the *National Bibliography* (Ottawa: National Library of Canada, 1998–).

Whether in print, microform, or CD-ROM, the National Library began its consecutive publication of this monthly list in 1952. It covers 'publications of Canadian interest received by National Library.' Since that time, one part has been devoted to federal government documents – English-language and French-language publications. Full bibliographic detail is given, and the arrangement is alphabetical by author. In print, there have been cumulative indexes issued for 1950–62; 1963–67; and 1968–72. The microform version was issued monthly, with annual cumulations from 1980 to 2000, with holdings going back to 1973. The CD-ROM version covers the period 1973 to date.

II Databases

The databases listed below are useful for searching scientific research literature published by federal scientists, issued as government documents, or in the open research literature. The most effective searching involves creating a search set representing the 'corporate' author, or author affiliation (this is one of the standard searchable fields in these databases), and combining that search set with any particular subject search set. For searching techniques involving databases, readers are referred to the author's previous work entitled *What Every Librarian Should Know About On-Line Searching.* 2nd Edition (Ottawa: Canadian Library Association, 1988. 380 p.). The databases are available for searching in a variety of formats – vendor on-line searching services, CD-ROM formats, or web-based versions, mounted locally in research libraries, or via consortial licensing by groups of research libraries.

Aerospace Database (New York: American Institute of Aeronautics and Astronautics, 1962–).

The Aerospace Database provides references, abstracts, and controlled vocabulary indexing of key scientific and technical documents, as well as books, reports, and conferences, which cover aerospace research and development in over 40 countries. The subject coverage is basic and applied research in aeronautics, astronautics, and space sciences, as well as technological developments and applications in complementary and supporting fields, such as chemistry, geosciences, physics, communications, and electronics. The database, which is the on-line equivalent of the print *International Aerospace Abstracts,* also published by the AIAA, contains reports issued by NASA, other US government agencies, international institutions, universities, and private firms. Over 1,600 periodicals from worldwide sources are scanned. The database, from 1962 to the present, contains over 2,200,000 records.

AGRICOLA (AGRICultural Online Access) (Beltsville, Maryland: National Agricultural Library, 1970–).

AGRICOLA is the database of the US National Agricultural Library, and consists of records for literature citations of journal articles, monographs, theses, patents, translations, microforms, audiovisuals, software, and technical reports. AGRICOLA records are used to produce the printed *Bibliography of Agriculture*, and other products of the NAC. This database provides selective worldwide coverage of primary information sources in agriculture and related fields. Library of Congress Subject Headings (and from 1985 on, the CAB – Commonwealth Agricultural Bureaux – Thesaurus) are used to select controlled vocabulary terms for subject indexing. Documents catalogued and indexed in the AGRICOLA database are selected from current acquisitions of the NAL and cooperating institutions. More than 2,000 serial titles are reviewed for indexing. The database, from 1970 to the present, contains over 3,550,000 records.

Aquatic Sciences and Fisheries Abstracts (Bethesda, Maryland: Cambridge Scientific Abstracts, 1978–).

Aquatic Sciences and Fisheries Abstracts (ASFA) is a comprehensive database covering the science, technology, and management of marine, brackish water, and freshwater environments and resources. The database corresponds to the printed publications *Aquatic Sciences and Fisheries Abstracts, Part 1: Biological Sciences and Living Resources*; *Part 2: Ocean Technology, Policy, and Non-Living Resources*; *Part 3: Aquatic Pollution, and Environmental Quality*; *ASFA Aquaculture Abstracts; and ASFA Marine Biotechnology Abstracts.* The database, which also contains records that have not appeared in the print publication, is a collaborative effort of the Food and Agricultural Organization (FAO) of the United Nations, as well as other international organizations and the national agencies in 23 countries, including Canada. Over 5,000 primary journals and a wide variety of other source documents, including books, monographic series, conference proceedings, and technical research reports, are monitored for inclusion in the database. The database, from 1978 to the present, contains approximately 700,000 records.

BIOSIS Previews (Philadelphia, Pennsylvania: BIOSIS, 1969–).

BIOSIS Previews contains citations from *Biological Abstracts* (BA), and *Biological Abstracts/ Reports, Reviews, and Meetings* (BA/RRM) (formerly *Bioresearch Index*), the major print publications of BIOSIS. Together, these publications constitute the major English-language service providing comprehensive worldwide coverage of research in the biological and biomedical sciences. The *Biological Abstracts* portion of the database includes approximately 350,000 accounts of original research yearly, from nearly 6,000 primary journal and monograph titles. The *Biological Abstracts/RRM* portion includes an additional 200,000-plus citations a year from meeting abstracts, reviews, books, book chapters, notes, letters, selected institutional and government reports, and research communications. US patents are included from 1986 through 1989. Abstracts are available for records from the *Biological Abstracts* portion of the database, starting in mid-1976, and for book synopses in BA/RRM, starting in 1985. Most BA/RRM records do not contain abstracts. For local web-based loading of the database from CD-ROMs, research libraries may subscribe to *Biological Abstracts* alone, or in combination with BA/RRM. The total database, from 1969 to the present, contains over 13,000,000 records.

CA SEARCH (Columbus, Ohio: Chemical Abstracts Service 1967–).

The CA SEARCH database is the major English-language service providing comprehensive worldwide coverage of the literature of chemistry and its applications. CA SEARCH corresponds to the bibliographic information and complete indexing found in the print *Chemical Abstracts*. The controlled vocabulary CA General Subject Index Headings, related general subject terminology from the CA Index Guide, and CAS Registry Numbers, each with its modifying phrase, are included. Chemical substances are represented by CAS Registry Numbers, unique numbers assigned to each specific chemical compound. Sources for the information are journal articles, patents, reviews, technical reports, monographs, conference and symposium proceedings, and dissertations. The database, from 1967 to the present, contains over 15 million records. Research libraries, with consortial arrangements to Chemical Abstracts Service, may offer a web-based version of the complete database called SciFinder Scholar.

Food Science and Technology Abstracts (Reading, UK: IFIS Publishing, 1969–).

Food Science and Technology Abstracts (FSTA) provides comprehensive coverage of research and new development in the areas related to food science and technology. It corresponds to the printed publication *Food Science and Technology Abstracts*. FSTA records include evaluated abstracts. Primary source material for FSTA comes from over 1,800 journals, which are published in over 90 countries. Additional sources of information include patents, reviews, theses, reports, books, conference proceedings, legislation, and standards. An extensive on-line thesaurus is included in the file. The database, from 1969 to the present, contains over 583,000 records.

GeoRef (Alexandria, Virginia: American Geological Institute, 1785–).

GeoRef, the database of the American Geological Institute (AGI), covers worldwide technical literature on geology and geophysics. GeoRef corresponds to the print publications *Bibliography and Index of North American Geology*; *Bibliography of Theses in Geology*; *Geophysical Abstracts*; *Bibliography and Index of Geology Exclusive of North America*; and the *Bibliography and Index of Geology*. GeoRef organizes and indexes papers from over 13,000 serials and other publications representative of the interests of 29 professional geological and earth science societies which are members of the AGI. GeoRef is also international and its current coverage includes over 3,500 journals, as well as books and book chapters, conference papers, government publications, theses, dissertations, reports, maps, and meeting papers. The database contains over 2,300,000 records (since 1785 to the present for North American Material and 1933 to the present for worldwide material).

Meteorological and Geoastrophysical Abstracts (Boston, Massachusetts: American Meteorological Society, 1970–).

Meteorological and Geoastrophysical Abstracts provides current citations in English for the most important meteorological and geoastrophysical research published in worldwide literature sources. Subject coverage includes meteorology (weather and climate), astrophysics, physical oceanography, hydrology, environmental sciences, and glaciology. Over 200 sources, including technical journals, monographs, proceedings, reviews, and annual publications, are scanned for relevant literature. Abstracts are included for records from 1972 to 1973, and from 1976 to the present. The database, from 1970 to the present, contains over 260,000 records.

INSPEC (Edison, NJ: INSPEC, 1969–).

INSPEC corresponds to the three Science Abstracts print publications: *Physics Abstracts, Electrical and Electronic Abstracts*, and *Computer and Control Abstracts*. The printed abstracting journals began as *Science Abstracts*, in 1898, and today approximately 16% of the databases' source publications are in languages other than English. However, all articles are abstracted and indexed in English. Author-prepared abstracts are used when available. The INSPEC database utilizes controlled vocabulary from the INSPEC Thesaurus, and beginning in January, 1987, INSPEC records also include chemical substance indexing and numerical index terms. As of November, 1990, over 4,100 journals and serials are scanned, of which 750 are abstracted cover to cover. These contribute 82% of the database, including 6% which are conference papers reported in journals. A further 16% comes from conference proceedings. Other source materials include books, reports, and dissertations. The database, from 1969 to the present, contains over 5,000,000 records.

Oceanic Abstracts (Bethesda, Maryland: Cambridge Scientific Abstracts, 1964–).

Oceanic Abstracts organizes and indexes technical literature published worldwide on marine-related subjects. Over 2,000 source publications – journals, books, technical reports, conference proceedings, government and trade publications – are scanned for references to major subject areas ranging from oceanography, marine biology, and marine pollution, to legal aspects of marine resources. The database, from 1964 to the present, contains over 330,000 records.

SciSearch (Philadelphia, Pennsylvania: Institute for Scientific Information, 1974–).

SciSearch is an international, multi-disciplinary index to the literature of science, technology, biomedicine, and related disciplines, produced by the Institute for Scientific Information. It contains all the records published in the print *Science Citation Index*, plus additional records from the Current Contents publications. SciSearch is unique because, first, the journals indexed are selected on the basis of several criteria, including citation analysis. This results in coverage of the most significant publications in the scientific, technical, and biomedical literature. Second, in addition to the more conventional retrieval methods, SciSearch offers citation indexing, which allows searching by cited references. For records added since January, 1991, author abstracts, author keywords, plus SciSearch added keywords, may be searched. SciSearch indexes all significant items – articles, review papers, meeting, abstracts, letters, editorials, book reviews, correction notices, etc. – from approximately 4,500 scientific and technical journals. Some 3,800 of these journals are further indexed by the references cited within each article, allowing for citation searching. The additional 700 journals indexed have been drawn from the ISI Current Contents series of publications. The standard database, from 1974 to the present, contains over 16,000,000 records. Research libraries, under consortial arrangements, may offer a web-based version, called Web of Science, which contains Science Citation Expanded, from 1970 to date; Social Science Citation Index, from 1970 to the present; and Arts and Humanities Index, from 1975 to the present.

Water Resources Abstracts (Bethesda, Maryland: Cambridge Scientific Abstracts, 1968–).

Water Resources Abstracts provides a comprehensive range of water-related topics in the biological and physical sciences, as well as the engineering and legal aspects of the conservation, control, use, and management of water. From 1968 until 1994, the database was produced by the US Geological Survey. Coverage emphasizes English-language materials, and most records contain an informative abstract. Records are indexed using controlled vocabulary from the Water Resources Thesaurus. The database, from 1968 to the present, contains more than 350,000 records.

Zoological Record Online (Philadelphia, Pennsylvania: BIOSIS, 1978–).

Zoological Record Online provides extensive coverage of the world's zoological literature, with particular emphasis on systematic/taxonomic information. The database corresponds closely to the printed index *Zoological Record*. The database includes subject indexing, using both controlled and natural language formats. It also includes a unique systematics field, which gives complete taxonomic hierarchy information for most organisms discussed. Over 4,500 international serials, monographs, conference proceedings, and other non-serial publications are monitored for inclusion in the database. The database, from 1978 to the present, contains over 1,500,000 records.

III Other Sources

Canadian Institute for Historical Microreproductions <<http://www.nls-bnc.ca/cihm>>.

The Canadian Institute for Historical Microreproductions (CIHM) was established in 1978, to locate early printed Canadian materials (books, annuals, and periodicals), preserve their content on microfilm, and make the resulting Early Canadiana Research Collection available to libraries and archives in Canada and abroad. A guide to the complete collection is the microfiche catalogue entitled *Canada, The Printed Record.* A new edition of the COM catalogue was issued in 1997, and includes more than 70,000 titles. The catalogue consists of two parts: (1) a register which reproduces, in computer-generated numerical sequence, the complete bibliographic entry for each record, and (2) seven indexes. These indexes are: Author/Title/Series; English Subject Headings; French Subject Headings; Dewey Decimal Classification; Place of Publication; Publication Date; and CIHM Series Numbers. Subsets of the pre-1900 monographs collection are available in subject sets. Of interest to science historians and students is the Science and Technology subset. It is strong in the areas of geology, geography, zoology, botany, and transportation. Orders or inquiries pertaining to CIHM can be sent to:

Canadian Institute for Historical Microreproductions
395 Wellington Street, Room 468
Ottawa, Ontario K1A 0N4
Email: cihmicmh@nk-bnc.ca
Telephone: (613) 235–2628

Bibliography

Athuns, C. Donald. *Meteorology Today: An Introduction to Weather, Climate, and the Environment.* 5th Edition. Minneapolis/St Paul: West Publishing Co., 1994. 591 p.

Bruce, James P. *Water Resources Research in Canada.* Ottawa: Science Secretariat Privy Council Office, 1968. 169 p. (Science Council of Canada. Special Study No. 5).

Cambridge Encyclopedia of Earth Sciences. Edited by David G. Smith. New York: Cambridge University Press, 1981. 496 p.

Canada Inquiry on Federal Water Policy. *Currents of Change: Final Report. Inquiry on Federal Water Policy.* Ottawa: Environment Canada, 1985. 222 p.

Canada. *The Fisheries Research Board of Canada: What It Is and What It Does.* Ottawa: Fisheries Research Board of Canada, 1968. 18 pp.

Cathermole, P.J. *The Story of the Earth.* Cambridge: Cambridge University Press, 1985. 224 p.

Centennial Conference of the Canadian Hydrographic Service. *From Leadline to Laser: Centennial Conference of the Canadian Hydrographic Service, April 5–8, 1983, Government Conference Centre, Ottawa, Canada.* Ottawa: Supply and Services Canada, 1983. 223 p. Canadian Special Publication of Fisheries and Aquatic Sciences No. 67.

Cranstone, D.A. *Canadian Ore Discoveries, 1946–1978: A Continuing Story of Success.* Ottawa: Energy, Mines and Resources Canada, 1980. 21 p.

Department of Energy, Mines and Resources. *A Collection of Papers Prepared for a Study of the Earth Sciences in Canada.* Ottawa: Department of Energy, Mines and Resources, 1970. 318 p.

Doern, G. Bruce. *Science and Politics in Canada.* Montreal: McGill-Queen's University Press, 1972. 328 p.

Filmore, Stanley. *The Chartmakers: A History of Nautical Surveying in Canada.* Toronto: NC Press in Association with the Canadian Hydrographic Service, 1983. 255 p.

Geological Survey of Canada. *Geology and Economic Minerals of Canada.* 5th Edition. Edited by R.J.W. Douglas. Ottawa: Department of Energy, Mines and Resources, 1976–82. 3 v.

Geological Survey of Canada. *The Geological Survey of Canada: Past and Present.* Ottawa: Geological Survey of Canada, 1986. 40 p.

The Geology of North America: An Overview. Edited by A.W. Bally and A.R. Palmer. Boulder, Colorado: Geological Society of America, 1989. 619 p.

Geology of the Innuition Orogen and Arctic Platform of Canada and Greenland. Edited by H.P. Trettin. Ottawa: Geological Survey of Canada, 1991. 2 v. (Geology of Canada) No. 3. Geology of North America, vol. E).

Geology of the Continental Margin of Eastern Canada. Edited by M.J. Keen and G.L. Williams. Ottawa: Geological Survey of Canada, 1990. 855 p. (Geology of Canada No. 2. Geology of North America, vol. I-1).

Geology of the Cordilleran Orogen in Canada. Edited by H. Gabrielse and C.J. Yorath. Ottawa: Geological Survey of Canada, 1992, 2 v. (Geology of Canada No. 4. Geology of North America, vol. G-2).

Healey, M.C., and R.R. Wallace, editors. *Canadian Aquatic Resources.* Ottawa: Supply and Services Canada, 1987. 533 p. (*Canadian Bulletin of Fisheries and Aquatic Sciences*, No. 215).

Ignatieff, A. *A Canadian Research Heritage: An Historical Account of 75 Years of Federal Research and Development in Minerals, Metals and Fuels at the Mines Branch.* Ottawa: Ministry of Supply and Services Canada, 1981. 353 p.

Johnstone, Kenneth. *The Aquatic Explorers: A History of the Fisheries Research Board of Canada.* Toronto: University of Toronto Press, 1977. 342 pp.

Martineau, René. *Insects Harmful to Forest Trees.* Montreal: Multiscience Publications, 1984. 261 p.

McDuffie, K., and A. Henderson-Sellers. *A Climate Modelling Primer.* 2nd Edition. New York: Wiley, 1997. 253 p.

Meteorological Challenges: A History. Edited by D.P. McIntyre. Ottawa: Information Canada, 1972. 338 p.

National Wetlands Working Group, Canada Committee on Ecological Land Classification. *Wetlands of Canada.* Montreal: Polyscience Publications in Cooperation with Environment Canada and the Canadian Government Publications Centre, 1988. 452 p. (Ecological Land Classification Series No. 24).

Neale, E.R.W. *The Earth Sciences in Canada: A Centennial Appraisal and Forecast.* Toronto: University of Toronto Press, 1968. 259 p. (Royal Society of Canada Special Publication No. 11).

Osborne, F.F. *Geochronology in Canada.* Toronto: University of Toronto Press, 1964. 156 p. (Royal Society of Canada Special Publication No. 8).

Perspectives in Science and Technology: The Legacy of Omond Solandt: Proceedings of a Symposium Held at the Donald Gordon Centre, Queen's University at Kingston, Ontario, 8–10 May, 1994. Edited by C.E. Law, G.R. Lindsey, and D.M. Grenville. [Kingston, Ontario], Published by *Queen's Quarterly*, [1995]. 239 p.

Ricker, W.E. *The Fisheries Research Board of Canada: Seventy-Five Years of Achievement.* Reprinted from the *Journal of the Fisheries Research Board of Canada,* v. 32, 1975. pp. 1465–90.

Science and Technology in Canada. Edited by Paul Dufour and John de la Mothe. Harlow, England: Longman, 1993. 331 p. (Longman Guide to World Science and Technology).

Sadler, Barry. *An Evaluation of the Beaufort Sea Environmental Assessment Panel Review.* Ottawa: Supply and Services Canada, 1990. 82p.

Stanton, Charles R. *Canadian Forestry: The View beyond the Trees.* Toronto: Macmillan of Canada and Maclean Hunter Press. 1976. 70 p.

Stevenson, J.C. *Fisheries Research Board of Canada, 1898 to 1973.* Ottawa: Environment Canada, 1973. 18p. (Miscellaneous Special Publication No. 20).

Thomas, Morley. *The Beginnings of Canadian Meteorology.* Toronto: ECW Press, 1991. 308 p.

Zaslow, M. *Reading the Rocks: The Story of the Geological Survey of Canada, 1842–1972.* Toronto: Macmillan, 1975. 599 p.

Index

Abbreviations

CANMET	Canada Centre for Mineral and Energy Technology
CCR	Canada Centre for Remote Sensing
CFS	Canada Forest Service
CHS	Canadian Hydrographic Service
CWS	Canadian Wildlife Service
GSC	Geological Survey of Canada
MOSST	Ministry of State for Science and Technology
NHRI	National Hydrology Research Institute
NRC	National Research Council
NSERC	Natural Sciences and Engineering Research Council
NWRI	National Water Research Institute
SCC	Science Council of Canada